Calculus

Tom M. Apostol

CALCULUS

VOLUME II

Multi-Variable Calculus and Linear Algebra, with Applications to Differential Equations and Probability

SECOND EDITION

JOHN WILEY & SONS
New York • Chichester • Brisbane • Toronto • Singapore

CONSULTING EDITOR

George Springer, *Indiana University*

Library of Congress Catalog Card Number: 67-14605
Printed in the United States of America.
ISBN 0 471 00007 8
Printed and bound by the Hamilton Printing Company.

30 29 28

To

Jane and Stephen

PREFACE

This book is a continuation of the author's *Calculus, Volume I, Second Edition.* The present volume has been written with the same underlying philosophy that prevailed in the first. Sound training in technique is combined with a strong theoretical development. Every effort has been made to convey the spirit of modern mathematics without undue emphasis on formalization. As in Volume I, historical remarks are included to give the student a sense of participation in the evolution of ideas.

The second volume is divided into three parts, entitled *Linear Analysis*, *Nonlinear Analysis*, and *Special Topics.* The last two chapters of Volume I have been repeated as the first two chapters of Volume II so that all the material on linear algebra will be complete in one volume.

Part 1 contains an introduction to linear algebra, including linear transformations, matrices, determinants, eigenvalues, and quadratic forms. Applications are given to analysis, in particular to the study of linear differential equations. Systems of differential equations are treated with the help of matrix calculus. Existence and uniqueness theorems are proved by Picard's method of successive approximations, which is also cast in the language of contraction operators.

Part 2 discusses the calculus of functions of several variables. Differential calculus is unified and simplified with the aid of linear algebra. It includes chain rules for scalar and vector fields, and applications to partial differential equations and extremum problems. Integral calculus includes line integrals, multiple integrals, and surface integrals, with applications to vector analysis. Here the treatment is along more or less classical lines and does not include a formal development of differential forms.

The special topics treated in Part 3 are *Probability* and *Numerical Analysis.* The material on probability is divided into two chapters, one dealing with finite or countably infinite sample spaces; the other with uncountable sample spaces, random variables, and distribution functions. The use of the calculus is illustrated in the study of both one- and two-dimensional random variables.

The last chapter contains an introduction to numerical analysis, the chief emphasis being on different kinds of polynomial approximation. Here again the ideas are unified by the notation and terminology of linear algebra. The book concludes with a treatment of approximate integration formulas, such as Simpson's rule, and a discussion of Euler's summation formula.

There is ample material in this volume for a full year's course meeting three or four times per week. It presupposes a knowledge of one-variable calculus as covered in most first-year calculus courses. The author has taught this material in a course with two lectures and two recitation periods per week, allowing about ten weeks for each part and omitting the starred sections.

This second volume has been planned so that many chapters can be omitted for a variety of shorter courses. For example, the last chapter of each part can be skipped without disrupting the continuity of the presentation. Part 1 by itself provides material for a combined course in linear algebra and ordinary differential equations. The individual instructor can choose topics to suit his needs and preferences by consulting the diagram on the next page which shows the logical interdependence of the chapters.

Once again I acknowledge with pleasure the assistance of many friends and colleagues. In preparing the second edition I received valuable help from Professors Herbert S. Zuckerman of the University of Washington, and Basil Gordon of the University of California, Los Angeles, each of whom suggested a number of improvements. Thanks are also due to the staff of Blaisdell Publishing Company for their assistance and cooperation.

As before, it gives me special pleasure to express my gratitude to my wife for the many ways in which she has contributed. In grateful acknowledgement I happily dedicate this book to her.

T. M. A.

Pasadena, California
September 16, 1968

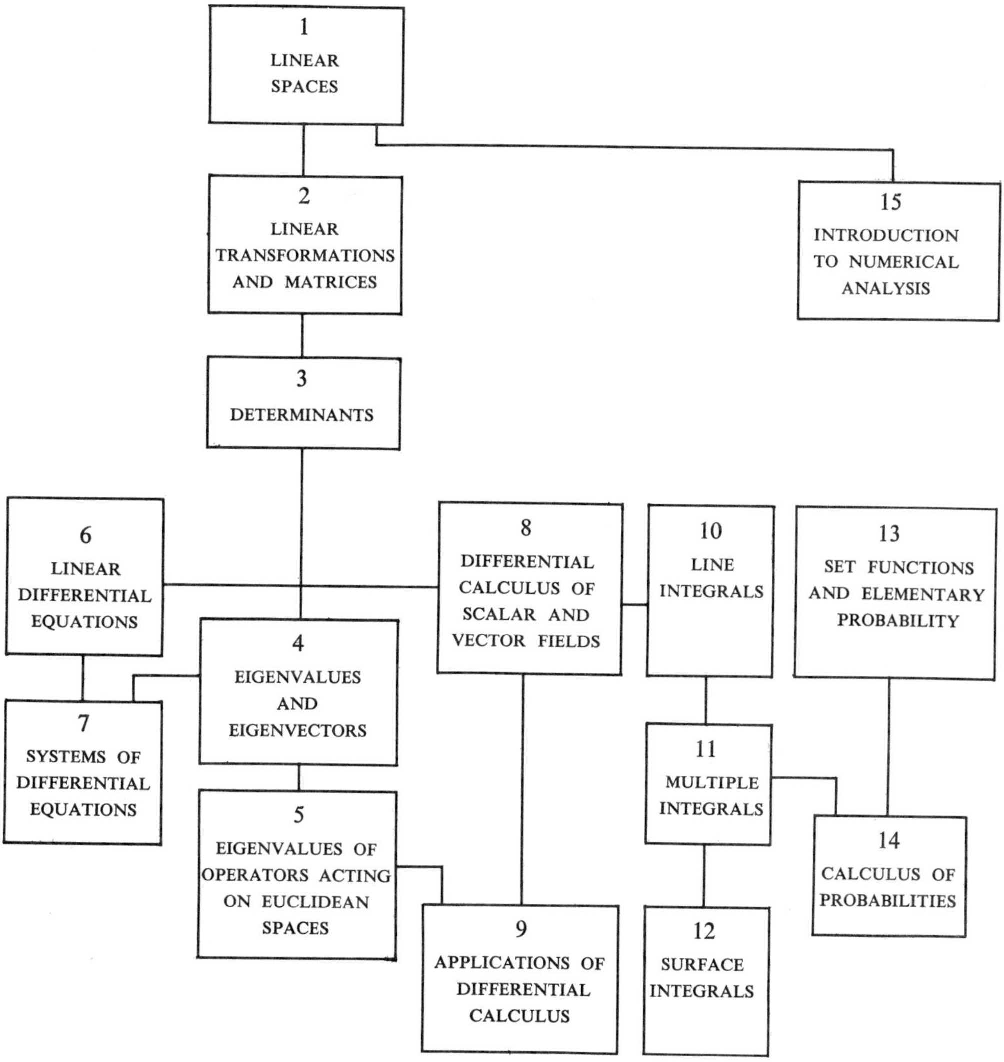
1
LINEAR SPACES
2
LINEAR TRANSFORMATIONS AND MATRICES
3
DETERMINANTS
4
EIGENVALUES AND EIGENVECTORS
5
EIGENVALUES OF OPERATORS ACTING ON EUCLIDEAN SPACES
6
LINEAR DIFFERENTIAL EQUATIONS
7
SYSTEMS OF DIFFERENTIAL EQUATIONS
8
DIFFERENTIAL CALCULUS OF SCALAR AND VECTOR FIELDS
9
APPLICATIONS OF DIFFERENTIAL CALCULUS
10
LINE INTEGRALS
11
MULTIPLE INTEGRALS
12
SURFACE INTEGRALS
13
SET FUNCTIONS AND ELEMENTARY PROBABILITY
14
CALCULUS OF PROBABILITIES
15
INTRODUCTION TO NUMERICAL ANALYSIS

CONTENTS

PART 1. LINEAR ANALYSIS

1. LINEAR SPACES

2. LINEAR TRANSFORMATIONS AND MATRICES

3. DETERMINANTS

4. EIGENVALUES AND EIGENVECTORS

5. EIGENVALUES OF OPERATORS ACTING ON EUCLIDEAN SPACES

6. LINEAR DIFFERENTIAL EQUATIONS

7. SYSTEMS OF DIFFERENTIAL EQUATIONS

9. APPLICATIONS OF THE DIFFERENTIAL CALCULUS

10. LINE INTEGRALS

11. MULTIPLE INTEGRALS

12. SURFACE INTEGRALS

PART 3. SPECIAL TOPICS

13. SET FUNCTIONS AND ELEMENTARY PROBABILITY

14. CALCULUS OF PROBABILITIES

15. INTRODUCTION TO NUMERICAL ANALYSIS

Calculus

PART 1

LINEAR ANALYSIS

1

LINEAR SPACES

1.1 Introduction

Throughout mathematics we encounter many examples of mathematical objects that can be added to each other and multiplied by real numbers. First of all, the real numbers themselves are such objects. Other examples are real-valued functions, the complex numbers, infinite series, vectors in n-space, and vector-valued functions. In this chapter we discuss a general mathematical concept, called a *linear space*, which includes all these examples and many others as special cases.

Briefly, a linear space is a set of elements of any kind on which certain operations (called *addition* and *multiplication by numbers*) can be performed. In defining a linear space, we do not specify the nature of the elements nor do we tell how the operations are to be performed on them. Instead, we require that the operations have certain properties which we take as axioms for a linear space. We turn now to a detailed description of these axioms.

1.2 The definition of a linear space

Let V denote a nonempty set of objects, called *elements*. The set V is called a linear space if it satisfies the following ten axioms which we list in three groups.

Closure axioms

AXIOM 1. CLOSURE UNDER ADDITION. *For every pair of elements x and y in V there corresponds a unique element in V called the sum of x and y, denoted by $x + y$.*

AXIOM 2. CLOSURE UNDER MULTIPLICATION BY REAL NUMBERS. *For every x in V and every real number a there corresponds an element in V called the product of a and x, denoted by ax.*

Axioms for addition

AXIOM 3. COMMUTATIVE LAW. *For all x and y in V, we have $x + y = y + x$.*

AXIOM 4. ASSOCIATIVE LAW. *For all x, y, and z in V, we have $(x + y) + z = x + (y + z)$.*

AXIOM 5. EXISTENCE OF ZERO ELEMENT. *There is an element in V, denoted by O, such that*

$$x + O = x \qquad \textit{for all } x \textit{ in } V.$$

AXIOM 6. EXISTENCE OF NEGATIVES. *For every x in V, the element $(-1)x$ has the property*

$$x + (-1)x = O.$$

Axioms for multiplication by numbers

AXIOM 7. ASSOCIATIVE LAW. *For every x in V and all real numbers a and b, we have*

$$a(bx) = (ab)x.$$

AXIOM 8. DISTRIBUTIVE LAW FOR ADDITION IN V. *For all x and y in V and all real a, we have*

$$a(x + y) = ax + ay.$$

AXIOM 9. DISTRIBUTIVE LAW FOR ADDITION OF NUMBERS. *For all x in V and all real a and b, we have*

$$(a + b)x = ax + bx.$$

AXIOM 10. EXISTENCE OF IDENTITY. *For every x in V, we have $1x = x$.*

Linear spaces, as defined above, are sometimes called *real* linear spaces to emphasize the fact that we are multiplying the elements of V by real numbers. If *real number* is replaced by *complex number* in Axioms 2, 7, 8, and 9, the resulting structure is called a *complex linear space*. Sometimes a linear space is referred to as a *linear vector space* or simply a *vector space;* the numbers used as multipliers are also called *scalars.* A real linear space has real numbers as scalars; a complex linear space has complex numbers as scalars. Although we shall deal primarily with examples of real linear spaces, all the theorems are valid for complex linear spaces as well. When we use the term linear space without further designation, it is to be understood that the space can be real or complex.

1.3 Examples of linear spaces

If we specify the set V and tell how to add its elements and how to multiply them by numbers, we get a concrete example of a linear space. The reader can easily verify that each of the following examples satisfies all the axioms for a real linear space.

EXAMPLE 1. Let $V = \mathbf{R}$, the set of all real numbers, and let $x + y$ and ax be ordinary addition and multiplication of real numbers.

EXAMPLE 2. Let $V = \mathbf{C}$, the set of all complex numbers, define $x + y$ to be ordinary addition of complex numbers, and define ax to be multiplication of the complex number x

by the real number a. Even though the elements of V are complex numbers, this is a real linear space because the scalars are real.

EXAMPLE 3. Let $V = V_n$, the vector space of all n-tuples of real numbers, with addition and multiplication by scalars defined in the usual way in terms of components.

EXAMPLE 4. Let V be the set of all vectors in V_n orthogonal to a given nonzero vector N. If $n = 2$, this linear space is a line through O with N as a normal vector. If $n = 3$, it is a plane through O with N as normal vector.

The following examples are called *function spaces*. The elements of V are real-valued functions, with addition of two functions f and g defined in the usual way:

$$(f + g)(x) = f(x) + g(x)$$

for every real x in the intersection of the domains of f and g. Multiplication of a function f by a real scalar a is defined as follows: af is that function whose value at each x in the domain of f is $af(x)$. The zero element is the function whose values are everywhere zero. The reader can easily verify that each of the following sets is a function space.

EXAMPLE 5. The set of all functions defined on a given interval.

EXAMPLE 6. The set of all polynomials.

EXAMPLE 7. The set of all polynomials of degree $\leq n$, where n is fixed. (Whenever we consider this set it is understood that the zero polynomial is also included.) The set of all polynomials of degree *equal* to n is not a linear space because the closure axioms are not satisfied. For example, the sum of two polynomials of degree n need not have degree n.

EXAMPLE 8. The set of all functions continuous on a given interval. If the interval is $[a, b]$, we denote this space by $C(a, b)$.

EXAMPLE 9. The set of all functions differentiable at a given point.

EXAMPLE 10. The set of all functions integrable on a given interval.

EXAMPLE 11. The set of all functions f defined at 1 with $f(1) = 0$. The number 0 is essential in this example. If we replace 0 by a nonzero number c, we violate the closure axioms.

EXAMPLE 12. The set of all solutions of a homogeneous linear differential equation $y'' + ay' + by = 0$, where a and b are given constants. Here again 0 is essential. The set of solutions of a nonhomogeneous differential equation does not satisfy the closure axioms.

These examples and many others illustrate how the linear space concept permeates algebra, geometry, and analysis. When a theorem is deduced from the axioms of a linear space, we obtain, in one stroke, a result valid for each concrete example. By unifying diverse examples in this way we gain a deeper insight into each. Sometimes special knowledge of one particular example helps to anticipate or interpret results valid for other examples and reveals relationships which might otherwise escape notice.

1.4 Elementary consequences of the axioms

The following theorems are easily deduced from the axioms for a linear space.

THEOREM 1.1. UNIQUENESS OF THE ZERO ELEMENT. *In any linear space there is one and only one zero element.*

Proof. Axiom 5 tells us that there is at least one zero element. Suppose there were two, say O_1 and O_2. Taking $x = O_1$ and $O = O_2$ in Axiom 5, we obtain $O_1 + O_2 = O_1$. Similarly, taking $x = O_2$ and $O = O_1$, we find $O_2 + O_1 = O_2$. But $O_1 + O_2 = O_2 + O_1$ because of the commutative law, so $O_1 = O_2$.

THEOREM 1.2. UNIQUENESS OF NEGATIVE ELEMENTS. *In any linear space every element has exactly one negative. That is, for every x there is one and only one y such that $x + y = O$.*

Proof. Axiom 6 tells us that each x has at least one negative, namely $(-1)x$. Suppose x has two negatives, say y_1 and y_2. Then $x + y_1 = O$ and $x + y_2 = O$. Adding y_2 to both members of the first equation and using Axioms 5, 4, and 3, we find that

$$y_2 + (x + y_1) = y_2 + O = y_2,$$

and

$$y_2 + (x + y_1) = (y_2 + x) + y_1 = O + y_1 = y_1 + O = y_1.$$

Therefore $y_1 = y_2$, so x has exactly one negative, the element $(-1)x$.

Notation. The negative of x is denoted by $-x$. The difference $y - x$ is defined to be the sum $y + (-x)$.

The next theorem describes a number of properties which govern elementary algebraic manipulations in a linear space.

THEOREM 1.3. *In a given linear space, let x and y denote arbitrary elements and let a and b denote arbitrary scalars. Then we have the following properties:*

(a) $0x = O$.
(b) $aO = O$.
(c) $(-a)x = -(ax) = a(-x)$.
(d) *If* $ax = O$, *then either* $a = 0$ *or* $x = O$.
(e) *If* $ax = ay$ *and* $a \neq 0$, *then* $x = y$.
(f) *If* $ax = bx$ *and* $x \neq O$, *then* $a = b$.
(g) $-(x + y) = (-x) + (-y) = -x - y$.
(h) $x + x = 2x$, $x + x + x = 3x$, *and in general*, $\sum_{i=1}^{n} x = nx$.

We shall prove (a), (b), and (c) and leave the proofs of the other properties as exercises.

Proof of (a). Let $z = 0x$. We wish to prove that $z = O$. Adding z to itself and using Axiom 9, we find that

$$z + z = 0x + 0x = (0 + 0)x = 0x = z.$$

Now add $-z$ to both members to get $z = O$.

Proof of (b). Let $z = aO$, add z to itself, and use Axiom 8.

Proof of (c). Let $z = (-a)x$. Adding z to ax and using Axiom 9, we find that

$$z + ax = (-a)x + ax = (-a + a)x = 0x = O,$$

so z is the negative of ax, $z = -(ax)$. Similarly, if we add $a(-x)$ to ax and use Axiom 8 and property (b), we find that $a(-x) = -(ax)$.

1.5 Exercises

In Exercises 1 through 28, determine whether each of the given sets is a real linear space, if addition and multiplication by real scalars are defined in the usual way. For those that are not, tell which axioms fail to hold. The functions in Exercises 1 through 17 are real-valued. In Exercises 3, 4, and 5, each function has domain containing 0 and 1. In Exercises 7 through 12, each domain contains all real numbers.

1. All rational functions.
2. All rational functions f/g, with the degree of $f \leq$ the degree of g (including $f = 0$).
3. All f with $f(0) = f(1)$.
4. All f with $2f(0) = f(1)$.
5. All f with $f(1) = 1 + f(0)$.
6. All step functions defined on [0, 1].
7. All f with $f(x) \to 0$ as $x \to +\infty$.
8. All even functions.
9. All odd functions.
10. All bounded functions.
11. All increasing functions.
12. All functions with period 2π.
13. All f integrable on [0, 1] with $\int_0^1 f(x)\,dx = 0$.
14. All f integrable on [0, 1] with $\int_0^1 f(x)\,dx \geq 0$.
15. All f satisfying $f(x) = f(1 - x)$ for all x.
16. All Taylor polynomials of degree $\leq n$ for a fixed n (including the zero polynomial).
17. All solutions of a linear second-order homogeneous differential equation $y'' + P(x)y' + Q(x)y = 0$, where P and Q are given functions, continuous everywhere.
18. All bounded real sequences.
19. All convergent real sequences.
20. All convergent real series.
21. All absolutely convergent real series.
22. All vectors (x, y, z) in V_3 with $z = 0$.
23. All vectors (x, y, z) in V_3 with $x = 0$ or $y = 0$.
24. All vectors (x, y, z) in V_3 with $y = 5x$.
25. All vectors (x, y, z) in V_3 with $3x + 4y = 1$, $z = 0$.
26. All vectors (x, y, z) in V_3 which are scalar multiples of (1, 2, 3).
27. All vectors (x, y, z) in V_3 whose components satisfy a system of three linear equations of the form:

$$a_{11}x + a_{12}y + a_{13}z = 0, \qquad a_{21}x + a_{22}y + a_{23}z = 0, \qquad a_{31}x + a_{32}y + a_{33}z = 0.$$

28. All vectors in V_n that are linear combinations of two given vectors A and B.
29. Let $V = \mathbf{R}^+$, the set of positive real numbers. Define the "sum" of two elements x and y in V to be their product $x \cdot y$ (in the usual sense), and define "multiplication" of an element x in V by a scalar c to be x^c. Prove that V is a real linear space with 1 as the zero element.
30. (a) Prove that Axiom 10 can be deduced from the other axioms.
    (b) Prove that Axiom 10 cannot be deduced from the other axioms if Axiom 6 is replaced by Axiom 6': For every x in V there is an element y in V such that $x + y = O$.
31. Let S be the set of all ordered pairs (x_1, x_2) of real numbers. In each case determine whether or not S is a linear space with the operations of addition and multiplication by scalars defined as indicated. If the set is not a linear space, indicate which axioms are violated.
    (a) $(x_1, x_2) + (y_1, y_2) = (x_1 + y_1, x_2 + y_2)$, $\quad a(x_1, x_2) = (ax_1, 0)$.
    (b) $(x_1, x_2) + (y_1, y_2) = (x_1 + y_1, 0)$, $\quad a(x_1, x_2) = (ax_1, ax_2)$.
    (c) $(x_1, x_2) + (y_1, y_2) = (x_1, x_2 + y_2)$, $\quad a(x_1, x_2) = (ax_1, ax_2)$.
    (d) $(x_1, x_2) + (y_1, y_2) = (|x_1 + x_2|, |y_1 + y_2|)$, $\quad a(x_1, x_2) = (|ax_1|, |ax_2|)$.
32. Prove parts (d) through (h) of Theorem 1.3.

1.6 Subspaces of a linear space

Given a linear space V, let S be a nonempty subset of V. If S is also a linear space, with the same operations of addition and multiplication by scalars, then S is called a *subspace* of V. The next theorem gives a simple criterion for determining whether or not a subset of a linear space is a subspace.

THEOREM 1.4. *Let S be a nonempty subset of a linear space V. Then S is a subspace if and only if S satisfies the closure axioms.*

Proof. If S is a subspace, it satisfies all the axioms for a linear space, and hence, in particular, it satisfies the closure axioms.

Now we show that if S satisfies the closure axioms it satisfies the others as well. The commutative and associative laws for addition (Axioms 3 and 4) and the axioms for multiplication by scalars (Axioms 7 through 10) are automatically satisfied in S because they hold for all elements of V. It remains to verify Axioms 5 and 6, the existence of a zero element in S, and the existence of a negative for each element in S.

Let x be any element of S. (S has at least one element since S is not empty.) By Axiom 2, ax is in S for every scalar a. Taking $a = 0$, it follows that $0x$ is in S. But $0x = O$, by Theorem 1.3(a), so $O \in S$, and Axiom 5 is satisfied. Taking $a = -1$, we see that $(-1)x$ is in S. But $x + (-1)x = O$ since both x and $(-1)x$ are in V, so Axiom 6 is satisfied in S. Therefore S is a subspace of V.

DEFINITION. *Let S be a nonempty subset of a linear space V. An element x in V of the form*

$$x = \sum_{i=1}^{k} c_i x_i,$$

where $x_1, \ldots, x_k$ are all in S and $c_1, \ldots, c_k$ are scalars, is called a finite linear combination of elements of S. The set of all finite linear combinations of elements of S satisfies the closure axioms and hence is a subspace of V. We call this the subspace spanned by S, or the linear span of S, and denote it by $L(S)$. If S is empty, we define $L(S)$ to be $\{O\}$, the set consisting of the zero element alone.

Different sets may span the same subspace. For example, the space V_2 is spanned by each of the following sets of vectors: $\{i, j\}$, $\{i, j, i + j\}$, $\{O, i, -i, j, -j, i + j\}$. The space of all polynomials $p(t)$ of degree $\leq n$ is spanned by the set of $n + 1$ polynomials

$$\{1, t, t^2, \ldots, t^n\}.$$

It is also spanned by the set $\{1, t/2, t^2/3, \ldots, t^n/(n + 1)\}$, and by $\{1, (1 + t), (1 + t)^2, \ldots, (1 + t)^n\}$. The space of all polynomials is spanned by the infinite set of polynomials $\{1, t, t^2, \ldots\}$.

A number of questions arise naturally at this point. For example, which spaces can be spanned by a finite set of elements? If a space can be spanned by a finite set of elements, what is the smallest number of elements required? To discuss these and related questions, we introduce the concepts of *dependence*, *independence*, *bases*, and *dimension*. These ideas were encountered in Volume I in our study of the vector space V_n. Now we extend them to general linear spaces.

1.7 Dependent and independent sets in a linear space

DEFINITION. *A set S of elements in a linear space V is called dependent if there is a finite set of distinct elements in S, say $x_1, \ldots, x_k$, and a corresponding set of scalars $c_1, \ldots, c_k$, not all zero, such that*

$$\sum_{i=1}^{k} c_i x_i = O .$$

An equation $\sum c_i x_i = O$ with not all $c_i = 0$ is said to be a nontrivial representation of O. The set S is called independent if it is not dependent. In this case, for all choices of distinct elements $x_1, \ldots, x_k$ in S and scalars $c_1, \ldots, c_k$,

$$\sum_{i=1}^{k} c_i x_i = O \qquad \textit{implies} \qquad c_1 = c_2 = \cdots = c_k = 0 .$$

Although dependence and independence are properties of sets of elements, we also apply these terms to the elements themselves. For example, the elements in an independent set are called independent elements.

If S is a finite set, the foregoing definition agrees with that given in Volume I for the space V_n. However, the present definition is not restricted to finite sets.

EXAMPLE 1. If a subset T of a set S is dependent, then S itself is dependent. This is logically equivalent to the statement that every subset of an independent set is independent.

EXAMPLE 2. If one element in S is a scalar multiple of another, then S is dependent.

EXAMPLE 3. If $O \in S$, then S is dependent.

EXAMPLE 4. The empty set is independent.

Many examples of dependent and independent sets of vectors in V_n were discussed in Volume I. The following examples illustrate these concepts in function spaces. In each case the underlying linear space V is the set of all real-valued functions defined on the real line.

EXAMPLE 5. Let $u_1(t) = \cos^2 t$, $u_2(t) = \sin^2 t$, $u_3(t) = 1$ for all real t. The Pythagorean identity shows that $u_1 + u_2 - u_3 = O$, so the three functions u_1, u_2, u_3 are dependent.

EXAMPLE 6. Let $u_k(t) = t^k$ for $k = 0, 1, 2, \ldots$, and t real. The set $S = \{u_0, u_1, u_2, \ldots\}$ is independent. To prove this, it suffices to show that for each n the $n + 1$ polynomials $u_0, u_1, \ldots, u_n$ are independent. A relation of the form $\sum c_k u_k = O$ means that

$$\sum_{k=0}^{n} c_k t^k = 0 \tag{1.1}$$

for all real t. When $t = 0$, this gives $c_0 = 0$. Differentiating (1.1) and setting $t = 0$, we find that $c_1 = 0$. Repeating the process, we find that each coefficient c_k is zero.

EXAMPLE 7. If $a_1, \ldots, a_n$ are distinct real numbers, the n exponential functions

$$u_1(x) = e^{a_1 x}, \ldots, u_n(x) = e^{a_n x}$$

are independent. We can prove this by induction on n. The result holds trivially when $n = 1$. Therefore, assume it is true for $n - 1$ exponential functions and consider scalars $c_1, \ldots, c_n$ such that

$$\sum_{k=1}^{n} c_k e^{a_k x} = 0. \tag{1.2}$$

Let a_M be the largest of the n numbers $a_1, \ldots, a_n$. Multiplying both members of (1.2) by $e^{-a_M x}$, we obtain

$$\sum_{k=1}^{n} c_k e^{(a_k - a_M)x} = 0. \tag{1.3}$$

If $k \neq M$, the number $a_k - a_M$ is negative. Therefore, when $x \to +\infty$ in Equation (1.3), each term with $k \neq M$ tends to zero and we find that $c_M = 0$. Deleting the Mth term from (1.2) and applying the induction hypothesis, we find that each of the remaining $n - 1$ coefficients c_k is zero.

THEOREM 1.5. *Let $S = \{x_1, \ldots, x_k\}$ be an independent set consisting of k elements in a linear space V and let $L(S)$ be the subspace spanned by S. Then every set of $k + 1$ elements in $L(S)$ is dependent.*

Proof. The proof is by induction on k, the number of elements in S. First suppose $k = 1$. Then, by hypothesis, S consists of one element x_1, where $x_1 \neq O$ since S is independent. Now take any two distinct elements y_1 and y_2 in $L(S)$. Then each is a scalar multiple of x_1, say $y_1 = c_1x_1$ and $y_2 = c_2x_1$, where c_1 and c_2 are not both 0. Multiplying y_1 by c_2 and y_2 by c_1 and subtracting, we find that

$$c_2y_1 - c_1y_2 = O.$$

This is a nontrivial representation of O so y_1 and y_2 are dependent. This proves the theorem when $k = 1$.

Now we assume the theorem is true for $k - 1$ and prove that it is also true for k. Take any set of $k + 1$ elements in $L(S)$, say $T = \{y_1, y_2, \ldots, y_{k+1}\}$. We wish to prove that T is dependent. Since each y_i is in $L(S)$ we may write

$$y_i = \sum_{j=1}^{k} a_{ij}x_j \tag{1.4}$$

for each $i = 1, 2, \ldots, k + 1$. We examine all the scalars a_{i1} that multiply x_1 and split the proof into two cases according to whether all these scalars are 0 or not.

CASE 1. $a_{i1} = 0$ for every $i = 1, 2, \ldots, k + 1$. In this case the sum in (1.4) does not involve x_1, so each y_i in T is in the linear span of the set $S' = \{x_2, \ldots, x_k\}$. But S' is independent and consists of $k - 1$ elements. By the induction hypothesis, the theorem is true for $k - 1$ so the set T is dependent. This proves the theorem in Case 1.

CASE 2. Not all the scalars a_{i1} are zero. Let us assume that $a_{11} \neq 0$. (If necessary, we can renumber the y's to achieve this.) Taking $i = 1$ in Equation (1.4) and multiplying both members by c_i, where $c_i = a_{i1}/a_{11}$, we get

$$c_iy_1 = a_{i1}x_1 + \sum_{j=2}^{k} c_ia_{1j}x_j.$$

From this we subtract Equation (1.4) to get

$$c_iy_1 - y_i = \sum_{j=2}^{k} (c_ia_{1j} - a_{ij})x_j,$$

for $i = 2, \ldots, k + 1$. This equation expresses each of the k elements $c_iy_1 - y_i$ as a linear combination of the $k - 1$ independent elements $x_2, \ldots, x_k$. By the induction hypothesis, the k elements $c_iy_1 - y_i$ must be dependent. Hence, for some choice of scalars $t_2, \ldots, t_{k+1}$, not all zero, we have

$$\sum_{i=2}^{k+1} t_i(c_iy_1 - y_i) = O,$$

from which we find

$$\left(\sum_{i=2}^{k+1} t_ic_i\right) y_1 - \sum_{i=2}^{k+1} t_iy_i = O.$$

But this is a nontrivial linear combination of $y_1, \ldots, y_{k+1}$ which represents the zero element, so the elements $y_1, \ldots, y_{k+1}$ must be dependent. This completes the proof.

1.8 Bases and dimension

DEFINITION. *A finite set S of elements in a linear space V is called a finite basis for V if S is independent and spans V. The space V is called finite-dimensional if it has a finite basis, or if V consists of O alone. Otherwise, V is called infinite-dimensional.*

THEOREM 1.6. *Let V be a finite-dimensional linear space. Then every finite basis for V has the same number of elements.*

Proof. Let S and T be two finite bases for V. Suppose S consists of k elements and T consists of m elements. Since S is independent and spans V, Theorem 1.5 tells us that every set of $k + 1$ elements in V is dependent. Therefore, every set of more than k elements in V is dependent. Since T is an independent set, we must have $m \le k$. The same argument with S and T interchanged shows that $k \le m$. Therefore $k = m$.

DEFINITION. *If a linear space V has a basis of n elements, the integer n is called the dimension of V. We write $n = \dim V$. If $V = \{O\}$, we say V has dimension* 0.

EXAMPLE 1. The space V_n has dimension n. One basis is the set of n unit coordinate vectors.

EXAMPLE 2. The space of all polynomials $p(t)$ of degree $\le n$ has dimension $n + 1$. One basis is the set of $n + 1$ polynomials $\{1, t, t^2, \ldots, t^n\}$. Every polynomial of degree $\le n$ is a linear combination of these $n + 1$ polynomials.

EXAMPLE 3. The space of solutions of the differential equation $y'' - 2y' - 3y = 0$ has dimension 2. One basis consists of the two functions $u_1(x) = e^{-x}$, $u_2(x) = e^{3x}$. Every solution is a linear combination of these two.

EXAMPLE 4. The space of all polynomials $p(t)$ is infinite-dimensional. Although the infinite set $\{1, t, t^2, \ldots\}$ spans this space, no *finite* set of polynomials spans the space.

THEOREM 1.7. *Let V be a finite-dimensional linear space with* $\dim V = n$. *Then we have the following:*

(a) *Any set of independent elements in V is a subset of some basis for V.*

(b) *Any set of n independent elements is a basis for V.*

Proof. To prove (a), let $S = \{x_1, \ldots, x_k\}$ be any independent set of elements in V. If $L(S) = V$, then S is a basis. If not, then there is some element y in V which is not in $L(S)$. Adjoin this element to S and consider the new set $S' = \{x_1, \ldots, x_k, y\}$. If this set were dependent there would be scalars $c_1, \ldots, c_{k+1}$, not all zero, such that

$$\sum_{i=1}^{k} c_i x_i + c_{k+1} y = O .$$

But $c_{k+1} \ne 0$ since $x_1, \ldots, x_k$ are independent. Hence, we could solve this equation for y and find that $y \in L(S)$, contradicting the fact that y is not in $L(S)$. Therefore, the set S'

is independent but contains $k + 1$ elements. If $L(S') = V$, then S' is a basis and, since S is a subset of S', part (a) is proved. If S' is not a basis, we can argue with S' as we did with S, getting a new set S'' which contains $k + 2$ elements and is independent. If S'' is a basis, then part (a) is proved. If not, we repeat the process. We must arrive at a basis in a finite number of steps, otherwise we would eventually obtain an independent set with $n + 1$ elements, contradicting Theorem 1.5. Therefore part (a) is proved.

To prove (b), let S be any independent set consisting of n elements. By part (a), S is a subset of some basis, say B. But by Theorem 1.6, the basis B has exactly n elements, so $S = B$.

1.9 Components

Let V be a linear space of dimension n and consider a basis whose elements $e_1, \ldots, e_n$ are taken in a given order. We denote such an ordered basis as an n-tuple $(e_1, \ldots, e_n)$. If $x \in V$, we can express x as a linear combination of these basis elements:

$$x = \sum_{i=1}^{n} c_i e_i. \tag{1.5}$$

The coefficients in this equation determine an n-tuple of numbers $(c_1, \ldots, c_n)$ that is uniquely determined by x. In fact, if we have another representation of x as a linear combination of $e_1, \ldots, e_n$, say $x = \sum_{i=1}^{n} d_i e_i$, then by subtraction from (1.5), we find that $\sum_{i=1}^{n} (c_i - d_i)e_i = O$. But since the basis elements are independent, this implies $c_i = d_i$ for each i, so we have $(c_1, \ldots, c_n) = (d_1, \ldots, d_n)$.

The components of the ordered n-tuple $(c_1, \ldots, c_n)$ determined by Equation (1.5) are called *the components of x relative to the ordered basis* $(e_1, \ldots, e_n)$.

1.10 Exercises

In each of Exercises 1 through 10, let S denote the set of all vectors (x, y, z) in V_3 whose components satisfy the condition given. Determine whether S is a subspace of V_3. If S is a subspace, compute dim S.

1. $x = 0$.
2. $x + y = 0$.
3. $x + y + z = 0$.
4. $x = y$.
5. $x = y = z$.
6. $x = y$ or $x = z$.
7. $x^2 - y^2 = 0$.
8. $x + y = 1$.
9. $y = 2x$ and $z = 3x$.
10. $x + y + z = 0$ and $x - y - z = 0$.

Let P_n denote the linear space of all real polynomials of degree $\le n$, where n is fixed. In each of Exercises 11 through 20, let S denote the set of all polynomials f in P_n satisfying the condition given. Determine whether or not S is a subspace of P_n. If S is a subspace, compute dim S.

11. $f(0) = 0$.
12. $f'(0) = 0$.
13. $f''(0) = 0$.
14. $f(0) + f'(0) = 0$.
15. $f(0) = f(1)$.
16. $f(0) = f(2)$.
17. f is even.
18. f is odd.
19. f has degree $\le k$, where $k < n$, or $f = 0$.
20. f has degree k, where $k < n$, or $f = 0$.
21. In the linear space of all real polynomials $p(t)$, describe the subspace spanned by each of the following subsets of polynomials and determine the dimension of this subspace.
(a) $\{1, t^2, t^4\}$; (b) $\{t, t^3, t^5\}$; (c) $\{t, t^2\}$; (d) $\{1 + t, (1 + t)^2\}$.

22. In this exercise, $L(S)$ denotes the subspace spanned by a subset S of a linear space V. Prove each of the statements (a) through (f).
    (a) $S \subseteq L(S)$.
    (b) If $S \subseteq T \subseteq V$ and if T is a subspace of V, then $L(S) \subseteq T$. This property is described by saying that $L(S)$ is the *smallest* subspace of V which contains S.
    (c) A subset S of V is a subspace of V if and only if $L(S) = S$.
    (d) If $S \subseteq T \subseteq V$, then $L(S) \subseteq L(T)$.
    (e) If S and T are subspaces of V, then so is $S \cap T$.
    (f) If S and T are subsets of V, then $L(S \cap T) \subseteq L(S) \cap L(T)$.
    (g) Give an example in which $L(S \cap T) \neq L(S) \cap L(T)$.
23. Let V be the linear space consisting of all real-valued functions defined on the real line. Determine whether each of the following subsets of V is dependent or independent. Compute the dimension of the subspace spanned by each set.
    (a) $\{1, e^{ax}, e^{bx}\}$, $a \neq b$.
    (b) $\{e^{ax}, xe^{ax}\}$.
    (c) $\{1, e^{ax}, xe^{ax}\}$.
    (d) $\{e^{ax}, xe^{ax}, x^2e^{ax}\}$.
    (e) $\{e^x, e^{-x}, \cosh x\}$.
    (f) $\{\cos x, \sin x\}$.
    (g) $\{\cos^2 x, \sin^2 x\}$.
    (h) $\{1, \cos 2x, \sin^2 x\}$.
    (i) $\{\sin x, \sin 2x\}$.
    (j) $\{e^x \cos x, e^{-x} \sin x\}$.
24. Let V be a finite-dimensional linear space, and let S be a subspace of V. Prove each of the following statements.
    (a) S is finite dimensional and $\dim S \leq \dim V$.
    (b) $\dim S = \dim V$ if and only if $S = V$.
    (c) Every basis for S is part of a basis for V.
    (d) A basis for V need not contain a basis for S.

1.11 Inner products, Euclidean spaces. Norms

In ordinary Euclidean geometry, those properties that rely on the possibility of measuring lengths of line segments and angles between lines are called *metric* properties. In our study of V_n, we defined lengths and angles in terms of the dot product. Now we wish to extend these ideas to more general linear spaces. We shall introduce first a generalization of the dot product, which we call an *inner product*, and then define length and angle in terms of the inner product.

The dot product $x \cdot y$ of two vectors $x = (x_1, \ldots, x_n)$ and $y = (y_1, \ldots, y_n)$ in V_n was defined in Volume I by the formula

$$x \cdot y = \sum_{i=1}^{n} x_i y_i . \tag{1.6}$$

In a general linear space, we write (x, y) instead of $x \cdot y$ for inner products, and we define the product axiomatically rather than by a specific formula. That is, we state a number of properties we wish inner products to satisfy and we regard these properties as *axioms*.

DEFINITION. *A real linear space V is said to have an inner product if for each pair of elements x and y in V there corresponds a unique real number (x, y) satisfying the following axioms for all choices of x, y, z in V and all real scalars c.*

(1) $(x, y) = (y, x)$ *(commutativity, or symmetry).*
(2) $(x, y + z) = (x, y) + (x, z)$ *(distributivity, or linearity).*
(3) $c(x, y) = (cx, y)$ *(associativity, or homogeneity).*
(4) $(x, x) > 0$ *if* $x \neq O$ *(positivity).*

A real linear space with an inner product is called a *real Euclidean space.*

Note: Taking $c = 0$ in (3), we find that $(O, y) = 0$ for all y.

In a complex linear space, an inner product (x, y) is a complex number satisfying the same axioms as those for a real inner product, except that the symmetry axiom is replaced by the relation

$$(1') \qquad (x, y) = \overline{(y, x)}, \qquad \textit{(Hermitian† symmetry)}$$

where $\overline{(y, x)}$ denotes the complex conjugate of (y, x). In the homogeneity axiom, the scalar multiplier c can be any complex number. From the homogeneity axiom and (1'), we get the companion relation

$$(3') \qquad (x, cy) = \overline{(cy, x)} = \bar{c}\overline{(y, x)} = \bar{c}(x, y).$$

A complex linear space with an inner product is called a *complex Euclidean space.* (Sometimes the term *unitary space* is also used.) One example is complex vector space $V_n(\mathbf{C})$ discussed briefly in Section 12.16 of Volume I.

Although we are interested primarily in examples of real Euclidean spaces, the theorems of this chapter are valid for complex Euclidean spaces as well. When we use the term Euclidean space without further designation, it is to be understood that the space can be real or complex.

The reader should verify that each of the following satisfies all the axioms for an inner product.

EXAMPLE 1. In V_n let $(x, y) = x \cdot y$, the usual dot product of x and y.

EXAMPLE 2. If $x = (x_1, x_2)$ and $y = (y_1, y_2)$ are any two vectors in V_2, define (x, y) by the formula

$$(x, y) = 2x_1y_1 + x_1y_2 + x_2y_1 + x_2y_2.$$

This example shows that there may be more than one inner product in a given linear space.

EXAMPLE 3. Let $C(a, b)$ denote the linear space of all real-valued functions continuous on an interval $[a, b]$. Define an inner product of two functions f and g by the formula

$$(f, g) = \int_a^b f(t)g(t)\,dt.$$

This formula is analogous to Equation (1.6) which defines the dot product of two vectors in V_n. The function values $f(t)$ and $g(t)$ play the role of the components x_i and y_i, and integration takes the place of summation.

† In honor of Charles Hermite (1822–1901), a French mathematician who made many contributions to algebra and analysis.

EXAMPLE 4. In the space $C(a, b)$, define

$$(f, g) = \int_a^b w(t)f(t)g(t)\,dt,$$

where w is a fixed positive function in $C(a, b)$. The function w is called a *weight function.* In Example 3 we have $w(t) = 1$ for all t.

EXAMPLE 5. In the linear space of all real polynomials, define

$$(f, g) = \int_0^\infty e^{-t}f(t)g(t)\,dt.$$

Because of the exponential factor, this improper integral converges for every choice of polynomials f and g.

THEOREM 1.8. *In a Euclidean space V, every inner product satisfies the Cauchy-Schwarz inequality:*

$$|(x, y)|^2 \le (x, x)(y, y) \qquad \textit{for all } x \textit{ and } y \textit{ in } V.$$

Moreover, the equality sign holds if and only if x and y are dependent.

Proof. If either $x = O$ or $y = O$ the result holds trivially, so we can assume that both x and y are nonzero. Let $z = ax + by$, where a and b are scalars to be specified later. We have the inequality $(z, z) \ge 0$ for all a and b. When we express this inequality in terms of x and y with an appropriate choice of a and b we will obtain the Cauchy-Schwarz inequality.

To express (z, z) in terms of x and y we use properties (1′), (2) and (3′) to obtain

$$\begin{aligned}(z, z) &= (ax + by, ax + by) = (ax, ax) + (ax, by) + (by, ax) + (by, by)\\ &= a\bar{a}(x, x) + a\bar{b}(x, y) + b\bar{a}(y, x) + b\bar{b}(y, y) \ge 0.\end{aligned}$$

Taking $a = (y, y)$ and cancelling the positive factor (y, y) in the inequality we obtain

$$(y, y)(x, x) + \bar{b}(x, y) + b(y, x) + b\bar{b} \ge 0.$$

Now we take $b = -(x, y)$. Then $\bar{b} = -(y, x)$ and the last inequality simplifies to

$$(y, y)(x, x) \ge (x, y)(y, x) = |(x, y)|^2$$

This proves the Cauchy-Schwarz inequality. The equality sign holds throughout the proof if and only if $z = O$. This holds, in turn, if and only if x and y are dependent.

EXAMPLE. Applying Theorem 1.8 to the space $C(a, b)$ with the inner product $(f, g) = \int_a^b f(t)g(t)\,dt$, we find that the Cauchy-Schwarz inequality becomes

$$\left(\int_a^b f(t)g(t)\,dt\right)^2 \le \left(\int_a^b f^2(t)\,dt\right)\left(\int_a^b g^2(t)\,dt\right).$$

The inner product can be used to introduce the metric concept of length in any Euclidean space.

DEFINITION. *In a Euclidean space V, the nonnegative number* $\|x\|$ *defined by the equation*

$$\|x\| = (x, x)^{1/2}$$

is called the norm of the element x.

When the Cauchy-Schwarz inequality is expressed in terms of norms, it becomes

$$|(x, y)| \le \|x\| \, \|y\| .$$

Since it may be possible to define an inner product in many different ways, the norm of an element will depend on the choice of inner product. This lack of uniqueness is to be expected. It is analogous to the fact that we can assign different numbers to measure the length of a given line segment, depending on the choice of scale or unit of measurement. The next theorem gives fundamental properties of norms that do not depend on the choice of inner product.

THEOREM 1.9. *In a Euclidean space, every norm has the following properties for all elements x and y and all scalars c:*

(a) $\|x\| = 0$ *if* $x = O$.
(b) $\|x\| > 0$ *if* $x \ne O$ *(positivity).*
(c) $\|cx\| = |c| \, \|x\|$ *(homogeneity).*
(d) $\|x + y\| \le \|x\| + \|y\|$ *(triangle inequality).*

The equality sign holds in (d) *if* $x = O$, *if* $y = O$, *or if* $y = cx$ *for some* $c > 0$.

Proof. Properties (a), (b) and (c) follow at once from the axioms for an inner product. To prove (d), we note that

$$\|x + y\|^2 = (x + y, x + y) = (x, x) + (y, y) + (x, y) + (y, x)$$
$$= \|x\|^2 + \|y\|^2 + (x, y) + \overline{(x, y)} .$$

The sum $(x, y) + \overline{(x, y)}$ is real. The Cauchy-Schwarz inequality shows that $|(x, y)| \le \|x\| \, \|y\|$ and $|\overline{(x, y)}| \le \|x\| \, \|y\|$, so we have

$$\|x + y\|^2 \le \|x\|^2 + \|y\|^2 + 2\|x\| \, \|y\| = (\|x\| + \|y\|)^2 .$$

This proves (d). When $y = cx$, where $c > 0$, we have

$$\|x + y\| = \|x + cx\| = (1 + c) \, \|x\| = \|x\| + \|cx\| = \|x\| + \|y\| .$$

DEFINITION. *In a real Euclidean space V, the angle between two nonzero elements x and y is defined to be that number θ in the interval $0 \le \theta \le \pi$ which satisfies the equation*

$$\cos\theta = \frac{(x, y)}{\|x\|\,\|y\|}. \tag{1.7}$$

Note: The Cauchy-Schwarz inequality shows that the quotient on the right of (1.7) lies in the interval $[-1, 1]$, so there is exactly one θ in $[0, \pi]$ whose cosine is equal to this quotient.

1.12 Orthogonality in a Euclidean space

DEFINITION. *In a Euclidean space V, two elements x and y are called orthogonal if their inner product is zero. A subset S of V is called an orthogonal set if $(x, y) = 0$ for every pair of distinct elements x and y in S. An orthogonal set is called orthonormal if each of its elements has norm 1.*

The zero element is orthogonal to every element of V; it is the only element orthogonal to itself. The next theorem shows a relation between orthogonality and independence.

THEOREM 1.10. *In a Euclidean space V, every orthogonal set of nonzero elements is independent. In particular, in a finite-dimensional Euclidean space with $\dim V = n$, every orthogonal set consisting of n nonzero elements is a basis for V.*

Proof. Let S be an orthogonal set of nonzero elements in V, and suppose some finite linear combination of elements of S is zero, say

$$\sum_{i=1}^{k} c_i x_i = O,$$

where each $x_i \in S$. Taking the inner product of each member with x_1 and using the fact that $(x_1, x_i) = 0$ if $i \ne 1$, we find that $c_1(x_1, x_1) = 0$. But $(x_1, x_1) \ne 0$ since $x_1 \ne O$ so $c_1 = 0$. Repeating the argument with x_1 replaced by x_j, we find that each $c_j = 0$. This proves that S is independent. If $\dim V = n$ and if S consists of n elements, Theorem 1.7(b) shows that S is a basis for V.

EXAMPLE. In the real linear space $C(0, 2\pi)$ with the inner product $(f, g) = \int_0^{2\pi} f(x)g(x)\,dx$, let S be the set of trigonometric functions $\{u_0, u_1, u_2, \ldots\}$ given by

$$u_0(x) = 1, \qquad u_{2n-1}(x) = \cos nx, \qquad u_{2n}(x) = \sin nx, \qquad \text{for } n = 1, 2, \ldots.$$

If $m \ne n$, we have the orthogonality relations

$$\int_0^{2\pi} u_n(x)u_m(x)\,dx = 0,$$

so S is an orthogonal set. Since no member of S is the zero element, S is independent. The norm of each element of S is easily calculated. We have $(u_0, u_0) = \int_0^{2\pi} dx = 2\pi$ and, for $n \geq 1$, we have

$$(u_{2n-1}, u_{2n-1}) = \int_0^{2\pi} \cos^2 nx\, dx = \pi, \qquad (u_{2n}, u_{2n}) = \int_0^{2\pi} \sin^2 nx\, dx = \pi.$$

Therefore, $\|u_0\| = \sqrt{2\pi}$ and $\|u_n\| = \sqrt{\pi}$ for $n \geq 1$. Dividing each u_n by its norm, we obtain an orthonormal set $\{\varphi_0, \varphi_1, \varphi_2, \ldots\}$ where $\varphi_n = u_n/\|u_n\|$. Thus, we have

$$\varphi_0(x) = \frac{1}{\sqrt{2\pi}}, \qquad \varphi_{2n-1}(x) = \frac{\cos nx}{\sqrt{\pi}}, \qquad \varphi_{2n}(x) = \frac{\sin nx}{\sqrt{\pi}}, \qquad \text{for} \quad n \geq 1.$$

In Section 1.14 we shall prove that every finite-dimensional Euclidean space has an orthogonal basis. The next theorem shows how to compute the components of an element relative to such a basis.

THEOREM 1.11. *Let V be a finite-dimensional Euclidean space with dimension n, and assume that $S = \{e_1, \ldots, e_n\}$ is an orthogonal basis for V. If an element x is expressed as a linear combination of the basis elements, say*

$$x = \sum_{i=1}^{n} c_i e_i, \tag{1.8}$$

then its components relative to the ordered basis $(e_1, \ldots, e_n)$ are given by the formulas

$$c_j = \frac{(x, e_j)}{(e_j, e_j)} \qquad \textit{for} \quad j = 1, 2, \ldots, n. \tag{1.9}$$

In particular, if S is an orthonormal basis, each c_j is given by

$$c_j = (x, e_j). \tag{1.10}$$

Proof. Taking the inner product of each member of (1.8) with e_j, we obtain

$$(x, e_j) = \sum_{i=1}^{n} c_i(e_i, e_j) = c_j(e_j, e_j)$$

since $(e_i, e_j) = 0$ if $i \neq j$. This implies (1.9), and when $(e_j, e_j) = 1$, we obtain (1.10).

If $\{e_1, \ldots, e_n\}$ is an orthonormal basis, Equation (1.9) can be written in the form

$$x = \sum_{i=1}^{n} (x, e_i)e_i. \tag{1.11}$$

The next theorem shows that in a finite-dimensional Euclidean space with an orthonormal basis the inner product of two elements can be computed in terms of their components.

THEOREM 1.12. *Let V be a finite-dimensional Euclidean space of dimension n, and assume that $\{e_1, \ldots, e_n\}$ is an orthonormal basis for V. Then for every pair of elements x and y in V, we have*

$$(1.12) \qquad (x, y) = \sum_{i=1}^{n} (x, e_i)\overline{(y, e_i)} \qquad \textit{(Parseval's formula)}.$$

In particular, when $x = y$, we have

$$(1.13) \qquad \|x\|^2 = \sum_{i=1}^{n} |(x, e_i)|^2.$$

Proof. Taking the inner product of both members of Equation (1.11) with y and using the linearity property of the inner product, we obtain (1.12). When $x = y$, Equation (1.12) reduces to (1.13).

Note: Equation (1.12) is named in honor of M. A. Parseval (circa 1776–1836), who obtained this type of formula in a special function space. Equation (1.13) is a generalization of the theorem of Pythagoras.

1.13 Exercises

1. Let $x = (x_1, \ldots, x_n)$ and $y = (y_1, \ldots, y_n)$ be arbitrary vectors in V_n. In each case, determine whether (x, y) is an inner product for V_n if (x, y) is defined by the formula given. In case (x, y) is not an inner product, tell which axioms are not satisfied.

(a) $(x, y) = \sum_{i=1}^{n} x_i |y_i|$.

(b) $(x, y) = \left|\sum_{i=1}^{n} x_i y_i\right|$.

(c) $(x, y) = \sum_{i=1}^{n} x_i \sum_{j=1}^{n} y_j$.

(d) $(x, y) = \left(\sum_{i=1}^{n} x_i^2 y_i^2\right)^{1/2}$.

(e) $(x, y) = \sum_{i=1}^{n} (x_i + y_i)^2 - \sum_{i=1}^{n} x_i^2 - \sum_{i=1}^{n} y_i^2$.

2. Suppose we retain the first three axioms for a real inner product (symmetry, linearity, and homogeneity but replace the fourth axiom by a new axiom (4′): $(x, x) = 0$ if and only if $x = O$. Prove that either $(x, x) > 0$ for all $x \neq O$ or else $(x, x) < 0$ for all $x \neq O$.

[*Hint:* Assume $(x, x) > 0$ for some $x \neq O$ and $(y, y) < 0$ for some $y \neq O$. In the space spanned by $\{x, y\}$, find an element $z \neq O$ with $(z, z) = 0$.]

Prove that each of the statements in Exercises 3 through 7 is valid for all elements x and y in a real Euclidean space.

3. $(x, y) = 0$ if and only if $\|x + y\| = \|x - y\|$.
4. $(x, y) = 0$ if and only if $\|x + y\|^2 = \|x\|^2 + \|y\|^2$.
5. $(x, y) = 0$ if and only if $\|x + cy\| \geq \|x\|$ for all real c.
6. $(x + y, x - y) = 0$ if and only if $\|x\| = \|y\|$.
7. If x and y are nonzero elements making an angle θ with each other, then

$$\|x - y\|^2 = \|x\|^2 + \|y\|^2 - 2\,\|x\|\,\|y\| \cos\theta.$$

8. In the real linear space $C(1, e)$, define an inner product by the equation

$$(f, g) = \int_1^e (\log x) f(x) g(x)\, dx .$$

(a) If $f(x) = \sqrt{x}$, compute $\|f\|$.
(b) Find a linear polynomial $g(x) = a + bx$ that is orthogonal to the constant function $f(x) = 1$.

9. In the real linear space $C(-1, 1)$, let $(f, g) = \int_{-1}^{1} f(t)g(t)\, dt$. Consider the three functions u_1, u_2, u_3 given by

$$u_1(t) = 1 , \qquad u_2(t) = t , \qquad u_3(t) = 1 + t .$$

Prove that two of them are orthogonal, two make an angle $\pi/3$ with each other, and two make an angle $\pi/6$ with each other.

10. In the linear space P_n of all real polynomials of degree $\leq n$, define

$$(f, g) = \sum_{k=0}^{n} f\left(\frac{k}{n}\right) g\left(\frac{k}{n}\right).$$

(a) Prove that (f, g) is an inner product for P_n.
(b) Compute (f, g) when $f(t) = t$ and $g(t) = at + b$.
(c) If $f(t) = t$, find all linear polynomials g orthogonal to f.

11. In the linear space of all real polynomials, define $(f, g) = \int_0^\infty e^{-t} f(t)g(t)\, dt$.
(a) Prove that this improper integral converges absolutely for all polynomials f and g.
(b) If $x_n(t) = t^n$ for $n = 0, 1, 2, \ldots$, prove that $(x_n, x_m) = (m + n)!$.
(c) Compute (f, g) when $f(t) = (t + 1)^2$ and $g(t) = t^2 + 1$.
(d) Find all linear polynomials $g(t) = a + bt$ orthogonal to $f(t) = 1 + t$.

12. In the linear space of all real polynomials, determine whether or not (f, g) is an inner product if (f, g) is defined by the formula given. In case (f, g) is not an inner product, indicate which axioms are violated. In (c), f' and g' denote derivatives.

(a) $(f, g) = f(1)g(1)$.
(b) $(f, g) = \left| \int_0^1 f(t)g(t)\, dt \right|$.
(c) $(f, g) = \int_0^1 f'(t)g'(t)\, dt$.
(d) $(f, g) = \left(\int_0^1 f(t)\, dt\right)\left(\int_0^1 g(t)\, dt\right)$.

13. Let V consist of all infinite sequences $\{x_n\}$ of real numbers for which the series Σx_n^2 converges. If $x = \{x_n\}$ and $y = \{y_n\}$ are two elements of V, define

$$(x, y) = \sum_{n=1}^{\infty} x_n y_n .$$

(a) Prove that this series converges absolutely.

[*Hint:* Use the Cauchy-Schwarz inequality to estimate the sum $\sum_{n=1}^{M} |x_n y_n|$.]

(b) Prove that V is a linear space with (x, y) as an inner product.
(c) Compute (x, y) if $x_n = 1/n$ and $y_n = 1/(n + 1)$ for $n \geq 1$.
(d) Compute (x, y) if $x_n = 2^n$ and $y_n = 1/n!$ for $n \geq 1$.

14. Let V be the set of all real functions f continuous on $[0, +\infty)$ and such that the integral $\int_0^\infty e^{-t} f^2(t)\, dt$ converges. Define $(f, g) = \int_0^\infty e^{-t} f(t)g(t)\, dt$.

(a) Prove that the integral for (f, g) converges absolutely for each pair of functions f and g in V.

[*Hint:* Use the Cauchy-Schwarz inequality to estimate the integral $\int_0^M e^{-t} |f(t)g(t)|\, dt$.]

(b) Prove that V is a linear space with (f, g) as an inner product.
(c) Compute (f, g) if $f(t) = e^{-t}$ and $g(t) = t^n$, where $n = 0, 1, 2, \ldots$.

15. In a complex Euclidean space, prove that the inner product has the following properties for all elements x, y and z, and all complex a and b.
(a) $(ax, by) = a\bar{b}(x, y)$. (b) $(x, ay + bz) = \bar{a}(x, y) + \bar{b}(x, z)$.
16. Prove that the following identities are valid in every Euclidean space.
(a) $\|x + y\|^2 = \|x\|^2 + \|y\|^2 + (x, y) + (y, x)$.
(b) $\|x + y\|^2 - \|x - y\|^2 = 2(x, y) + 2(y, x)$.
(c) $\|x + y\|^2 + \|x - y\|^2 = 2\|x\|^2 + 2\|y\|^2$.
17. Prove that the space of all complex-valued functions continuous on an interval $[a, b]$ becomes a unitary space if we define an inner product by the formula

$$(f, g) = \int_a^b w(t) f(t) \overline{g(t)}\, dt,$$

where w is a fixed positive function, continuous on $[a, b]$.

1.14 Construction of orthogonal sets. The Gram-Schmidt process

Every finite-dimensional linear space has a finite basis. If the space is Euclidean, we can always construct an *orthogonal* basis. This result will be deduced as a consequence of a general theorem whose proof shows how to construct orthogonal sets in any Euclidean space, finite or infinite dimensional. The construction is called the *Gram-Schmidt orthogonalization process*, in honor of J. P. Gram (1850–1916) and E. Schmidt (1845–1921).

THEOREM 1.13. ORTHOGONALIZATION THEOREM. *Let $x_1, x_2, \ldots,$ be a finite or infinite sequence of elements in a Euclidean space V, and let $L(x_1, \ldots, x_k)$ denote the subspace spanned by the first k of these elements. Then there is a corresponding sequence of elements $y_1, y_2, \ldots,$ in V which has the following properties for each integer k:*

(a) *The element y_k is orthogonal to every element in the subspace $L(y_1, \ldots, y_{k-1})$.*
(b) *The subspace spanned by $y_1, \ldots, y_k$ is the same as that spanned by $x_1, \ldots, x_k$:*

$$L(y_1, \ldots, y_k) = L(x_1, \ldots, x_k).$$

(c) *The sequence $y_1, y_2, \ldots,$ is unique, except for scalar factors. That is, if $y_1', y_2', \ldots,$ is another sequence of elements in V satisfying properties* (a) *and* (b) *for all k, then for each k there is a scalar c_k such that $y_k' = c_k y_k$.*

Proof. We construct the elements $y_1, y_2, \ldots,$ by induction. To start the process, we take $y_1 = x_1$. Now assume we have constructed $y_1, \ldots, y_r$ so that (a) and (b) are satisfied when $k = r$. Then we define y_{r+1} by the equation

$$y_{r+1} = x_{r+1} - \sum_{i=1}^{r} a_i y_i, \tag{1.14}$$

where the scalars $a_1, \ldots, a_r$ are to be determined. For $j \leq r$, the inner product of y_{r+1} with y_j is given by

$$(y_{r+1}, y_j) = (x_{r+1}, y_j) - \sum_{i=1}^{r} a_i(y_i, y_j) = (x_{r+1}, y_j) - a_j(y_j, y_j),$$

since $(y_i, y_j) = 0$ if $i \neq j$. If $y_j \neq O$, we can make y_{r+1} orthogonal to y_j by taking

$$a_j = \frac{(x_{r+1}, y_j)}{(y_j, y_j)}. \tag{1.15}$$

If $y_j = O$, then y_{r+1} is orthogonal to y_j for any choice of a_j, and in this case we choose $a_j = 0$. Thus, the element y_{r+1} is well defined and is orthogonal to each of the earlier elements $y_1, \ldots, y_r$. Therefore, it is orthogonal to every element in the subspace

$$L(y_1, \ldots, y_r).$$

This proves (a) when $k = r + 1$.

To prove (b) when $k = r + 1$, we must show that $L(y_1, \ldots, y_{r+1}) = L(x_1, \ldots, x_{r+1})$, given that $L(y_1, \ldots, y_r) = L(x_1, \ldots, x_r)$. The first r elements $y_1, \ldots, y_r$ are in

$$L(x_1, \ldots, x_r)$$

and hence they are in the larger subspace $L(x_1, \ldots, x_{r+1})$. The new element y_{r+1} given by (1.14) is a difference of two elements in $L(x_1, \ldots, x_{r+1})$ so it, too, is in $L(x_1, \ldots, x_{r+1})$. This proves that

$$L(y_1, \ldots, y_{r+1}) \subseteq L(x_1, \ldots, x_{r+1}).$$

Equation (1.14) shows that x_{r+1} is the sum of two elements in $L(y_1, \ldots, y_{r+1})$ so a similar argument gives the inclusion in the other direction:

$$L(x_1, \ldots, x_{r+1}) \subseteq L(y_1, \ldots, y_{r+1}).$$

This proves (b) when $k = r + 1$. Therefore both (a) and (b) are proved by induction on k.

Finally we prove (c) by induction on k. The case $k = 1$ is trivial. Therefore, assume (c) is true for $k = r$ and consider the element y'_{r+1}. Because of (b), this element is in

$$L(y_1, \ldots, y_{r+1}),$$

so we can write

$$y'_{r+1} = \sum_{i=1}^{r+1} c_i y_i = z_r + c_{r+1} y_{r+1},$$

where $z_r \in L(y_1, \ldots, y_r)$. We wish to prove that $z_r = O$. By property (a), both y'_{r+1} and $c_{r+1} y_{r+1}$ are orthogonal to z_r. Therefore, their difference, z_r, is orthogonal to z_r. In other words, z_r is orthogonal to itself, so $z_r = O$. This completes the proof of the orthogonalization theorem.

In the foregoing construction, suppose we have $y_{r+1} = O$ for some r. Then (1.14) shows that x_{r+1} is a linear combination of $y_1, \ldots, y_r$, and hence of $x_1, \ldots, x_r$, so the elements $x_1, \ldots, x_{r+1}$ are dependent. In other words, if the first k elements $x_1, \ldots, x_k$ are independent, then the corresponding elements $y_1, \ldots, y_k$ are *nonzero*. In this case the coefficients a_i in (1.14) are given by (1.15), and the formulas defining $y_1, \ldots, y_k$ become

$$(1.16) \quad y_1 = x_1, \qquad y_{r+1} = x_{r+1} - \sum_{i=1}^{r} \frac{(x_{r+1}, y_i)}{(y_i, y_i)} y_i \qquad \text{for} \quad r = 1, 2, \ldots, k-1.$$

These formulas describe the Gram-Schmidt process for constructing an orthogonal set of nonzero elements $y_1, \ldots, y_k$ which spans the same subspace as a given independent set $x_1, \ldots, x_k$. In particular, if $x_1, \ldots, x_k$ is a basis for a finite-dimensional Euclidean space, then $y_1, \ldots, y_k$ is an orthogonal basis for the same space. We can also convert this to an orthonormal basis by *normalizing* each element y_i, that is, by dividing it by its norm. Therefore, as a corollary of Theorem 1.13 we have the following.

THEOREM 1.14. *Every finite-dimensional Euclidean space has an orthonormal basis.*

If x and y are elements in a Euclidean space, with $y \neq O$, the element

$$\frac{(x, y)}{(y, y)} y$$

is called the *projection of x along y*. In the Gram-Schmidt process (1.16), we construct the element y_{r+1} by subtracting from x_{r+1} the projection of x_{r+1} along each of the earlier elements $y_1, \ldots, y_r$. Figure 1.1 illustrates the construction geometrically in the vector space V_3.

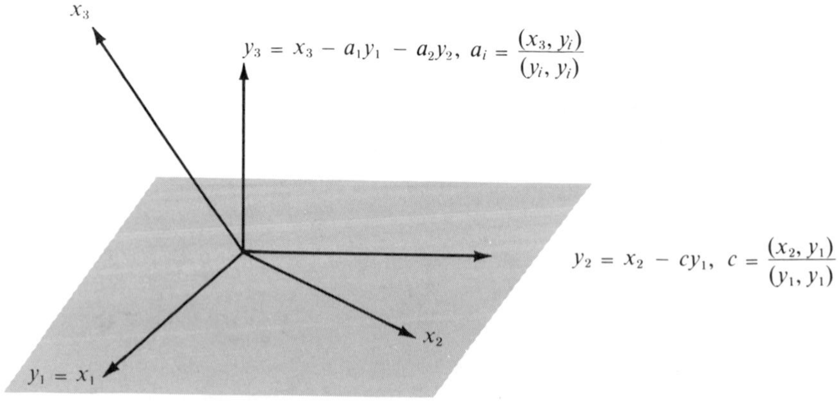

FIGURE 1.1 The Gram-Schmidt process in V_3. An orthogonal set $\{y_1, y_2, y_3\}$ is constructed from a given independent set $\{x_1, x_2, x_3\}$.

EXAMPLE 1. In V_4, find an orthonormal basis for the subspace spanned by the three vectors $x_1 = (1, -1, 1, -1)$, $x_2 = (5, 1, 1, 1)$, and $x_3 = (-3, -3, 1, -3)$.

Solution. Applying the Gram-Schmidt process, we find

$$y_1 = x_1 = (1, -1, 1, -1),$$

$$y_2 = x_2 - \frac{(x_2, y_1)}{(y_1, y_1)} y_1 = x_2 - y_1 = (4, 2, 0, 2),$$

$$y_3 = x_3 - \frac{(x_3, y_1)}{(y_1, y_1)} y_1 - \frac{(x_3, y_2)}{(y_2, y_2)} y_2 = x_3 - y_1 + y_2 = (0, 0, 0, 0).$$

Since $y_3 = O$, the three vectors x_1, x_2, x_3 must be dependent. But since y_1 and y_2 are nonzero, the vectors x_1 and x_2 are independent. Therefore $L(x_1, x_2, x_3)$ is a subspace of dimension 2. The set $\{y_1, y_2\}$ is an orthogonal basis for this subspace. Dividing each of y_1 and y_2 by its norm we get an orthonormal basis consisting of the two vectors

$$\frac{y_1}{\|y_1\|} = \tfrac{1}{2}(1, -1, 1, -1) \quad \text{and} \quad \frac{y_2}{\|y_2\|} = \frac{1}{\sqrt{6}}(2, 1, 0, 1).$$

EXAMPLE 2. *The Legendre polynomials.* In the linear space of all polynomials, with the inner product $(x, y) = \int_{-1}^{1} x(t)y(t)\,dt$, consider the infinite sequence $x_0, x_1, x_2, \ldots$, where $x_n(t) = t^n$. When the orthogonalization theorem is applied to this sequence it yields another sequence of polynomials $y_0, y_1, y_2, \ldots$, first encountered by the French mathematician A. M. Legendre (1752–1833) in his work on potential theory. The first few polynomials are easily calculated by the Gram-Schmidt process. First of all, we have $y_0(t) = x_0(t) = 1$. Since

$$(y_0, y_0) = \int_{-1}^{1} dt = 2 \quad \text{and} \quad (x_1, y_0) = \int_{-1}^{1} t\,dt = 0,$$

we find that

$$y_1(t) = x_1(t) - \frac{(x_1, y_0)}{(y_0, y_0)} y_0(t) = x_1(t) = t.$$

Next, we use the relations

$$(x_2, y_0) = \int_{-1}^{1} t^2\,dt = \tfrac{2}{3}, \qquad (x_2, y_1) = \int_{-1}^{1} t^3\,dt = 0, \qquad (y_1, y_1) = \int_{-1}^{1} t^2\,dt = \tfrac{2}{3},$$

to obtain

$$y_2(t) = x_2(t) - \frac{(x_2, y_0)}{(y_0, y_0)} y_0(t) - \frac{(x_2, y_1)}{(y_1, y_1)} y_1(t) = t^2 - \tfrac{1}{3}.$$

Similarly, we find that

$$y_3(t) = t^3 - \tfrac{3}{5}t, \qquad y_4(t) = t^4 - \tfrac{6}{7}t^2 + \tfrac{3}{35}, \qquad y_5(t) = t^5 - \tfrac{10}{9}t^3 + \tfrac{5}{21}t.$$

We shall encounter these polynomials again in Chapter 6 in our further study of differential equations, and we shall prove that

$$y_n(t) = \frac{n!}{(2n)!} \frac{d^n}{dt^n} (t^2 - 1)^n .$$

The polynomials P_n given by

$$P_n(t) = \frac{(2n)!}{2^n(n!)^2} y_n(t) = \frac{1}{2^n n!} \frac{d^n}{dt^n} (t^2 - 1)^n$$

are known as the *Legendre polynomials*. The polynomials in the corresponding orthonormal sequence $\varphi_0, \varphi_1, \varphi_2, \ldots$, given by $\varphi_n = y_n / \|y_n\|$ are called the *normalized Legendre polynomials*. From the formulas for $y_0, \ldots, y_5$ given above, we find that

$$\varphi_0(t) = \sqrt{\tfrac{1}{2}}\,, \qquad \varphi_1(t) = \sqrt{\tfrac{3}{2}}\, t\,, \qquad \varphi_2(t) = \tfrac{1}{2}\sqrt{\tfrac{5}{2}}\,(3t^2 - 1)\,, \qquad \varphi_3(t) = \tfrac{1}{2}\sqrt{\tfrac{7}{2}}\,(5t^3 - 3t)\,,$$

$$\varphi_4(t) = \tfrac{1}{8}\sqrt{\tfrac{9}{2}}\,(35t^4 - 30t^2 + 3)\,, \qquad \varphi_5(t) = \tfrac{1}{8}\sqrt{\tfrac{11}{2}}\,(63t^5 - 70t^3 + 15t)\,.$$

1.15. Orthogonal complements. Projections

Let V be a Euclidean space and let S be a finite-dimensional subspace. We wish to consider the following type of approximation problem: *Given an element x in V, to determine an element in S whose distance from x is as small as possible.* The distance between two elements x and y is defined to be the norm $\|x - y\|$.

Before discussing this problem in its general form, we consider a special case, illustrated in Figure 1.2. Here V is the vector space V_3 and S is a two-dimensional subspace, a plane through the origin. Given x in V, the problem is to find, in the plane S, that point s nearest to x.

If $x \in S$, then clearly $s = x$ is the solution. If x is not in S, then the nearest point s is obtained by dropping a perpendicular from x to the plane. This simple example suggests an approach to the general approximation problem and motivates the discussion that follows.

DEFINITION. *Let S be a subset of a Euclidean space V. An element in V is said to be orthogonal to S if it is orthogonal to every element of S. The set of all elements orthogonal to S is denoted by $S^\perp$ and is called "S perpendicular."*

It is a simple exercise to verify that $S^\perp$ is a subspace of V, whether or not S itself is one. In case S is a subspace, then $S^\perp$ is called the *orthogonal complement* of S.

EXAMPLE. If S is a plane through the origin, as shown in Figure 1.2, then $S^\perp$ is a line through the origin perpendicular to this plane. This example also gives a geometric interpretation for the next theorem.

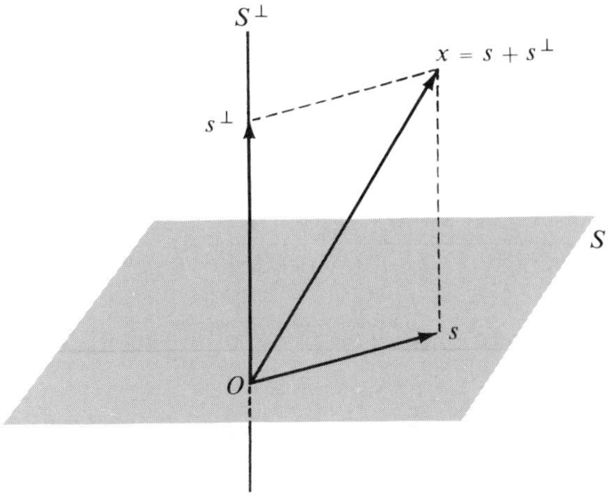

FIGURE 1.2 Geometric interpretation of the orthogonal decomposition theorem in V_3.

THEOREM 1.15. ORTHOGONAL DECOMPOSITION THEOREM. *Let V be a Euclidean space and let S be a finite-dimensional subspace of V. Then every element x in V can be represented uniquely as a sum of two elements, one in S and one in $S^{\perp}$. That is, we have*

$$x = s + s^{\perp}, \qquad \text{where} \quad s \in S \quad \text{and} \quad s^{\perp} \in S^{\perp}. \tag{1.17}$$

Moreover, the norm of x is given by the Pythagorean formula

$$\|x\|^2 = \|s\|^2 + \|s^{\perp}\|^2. \tag{1.18}$$

Proof. First we prove that an orthogonal decomposition (1.17) actually exists. Since S is finite-dimensional, it has a finite orthonormal basis, say $\{e_1, \ldots, e_n\}$. Given x, define the elements s and $s^{\perp}$ as follows:

$$s = \sum_{i=1}^{n} (x, e_i)e_i, \qquad s^{\perp} = x - s. \tag{1.19}$$

Note that each term $(x, e_i)e_i$ is the projection of x along e_i. The element s is the sum of the projections of x along each basis element. Since s is a linear combination of the basis elements, s lies in S. The definition of $s^{\perp}$ shows that Equation (1.17) holds. To prove that $s^{\perp}$ lies in $S^{\perp}$, we consider the inner product of $s^{\perp}$ and any basis element e_j. We have

$$(s^{\perp}, e_j) = (x - s, e_j) = (x, e_j) - (s, e_j).$$

But from (1.19), we find that $(s, e_j) = (x, e_j)$, so $s^{\perp}$ is orthogonal to e_j. Therefore $s^{\perp}$ is orthogonal to every element in S, which means that $s^{\perp} \in S^{\perp}$.

Next we prove that the orthogonal decomposition (1.17) is unique. Suppose that x has two such representations, say

$$x = s + s^{\perp} \qquad \text{and} \qquad x = t + t^{\perp}, \tag{1.20}$$

where s and t are in S, and $s^\perp$ and $t^\perp$ are in $S^\perp$. We wish to prove that $s = t$ and $s^\perp = t^\perp$. From (1.20), we have $s - t = t^\perp - s^\perp$, so we need only prove that $s - t = O$. But $s - t \in S$ and $t^\perp - s^\perp \in S^\perp$ so $s - t$ is both orthogonal to $t^\perp - s^\perp$ and equal to $t^\perp - s^\perp$. Since the zero element is the only element orthogonal to itself, we must have $s - t = O$. This shows that the decomposition is unique.

Finally, we prove that the norm of x is given by the Pythagorean formula. We have

$$\|x\|^2 = (x, x) = (s + s^\perp, s + s^\perp) = (s, s) + (s^\perp, s^\perp),$$

the remaining terms being zero since s and $s^\perp$ are orthogonal. This proves (1.18).

DEFINITION. *Let S be a finite-dimensional subspace of a Euclidean space V, and let $\{e_1, \ldots, e_n\}$ be an orthonormal basis for S. If $x \in V$, the element s defined by the equation*

$$s = \sum_{i=1}^{n} (x, e_i)e_i$$

is called the projection of x on the subspace S.

We prove next that the projection of x on S is the solution to the approximation problem stated at the beginning of this section.

1.16 Best approximation of elements in a Euclidean space by elements in a finite-dimensional subspace

THEOREM 1.16. APPROXIMATION THEOREM. *Let S be a finite-dimensional subspace of a Euclidean space V, and let x be any element of V. Then the projection of x on S is nearer to x than any other element of S. That is, if s is the projection of x on S, we have*

$$\|x - s\| \le \|x - t\|$$

for all t in S; the equality sign holds if and only if $t = s$.

Proof. By Theorem 1.15 we can write $x = s + s^\perp$, where $s \in S$ and $s^\perp \in S^\perp$. Then, for any t in S, we have

$$x - t = (x - s) + (s - t).$$

Since $s - t \in S$ and $x - s = s^\perp \in S^\perp$, this is an orthogonal decomposition of $x - t$, so its norm is given by the Pythagorean formula

$$\|x - t\|^2 = \|x - s\|^2 + \|s - t\|^2.$$

But $\|s - t\|^2 \ge 0$, so we have $\|x - t\|^2 \ge \|x - s\|^2$, with equality holding if and only if $s = t$. This completes the proof.

EXAMPLE 1. *Approximation of continuous functions on* $[0, 2\pi]$ *by trigonometric polynomials.* Let $V = C(0, 2\pi)$, the linear space of all real functions continuous on the interval $[0, 2\pi]$, and define an inner product by the equation $(f, g) = \int_0^{2\pi} f(x)g(x)\,dx$. In Section 1.12 we exhibited an orthonormal set of trigonometric functions $\varphi_0, \varphi_1, \varphi_2, \ldots$, where

$$(1.21)\qquad \varphi_0(x) = \frac{1}{\sqrt{2\pi}}, \qquad \varphi_{2k-1}(x) = \frac{\cos kx}{\sqrt{\pi}}, \qquad \varphi_{2k}(x) = \frac{\sin kx}{\sqrt{\pi}}, \qquad \text{for} \quad k \geq 1.$$

The $2n + 1$ elements $\varphi_0, \varphi_1, \ldots, \varphi_{2n}$ span a subspace S of dimension $2n + 1$. The elements of S are called *trigonometric polynomials*.

If $f \in C(0, 2\pi)$, let f_n denote the projection of f on the subspace S. Then we have

$$(1.22)\qquad f_n = \sum_{k=0}^{2n} (f, \varphi_k)\varphi_k, \qquad \text{where} \quad (f, \varphi_k) = \int_0^{2\pi} f(x)\varphi_k(x)\,dx.$$

The numbers (f, φ_k) are called *Fourier coefficients* of f. Using the formulas in (1.21), we can rewrite (1.22) in the form

$$(1.23)\qquad f_n(x) = \tfrac{1}{2}a_0 + \sum_{k=1}^{n} (a_k \cos kx + b_k \sin kx),$$

where

$$a_k = \frac{1}{\pi}\int_0^{2\pi} f(x)\cos kx\,dx, \qquad b_k = \frac{1}{\pi}\int_0^{2\pi} f(x)\sin kx\,dx$$

for $k = 0, 1, 2, \ldots, n$. The approximation theorem tells us that the trigonometric polynomial in (1.23) approximates f better than any other trigonometric polynomial in S, in the sense that the norm $\|f - f_n\|$ is as small as possible.

EXAMPLE 2. *Approximation of continuous functions on* $[-1, 1]$ *by polynomials of degree* $\leq n$. Let $V = C(-1, 1)$, the space of real continuous functions on $[-1, 1]$, and let $(f, g) = \int_{-1}^{1} f(x)g(x)\,dx$. The $n + 1$ normalized Legendre polynomials $\varphi_0, \varphi_1, \ldots, \varphi_n$, introduced in Section 1.14, span a subspace S of dimension $n + 1$ consisting of all polynomials of degree $\leq n$. If $f \in C(-1, 1)$, let f_n denote the projection of f on S. Then we have

$$f_n = \sum_{k=0}^{n} (f, \varphi_k)\varphi_k, \qquad \text{where} \quad (f, \varphi_k) = \int_{-1}^{1} f(t)\varphi_k(t)\,dt.$$

This is the polynomial of degree $\leq n$ for which the norm $\|f - f_n\|$ is smallest. For example, when $f(x) = \sin \pi x$, the coefficients (f, φ_k) are given by

$$(f, \varphi_k) = \int_{-1}^{1} \sin \pi t\, \varphi_k(t)\,dt.$$

In particular, we have $(f, \varphi_0) = 0$ and

$$(f, \varphi_1) = \int_{-1}^{1} \sqrt{\frac{3}{2}}\, t \sin \pi t\,dt = \sqrt{\frac{3}{2}}\,\frac{2}{\pi}.$$

Therefore the linear polynomial $f_1(t)$ which is nearest to $\sin \pi t$ on $[-1, 1]$ is

$$f_1(t) = \sqrt{\frac{3}{2}}\frac{2}{\pi}\varphi_1(t) = \frac{3}{\pi}t.$$

Since $(f, \varphi_2) = 0$, this is also the nearest quadratic approximation.

1.17 Exercises

1. In each case, find an orthonormal basis for the subspace of V_3 spanned by the given vectors.
(a) $x_1 = (1, 1, 1)$, $x_2 = (1, 0, 1)$, $x_3 = (3, 2, 3)$.
(b) $x_1 = (1, 1, 1)$, $x_2 = (-1, 1, -1)$, $x_3 = (1, 0, 1)$.
2. In each case, find an orthonormal basis for the subspace of V_4 spanned by the given vectors.
(a) $x_1 = (1, 1, 0, 0)$, $x_2 = (0, 1, 1, 0)$, $x_3 = (0, 0, 1, 1)$, $x_4 = (1, 0, 0, 1)$.
(b) $x_1 = (1, 1, 0, 1)$, $x_2 = (1, 0, 2, 1)$, $x_3 = (1, 2, -2, 1)$.
3. In the real linear space $C(0, \pi)$, with inner product $(x, y) = \int_0^{\pi} x(t)y(t)\,dt$, let $x_n(t) = \cos nt$ for $n = 0, 1, 2, \ldots$. Prove that the functions $y_0, y_1, y_2, \ldots$, given by

$$y_0(t) = \frac{1}{\sqrt{\pi}} \quad \text{and} \quad y_n(t) = \sqrt{\frac{2}{\pi}}\cos nt \quad \text{for} \quad n \geq 1,$$

form an orthonormal set spanning the same subspace as $x_0, x_1, x_2, \ldots$.
4. In the linear space of all real polynomials, with inner product $(x, y) = \int_0^1 x(t)y(t)\,dt$, let $x_n(t) = t^n$ for $n = 0, 1, 2, \ldots$. Prove that the functions

$$y_0(t) = 1, \quad y_1(t) = \sqrt{3}\,(2t - 1), \quad y_2(t) = \sqrt{5}\,(6t^2 - 6t + 1)$$

form an orthonormal set spanning the same subspace as $\{x_0, x_1, x_2\}$.
5. Let V be the linear space of all real functions f continuous on $[0, +\infty)$ and such that the integral $\int_0^{\infty} e^{-t}f^2(t)\,dt$ converges. Define $(f, g) = \int_0^{\infty} e^{-t}f(t)g(t)\,dt$, and let $y_0, y_1, y_2, \ldots$, be the set obtained by applying the Gram-Schmidt process to $x_0, x_1, x_2, \ldots$, where $x_n(t) = t^n$ for $n \geq 0$. Prove that $y_0(t) = 1$, $y_1(t) = t - 1$, $y_2(t) = t^2 - 4t + 2$, $y_3(t) = t^3 - 9t^2 + 18t - 6$.
6. In the real linear space $C(1, 3)$ with inner product $(f, g) = \int_1^3 f(x)g(x)\,dx$, let $f(x) = 1/x$ and show that the constant polynomial g nearest to f is $g = \frac{1}{2}\log 3$. Compute $\|g - f\|^2$ for this g.
7. In the real linear space $C(0, 2)$ with inner product $(f, g) = \int_0^2 f(x)g(x)\,dx$, let $f(x) = e^x$ and show that the constant polynomial g nearest to f is $g = \frac{1}{2}(e^2 - 1)$. Compute $\|g - f\|^2$ for this g.
8. In the real linear space $C(-1, 1)$ with inner product $(f, g) = \int_{-1}^1 f(x)g(x)\,dx$, let $f(x) = e^x$ and find the linear polynomial g nearest to f. Compute $\|g - f\|^2$ for this g.
9. In the real linear space $C(0, 2\pi)$ with inner product $(f, g) = \int_0^{2\pi} f(x)g(x)\,dx$, let $f(x) = x$. In the subspace spanned by $u_0(x) = 1$, $u_1(x) = \cos x$, $u_2(x) = \sin x$, find the trigonometric polynomial nearest to f.
10. In the linear space V of Exercise 5, let $f(x) = e^{-x}$ and find the linear polynomial that is nearest to f.

2

LINEAR TRANSFORMATIONS AND MATRICES

2.1 Linear transformations

One of the ultimate goals of analysis is a comprehensive study of functions whose domains and ranges are subsets of linear spaces. Such functions are called *transformations*, *mappings*, or *operators*. This chapter treats the simplest examples, called *linear* transformations, which occur in all branches of mathematics. Properties of more general transformations are often obtained by approximating them by linear transformations.

First we introduce some notation and terminology concerning arbitrary functions. Let V and W be two sets. The symbol

$$T: V \to W$$

will be used to indicate that T is a function whose domain is V and whose values are in W. For each x in V, the element $T(x)$ in W is called the *image of x under T*, and we say that T *maps x onto $T(x)$*. If A is any subset of V, the set of all images $T(x)$ for x in A is called the *image of A under T* and is denoted by $T(A)$. The image of the domain V, $T(V)$, is the range of T.

Now we assume that V and W are linear spaces having the same set of scalars, and we define a linear transformation as follows.

DEFINITION. *If V and W are linear spaces, a function $T: V \to W$ is called a linear transformation of V into W if it has the following two properties:*

(a) $T(x + y) = T(x) + T(y)$ *for all x and y in V,*

(b) $T(cx) = cT(x)$ *for all x in V and all scalars c.*

These properties are verbalized by saying that T preserves addition and multiplication by scalars. The two properties can be combined into one formula which states that

$$T(ax + by) = aT(x) + bT(y)$$

for all x, y in V and all scalars a and b. By induction, we also have the more general relation

$$T\left(\sum_{i=1}^{n} a_i x_i\right) = \sum_{i=1}^{n} a_i T(x_i)$$

for any n elements $x_1, \ldots, x_n$ in V and any n scalars $a_1, \ldots, a_n$.

The reader can easily verify that the following examples are linear transformations.

EXAMPLE 1. *The identity transformation.* The transformation $T: V \to V$, where $T(x) = x$ for each x in V, is called the identity transformation and is denoted by I or by I_V.

EXAMPLE 2. *The zero transformation.* The transformation $T: V \to V$ which maps each element of V onto O is called the zero transformation and is denoted by O.

EXAMPLE 3. *Multiplication by a fixed scalar c.* Here we have $T: V \to V$, where $T(x) = cx$ for all x in V. When $c = 1$, this is the identity transformation. When $c = 0$, it is the zero transformation.

EXAMPLE 4. *Linear equations.* Let $V = V_n$ and $W = V_m$. Given mn real numbers a_{ik}, where $i = 1, 2, \ldots, m$ and $k = 1, 2, \ldots, n$, define $T: V_n \to V_m$ as follows: T maps each vector $x = (x_1, \ldots, x_n)$ in V_n onto the vector $y = (y_1, \ldots, y_m)$ in V_m according to the equations

$$y_i = \sum_{k=1}^{n} a_{ik} x_k \qquad \text{for} \quad i = 1, 2, \ldots, m.$$

EXAMPLE 5. *Inner product with a fixed element.* Let V be a real Euclidean space. For a fixed element z in V, define $T: V \to \mathbf{R}$ as follows: If $x \in V$, then $T(x) = (x, z)$, the inner product of x with z.

EXAMPLE 6. *Projection on a subspace.* Let V be a Euclidean space and let S be a finite-dimensional subspace of V. Define $T: V \to S$ as follows: If $x \in V$, then $T(x)$ is the projection of x on S.

EXAMPLE 7. *The differentiation operator.* Let V be the linear space of all real functions f differentiable on an open interval (a, b). The linear transformation which maps each function f in V onto its derivative f' is called the differentiation operator and is denoted by D. Thus, we have $D: V \to W$, where $D(f) = f'$ for each f in V. The space W consists of all derivatives f'.

EXAMPLE 8. *The integration operator.* Let V be the linear space of all real functions continuous on an interval $[a, b]$. If $f \in V$, define $g = T(f)$ to be that function in V given by

$$g(x) = \int_a^x f(t)\,dt \qquad \text{if} \quad a \le x \le b.$$

This transformation T is called the integration operator.

2.2 Null space and range

In this section, T denotes a linear transformation of a linear space V into a linear space W.

THEOREM 2.1. *The set $T(V)$ (the range of T) is a subspace of W. Moreover, T maps the zero element of V onto the zero element of W.*

Proof. To prove that $T(V)$ is a subspace of W, we need only verify the closure axioms. Take any two elements of $T(V)$, say $T(x)$ and $T(y)$. Then $T(x) + T(y) = T(x + y)$, so $T(x) + T(y)$ is in $T(V)$. Also, for any scalar c we have $cT(x) = T(cx)$, so $cT(x)$ is in $T(V)$. Therefore, $T(V)$ is a subspace of W. Taking $c = 0$ in the relation $T(cx) = cT(x)$, we find that $T(O) = O$.

DEFINITION. *The set of all elements in V that T maps onto O is called the null space of T and is denoted by $N(T)$. Thus, we have*

$$N(T) = \{x \mid x \in V \quad \textit{and} \quad T(x) = O\}.$$

The null space is sometimes called the kernel of T.

THEOREM 2.2. *The null space of T is a subspace of V.*

Proof. If x and y are in $N(T)$, then so are $x + y$ and cx for all scalars c, since

$$T(x + y) = T(x) + T(y) = O \qquad \text{and} \qquad T(cx) = cT(x) = O.$$

The following examples describe the null spaces of the linear transformations given in Section 2.1.

EXAMPLE 1. *Identity transformation.* The null space is $\{O\}$, the subspace consisting of the zero element alone.

EXAMPLE 2. *Zero transformation.* Since every element of V is mapped onto zero, the null space is V itself.

EXAMPLE 3. *Multiplication by a fixed scalar c.* If $c \neq 0$, the null space contains only O. If $c = 0$, the null space is V.

EXAMPLE 4. *Linear equations.* The null space consists of all vectors $(x_1, \ldots, x_n)$ in V_n for which

$$\sum_{k=1}^{n} a_{ik} x_k = 0 \qquad \text{for} \quad i = 1, 2, \ldots, m.$$

EXAMPLE 5. *Inner product with a fixed element z.* The null space consists of all elements in V orthogonal to z.

EXAMPLE 6. *Projection on a subspace S.* If $x \in V$, we have the unique orthogonal decomposition $x = s + s^{\perp}$ (by Theorem 1.15). Since $T(x) = s$, we have $T(x) = O$ if and only if $x = s^{\perp}$. Therefore, the null space is $S^{\perp}$, the orthogonal complement of S.

EXAMPLE 7. *Differentiation operator.* The null space consists of all functions that are constant on the given interval.

EXAMPLE 8. *Integration operator.* The null space contains only the zero function.

2.3 Nullity and rank

Again in this section T denotes a linear transformation of a linear space V into a linear space W. We are interested in the relation between the dimensionality of V, of the null space $N(T)$, and of the range $T(V)$. If V is finite-dimensional, then the null space is also finite-dimensional since it is a subspace of V. The dimension of $N(T)$ is called the *nullity* of T. In the next theorem, we prove that the range is also finite-dimensional; its dimension is called the *rank* of T.

THEOREM 2.3. NULLITY PLUS RANK THEOREM. *If V is finite-dimensional, then $T(V)$ is also finite-dimensional, and we have*

$$\dim N(T) + \dim T(V) = \dim V . \tag{2.1}$$

In other words, the nullity plus the rank of a linear transformation is equal to the dimension of its domain.

Proof. Let $n = \dim V$ and let $e_1, \ldots, e_k$ be a basis for $N(T)$, where $k = \dim N(T) \leq n$. By Theorem 1.7, these elements are part of some basis for V, say the basis

$$e_1, \ldots, e_k, e_{k+1}, \ldots, e_{k+r}, \tag{2.2}$$

where $k + r = n$. We shall prove that the r elements

$$T(e_{k+1}), \ldots, T(e_{k+r}) \tag{2.3}$$

form a basis for $T(V)$, thus proving that $\dim T(V) = r$. Since $k + r = n$, this also proves (2.1).

First we show that the r elements in (2.3) span $T(V)$. If $y \in T(V)$, we have $y = T(x)$ for some x in V, and we can write $x = c_1 e_1 + \cdots + c_{k+r} e_{k+r}$. Hence, we have

$$y = T(x) = \sum_{i=1}^{k+r} c_i T(e_i) = \sum_{i=1}^{k} c_i T(e_i) + \sum_{i=k+1}^{k+r} c_i T(e_i) = \sum_{i=k+1}^{k+r} c_i T(e_i)$$

since $T(e_1) = \cdots = T(e_k) = O$. This shows that the elements in (2.3) span $T(V)$.

Now we show that these elements are independent. Suppose that there are scalars $c_{k+1}, \ldots, c_{k+r}$ such that

$$\sum_{i=k+1}^{k+r} c_i T(e_i) = O .$$

This implies that

$$T\left(\sum_{i=k+1}^{k+r} c_i e_i\right) = O$$

so the element $x = c_{k+1} e_{k+1} + \cdots + c_{k+r} e_{k+r}$ is in the null space $N(T)$. This means there are scalars $c_1, \ldots, c_k$ such that $x = c_1e_1 + \cdots + c_ke_k$, so we have

$$x - x = \sum_{i=1}^{k} c_ie_i - \sum_{i=k+1}^{k+r} c_ie_i = O .$$

But since the elements in (2.2) are independent, this implies that all the scalars c_i are zero. Therefore, the elements in (2.3) are independent.

Note: If V is infinite-dimensional, then at least one of $N(T)$ or $T(V)$ is infinite-dimensional. A proof of of this fact is outlined in Exercise 30 of Section 2.4.

2.4 Exercises

In each of Exercises 1 through 10, a transformation $T: V_2 \to V_2$ is defined by the formula given for $T(x, y)$, where (x, y) is an arbitrary point in V_2. In each case determine whether T is linear. If T is linear, describe its null space and range, and compute its nullity and rank.

1. $T(x, y) = (y, x)$.
2. $T(x, y) = (x, -y)$.
3. $T(x, y) = (x, 0)$.
4. $T(x, y) = (x, x)$.
5. $T(x, y) = (x^2, y^2)$.
6. $T(x, y) = (e^x, e^y)$.
7. $T(x, y) = (x, 1)$.
8. $T(x, y) = (x + 1, y + 1)$.
9. $T(x, y) = (x - y, x + y)$.
10. $T(x, y) = (2x - y, x + y)$.

Do the same as above for each of Exercises 11 through 15 if the transformation $T: V_2 \to V_2$ is described as indicated.

11. T rotates every point through the same angle φ about the origin. That is, T maps a point with polar coordinates (r, θ) onto the point with polar coordinates $(r, \theta + \varphi)$, where φ is fixed. Also, T maps O onto itself.
12. T maps each point onto its reflection with respect to a fixed line through the origin.
13. T maps every point onto the point $(1, 1)$.
14. T maps each point with polar coordinates (r, θ) onto the point with polar coordinates $(2r, \theta)$. Also, T maps O onto itself.
15. T maps each point with polar coordinates (r, θ) onto the point with polar coordinates $(r, 2\theta)$. Also, T maps O onto itself.

Do the same as above in each of Exercises 16 through 23 if a transformation $T: V_3 \to V_3$ is defined by the formula given for $T(x, y, z)$, where (x, y, z) is an arbitrary point of V_3.

16. $T(x, y, z) = (z, y, x)$.
17. $T(x, y, z) = (x, y, 0)$.
18. $T(x, y, z) = (x, 2y, 3z)$.
19. $T(x, y, z) = (x, y, 1)$.
20. $T(x, y, z) = (x + 1, y + 1, z - 1)$.
21. $T(x, y, z) = (x + 1, y + 2, z + 3)$.
22. $T(x, y, z) = (x, y^2, z^3)$.
23. $T(x, y, z) = (x + z, 0, x + y)$.

In each of Exercises 24 through 27, a transformation $T: V \to V$ is described as indicated. In each case, determine whether T is linear. If T is linear, describe its null space and range, and compute the nullity and rank when they are finite.

24. Let V be the linear space of all real polynomials $p(x)$ of degree $\le n$. If $p \in V$, $q = T(p)$ means that $q(x) = p(x + 1)$ for all real x.
25. Let V be the linear space of all real functions differentiable on the open interval $(-1, 1)$. If $f \in V$, $g = T(f)$ means that $g(x) = xf'(x)$ for all x in $(-1, 1)$.

26. Let V be the linear space of all real functions continuous on $[a, b]$. If $f \in V$, $g = T(f)$ means that

$$g(x) = \int_a^b f(t) \sin (x - t)\, dt \qquad \text{for} \quad a \le x \le b\,.$$

27. Let V be the space of all real functions twice differentiable on an open interval (a, b). If $y \in V$, define $T(y) = y'' + Py' + Qy$, where P and Q are fixed constants.
28. Let V be the linear space of all real convergent sequences $\{x_n\}$. Define a transformation $T: V \to V$ as follows: If $x = \{x_n\}$ is a convergent sequence with limit a, let $T(x) = \{y_n\}$, where $y_n = a - x_n$ for $n \ge 1$. Prove that T is linear and describe the null space and range of T.
29. Let V denote the linear space of all real functions continuous on the interval $[-\pi, \pi]$. Let S be that subset of V consisting of all f satisfying the three equations

$$\int_{-\pi}^{\pi} f(t)\, dt = 0\,, \qquad \int_{-\pi}^{\pi} f(t) \cos t\, dt = 0\,, \qquad \int_{-\pi}^{\pi} f(t) \sin t\, dt = 0\,.$$

(a) Prove that S is a subspace of V.
(b) Prove that S contains the functions $f(x) = \cos nx$ and $f(x) = \sin nx$ for each $n = 2, 3, \ldots$.
(c) Prove that S is infinite-dimensional.

Let $T: V \to V$ be the linear transformation defined as follows: If $f \in V$, $g = T(f)$ means that

$$g(x) = \int_{-\pi}^{\pi} \{1 + \cos (x - t)\} f(t)\, dt\,.$$

(d) Prove that $T(V)$, the range of T, is finite-dimensional and find a basis for $T(V)$.
(e) Determine the null space of T.
(f) Find all real $c \ne 0$ and all nonzero f in V such that $T(f) = cf$. (Note that such an f lies in the range of T.)

30. Let $T: V \to W$ be a linear transformation of a linear space V into a linear space W. If V is infinite-dimensional, prove that at least one of $T(V)$ or $N(T)$ is infinite-dimensional.

[*Hint:* Assume $\dim N(T) = k$, $\dim T(V) = r$, let $e_1, \ldots, e_k$ be a basis for $N(T)$ and let $e_1, \ldots, e_k, e_{k+1}, \ldots, e_{k+n}$ be independent elements in V, where $n > r$. The elements $T(e_{k+1}), \ldots, T(e_{k+n})$ are dependent since $n > r$. Use this fact to obtain a contradiction.]

2.5 Algebraic operations on linear transformations

Functions whose values lie in a given linear space W can be added to each other and can be multiplied by the scalars in W according to the following definition.

DEFINITION. *Let $S: V \to W$ and $T: V \to W$ be two functions with a common domain V and with values in a linear space W. If c is any scalar in W, we define the sum $S + T$ and the product cT by the equations*

$$(2.4) \qquad (S + T)(x) = S(x) + T(x)\,, \qquad (cT)(x) = cT(x)$$

for all x in V.

We are especially interested in the case where V is also a linear space having the same scalars as W. In this case we denote by $\mathscr{L}(V, W)$ the set of all linear transformations of V into W.

If S and T are two linear transformations in $\mathscr{L}(V, W)$, it is an easy exercise to verify that $S + T$ and cT are also linear transformations in $\mathscr{L}(V, W)$. More than this is true. With the operations just defined, the set $\mathscr{L}(V, W)$ itself becomes a new linear space. The zero transformation serves as the zero element of this space, and the transformation $(-1)T$ is the negative of T. It is a straightforward matter to verify that all ten axioms for a linear space are satisfied. Therefore, we have the following.

THEOREM 2.4. *The set $\mathscr{L}(V, W)$ of all linear transformations of V into W is a linear space with the operations of addition and multiplication by scalars defined as in* (2.4).

A more interesting algebraic operation on linear transformations is *composition* or *multiplication* of transformations. This operation makes no use of the algebraic structure of a linear space and can be defined quite generally as follows.

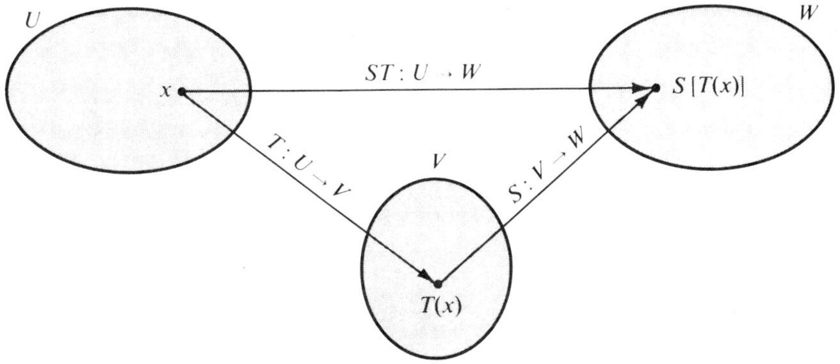

FIGURE 2.1 Illustrating the composition of two transformations.

DEFINITION. *Let U, V, W be sets. Let $T: U \to V$ be a function with domain U and values in V, and let $S: V \to W$ be another function with domain V and values in W. Then the composition ST is the function $ST: U \to W$ defined by the equation*

$$(ST)(x) = S[T(x)] \qquad \textit{for every } x \textit{ in } U.$$

Thus, to map x by the composition ST, we first map x by T and then map $T(x)$ by S. This is illustrated in Figure 2.1.

Composition of real-valued functions has been encountered repeatedly in our study of calculus, and we have seen that the operation is, in general, not commutative. However, as in the case of real-valued functions, composition does satisfy an associative law.

THEOREM 2.5. *If $T: U \to V$, $S: V \to W$, and $R: W \to X$ are three functions, then we have*

$$R(ST) = (RS)T.$$

Proof. Both functions $R(ST)$ and $(RS)T$ have domain U and values in X. For each x in U, we have

$$[R(ST)](x) = R[(ST)(x)] = R[S[T(x)]] \quad \text{and} \quad [(RS)T](x) = (RS)[T(x)] = R[S[T(x)]],$$

which proves that $R(ST) = (RS)T$.

DEFINITION. *Let $T: V \to V$ be a function which maps V into itself. We define integral powers of T inductively as follows:*

$$T^0 = I, \qquad T^n = TT^{n-1} \quad \textit{for} \quad n \geq 1.$$

Here I is the identity transformation. The reader may verify that the associative law implies the law of exponents $T^mT^n = T^{m+n}$ for all nonnegative integers m and n.

The next theorem shows that the composition of *linear* transformations is again linear.

THEOREM 2.6. *If U, V, W are linear spaces with the same scalars, and if $T: U \to V$ and $S: V \to W$ are linear transformations, then the composition $ST: U \to W$ is linear.*

Proof. For all x, y in U and all scalars a and b, we have

$$(ST)(ax + by) = S[T(ax + by)] = S[aT(x) + bT(y)] = aST(x) + bST(y).$$

Composition can be combined with the algebraic operations of addition and multiplication of scalars in $\mathscr{L}(V, W)$ to give us the following.

THEOREM 2.7. *Let U, V, W be linear spaces with the same scalars, assume S and T are in $\mathscr{L}(V, W)$, and let c be any scalar.*

(a) *For any function R with values in V, we have*

$$(S + T)R = SR + TR \quad \textit{and} \quad (cS)R = c(SR).$$

(b) *For any linear transformation $R: W \to U$, we have*

$$R(S + T) = RS + RT \quad \textit{and} \quad R(cS) = c(RS).$$

The proof is a straightforward application of the definition of composition and is left as an exercise.

2.6 Inverses

In our study of real-valued functions we learned how to construct new functions by inversion of monotonic functions. Now we wish to extend the process of inversion to a more general class of functions.

Given a function T, our goal is to find, if possible, another function S whose composition with T is the identity transformation. Since composition is in general not commutative, we have to distinguish between ST and TS. Therefore we introduce two kinds of inverses which we call left and right inverses.

DEFINITION. *Given two sets V and W and a function $T: V \to W$. A function $S: T(V) \to V$ is called a left inverse of T if $S[T(x)] = x$ for all x in V, that is, if*

$$ST = I_V ,$$

where I_V is the identity transformation on V. A function $R: T(V) \to V$ is called a right inverse of T if $T[R(y)] = y$ for all y in $T(V)$, that is, if

$$TR = I_{T(V)} ,$$

where $I_{T(V)}$ is the identity transformation on $T(V)$.

EXAMPLE. *A function with no left inverse but with two right inverses.* Let $V = \{1, 2\}$ and let $W = \{0\}$. Define $T: V \to W$ as follows: $T(1) = T(2) = 0$. This function has two right inverses $R: W \to V$ and $R': W \to V$ given by

$$R(0) = 1 , \qquad R'(0) = 2 .$$

It cannot have a left inverse S since this would require

$$1 = S[T(1)] = S(0) \qquad \text{and} \qquad 2 = S[T(2)] = S(0) .$$

This simple example shows that left inverses need not exist and that right inverses need not be unique.

Every function $T: V \to W$ has at least one right inveise. In fact, each y in $T(V)$ has the form $y = T(x)$ for at least one x in V. If we select one such x and define $R(y) = x$, then $T[R(y)] = T(x) = y$ for each y in $T(V)$, so R is a right inverse. Nonuniqueness may occur because there may be more than one x in V which maps onto a given y in $T(V)$. We shall prove presently (in Theorem 2.9) that if each y in $T(V)$ is the image of *exactly one* x in V, then right inverses are unique.

First we prove that if a left inverse exists it is unique and, at the same time, is a right inverse.

THEOREM 2.8. *A function $T: V \to W$ can have at most one left inverse. If T has a left inverse S, then S is also a right inverse.*

Proof. Assume T has two left inverses, $S: T(V) \to V$ and $S': T(V) \to V$. Choose any y in $T(V)$. We shall prove that $S(y) = S'(y)$. Now $y = T(x)$ for some x in V, so we have

$$S[T(x)] = x \qquad \text{and} \qquad S'[T(x)] = x ,$$

since both S and S' are left inverses. Therefore $S(y) = x$ and $S'(y) = x$, so $S(y) = S'(y)$ for all y in $T(V)$. Therefore $S = S'$ which proves that left inverses are unique.

Now we prove that every left inverse S is also a right inverse. Choose any element y in $T(V)$. We shall prove that $T[S(y)] = y$. Since $y \in T(V)$, we have $y = T(x)$ for some x in V. But S is a left inverse, so

$$x = S[T(x)] = S(y).$$

Applying T, we get $T(x) = T[S(y)]$. But $y = T(x)$, so $y = T[S(y)]$, which completes the proof.

The next theorem characterizes all functions having left inverses.

THEOREM 2.9. *A function $T: V \to W$ has a left inverse if and only if T maps distinct elements of V onto distinct elements of W; that is, if and only if, for all x and y in V,*

(2.5) $$x \neq y \quad \textit{implies} \quad T(x) \neq T(y).$$

Note: Condition (2.5) is equivalent to the statement

(2.6) $$T(x) = T(y) \quad \text{implies} \quad x = y.$$

A function T satisfying (2.5) or (2.6) for all x and y in V is said to be *one-to-one* on V.

Proof. Assume T has a left inverse S, and assume that $T(x) = T(y)$. We wish to prove that $x = y$. Applying S, we find $S[T(x)] = S[T(y)]$. Since $S[T(x)] = x$ and $S[T(y)] = y$, this implies $x = y$. This proves that a function with a left inverse is one-to-one on its domain.

Now we prove the converse. Assume T is one-to-one on V. We shall exhibit a function $S: T(V) \to V$ which is a left inverse of T. If $y \in T(V)$, then $y = T(x)$ for some x in V. By (2.6), there is *exactly one* x in V for which $y = T(x)$. Define $S(y)$ to be this x. That is, we define S on $T(V)$ as follows:

$$S(y) = x \quad \text{means that} \quad T(x) = y.$$

Then we have $S[T(x)] = x$ for each x in V, so $ST = I_V$. Therefore, the function S so defined is a left inverse of T.

DEFINITION. *Let $T: V \to W$ be one-to-one on V. The unique left inverse of T (which we know is also a right inverse) is denoted by T^{-1}. We say that T is invertible, and we call T^{-1} the inverse of T.*

The results of this section refer to arbitrary functions. Now we apply these ideas to linear transformations.

2.7 One-to-one linear transformations

In this section, V and W denote linear spaces with the same scalars, and $T: V \to W$ denotes a linear transformation in $\mathscr{L}(V, W)$. The linearity of T enables us to express the one-to-one property in several equivalent forms.

THEOREM 2.10. *Let $T: V \to W$ be a linear transformation in $\mathscr{L}(V, W)$. Then the following statements are equivalent.*

(a) *T is one-to-one on V.*
(b) *T is invertible and its inverse $T^{-1}: T(V) \to V$ is linear.*
(c) *For all x in V, $T(x) = O$ implies $x = O$. That is, the null space $N(T)$ contains only the zero element of V.*

Proof. We shall prove that (a) implies (b), (b) implies (c), and (c) implies (a). First assume (a) holds. Then T has an inverse (by Theorem 2.9), and we must show that T^{-1} is linear. Take any two elements u and v in $T(V)$. Then $u = T(x)$ and $v = T(y)$ for some x and y in V. For any scalars a and b, we have

$$au + bv = aT(x) + bT(y) = T(ax + by),$$

since T is linear. Hence, applying T^{-1}, we have

$$T^{-1}(au + bv) = ax + by = aT^{-1}(u) + bT^{-1}(v),$$

so T^{-1} is linear. Therefore (a) implies (b).

Next assume that (b) holds. Take any x in V for which $T(x) = O$. Applying T^{-1}, we find that $x = T^{-1}(O) = O$, since T^{-1} is linear. Therefore, (b) implies (c).

Finally, assume (c) holds. Take any two elements u and v in V with $T(u) = T(v)$. By linearity, we have $T(u - v) = T(u) - T(v) = O$, so $u - v = O$. Therefore, T is one-to-one on V, and the proof of the theorem is complete.

When V is finite-dimensional, the one-to-one property can be formulated in terms of independence and dimensionality, as indicated by the next theorem.

THEOREM 2.11. *Let $T: V \to W$ be a linear transformation in $\mathscr{L}(V, W)$ and assume that V is finite-dimensional, say* $\dim V = n$. *Then the following statements are equivalent.*

(a) *T is one-to-one on V.*
(b) *If $e_1, \ldots, e_p$ are independent elements in V, then $T(e_1), \ldots, T(e_p)$ are independent elements in $T(V)$.*
(c) $\dim T(V) = n$.
(d) *If $\{e_1, \ldots, e_n\}$ is a basis for V, then $\{T(e_1), \ldots, T(e_n)\}$ is a basis for $T(V)$.*

Proof. We shall prove that (a) implies (b), (b) implies (c), (c) implies (d), and (d) implies (a). Assume (a) holds. Let $e_1, \ldots, e_p$ be independent elements of V and consider the elements $T(e_1), \ldots, T(e_p)$ in $T(V)$. Suppose that

$$\sum_{i=1}^{p} c_i T(e_i) = O$$

for certain scalars $c_1, \ldots, c_p$. By linearity, we obtain

$$T\left(\sum_{i=1}^{p} c_i e_i\right) = O, \qquad \text{and hence} \qquad \sum_{i=1}^{p} c_i e_i = O$$

since T is one-to-one. But $e_1, \ldots, e_p$ are independent, so $c_1 = \cdots = c_p = 0$. Therefore (a) implies (b).

Now assume (b) holds. Let $\{e_1, \ldots, e_n\}$ be a basis for V. By (b), the n elements $T(e_1), \ldots, T(e_n)$ in $T(V)$ are independent. Therefore, $\dim T(V) \geq n$. But, by Theorem 2.3, we have $\dim T(V) \leq n$. Therefore $\dim T(V) = n$, so (b) implies (c).

Next, assume (c) holds and let $\{e_1, \ldots, e_n\}$ be a basis for V. Take any element y in $T(V)$. Then $y = T(x)$ for some x in V, so we have

$$x = \sum_{i=1}^{n} c_i e_i, \qquad \text{and hence} \qquad y = T(x) = \sum_{i=1}^{n} c_i T(e_i).$$

Therefore $\{T(e_1), \ldots, T(e_n)\}$ spans $T(V)$. But we are assuming $\dim T(V) = n$, so $\{T(e_1), \ldots, T(e_n)\}$ is a basis for $T(V)$. Therefore (c) implies (d).

Finally, assume (d) holds. We will prove that $T(x) = O$ implies $x = O$. Let $\{e_1, \ldots, e_n\}$ be a basis for V. If $x \in V$, we may write

$$x = \sum_{i=1}^{n} c_i e_i, \qquad \text{and hence} \qquad T(x) = \sum_{i=1}^{n} c_i T(e_i).$$

If $T(x) = O$, then $c_1 = \cdots = c_n = 0$, since the elements $T(e_1), \ldots, T(e_n)$ are independent. Therefore $x = O$, so T is one-to-one on V. Thus, (d) implies (a) and the proof is complete.

2.8 Exercises

1. Let $V = \{0, 1\}$. Describe all functions $T: V \to V$. There are four altogether. Label them as T_1, T_2, T_3, T_4 and make a multiplication table showing the composition of each pair. Indicate which functions are one-to-one on V and give their inverses.
2. Let $V = \{0, 1, 2\}$. Describe all functions $T: V \to V$ for which $T(V) = V$. There are six altogether. Label them as $T_1, \ldots, T_6$ and make a multiplication table showing the composition of each pair. Indicate which functions are one-to-one on V, and give their inverses.

In each of Exercises 3 through 12, a function $T: V_2 \to V_2$ is defined by the formula given for $T(x, y)$, where (x, y) is an arbitrary point in V_2. In each case determine whether T is one-to-one on V_2. If it is, describe its range $T(V_2)$; for each point (u, v) in $T(V_2)$, let $(x, y) = T^{-1}(u, v)$ and give formulas for determining x and y in terms of u and v.

3. $T(x, y) = (y, x)$.
4. $T(x, y) = (x, -y)$.
5. $T(x, y) = (x, 0)$.
6. $T(x, y) = (x, x)$.
7. $T(x, y) = (x^2, y^2)$.
8. $T(x, y) = (e^x, e^y)$.
9. $T(x, y) = (x, 1)$.
10. $T(x, y) = (x + 1, y + 1)$.
11. $T(x, y) = (x - y, x + y)$.
12. $T(x, y) = (2x - y, x + y)$.

In each of Exercises 13 through 20, a function $T: V_3 \to V_3$ is defined by the formula given for $T(x, y, z)$, where (x, y, z) is an arbitrary point in V_3. In each case, determine whether T is one-to-one on V_3. If it is, describe its range $T(V_3)$; for each point (u, v, w) in $T(V_3)$, let $(x, y, z) = T^{-1}(u, v, w)$ and give formulas for determining x, y, and z in terms of u, v, and w.

13. $T(x, y, z) = (z, y, x)$.
14. $T(x, y, z) = (x, y, 0)$.
15. $T(x, y, z) = (x, 2y, 3z)$.
16. $T(x, y, z) = (x, y, x + y + z)$.
17. $T(x, y, z) = (x + 1, y + 1, z - 1)$.
18. $T(x, y, z) = (x + 1, y + 2, z + 3)$.
19. $T(x, y, z) = (x, x + y, x + y + z)$.
20. $T(x, y, z) = (x + y, y + z, x + z)$.

21. Let $T: V \to V$ be a function which maps V into itself. Powers are defined inductively by the formulas $T^0 = I$, $T^n = TT^{n-1}$ for $n \geq 1$. Prove that the associative law for composition implies the law of exponents: $T^mT^n = T^{m+n}$. If T is invertible, prove that T^n is also invertible and that $(T^n)^{-1} = (T^{-1})^n$.

In Exercises 22 through 25, S and T denote functions with domain V and values in V. In general, $ST \neq TS$. If $ST = TS$, we say that S and T *commute*.

22. If S and T commute, prove that $(ST)^n = S^nT^n$ for all integers $n \geq 0$.
23. If S and T are invertible, prove that ST is also invertible and that $(ST)^{-1} = T^{-1}S^{-1}$. In other words, the inverse of ST is the composition of inverses, taken in reverse order.
24. If S and T are invertible and commute, prove that their inverses also commute.
25. Let V be a linear space. If S and T commute, prove that

$$(S + T)^2 = S^2 + 2ST + T^2 \quad \text{and} \quad (S + T)^3 = S^3 + 3S^2T + 3ST^2 + T^3.$$

Indicate how these formulas must be altered if $ST \neq TS$.

26. Let S and T be the linear transformations of V_3 into V_3 defined by the formulas $S(x, y, z) = (z, y, x)$ and $T(x, y, z) = (x, x + y, x + y + z)$, where (x, y, z) is an arbitrary point of V_3.
(a) Determine the image of (x, y, z) under each of the following transformations: ST, TS, $ST - TS$, S^2, T^2, $(ST)^2$, $(TS)^2$, $(ST - TS)^2$.
(b) Prove that S and T are one-to-one on V_3 and find the image of (u, v, w) under each of the following transformations: S^{-1}, T^{-1}, $(ST)^{-1}$, $(TS)^{-1}$.
(c) Find the image of (x, y, z) under $(T - I)^n$ for each $n \geq 1$.
27. Let V be the linear space of all real polynomials $p(x)$. Let D denote the differentiation operator and let T denote the integration operator which maps each polynomial p onto the polynomial q given by $q(x) = \int_0^x p(t)\, dt$. Prove that $DT = I_V$ but that $TD \neq I_V$. Describe the null space and range of TD.
28. Let V be the linear space of all real polynomials $p(x)$. Let D denote the differentiation operator and let T be the linear transformation that maps $p(x)$ onto $xp'(x)$.
(a) Let $p(x) = 2 + 3x - x^2 + 4x^3$ and determine the image of p under each of the following transformations: D, T, DT, TD, $DT - TD$, $T^2D^2 - D^2T^2$.
(b) Determine those p in V for which $T(p) = p$.
(c) Determine those p in V for which $(DT - 2D)(p) = O$.
(d) Determine those p in V for which $(DT - TD)^n(p) = D^n(p)$.
29. Let V and D be as in Exercise 28 but let T be the linear transformation that maps $p(x)$ onto $xp(x)$. Prove that $DT - TD = I$ and that $DT^n - T^nD = nT^{n-1}$ for $n \geq 2$.
30. Let S and T be in $\mathscr{L}(V, V)$ and assume that $ST - TS = I$. Prove that $ST^n - T^nS = nT^{n-1}$ for all $n \geq 1$.
31. Let V be the linear space of all real polynomials $p(x)$. Let R, S, T be the functions which map an arbitrary polynomial $p(x) = c_0 + c_1x + \cdots + c_nx^n$ in V onto the polynomials $r(x)$, $s(x)$,

and $t(x)$, respectively, where

$$r(x) = p(0), \qquad s(x) = \sum_{k=1}^{n} c_k x^{k-1}, \qquad t(x) = \sum_{k=0}^{n} c_k x^{k+1}.$$

(a) Let $p(x) = 2 + 3x - x^2 + x^3$ and determine the image of p under each of the following transformations: $R, S, T, ST, TS, (TS)^2, T^2S^2, S^2T^2, TRS, RST$.
(b) Prove that R, S, and T are linear and determine the null space and range of each.
(c) Prove that T is one-to-one on V and determine its inverse.
(d) If $n \geq 1$, express $(TS)^n$ and S^nT^n in terms of I and R.

32. Refer to Exercise 28 of Section 2.4. Determine whether T is one-to-one on V. If it is, describe its inverse.

2.9 Linear transformations with prescribed values

If V is finite-dimensional, we can always construct a linear transformation $T: V \to W$ with prescribed values at the basis elements of V, as described in the next theorem.

THEOREM 2.12. *Let $e_1, \ldots, e_n$ be a basis for an n-dimensional linear space V. Let $u_1, \ldots, u_n$ be n arbitrary elements in a linear space W. Then there is one and only one linear transformation $T: V \to W$ such that*

$$(2.7) \qquad T(e_k) = u_k \quad \textit{for} \quad k = 1, 2, \ldots, n.$$

This T maps an arbitrary element x in V as follows:

$$(2.8) \qquad \textit{If} \quad x = \sum_{k=1}^{n} x_k e_k, \qquad \textit{then} \quad T(x) = \sum_{k=1}^{n} x_k u_k.$$

Proof. Every x in V can be expressed uniquely as a linear combination of $e_1, \ldots, e_n$, the multipliers $x_1, \ldots, x_n$ being the components of x relative to the ordered basis $(e_1, \ldots, e_n)$. If we define T by (2.8), it is a straightforward matter to verify that T is linear. If $x = e_k$ for some k, then all components of x are 0 except the kth, which is 1, so (2.8) gives $T(e_k) = u_k$, are required.

To prove that there is only one linear transformation satisfying (2.7), let T' be another and compute $T'(x)$. We find that

$$T'(x) = T'\left(\sum_{k=1}^{n} x_k e_k\right) = \sum_{k=1}^{n} x_k T'(e_k) = \sum_{k=1}^{n} x_k u_k = T(x).$$

Since $T'(x) = T(x)$ for all x in V, we have $T' = T$, which completes the proof.

EXAMPLE. Determine the linear transformation $T: V_2 \to V_2$ which maps the basis elements $\boldsymbol{i} = (1, 0)$ and $\boldsymbol{j} = (0, 1)$ as follows:

$$T(\boldsymbol{i}) = \boldsymbol{i} + \boldsymbol{j}, \qquad T(\boldsymbol{j}) = 2\boldsymbol{i} - \boldsymbol{j}.$$

Solution. If $x = x_1 i + x_2 j$ is an arbitrary element of V_2, then $T(x)$ is given by

$$T(x) = x_1 T(i) + x_2 T(j) = x_1(i + j) + x_2(2i - j) = (x_1 + 2x_2)i + (x_1 - x_2)j.$$

2.10 Matrix representations of linear transformations

Theorem 2.12 shows that a linear transformation $T: V \to W$ of a finite-dimensional linear space V is completely determined by its action on a given set of basis elements $e_1, \ldots, e_n$. Now, suppose the space W is also finite-dimensional, say $\dim W = m$, and let $w_1, \ldots, w_m$ be a basis for W. (The dimensions n and m may or may not be equal.) Since T has values in W, each element $T(e_k)$ can be expressed uniquely as a linear combination of the basis elements $w_1, \ldots, w_m$, say

$$T(e_k) = \sum_{i=1}^{m} t_{ik} w_i ,$$

where $t_{1k}, \ldots, t_{mk}$ are the components of $T(e_k)$ relative to the ordered basis $(w_1, \ldots, w_m)$. We shall display the m-tuple $(t_{1k}, \ldots, t_{mk})$ vertically, as follows:

$$\text{(2.9)} \qquad \begin{bmatrix} t_{1k} \\ t_{2k} \\ \cdot \\ \cdot \\ \cdot \\ t_{mk} \end{bmatrix} .$$

This array is called a *column vector* or a *column matrix*. We have such a column vector for each of the n elements $T(e_1), \ldots, T(e_n)$. We place them side by side and enclose them in one pair of brackets to obtain the following rectangular array:

$$\begin{bmatrix} t_{11} & t_{12} & \cdots & t_{1n} \\ t_{21} & t_{22} & \cdots & t_{2n} \\ \cdot & \cdot & & \cdot \\ \cdot & \cdot & & \cdot \\ \cdot & \cdot & & \cdot \\ t_{m1} & t_{m2} & \cdots & t_{mn} \end{bmatrix} .$$

This array is called a *matrix* consisting of m *rows* and n *columns*. We call it an m by n matrix, or an $m \times n$ matrix. The first row is the $1 \times n$ matrix $(t_{11}, t_{12}, \ldots, t_{1n})$. The $m \times 1$ matrix displayed in (2.9) is the kth column. The scalars t_{ik} are indexed so the first subscript i indicates the *row*, and the second subscript k indicates the *column* in which t_{ik} occurs. We call t_{ik} the *ik-entry* or the *ik-element* of the matrix. The more compact notation

$$(t_{ik}), \qquad \text{or} \qquad (t_{ik})_{i,k=1}^{m,n},$$

is also used to denote the matrix whose ik-entry is t_{ik}.

Thus, every linear transformation T of an n-dimensional space V into an m-dimensional space W gives rise to an $m \times n$ matrix (t_{ik}) whose columns consist of the components of $T(e_1), \ldots, T(e_n)$ relative to the basis $(w_1, \ldots, w_m)$. We call this the *matrix representation* of T relative to the given choice of ordered bases $(e_1, \ldots, e_n)$ for V and $(w_1, \ldots, w_m)$ for W. Once we know the matrix (t_{ik}), the components of any element $T(x)$ relative to the basis $(w_1, \ldots, w_m)$ can be determined as described in the next theorem.

THEOREM 2.13. *Let T be a linear transformation in $\mathscr{L}(V, W)$, where* $\dim V = n$ *and* $\dim W = m$. *Let $(e_1, \ldots, e_n)$ and $(w_1, \ldots, w_m)$ be ordered bases for V and W, respectively, and let (t_{ik}) be the $m \times n$ matrix whose entries are determined by the equations*

$$T(e_k) = \sum_{i=1}^{m} t_{ik} w_i, \qquad \textit{for} \quad k = 1, 2, \ldots, n. \tag{2.10}$$

Then an arbitrary element

$$x = \sum_{k=1}^{n} x_k e_k \tag{2.11}$$

in V with components $(x_1, \ldots, x_n)$ relative to $(e_1, \ldots, e_n)$ is mapped by T onto the element

$$T(x) = \sum_{i=1}^{m} y_i w_i \tag{2.12}$$

in W with components $(y_1, \ldots, y_m)$ relative to $(w_1, \ldots, w_m)$. The y_i are related to the components of x by the linear equations

$$y_i = \sum_{k=1}^{n} t_{ik} x_k \qquad \textit{for} \quad i = 1, 2, \ldots, m. \tag{2.13}$$

Proof. Applying T to each member of (2.11) and using (2.10), we obtain

$$T(x) = \sum_{k=1}^{n} x_k T(e_k) = \sum_{k=1}^{n} x_k \sum_{i=1}^{m} t_{ik} w_i = \sum_{i=1}^{m} \left(\sum_{k=1}^{n} t_{ik} x_k \right) w_i = \sum_{i=1}^{m} y_i w_i ,$$

where each y_i is given by (2.13). This completes the proof.

Having chosen a pair of bases $(e_1, \ldots, e_n)$ and $(w_1, \ldots, w_m)$ for V and W, respectively, every linear transformation $T: V \to W$ has a matrix representation (t_{ik}). Conversely, if we start with any mn scalars arranged as a rectangular matrix (t_{ik}) and choose a pair of ordered bases for V and W, then it is easy to prove that there is exactly one linear transformation $T: V \to W$ having this matrix representation. We simply define T at the basis elements of V by the equations in (2.10). Then, by Theorem 2.12, there is one and only one linear transformation $T: V \to W$ with these prescribed values. The image $T(x)$ of an arbitrary point x in V is then given by Equations (2.12) and (2.13).

EXAMPLE 1. *Construction of a linear transformation from a given matrix.* Suppose we start with the 2×3 matrix

$$\begin{bmatrix} 3 & 1 & -2 \\ 1 & 0 & 4 \end{bmatrix}.$$

Choose the usual bases of unit coordinate vectors for V_3 and V_2. Then the given matrix represents a linear transformation $T: V_3 \to V_2$ which maps an arbitrary vector (x_1, x_2, x_3) in V_3 onto the vector (y_1, y_2) in V_2 according to the linear equations

$$y_1 = 3x_1 + x_2 - 2x_3,$$

$$y_2 = x_1 + 0x_2 + 4x_3.$$

EXAMPLE 2. *Construction of a matrix representation of a given linear transformation.* Let V be the linear space of all real polynomials $p(x)$ of degree ≤ 3. This space has dimension 4, and we choose the basis $(1, x, x^2, x^3)$. Let D be the differentiation operator which maps each polynomial $p(x)$ in V onto its derivative $p'(x)$. We can regard D as a linear transformation of V into W, where W is the 3-dimensional space of all real polynomials of degree ≤ 2. In W we choose the basis $(1, x, x^2)$. To find the matrix representation of D relative to this choice of bases, we transform (differentiate) each basis element of V and express it as a linear combination of the basis elements of W. Thus, we find that

$$D(1) = 0 = 0 + 0x + 0x^2, \qquad D(x) = 1 = 1 + 0x + 0x^2,$$

$$D(x^2) = 2x = 0 + 2x + 0x^2, \qquad D(x^3) = 3x^2 = 0 + 0x + 3x^2.$$

The coefficients of these polynomials determine the *columns* of the matrix representation of D. Therefore, the required representation is given by the following 3×4 matrix:

$$\begin{bmatrix} 0 & 1 & 0 & 0 \\ 0 & 0 & 2 & 0 \\ 0 & 0 & 0 & 3 \end{bmatrix}.$$

To emphasize that the matrix representation depends not only on the basis elements but also on their order, let us reverse the order of the basis elements in W and use, instead, the ordered basis $(x^2, x, 1)$. Then the basis elements of V are transformed into the same polynomials obtained above, but the components of these polynomials relative to the new basis $(x^2, x, 1)$ appear in reversed order. Therefore, the matrix representation of D now becomes

$$\begin{bmatrix} 0 & 0 & 0 & 3 \\ 0 & 0 & 2 & 0 \\ 0 & 1 & 0 & 0 \end{bmatrix}.$$

Let us compute a third matrix representation for D, using the basis $(1, 1 + x, 1 + x + x^2, 1 + x + x^2 + x^3)$ for V, and the basis $(1, x, x^2)$ for W. The basis elements of V are transformed as follows:

$$D(1) = 0, \qquad D(1 + x) = 1, \qquad D(1 + x + x^2) = 1 + 2x,$$

$$D(1 + x + x^2 + x^3) = 1 + 2x + 3x^2,$$

so the matrix representation in this case is

$$\begin{bmatrix} 0 & 1 & 1 & 1 \\ 0 & 0 & 2 & 2 \\ 0 & 0 & 0 & 3 \end{bmatrix}.$$

2.11 Construction of a matrix representation in diagonal form

Since it is possible to obtain different matrix representations of a given linear transformation by different choices of bases, it is natural to try to choose the bases so that the resulting matrix will have a particularly simple form. The next theorem shows that we can make all the entries 0 except possibly along the diagonal starting from the upper left-hand corner of the matrix. Along this diagonal there will be a string of ones followed by zeros, the number of ones being equal to the rank of the transformation. A matrix (t_{ik}) with all entries $t_{ik} = 0$ when $i \neq k$ is said to be a *diagonal matrix*.

THEOREM 2.14. *Let V and W be finite-dimensional linear spaces, with* $\dim V = n$ *and* $\dim W = m$. *Assume $T \in \mathscr{L}(V, W)$ and let $r = \dim T(V)$ denote the rank of T. Then there exists a basis $(e_1, \ldots, e_n)$ for V and a basis $(w_1, \ldots, w_m)$ for W such that*

(2.14) $$T(e_i) = w_i \quad \text{for} \quad i = 1, 2, \ldots, r,$$

and

(2.15) $$T(e_i) = O \quad \text{for} \quad i = r + 1, \ldots, n.$$

Therefore, the matrix (t_{ik}) of T relative to these bases has all entries zero except for the r diagonal entries

$$t_{11} = t_{22} = \cdots = t_{rr} = 1.$$

Proof. First we construct a basis for W. Since $T(V)$ is a subspace of W with $\dim T(V) = r$, the space $T(V)$ has a basis of r elements in W, say $w_1, \ldots, w_r$. By Theorem 1.7, these elements form a subset of some basis for W. Therefore we can adjoin elements $w_{r+1}, \ldots, w_m$ so that

(2.16) $$(w_1, \ldots, w_r, w_{r+1}, \ldots, w_m)$$

is a basis for W.

Now we construct a basis for V. Each of the first r elements w_i in (2.16) is the image of at least one element in V. Choose one such element in V and call it e_i. Then $T(e_i) = w_i$ for $i = 1, 2, \ldots, r$ so (2.14) is satisfied. Now let k be the dimension of the null space $N(T)$. By Theorem 2.3 we have $n = k + r$. Since $\dim N(T) = k$, the space $N(T)$ has a basis consisting of k elements in V which we designate as $e_{r+1}, \ldots, e_{r+k}$. For each of these elements, Equation (2.15) is satisfied. Therefore, to complete the proof, we must show that the ordered set

$$(e_1, \ldots, e_r, e_{r+1}, \ldots, e_{r+k}) \tag{2.17}$$

is a basis for V. Since $\dim V = n = r + k$, we need only show that these elements are independent. Suppose that some linear combination of them is zero, say

$$\sum_{i=1}^{r+k} c_i e_i = O. \tag{2.18}$$

Applying T and using Equations (2.14) and (2.15), we find that

$$\sum_{i=1}^{r+k} c_i T(e_i) = \sum_{i=1}^{r} c_i w_i = O.$$

But $w_1, \ldots, w_r$ are independent, and hence $c_1 = \cdots = c_r = 0$. Therefore, the first r terms in (2.18) are zero, so (2.18) reduces to

$$\sum_{i=r+1}^{r+k} c_i e_i = O.$$

But $e_{r+1}, \ldots, e_{r+k}$ are independent since they form a basis for $N(T)$, and hence $c_{r+1} = \cdots = c_{r+k} = 0$. Therefore, all the c_i in (2.18) are zero, so the elements in (2.17) form a basis for V. This completes the proof.

EXAMPLE. We refer to Example 2 of Section 2.10, where D is the differentiation operator which maps the space V of polynomials of degree ≤ 3 into the space W of polynomials of degree ≤ 2. In this example, the range $T(V) = W$, so T has rank 3. Applying the method used to prove Theorem 2.14, we choose any basis for W, for example the basis $(1, x, x^2)$. A set of polynomials in V which map onto these elements is given by $(x, \frac{1}{2}x^2, \frac{1}{3}x^3)$. We extend this set to get a basis for V by adjoining the constant polynomial 1, which is a basis for the null space of D. Therefore, if we use the basis $(x, \frac{1}{2}x^2, \frac{1}{3}x^3, 1)$ for V and the basis $(1, x, x^2)$ for W, the corresponding matrix representation for D has the diagonal form

$$\begin{bmatrix} 1 & 0 & 0 & 0 \\ 0 & 1 & 0 & 0 \\ 0 & 0 & 1 & 0 \end{bmatrix}.$$

2.12 Exercises

In all exercises involving the vector space V_n, the usual basis of unit coordinate vectors is to be chosen unless another basis is specifically mentioned. In exercises concerned with the matrix of a linear transformation $T: V \to W$ where $V = W$, we take the same basis in both V and W unless another choice is indicated.

1. Determine the matrix of each of the following linear transformations of V_n into V_n:
   (a) the identity transformation,
   (b) the zero transformation,
   (c) multiplication by a fixed scalar c.
2. Determine the matrix for each of the following projections.
   (a) $T: V_3 \to V_2$, where $T(x_1, x_2, x_3) = (x_1, x_2)$.
   (b) $T: V_3 \to V_2$, where $T(x_1, x_2, x_3) = (x_2, x_3)$.
   (c) $T: V_5 \to V_3$, where $T(x_1, x_2, x_3, x_4, x_5) = (x_2, x_3, x_4)$.
3. A linear transformation $T: V_2 \to V_2$ maps the basis vectors $\boldsymbol{i}$ and $\boldsymbol{j}$ as follows:

$$T(\boldsymbol{i}) = \boldsymbol{i} + \boldsymbol{j}, \qquad T(\boldsymbol{j}) = 2\boldsymbol{i} - \boldsymbol{j}.$$

   (a) Compute $T(3\boldsymbol{i} - 4\boldsymbol{j})$ and $T^2(3\boldsymbol{i} - 4\boldsymbol{j})$ in terms of $\boldsymbol{i}$ and $\boldsymbol{j}$.
   (b) Determine the matrix of T and of T^2.
   (c) Solve part (b) if the basis $(\boldsymbol{i}, \boldsymbol{j})$ is replaced by (e_1, e_2), where $e_1 = \boldsymbol{i} - \boldsymbol{j}$, $e_2 = 3\boldsymbol{i} + \boldsymbol{j}$.
4. A linear transformation $T: V_2 \to V_2$ is defined as follows: Each vector (x, y) is reflected in the y-axis and then doubled in length to yield $T(x, y)$. Determine the matrix of T and of T^2.
5. Let $T: V_3 \to V_3$ be a linear transformation such that

$$T(\boldsymbol{k}) = 2\boldsymbol{i} + 3\boldsymbol{j} + 5\boldsymbol{k}, \qquad T(\boldsymbol{j} + \boldsymbol{k}) = \boldsymbol{i}, \qquad T(\boldsymbol{i} + \boldsymbol{j} + \boldsymbol{k}) = \boldsymbol{j} - \boldsymbol{k}.$$

   (a) Compute $T(\boldsymbol{i} + 2\boldsymbol{j} + 3\boldsymbol{k})$ and determine the nullity and rank of T.
   (b) Determine the matrix of T.
6. For the linear transformation in Exercise 5, choose both bases to be (e_1, e_2, e_3), where $e_1 = (2, 3, 5)$, $e_2 = (1, 0, 0)$, $e_3 = (0, 1, -1)$, and determine the matrix of T relative to the new bases.
7. A linear transformation $T: V_3 \to V_2$ maps the basis vectors as follows: $T(\boldsymbol{i}) = (0, 0)$, $T(\boldsymbol{j}) = (1, 1)$, $T(\boldsymbol{k}) = (1, -1)$.
   (a) Compute $T(4\boldsymbol{i} - \boldsymbol{j} + \boldsymbol{k})$ and determine the nullity and rank of T.
   (b) Determine the matrix of T.
   (c) Use the basis $(\boldsymbol{i}, \boldsymbol{j}, \boldsymbol{k})$ in V_3 and the basis (w_1, w_2) in V_2, where $w_1 = (1, 1)$, $w_2 = (1, 2)$. Determine the matrix of T relative to these bases.
   (d) Find bases (e_1, e_2, e_3) for V_3 and (w_1, w_2) for V_2 relative to which the matrix of T will be in diagonal form.
8. A linear transformation $T: V_2 \to V_3$ maps the basis vectors as follows: $T(\boldsymbol{i}) = (1, 0, 1)$, $T(\boldsymbol{j}) = (-1, 0, 1)$.
   (a) Compute $T(2\boldsymbol{i} - 3\boldsymbol{j})$ and determine the nullity and rank of T.
   (b) Determine the matrix of T.
   (c) Find bases (e_1, e_2) for V_2 and (w_1, w_2, w_3) for V_3 relative to which the matrix of T will be in diagonal form.
9. Solve Exercise 8 if $T(\boldsymbol{i}) = (1, 0, 1)$ and $T(\boldsymbol{j}) = (1, 1, 1)$.
10. Let V and W be linear spaces, each with dimension 2 and each with basis (e_1, e_2). Let $T: V \to W$ be a linear transformation such that $T(e_1 + e_2) = 3e_1 + 9e_2$, $T(3e_1 + 2e_2) = 7e_1 + 23e_2$.
   (a) Compute $T(e_2 - e_1)$ and determine the nullity and rank of T.
   (b) Determine the matrix of T relative to the given basis.

(c) Use the basis (e_1, e_2) for V and find a new basis of the form $(e_1 + ae_2, 2e_1 + be_2)$ for W, relative to which the matrix of T will be in diagonal form.

In the linear space of all real-valued functions, each of the following sets is independent and spans a finite-dimensional subspace V. Use the given set as a basis for V and let $D: V \to V$ be the differentiation operator. In each case, find the matrix of D and of D^2 relative to this choice of basis.

11. $(\sin x, \cos x)$.
12. $(1, x, e^x)$.
13. $(1, 1 + x, 1 + x + e^x)$.
14. (e^x, xe^x).
15. $(-\cos x, \sin x)$.
16. $(\sin x, \cos x, x \sin x, x \cos x)$.
17. $(e^x \sin x, e^x \cos x)$.
18. $(e^{2x} \sin 3x, e^{2x} \cos 3x)$.
19. Choose the basis $(1, x, x^2, x^3)$ in the linear space V of all real polynomials of degree ≤ 3. Let D denote the differentiation operator and let $T: V \to V$ be the linear transformation which maps $p(x)$ onto $xp'(x)$. Relative to the given basis, determine the matrix of each of the following transformations: (a) T; (b) DT; (c) TD; (d) $TD - DT$; (e) T^2; (f) $T^2D^2 - D^2T^2$.
20. Refer to Exercise 19. Let W be the image of V under TD. Find bases for V and for W relative to which the matrix of TD is in diagonal form.

2.13 Linear spaces of matrices

We have seen how matrices arise in a natural way as representations of linear transformations. Matrices can also be considered as objects existing in their own right, without necessarily being connected to linear transformations. As such, they form another class of mathematical objects on which algebraic operations can be defined. The connection with linear transformations serves as motivation for these definitions, but this connection will be ignored for the moment.

Let m and n be two positive integers, and let $I_{m,n}$ be the set of all pairs of integers (i, j) such that $1 \leq i \leq m, 1 \leq j \leq n$. Any function A whose domain is $I_{m,n}$ is called an $m \times n$ *matrix*. The function value $A(i, j)$ is called the *ij-entry* or *ij-element* of the matrix and will be denoted also by a_{ij}. It is customary to display all the function values in a rectangular array consisting of m rows and n columns, as follows:

$$\begin{bmatrix} a_{11} & a_{12} & \cdots & a_{1n} \\ a_{21} & a_{22} & \cdots & a_{2n} \\ \cdot & \cdot & & \cdot \\ \cdot & \cdot & & \cdot \\ \cdot & \cdot & & \cdot \\ a_{m1} & a_{m2} & \cdots & a_{mn} \end{bmatrix}.$$

The elements a_{ij} may be arbitrary objects of any kind. Usually they will be real or complex numbers, but sometimes it is convenient to consider matrices whose elements are other objects, for example, functions. We also denote matrices by the more compact notation

$$A = (a_{ij})_{i,j=1}^{m,n} \qquad \text{or} \qquad A = (a_{ij}).$$

If $m = n$, the matrix is said to be a *square matrix*. A $1 \times n$ matrix is called a *row matrix;* an $m \times 1$ matrix is called a *column matrix.*

Two functions are equal if and only if they have the same domain and take the same function value at each element in the domain. Since matrices are functions, two matrices $A = (a_{ij})$ and $B = (b_{ij})$ are equal if and only if they have the same number of rows, the same number of columns, and equal entries $a_{ij} = b_{ij}$ for each pair (i, j).

Now we assume the entries are numbers (real or complex) and we define addition of matrices and multiplication by scalars by the same method used for any real- or complex-valued functions.

DEFINITION. *If $A = (a_{ij})$ and $B = (b_{ij})$ are two $m \times n$ matrices and if c is any scalar, we define matrices $A + B$ and cA as follows:*

$$A + B = (a_{ij} + b_{ij}), \qquad cA = (ca_{ij}).$$

The sum is defined only when A and B have the same size.

EXAMPLE. If

$$A = \begin{bmatrix} 1 & 2 & -3 \\ -1 & 0 & 4 \end{bmatrix} \quad \text{and} \quad B = \begin{bmatrix} 5 & 0 & 1 \\ 1 & -2 & 3 \end{bmatrix},$$

then we have

$$A + B = \begin{bmatrix} 6 & 2 & -2 \\ 0 & -2 & 7 \end{bmatrix}, \quad 2A = \begin{bmatrix} 2 & 4 & -6 \\ -2 & 0 & 8 \end{bmatrix}, \quad (-1)B = \begin{bmatrix} -5 & 0 & -1 \\ -1 & 2 & -3 \end{bmatrix}.$$

We define the zero matrix O to be the $m \times n$ matrix all of whose elements are 0. With these definitions, it is a straightforward exercise to verify that the collection of all $m \times n$ matrices is a linear space. We denote this linear space by $M_{m,n}$. If the entries are real numbers, the space $M_{m,n}$ is a real linear space. If the entries are complex, $M_{m,n}$ is a complex linear space. It is also easy to prove that this space has dimension mn. In fact, a basis for $M_{m,n}$ consists of the mn matrices having one entry equal to 1 and all others equal to 0. For example, the six matrices

$$\begin{bmatrix} 1 & 0 & 0 \\ 0 & 0 & 0 \end{bmatrix}, \quad \begin{bmatrix} 0 & 1 & 0 \\ 0 & 0 & 0 \end{bmatrix}, \quad \begin{bmatrix} 0 & 0 & 1 \\ 0 & 0 & 0 \end{bmatrix}, \quad \begin{bmatrix} 0 & 0 & 0 \\ 1 & 0 & 0 \end{bmatrix}, \quad \begin{bmatrix} 0 & 0 & 0 \\ 0 & 1 & 0 \end{bmatrix}, \quad \begin{bmatrix} 0 & 0 & 0 \\ 0 & 0 & 1 \end{bmatrix},$$

form a basis for the set of all 2×3 matrices.

2.14 Isomorphism between linear transformations and matrices

We return now to the connection between matrices and linear transformations. Let V and W be finite-dimensional linear spaces with $\dim V = n$ and $\dim W = m$. Choose a basis $(e_1, \ldots, e_n)$ for V and a basis $(w_1, \ldots, w_m)$ for W. In this discussion, these bases are kept fixed. Let $\mathscr{L}(V, W)$ denote the linear space of all linear transformations of V into W. If $T \in \mathscr{L}(V, W)$, let $m(T)$ denote the matrix of T relative to the given bases. We recall that $m(T)$ is defined as follows.

The image of each basis element e_k is expressed as a linear combination of the basis elements in W:

(2.19) $$T(e_k) = \sum_{i=1}^{m} t_{ik} w_i \qquad \text{for} \quad k = 1, 2, \ldots, n.$$

The scalar multipliers t_{ik} are the ik-entries of $m(T)$. Thus, we have

(2.20) $$m(T) = (t_{ik})_{i,k=1}^{m,n}.$$

Equation (2.20) defines a new function m whose domain is $\mathscr{L}(V, W)$ and whose values are matrices in $M_{m,n}$. Since every $m \times n$ matrix is the matrix $m(T)$ for some T in $\mathscr{L}(V, W)$, the range of m is $M_{m,n}$. The next theorem shows that the transformation $m: \mathscr{L}(V, W) \to M_{m,n}$ is linear and one-to-one on $\mathscr{L}(V, W)$.

THEOREM 2.15. ISOMORPHISM THEOREM. *For all S and T in $\mathscr{L}(V, W)$ and all scalars c, we have*

$$m(S + T) = m(S) + m(T) \qquad \textit{and} \qquad m(cT) = cm(T).$$

Moreover,

$$m(S) = m(T) \qquad \textit{implies} \qquad S = T,$$

so m is one-to-one on $\mathscr{L}(V, W)$.

Proof. The matrix $m(T)$ is formed from the multipliers t_{ik} in (2.19). Similarly, the matrix $m(S)$ is formed from the multipliers s_{ik} in the equations

(2.21) $$S(e_k) = \sum_{i=1}^{m} s_{ik} w_i \qquad \text{for} \quad k = 1, 2, \ldots, n.$$

Since we have

$$(S + T)(e_k) = \sum_{i=1}^{m} (s_{ik} + t_{ik}) w_i \qquad \text{and} \qquad (cT)(e_k) = \sum_{i=1}^{m} (ct_{ik}) w_i,$$

we obtain $m(S + T) = (s_{ik} + t_{ik}) = m(S) + m(T)$ and $m(cT) = (ct_{ik}) = cm(T)$. This proves that m is linear.

To prove that m is one-to-one, suppose that $m(S) = m(T)$, where $S = (s_{ik})$ and $T = (t_{ik})$. Equations (2.19) and (2.21) show that $S(e_k) = T(e_k)$ for each basis element e_k, so $S(x) = T(x)$ for all x in V, and hence $S = T$.

Note: The function m is called an *isomorphism*. For a given choice of bases, m establishes a one-to-one correspondence between the set of linear transformations $\mathscr{L}(V, W)$ and the set of $m \times n$ matrices $M_{m,n}$. The operations of addition and multiplication by scalars are preserved under this correspondence. The linear spaces $\mathscr{L}(V, W)$ and $M_{m,n}$ are said to be *isomorphic*. Incidentally, Theorem 2.11 shows that the domain of a one-to-one linear transformation has the same dimension as its range. Therefore, $\dim \mathscr{L}(V, W) = \dim M_{m,n} = mn$.

If $V = W$ and if we choose the same basis in both V and W, then the matrix $m(I)$ which corresponds to the identity transformation $I: V \to V$ is an $n \times n$ diagonal matrix with each diagonal entry equal to 1 and all others equal to 0. This is called the *identity* or *unit matrix* and is denoted by I or by I_n.

2.15 Multiplication of matrices

Some linear transformations can be multiplied by means of composition. Now we shall define multiplication of matrices in such a way that the product of two matrices corresponds to the composition of the linear transformations they represent.

We recall that if $T: U \to V$ and $S: V \to W$ are linear transformations, their composition $ST: U \to W$ is a linear transformation given by

$$ST(x) = S[T(x)] \qquad \text{for all } x \text{ in } U.$$

Suppose that U, V, and W are finite-dimensional, say

$$\dim U = n, \qquad \dim V = p, \qquad \dim W = m.$$

Choose bases for U, V, and W. Relative to these bases, the matrix $m(S)$ is an $m \times p$ matrix, the matrix T is a $p \times n$ matrix, and the matrix of ST is an $m \times n$ matrix. The following definition of matrix multiplication will enable us to deduce the relation $m(ST) = m(S)m(T)$. This extends the isomorphism property to products.

DEFINITION. *Let A be any $m \times p$ matrix, and let B be any $p \times n$ matrix, say*

$$A = (a_{ij})_{i,j=1}^{m,p} \qquad \text{and} \qquad B = (b_{ij})_{i,j=1}^{p,n}.$$

Then the product AB is defined to be the $m \times n$ matrix $C = (c_{ij})$ whose ij-entry is given by

$$c_{ij} = \sum_{k=1}^{p} a_{ik} b_{kj}. \tag{2.22}$$

Note: The product AB is not defined unless the number of columns of A is equal to the number of rows of B.

If we write A_i for the ith row of A, and B^j for the jth column of B, and think of these as p-dimensional vectors, then the sum in (2.22) is simply the dot product $A_i \cdot B^j$. In other words, the ij-entry of AB is the dot product of the ith row of A with the jth column of B:

$$AB = (A_i \cdot B^j)_{i,j=1}^{m,n}.$$

Thus, matrix multiplication can be regarded as a generalization of the dot product.

EXAMPLE 1. Let $A = \begin{bmatrix} 3 & 1 & 2 \\ -1 & 1 & 0 \end{bmatrix}$ and $B = \begin{bmatrix} 4 & 6 \\ 5 & -1 \\ 0 & 2 \end{bmatrix}$. Since A is 2×3 and B is 3×2, the product AB is the 2×2 matrix

$$AB = \begin{bmatrix} A_1 \cdot B^1 & A_1 \cdot B^2 \\ A_2 \cdot B^1 & A_2 \cdot B^2 \end{bmatrix} = \begin{bmatrix} 17 & 21 \\ 1 & -7 \end{bmatrix}.$$

The entries of AB are computed as follows:

$$A_1 \cdot B^1 = 3 \cdot 4 + 1 \cdot 5 + 2 \cdot 0 = 17, \qquad A_1 \cdot B^2 = 3 \cdot 6 + 1 \cdot (-1) + 2 \cdot 2 = 21,$$

$$A_2 \cdot B^1 = (-1) \cdot 4 + 1 \cdot 5 + 0 \cdot 0 = 1, \qquad A_2 \cdot B^2 = (-1) \cdot 6 + 1 \cdot (-1) + 0 \cdot 2 = -7.$$

EXAMPLE 2. Let

$$A = \begin{bmatrix} 2 & 1 & -3 \\ 1 & 2 & 4 \end{bmatrix} \quad \text{and} \quad B = \begin{bmatrix} -2 \\ 1 \\ 2 \end{bmatrix}.$$

Here A is 2×3 and B is 3×1, so AB is the 2×1 matrix given by

$$AB = \begin{bmatrix} A_1 \cdot B^1 \\ A_2 \cdot B^1 \end{bmatrix} = \begin{bmatrix} -9 \\ 8 \end{bmatrix},$$

since $A_1 \cdot B^1 = 2 \cdot (-2) + 1 \cdot 1 + (-3) \cdot 2 = -9$ and $A_2 \cdot B^1 = 1 \cdot (-2) + 2 \cdot 1 + 4 \cdot 2 = 8$.

EXAMPLE 3. If A and B are both square matrices of the same size, then both AB and BA are defined. For example, if

$$A = \begin{bmatrix} 1 & 2 \\ -1 & 1 \end{bmatrix} \quad \text{and} \quad B = \begin{bmatrix} 3 & 4 \\ 5 & 2 \end{bmatrix},$$

we find that

$$AB = \begin{bmatrix} 13 & 8 \\ 2 & -2 \end{bmatrix}, \qquad BA = \begin{bmatrix} -1 & 10 \\ 3 & 12 \end{bmatrix}.$$

This example shows that in general $AB \neq BA$. If $AB = BA$, we say A and B *commute.*

EXAMPLE 4. If I_p is the $p \times p$ identity matrix, then $I_pA = A$ for every $p \times n$ matrix A, and $BI_p = B$ for every $m \times p$ matrix B. For example,

$$\begin{bmatrix} 1 & 0 & 0 \\ 0 & 1 & 0 \\ 0 & 0 & 1 \end{bmatrix}\begin{bmatrix} 2 \\ 3 \\ 4 \end{bmatrix} = \begin{bmatrix} 2 \\ 3 \\ 4 \end{bmatrix}, \qquad \begin{bmatrix} 1 & 2 & 3 \\ 4 & 5 & 6 \end{bmatrix}\begin{bmatrix} 1 & 0 & 0 \\ 0 & 1 & 0 \\ 0 & 0 & 1 \end{bmatrix} = \begin{bmatrix} 1 & 2 & 3 \\ 4 & 5 & 6 \end{bmatrix}.$$

Now we prove that the matrix of a composition ST is the product of the matrices $m(S)$ and $m(T)$.

THEOREM 2.16. *Let $T: U \to V$ and $S: V \to W$ be linear transformations, where U, V, W are finite-dimensional linear spaces. Then, for a fixed choice of bases, the matrices of S, T, and ST are related by the equation*

$$m(ST) = m(S)m(T).$$

Proof. Assume $\dim U = n$, $\dim V = p$, $\dim W = m$. Let $(u_1, \ldots, u_n)$ be a basis for U, $(v_1, \ldots, v_p)$ a basis for V, and $(w_1, \ldots, w_m)$ a basis for W. Relative to these bases, we have

$$m(S) = (s_{ij})_{i,j=1}^{m,p}, \qquad \text{where} \quad S(v_k) = \sum_{i=1}^{m} s_{ik} w_i \qquad \text{for} \quad k = 1, 2, \ldots, p,$$

and

$$m(T) = (t_{ij})_{i,j=1}^{p,n}, \qquad \text{where} \quad T(u_j) = \sum_{k=1}^{p} t_{kj} v_k \qquad \text{for} \quad j = 1, 2, \ldots, n.$$

Therefore, we have

$$ST(u_j) = S[T(u_j)] = \sum_{k=1}^{p} t_{kj} S(v_k) = \sum_{k=1}^{p} t_{kj} \sum_{i=1}^{m} s_{ik} w_i = \sum_{i=1}^{m} \left(\sum_{k=1}^{p} s_{ik} t_{kj} \right) w_i,$$

so we find that

$$m(ST) = \left(\sum_{k=1}^{p} s_{ik} t_{kj} \right)_{i,j=1}^{m,n} = m(S)m(T).$$

We have already noted that matrix multiplication does not always satisfy the commutative law. The next theorem shows that it does satisfy the associative and distributive laws.

THEOREM 2.17. ASSOCIATIVE AND DISTRIBUTIVE LAWS FOR MATRIX MULTIPLICATION. *Given matrices A, B, C.*

(a) *If the products $A(BC)$ and $(AB)C$ are meaningful, we have*

$$A(BC) = (AB)C \qquad (\textit{associative law}).$$

(b) *Assume A and B are of the same size. If AC and BC are meaningful, we have*

$$(A + B)C = AC + BC \qquad (\textit{right distributive law}),$$

whereas if CA and CB are meaningful, we have

$$C(A + B) = CA + CB \qquad (\textit{left distributive law}).$$

Proof. These properties can be deduced directly from the definition of matrix multiplication, but we prefer the following type of argument. Introduce finite-dimensional linear spaces U, V, W, X and linear transformations $T: U \to V$, $S: V \to W$, $R: W \to X$ such that, for a fixed choice of bases, we have

$$A = m(R), \qquad B = m(S), \qquad C = m(T).$$

By Theorem 2.16, we have $m(RS) = AB$ and $m(ST) = BC$. From the associative law for composition, we find that $R(ST) = (RS)T$. Applying Theorem 2.16 once more to this equation, we obtain $m(R)m(ST) = m(RS)m(T)$ or $A(BC) = (AB)C$, which proves (a). The proof of (b) can be given by a similar type of argument.

DEFINITION. *If A is a square matrix, we define integral powers of A inductively as follows:*

$$A^0 = I, \qquad A^n = AA^{n-1} \quad \textit{for} \quad n \geq 1.$$

2.16 Exercises

1. If $A = \begin{bmatrix} 1 & -4 & 2 \\ -1 & 4 & -2 \end{bmatrix}$, $B = \begin{bmatrix} 1 & 2 \\ -1 & 3 \\ 5 & -2 \end{bmatrix}$, $C = \begin{bmatrix} 2 & 2 \\ 1 & -1 \\ 1 & -3 \end{bmatrix}$, compute $B + C$, AB, BA, AC, CA, $A(2B - 3C)$.

2. Let $A = \begin{bmatrix} 0 & 1 \\ 0 & 2 \end{bmatrix}$. Find all 2×2 matrices B such that (a) $AB = O$; (b) $BA = O$.

3. In each case find a, b, c, d to satisfy the given equation.

(a) $\begin{bmatrix} 0 & 0 & 1 & 0 \\ 1 & 0 & 0 & 0 \\ 0 & 1 & 0 & 0 \\ 0 & 0 & 0 & 1 \end{bmatrix} \begin{bmatrix} a \\ b \\ c \\ d \end{bmatrix} = \begin{bmatrix} 1 \\ 9 \\ 6 \\ 5 \end{bmatrix}$; (b) $\begin{bmatrix} a & b & c & d \\ 1 & 4 & 9 & 2 \end{bmatrix} \begin{bmatrix} 1 & 0 & 2 & 0 \\ 0 & 0 & 1 & 1 \\ 0 & 1 & 0 & 0 \\ 0 & 0 & 1 & 0 \end{bmatrix} = \begin{bmatrix} 1 & 0 & 6 & 6 \\ 1 & 9 & 8 & 4 \end{bmatrix}$.

4. Calculate $AB - BA$ in each case.

(a) $A = \begin{bmatrix} 1 & 2 & 2 \\ 2 & 1 & 2 \\ 1 & 2 & 3 \end{bmatrix}$, $B = \begin{bmatrix} 4 & 1 & 1 \\ -4 & 2 & 0 \\ 1 & 2 & 1 \end{bmatrix}$;

(b) $A = \begin{bmatrix} 2 & 0 & 0 \\ 1 & 1 & 2 \\ -1 & 2 & 1 \end{bmatrix}$, $B = \begin{bmatrix} 3 & 1 & -2 \\ 3 & -2 & 4 \\ -3 & 5 & 11 \end{bmatrix}$.

5. If A is a square matrix, prove that $A^nA^m = A^{m+n}$ for all integers $m \geq 0$, $n \geq 0$.

6. Let $A = \begin{bmatrix} 1 & 1 \\ 0 & 1 \end{bmatrix}$. Verify that $A^2 = \begin{bmatrix} 1 & 2 \\ 0 & 1 \end{bmatrix}$ and compute A^n.

7. Let $A = \begin{bmatrix} \cos\theta & -\sin\theta \\ \sin\theta & \cos\theta \end{bmatrix}$. Verify that $A^2 = \begin{bmatrix} \cos 2\theta & -\sin 2\theta \\ \sin 2\theta & \cos 2\theta \end{bmatrix}$ and compute A^n.

8. Let $A = \begin{bmatrix} 1 & 1 & 1 \\ 0 & 1 & 1 \\ 0 & 0 & 1 \end{bmatrix}$. Verify that $A^2 = \begin{bmatrix} 1 & 2 & 3 \\ 0 & 1 & 2 \\ 0 & 0 & 1 \end{bmatrix}$. Compute A^3 and A^4. Guess a general formula for A^n and prove it by induction.

9. Let $A = \begin{bmatrix} 1 & 0 \\ -1 & 1 \end{bmatrix}$. Prove that $A^2 = 2A - I$ and compute A^{100}.

10. Find all 2×2 matrices A such that $A^2 = O$.
11. (a) Prove that a 2×2 matrix A commutes with every 2×2 matrix if and only if A commutes with each of the four matrices

$$\begin{bmatrix} 1 & 0 \\ 0 & 0 \end{bmatrix}, \quad \begin{bmatrix} 0 & 1 \\ 0 & 0 \end{bmatrix}, \quad \begin{bmatrix} 0 & 0 \\ 1 & 0 \end{bmatrix}, \quad \begin{bmatrix} 0 & 0 \\ 0 & 1 \end{bmatrix}.$$

(b) Find all such matrices A.
12. The equation $A^2 = I$ is satisfied by each of the 2×2 matrices

$$\begin{bmatrix} 1 & 0 \\ 0 & 1 \end{bmatrix}, \quad \begin{bmatrix} 1 & 0 \\ c & -1 \end{bmatrix}, \quad \begin{bmatrix} 1 & b \\ 0 & -1 \end{bmatrix},$$

where b and c are arbitrary real numbers. Find all 2×2 matrices A such that $A^2 = I$.

13. If $A = \begin{bmatrix} 2 & -1 \\ -2 & 3 \end{bmatrix}$ and $B = \begin{bmatrix} 7 & 6 \\ 9 & 8 \end{bmatrix}$, find 2×2 matrices C and D such that $AC = B$ and $DA = B$.
14. (a) Verify that the algebraic identities

$$(A + B)^2 = A^2 + 2AB + B^2 \quad \text{and} \quad (A + B)(A - B) = A^2 - B^2$$

do not hold for the 2×2 matrices $A = \begin{bmatrix} 1 & -1 \\ 0 & 2 \end{bmatrix}$ and $B = \begin{bmatrix} 1 & 0 \\ 1 & 2 \end{bmatrix}$.

(b) Amend the right-hand members of these identities to obtain formulas valid for all square matrices A and B.
(c) For which matrices A and B are the identities valid as stated in (a)?

2.17 Systems of linear equations

Let $A = (a_{ij})$ be a given $m \times n$ matrix of numbers, and let $c_1, \ldots, c_m$ be m further numbers. A set of m equations of the form

$$\sum_{k=1}^{n} a_{ik}x_k = c_i \quad \text{for} \quad i = 1, 2, \ldots, m, \tag{2.23}$$

is called a system of m linear equations in n unknowns. Here $x_1, \ldots, x_n$ are regarded as unknown. A *solution* of the system is any n-tuple of numbers $(x_1, \ldots, x_n)$ for which all the equations are satisfied. The matrix A is called the *coefficient-matrix* of the system.

Linear systems can be studied with the help of linear transformations. Choose the usual bases of unit coordinate vectors in V_n and in V_m. The coefficient-matrix A determines a linear transformation, $T: V_n \to V_m$, which maps an arbitrary vector $x = (x_1, \ldots, x_n)$ in V_n onto the vector $y = (y_1, \ldots, y_m)$ in V_m given by the m linear equations

$$y_i = \sum_{k=1}^{n} a_{ik} x_k \qquad \text{for} \quad i = 1, 2, \ldots, m.$$

Let $c = (c_1, \ldots, c_m)$ be the vector in V_m whose components are the numbers appearing in system (2.23). This system can be written more simply as

$$T(x) = c.$$

The system has a solution if and only if c is in the range of T. If exactly one x in V_n maps onto c, the system has exactly one solution. If more than one x maps onto c, the system has more than one solution.

EXAMPLE 1. *A system with no solution.* The system $x + y = 1$, $x + y = 2$ has no solution. The sum of two numbers cannot be both 1 and 2.

EXAMPLE 2. *A system with exactly one solution.* The system $x + y = 1$, $x - y = 0$ has exactly one solution: $(x, y) = (\frac{1}{2}, \frac{1}{2})$.

EXAMPLE 3. *A system with more than one solution.* The system $x + y = 1$, consisting of one equation in two unknowns, has more than one solution. Any two numbers whose sum is 1 gives a solution.

With each linear system (2.23), we can associate another system

$$\sum_{k=1}^{n} a_{ik} x_k = 0 \qquad \text{for} \quad i = 1, 2, \ldots, m,$$

obtained by replacing each c_i in (2.23) by 0. This is called the *homogeneous system* corresponding to (2.23). If $c \neq O$, system (2.23) is called a *nonhomogeneous system.* A vector x in V_n will satisfy the homogeneous system if and only if

$$T(x) = O,$$

where T is the linear transformation determined by the coefficient-matrix. The homogeneous system always has one solution, namely $x = O$, but it may have others. The set of solutions of the homogeneous system is the null space of T. The next theorem describes the relation between solutions of the homogeneous system and those of the nonhomogeneous system.

THEOREM 2.18. *Assume the nonhomogeneous system* (2.23) *has a solution, say b.*

(a) *If a vector x is a solution of the nonhomogeneous system, then the vector $v = x - b$ is a solution of the corresponding homogeneous system.*

(b) *If a vector v is a solution of the homogeneous system, then the vector $x = v + b$ is a solution of the nonhomogeneous system.*

Proof. Let $T: V_n \to V_m$ be the linear transformation determined by the coefficient-matrix, as described above. Since b is a solution of the nonhomogeneous system we have $T(b) = c$. Let x and v be two vectors in V_n such that $v = x - b$. Then we have

$$T(v) = T(x - b) = T(x) - T(b) = T(x) - c .$$

Therefore $T(x) = c$ if and only if $T(v) = O$. This proves both (a) and (b).

This theorem shows that the problem of finding all solutions of a nonhomogeneous system splits naturally into two parts: (1) Finding all solutions v of the homogeneous system, that is, determining the null space of T; and (2) finding one particular solution b of the nonhomogeneous system. By adding b to each vector v in the null space of T, we thereby obtain all solutions $x = v + b$ of the nonhomogeneous system.

Let k denote the dimension of $N(T)$ (the nullity of T). If we can find k *independent* solutions $v_1, \ldots, v_k$ of the homogeneous system, they will form a basis for $N(T)$, and we can obtain every v in $N(T)$ by forming all possible linear combinations

$$v = t_1v_1 + \cdots + t_kv_k ,$$

where $t_1, \ldots, t_k$ are arbitrary scalars. This linear combination is called the *general solution of the homogeneous system.* If b is one particular solution of the nonhomogeneous system, then all solutions x are given by

$$x = b + t_1v_1 + \cdots + t_kv_k .$$

This linear combination is called the *general solution of the nonhomogeneous system.* Theorem 2.18 can now be restated as follows.

THEOREM 2.19. *Let $T: V_n \to V_m$ be the linear transformation such that $T(x) = y$, where $x = (x_1, \ldots, x_n)$, $y = (y_1, \ldots, y_m)$ and*

$$y_i = \sum_{k=1}^{n} a_{ik}x_k \qquad \textit{for} \quad i = 1, 2, \ldots, m .$$

Let k denote the nullity of T. If $v_1, \ldots, v_k$ are k independent solutions of the homogeneous system $T(x) = O$, and if b is one particular solution of the nonhomogeneous system $T(x) = c$, then the general solution of the nonhomogeneous system is

$$x = b + t_1v_1 + \cdots + t_kv_k ,$$

where $t_1, \ldots, t_k$ are arbitrary scalars.

This theorem does not tell us how to decide if a nonhomogeneous system has a particular solution b, nor does it tell us how to determine solutions $v_1, \ldots, v_k$ of the homogeneous system. It does tell us what to expect when the nonhomogeneous system has a solution. The following example, although very simple, illustrates the theorem.

EXAMPLE. The system $x + y = 2$ has for its associated homogeneous system the equation $x + y = 0$. Therefore, the null space consists of all vectors in V_2 of the form $(t, -t)$, where t is arbitrary. Since $(t, -t) = t(1, -1)$, this is a one-dimensional subspace of V_2 with basis $(1, -1)$. A particular solution of the nonhomogeneous system is $(0, 2)$. Therefore the general solution of the nonhomogeneous system is given by

$$(x, y) = (0, 2) + t(1, -1) \qquad \text{or} \qquad x = t, \qquad y = 2 - t,$$

where t is arbitrary.

2.18 Computation techniques

We turn now to the problem of actually computing the solutions of a nonhomogeneous linear system. Although many methods have been developed for attacking this problem, all of them require considerable computation if the system is large. For example, to solve a system of ten equations in as many unknowns can require several hours of hand computation, even with the aid of a desk calculator.

We shall discuss a widely-used method, known as the *Gauss-Jordan elimination method*, which is relatively simple and can be easily programmed for high-speed electronic computing machines. The method consists of applying three basic types of operations on the equations of a linear system:

(1) *Interchanging two equations;*
(2) *Multiplying all the terms of an equation by a nonzero scalar;*
(3) *Adding to one equation a multiple of another.*

Each time we perform one of these operations on the system we obtain a new system having exactly the same solutions. Two such systems are called *equivalent*. By performing these operations over and over again in a systematic fashion we finally arrive at an equivalent system which can be solved by inspection.

We shall illustrate the method with some specific examples. It will then be clear how the method is to be applied in general.

EXAMPLE 1. *A system with a unique solution.* Consider the system

$$\begin{aligned} 2x - 5y + 4z &= -3 \\ x - 2y + z &= 5 \\ x - 4y + 6z &= 10. \end{aligned}$$

This particular system has a unique solution, $x = 124$, $y = 75$, $z = 31$, which we shall obtain by the Gauss-Jordan elimination process. To save labor we do not bother to copy the letters x, y, z and the equals sign over and over again, but work instead with the *augmented matrix*

$$\left[\begin{array}{ccc|c} 2 & -5 & 4 & -3 \\ 1 & -2 & 1 & 5 \\ 1 & -4 & 6 & 10 \end{array}\right] \tag{2.24}$$

obtained by adjoining the right-hand members of the system to the coefficient matrix. The three basic types of operation mentioned above are performed on the rows of the augmented matrix and are called *row operations.* At any stage of the process we can put the letters x, y, z back again and insert equals signs along the vertical line to obtain equations. Our ultimate goal is to arrive at the augmented matrix

$$\left[\begin{array}{ccc|c} 1 & 0 & 0 & 124 \\ 0 & 1 & 0 & 75 \\ 0 & 0 & 1 & 31 \end{array}\right] \tag{2.25}$$

after a succession of row operations. The corresponding system of equations is $x = 124$, $y = 75$, $z = 31$, which gives the desired solution.

The first step is to obtain a 1 in the upper left-hand corner of the matrix. We can do this by interchanging the first row of the given matrix (2.24) with either the second or third row. Or, we can multiply the first row by $\frac{1}{2}$. Interchanging the first and second rows, we get

$$\left[\begin{array}{ccc|c} 1 & -2 & 1 & 5 \\ 2 & -5 & 4 & -3 \\ 1 & -4 & 6 & 10 \end{array}\right].$$

The next step is to make all the remaining entries in the first column equal to zero, leaving the first row intact. To do this we multiply the first row by -2 and add the result to the second row. Then we multiply the first row by -1 and add the result to the third row. After these two operations, we obtain

$$\left[\begin{array}{ccc|c} 1 & -2 & 1 & 5 \\ 0 & -1 & 2 & -13 \\ 0 & -2 & 5 & 5 \end{array}\right]. \tag{2.26}$$

Now we repeat the process on the smaller matrix $\left[\begin{array}{cc|c} -1 & 2 & -13 \\ -2 & 5 & 5 \end{array}\right]$ which appears adjacent to the two zeros. We can obtain a 1 in *its* upper left-hand corner by multiplying the second row of (2.26) by -1. This gives us the matrix

$$\left[\begin{array}{ccc|c} 1 & -2 & 1 & 5 \\ 0 & 1 & -2 & 13 \\ 0 & -2 & 5 & 5 \end{array}\right].$$

Multiplying the second row by 2 and adding the result to the third, we get

$$\left[\begin{array}{ccc|c} 1 & -2 & 1 & 5 \\ 0 & 1 & -2 & 13 \\ 0 & 0 & 1 & 31 \end{array}\right]. \tag{2.27}$$

At this stage, the corresponding system of equations is given by

$$\begin{aligned} x - 2y + z &= 5 \\ y - 2z &= 13 \\ z &= 31\,. \end{aligned}$$

These equations can be solved in succession, starting with the third one and working backwards, to give us

$$z = 31\,, \qquad y = 13 + 2z = 13 + 62 = 75\,, \qquad x = 5 + 2y - z = 5 + 150 - 31 = 124\,.$$

Or, we can continue the Gauss–Jordan process by making all the entries zero above the diagonal elements in the second and third columns. Multiplying the second row of (2.27) by 2 and adding the result to the first row, we obtain

$$\left[\begin{array}{ccc|c} 1 & 0 & -3 & 31 \\ 0 & 1 & -2 & 13 \\ 0 & 0 & 1 & 31 \end{array}\right].$$

Finally, we multiply the third row by 3 and add the result to the first row, and then multiply the third row by 2 and add the result to the second row to get the matrix in (2.25).

EXAMPLE 2. *A system with more than one solution.* Consider the following system of 3 equations in 5 unknowns:

$$\begin{aligned} 2x - 5y + 4z + u - v &= -3 \\ x - 2y + z - u + v &= 5 \\ x - 4y + 6z + 2u - v &= 10\,. \end{aligned} \tag{2.28}$$

The corresponding augmented matrix is

$$\left[\begin{array}{ccccc|c} 2 & -5 & 4 & 1 & -1 & -3 \\ 1 & -2 & 1 & -1 & 1 & 5 \\ 1 & -4 & 6 & 2 & -1 & 10 \end{array}\right].$$

The coefficients of x, y, z and the right-hand members are the same as those in Example 1. If we perform the same row operations used in Example 1, we finally arrive at the augmented matrix

$$\left[\begin{array}{ccccc|c} 1 & 0 & 0 & -16 & 19 & 124 \\ 0 & 1 & 0 & -9 & 11 & 75 \\ 0 & 0 & 1 & -3 & 4 & 31 \end{array}\right].$$

The corresponding system of equations can be solved for x, y, and z in terms of u and v, giving us

$$\begin{aligned} x &= 124 + 16u - 19v \\ y &= 75 + 9u - 11v \\ z &= 31 + 3u - 4v\,. \end{aligned}$$

If we let $u = t_1$ and $v = t_2$, where t_1 and t_2 are arbitrary real numbers, and determine x, y, z by these equations, the vector (x, y, z, u, v) in V_5 given by

$$(x, y, z, u, v) = (124 + 16t_1 - 19t_2, 75 + 9t_1 - 11t_2, 31 + 3t_1 - 4t_2, t_1, t_2)$$

is a solution. By separating the parts involving t_1 and t_2, we can rewrite this as follows:

$$(x, y, z, u, v) = (124, 75, 31, 0, 0) + t_1(16, 9, 3, 1, 0) + t_2(-19, -11, -4, 0, 1)\,.$$

This equation gives the general solution of the system. The vector $(124, 75, 31, 0, 0)$ is a particular solution of the nonhomogeneous system (2.28). The two vectors $(16, 9, 3, 1, 0)$ and $(-19, -11, -4, 0, 1)$ are solutions of the corresponding homogeneous system. Since they are independent, they form a basis for the space of all solutions of the homogeneous system.

EXAMPLE 3. *A system with no solution.* Consider the system

$$\begin{aligned} 2x - 5y + 4z &= -3 \\ x - 2y + z &= 5 \\ x - 4y + 5z &= 10\,. \end{aligned} \tag{2.29}$$

This system is almost identical to that of Example 1 except that the coefficient of z in the third equation has been changed from 6 to 5. The corresponding augmented matrix is

$$\left[\begin{array}{ccc|c} 2 & -5 & 4 & -3 \\ 1 & -2 & 1 & 5 \\ 1 & -4 & 5 & 10 \end{array}\right].$$

Applying the same row operations used in Example 1 to transform (2.24) into (2.27), we arrive at the augmented matrix

$$\left[\begin{array}{ccc|c} 1 & -2 & 1 & 5 \\ 0 & 1 & -2 & 13 \\ 0 & 0 & 0 & 31 \end{array}\right]. \tag{2.30}$$

When the bottom row is expressed as an equation, it states that $0 = 31$. Therefore the original system has no solution since the two systems (2.29) and (2.30) are equivalent.

In each of the foregoing examples, the number of equations did not exceed the number of unknowns. If there are more equations than unknowns, the Gauss-Jordan process is still applicable. For example, suppose we consider the system of Example 1, which has the solution $x = 124, y = 75, z = 31$. If we adjoin a new equation to this system which is also satisfied by the same triple, for example, the equation $2x - 3y + z = 54$, then the elimination process leads to the agumented matrix

$$\left[\begin{array}{ccc|c} 1 & 0 & 0 & 124 \\ 0 & 1 & 0 & 75 \\ 0 & 0 & 1 & 31 \\ 0 & 0 & 0 & 0 \end{array}\right]$$

with a row of zeros along the bottom. But if we adjoin a new equation which is not satisfied by the triple (124, 75, 31), for example the equation $x + y + z = 1$, then the elimination process leads to an augmented matrix of the form

$$\left[\begin{array}{ccc|c} 1 & 0 & 0 & 124 \\ 0 & 1 & 0 & 75 \\ 0 & 0 & 1 & 31 \\ 0 & 0 & 0 & a \end{array}\right],$$

where $a \neq 0$. The last row now gives a contradictory equation $0 = a$ which shows that the system has no solution.

2.19 Inverses of square matrices

Let $A = (a_{ij})$ be a square $n \times n$ matrix. If there is another $n \times n$ matrix B such that $BA = I$, where I is the $n \times n$ identity matrix, then A is called *nonsingular* and B is called a *left inverse* of A.

Choose the usual basis of unit coordinate vectors in V_n and let $T: V_n \to V_n$ be the linear transformation with matrix $m(T) = A$. Then we have the following.

THEOREM 2.20. *The matrix A is nonsingular if and only if T is invertible. If $BA = I$ then $B = m(T^{-1})$.*

Proof. Assume that A is nonsingular and that $BA = I$. We shall prove that $T(x) = O$ implies $x = O$. Given x such that $T(x) = O$, let X be the $n \times 1$ column matrix formed from the components of x. Since $T(x) = O$, the matrix product AX is an $n \times 1$ column matrix consisting of zeros, so $B(AX)$ is also a column matrix of zeros. But $B(AX) = (BA)X = IX = X$, so every component of x is 0. Therefore, T is invertible, and the equation $TT^{-1} = I$ implies that $m(T)m(T^{-1}) = I$ or $Am(T^{-1}) = I$. Multiplying on the left by B, we find $m(T^{-1}) = B$. Conversely, if T is invertible, then $T^{-1}T$ is the identity transformation so $m(T^{-1})m(T)$ is the identity matrix. Therefore A is nonsingular and $m(T^{-1})A = I$.

All the properties of invertible linear transformations have their counterparts for nonsingular matrices. In particular, left inverses (if they exist) are unique, and every left inverse is also a right inverse. In other words, if A is nonsingular and $BA = I$, then $AB = I$. We call B the *inverse* of A and denote it by A^{-1}. The inverse A^{-1} is also nonsingular and *its* inverse is A.

Now we show that the problem of actually determining the entries of the inverse of a nonsingular matrix is equivalent to solving n separate nonhomogeneous linear systems.

Let $A = (a_{ij})$ be nonsingular and let $A^{-1} = (b_{ij})$ be its inverse. The entries of A and A^{-1} are related by the n^2 equations

$$\sum_{k=1}^{n} a_{ik}b_{kj} = \delta_{ij}, \tag{2.31}$$

where $\delta_{ij} = 1$ if $i = j$, and $\delta_{ij} = 0$ if $i \neq j$. For each fixed choice of j, we can regard this as a nonhomogeneous system of n linear equations in n unknowns $b_{1j}, b_{2j}, \ldots, b_{nj}$. Since A is nonsingular, each of these systems has a unique solution, the jth column of B. All these systems have the same coefficient-matrix A and differ only in their right members. For example, if A is a 3×3 matrix, there are 9 equations in (2.31) which can be expressed as 3 separate linear systems having the following augmented matrices:

$$\left[\begin{array}{ccc|c} a_{11} & a_{12} & a_{13} & 1 \\ a_{21} & a_{22} & a_{23} & 0 \\ a_{31} & a_{32} & a_{33} & 0 \end{array}\right], \quad \left[\begin{array}{ccc|c} a_{11} & a_{12} & a_{13} & 0 \\ a_{21} & a_{22} & a_{23} & 1 \\ a_{31} & a_{32} & a_{33} & 0 \end{array}\right], \quad \left[\begin{array}{ccc|c} a_{11} & a_{12} & a_{13} & 0 \\ a_{21} & a_{22} & a_{23} & 0 \\ a_{31} & a_{32} & a_{33} & 1 \end{array}\right].$$

If we apply the Gauss-Jordan process, we arrive at the respective augmented matrices

$$\left[\begin{array}{ccc|c} 1 & 0 & 0 & b_{11} \\ 0 & 1 & 0 & b_{21} \\ 0 & 0 & 1 & b_{31} \end{array}\right], \quad \left[\begin{array}{ccc|c} 1 & 0 & 0 & b_{12} \\ 0 & 1 & 0 & b_{22} \\ 0 & 0 & 1 & b_{32} \end{array}\right], \quad \left[\begin{array}{ccc|c} 1 & 0 & 0 & b_{13} \\ 0 & 1 & 0 & b_{23} \\ 0 & 0 & 1 & b_{33} \end{array}\right].$$

In actual practice we exploit the fact that all three systems have the same coefficient-matrix and solve all three systems at once by working with the enlarged matrix

$$\left[\begin{array}{ccc|ccc} a_{11} & a_{12} & a_{13} & 1 & 0 & 0 \\ a_{21} & a_{22} & a_{23} & 0 & 1 & 0 \\ a_{31} & a_{32} & a_{33} & 0 & 0 & 1 \end{array}\right].$$

The elimination process then leads to

$$\left[\begin{array}{ccc|ccc} 1 & 0 & 0 & b_{11} & b_{12} & b_{13} \\ 0 & 1 & 0 & b_{21} & b_{22} & b_{23} \\ 0 & 0 & 1 & b_{31} & b_{32} & b_{33} \end{array}\right].$$

The matrix on the right of the vertical line is the required inverse. The matrix on the left of the line is the 3×3 identity matrix.

It is not necessary to know in advance whether A is nonsingular. If A is *singular* (not nonsingular), we can still apply the Gauss-Jordan method, but somewhere in the process one of the diagonal elements will become zero, and it will not be possible to transform A to the identity matrix.

A system of n linear equations in n unknowns, say

$$\sum_{k=1}^{n} a_{ik}x_k = c_i, \qquad i = 1, 2, \ldots, n,$$

can be written more simply as a matrix equation,

$$AX = C,$$

where $A = (a_{ij})$ is the coefficient matrix, and X and C are column matrices,

$$X = \begin{bmatrix} x_1 \\ x_2 \\ \cdot \\ \cdot \\ \cdot \\ x_n \end{bmatrix}, \qquad C = \begin{bmatrix} c_1 \\ c_2 \\ \cdot \\ \cdot \\ \cdot \\ c_n \end{bmatrix}.$$

If A is nonsingular there is a unique solution of the system given by $X = A^{-1}C$.

2.20 Exercises

Apply the Gauss-Jordan elimination process to each of the following systems. If a solution exists, determine the general solution.

1. $\begin{aligned} x + y + 3z &= 5 \\ 2x - y + 4z &= 11 \\ -y + z &= 3. \end{aligned}$

2. $\begin{aligned} 3x + 2y + z &= 1 \\ 5x + 3y + 3z &= 2 \\ x + y - z &= 1. \end{aligned}$

3. $\begin{aligned} 3x + 2y + z &= 1 \\ 5x + 3y + 3z &= 2 \\ 7x + 4y + 5z &= 3. \end{aligned}$

4. $\begin{aligned} 3x + 2y + z &= 1 \\ 5x + 3y + 3z &= 2 \\ 7x + 4y + 5z &= 3 \\ x + y - z &= 0. \end{aligned}$

5. $\begin{aligned} 3x - 2y + 5z + u &= 1 \\ x + y - 3z + 2u &= 2 \\ 6x + y - 4z + 3u &= 7. \end{aligned}$

6. $\begin{aligned} x + y - 3z + u &= 5 \\ 2x - y + z - 2u &= 2 \\ 7x + y - 7z + 3u &= 3. \end{aligned}$

7. $\begin{aligned} x + y + 2z + 3u + 4v &= 0 \\ 2x + 2y + 7z + 11u + 14v &= 0 \\ 3x + 3y + 6z + 10u + 15v &= 0. \end{aligned}$

8. $\begin{aligned} x - 2y + z + 2u &= -2 \\ 2x + 3y - z - 5u &= 9 \\ 4x - y + z - u &= 5 \\ 5x - 3y + 2z + u &= 3. \end{aligned}$

9. Prove that the system $x + y + 2z = 2$, $2x - y + 3z = 2$, $5x - y + az = 6$, has a unique solution if $a \neq 8$. Find all solutions when $a = 8$.

10. (a) Determine all solutions of the system

$$\begin{aligned} 5x + 2y - 6z + 2u &= -1 \\ x - y + z - u &= -2. \end{aligned}$$

(b) Determine all solutions of the system

$$\begin{aligned} 5x + 2y - 6z + 2u &= -1 \\ x - y + z - u &= -2 \\ x + y + z \quad\quad &= 6. \end{aligned}$$

11. This exercise tells how to determine all nonsingular 2×2 matrices. Prove that

$$\begin{bmatrix} a & b \\ c & d \end{bmatrix} \begin{bmatrix} d & -b \\ -c & a \end{bmatrix} = (ad - bc)I.$$

Deduce that $\begin{bmatrix} a & b \\ c & d \end{bmatrix}$ is nonsingular if and only if $ad - bc \neq 0$, in which case its inverse is

$$\frac{1}{ad - bc} \begin{bmatrix} d & -b \\ -c & a \end{bmatrix}.$$

Determine the inverse of each of the matrices in Exercises 12 through 16.

12. $\begin{bmatrix} 2 & 3 & 4 \\ 2 & 1 & 1 \\ -1 & 1 & 2 \end{bmatrix}.$

13. $\begin{bmatrix} 1 & 2 & 2 \\ 2 & -1 & 1 \\ 1 & 3 & 2 \end{bmatrix}.$

14. $\begin{bmatrix} 1 & -2 & 1 \\ -2 & 5 & -4 \\ 1 & -4 & 6 \end{bmatrix}.$

15. $\begin{bmatrix} 1 & 2 & 3 & 4 \\ 0 & 1 & 2 & 3 \\ 0 & 0 & 1 & 2 \\ 0 & 0 & 0 & 1 \end{bmatrix}.$

16. $\begin{bmatrix} 0 & 1 & 0 & 0 & 0 & 0 \\ 2 & 0 & 2 & 0 & 0 & 0 \\ 0 & 3 & 0 & 1 & 0 & 0 \\ 0 & 0 & 1 & 0 & 2 & 0 \\ 0 & 0 & 0 & 3 & 0 & 1 \\ 0 & 0 & 0 & 0 & 2 & 0 \end{bmatrix}.$

2.21 Miscellaneous exercises on matrices

1. If a square matrix has a row of zeros or a column of zeros, prove that it is singular.
2. For each of the following statements about $n \times n$ matrices, give a proof or exhibit a counter example.
   (a) If $AB + BA = O$, then $A^2B^3 = B^3A^2$.
   (b) If A and B are nonsingular, then $A + B$ is nonsingular.
   (c) If A and B are nonsingular, then AB is nonsingular.
   (d) If A, B, and $A + B$ are nonsingular, then $A - B$ is nonsingular.
   (e) If $A^3 = O$, then $A - I$ is nonsingular.
   (f) If the product of k matrices $A_1 \cdots A_k$ is nonsingular, then each matrix A_i is nonsingular.

3. If $A = \begin{bmatrix} 1 & 2 \\ 5 & 4 \end{bmatrix}$, find a nonsingular matrix P such that $P^{-1}AP = \begin{bmatrix} 6 & 0 \\ 0 & -1 \end{bmatrix}$.
4. The matrix $A = \begin{bmatrix} a & i \\ i & b \end{bmatrix}$, where $i^2 = -1$, $a = \frac{1}{2}(1 + \sqrt{5})$, and $b = \frac{1}{2}(1 - \sqrt{5})$, has the property that $A^2 = A$. Describe completely all 2×2 matrices A with complex entries such that $A^2 = A$.
5. If $A^2 = A$, prove that $(A + I)^k = I + (2^k - 1)A$.
6. The special theory of relativity makes use of a set of equations of the form $x' = a(x - vt)$, $y' = y$, $z' = z$, $t' = a(t - vx/c^2)$. Here v represents the velocity of a moving object, c the speed of light, and $a = c/\sqrt{c^2 - v^2}$, where $|v| < c$. The linear transformation which maps the two-dimensional vector (x, t) onto (x', t') is called a *Lorentz transformation.* Its matrix relative to the usual bases is denoted by $L(v)$ and is given by

$$L(v) = a\begin{bmatrix} 1 & -v \\ -vc^{-2} & 1 \end{bmatrix}.$$

Note that $L(v)$ is nonsingular and that $L(0) = I$. Prove that $L(v)L(u) = L(w)$, where $w = (u + v)c^2/(uv + c^2)$. In other words, the product of two Lorentz transformations is another Lorentz transformation.
7. If we interchange the rows and columns of a rectangular matrix A, the new matrix so obtained is called the *transpose* of A and is denoted by A^t. For example, if we have

$$A = \begin{bmatrix} 1 & 2 & 3 \\ 4 & 5 & 6 \end{bmatrix}, \qquad \text{then} \quad A^t = \begin{bmatrix} 1 & 4 \\ 2 & 5 \\ 3 & 6 \end{bmatrix}.$$

Prove that transposes have the following properties:
(a) $(A^t)^t = A$. (b) $(A + B)^t = A^t + B^t$. (c) $(cA)^t = cA^t$.
(d) $(AB)^t = B^tA^t$. (e) $(A^t)^{-1} = (A^{-1})^t$ if A is nonsingular.
8. A square matrix A is called an orthogonal matrix if $AA^t = I$. Verify that the 2×2 matrix $\begin{bmatrix} \cos\theta & -\sin\theta \\ \sin\theta & \cos\theta \end{bmatrix}$ is orthogonal for each real θ. If A is any $n \times n$ orthogonal matrix, prove that its rows, considered as vectors in V_n, form an orthonormal set.
9. For each of the following statements about $n \times n$ matrices, give a proof or else exhibit a counter example.
(a) If A and B are orthogonal, then $A + B$ is orthogonal.
(b) If A and B are orthogonal, then AB is orthogonal.
(c) If A and B are orthogonal, then B is orthogonal.
10. *Hadamard matrices*, named for Jacques Hadamard (1865–1963), are those $n \times n$ matrices with the following properties:
I. Each entry is 1 or -1.
II. Each row, considered as a vector in V_n, has length $\sqrt{n}$.
III. The dot product of any two distinct rows is 0.
Hadamard matrices arise in certain problems in geometry and the theory of numbers, and they have been applied recently to the construction of optimum code words in space communication. In spite of their apparent simplicity, they present many unsolved problems. The

main unsolved problem at this time is to determine all n for which an $n \times n$ Hadamard matrix exists. This exercise outlines a partial solution.
(a) Determine all 2×2 Hadamard matrices (there are exactly 8).
(b) This part of the exercise outlines a simple proof of the following theorem: *If A is an $n \times n$ Hadamard matrix, where $n > 2$, then n is a multiple of* 4. The proof is based on two very simple lemmas concerning vectors in n-space. Prove each of these lemmas and apply them to the rows of Hadamard matrix to prove the theorem.

LEMMA 1. *If X, Y, Z are orthogonal vectors in V_n, then we have*

$$(X + Y) \cdot (X + Z) = \|X\|^2 .$$

LEMMA 2. *Write $X = (x_1, \ldots, x_n)$, $Y = (y_1, \ldots, y_n)$, $Z = (z_1, \ldots, z_n)$. If each component x_i, y_i, z_i is either* 1 *or* -1 , *then the product $(x_i + y_i)(x_i + z_i)$ is either* 0 *or* 4.

3

DETERMINANTS

3.1 Introduction

In many applications of linear algebra to geometry and analysis the concept of a determinant plays an important part. This chapter studies the basic properties of determinants and some of their applications.

Determinants of order two and three were intoduced in Volume I as a useful notation for expressing certain formulas in a compact form. We recall that a determinant of order two was defined by the formula

$$\begin{vmatrix} a_{11} & a_{12} \\ a_{21} & a_{22} \end{vmatrix} = a_{11}a_{22} - a_{12}a_{21}. \tag{3.1}$$

Despite similarity in notation, the determinant $\begin{vmatrix} a_{11} & a_{12} \\ a_{21} & a_{22} \end{vmatrix}$ (written with vertical bars) is conceptually distinct from the matrix $\begin{bmatrix} a_{11} & a_{12} \\ a_{21} & a_{22} \end{bmatrix}$ (written with square brackets). The determinant is a *number* assigned to the matrix according to Formula (3.1). To emphasize this connection we also write

$$\begin{vmatrix} a_{11} & a_{12} \\ a_{21} & a_{22} \end{vmatrix} = \det \begin{bmatrix} a_{11} & a_{12} \\ a_{21} & a_{22} \end{bmatrix}.$$

Determinants of order three were defined in Volume I in terms of second-order determinants by the formula

$$\det \begin{bmatrix} a_{11} & a_{12} & a_{13} \\ a_{21} & a_{22} & a_{23} \\ a_{31} & a_{32} & a_{33} \end{bmatrix} = a_{11} \begin{vmatrix} a_{22} & a_{23} \\ a_{32} & a_{33} \end{vmatrix} - a_{12} \begin{vmatrix} a_{21} & a_{23} \\ a_{31} & a_{33} \end{vmatrix} + a_{13} \begin{vmatrix} a_{21} & a_{22} \\ a_{31} & a_{32} \end{vmatrix}. \tag{3.2}$$

This chapter treats the more general case, the determinant of a square matrix of order n for any integer $n \geq 1$. Our point of view is to treat the determinant as a function which assigns to each square matrix A a number called the determinant of A and denoted by det A. It is possible to define this function by an explicit formula generalizing (3.1) and (3.2). This formula is a sum containing $n!$ products of entries of A. For large n the formula is unwieldy and is rarely used in practice. It seems preferable to study determinants from another point of view which emphasizes more clearly their essential properties. These properties, which are important in the applications, will be taken as *axioms* for a determinant function. Initially, our program will consist of three parts: (1) To motivate the choice of axioms. (2) To deduce further properties of determinants from the axioms. (3) To prove that there is one and only one function which satisfies the axioms.

3.2 Motivation for the choice of axioms for a determinant function

In Volume I we proved that the scalar triple product of three vectors A_1, A_2, A_3 in 3-space can be expressed as the determinant of a matrix whose rows are the given vectors. Thus we have

$$A_1 \times A_2 \cdot A_3 = \det \begin{bmatrix} a_{11} & a_{12} & a_{13} \\ a_{21} & a_{22} & a_{23} \\ a_{31} & a_{32} & a_{33} \end{bmatrix},$$

where $A_1 = (a_{11}, a_{12}, a_{13})$, $A_2 = (a_{21}, a_{22}, a_{23})$, and $A_3 = (a_{31}, a_{32}, a_{33})$.

If the rows are linearly independent the scalar triple product is nonzero; the absolute value of the product is equal to the volume of the parallelepiped determined by the three vectors A_1, A_2, A_3. If the rows are dependent the scalar triple product is zero. In this case the vectors A_1, A_2, A_3 are coplanar and the parallelepiped degenerates to a plane figure of zero volume.

Some of the properties of the scalar triple product will serve as motivation for the choice of axioms for a determinant function in the higher-dimensional case. To state these properties in a form suitable for generalization, we consider the scalar triple product as a function of the three row-vectors A_1, A_2, A_3. We denote this function by d; thus,

$$d(A_1, A_2, A_3) = A_1 \times A_2 \cdot A_3 .$$

We focus our attention on the following properties:

(a) *Homogeneity in each row.* For example, homogeneity in the first row states that

$$d(tA_1, A_2, A_3) = t\, d(A_1, A_2, A_3) \qquad \text{for every scalar } t .$$

(b) *Additivity in each row.* For example, additivity in the second row states that

$$d(A_1, A_2 + C, A_3) = d(A_1, A_2, A_3) + d(A_1, C, A_3)$$

for every vector C.

(c) *The scalar triple product is zero if two of the rows are equal.*

(d) *Normalization:*

$$d(\boldsymbol{i}, \boldsymbol{j}, \boldsymbol{k}) = 1, \qquad \text{where } \boldsymbol{i} = (1, 0, 0), \quad \boldsymbol{j} = (0, 1, 0), \quad \boldsymbol{k} = (0, 0, 1).$$

Each of these properties can be easily verified from properties of the dot and cross product. Some of these properties are suggested by the geometric relation between the scalar triple product and the volume of the parallelepiped determined by the geometric vectors A_1, A_2, A_3. The geometric meaning of the additive property (b) in a special case is of particular interest. If we take $C = A_1$ in (b) the second term on the right is zero because of (c), and relation (b) becomes

$$d(A_1, A_2 + A_1, A_3) = d(A_1, A_2, A_3). \tag{3.3}$$

This property is illustrated in Figure 3.1 which shows a parallelepiped determined by A_1, A_2, A_3, and another parallelepiped determined by $A_1, A_1 + A_2, A_3$. Equation (3.3) merely states that these two parallelepipeds have equal volumes. This is evident geometrically because the parallelepipeds have equal altitudes and bases of equal area.

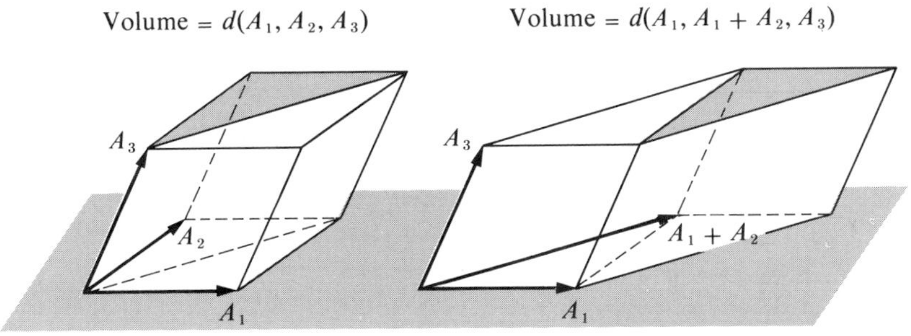

FIGURE 3.1

Geometric interpretation of the property $d(A_1, A_2, A_3) = d(A_1, A_1 + A_2, A_3)$. The two parallelepipeds have equal volumes.

3.3 A set of axioms for a determinant function

The properties of the scalar triple product mentioned in the foregoing section can be suitably generalized and used as axioms for determinants of order n. If $A = (a_{ij})$ is an $n \times n$ matrix with real or complex entries, we denote its rows by $A_1, \ldots, A_n$. Thus, the ith row of A is a vector in n-space given by

$$A_i = (a_{i1}, a_{i2}, \ldots, a_{in}).$$

We regard the determinant as a function of the n rows $A_1, \ldots, A_n$ and denote its values by $d(A_1, \ldots, A_n)$ or by det A.

AXIOMATIC DEFINITION OF A DETERMINANT FUNCTION. *A real- or complex-valued function d, defined for each ordered n-tuple of vectors $A_1, \ldots, A_n$ in n-space, is called a determinant function of order n if it satisfies the following axioms for all choices of vectors $A_1, \ldots, A_n$ and C in n-space:*

AXIOM 1. HOMOGENEITY IN EACH ROW. *If the kth row A_k is multiplied by a scalar t, then the determinant is also multiplied by t:*

$$d(\ldots, tA_k, \ldots) = t\, d(\ldots, A_k, \ldots).$$

AXIOM 2. ADDITIVITY IN EACH ROW. *For each k we have*

$$d(A_1, \ldots, A_k + C, \ldots, A_n) = d(A_1, \ldots, A_k, \ldots, A_n) + d(A_1, \ldots, C, \ldots, A_n).$$

AXIOM 3. THE DETERMINANT VANISHES IF ANY TWO ROWS ARE EQUAL:

$$d(A_1, \ldots, A_n) = 0 \quad \text{if} \quad A_i = A_j \quad \text{for some } i \text{ and } j \text{ with } i \neq j.$$

AXIOM 4. THE DETERMINANT OF THE IDENTITY MATRIX IS EQUAL TO 1:

$$d(I_1, \ldots, I_n) = 1, \qquad \text{where } I_k \text{ is the kth unit coordinate vector}.$$

The first two axioms state that the determinant of a matrix is a linear function of each of its rows. This is often described by saying that the determinant is a *multilinear* function of its rows. By repeated application of linearity in the first row we can write

$$d\left(\sum_{k=1}^{p} t_k C_k, A_2, \ldots, A_n\right) = \sum_{k=1}^{p} t_k\, d(C_k, A_2, \ldots, A_n),$$

where $t_1, \ldots, t_p$ are scalars and $C_1, \ldots, C_p$ are any vectors in n-space.

Sometimes a weaker version of Axiom 3 is used:

AXIOM 3′. THE DETERMINANT VANISHES IF TWO ADJACENT ROWS ARE EQUAL:

$$d(A_1, \ldots, A_n) = 0 \quad \text{if} \quad A_k = A_{k+1} \quad \text{for some } k = 1, 2, \ldots, n-1.$$

It is a remarkable fact that for a given n there is one and only one function d which satisfies Axioms 1, 2, 3′ and 4. The proof of this fact, one of the principal results of this chapter, will be given later. The next theorem gives properties of determinants deduced from Axioms 1, 2, and 3′ alone. One of these properties is Axiom 3. It should be noted that Axiom 4 is not used in the proof of this theorem. This observation will be useful later when we prove uniqueness of the determinant function.

THEOREM 3.1. *A determinant function satisfying Axioms* 1, 2, *and* 3′ *has the following further properties:*

(a) *The determinant vanishes if some row is O:*

$$d(A_1, \ldots, A_n) = 0 \quad \text{if} \quad A_k = O \quad \text{for some } k.$$

(b) *The determinant changes sign if two adjacent rows are interchanged:*

$$d(\ldots, A_k, A_{k+1}, \ldots) = -d(\ldots, A_{k+1}, A_k, \ldots).$$

(c) *The determinant changes sign if any two rows A_i and A_j with $i \neq j$ are interchanged.*
(d) *The determinant vanishes if any two rows are equal:*

$$d(A_1, \ldots, A_n) = 0 \qquad \text{if} \quad A_i = A_j \qquad \text{for some } i \text{ and } j \text{ with } i \neq j.$$

(e) *The determinant vanishes if its rows are dependent.*

Proof. To prove (a) we simply take $t = 0$ in Axiom 1. To prove (b), let B be a matrix having the same rows as A except for row k and row $k + 1$. Let both rows B_k and B_{k+1} be equal to $A_k + A_{k+1}$. Then $\det B = 0$ by Axiom 3′. Thus we may write

$$d(\ldots, A_k + A_{k+1}, A_k + A_{k+1}, \ldots) = 0.$$

Applying the additive property to row k and to row $k + 1$ we can rewrite this equation as follows:

$$d(\ldots, A_k, A_k, \ldots) + d(\ldots, A_k, A_{k+1}, \ldots) + d(\ldots, A_{k+1}, A_k, \ldots) + d(\ldots, A_{k+1}, A_{k+1}, \ldots) = 0.$$

The first and fourth terms are zero by Axiom 3′. Hence the second and third terms are negatives of each other, which proves (b).

To prove (c) we can assume that $i < j$. We can interchange rows A_i and A_j by performing an odd number of interchanges of *adjacent* rows. First we interchange row A_j successively with the earlier adjacent rows $A_{j-1}, A_{j-2}, \ldots, A_i$. This requires $j - i$ interchanges. Then we interchange row A_i successively with the later adjacent rows $A_{i+1}, A_{i+2}, \ldots, A_{j-1}$. This requires $j - i - 1$ further interchanges. Each interchange of adjacent rows reverses the sign of the determinant. Since there are $(j - i) + (j - i - 1) = 2(j - i) - 1$ interchanges altogether (an *odd* number), the determinant changes sign an odd number of times, which proves (c).

To prove (d), let B be the matrix obtained from A by interchanging rows A_i and A_j. Since $A_i = A_j$ we have $B = A$ and hence $\det B = \det A$. But by (c), $\det B = -\det A$. Therefore $\det A = 0$.

To prove (e) suppose scalars $c_1, \ldots, c_n$ exist, not all zero, such that $\sum_{k=1}^{n} c_k A_k = O$. Then any row A_k with $c_k \neq 0$ can be expressed as a linear combination of the other rows. For simplicity, suppose that A_1 is a linear combination of the others, say $A_1 = \sum_{k=2}^{n} t_k A_k$. By linearity of the first row we have

$$d(A_1, A_2, \ldots, A_n) = d\left(\sum_{k=2}^{n} t_k A_k, A_2, \ldots, A_n\right) = \sum_{k=2}^{n} t_k \, d(A_k, A_2, \ldots, A_n).$$

But each term $d(A_k, A_2, \ldots, A_n)$ in the last sum is zero since A_k is equal to at least one of $A_2, \ldots, A_n$. Hence the whole sum is zero. If row A_i is a linear combination of the other rows we argue the same way, using linearity in the ith row. This proves (e).

3.4 Computation of determinants

At this stage it may be instructive to compute some determinants, using only the axioms and the properties in Theorem 3.1, assuming throughout that determinant functions exist. In each of the following examples we do not use Axiom 4 until the very last step in the computation.

EXAMPLE 1. *Determinant of a* 2×2 *matrix.* We shall prove that

$$\det \begin{bmatrix} a_{11} & a_{12} \\ a_{21} & a_{22} \end{bmatrix} = a_{11}a_{22} - a_{12}a_{21}. \tag{3.4}$$

Write the row vectors as linear combinations of the unit coordinate vectors $\boldsymbol{i} = (1, 0)$ and $\boldsymbol{j} = (0, 1)$:

$$A_1 = (a_{11}, a_{12}) = a_{11}\boldsymbol{i} + a_{12}\boldsymbol{j}, \qquad A_2 = (a_{21}, a_{22}) = a_{21}\boldsymbol{i} + a_{22}\boldsymbol{j}.$$

Using linearity in the first row we have

$$d(A_1, A_2) = d(a_{11}\boldsymbol{i} + a_{12}\boldsymbol{j}, A_2) = a_{11}\, d(\boldsymbol{i}, A_2) + a_{12}\, d(\boldsymbol{j}, A_2).$$

Now we use linearity in the second row to obtain

$$d(\boldsymbol{i}, A_2) = d(\boldsymbol{i}, a_{21}\boldsymbol{i} + a_{22}\boldsymbol{j}) = a_{21}\, d(\boldsymbol{i}, \boldsymbol{i}) + a_{22}\, d(\boldsymbol{i}, \boldsymbol{j}) = a_{22}\, d(\boldsymbol{i}, \boldsymbol{j}),$$

since $d(\boldsymbol{i}, \boldsymbol{i}) = 0$. Similarly we find

$$d(\boldsymbol{j}, A_2) = d(\boldsymbol{j}, a_{21}\boldsymbol{i} + a_{22}\boldsymbol{j}) = a_{21}\, d(\boldsymbol{j}, \boldsymbol{i}) = -a_{21}\, d(\boldsymbol{i}, \boldsymbol{j}).$$

Hence we obtain

$$d(A_1, A_2) = (a_{11}a_{22} - a_{12}a_{21})\, d(\boldsymbol{i}, \boldsymbol{j}).$$

But $d(\boldsymbol{i}, \boldsymbol{j}) = 1$ by Axiom 4, so $d(A_1, A_2) = a_{11}a_{22} - a_{12}a_{21}$, as asserted.

This argument shows that if a determinant function exists for 2×2 matrices, then it must necessarily have the form (3.4). Conversely, it is easy to verify that this formula does, indeed, define a determinant function of order 2. Therefore we have shown that there is one and only one determinant function of order 2.

EXAMPLE 2. *Determinant of a diagonal matrix.* A square matrix of the form

$$A = \begin{bmatrix} a_{11} & 0 & \cdots & 0 \\ 0 & a_{22} & & 0 \\ \vdots & & \ddots & \vdots \\ 0 & 0 & \cdots & a_{nn} \end{bmatrix}$$

is called a *diagonal matrix*. Each entry a_{ij} off the main diagonal ($i \neq j$) is zero. We shall prove that the determinant of A is equal to the product of its diagonal elements,

$$\det A = a_{11}a_{22}\cdots a_{nn}. \tag{3.5}$$

The kth row of A is simply a scalar multiple of the kth unit coordinate vector, $A_k = a_{kk}I_k$. Applying the homogeneity property repeatedly to factor out the scalars one at a time we get

$$\det A = d(A_1, \ldots, A_n) = d(a_{11}I_1, \ldots, a_{nn}I_n) = a_{11}\cdots a_{nn}\, d(I_1, \ldots, I_n).$$

This formula can be written in the form

$$\det A = a_{11}\cdots a_{nn} \det I,$$

where I is the identity matrix. Axiom 4 tells us that $\det I = 1$ so we obtain (3.5).

EXAMPLE 3. *Determinant of an upper triangular matrix.* A square matrix of the form

$$U = \begin{bmatrix} u_{11} & u_{12} & \cdots & u_{1n} \\ 0 & u_{22} & \cdots & u_{2n} \\ \cdot & \cdot & & \cdot \\ \cdot & \cdot & & \cdot \\ \cdot & \cdot & & \cdot \\ 0 & 0 & \cdots & u_{nn} \end{bmatrix}$$

is called an upper triangular matrix. All the entries below the main diagonal are zero. We shall prove that the determinant of such a matrix is equal to the product of its diagonal elements,

$$\det U = u_{11}u_{22}\cdots u_{nn}.$$

First we prove that $\det U = 0$ if some diagonal element $u_{ii} = 0$. If the last diagonal element u_{nn} is zero, then the last row is O and $\det U = 0$ by Theorem 3.1(a). Suppose, then, that some earlier diagonal element u_{ii} is zero. To be specific, say $u_{22} = 0$. Then each of the $n - 1$ row-vectors $U_2, \ldots, U_n$ has its first two components zero. Hence these vectors span a subspace of dimension at most $n - 2$. Therefore these $n - 1$ rows (and hence *all* the rows) are dependent. By Theorem 3.1(e), $\det U = 0$. In the same way we find that $\det U = 0$ if any diagonal element is zero.

Now we treat the general case. First we write the first row U_1 as a sum of two row-vectors,

$$U_1 = V_1 + V_1',$$

where $V_1 = [u_{11}, 0, \ldots, 0]$ and $V_1' = [0, u_{12}, u_{13}, \ldots, u_{1n}]$. By linearity in the first row we have

$$\det U = \det(V_1, U_2, \ldots, U_n) + \det(V_1', U_2, \ldots, U_n).$$

But $\det(V_1', U_2, \ldots, U_n) = 0$ since this is the determinant of an upper triangular matrix with a diagonal element equal to 0. Hence we have

$$\det U = \det(V_1, U_2, \ldots, U_n). \tag{3.6}$$

Now we treat row-vector U_2 in a similar way, expressing it as a sum,

$$U_2 = V_2 + V_2',$$

where

$$V_2 = [0, u_{22}, 0, \ldots, 0] \quad \text{and} \quad V_2' = [0, 0, u_{23}, \ldots, u_{2n}].$$

We use this on the right of (3.6) and apply linearity in the second row to obtain

$$\det U = \det(V_1, V_2, U_3, \ldots, U_n), \tag{3.7}$$

since $\det(V_1, V_2', U_3, \ldots, U_n) = 0$. Repeating the argument on each of the succeeding rows in the right member of (3.7) we finally obtain

$$\det U = \det(V_1, V_2, \ldots, V_n),$$

where $(V_1, V_2, \ldots, V_n)$ is a diagonal matrix with the same diagonal elements as U. Therefore, by Example 2 we have

$$\det U = u_{11}u_{22}\cdots u_{nn},$$

as required.

EXAMPLE 4. *Computation by the Gauss-Jordan process.* The Gauss-Jordan elimination process for solving systems of linear equations is also one of the best methods for computing determinants. We recall that the method consists in applying three types of operations to the rows of a matrix:

(1) *Interchanging two rows.*
(2) *Multiplying all the elements in a row by a nonzero scalar.*
(3) *Adding to one row a scalar multiple of another.*

By performing these operations over and over again in a systematic fashion we can transform any square matrix A to an upper triangular matrix U whose determinant we now know how to compute. It is easy to determine the relation between $\det A$ and $\det U$. Each time operation (1) is performed the determinant changes sign. Each time (2) is performed with a scalar $c \neq 0$, the determinant is multiplied by c. Each time (3) is performed the determinant is unaltered. Therefore, if operation (1) is performed p times and if $c_1, \ldots, c_q$ are the nonzero scalar multipliers used in connection with operation (2), then we have

$$\det A = (-1)^p(c_1c_2\cdots c_q)^{-1}\det U. \tag{3.8}$$

Again we note that this formula is a consequence of the first three axioms alone. Its proof does not depend on Axiom 4.

3.5 The uniqueness theorem

In Example 3 of the foregoing section we showed that Axioms 1, 2, and 3 imply the formula $\det U = u_{11}u_{22}\cdots u_{nn} \det I$. Combining this with (3.8) we see that for every $n \times n$ matrix A there is scalar c (depending on A) such that

$$d(A_1, \ldots, A_n) = c\, d(I_1, \ldots, I_n). \tag{3.9}$$

Moreover, this formula is a consequence of Axioms 1, 2, and 3 alone. From this we can easily prove that there cannot be more than one determinant function.

THEOREM 3.2. UNIQUENESS THEOREM FOR DETERMINANTS. *Let d be a function satisfying all four axioms for a determinant function of order n, and let f be another function satisfying Axioms 1, 2, and 3. Then for every choice of vectors $A_1, \ldots, A_n$ in n-space we have*

$$f(A_1, \ldots, A_n) = d(A_1, \ldots, A_n) f(I_1, \ldots, I_n). \tag{3.10}$$

In particular, if f also satisfies Axiom 4 we have $f(A_1, \ldots, A_n) = d(A_1, \ldots, A_n)$.

Proof. Let $g(A_1, \ldots, A_n) = f(A_1, \ldots, A_n) - d(A_1, \ldots, A_n) f(I_1, \ldots, I_n)$. We will prove that $g(A_1, \ldots, A_n) = 0$ for every choice of $A_1, \ldots, A_n$. Since both d and f satisfy Axioms 1, 2, and 3 the same is true of g. Hence g also satisfies Equation (3.9) since this was deduced from the first three axioms alone. Therefore we can write

$$g(A_1, \ldots, A_n) = c\, g(I_1, \ldots, I_n), \tag{3.11}$$

where c is a scalar depending on A. Taking $A = I$ in the definition of g and noting that d satisfies Axiom 4 we find

$$g(I_1, \ldots, I_n) = f(I_1, \ldots, I_n) - f(I_1, \ldots, I_n) = 0.$$

Therefore Equation (3.11) becomes $g(A_1, \ldots, A_n) = 0$. This completes the proof.

3.6 Exercises

In this set of exercises you may assume existence of a determinant function. Determinants of order 3 may be computed by Equation (3.2).

1. Compute each of the following determinants.

$$\text{(a)} \begin{vmatrix} 2 & 1 & 1 \\ 1 & 4 & -4 \\ 1 & 0 & 2 \end{vmatrix}, \qquad \text{(b)} \begin{vmatrix} 3 & 0 & 8 \\ 5 & 0 & 7 \\ -1 & 4 & 2 \end{vmatrix}, \qquad \text{(c)} \begin{vmatrix} a & 1 & 0 \\ 2 & a & 2 \\ 0 & 1 & a \end{vmatrix}.$$

2. If $\det \begin{bmatrix} x & y & z \\ 3 & 0 & 2 \\ 1 & 1 & 1 \end{bmatrix} = 1$, compute the determinant of each of the following matrices:

(a) $\begin{bmatrix} 2x & 2y & 2z \\ \frac{3}{2} & 0 & 1 \\ 1 & 1 & 1 \end{bmatrix}$, (b) $\begin{bmatrix} x & y & z \\ 3x+3 & 3y & 3z+2 \\ x+1 & y+1 & z+1 \end{bmatrix}$, (c) $\begin{bmatrix} x-1 & y-1 & z-1 \\ 4 & 1 & 3 \\ 1 & 1 & 1 \end{bmatrix}$.

3. (a) Prove that $\begin{vmatrix} 1 & 1 & 1 \\ a & b & c \\ a^2 & b^2 & c^2 \end{vmatrix} = (b-a)(c-a)(c-b)$.

(b) Find corresponding formulas for the determinants

$$\begin{vmatrix} 1 & 1 & 1 \\ a & b & c \\ a^3 & b^3 & c^3 \end{vmatrix} \quad \text{and} \quad \begin{vmatrix} 1 & 1 & 1 \\ a^2 & b^2 & c^2 \\ a^3 & b^3 & c^3 \end{vmatrix}.$$

4. Compute the determinant of each of the following matrices by transforming each of them to an upper triangular matrix.

(a) $\begin{bmatrix} 1 & -1 & 1 & 1 \\ 1 & -1 & -1 & -1 \\ 1 & 1 & -1 & -1 \\ 1 & 1 & 1 & -1 \end{bmatrix}$, (b) $\begin{bmatrix} 1 & 1 & 1 & 1 \\ a & b & c & d \\ a^2 & b^2 & c^2 & d^2 \\ a^3 & b^3 & c^3 & d^3 \end{bmatrix}$, (c) $\begin{bmatrix} 1 & 1 & 1 & 1 \\ a & b & c & d \\ a^2 & b^2 & c^2 & d^2 \\ a^4 & b^4 & c^4 & d^4 \end{bmatrix}$,

(d) $\begin{bmatrix} a & 1 & 0 & 0 & 0 \\ 4 & a & 2 & 0 & 0 \\ 0 & 3 & a & 3 & 0 \\ 0 & 0 & 2 & a & 4 \\ 0 & 0 & 0 & 1 & a \end{bmatrix}$, (e) $\begin{bmatrix} 1 & 1 & 1 & 1 & 1 & 1 \\ 1 & 1 & 1 & -1 & -1 & -1 \\ 1 & 1 & -1 & -1 & 1 & 1 \\ 1 & -1 & -1 & 1 & -1 & 1 \\ 1 & -1 & 1 & -1 & 1 & 1 \\ 1 & -1 & -1 & 1 & 1 & -1 \end{bmatrix}$.

5. A lower triangular matrix $A = (a_{ij})$ is a square matrix with all entries above the main diagonal equal to 0; that is, $a_{ij} = 0$ whenever $i < j$. Prove that the determinant of such a matrix is equal to the product of its diagonal entries: $\det A = a_{11}a_{22}\cdots a_{nn}$.

6. Let f_1, f_2, g_1, g_2 be four functions differentiable on an interval (a, b). Define

$$F(x) = \begin{vmatrix} f_1(x) & f_2(x) \\ g_1(x) & g_2(x) \end{vmatrix}$$

for each x in (a, b). Prove that

$$F'(x) = \begin{vmatrix} f_1'(x) & f_2'(x) \\ g_1(x) & g_2(x) \end{vmatrix} + \begin{vmatrix} f_1(x) & f_2(x) \\ g_1'(x) & g_2'(x) \end{vmatrix}.$$

7. State and prove a generalization of Exercise 6 for the determinant

$$F(x) = \begin{vmatrix} f_1(x) & f_2(x) & f_3(x) \\ g_1(x) & g_2(x) & g_3(x) \\ h_1(x) & h_2(x) & h_3(x) \end{vmatrix}.$$

8. (a) If $F(x) = \begin{vmatrix} f_1(x) & f_2(x) \\ f_1'(x) & f_2'(x) \end{vmatrix}$, prove that $F'(x) = \begin{vmatrix} f_1(x) & f_2(x) \\ f_1''(x) & f_2''(x) \end{vmatrix}$.

   (b) State and prove a corresponding result for 3×3 determinants, assuming the validity of Equation (3.2).
9. Let U and V be two $n \times n$ upper triangular matrices.
   (a) Prove that each of $U + V$ and UV is an upper triangular matrix.
   (b) Prove that $\det(UV) = (\det U)(\det V)$.
   (c) If $\det U \neq 0$ prove that there is an upper triangular matrix U^{-1} such that $UU^{-1} = I$, and deduce that $\det(U^{-1}) = 1/\det U$.
10. Calculate $\det A$, $\det(A^{-1})$, and A^{-1} for the following upper triangular matrix:

$$A = \begin{bmatrix} 2 & 3 & 4 & 5 \\ 0 & 2 & 3 & 4 \\ 0 & 0 & 2 & 3 \\ 0 & 0 & 0 & 2 \end{bmatrix}.$$

3.7 The product formula for determinants

In this section we use the uniqueness theorem to prove that the determinant of a product of two square matrices is equal to the product of their determinants,

$$\det(AB) = (\det A)(\det B),$$

assuming that a determinant function exists.

We recall that the product AB of two matrices $A = (a_{ij})$ and $B = (b_{ij})$ is the matrix $C = (c_{ij})$ whose i, j entry is given by the formula

$$c_{ij} = \sum_{k=1}^{n} a_{ik} b_{kj}. \tag{3.12}$$

The product is defined only if the number of columns of the left-hand factor A is equal to the number of rows of the right-hand factor B. This is always the case if both A and B are square matrices of the same size.

The proof of the product formula will make use of a simple relation which holds between the rows of AB and the rows of A. We state this as a lemma. As usual, we let A_i denote the ith row of matrix A.

LEMMA 3.3. *If A is an $m \times n$ matrix and B an $n \times p$ matrix, then we have*

$$(AB)_i = A_i B .$$

That is, the ith row of the product AB is equal to the product of row matrix A_i with B.

Proof. Let B^j denote the jth column of B and let $C = AB$. Then the sum in (3.12) can be regarded as the dot product of the ith row of A with the jth column of B,

$$c_{ij} = A_i \cdot B^j .$$

Therefore the ith row C_i is the row matrix

$$C_i = [A_i \cdot B^1, A_i \cdot B^2, \ldots, A_i \cdot B^p] .$$

But this is also the result of matrix multiplication of row matrix A_i with B, since

$$A_i B = [a_{i1}, \ldots, a_{in}] \begin{bmatrix} b_{11} & b_{12} & \cdots & b_{1p} \\ \cdot & & & \cdot \\ \cdot & & & \cdot \\ \cdot & & & \cdot \\ b_{n1} & b_{n2} & \cdots & b_{np} \end{bmatrix} = [A_i \cdot B^1, \ldots, A_i \cdot B^p] .$$

Therefore $C_i = A_i B$, which proves the lemma.

THEOREM 3.4. PRODUCT FORMULA FOR DETERMINANTS. *For any $n \times n$ matrices A and B we have*

$$\det (AB) = (\det A)(\det B) .$$

Proof. Since $(AB)_i = A_i B$, we are to prove that

$$d(A_1 B, \ldots, A_n B) = d(A_1, \ldots, A_n)\, d(B_1, \ldots, B_n) .$$

Using the lemma again we also have $B_i = (IB)_i = I_i B$, where I is the $n \times n$ identity matrix. Therefore $d(B_1, \ldots, B_n) = d(I_1 B, \ldots, I_n B)$, and we are to prove that

$$d(A_1 B, \ldots, A_n B) = d(A_1, \ldots, A_n)\, d(I_1 B, \ldots, I_n B) .$$

We keep B fixed and introduce a function f defined by the formula

$$f(A_1, \ldots, A_n) = d(A_1 B, \ldots, A_n B) .$$

The equation we wish to prove states that

$$f(A_1, \ldots, A_n) = d(A_1, \ldots, A_n) f(I_1, \ldots, I_n) . \tag{3.13}$$

But now it is a simple matter to verify that f satisfies Axioms 1, 2, and 3 for a determinant function so, by the uniqueness theorem, Equation (3.13) holds for every matrix A. This completes the proof.

Applications of the product formula are given in the next two sections.

3.8 The determinant of the inverse of a nonsingular matrix

We recall that a square matrix A is called nonsingular if it has a left inverse B such that $BA = I$. If a left inverse exists it is unique and is also a right inverse, $AB = I$. We denote the inverse by A^{-1}. The relation between $\det A$ and $\det A^{-1}$ is as natural as one could expect.

THEOREM 3.5. *If matrix A is nonsingular, then* $\det A \neq 0$ *and we have*

$$\det A^{-1} = \frac{1}{\det A}. \tag{3.14}$$

Proof. From the product formula we have

$$(\det A)(\det A^{-1}) = \det (AA^{-1}) = \det I = 1.$$

Hence $\det A \neq 0$ and (3.14) holds.

Theorem 3.5 shows that nonvanishing of $\det A$ is a necessary condition for A to be nonsingular. Later we will prove that this condition is also sufficient. That is, if $\det A \neq 0$ then A^{-1} exists.

3.9 Determinants and independence of vectors

A simple criterion for testing independence of vectors can be deduced from Theorem 3.5.

THEOREM 3.6. *A set of n vectors $A_1, \ldots, A_n$ in n-space is independent if and only if* $d(A_1, \ldots, A_n) \neq 0$.

Proof. We already proved in Theorem 3.2(e) that dependence implies $d(A_1, \ldots, A_n) = 0$. To prove the converse, we assume that $A_1, \ldots, A_n$ are independent and prove that $d(A_1, \ldots, A_n) \neq 0$.

Let V_n denote the linear space of n-tuples of scalars. Since $A_1, \ldots, A_n$ are n independent elements in an n-dimensional space they form a basis for V_n. By Theorem 2.12 there is a linear transformation $T: V_n \to V_n$ which maps these n vectors onto the unit coordinate vectors,

$$T(A_k) = I_k \qquad \text{for} \quad k = 1, 2, \ldots, n.$$

Therefore there is an $n \times n$ matrix B such that

$$A_k B = I_k \qquad \text{for} \quad k = 1, 2, \ldots, n.$$

But by Lemma 3.3 we have $A_k B = (AB)_k$, where A is the matrix with rows $A_1, \ldots, A_n$. Hence $AB = I$, so A is nonsingular and $\det A \neq 0$.

3.10 The determinant of a block-diagonal matrix

A square matrix C of the form

$$C = \begin{bmatrix} A & O \\ O & B \end{bmatrix},$$

where A and B are square matrices and each O denotes a matrix of zeros, is called a *block-diagonal matrix* with two diagonal blocks A and B. An example is the 5×5 matrix

$$C = \begin{bmatrix} 1 & 0 & 0 & 0 & 0 \\ 0 & 1 & 0 & 0 & 0 \\ 0 & 0 & 1 & 2 & 3 \\ 0 & 0 & 4 & 5 & 6 \\ 0 & 0 & 7 & 8 & 9 \end{bmatrix}.$$

The diagonal blocks in this case are

$$A = \begin{bmatrix} 1 & 0 \\ 0 & 1 \end{bmatrix} \quad \text{and} \quad B = \begin{bmatrix} 1 & 2 & 3 \\ 4 & 5 & 6 \\ 7 & 8 & 9 \end{bmatrix}.$$

The next theorem shows that the determinant of a block-diagonal matrix is equal to the product of the determinants of its diagonal blocks.

THEOREM 3.7. *For any two square matrices A and B we have*

$$\det \begin{bmatrix} A & O \\ O & B \end{bmatrix} = (\det A)(\det B). \tag{3.15}$$

Proof. Assume A is $n \times n$ and B is $m \times m$. We note that the given block-diagonal matrix can be expressed as a product of the form

$$\begin{bmatrix} A & O \\ O & B \end{bmatrix} = \begin{bmatrix} A & O \\ O & I_m \end{bmatrix} \begin{bmatrix} I_n & O \\ O & B \end{bmatrix}$$

where I_n and I_m are identity matrices of orders n and m, respectively. Therefore, by the product formula for determinants we have

$$\det \begin{bmatrix} A & O \\ O & B \end{bmatrix} = \det \begin{bmatrix} A & O \\ O & I_m \end{bmatrix} \det \begin{bmatrix} I_n & O \\ O & B \end{bmatrix}. \tag{3.16}$$

Now we regard the determinant $\det \begin{bmatrix} A & O \\ O & I_m \end{bmatrix}$ as a function of the n rows of A. This is possible because of the block of zeros in the upper right-hand corner. It is easily verified that this function satisfies all four axioms for a determinant function of order n. Therefore, by the uniqueness theorem, we must have

$$\det \begin{bmatrix} A & O \\ O & I_m \end{bmatrix} = \det A .$$

A similar argument shows that $\det \begin{bmatrix} I_n & O \\ O & B \end{bmatrix} = \det B$. Hence (3.16) implies (3.15).

3.11 Exercises

1. For each of the following statements about square matrices, give a proof or exhibit a counter example.
   (a) $\det (A + B) = \det A + \det B$.
   (b) $\det \{(A + B)^2\} = \{\det (A + B)\}^2$
   (c) $\det \{(A + B)^2\} = \det (A^2 + 2AB + B^2)$
   (d) $\det \{(A + B)^2\} = \det (A^2 + B^2)$.
2. (a) Extend Theorem 3.7 to block-diagonal matrices with three diagonal blocks:

$$\det \begin{bmatrix} A & O & O \\ O & B & O \\ O & O & C \end{bmatrix} = (\det A)(\det B)(\det C).$$

   (b) State and prove a generalization for block-diagonal matrices with any number of diagonal blocks.
3. Let $A = \begin{bmatrix} 1 & 0 & 0 & 0 \\ 0 & 1 & 0 & 0 \\ a & b & c & d \\ e & f & g & h \end{bmatrix}$, $B = \begin{bmatrix} a & b & c & d \\ e & f & g & h \\ 0 & 0 & 1 & 0 \\ 0 & 0 & 0 & 1 \end{bmatrix}$. Prove that $\det A = \det \begin{bmatrix} c & d \\ g & h \end{bmatrix}$ and that $\det B = \det \begin{bmatrix} a & b \\ e & f \end{bmatrix}$.
4. State and prove a generalization of Exercise 3 for $n \times n$ matrices.
5. Let $A = \begin{bmatrix} a & b & 0 & 0 \\ c & d & 0 & 0 \\ e & f & g & h \\ x & y & z & w \end{bmatrix}$. Prove that $\det A = \det \begin{bmatrix} a & b \\ c & d \end{bmatrix} \det \begin{bmatrix} g & h \\ z & w \end{bmatrix}$.
6. State and prove a generalization of Exercise 5 for $n \times n$ matrices of the form

$$A = \begin{bmatrix} B & O \\ C & D \end{bmatrix},$$

where B, C, D denote square matrices and O denotes a matrix of zeros.

7. Use Theorem 3.6 to determine whether the following sets of vectors are linearly dependent or independent.
   (a) $A_1 = (1, -1, 0)$, $A_2 = (0, 1, -1)$, $A_3 = (2, 3, -1)$.
   (b) $A_1 = (1, -1, 2, 1)$, $A_2 = (-1, 2, -1, 0)$, $A_3 = (3, -1, 1, 0)$, $A_4 = (1, 0, 0, 1)$.
   (c) $A_1 = (1, 0, 0, 0, 1)$, $A_2 = (1, 1, 0, 0, 0)$, $A_3 = (1, 0, 1, 0, 1)$, $A_4 = (1, 1, 0, 1, 1)$, $A_5 = (0, 1, 0, 1, 0)$.

3.12 Expansion formulas for determinants. Minors and cofactors

We still have not shown that a determinant function actually exists, except in the 2×2 case. In this section we exploit the linearity property and the uniqueness theorem to show that if determinants exist they can be computed by a formula which expresses every determinant of order n as a linear combination of determinants of order $n - 1$. Equation (3.2) in Section 3.1 is an example of this formula in the 3×3 case. The general formula will suggest a method for proving existence of determinant functions by induction.

Every row of an $n \times n$ matrix A can be expressed as a linear combination of the n unit coordinate vectors $I_1, \ldots, I_n$. For example, the first row A_1 can be written as follows:

$$A_1 = \sum_{j=1}^{n} a_{1j} I_j .$$

Since determinants are linear in the first row we have

$$(3.17) \quad d(A_1, A_2, \ldots, A_n) = d\left(\sum_{j=1}^{n} a_{1j} I_j, A_2, \ldots, A_n\right) = \sum_{j=1}^{n} a_{1j}\, d(I_j, A_2, \ldots, A_n).$$

Therefore to compute $\det A$ it suffices to compute $d(I_j, A_2, \ldots, A_n)$ for each unit coordinate vector I_j.

Let us use the notation A'_{1j} to denote the matrix obtained from A by replacing the first row A_1 by the unit vector I_j. For example, if $n = 3$ there are three such matrices:

$$A'_{11} = \begin{bmatrix} 1 & 0 & 0 \\ a_{21} & a_{22} & a_{23} \\ a_{31} & a_{32} & a_{33} \end{bmatrix}, \quad A'_{12} = \begin{bmatrix} 0 & 1 & 0 \\ a_{21} & a_{22} & a_{23} \\ a_{31} & a_{32} & a_{33} \end{bmatrix}, \quad A'_{13} = \begin{bmatrix} 0 & 0 & 1 \\ a_{21} & a_{22} & a_{23} \\ a_{31} & a_{32} & a_{33} \end{bmatrix}.$$

Note that $\det A'_{1j} = d(I_j, A_2, \ldots, A_n)$. Equation (3.17) can now be written in the form

$$(3.18) \qquad \det A = \sum_{j=1}^{n} a_{1j} \det A'_{1j} .$$

This is called an *expansion formula;* it expresses the determinant of A as a linear combination of the elements in its first row.

The argument used to derive (3.18) can be applied to the kth row instead of the first row. The result is an expansion formula in terms of elements of the kth row.

THEOREM 3.8. EXPANSION BY COFACTORS. *Let A'_{kj} denote the matrix obtained from A by replacing the kth row A_k by the unit coordinate vector I_j. Then we have the expansion formula*

$$\det A = \sum_{j=1}^{n} a_{kj} \det A'_{kj} \tag{3.19}$$

which expresses the determinant of A as a linear combination of the elements of the kth row. The number $\det A'_{kj}$ *is called the cofactor of entry a_{kj}.*

In the next theorem we shall prove that each cofactor is, except for a plus or minus sign, equal to a determinant of a matrix of order $n - 1$. These smaller matrices are called *minors.*

DEFINITION. *Given a square matrix A of order $n \geq 2$, the square matrix of order $n - 1$ obtained by deleting the kth row and the jth column of A is called the k, j minor of A and is denoted by A_{kj}.*

EXAMPLE. A matrix $A = (a_{kj})$ of order 3 has nine minors. Three of them are

$$A_{11} = \begin{bmatrix} a_{22} & a_{23} \\ a_{32} & a_{33} \end{bmatrix}, \qquad A_{12} = \begin{bmatrix} a_{21} & a_{23} \\ a_{31} & a_{33} \end{bmatrix}, \qquad A_{13} = \begin{bmatrix} a_{21} & a_{22} \\ a_{31} & a_{32} \end{bmatrix}.$$

Equation (3.2) expresses the determinant of a 3×3 matrix as a linear combination of determinants of these three minors. The formula can be written as follows:

$$\det A = a_{11} \det A_{11} - a_{12} \det A_{12} + a_{13} \det A_{13}.$$

The next theorem extends this formula to the $n \times n$ case for any $n \geq 2$.

THEOREM 3.9. EXPANSION BY kTH-ROW MINORS. *For any $n \times n$ matrix A, $n \geq 2$, the cofactor of a_{kj} is related to the minor A_{kj} by the formula*

$$\det A'_{kj} = (-1)^{k+j} \det A_{kj}. \tag{3.20}$$

Therefore the expansion of $\det A$ *in terms of elements of the kth row is given by*

$$\det A = \sum_{j=1}^{n} (-1)^{k+j} a_{kj} \det A_{kj}. \tag{3.21}$$

Proof. We illustrate the idea of the proof by considering first the special case $k = j = 1$. The matrix A'_{11} has the form

$$A'_{11} = \begin{bmatrix} 1 & 0 & 0 & \cdots & 0 \\ a_{21} & a_{22} & a_{23} & \cdots & a_{2n} \\ a_{31} & a_{32} & a_{33} & \cdots & a_{3n} \\ \cdot & & & & \cdot \\ \cdot & & & & \cdot \\ \cdot & & & & \cdot \\ a_{n1} & a_{n2} & a_{n3} & \cdots & a_{nn} \end{bmatrix}.$$

By applying elementary row operations of type (3) we can make every entry below the 1 in the first column equal to zero, leaving all the remaining entries intact. For example, if we multiply the first row of A'_{11} by $-a_{21}$ and add the result to the second row, the new second row becomes $(0, a_{22}, a_{23}, \ldots, a_{2n})$. By a succession of such elementary row operations we obtain a new matrix which we shall denote by A^0_{11} and which has the form

$$A^0_{11} = \begin{bmatrix} 1 & 0 & \cdots & 0 \\ 0 & a_{22} & \cdots & a_{2n} \\ 0 & a_{32} & \cdots & a_{3n} \\ \cdot & \cdot & & \cdot \\ \cdot & \cdot & & \cdot \\ \cdot & \cdot & & \cdot \\ 0 & a_{n2} & \cdots & a_{nn} \end{bmatrix}.$$

Since row operations of type (3) leave the determinant unchanged we have

$$\det A^0_{11} = \det A'_{11}. \tag{3.22}$$

But A^0_{11} is a block-diagonal matrix so, by Theorem 3.7, we have

$$\det A^0_{11} = \det A_{11},$$

where A_{11} is the 1, 1 minor of A,

$$A_{11} = \begin{bmatrix} a_{22} & \cdots & a_{2n} \\ a_{32} & \cdots & a_{3n} \\ \cdot & & \cdot \\ \cdot & & \cdot \\ \cdot & & \cdot \\ a_{n2} & \cdots & a_{nn} \end{bmatrix}.$$

Therefore $\det A'_{11} = \det A_{11}$, which proves (3.20) for $k = j = 1$.

We consider next the special case $k = 1, j$ arbitrary, and prove that

$$\det A'_{1j} = (-1)^{j-1} \det A_{1j}. \tag{3.23}$$

Once we prove (3.23) the more general formula (3.20) follows at once because matrix A'_{kj} can be transformed to a matrix of the form B'_{1j} by $k - 1$ successive interchanges of adjacent rows. The determinant changes sign at each interchange so we have

$$\det A'_{kj} = (-1)^{k-1} \det B'_{1j}, \tag{3.24}$$

where B is an $n \times n$ matrix whose first row is I_j and whose $1, j$ minor B_{1j} is A_{kj}. By (3.23), we have

$$\det B'_{1j} = (-1)^{j-1} \det B_{1j} = (-1)^{j-1} \det A_{kj},$$

so (3.24) gives us

$$\det A'_{kj} = (-1)^{k-1}(-1)^{j-1} \det A_{kj} = (-1)^{k+j} \det A_{kj}.$$

Therefore if we prove (3.23) we also prove (3.20).

We turn now to the proof of (3.23). The matrix A'_{1j} has the form

$$A'_{1j} = \begin{bmatrix} 0 & \cdots & 1 & \cdots & 0 \\ a_{21} & \cdots & a_{2j} & \cdots & a_{2n} \\ \cdot & & \cdot & & \cdot \\ \cdot & & \cdot & & \cdot \\ \cdot & & \cdot & & \cdot \\ a_{n1} & \cdots & a_{nj} & \cdots & a_{nn} \end{bmatrix}.$$

By elementary row operations of type (3) we introduce a column of zeros below the 1 and transform this to

$$A^0_{1j} = \begin{bmatrix} 0 & \cdots & 0 & 1 & 0 & \cdots & 0 \\ a_{21} & \cdots & a_{2,j-1} & 0 & a_{2,j+1} & \cdots & a_{2n} \\ \cdot & & \cdot & & \cdot & & \cdot \\ \cdot & & \cdot & & \cdot & & \cdot \\ \cdot & & \cdot & & \cdot & & \cdot \\ a_{n1} & \cdots & a_{n,j-1} & 0 & a_{n,j+1} & \cdots & a_{nn} \end{bmatrix}.$$

As before, the determinant is unchanged so $\det A^0_{1j} = \det A'_{1j}$. The $1, j$ minor A_{1j} has the form

$$A_{1j} = \begin{bmatrix} a_{21} & \cdots & a_{2,j-1} & a_{2,j+1} & \cdots & a_{2n} \\ \cdot & & \cdot & \cdot & & \cdot \\ \cdot & & \cdot & \cdot & & \cdot \\ \cdot & & \cdot & \cdot & & \cdot \\ a_{n1} & \cdots & a_{n,j-1} & a_{n,j+1} & \cdots & a_{nn} \end{bmatrix}.$$

Now we regard $\det A^0_{1j}$ as a function of the $n-1$ rows of A_{1j}, say $\det A^0_{1j} = f(A_{1j})$. The function f satisfies the first three axioms for a determinant function of order $n-1$. Therefore, by the uniqueness theorem we can write

(3.25) $$f(A_{1j}) = f(J) \det A_{1j},$$

where J is the identity matrix of order $n-1$. Therefore, to prove (3.23) we must show that $f(J) = (-1)^{j-1}$. Now $f(J)$ is by definition the determinant of the matrix

$$C = \begin{bmatrix} 0 & \cdots & 0 & 1 & 0 & \cdots & 0 \\ 1 & & 0 & 0 & 0 & & 0 \\ \cdot & \diagdown & \cdot & \cdot & \cdot & & \cdot \\ \cdot & & \cdot & \cdot & \cdot & & \cdot \\ \cdot & & \cdot & \cdot & \cdot & & \cdot \\ 0 & \cdots & 1 & 0 & 0 & \cdots & 0 \\ 0 & \cdots & 0 & 0 & 1 & \cdots & 0 \\ \cdot & & \cdot & \cdot & \diagdown & & \cdot \\ \cdot & & \cdot & \cdot & & \diagdown & \cdot \\ 0 & \cdots & 0 & 0 & 0 & \cdots & 1 \end{bmatrix} \leftarrow j\text{th row}$$

$$\underset{j\text{th column}}{\uparrow}$$

The entries along the sloping lines are all 1. The remaining entries not shown are all 0. By interchanging the first row of C successively with rows $2, 3, \ldots, j$ we arrive at the $n \times n$ identity matrix I after $j - 1$ interchanges. The determinant changes sign at each interchange, so $\det C = (-1)^{j-1}$. Hence $f(J) = (-1)^{j-1}$, which proves (3.23) and hence (3.20).

3.13 Existence of the determinant function

In this section we use induction on n, the size of a matrix, to prove that determinant functions of every order exist. For $n = 2$ we have already shown that a determinant function exists. We can also dispense with the case $n = 1$ by defining $\det [a_{11}] = a_{11}$.

Assuming a determinant function exists of order $n - 1$, a logical candidate for a determinant function of order n would be one of the expansion formulas of Theorem 3.9, for example, the expansion in terms of the first-row minors. However, it is easier to verify the axioms if we use a different but analogous formula expressed in terms of the first-*column* minors.

THEOREM 3.10. *Assume determinants of order $n - 1$ exist. For any $n \times n$ matrix $A = (a_{jk})$, let f be the function defined by the formula*

$$f(A_1, \ldots, A_n) = \sum_{j=1}^{n} (-1)^{j+1} a_{j1} \det A_{j1}. \tag{3.26}$$

Then f satisfies all four axioms for a determinant function of order n. Therefore, by induction, determinants of order n exist for every n.

Proof. We regard each term of the sum in (3.26) as a function of the rows of A and we write

$$f_j(A_1, \ldots, A_n) = (-1)^{j+1} a_{j1} \det A_{j1}.$$

If we verify that each f_j satisfies Axioms 1 and 2 the same will be true for f.

Consider the effect of multiplying the first row of A by a scalar t. The minor A_{11} is not affected since it does not involve the first row. The coefficient a_{11} is multiplied by t, so we have

$$f_1(tA_1, A_2, \ldots, A_n) = ta_{11} \det A_{11} = tf_1(A_1, \ldots, A_n).$$

If $j > 1$ the first row of each minor A_{j1} gets multiplied by t and the coefficient a_{j1} is not affected, so again we have

$$f_j(tA_1, A_2, \ldots, A_n) = tf_j(A_1, A_2, \ldots, A_n).$$

Therefore each f_j is homogeneous in the first row.

If the kth row of A is multiplied by t, where $k > 1$, the minor A_{k1} is not affected but a_{k1} is multiplied by t, so f_k is homogeneous in the kth row. If $j \neq k$, the coefficient a_{j1} is not affected but some row of A_{j1} gets multiplied by t. Hence every f_j is homogeneous in the kth row.

A similar argument shows that each f_j is additive in every row, so f satisfies Axioms 1 and 2. We prove next that f satisfies Axiom 3′, the weak version of Axiom 3. From Theorem 3.1, it then follows that f satisfies Axiom 3.

To verify that f satisfies Axiom 3′, assume two adjacent rows of A are equal, say $A_k = A_{k+1}$. Then, except for minors A_{k1} and $A_{k+1,1}$, each minor A_{j1} has two equal rows so $\det A_{j1} = 0$. Therefore the sum in (3.26) consists only of the two terms corresponding to $j = k$ and $j = k + 1$,

$$f(A_1, \ldots, A_n) = (-1)^{k+1} a_{k1} \det A_{k1} + (-1)^{k+2} a_{k+1,1} \det A_{k+1,1}. \tag{3.27}$$

But $A_{k1} = A_{k+1,1}$ and $a_{k1} = a_{k+1,1}$ since $A_k = A_{k+1}$. Therefore the two terms in (3.27) differ only in sign, so $f(A_1, \ldots, A_n) = 0$. Thus, f satisfies Axiom 3′.

Finally, we verify that f satisfies Axiom 4. When $A = I$ we have $a_{11} = 1$ and $a_{j1} = 0$ for $j > 1$. Also, A_{11} is the identity matrix of order $n - 1$, so each term in (3.26) is zero except the first, which is equal to 1. Hence $f(I_1, \ldots, I_n) = 1$ so f satisfies Axiom 4.

In the foregoing proof we could just as well have used a function f defined in terms of the kth-column minors A_{jk} instead of the first-column minors A_{j1}. In fact, if we let

$$f(A_1, \ldots, A_n) = \sum_{j=1}^{n} (-1)^{j+k} a_{jk} \det A_{jk}, \tag{3.28}$$

exactly the same type of proof shows that this f satisfies all four axioms for a determinant function. Since determinant functions are unique, the expansion formulas in (3.28) and those in (3.21) are all equal to $\det A$.

The expansion formulas (3.28) not only establish the existence of determinant functions but also reveal a new aspect of the theory of determinants—a connection between row-properties and column-properties. This connection is discussed further in the next section.

3.14 The determinant of a transpose

Associated with each matrix A is another matrix called the *transpose* of A and denoted by A^t. The rows of A^t are the columns of A. For example, if

$$A = \begin{bmatrix} 1 & 2 & 3 \\ 4 & 5 & 6 \end{bmatrix}, \qquad \text{then} \quad A^t = \begin{bmatrix} 1 & 4 \\ 2 & 5 \\ 3 & 6 \end{bmatrix}.$$

A formal definition may be given as follows.

DEFINITION OF TRANSPOSE. *The transpose of an $m \times n$ matrix $A = (a_{ij})_{i,j=1}^{m,n}$ is the $n \times m$ matrix A^t whose i, j entry is a_{ji}.*

Although transposition can be applied to any rectangular matrix we shall be concerned primarily with square matrices. We prove next that transposition of a square matrix does not alter its determinant.

THEOREM 3.11. *For any $n \times n$ matrix A we have*

$$\det A = \det A^t.$$

Proof. The proof is by induction on n. For $n = 1$ and $n = 2$ the result is easily verified. Assume, then, that the theorem is true for matrices of order $n - 1$. Let $A = (a_{ij})$ and let $B = A^t = (b_{ij})$. Expanding $\det A$ by its first-column minors and $\det B$ by its first-row minors we have

$$\det A = \sum_{j=1}^{n} (-1)^{j+1} a_{j1} \det A_{j1}, \qquad \det B = \sum_{j=1}^{n} (-1)^{j+1} b_{1j} \det B_{1j}.$$

But from the definition of transpose we have $b_{1j} = a_{j1}$ and $B_{1j} = (A_{j1})^t$. Since we are assuming the theorem is true for matrices of order $n - 1$ we have $\det B_{1j} = \det A_{j1}$. Hence the foregoing sums are equal term by term, so $\det A = \det B$.

3.15 The cofactor matrix

Theorem 3.5 showed that if A is nonsingular then $\det A \neq 0$. The next theorem proves the converse. That is, if $\det A \neq 0$, then A^{-1} exists. Moreover, it gives an explicit formula for expressing A^{-1} in terms of a matrix formed from the cofactors of the entries of A.

In Theorem 3.9 we proved that the i, j cofactor of a_{ij} is equal to $(-1)^{i+j} \det A_{ij}$, where A_{ij} is the i, j minor of A. Let us denote this cofactor by $\operatorname{cof} a_{ij}$. Thus, by definition,

$$\operatorname{cof} a_{ij} = (-1)^{i+j} \det A_{ij}.$$

DEFINITION OF THE COFACTOR MATRIX. *The matrix whose i, j entry is $\operatorname{cof} a_{ij}$ is called the cofactor matrix*† *of A and is denoted by $\operatorname{cof} A$. Thus, we have*

$$\operatorname{cof} A = (\operatorname{cof} a_{ij})_{i,j=1}^{n} = ((-1)^{i+j} \det A_{ij})_{i,j=1}^{n}.$$

The next theorem shows that the product of A with the transpose of its cofactor matrix is, apart from a scalar factor, the identity matrix I.

THEOREM 3.12. *For any $n \times n$ matrix A with $n \geq 2$ we have*

$$A(\operatorname{cof} A)^t = (\det A)I. \tag{3.29}$$

In particular, if $\det A \neq 0$ the inverse of A exists and is given by

$$A^{-1} = \frac{1}{\det A} (\operatorname{cof} A)^t.$$

† In much of the matrix literature the transpose of the cofactor matrix is called the *adjugate* of A. Some of the older literature calls it the *adjoint* of A. However, current nomenclature reserves the term *adjoint* for an entirely different object, discussed in Section 5.8.

Proof. Using Theorem 3.9 we express $\det A$ in terms of its kth-row cofactors by the formula

$$\det A = \sum_{j=1}^{n} a_{kj} \operatorname{cof} a_{kj}. \tag{3.30}$$

Keep k fixed and apply this relation to a new matrix B whose ith row is equal to the kth row of A for some $i \neq k$, and whose remaining rows are the same as those of A. Then $\det B = 0$ because the ith and kth rows of B are equal. Expressing $\det B$ in terms of its ith-row cofactors we have

$$\det B = \sum_{j=1}^{n} b_{ij} \operatorname{cof} b_{ij} = 0. \tag{3.31}$$

But since the ith row of B is equal to the kth row of A we have

$$b_{ij} = a_{kj} \qquad \text{and} \qquad \operatorname{cof} b_{ij} = \operatorname{cof} a_{ij} \qquad \text{for every } j.$$

Hence (3.31) states that

$$\sum_{j=1}^{n} a_{kj} \operatorname{cof} a_{ij} = 0 \qquad \text{if} \quad k \neq i. \tag{3.32}$$

Equations (3.30) and (3.32) together can be written as follows:

$$\sum_{j=1}^{n} a_{kj} \operatorname{cof} a_{ij} = \begin{cases} \det A & \text{if} \quad i = k \\ 0 & \text{if} \quad i \neq k. \end{cases} \tag{3.33}$$

But the sum appearing on the left of (3.33) is the k, i entry of the product $A(\operatorname{cof} A)^t$. Therefore (3.33) implies (3.29).

As a direct corollary of Theorems 3.5 and 3.12 we have the following necessary and sufficient condition for a square matrix to be nonsingular.

THEOREM 3.13. *A square matrix A is nonsingular if and only if* $\det A \neq 0$.

3.16 Cramer's rule

Theorem 3.12 can also be used to give explicit formulas for the solutions of a system of linear equations with a nonsingular coefficient matrix. The formulas are called *Cramer's rule*, in honor of the Swiss mathematician Gabriel Cramer (1704–1752).

THEOREM 3.14. CRAMER'S RULE. *If a system of n linear equations in n unknowns* $x_1, \ldots, x_n$,

$$\sum_{j=1}^{n} a_{ij} x_j = b_i \qquad (i = 1, 2, \ldots, n)$$

has a nonsingular coefficient-matrix $A = (a_{ij})$, *then there is a unique solution for the system given by the formulas*

(3.34) $$x_j = \frac{1}{\det A}\sum_{k=1}^{n} b_k \operatorname{cof} a_{kj}, \qquad \text{for } j = 1, 2, \ldots, n.$$

Proof. The system can be written as a matrix equation,

$$AX = B,$$

where X and B are column matrices, $X = \begin{bmatrix} x_1 \\ \cdot \\ \cdot \\ \cdot \\ x_n \end{bmatrix}$, $B = \begin{bmatrix} b_1 \\ \cdot \\ \cdot \\ \cdot \\ b_n \end{bmatrix}$. Since A is nonsingular there is a unique solution X given by

(3.35) $$X = A^{-1}B = \frac{1}{\det A}(\operatorname{cof} A)^t B.$$

The formulas in (3.34) follow by equating components in (3.35).

It should be noted that the formula for x_j in (3.34) can be expressed as the quotient of two determinants,

$$x_j = \frac{\det C_j}{\det A},$$

where C_j is the matrix obtained from A by replacing the jth column of A by the column matrix B.

3.17 Exercises

1. Determine the cofactor matrix of each of the following matrices:

(a) $\begin{bmatrix} 1 & 2 \\ 3 & 4 \end{bmatrix}$, (b) $\begin{bmatrix} 2 & -1 & 3 \\ 0 & 1 & 1 \\ -1 & -2 & 0 \end{bmatrix}$, (c) $\begin{bmatrix} 3 & 1 & 2 & 4 \\ 2 & 0 & 5 & 1 \\ 1 & -1 & -2 & 6 \\ -2 & 3 & 2 & 3 \end{bmatrix}$.

2. Determine the inverse of each of the nonsingular matrices in Exercise 1.
3. Find all values of the scalar λ for which the matrix $\lambda I - A$ is singular, if A is equal to

(a) $\begin{bmatrix} 0 & 3 \\ 2 & -1 \end{bmatrix}$, (b) $\begin{bmatrix} 1 & 0 & 2 \\ 0 & -1 & -2 \\ 2 & -2 & 0 \end{bmatrix}$, (c) $\begin{bmatrix} 11 & -2 & 8 \\ 19 & -3 & 14 \\ -8 & 2 & -5 \end{bmatrix}$.

4. If A is an $n \times n$ matrix with $n \geq 2$, prove each of the following properties of its cofactor matrix:
(a) $\text{cof}\,(A^t) = (\text{cof}\,A)^t$. (b) $(\text{cof}\,A)^t A = (\det A)I$.
(c) $A(\text{cof}\,A)^t = (\text{cof}\,A)^t A$ (A commutes with the transpose of its cofactor matrix).
5. Use Cramer's rule to solve each of the following systems:
(a) $x + 2y + 3z = 8$, $2x - y + 4z = 7$, $-y + z = 1$.
(b) $x + y + 2z = 0$, $3x - y - z = 3$, $2x + 5y + 3z = 4$.
6. (a) Explain why each of the following is a Cartesian equation for a straight line in the xy-plane passing through two distinct points (x_1, y_1) and (x_2, y_2).

$$\det\begin{bmatrix} x - x_1 & y - y_1 \\ x_2 - x_1 & y_2 - y_1 \end{bmatrix} = 0; \qquad \det\begin{bmatrix} x & y & 1 \\ x_1 & y_1 & 1 \\ x_2 & y_2 & 1 \end{bmatrix} = 0.$$

(b) State and prove corresponding relations for a plane in 3-space passing through three distinct points.
(c) State and prove corresponding relations for a circle in the xy-plane passing through three noncolinear points.
7. Given n^2 functions f_{ij}, each differentiable on an interval (a, b), define $F(x) = \det\,[f_{ij}(x)]$ for each x in (a, b). Prove that the derivative $F'(x)$ is a sum of n determinants,

$$F'(x) = \sum_{i=1}^{n} \det A_i(x),$$

where $A_i(x)$ is the matrix obtained by differentiating the functions in the ith row of $[f_{ij}(x)]$.
8. An $n \times n$ matrix of functions of the form $W(x) = [u_j^{(i-1)}(x)]$, in which each row after the first is the derivative of the previous row, is called a *Wronskian matrix* in honor of the Polish mathematician J. M. H. Wronski (1778–1853). Prove that the derivative of the determinant of $W(x)$ is the determinant of the matrix obtained by differentiating each entry in the last row of $W(x)$.

[*Hint:* Use Exercise 7.]

4

EIGENVALUES AND EIGENVECTORS

4.1 Linear transformations with diagonal matrix representations

Let $T: V \to V$ be a linear transformation on a finite-dimensional linear space V. Those properties of T which are independent of any coordinate system (basis) for V are called *intrinsic properties* of T. They are shared by all the matrix representations of T. If a basis can be chosen so that the resulting matrix has a particularly simple form it may be possible to detect some of the intrinsic properties directly from the matrix representation.

Among the simplest types of matrices are the diagonal matrices. Therefore we might ask whether every linear transformation has a diagonal matrix representation. In Chapter 2 we treated the problem of finding a diagonal matrix representation of a linear transformation $T: V \to W$, where $\dim V = n$ and $\dim W = m$. In Theorem 2.14 we proved that there always exists a basis $(e_1, \ldots, e_n)$ for V and a basis $(w_1, \ldots, w_m)$ for W such that the matrix of T relative to this pair of bases is a diagonal matrix. In particular, if $W = V$ the matrix will be a square diagonal matrix. The new feature now is that we want to use the *same basis* for both V and W. With this restriction it is not always possible to find a diagonal matrix representation for T. We turn, then, to the problem of determining which transformations do have a diagonal matrix representation.

Notation: If $A = (a_{ij})$ is a diagonal matrix, we write $A = \operatorname{diag}(a_{11}, a_{22}, \ldots, a_{nn})$.

It is easy to give a necessary and sufficient condition for a linear transformation to have a diagonal matrix representation.

THEOREM 4.1. *Given a linear transformation $T: V \to V$, where $\dim V = n$. If T has a diagonal matrix representation, then there exists an independent set of elements $u_1, \ldots, u_n$ in V and a corresponding set of scalars $\lambda_1, \ldots, \lambda_n$ such that*

$$T(u_k) = \lambda_k u_k \quad \text{for} \quad k = 1, 2, \ldots, n. \tag{4.1}$$

Conversely, if there is an independent set $u_1, \ldots, u_n$ in V and a corresponding set of scalars $\lambda_1, \ldots, \lambda_n$ satisfying (4.1), *then the matrix*

$$A = \operatorname{diag}(\lambda_1, \ldots, \lambda_n)$$

is a representation of T relative to the basis $(u_1, \ldots, u_n)$.

Proof. Assume first that T has a diagonal matrix representation $A = (a_{ik})$ relative to some basis $(e_1, \ldots, e_n)$. The action of T on the basis elements is given by the formula

$$T(e_k) = \sum_{i=1}^{n} a_{ik}e_i = a_{kk}e_k$$

since $a_{ik} = 0$ for $i \neq k$. This proves (4.1) with $u_k = e_k$ and $\lambda_k = a_{kk}$.

Now suppose independent elements $u_1, \ldots, u_n$ and scalars $\lambda_1, \ldots, \lambda_n$ exist satisfying (4.1). Since $u_1, \ldots, u_n$ are independent they form a basis for V. If we define $a_{kk} = \lambda_k$ and $a_{ik} = 0$ for $i \neq k$, then the matrix $A = (a_{ik})$ is a diagonal matrix which represents T relative to the basis $(u_1, \ldots, u_n)$.

Thus the problem of finding a diagonal matrix representation of a linear transformation has been transformed to another problem, that of finding independent elements $u_1, \ldots, u_n$ and scalars $\lambda_1, \ldots, \lambda_n$ to satisfy (4.1). Elements u_k and scalars λ_k satisfying (4.1) are called *eigenvectors* and *eigenvalues* of T, respectively. In the next section we study eigenvectors and eigenvalues† in a more general setting.

4.2 Eigenvectors and eigenvalues of a linear transformation

In this discussion V denotes a linear space and S denotes a subspace of V. The spaces S and V are not required to be finite dimensional.

DEFINITION. *Let $T: S \to V$ be a linear transformation of S into V. A scalar λ is called an eigenvalue of T if there is a nonzero element x in S such that*

$$T(x) = \lambda x. \tag{4.2}$$

The element x is called an eigenvector of T belonging to λ. The scalar λ is called an eigenvalue corresponding to x.

There is exactly one eigenvalue corresponding to a given eigenvector x. In fact, if we have $T(x) = \lambda x$ and $T(x) = \mu x$ for some $x \neq O$, then $\lambda x = \mu x$ so $\lambda = \mu$.

Note: Although Equation (4.2) always holds for $x = O$ and any scalar λ, the definition excludes O as an eigenvector. One reason for this prejudice against O is to have exactly one eigenvalue λ associated with a given eigenvector x.

The following examples illustrate the meaning of these concepts.

EXAMPLE 1. *Multiplication by a fixed scalar.* Let $T: S \to V$ be the linear transformation defined by the equation $T(x) = cx$ for each x in S, where c is a fixed scalar. In this example every nonzero element of S is an eigenvector belonging to the scalar c.

† The words *eigenvector* and *eigenvalue* are partial translations of the German words *Eigenvektor* and *Eigenwert*, respectively. Some authors use the terms *characteristic vector*, or *proper vector* as synonyms for eigenvector. Eigenvalues are also called *characteristic values, proper values*, or *latent roots*.

EXAMPLE 2. *The eigenspace $E(\lambda)$ consisting of all x such that $T(x) = \lambda x$.* Let $T: S \to V$ be a linear transformation having an eigenvalue λ. Let $E(\lambda)$ be the set of *all* elements x in S such that $T(x) = \lambda x$. This set contains the zero element O and all eigenvectors belonging to λ. It is easy to prove that $E(\lambda)$ is a subspace of S, because if x and y are in $E(\lambda)$ we have

$$T(ax + by) = aT(x) + bT(y) = a\lambda x + b\lambda y = \lambda(ax + by)$$

for all scalars a and b. Hence $(ax + by) \in E(\lambda)$ so $E(\lambda)$ is a subspace. The space $E(\lambda)$ is called the *eigenspace* corresponding to λ. It may be finite- or infinite-dimensional. If $E(\lambda)$ is finite-dimensional then $\dim E(\lambda) \geq 1$, since $E(\lambda)$ contains at least one nonzero element x corresponding to λ.

EXAMPLE 3. *Existence of zero eigenvalues.* If an eigenvector exists it cannot be zero, by definition. However, the zero scalar *can* be an eigenvalue. In fact, if 0 is an eigenvalue for x then $T(x) = 0x = O$, so x is in the null space of T. Conversely, if the null space of T contains any nonzero elements then each of these is an eigenvector with eigenvalue 0. In general, $E(\lambda)$ is the null space of $T - \lambda I$.

EXAMPLE 4. *Reflection in the xy-plane.* Let $S = V = V_3(\mathbf{R})$ and let T be a reflection in the xy-plane. That is, let T act on the basis vectors $\boldsymbol{i}, \boldsymbol{j}, \boldsymbol{k}$ as follows: $T(\boldsymbol{i}) = \boldsymbol{i}$, $T(\boldsymbol{j}) = \boldsymbol{j}$, $T(\boldsymbol{k}) = -\boldsymbol{k}$. Every nonzero vector in the xy-plane is an eigenvector with eigenvalue 1. The remaining eigenvectors are those of the form $c\boldsymbol{k}$, where $c \neq 0$; each of them has eigenvalue -1.

EXAMPLE 5. *Rotation of the plane through a fixed angle α.* This example is of special interest because it shows that the existence of eigenvectors may depend on the underlying field of scalars. The plane can be regarded as a linear space in two different ways: (1) As a 2-dimensional *real* linear space, $V = V_2(\mathbf{R})$, with two basis elements (1, 0) and (0, 1), and with real numbers as scalars; or (2) as a 1-dimensional *complex* linear space, $V = V_1(\mathbf{C})$, with one basis element 1, and complex numbers as scalars.

Consider the second interpretation first. Each element $z \neq 0$ of $V_1(\mathbf{C})$ can be expressed in polar form, $z = re^{i\theta}$. If T rotates z through an angle α then $T(z) = re^{i(\theta+\alpha)} = e^{i\alpha}z$. Thus, each $z \neq 0$ is an eigenvector with eigenvalue $\lambda = e^{i\alpha}$. Note that the eigenvalue $e^{i\alpha}$ is not real unless α is an integer multiple of π.

Now consider the plane as a *real* linear space, $V_2(\mathbf{R})$. Since the scalars of $V_2(\mathbf{R})$ are real numbers the rotation T has real eigenvalues only if α is an integer multiple of π. In other words, if α is not an integer multiple of π then T has no real eigenvalues and hence no eigenvectors. Thus the existence of eigenvectors and eigenvalues may depend on the choice of scalars for V.

EXAMPLE 6. *The differentiation operator.* Let V be the linear space of all real functions f having derivatives of every order on a given open interval. Let D be the linear transformation which maps each f onto its derivative, $D(f) = f'$. The eigenvectors of D are those nonzero functions f satisfying an equation of the form

$$f' = \lambda f$$

for some real λ. This is a first order linear differential equation. All its solutions are given by the formula

$$f(x) = ce^{\lambda x},$$

where c is an arbitrary real constant. Therefore the eigenvectors of D are all exponential functions $f(x) = ce^{\lambda x}$ with $c \neq 0$. The eigenvalue corresponding to $f(x) = ce^{\lambda x}$ is λ. In examples like this one where V is a function space the eigenvectors are called *eigenfunctions.*

EXAMPLE 7. *The integration operator.* Let V be the linear space of all real functions continuous on a finite interval $[a, b]$. If $f \in V$ define $g = T(f)$ to be that function given by

$$g(x) = \int_a^x f(t)\, dt \qquad \text{if} \quad a \leq x \leq b\,.$$

The eigenfunctions of T (if any exist) are those nonzero f satisfying an equation of the form

$$\int_a^x f(t)\, dt = \lambda f(x) \tag{4.3}$$

for some real λ. If an eigenfunction exists we may differentiate this equation to obtain the relation $f(x) = \lambda f'(x)$, from which we find $f(x) = ce^{x/\lambda}$, provided $\lambda \neq 0$. In other words, the only candidates for eigenfunctions are those exponential functions of the form $f(x) = ce^{x/\lambda}$ with $c \neq 0$ and $\lambda \neq 0$. However, if we put $x = a$ in (4.3) we obtain

$$0 = \lambda f(a) = \lambda c e^{a/\lambda}.$$

Since $e^{a/\lambda}$ is never zero we see that the equation $T(f) = \lambda f$ cannot be satisfied with a nonzero f, so T has no eigenfunctions and no eigenvalues.

EXAMPLE 8. *The subspace spanned by an eigenvector.* Let $T: S \to V$ be a linear transformation having an eigenvalue λ. Let x be an eigenvector belonging to λ and let $L(x)$ be the subspace spanned by x. That is, $L(x)$ is the set of all scalar multiples of x. It is easy to show that T maps $L(x)$ into itself. In fact, if $y = cx$ we have

$$T(y) = T(cx) = cT(x) = c(\lambda x) = \lambda(cx) = \lambda y\,.$$

If $c \neq 0$ then $y \neq O$ so every nonzero element y of $L(x)$ is also an eigenvector belonging to λ.

A subspace U of S is called *invariant* under T if T maps each element of U onto an element of U. We have just shown that the subspace spanned by an eigenvector is invariant under T.

4.3 Linear independence of eigenvectors corresponding to distinct eigenvalues

One of the most important properties of eigenvalues is described in the next theorem. As before, S denotes a subspace of a linear space V.

THEOREM 4.2. *Let $u_1, \ldots, u_k$ be eigenvectors of a linear transformation $T: S \to V$, and assume that the corresponding eigenvalues $\lambda_1, \ldots, \lambda_k$ are distinct. Then the eigenvectors $u_1, \ldots, u_k$ are independent.*

Proof. The proof is by induction on k. The result is trivial when $k = 1$. Assume, then, that it has been proved for every set of $k - 1$ eigenvectors. Let $u_1, \ldots, u_k$ be k eigenvectors belonging to distinct eigenvalues, and assume that scalars c_i exist such that

$$\sum_{i=1}^{k} c_i u_i = O. \tag{4.4}$$

Applying T to both members of (4.4) and using the fact that $T(u_i) = \lambda_i u_i$ we find

$$\sum_{i=1}^{k} c_i \lambda_i u_i = O. \tag{4.5}$$

Multiplying (4.4) by λ_k and subtracting from (4.5) we obtain the equation

$$\sum_{i=1}^{k-1} c_i(\lambda_i - \lambda_k) u_i = O.$$

But since $u_1, \ldots, u_{k-1}$ are independent we must have $c_i(\lambda_i - \lambda_k) = 0$ for each $i = 1, 2, \ldots, k - 1$. Since the eigenvalues are distinct we have $\lambda_i \neq \lambda_k$ for $i \neq k$ so $c_i = 0$ for $i = 1, 2, \ldots, k - 1$. From (4.4) we see that c_k is also 0 so the eigenvectors $u_1, \ldots, u_k$ are independent.

Note that Theorem 4.2 would not be true if the zero element were allowed to be an eigenvector. This is another reason for excluding O as an eigenvector.

> *Warning:* The converse of Theorem 4.2 does not hold. That is, if T has independent eigenvectors $u_1, \ldots, u_k$, then the corresponding eigenvalues $\lambda_1, \ldots, \lambda_k$ need not be distinct. For example, if T is the identity transformation, $T(x) = x$ for all x, then every $x \neq O$ is an eigenvector but there is only one eigenvalue, $\lambda = 1$.

Theorem 4.2. has important consequences for the finite-dimensional case.

THEOREM 4.3. *If* $\dim V = n$, *every linear transformation $T: V \to V$ has at most n distinct eigenvalues. If T has exactly n distinct eigenvalues, then the corresponding eigenvectors form a basis for V and the matrix of T relative to this basis is a diagonal matrix with the eigenvalues as diagonal entries.*

Proof. If there were $n + 1$ distinct eigenvalues then, by Theorem 4.2, V would contain $n + 1$ independent elements. This is not possible since $\dim V = n$. The second assertion follows from Theorems 4.1 and 4.2.

Note: Theorem 4.3 tells us that the existence of n distinct eigenvalues is a *sufficient* condition for T to have a diagonal matrix representation. This condition is not necessary. There exist linear transformations with less than n distinct eigenvalues that can be represented by diagonal matrices. The identity transformation is an example. All its eigenvalues are equal to 1 but it can be represented by the identity matrix. Theorem 4.1 tells us that the existence of n independent eigenvectors is *necessary and sufficient* for T to have a diagonal matrix representation.

4.4 Exercises

1. (a) If T has an eigenvalue λ, prove that aT has the eigenvalue $a\lambda$.
(b) If x is an eigenvector for both T_1 and T_2, prove that x is also an eigenvector for $aT_1 + bT_2$. How are the eigenvalues related?
2. Assume $T: V \to V$ has an eigenvector x belonging to an eigenvalue λ. Prove that x is an eigenvector of T^2 belonging to λ^2 and, more generally, x is an eigenvector of T^n belonging to λ^n. Then use the result of Exercise 1 to show that if P is a polynomial, then x is an eigenvector of $P(T)$ belonging to $P(\lambda)$.
3. Consider the plane as a real linear space, $V = V_2(\mathbf{R})$, and let T be a rotation of V through an angle of $\pi/2$ radians. Although T has no eigenvectors, prove that every nonzero vector is an eigenvector for T^2.
4. If $T: V \to V$ has the property that T^2 has a nonnegative eigenvalue λ^2, prove that at least one of λ or $-\lambda$ is an eigenvalue for T. [*Hint:* $T^2 - \lambda^2 I = (T + \lambda I)(T - \lambda I)$.]
5. Let V be the linear space of all real functions differentiable on (0, 1). If $f \in V$, define $g = T(f)$ to mean that $g(t) = tf'(t)$ for all t in (0, 1). Prove that every real λ is an eigenvalue for T, and determine the eigenfunctions corresponding to λ.
6. Let V be the linear space of all real polynomials $p(x)$ of degree $\le n$. If $p \in V$, define $q = T(p)$ to mean that $q(t) = p(t + 1)$ for all real t. Prove that T has only the eigenvalue 1. What are the eigenfunctions belonging to this eigenvalue?
7. Let V be the linear space of all functions continuous on $(-\infty, +\infty)$ and such that the integral $\int_{-\infty}^{x} f(t)\,dt$ exists for all real x. If $f \in V$, let $g = T(f)$ be defined by the equation $g(x) = \int_{-\infty}^{x} f(t)\,dt$. Prove that every positive λ is an eigenvalue for T and determine the eigenfunctions corresponding to λ.
8. Let V be the linear space of all functions continuous on $(-\infty, +\infty)$ and such that the integral $\int_{-\infty}^{x} t\, f(t)\,dt$ exists for all real x. If $f \in V$ let $g = T(f)$ be defined by the equation $g(x) = \int_{-\infty}^{x} t\, f(t)\,dt$. Prove that every negative λ is an eigenvalue for T and determine the eigenfunctions corresponding to λ.
9. Let $V = C(0, \pi)$ be the real linear space of all real functions continuous on the interval $[0, \pi]$. Let S be the subspace of all functions f which have a continuous second derivative in linear and which also satisfy the boundary conditions $f(0) = f(\pi) = 0$. Let $T: S \to V$ be the linear transformation which maps each f onto its second derivative, $T(f) = f''$. Prove that the eigenvalues of T are the numbers of the form $-n^2$, where $n = 1, 2, \ldots$, and that the eigenfunctions corresponding to $-n^2$ are $f(t) = c_n \sin nt$, where $c_n \neq 0$.
10. Let V be the linear space of all real convergent sequences $\{x_n\}$. Define $T: V \to V$ as follows: If $x = \{x_n\}$ is a convergent sequence with limit a, let $T(x) = \{y_n\}$, where $y_n = a - x_n$ for $n \ge 1$. Prove that T has only two eigenvalues, $\lambda = 0$ and $\lambda = -1$, and determine the eigenvectors belonging to each such λ.
11. Assume that a linear transformation T has two eigenvectors x and y belonging to distinct eigenvalues λ and μ. If $ax + by$ is an eigenvector of T, prove that $a = 0$ or $b = 0$.
12. Let $T: S \to V$ be a linear transformation such that every nonzero element of S is an eigenvector. Prove that there exists a scalar c such that $T(x) = cx$. In other words, the only transformation with this property is a scalar times the identity. [*Hint:* Use Exercise 11.]

4.5 The finite-dimensional case. Characteristic polynomials

If dim $V = n$, the problem of finding the eigenvalues of a linear transformation $T: V \to V$ can be solved with the help of determinants. We wish to find those scalars λ such that the equation $T(x) = \lambda x$ has a solution x with $x \neq O$. The equation $T(x) = \lambda x$ can be written in the form

$$(\lambda I - T)(x) = O,$$

where I is the identity transformation. If we let $T_\lambda = \lambda I - T$, then λ is an eigenvalue if and only if the equation

$$T_\lambda(x) = O \tag{4.6}$$

has a nonzero solution x, in which case T_λ is not invertible (because of Theorem 2.10). Therefore, by Theorem 2.20, a nonzero solution of (4.6) exists if and only if the matrix of T_λ is singular. If A is a matrix representation for T, then $\lambda I - A$ is a matrix representation for T_λ. By Theorem 3.13, the matrix $\lambda I - A$ is singular if and only if $\det(\lambda I - A) = 0$. Thus, if λ is an eigenvalue for T it satisfies the equation

$$\det(\lambda I - A) = 0. \tag{4.7}$$

Conversely, any λ *in the underlying field of scalars* which satisfies (4.7) is an eigenvalue. This suggests that we should study the determinant $\det(\lambda I - A)$ as a function of λ.

THEOREM 4.4. *If A is any $n \times n$ matrix and if I is the $n \times n$ identity matrix, the function f defined by the equation*

$$f(\lambda) = \det(\lambda I - A)$$

is a polynomial in λ of degree n. Moreover, the term of highest degree is λ^n, and the constant term is $f(0) = \det(-A) = (-1)^n \det A$.

Proof. The statement $f(0) = \det(-A)$ follows at once from the definition of f. We prove that f is a polynomial of degree n only for the case $n \leq 3$. The proof in the general case can be given by induction and is left as an exercise. (See Exercise 9 in Section 4.8.)

For $n = 1$ the determinant is the linear polynomial $f(\lambda) = \lambda - a_{11}$. For $n = 2$ we have

$$\det(\lambda I - A) = \begin{vmatrix} \lambda - a_{11} & -a_{12} \\ -a_{21} & \lambda - a_{22} \end{vmatrix} = (\lambda - a_{11})(\lambda - a_{22}) - a_{12}a_{21}$$

$$= \lambda^2 - (a_{11} + a_{22})\lambda + (a_{11}a_{22} - a_{12}a_{21}),$$

a quadratic polynomial in λ. For $n = 3$ we have

$$\det(\lambda I - A) = \begin{vmatrix} \lambda - a_{11} & -a_{12} & -a_{13} \\ -a_{21} & \lambda - a_{22} & -a_{23} \\ -a_{31} & -a_{32} & \lambda - a_{33} \end{vmatrix}$$

$$= (\lambda - a_{11})\begin{vmatrix} \lambda - a_{22} & -a_{23} \\ -a_{32} & \lambda - a_{33} \end{vmatrix} + a_{12}\begin{vmatrix} -a_{21} & -a_{23} \\ -a_{31} & \lambda - a_{33} \end{vmatrix} - a_{13}\begin{vmatrix} -a_{21} & \lambda - a_{22} \\ -a_{31} & -a_{32} \end{vmatrix}.$$

The last two terms are linear polynomials in λ. The first term is a cubic polynomial, the term of highest degree being λ^3.

DEFINITION. *If A is an $n \times n$ matrix the determinant*

$$f(\lambda) = \det (\lambda I - A)$$

is called the characteristic polynomial of A.

The roots of the characteristic polynomial of A are complex numbers, some of which may be real. If we let F denote either the real field $\mathbf{R}$ or the complex field $\mathbf{C}$, we have the following theorem.

THEOREM 4.5. *Let $T: V \to V$ be a linear transformation, where V has scalars in F, and* $\dim V = n$. *Let A be a matrix representation of T. Then the set of eigenvalues of T consists of those roots of the characteristic polynomial of A which lie in F.*

Proof. The discussion preceding Theorem 4.4 shows that every eigenvalue of T satisfies the equation $\det (\lambda I - A) = 0$ and that any root of the characteristic polynomial of A which lies in F is an eigenvalue of T.

The matrix A depends on the choice of basis for V, but the eigenvalues of T were defined without reference to a basis. Therefore, the *set* of roots of the characteristic polynomial of A must be independent of the choice of basis. More than this is true. In a later section we shall prove that the characteristic polynomial itself is independent of the choice of basis. We turn now to the problem of actually calculating the eigenvalues and eigenvectors in the finite-dimensional case.

4.6 Calculation of eigenvalues and eigenvectors in the finite-dimensional case

In the finite-dimensional case the eigenvalues and eigenvectors of a linear transformation T are also called eigenvalues and eigenvectors of each matrix representation of T. Thus, the eigenvalues of a square matrix A are the roots of the characteristic polynomial $f(\lambda) = \det (\lambda I - A)$. The eigenvectors corresponding to an eigenvalue λ are those nonzero vectors $X = (x_1, \ldots, x_n)$ regarded as $n \times 1$ column matrices satisfying the matrix equation

$$AX = \lambda X, \qquad \text{or} \qquad (\lambda I - A)X = O.$$

This is a system of n linear equations for the components $x_1, \ldots, x_n$. Once we know λ we can obtain the eigenvectors by solving this system. We illustrate with three examples that exhibit different features.

EXAMPLE 1. *A matrix with all its eigenvalues distinct.* The matrix

$$A = \begin{bmatrix} 2 & 1 & 1 \\ 2 & 3 & 4 \\ -1 & -1 & -2 \end{bmatrix}$$

has the characteristic polynomial

$$\det(\lambda I - A) = \det \begin{bmatrix} \lambda - 2 & -1 & -1 \\ -2 & \lambda - 3 & -4 \\ 1 & 1 & \lambda + 2 \end{bmatrix} = (\lambda - 1)(\lambda + 1)(\lambda - 3),$$

so there are three distinct eigenvalues: $1, -1$, and 3. To find the eigenvectors corresponding to $\lambda = 1$ we solve the system $AX = X$, or

$$\begin{bmatrix} 2 & 1 & 1 \\ 2 & 3 & 4 \\ -1 & -1 & -2 \end{bmatrix} \begin{bmatrix} x_1 \\ x_2 \\ x_3 \end{bmatrix} = \begin{bmatrix} x_1 \\ x_2 \\ x_3 \end{bmatrix}.$$

This gives us

$$\begin{aligned} 2x_1 + x_2 + x_3 &= x_1 \\ 2x_1 + 3x_2 + 4x_3 &= x_2 \\ -x_1 - x_2 - 2x_3 &= x_3, \end{aligned}$$

which can be rewritten as

$$\begin{aligned} x_1 + x_2 + x_3 &= 0 \\ 2x_1 + 2x_2 + 4x_3 &= 0 \\ -x_1 - x_2 - 3x_3 &= 0. \end{aligned}$$

Adding the first and third equations we find $x_3 = 0$, and all three equations then reduce to $x_1 + x_2 = 0$. Thus the eigenvectors corresponding to $\lambda = 1$ are $X = t(1, -1, 0)$, where t is any nonzero scalar.

By similar calculations we find the eigenvectors $X = t(0, 1, -1)$ corresponding to $\lambda = -1$, and $X = t(2, 3, -1)$ corresponding to $\lambda = 3$, with t any nonzero scalar. Since the eigenvalues are distinct the corresponding eigenvectors $(1, -1, 0)$, $(0, 1, -1)$, and $(2, 3, -1)$ are independent. The results can be summarized in tabular form as follows. In the third column we have listed the dimension of the eigenspace $E(\lambda)$.

Eigenvalue λ	Eigenvectors	dim $E(\lambda)$
1	$t(1, -1, 0), \quad t \neq 0$	1
-1	$t(0, 1, -1), \quad t \neq 0$	1
3	$t(2, 3, -1), \quad t \neq 0$	1

EXAMPLE 2. *A matrix with repeated eigenvalues.* The matrix

$$A = \begin{bmatrix} 2 & -1 & 1 \\ 0 & 3 & -1 \\ 2 & 1 & 3 \end{bmatrix}$$

has the characteristic polynomial

$$\det(\lambda I - A) = \det \begin{bmatrix} \lambda - 2 & 1 & -1 \\ 0 & \lambda - 3 & 1 \\ -2 & -1 & \lambda - 3 \end{bmatrix} = (\lambda - 2)(\lambda - 2)(\lambda - 4).$$

The eigenvalues are 2, 2, and 4. (We list the eigenvalue 2 twice to emphasize that it is a double root of the characteristic polynomial.) To find the eigenvectors corresponding to $\lambda = 2$ we solve the system $AX = 2X$, which reduces to

$$\begin{aligned} -x_2 + x_3 &= 0 \\ x_2 - x_3 &= 0 \\ 2x_1 + x_2 + x_3 &= 0 . \end{aligned}$$

This has the solution $x_2 = x_3 = -x_1$ so the eigenvectors corresponding to $\lambda = 2$ are $t(-1, 1, 1)$, where $t \neq 0$. Similarly we find the eigenvectors $t(1, -1, 1)$ corresponding to the eigenvalue $\lambda = 4$. The results can be summarized as follows:

Eigenvalue	Eigenvectors	dim $E(\lambda)$
2, 2	$t(-1, 1, 1), \quad t \neq 0$	1
4	$t(1, -1, 1), \quad t \neq 0$	1

EXAMPLE 3. *Another matrix with repeated eigenvalues.* The matrix

$$A = \begin{bmatrix} 2 & 1 & 1 \\ 2 & 3 & 2 \\ 3 & 3 & 4 \end{bmatrix}$$

has the characteristic polynomial $(\lambda - 1)(\lambda - 1)(\lambda - 7)$. When $\lambda = 7$ the system $AX = 7X$ becomes

$$\begin{aligned} 5x_1 - x_2 - x_3 &= 0 \\ -2x_1 + 4x_2 - 2x_3 &= 0 \\ -3x_1 - 3x_2 + 3x_3 &= 0 . \end{aligned}$$

This has the solution $x_2 = 2x_1$, $x_3 = 3x_1$, so the eigenvectors corresponding to $\lambda = 7$ are $t(1, 2, 3)$, where $t \neq 0$. For the eigenvalue $\lambda = 1$, the system $AX = X$ consists of the equation

$$x_1 + x_2 + x_3 = 0$$

repeated three times. To solve this equation we may take $x_1 = a$, $x_2 = b$, where a and b are arbitrary, and then take $x_3 = -a - b$. Thus every eigenvector corresponding to $\lambda = 1$ has the form

$$(a, b, -a - b) = a(1, 0, -1) + b(0, 1, -1),$$

where $a \neq 0$ or $b \neq 0$. This means that the vectors $(1, 0, -1)$ and $(0, 1, -1)$ form a basis for $E(1)$. Hence $\dim E(\lambda) = 2$ when $\lambda = 1$. The results can be summarized as follows:

Eigenvalue	Eigenvectors	$\dim E(\lambda)$
7	$t(1, 2, 3),\quad t \neq 0$	1
1, 1	$a(1, 0, -1) + b(0, 1, -1),\quad a, b$ not both 0.	2

Note that in this example there are three independent eigenvectors but only two distinct eigenvalues.

4.7 Trace of a matrix

Let $f(\lambda)$ be the characteristic polynomial of an $n \times n$ matrix A. We denote the n roots of $f(\lambda)$ by $\lambda_1, \ldots, \lambda_n$, with each root written as often as its multiplicity indicates. Then we have the factorization

$$f(\lambda) = (\lambda - \lambda_1) \cdots (\lambda - \lambda_n).$$

We can also write $f(\lambda)$ in decreasing powers of λ as follows,

$$f(\lambda) = \lambda^n + c_{n-1}\lambda^{n-1} + \cdots + c_1\lambda + c_0.$$

Comparing this with the factored form we find that the constant term c_0 and the coefficient of λ^{n-1} are given by the formulas

$$c_0 = (-1)^n \lambda_1 \cdots \lambda_n \qquad \text{and} \qquad c_{n-1} = -(\lambda_1 + \cdots + \lambda_n).$$

Since we also have $c_0 = (-1)^n \det A$, we see that

$$\lambda_1 \cdots \lambda_n = \det A.$$

That is, *the product of the roots of the characteristic polynomial of A is equal to the determinant of A.*

The *sum* of the roots of $f(\lambda)$ is called the *trace* of A, denoted by $\operatorname{tr} A$. Thus, by definition,

$$\operatorname{tr} A = \sum_{i=1}^{n} \lambda_i.$$

The coefficient of λ^{n-1} is given by $c_{n-1} = -\operatorname{tr} A$. We can also compute this coefficient from the determinant form for $f(\lambda)$ and we find that

$$c_{n-1} = -(a_{11} + \cdots + a_{nn}).$$

(A proof of this formula is requested in Exercise 12 of Section 4.8.) The two formulas for c_{n-1} show that

$$\operatorname{tr} A = \sum_{i=1}^{n} a_{ii}.$$

That is, *the trace of A is also equal to the sum of the diagonal elements of A.*

Since the sum of the diagonal elements is easy to compute it can be used as a numerical check in calculations of eigenvalues. Further properties of the trace are described in the next set of exercises.

4.8 Exercises

Determine the eigenvalues and eigenvectors of each of the matrices in Exercises 1 through 3. Also, for each eigenvalue λ compute the dimension of the eigenspace $E(\lambda)$.

1. (a) $\begin{bmatrix} 1 & 0 \\ 0 & 1 \end{bmatrix}$, (b) $\begin{bmatrix} 1 & 1 \\ 0 & 1 \end{bmatrix}$, (c) $\begin{bmatrix} 1 & 0 \\ 1 & 1 \end{bmatrix}$, (d) $\begin{bmatrix} 1 & 1 \\ 1 & 1 \end{bmatrix}$.

2. $\begin{bmatrix} 1 & a \\ b & 1 \end{bmatrix}$, $a > 0, b > 0$.

3. $\begin{bmatrix} \cos\theta & -\sin\theta \\ \sin\theta & \cos\theta \end{bmatrix}$.

4. The matrices $P_1 = \begin{bmatrix} 0 & 1 \\ 1 & 0 \end{bmatrix}$, $P_2 = \begin{bmatrix} 0 & -i \\ i & 0 \end{bmatrix}$, $P_3 = \begin{bmatrix} 1 & 0 \\ 0 & -1 \end{bmatrix}$ occur in the quantum mechanical theory of electron spin and are called *Pauli spin matrices*, in honor of the physicist Wolfgang Pauli (1900–1958). Verify that they all have eigenvalues 1 and -1. Then determine all 2×2 matrices with complex entries having the two eigenvalues 1 and -1.
5. Determine all 2×2 matrices with real entries whose eigenvalues are (a) real and distinct, (b) real and equal, (c) complex conjugates.
6. Determine a, b, c, d, e, f, given that the vectors (1, 1, 1), (1, 0, -1), and (1, -1, 0) are eigenvectors of the matrix

$$\begin{bmatrix} 1 & 1 & 1 \\ a & b & c \\ d & e & f \end{bmatrix}.$$

7. Calculate the eigenvalues and eigenvectors of each of the following matrices. Also, compute the dimension of the eigenspace $E(\lambda)$ for each eigenvalue λ.

(a) $\begin{bmatrix} 1 & 0 & 0 \\ -3 & 1 & 0 \\ 4 & -7 & 1 \end{bmatrix}$, (b) $\begin{bmatrix} 2 & 1 & 3 \\ 1 & 2 & 3 \\ 3 & 3 & 20 \end{bmatrix}$, (c) $\begin{bmatrix} 5 & -6 & -6 \\ -1 & 4 & 2 \\ 3 & -6 & -4 \end{bmatrix}$.

8. Calculate the eigenvalues of each of the five matrices

(a) $\begin{bmatrix} 0 & 0 & 1 & 0 \\ 0 & 0 & 0 & 1 \\ 1 & 0 & 0 & 0 \\ 0 & 1 & 0 & 0 \end{bmatrix}$, (b) $\begin{bmatrix} 1 & 0 & 0 & 0 \\ 0 & 1 & 0 & 0 \\ 0 & 0 & -1 & 0 \\ 0 & 0 & 0 & -1 \end{bmatrix}$, (c) $\begin{bmatrix} 0 & 1 & 0 & 0 \\ 1 & 0 & 0 & 0 \\ 0 & 0 & 0 & 1 \\ 0 & 0 & 1 & 0 \end{bmatrix}$,

$$\text{(d)}\begin{bmatrix} 0 & -i & 0 & 0 \\ i & 0 & 0 & 0 \\ 0 & 0 & 0 & -i \\ 0 & 0 & i & 0 \end{bmatrix}, \qquad \text{(e)}\begin{bmatrix} 1 & 0 & 0 & 0 \\ 0 & -1 & 0 & 0 \\ 0 & 0 & 1 & 0 \\ 0 & 0 & 0 & -1 \end{bmatrix}.$$

These are called *Dirac matrices* in honor of Paul A. M. Dirac (1902–), the English physicist. They occur in the solution of the relativistic wave equation in quantum mechanics.

9. If A and B are $n \times n$ matrices, with B a diagonal matrix, prove (by induction) that the determinant $f(\lambda) = \det(\lambda B - A)$ is a polynomial in λ with $f(0) = (-1)^n \det A$, and with the coefficient of λ^n equal to the product of the diagonal entries of B.
10. Prove that a square matrix A and its transpose A^t have the same characteristic polynomial.
11. If A and B are $n \times n$ matrices, with A nonsingular, prove that AB and BA have the same set of eigenvalues. *Note:* It can be shown that AB and BA have the same characteristic polynomial, even if A is singular, but you are not required to prove this.
12. Let A be an $n \times n$ matrix with characteristic polynomial $f(\lambda)$. Prove (by induction) that the coefficient of λ^{n-1} in $f(\lambda)$ is $-\text{tr } A$.
13. Let A and B be $n \times n$ matrices with $\det A = \det B$ and $\text{tr } A = \text{tr } B$. Prove that A and B have the same characteristic polynomial if $n = 2$ but that this need not be the case if $n > 2$.
14. Prove each of the following statements about the trace.

    (a) $\text{tr}(A + B) = \text{tr } A + \text{tr } B$.
    (b) $\text{tr}(cA) = c \text{ tr } A$.
    (c) $\text{tr}(AB) = \text{tr}(BA)$.
    (d) $\text{tr } A^t = \text{tr } A$.

4.9 Matrices representing the same linear transformation. Similar matrices

In this section we prove that two different matrix representations of a linear transformation have the same characteristic polynomial. To do this we investigate more closely the relation between matrices which represent the same transformation.

Let us recall how matrix representations are defined. Suppose $T: V \to W$ is a linear mapping of an n-dimensional space V into an m-dimensional space W. Let $(e_1, \ldots, e_n)$ and $(w_1, \ldots, w_m)$ be ordered bases for V and W respectively. The matrix representation of T relative to this choice of bases is the $m \times n$ matrix whose columns consist of the components of $T(e_1), \ldots, T(e_n)$ relative to the basis $(w_1, \ldots, w_m)$. Different matrix representations arise from different choices of the bases.

We consider now the case in which $V = W$, and we assume that the same ordered basis $(e_1, \ldots, e_n)$ is used for both V and W. Let $A = (a_{ik})$ be the matrix of T relative to this basis. This means that we have

$$T(e_k) = \sum_{i=1}^{n} a_{ik} e_i \quad \text{for} \quad k = 1, 2, \ldots, n. \tag{4.8}$$

Now choose another ordered basis $(u_1, \ldots, u_n)$ for both V and W and let $B = (b_{kj})$ be the matrix of T relative to this new basis. Then we have

(4.9) $$T(u_j) = \sum_{k=1}^{n} b_{kj} u_k \qquad \text{for} \quad j = 1, 2, \ldots, n.$$

Since each u_j is in the space spanned by $e_1, \ldots, e_n$ we can write

(4.10) $$u_j = \sum_{k=1}^{n} c_{kj} e_k \qquad \text{for} \quad j = 1, 2, \ldots, n,$$

for some set of scalars c_{kj}. The $n \times n$ matrix $C = (c_{kj})$ determined by these scalars is nonsingular because it represents a linear transformation which maps a basis of V onto another basis of V. Applying T to both members of (4.10) we also have the equations

(4.11) $$T(u_j) = \sum_{k=1}^{n} c_{kj} T(e_k) \qquad \text{for} \quad j = 1, 2, \ldots, n.$$

The systems of equations in (4.8) through (4.11) can be written more simply in matrix form by introducing matrices with vector entries. Let

$$E = [e_1, \ldots, e_n] \qquad \text{and} \qquad U = [u_1, \ldots, u_n]$$

be $1 \times n$ row matrices whose entries are the basis elements in question. Then the set of equations in (4.10) can be written as a single matrix equation,

(4.12) $$U = EC.$$

Similarly, if we introduce

$$E' = [T(e_1), \ldots, T(e_n)] \qquad \text{and} \qquad U' = [T(u_1), \ldots, T(u_n)],$$

Equations (4.8), (4.9), and (4.11) become, respectively,

(4.13) $$E' = EA, \qquad U' = UB, \qquad U' = E'C.$$

From (4.12) we also have

$$E = UC^{-1}.$$

To find the relation between A and B we express U' in two ways in terms of U. From (4.13) we have

$$U' = UB$$

and

$$U' = E'C = EAC = UC^{-1}AC.$$

Therefore $UB = UC^{-1}AC$. But each entry in this matrix equation is a linear combination of the basis vectors $u_1, \ldots, u_n$. Since the u_i are independent we must have

$$B = C^{-1}AC.$$

Thus, we have proved the following theorem.

THEOREM 4.6. *If two $n \times n$ matrices A and B represent the same linear transformation T, then there is a nonsingular matrix C such that*

$$B = C^{-1}AC.$$

Moreover, if A is the matrix of T relative to a basis $E = [e_1, \ldots, e_n]$ and if B is the matrix of T relative to a basis $U = [u_1, \ldots, u_n]$, then for C we can take the nonsingular matrix relating the two bases according to the matrix equation $U = EC$.

The converse of Theorem 4.6 is also true.

THEOREM 4.7. *Let A and B be two $n \times n$ matrices related by an equation of the form $B = C^{-1}AC$, where C is a nonsingular $n \times n$ matrix. Then A and B represent the same linear transformation.*

Proof. Choose a basis $E = [e_1, \ldots, e_n]$ for an n-dimensional space V. Let $u_1, \ldots, u_n$ be the vectors determined by the equations

$$u_j = \sum_{k=1}^{n} c_{kj} e_k \qquad \text{for } j = 1, 2, \ldots, n, \tag{4.14}$$

where the scalars c_{kj} are the entries of C. Since C is nonsingular it represents an invertible linear transformation, so $U = [u_1, \ldots, u_n]$ is also a basis for V, and we have $U = EC$.

Let T be the linear transformation having the matrix representation A relative to the basis E, and let S be the transformation having the matrix representation B relative to the basis U. Then we have

$$T(e_k) = \sum_{i=1}^{n} a_{ik} e_i \qquad \text{for } k = 1, 2, \ldots, n \tag{4.15}$$

and

$$S(u_j) = \sum_{k=1}^{n} b_{kj} u_k \qquad \text{for } j = 1, 2, \ldots, n. \tag{4.16}$$

We shall prove that $S = T$ by showing that $T(u_j) = S(u_j)$ for each j.

Equations (4.15) and (4.16) can be written more simply in matrix form as follows,

$$[T(e_1), \ldots, T(e_n)] = EA, \qquad [S(u_1), \ldots, S(u_n)] = UB.$$

Applying T to (4.14) we also obtain the relation $T(u_j) = \sum c_{kj} T(e_k)$, or

$$[T(u_1), \ldots, T(u_n)] = EAC.$$

But we have

$$UB = ECB = EC(C^{-1}AC) = EAC,$$

which shows that $T(u_j) = S(u_j)$ for each j. Therefore $T(x) = S(x)$ for each x in V, so $T = S$. In other words, the matrices A and B represent the same linear transformation.

DEFINITION. *Two $n \times n$ matrices A and B are called similar if there is a nonsingular matrix C such that $B = C^{-1}AC$.*

Theorems 4.6 and 4.7 can be combined to give us

THEOREM 4.8. *Two $n \times n$ matrices are similar if and only if they represent the same linear transformation.*

Similar matrices share many properties. For example, they have the same determinant since

$$\det(C^{-1}AC) = \det(C^{-1})(\det A)(\det C) = \det A.$$

This property gives us the following theorem.

THEOREM 4.9. *Similar matrices have the same characteristic polynomial and therefore the same eigenvalues.*

Proof. If A and B are similar there is a nonsingular matrix C such that $B = C^{-1}AC$. Therefore we have

$$\lambda I - B = \lambda I - C^{-1}AC = \lambda C^{-1}IC - C^{-1}AC = C^{-1}(\lambda I - A)C.$$

This shows that $\lambda I - B$ and $\lambda I - A$ are similar, so $\det(\lambda I - B) = \det(\lambda I - A)$.

Theorems 4.8 and 4.9 together show that all matrix representations of a given linear transformation T have the same characteristic polynomial. This polynomial is also called the characteristic polynomial of T.

The next theorem is a combination of Theorems 4.5, 4.2, and 4.6. In Theorem 4.10, F denotes either the real field $\mathbf{R}$ or the complex field $\mathbf{C}$.

THEOREM 4.10. *Let $T: V \to V$ be a linear transformation, where V has scalars in F, and $\dim V = n$. Assume that the characteristic polynomial of T has n distinct roots $\lambda_1, \ldots, \lambda_n$ in F. Then we have:*

(a) *The corresponding eigenvectors $u_1, \ldots, u_n$ form a basis for V.*

(b) *The matrix of T relative to the ordered basis $U = [u_1, \ldots, u_n]$ is the diagonal matrix Λ having the eigenvalues as diagonal entries:*

$$\Lambda = \operatorname{diag}(\lambda_1, \ldots, \lambda_n).$$

(c) *If A is the matrix of T relative to another basis $E = [e_1, \ldots, e_n]$, then*

$$\Lambda = C^{-1}AC,$$

where C is the nonsingular matrix relating the two bases by the equation

$$U = EC.$$

Proof. By Theorem 4.5 each root λ_i is an eigenvalue. Since there are n distinct roots, Theorem 4.2 tells us that the corresponding eigenvectors $u_1, \ldots, u_n$ are independent. Hence they form a basis for V. This proves (a). Since $T(u_i) = \lambda_i u_i$ the matrix of T relative to U is the diagonal matrix Λ, which proves (b). To prove (c) we use Theorem 4.6.

Note: The nonsingular matrix C in Theorem 4.10 is called a *diagonalizing matrix.* If $(e_1, \ldots, e_n)$ is the basis of unit coordinate vectors $(I_1, \ldots, I_n)$, then the equation $U = EC$ in Theorem 4.10 shows that the kth column of C consists of the components of the eigenvector u_k relative to $(I_1, \ldots, I_n)$.

If the eigenvalues of A are distinct then A is similar to a diagonal matrix. If the eigenvalues are not distinct then A still might be similar to a diagonal matrix. This will happen if and only if there are k independent eigenvectors corresponding to each eigenvalue of multiplicity k. Examples occur in the next set of exercises.

4.10 Exercises

1. Prove that the matrices $\begin{bmatrix} 1 & 1 \\ 0 & 1 \end{bmatrix}$ and $\begin{bmatrix} 1 & 0 \\ 0 & 1 \end{bmatrix}$ have the same eigenvalues but are not similar.

2. In each case find a nonsingular matrix C such that $C^{-1}AC$ is a diagonal matrix or explain why no such C exists.

(a) $A = \begin{bmatrix} 1 & 0 \\ 1 & 3 \end{bmatrix}$, (b) $A = \begin{bmatrix} 1 & 2 \\ 5 & 4 \end{bmatrix}$, (c) $A = \begin{bmatrix} 2 & 1 \\ -1 & 4 \end{bmatrix}$, (d) $A = \begin{bmatrix} 2 & 1 \\ -1 & 0 \end{bmatrix}$.

3. Three bases in the plane are given. With respect to these bases a point has components (x_1, x_2), (y_1, y_2), and (z_1, z_2), respectively. Suppose that $[y_1, y_2] = [x_1, x_2]A$, $[z_1, z_2] = [x_1, x_2]B$, and $[z_1, z_2] = [y_1, y_2]C$, where A, B, C are 2×2 matrices. Express C in terms of A and B.

4. In each case, show that the eigenvalues of A are not distinct but that A has three independent eigenvectors. Find a nonsingular matrix C such that $C^{-1}AC$ is a diagonal matrix.

$$\text{(a) } A = \begin{bmatrix} 0 & 0 & 1 \\ 0 & 1 & 0 \\ 1 & 0 & 0 \end{bmatrix}, \qquad \text{(b) } A = \begin{bmatrix} 1 & -1 & -1 \\ 1 & 3 & 1 \\ -1 & -1 & 1 \end{bmatrix}.$$

5. Show that none of the following matrices is similar to a diagonal matrix, but that each is similar to a triangular matrix of the form $\begin{bmatrix} \lambda & 0 \\ 1 & \lambda \end{bmatrix}$, where λ is an eigenvalue.

$$\text{(a) } \begin{bmatrix} 2 & -1 \\ 0 & 2 \end{bmatrix}, \qquad \text{(b) } \begin{bmatrix} 2 & 1 \\ -1 & 4 \end{bmatrix}.$$

6. Determine the eigenvalues and eigenvectors of the matrix $\begin{bmatrix} 0 & -1 & 0 \\ 0 & 0 & 1 \\ -1 & -3 & 3 \end{bmatrix}$ and thereby show that it is not similar to a diagonal matrix.

7. (a) Prove that a square matrix A is nonsingular if and only if 0 is not an eigenvalue of A.
(b) If A is nonsingular, prove that the eigenvalues of A^{-1} are the reciprocals of the eigenvalues of A.

8. Given an $n \times n$ matrix A with real entries such that $A^2 = -I$. Prove the following statements about A.
(a) A is nonsingular.
(b) n is even.
(c) A has no real eigenvalues.
(d) $\det A = 1$.

5

EIGENVALUES OF OPERATORS ACTING ON EUCLIDEAN SPACES

5.1 Eigenvalues and inner products

This chapter describes some properties of eigenvalues and eigenvectors of linear transformations that operate on Euclidean spaces, that is, on linear spaces having an inner product. We recall the fundamental properties of inner products.

In a *real* Euclidean space an inner product (x, y) of two elements x and y is a real number satisfying the following properties:

(1) $(x, y) = (y, x)$ (symmetry)
(2) $(x + z, y) = (x, y) + (z, y)$ (linearity)
(3) $(cx, y) = c(x, y)$ (homogeneity)
(4) $(x, x) > 0$ if $x \neq O$ (positivity).

In a *complex* Euclidean space the inner product is a complex number satisfying the same properties, with the exception that symmetry is replaced by *Hermitian symmetry*,

$$(1') \qquad (x, y) = \overline{(y, x)},$$

where the bar denotes the complex conjugate. In (3) the scalar c is complex. From (1′) and (3) we obtain

$$(3') \qquad (x, cy) = \bar{c}(x, y),$$

which tells us that scalars are conjugated when taken out of the second factor. Taking $x = y$ in (1′) we see that (x, x) is real so property (4) is meaningful if the space is complex.

When we use the term Euclidean space without further designation it is to be understood that the space can be real or complex. Although most of our applications will be to finite-dimensional spaces, we do not require this restriction at the outset.

The first theorem shows that eigenvalues (if they exist) can be expressed in terms of the inner product.

THEOREM 5.1. *Let E be a Euclidean space, let V be a subspace of E, and let T: $V \to E$ be a linear transformation having an eigenvalue λ with a corresponding eigenvector x. Then we have*

$$(5.1) \qquad \lambda = \frac{(T(x), x)}{(x, x)}.$$

Proof. Since $T(x) = \lambda x$ we have

$$(T(x), x) = (\lambda x, x) = \lambda(x, x).$$

Since $x \neq O$ we can divide by (x, x) to get (5.1).

Several properties of eigenvalues are easily deduced from Equation (5.1). For example, from the Hermitian symmetry of the inner product we have the companion formula

$$\bar{\lambda} = \frac{(x, T(x))}{(x, x)} \tag{5.2}$$

for the complex conjugate of λ. From (5.1) and (5.2) we see that λ is real ($\lambda = \bar{\lambda}$) if and only if $(T(x), x)$ is real, that is, if and only if

$$(T(x), x) = (x, T(x)) \qquad \text{for the eigenvector } x.$$

(This condition is trivially satisfied in a real Euclidean space.) Also, λ is pure imaginary ($\lambda = -\bar{\lambda}$) if and only if $(T(x), x)$ is pure imaginary, that is, if and only if

$$(T(x), x) = -(x, T(x)) \qquad \text{for the eigenvector } x.$$

5.2 Hermitian and skew-Hermitian transformations

In this section we introduce two important types of linear operators which act on Euclidean spaces. These operators have two categories of names, depending on whether the underlying Euclidean space has a real or complex inner product. In the real case the transformations are called *symmetric* and *skew-symmetric.* In the complex case they are called *Hermitian* and *skew-Hermitian.* These transformations occur in many different applications. For example, Hermitian operators on infinite-dimensional spaces play an important role in quantum mechanics. We shall discuss primarily the complex case since it presents no added difficulties.

DEFINITION. *Let E be a Euclidean space and let V be a subspace of E. A linear transformation $T: V \to E$ is called Hermitian on V if*

$$(T(x), y) = (x, T(y)) \qquad \textit{for all } x \textit{ and } y \textit{ in } V.$$

Operator T is called skew-Hermitian on V if

$$(T(x), y) = -(x, T(y)) \qquad \textit{for all } x \textit{ and } y \textit{ in } V.$$

In other words, a Hermitian operator T can be shifted from one factor of an inner product to the other without changing the value of the product. Shifting a skew-Hermitian operator changes the sign of the product.

Note: As already mentioned, if E is a *real* Euclidean space, Hermitian transformations are also called *symmetric*; skew-Hermitian transformations are called *skew-symmetric.*

EXAMPLE 1. *Symmetry and skew-symmetry in the space $C(a, b)$.* Let $C(a, b)$ denote the space of all real functions continuous on a closed interval $[a, b]$, with the real inner product

$$(f, g) = \int_a^b f(t)g(t)\,dt .$$

Let V be a subspace of $C(a, b)$. If $T: V \to C(a, b)$ is a linear transformation then $(f, T(g)) = \int_a^b f(t)Tg(t)\,dt$, where we have written $Tg(t)$ for $T(g)(t)$. Therefore the conditions for symmetry and skew-symmetry become

$$(5.3) \qquad \int_a^b \{f(t)Tg(t) - g(t)Tf(t)\}\,dt = 0 \qquad \text{if } T \text{ is symmetric},$$

and

$$(5.4) \qquad \int_a^b \{f(t)Tg(t) + g(t)Tf(t)\}\,dt = 0 \qquad \text{if } T \text{ is skew-symmetric}.$$

EXAMPLE 2. *Multiplication by a fixed function.* In the space $C(a, b)$ of Example 1, choose a fixed function p and define $T(f) = pf$, the product of p and f. For this T, Equation (5.3) is satisfied for all f and g in $C(a, b)$ since the integrand is zero. Therefore, multiplication by a fixed function is a symmetric operator.

EXAMPLE 3. *The differentiation operator.* In the space $C(a, b)$ of Example 1, let V be the subspace consisting of all functions f which have a continuous derivative in the open interval (a, b) and which also satisfy the *boundary condition* $f(a) = f(b)$. Let $D: V \to C(a, b)$ be the differentiation operator given by $D(f) = f'$. It is easy to prove that D is skew-symmetric. In this case the integrand in (5.4) is the derivative of the product fg, so the integral is equal to

$$\int_a^b (fg)'(t)\,dt = f(b)g(b) - f(a)g(a) .$$

Since both f and g satisfy the boundary condition, we have $f(b)g(b) - f(a)g(a) = 0$. Thus, the boundary condition implies skew-symmetry of D. The only eigenfunctions in the subspace V are the constant functions. They belong to the eigenvalue 0.

EXAMPLE 4. *Sturm-Liouville operators.* This example is important in the theory of linear second-order differential equations. We use the space $C(a, b)$ of Example 1 once more and let V be the subspace consisting of all f which have a continuous second derivative in $[a, b]$ and which also satisfy the two boundary conditions

$$(5.5) \qquad p(a)f(a) = 0, \qquad p(b)f(b) = 0,$$

where p is a fixed function in $C(a, b)$ with a continuous derivative on $[a, b]$. Let q be another fixed function in $C(a, b)$ and let $T: V \to C(a, b)$ be the operator defined by the equation

$$T(f) = (pf')' + qf .$$

This is called a *Sturm-Liouville operator.* To test for symmetry we note that $fT(g) - gT(f) = f(pg')' - g(pf')'$. Using this in (5.3) and integrating both $\int_a^b f \cdot (pg')'\, dt$ and $\int_a^b g \cdot (pf')'\, dt$ by parts, we find

$$\int_a^b \{fT(g) - gT(f)\}\, dt = fpg' \Big|_a^b - \int_a^b pg'f'\, dt - gpf' \Big|_a^b + \int_a^b pf'g'\, dt = 0,$$

since both f and g satisfy the boundary conditions (5.5). Hence T is symmetric on V. The eigenfunctions of T are those nonzero f which satisfy, for some real λ, a differential equation of the form

$$(pf')' + qf = \lambda f$$

on $[a, b]$, and also satisfy the boundary conditions (5.5).

5.3 Eigenvalues and eigenvectors of Hermitian and skew-Hermitian operators

Regarding eigenvalues we have the following theorem.

THEOREM 5.2. *Assume T has an eigenvalue λ. Then we have:*
(a) *If T is Hermitian, λ is real: $\lambda = \bar{\lambda}$.*
(b) *If T is skew-Hermitian, λ is pure imaginary: $\lambda = -\bar{\lambda}$.*

Proof. Let x be an eigenvector corresponding to λ. Then we have

$$\lambda = \frac{(T(x), x)}{(x, x)} \qquad \text{and} \qquad \bar{\lambda} = \frac{(x, T(x))}{(x, x)}.$$

If T is Hermitian we have $(T(x), x) = (x, T(x))$ so $\lambda = \bar{\lambda}$. If T is skew-Hermitian we have $(T(x), x) = -(x, T(x))$ so $\lambda = -\bar{\lambda}$.

Note: If T is *symmetric*, Theorem 5.2 tells us nothing new about the eigenvalues of T since all the eigenvalues must be real if the inner product is real. If T is *skew-symmetric*, the eigenvalues of T must be both real and pure imaginary. Hence all the eigenvalues of a skew-symmetric operator must be zero (if any exist).

5.4 Orthogonality of eigenvectors corresponding to distinct eigenvalues

Distinct eigenvalues of any linear transformation correspond to independent eigenvectors (by Theorem 4.2). For Hermitian and skew-Hermitian transformations more is true.

THEOREM 5.3. *Let T be a Hermitian or skew-Hermitian transformation, and let λ and μ be distinct eigenvalues of T with corresponding eigenvectors x and y. Then x and y are orthogonal; that is, $(x, y) = 0$.*

Proof. We write $T(x) = \lambda x$, $T(y) = \mu y$ and compare the two inner products $(T(x), y)$ and $(x, T(y))$. We have

$$(T(x), y) = (\lambda x, y) = \lambda(x, y) \qquad \text{and} \qquad (x, T(y)) = (x, \mu y) = \bar{\mu}(x, y).$$

If T is Hermitian this gives us $\lambda(x, y) = \bar{\mu}(x, y) = \mu(x, y)$ since $\mu = \bar{\mu}$. Therefore $(x, y) = 0$ since $\lambda \neq \mu$. If T is skew-Hermitian we obtain $\lambda(x, y) = -\bar{\mu}(x, y) = \mu(x, y)$ which again implies $(x, y) = 0$.

EXAMPLE. We apply Theorem 5.3 to those nonzero functions which satisfy a differential equation of the form

$$(pf')' + qf = \lambda f \tag{5.6}$$

on an interval $[a, b]$, and which also satisfy the boundary conditions $p(a)f(a) = p(b)f(b) = 0$. The conclusion is that any two solutions f and g corresponding to two distinct values of λ are orthogonal. For example, consider the differential equation of simple harmonic motion,

$$f'' + k^2 f = 0$$

on the interval $[0, \pi]$, where $k \neq 0$. This has the form (5.6) with $p = 1$, $q = 0$, and $\lambda = -k^2$. All solutions are given by $f(t) = c_1 \cos kt + c_2 \sin kt$. The boundary condition $f(0) = 0$ implies $c_1 = 0$. The second boundary condition, $f(\pi) = 0$, implies $c_2 \sin k\pi = 0$. Since $c_2 \neq 0$ for a nonzero solution, we must have $\sin k\pi = 0$, which means that k is an integer. In other words, nonzero solutions which satisfy the boundary conditions exist if and only if k is an integer. These solutions are $f(t) = \sin nt$, $n = \pm 1, \pm 2, \ldots$. The orthogonality condition implied by Theorem 5.3 now becomes the familiar relation

$$\int_0^{\pi} \sin nt \sin mt \, dt = 0$$

if m^2 and n^2 are distinct integers.

5.5 Exercises

1. Let E be a Euclidean space, let V be a subspace, and let $T: V \to E$ be a given linear transformation. Let λ be a scalar and x a nonzero element of V. Prove that λ is an eigenvalue of T with x as an eigenvector if and only if

$$(T(x), y) = \lambda(x, y) \qquad \text{for every } y \text{ in } E.$$

2. Let $T(x) = cx$ for every x in a linear space V, where c is a fixed scalar. Prove that T is symmetric if V is a real Euclidean space.
3. Assume $T: V \to V$ is a Hermitian transformation.
   (a) Prove that T^n is Hermitian for every positive integer n, and that T^{-1} is Hermitian if T is invertible.
   (b) What can you conclude about T^n and T^{-1} if T is skew-Hermitian?
4. Let $T_1: V \to E$ and $T_2: V \to E$ be two Hermitian transformations.
   (a) Prove that $aT_1 + bT_2$ is Hermitian for all real scalars a and b.
   (b) Prove that the product (composition) T_1T_2 is Hermitian if T_1 and T_2 commute, that is, if $T_1T_2 = T_2T_1$.
5. Let $V = V_3(\mathbf{R})$ with the usual dot product as inner product. Let T be a reflection in the xy-plane; that is, let $T(\boldsymbol{i}) = \boldsymbol{i}$, $T(\boldsymbol{j}) = \boldsymbol{j}$, and $T(\boldsymbol{k}) = -\boldsymbol{k}$. Prove that T is symmetric.

6. Let $C(0, 1)$ be the real linear space of all real functions continuous on $[0, 1]$ with inner product $(f, g) = \int_0^1 f(t)g(t)\,dt$. Let V be the subspace of all f such that $\int_0^1 f(t)\,dt = 0$. Let $T\colon V \to C(0, 1)$ be the integration operator defined by $Tf(x) = \int_0^x f(t)\,dt$. Prove that T is skew-symmetric.
7. Let V be the real Euclidean space of all real polynomials with the inner product $(f, g) = \int_{-1}^1 f(t)g(t)\,dt$. Determine which of the following transformations $T\colon V \to V$ is symmetric or skew-symmetric:
   (a) $Tf(x) = f(-x)$.
   (b) $Tf(x) = f(x)f(-x)$.
   (c) $Tf(x) = f(x) + f(-x)$.
   (d) $Tf(x) = f(x) - f(-x)$.
8. Refer to Example 4 of Section 5.2. Modify the inner product as follows:

$$(f, g) = \int_a^b f(t)g(t)w(t)\,dt,$$

where w is a fixed positive function in $C(a, b)$. Modify the Sturm-Liouville operator T by writing

$$T(f) = \frac{(pf')' + qf}{w}.$$

Prove that the modified operator is symmetric on the subspace V.
9. Let V be a subspace of a complex Euclidean space E. Let $T\colon V \to E$ be a linear transformation and define a scalar-valued function Q on V as follows:

$$Q(x) = (T(x), x) \qquad \text{for all } x \text{ in } V.$$

(a) If T is Hermitian on V, prove that $Q(x)$ is real for all x.
(b) If T is skew-Hermitian, prove that $Q(x)$ is pure imaginary for all x.
(c) Prove that $Q(tx) = t\bar{t}Q(x)$ for every scalar t.
(d) Prove that $Q(x + y) = Q(x) + Q(y) + (T(x), y) + (T(y), x)$, and find a corresponding formula for $Q(x + ty)$.
(e) If $Q(x) = 0$ for all x prove that $T(x) = O$ for all x.
(f) If $Q(x)$ is real for all x prove that T is Hermitian. [*Hint:* Use the fact that $Q(x + ty)$ equals its conjugate for every scalar t.]
10. This exercise shows that the Legendre polynomials (introduced in Section 1.14) are eigenfunctions of a Sturm-Liouville operator. The Legendre polynomials are defined by the equation

$$P_n(t) = \frac{1}{2^n n!} f_n^{(n)}(t), \qquad \text{where } f_n(t) = (t^2 - 1)^n.$$

(a) Verify that $(t^2 - 1)f_n'(t) = 2nt f_n(t)$.
(b) Differentiate the equation in (a) $n + 1$ times, using Leibniz's formula (see p. 222 of Volume I) to obtain

$$(t^2 - 1)f_n^{(n+2)}(t) + 2t(n + 1)f_n^{(n+1)}(t) + n(n + 1)f_n^{(n)}(t) = 2ntf_n^{(n+1)}(t) + 2n(n + 1)f_n^{(n)}(t).$$

(c) Show that the equation in (b) can be rewritten in the form

$$[(t^2 - 1)P_n'(t)]' = n(n + 1)P_n(t).$$

This shows that $P_n(t)$ is an eigenfunction of the Sturm-Liouville operator T given on the interval $[-1, 1]$ by $T(f) = (pf')'$, where $p(t) = t^2 - 1$. The eigenfunction $P_n(t)$ belongs to the eigenvalue $\lambda = n(n + 1)$. In this example the boundary conditions for symmetry are automatically satisfied since $p(1) = p(-1) = 0$.

5.6 Existence of an orthonormal set of eigenvectors for Hermitian and skew-Hermitian operators acting on finite-dimensional spaces

Both Theorems 5.2 and 5.3 are based on the assumption that T has an eigenvalue. As we know, eigenvalues need not exist. However, if T acts on a *finite*-dimensional *complex* space, then eigenvalues always exist since they are the roots of the characteristic polynomial. If T is Hermitian, all the eigenvalues are real. If T is skew-Hermitian, all the eigenvalues are pure imaginary.

We also know that two distinct eigenvalues belong to orthogonal eigenvectors if T is Hermitian or skew-Hermitian. Using this property we can prove that T has an orthonormal set of eigenvectors which spans the whole space. (We recall that an orthogonal set is called orthonormal if each of its elements has norm 1.)

THEOREM 5.4. *Assume* $\dim V = n$ *and let* $T: V \to V$ *be Hermitian or skew-Hermitian. Then there exist n eigenvectors* $u_1, \ldots, u_n$ *of T which form an orthonormal basis for V. Hence the matrix of T relative to this basis is the diagonal matrix* $\Lambda = \operatorname{diag}(\lambda_1, \ldots, \lambda_n)$, *where* λ_k *is the eigenvalue belonging to* u_k.

Proof. We use induction on the dimension n. If $n = 1$, then T has exactly one eigenvalue. Any eigenvector u_1 of norm 1 is an orthonormal basis for V.

Now assume the theorem is true for every Euclidean space of dimension $n - 1$. To prove it is also true for V we choose an eigenvalue λ_1 for T and a corresponding eigenvector u_1 of norm 1. Then $T(u_1) = \lambda_1 u_1$ and $\|u_1\| = 1$. Let S be the subspace spanned by u_1. We shall apply the induction hypothesis to the subspace $S^\perp$ consisting of all elements in V which are orthogonal to u_1,

$$S^\perp = \{x \mid x \in V, (x, u_1) = 0\}.$$

To do this we need to know that $\dim S^\perp = n - 1$ and that T maps $S^\perp$ into itself.

From Theorem 1.7(a) we know that u_1 is part of a basis for V, say the basis $(u_1, v_2, \ldots, v_n)$. We can assume, without loss in generality, that this is an orthonormal basis. (If not, we apply the Gram-Schmidt process to convert it into an orthonormal basis, keeping u_1 as the first basis element.) Now take any x in $S^\perp$ and write

$$x = x_1 u_1 + x_2 v_2 + \cdots + x_n v_n.$$

Then $x_1 = (x, u_1) = 0$ since the basis is orthonormal, so x is in the space spanned by $v_2, \ldots, v_n$. Hence $\dim S^\perp = n - 1$.

Next we show that T maps $S^\perp$ into itself. Assume T is Hermitian. If $x \in S^\perp$ we have

$$(T(x), u_1) = (x, T(u_1)) = (x, \lambda_1 u_1) = \bar{\lambda}_1 (x, u_1) = 0,$$

so $T(x) \in S^{\perp}$. Since T is Hermitian on $S^{\perp}$ we can apply the induction hypothesis to find that T has $n-1$ eigenvectors $u_2, \ldots, u_n$ which form an orthonormal basis for $S^{\perp}$. Therefore the orthogonal set $u_1, \ldots, u_n$ is an orthonormal basis for V. This proves the theorem if T is Hermitian. A similar argument works if T is skew-Hermitian.

5.7 Matrix representations for Hermitian and skew-Hermitian operators

In this section we assume that V is a finite-dimensional Euclidean space. A Hermitian or skew-Hermitian transformation can be characterized in terms of its action on the elements of any basis.

THEOREM 5.5. *Let $(e_1, \ldots, e_n)$ be a basis for V and let $T: V \to V$ be a linear transformation. Then we have:*

(a) *T is Hermitian if and only if $(T(e_j), e_i) = (e_j, T(e_i))$ for all i and j.*

(b) *T is skew-Hermitian if and only if $(T(e_j), e_i) = -(e_j, T(e_i))$ for all i and j.*

Proof. Take any two elements x and y in V and express each in terms of the basis elements, say $x = \sum x_j e_j$ and $y = \sum y_i e_i$. Then we have

$$(T(x), y) = \left(\sum_{j=1}^{n} x_j T(e_j), y\right) = \sum_{j=1}^{n} x_j \left(T(e_j), \sum_{i=1}^{n} y_i e_i\right) = \sum_{j=1}^{n} \sum_{i=1}^{n} x_j \bar{y}_i (T(e_j), e_i).$$

Similarly we find

$$(x, T(y)) = \sum_{j=1}^{n} \sum_{i=1}^{n} x_j \bar{y}_i (e_j, T(e_i)).$$

Statements (a) and (b) following immediately from these equations.

Now we characterize these concepts in terms of a matrix representation of T.

THEOREM 5.6. *Let $(e_1, \ldots, e_n)$ be an orthonormal basis for V, and let $A = (a_{ij})$ be the matrix representation of a linear transformation $T: V \to V$ relative to this basis. Then we have:*

(a) *T is Hermitian if and only if $a_{ij} = \bar{a}_{ji}$ for all i and j.*

(b) *T is skew-Hermitian if and only if $a_{ij} = -\bar{a}_{ji}$ for all i and j.*

Proof. Since A is the matrix of T we have $T(e_j) = \sum_{k=1}^{n} a_{kj} e_k$. Taking the inner product of $T(e_j)$ with e_i and using the linearity of the inner product we obtain

$$(T(e_j), e_i) = \left(\sum_{k=1}^{n} a_{kj} e_k, e_i\right) = \sum_{k=1}^{n} a_{kj} (e_k, e_i).$$

But $(e_k, e_i) = 0$ unless $k = i$, so the last sum simplifies to $a_{ij}(e_i, e_i) = a_{ij}$ since $(e_i, e_i) = 1$. Hence we have

$$a_{ij} = (T(e_j), e_i) \qquad \text{for all } i, j.$$

Interchanging i and j, taking conjugates, and using the Hermitian symmetry of the inner product, we find

$$\bar{a}_{ji} = (e_j, T(e_i)) \qquad \text{for all } i, j.$$

Now we apply Theorem 5.5 to complete the proof.

5.8 Hermitian and skew-Hermitian matrices. The adjoint of a matrix

The following definition is suggested by Theorem 5.6.

DEFINITION. *A square matrix $A = (a_{ij})$ is called Hermitian if $a_{ij} = \bar{a}_{ji}$ for all i and j. Matrix A is called skew-Hermitian if $a_{ij} = -\bar{a}_{ji}$ for all i and j.*

Theorem 5.6 states that a transformation T on a finite-dimensional space V is Hermitian or skew-Hermitian according as its matrix relative to an orthonormal basis is Hermitian or skew-Hermitian.

These matrices can be described in another way. Let $\bar{A}$ denote the matrix obtained by replacing each entry of A by its complex conjugate. Matrix $\bar{A}$ is called the *conjugate* of A. Matrix A is Hermitian if and only if it is equal to the transpose of its conjugate, $A = \bar{A}^t$. It is skew-Hermitian if $A = -\bar{A}^t$.

The transpose of the conjugate is given a special name.

DEFINITION OF THE ADJOINT OF A MATRIX. *For any matrix A, the transpose of the conjugate, $\bar{A}^t$, is also called the adjoint of A and is denoted by A^*.*

Thus, a square matrix A is Hermitian if $A = A^*$, and skew-Hermitian if $A = -A^*$. A Hermitian matrix is also called *self-adjoint*.

Note: Much of the older matrix literature uses the term *adjoint* for the transpose of the cofactor matrix, an entirely different object. The definition given here conforms to the current nomenclature in the theory of linear operators.

5.9 Diagonalization of a Hermitian or skew-Hermitian matrix

THEOREM 5.7. *Every $n \times n$ Hermitian or skew-Hermitian matrix A is similar to the diagonal matrix $\Lambda = \text{diag}(\lambda_1, \ldots, \lambda_n)$ of its eigenvalues. Moreover, we have*

$$\Lambda = C^{-1}AC,$$

where C is a nonsingular matrix whose inverse is its adjoint, $C^{-1} = C^*$.

Proof. Let V be the space of n-tuples of complex numbers, and let $(e_1, \ldots, e_n)$ be the orthonormal basis of unit coordinate vectors. If $x = \sum x_i e_i$ and $y = \sum y_i e_i$ let the inner product be given by $(x, y) = \sum x_i \bar{y}_i$. For the given matrix A, let T be the transformation represented by A relative to the chosen basis. Then Theorem 5.4 tells us that V has an orthonormal basis of eigenvectors $(u_1, \ldots, u_n)$, relative to which T has the diagonal matrix representation $\Lambda = \operatorname{diag}(\lambda_1, \ldots, \lambda_n)$, where λ_k is the eigenvalue belonging to u_k. Since both A and Λ represent T they are similar, so we have $\Lambda = C^{-1}AC$, where $C = (c_{ij})$ is the nonsingular matrix relating the two bases:

$$[u_1, \ldots, u_n] = [e_1, \ldots, e_n]C.$$

This equation shows that the jth column of C consists of the components of u_j relative to $(e_1, \ldots, e_n)$. Therefore c_{ij} is the ith component of u_j. The inner product of u_j and u_i is given by

$$(u_j, u_i) = \sum_{k=1}^{n} c_{kj}\bar{c}_{ki}.$$

Since $\{u_1, \ldots, u_n\}$ is an orthonormal set, this shows that $CC^* = I$, so $C^{-1} = C^*$.

Note: The proof of Theorem 5.7 also tells us how to determine the diagonalizing matrix C. We find an orthonormal set of eigenvectors $u_1, \ldots, u_n$ and then use the components of u_j (relative to the basis of unit coordinate vectors) as the entries of the jth column of C.

EXAMPLE 1. The real Hermitian matrix $A = \begin{bmatrix} 2 & 2 \\ 2 & 5 \end{bmatrix}$ has eigenvalues $\lambda_1 = 1$ and $\lambda_2 = 6$. The eigenvectors belonging to 1 are $t(2, -1)$, $t \neq 0$. Those belonging to 6 are $t(1, 2)$, $t \neq 0$. The two eigenvectors $u_1 = t(2, -1)$ and $u_2 = t(1, 2)$ with $t = 1/\sqrt{5}$ form an orthonormal set. Therefore the matrix

$$C = \frac{1}{\sqrt{5}}\begin{bmatrix} 2 & 1 \\ -1 & 2 \end{bmatrix}$$

is a diagonalizing matrix for A. In this case $C^* = C^t$ since C is real. It is easily verified that $C^tAC = \begin{bmatrix} 1 & 0 \\ 0 & 6 \end{bmatrix}$.

EXAMPLE 2. If A is already a diagonal matrix, then the diagonalizing matrix C of Theorem 5.7 either leaves A unchanged or merely rearranges the diagonal entries.

5.10 Unitary matrices. Orthogonal matrices

DEFINITION. *A square matrix A is called unitary if $AA^* = I$. It is called orthogonal if $AA^t = I$.*

Note: Every real unitary matrix is orthogonal since $A^* = A^t$.

Theorem 5.7 tells us that a Hermitian or skew-Hermitian matrix can always be diagonalized by a unitary matrix. A real Hermitian matrix has real eigenvalues and the corresponding eigenvectors can be taken real. Therefore a real Hermitian matrix can be diagonalized by a real orthogonal matrix. This is *not* true for real skew-Hermitian matrices. (See Exercise 11 in Section 5.11.)

We also have the following related concepts.

DEFINITION. *A square matrix A with real or complex entries is called symmetric if $A = A^t$; it is called skew-symmetric if $A = -A^t$.*

EXAMPLE 3. If A is real, its adjoint is equal to its transpose, $A^* = A^t$. Thus, every *real* Hermitian matrix is symmetric, but a symmetric matrix need not be Hermitian.

EXAMPLE 4. If $A = \begin{bmatrix} 1+i & 2 \\ 3-i & 4i \end{bmatrix}$, then $\bar{A} = \begin{bmatrix} 1-i & 2 \\ 3+i & -4i \end{bmatrix}$, $A^t = \begin{bmatrix} 1+i & 3-i \\ 2 & 4i \end{bmatrix}$ and $A^* = \begin{bmatrix} 1-i & 3+i \\ 2 & -4i \end{bmatrix}$.

EXAMPLE 5. Both matrices $\begin{bmatrix} 1 & 2 \\ 2 & 3 \end{bmatrix}$ and $\begin{bmatrix} 1 & 2+i \\ 2-i & 3 \end{bmatrix}$ are Hermitian. The first is symmetric, the second is not.

EXAMPLE 6. Both matrices $\begin{bmatrix} 0 & -2 \\ 2 & 0 \end{bmatrix}$ and $\begin{bmatrix} i & -2 \\ 2 & 3i \end{bmatrix}$ are skew-Hermitian. The first is skew-symmetric, the second is not.

EXAMPLE 7. All the diagonal elements of a Hermitian matrix are real. All the diagonal elements of a skew-Hermitian matrix are pure imaginary. All the diagonal elements of a skew-symmetric matrix are zero.

EXAMPLE 8. For any square matrix A, the matrix $B = \frac{1}{2}(A + A^*)$ is Hermitian, and the matrix $C = \frac{1}{2}(A - A^*)$ is skew-Hermitian. Their sum is A. Thus, every square matrix A can be expressed as a sum $A = B + C$, where B is Hermitian and C is skew-Hermitian. It is an easy exercise to verify that this decomposition is unique. Also every square matrix A can be expressed uniquely as the sum of a symmetric matrix, $\frac{1}{2}(A + A^t)$, and a skew-symmetric matrix, $\frac{1}{2}(A - A^t)$.

EXAMPLE 9. If A is orthogonal we have $1 = \det(AA^t) = (\det A)(\det A^t) = (\det A)^2$, so $\det A = \pm 1$.

5.11 Exercises

1. Determine which of the following matrices are symmetric, skew-symmetric, Hermitian, skew-Hermitian.

$$\text{(a)} \begin{bmatrix} 0 & 1 & 2 \\ 1 & 0 & 3 \\ 2 & 3 & 4 \end{bmatrix}, \quad \text{(b)} \begin{bmatrix} 0 & i & 2 \\ i & 0 & 3 \\ -2 & -3 & 4i \end{bmatrix}, \quad \text{(c)} \begin{bmatrix} 0 & i & 2 \\ -i & 0 & 3 \\ -2 & -3 & 0 \end{bmatrix}, \quad \text{(d)} \begin{bmatrix} 0 & 1 & 2 \\ -1 & 0 & 3 \\ -2 & -3 & 0 \end{bmatrix}.$$

2. (a) Verify that the 2×2 matrix $A = \begin{bmatrix} \cos\theta & -\sin\theta \\ \sin\theta & \cos\theta \end{bmatrix}$ is an orthogonal matrix.

   (b) Let T be the linear transformation with the above matrix A relative to the usual basis $\{\boldsymbol{i}, \boldsymbol{j}\}$. Prove that T maps each point in the plane with polar coordinates (r, α) onto the point with polar coordinates $(r, \alpha + \theta)$. Thus, T is a rotation of the plane about the origin, θ being the angle of rotation.

3. Let V be real 3-space with the usual basis vectors $\boldsymbol{i}, \boldsymbol{j}, \boldsymbol{k}$. Prove that each of the following matrices is orthogonal and represents the transformation indicated.

   (a) $\begin{bmatrix} 1 & 0 & 0 \\ 0 & 1 & 0 \\ 0 & 0 & -1 \end{bmatrix}$ (reflection in the xy-plane).

   (b) $\begin{bmatrix} 1 & 0 & 0 \\ 0 & -1 & 0 \\ 0 & 0 & -1 \end{bmatrix}$ (reflection through the x-axis).

   (c) $\begin{bmatrix} -1 & 0 & 0 \\ 0 & -1 & 0 \\ 0 & 0 & -1 \end{bmatrix}$ (reflection through the origin).

   (d) $\begin{bmatrix} 1 & 0 & 0 \\ 0 & \cos\theta & -\sin\theta \\ 0 & \sin\theta & \cos\theta \end{bmatrix}$ (rotation about the x-axis).

   (e) $\begin{bmatrix} -1 & 0 & 0 \\ 0 & \cos\theta & -\sin\theta \\ 0 & \sin\theta & \cos\theta \end{bmatrix}$ (rotation about x-axis followed by reflection in the yz-plane).

4. A real orthogonal matrix A is called *proper* if $\det A = 1$, and *improper* if $\det A = -1$.

   (a) If A is a proper 2×2 matrix, prove that $A = \begin{bmatrix} \cos\theta & -\sin\theta \\ \sin\theta & \cos\theta \end{bmatrix}$ for some θ. This represents a rotation through an angle θ.

   (b) Prove that $\begin{bmatrix} 1 & 0 \\ 0 & -1 \end{bmatrix}$ and $\begin{bmatrix} -1 & 0 \\ 0 & 1 \end{bmatrix}$ are improper matrices. The first matrix represents a reflection of the xy-plane through the x-axis; the second represents a reflection through the y-axis. Find all improper 2×2 matrices.

In each of Exercises 5 through 8, find (a) an orthogonal set of eigenvectors for A, and (b) a unitary matrix C such that $C^{-1}AC$ is a diagonal matrix.

5. $A = \begin{bmatrix} 9 & 12 \\ 12 & 16 \end{bmatrix}$.

6. $A = \begin{bmatrix} 0 & -2 \\ 2 & 0 \end{bmatrix}$.

7. $A = \begin{bmatrix} 1 & 3 & 0 \\ 3 & -2 & -1 \\ 0 & -1 & 1 \end{bmatrix}$.

8. $A = \begin{bmatrix} 1 & 3 & 4 \\ 3 & 1 & 0 \\ 4 & 0 & 1 \end{bmatrix}$.

9. Determine which of the following matrices are unitary, and which are orthogonal (a, b, θ real).

$$\text{(a)}\ \begin{bmatrix} e^{ia} & 0 \\ 0 & e^{ib} \end{bmatrix}, \qquad \text{(b)}\ \begin{bmatrix} \cos\theta & 0 & \sin\theta \\ 0 & 1 & 0 \\ -\sin\theta & 0 & \cos\theta \end{bmatrix}, \qquad \text{(c)}\ \begin{bmatrix} \frac{1}{2}\sqrt{2} & -\frac{1}{3}\sqrt{3} & \frac{1}{6}\sqrt{6} \\ 0 & \frac{1}{3}\sqrt{3} & \frac{1}{3}\sqrt{6} \\ \frac{1}{2}\sqrt{2} & \frac{1}{3}\sqrt{3} & -\frac{1}{6}\sqrt{6} \end{bmatrix}.$$

10. The special theory of relativity makes use of the equations

$$x' = a(x - vt), \qquad y' = y, \qquad z' = z, \qquad t' = a(t - vx/c^2).$$

Here v is the velocity of a moving object, c the speed of light, and $a = c/\sqrt{c^2 - v^2}$. The linear transformation which maps (x, y, z, t) onto (x', y', z', t') is called a *Lorentz transformation.*
(a) Let $(x_1, x_2, x_3, x_4) = (x, y, z, ict)$ and $(x_1', x_2', x_3', x_4') = (x', y', z', ict')$. Show that the four equations can be written as one matrix equation,

$$[x_1', x_2', x_3', x_4'] = [x_1, x_2, x_3, x_4]\begin{bmatrix} a & 0 & 0 & -iav/c \\ 0 & 1 & 0 & 0 \\ 0 & 0 & 1 & 0 \\ iav/c & 0 & 0 & a \end{bmatrix}.$$

(b) Prove that the 4×4 matrix in (a) is orthogonal but not unitary.

11. Let a be a nonzero real number and let A be the skew-symmetric matrix $A = \begin{bmatrix} 0 & a \\ -a & 0 \end{bmatrix}$.
(a) Find an orthonormal set of eigenvectors for A.
(b) Find a unitary matrix C such that $C^{-1}AC$ is a diagonal matrix.
(c) Prove that there is no real orthogonal matrix C such that $C^{-1}AC$ is a diagonal matrix.
12. If the eigenvalues of a Hermitian or skew-Hermitian matrix A are all equal to c, prove that $A = cI$.
13. If A is a real skew-symmetric matrix, prove that both $I - A$ and $I + A$ are nonsingular and that $(I - A)(I + A)^{-1}$ is orthogonal.
14. For each of the following statements about $n \times n$ matrices, give a proof or exhibit a counter example.
(a) If A and B are unitary, then $A + B$ is unitary.
(b) If A and B are unitary, then AB is unitary.
(c) If A and AB are unitary, then B is unitary.
(d) If A and B are unitary, then $A + B$ is not unitary.

5.12 Quadratic forms

Let V be a *real* Euclidean space and let $T: V \to V$ be a symmetric operator. This means that T can be shifted from one factor of an inner product to the other,

$$(T(x), y) = (x, T(y)) \qquad \text{for all } x \text{ and } y \text{ in } V.$$

Given T, we define a real-valued function Q on V by the equation

$$Q(x) = (T(x), x).$$

The function Q is called the *quadratic form* associated with T. The term "quadratic" is suggested by the following theorem which shows that in the finite-dimensional case $Q(x)$ is a quadratic polynomial in the components of x.

THEOREM 5.8. *Let $(e_1, \ldots, e_n)$ be an orthonormal basis for a real Euclidean space V. Let $T: V \to V$ be a symmetric transformation, and let $A = (a_{ij})$ be the matrix of T relative to this basis. Then the quadratic form $Q(x) = (T(x), x)$ is related to A as follows:*

$$Q(x) = \sum_{i=1}^{n} \sum_{j=1}^{n} a_{ij} x_i x_j \quad \text{if} \quad x = \sum_{i=1}^{n} x_i e_i . \tag{5.7}$$

Proof. By linearity we have $T(x) = \sum x_i T(e_i)$. Therefore

$$Q(x) = \left(\sum_{i=1}^{n} x_i T(e_i), \sum_{j=1}^{n} x_j e_j \right) = \sum_{i=1}^{n} \sum_{j=1}^{n} x_i x_j (T(e_i), e_j) .$$

This proves (5.7) since $a_{ij} = a_{ji} = (T(e_i), e_j)$.

The sum appearing in (5.7) is meaningful even if the matrix A is not symmetric.

DEFINITION. *Let V be any real Euclidean space with an orthonormal basis $(e_1, \ldots, e_n)$, and let $A = (a_{ij})$ by any $n \times n$ matrix of scalars. The scalar-valued function Q defined at each element $x = \sum x_i e_i$ of V by the double sum*

$$Q(x) = \sum_{i=1}^{n} \sum_{j=1}^{n} a_{ij} x_i x_j \tag{5.8}$$

is called the quadratic form associated with A.

If A is a diagonal matrix, then $a_{ij} = 0$ if $i \neq j$ so the sum in (5.8) contains only squared terms and can be written more simply as

$$Q(x) = \sum_{i=1}^{n} a_{ii} x_i^2 .$$

In this case the quadratic form is called a *diagonal form*.

The double sum appearing in (5.8) can also be expressed as a product of three matrices.

THEOREM 5.9. *Let $X = [x_1, \ldots, x_n]$ be a $1 \times n$ row matrix, and let $A = (a_{ij})$ be an $n \times n$ matrix. Then XAX^t is a 1×1 matrix with entry*

$$\sum_{i=1}^{n} \sum_{j=1}^{n} a_{ij} x_i x_j . \tag{5.9}$$

Proof. The product XA is a $1 \times n$ matrix, $XA = [y_1, \ldots, y_n]$, where entry y_j is the dot product of X with the jth column of A,

$$y_j = \sum_{i=1}^{n} x_i a_{ij} .$$

Therefore the product XAX^t is a 1×1 matrix whose single entry is the dot product

$$\sum_{j=1}^{n} y_j x_j = \sum_{j=1}^{n} \left(\sum_{i=1}^{n} x_i a_{ij} \right) x_j = \sum_{i=1}^{n} \sum_{j=1}^{n} a_{ij} x_i x_j .$$

Note: It is customary to identify the 1×1 matrix XAX^t with the sum in (5.9) and to call the product XAX^t a quadratic form. Equation (5.8) is written more simply as follows:

$$Q(x) = XAX^t .$$

EXAMPLE 1. Let $A = \begin{bmatrix} 1 & -1 \\ -3 & 5 \end{bmatrix}$, $X = [x_1, x_2]$. Then we have

$$XA = [x_1, x_2] \begin{bmatrix} 1 & -1 \\ -3 & 5 \end{bmatrix} = [x_1 - 3x_2, -x_1 + 5x_2],$$

and hence

$$XAX^t = [x_1 - 3x_2, -x_1 + 5x_2] \begin{bmatrix} x_1 \\ x_2 \end{bmatrix} = x_1^2 - 3x_2x_1 - x_1x_2 + 5x_2^2 .$$

EXAMPLE 2. Let $B = \begin{bmatrix} 1 & -2 \\ -2 & 5 \end{bmatrix}$, $X = [x_1, x_2]$. Then we have

$$XBX^t = [x_1, x_2] \begin{bmatrix} 1 & -2 \\ -2 & 5 \end{bmatrix} \begin{bmatrix} x_1 \\ x_2 \end{bmatrix} = x_1^2 - 2x_2x_1 - 2x_1x_2 + 5x_2^2 .$$

In both Examples 1 and 2 the two mixed product terms add up to $-4x_1x_2$ so $XAX^t = XBX^t$. These examples show that different matrices can lead to the same quadratic form. Note that one of these matrices is symmetric. This illustrates the next theorem.

THEOREM 5.10. *For any $n \times n$ matrix A and any $1 \times n$ row matrix X we have $XAX^t = XBX^t$ where B is the symmetric matrix $B = \frac{1}{2}(A + A^t)$.*

Proof. Since XAX^t is a 1×1 matrix it is equal to its transpose, $XAX^t = (XAX^t)^t$. But the transpose of a product is the product of transposes in reversed order, so we have $(XAX^t)^t = XA^tX^t$. Therefore $XAX^t = \frac{1}{2}XAX^t + \frac{1}{2}XA^tX^t = XBX^t$.

5.13 Reduction of a real quadratic form to a diagonal form

A real symmetric matrix A is Hermitian. Therefore, by Theorem 5.7 it is similar to the diagonal matrix $\Lambda = \text{diag}(\lambda_1, \ldots, \lambda_n)$ of its eigenvalues. Moreover, we have $\Lambda = C^tAC$, where C is an orthogonal matrix. Now we show that C can be used to convert the quadratic form XAX^t to a diagonal form.

THEOREM 5.11. *Let XAX^t be the quadratic form associated with a real symmetric matrix A, and let C be an orthogonal matrix that converts A to a diagonal matrix $\Lambda = C^tAC$. Then we have*

$$XAX^t = Y\Lambda Y^t = \sum_{i=1}^{n} \lambda_i y_i^2,$$

where $Y = [y_1, \ldots, y_n]$ is the row matrix $Y = XC$, and $\lambda_1, \ldots, \lambda_n$ are the eigenvalues of A.

Proof. Since C is orthogonal we have $C^{-1} = C^t$. Therefore the equation $Y = XC$ implies $X = YC^t$, and we obtain

$$XAX^t = (YC^t)A(YC^t)^t = Y(C^tAC)Y^t = Y\Lambda Y^t.$$

Note: Theorem 5.11 is described by saying that the linear transformation $Y = XC$ reduces the quadratic form XAX^t to a diagonal form $Y\Lambda Y^t$.

EXAMPLE 1. The quadratic form belonging to the identity matrix is

$$XIX^t = \sum_{i=1}^{n} x_i^2 = \|X\|^2,$$

the square of the length of the vector $X = (x_1, \ldots, x_n)$. A linear transformation $Y = XC$, where C is an orthogonal matrix, gives a new quadratic form $Y\Lambda Y^t$ with $\Lambda = CIC^t = CC^t = I$. Since $XIX^t = YIY^t$ we have $\|X\|^2 = \|Y\|^2$, so Y has the same length as X. A linear transformation which preserves the length of each vector is called an *isometry*. These transformations are discussed in more detail in Section 5.19.

EXAMPLE 2. Determine an orthogonal matrix C which reduces the quadratic form $Q(x) = 2x_1^2 + 4x_1x_2 + 5x_2^2$ to a diagonal form.

Solution. We write $Q(x) = XAX^t$, where $A = \begin{bmatrix} 2 & 2 \\ 2 & 5 \end{bmatrix}$. This symmetric matrix was diagonalized in Example 1 following Theorem 5.7. It has the eigenvalues $\lambda_1 = 1$, $\lambda_2 = 6$, and an orthonormal set of eigenvectors u_1, u_2, where $u_1 = t(2, -1)$, $u_2 = t(1, 2)$, $t = 1/\sqrt{5}$. An orthogonal diagonalizing matrix is $C = t\begin{bmatrix} 2 & 1 \\ -1 & 2 \end{bmatrix}$. The corresponding diagonal form is

$$Y\Lambda Y^t = \lambda_1 y_1^2 + \lambda_2 y_2^2 = y_1^2 + 6y_2^2.$$

The result of Example 2 has a simple geometric interpretation, illustrated in Figure 5.1. The linear transformation $Y = XC$ can be regarded as a rotation which maps the basis $\boldsymbol{i}$, $\boldsymbol{j}$ onto the new basis u_1, u_2. A point with coordinates (x_1, x_2) relative to the first basis has new coordinates (y_1, y_2) relative to the second basis. Since $XAX^t = Y\Lambda Y^t$, the set of points (x_1, x_2) satisfying the equation $XAX^t = c$ for some c is identical with the set of points (y_1, y_2) satisfying $Y\Lambda Y^t = c$. The second equation, written as $y_1^2 + 6y_2^2 = c$, is

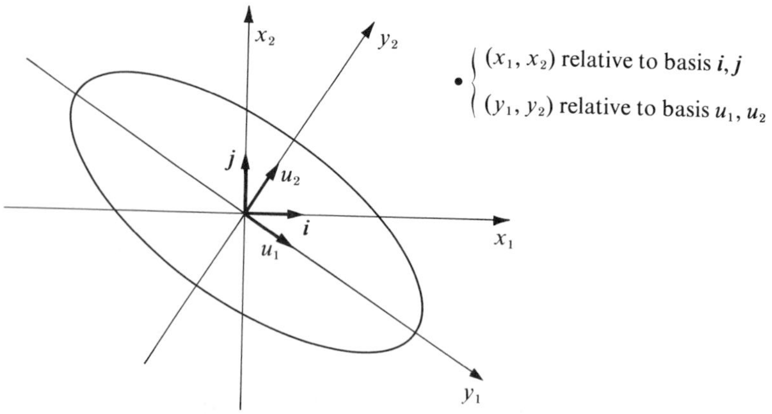

FIGURE 5.1 Rotation of axes by an orthogonal matrix. The ellipse has Cartesian equation $XAX^t = 9$ in the x_1x_2-system, and equation $Y\Lambda Y^t = 9$ in the y_1y_2-system.

the Cartesian equation of an ellipse if $c > 0$. Therefore the equation $XAX^t = c$, written as $2x_1^2 + 4x_1x_2 + 5x_2^2 = c$, represents the same ellipse in the original coordinate system. Figure 5.1 shows the ellipse corresponding to $c = 9$.

5.14 Applications to analytic geometry

The reduction of a quadratic form to a diagonal form can be used to identify the set of all points (x, y) in the plane which satisfy a Cartesian equation of the form

$$ax^2 + bxy + cy^2 + dx + ey + f = 0. \tag{5.10}$$

We shall find that this set is always a conic section, that is, an ellipse, hyperbola, parabola, or one of the degenerate cases (the empty set, a single point, or one or two straight lines). The type of conic is governed by the second-degree terms, that is, by the quadratic form $ax^2 + bxy + cy^2$. To conform with the notation used earlier, we write x_1 for x, x_2 for y, and express this quadratic form as a matrix product,

$$XAX^t = ax_1^2 + bx_1x_2 + cx_2^2,$$

where $X = [x_1, x_2]$ and $A = \begin{bmatrix} a & b/2 \\ b/2 & c \end{bmatrix}$. By a rotation $Y = XC$ we reduce this form to a diagonal form $\lambda_1 y_1^2 + \lambda_2 y_2^2$, where λ_1, λ_2 are the eigenvalues of A. An orthonormal set of eigenvectors u_1, u_2 determines a new set of coordinate axes, relative to which the Cartesian equation (5.10) becomes

$$\lambda_1 y_1^2 + \lambda_2 y_2^2 + d'y_1 + e'y_2 + f = 0, \tag{5.11}$$

with new coefficients d' and e' in the linear terms. In this equation there is no mixed product term y_1y_2, so the type of conic is easily identified by examining the eigenvalues λ_1 and λ_2.

If the conic is not degenerate, Equation (5.11) represents an *ellipse* if λ_1, λ_2 have the same sign, a *hyperbola* if λ_1, λ_2 have opposite signs, and a *parabola* if either λ_1 or λ_2 is zero. The three cases correspond to $\lambda_1\lambda_2 > 0$, $\lambda_1\lambda_2 < 0$, and $\lambda_1\lambda_2 = 0$. We illustrate with some specific examples.

EXAMPLE 1. $2x^2 + 4xy + 5y^2 + 4x + 13y - \frac{1}{4} = 0$. We rewrite this as

$$2x_1^2 + 4x_1x_2 + 5x_2^2 + 4x_1 + 13x_2 - \tfrac{1}{4} = 0. \tag{5.12}$$

The quadratic form $2x_1^2 + 4x_1x_2 + 5x_2^2$ is the one treated in Example 2 of the foregoing section. Its matrix has eigenvalues $\lambda_1 = 1$, $\lambda_2 = 6$, and an orthonormal set of eigenvectors $u_1 = t(2, -1)$, $u_2 = t(1, 2)$, where $t = 1/\sqrt{5}$. An orthogonal diagonalizing matrix is $C = t\begin{bmatrix} 2 & 1 \\ -1 & 2 \end{bmatrix}$. This reduces the quadratic part of (5.12) to the form $y_1^2 + 6y_2^2$. To determine the effect on the linear part we write the equation of rotation $Y = XC$ in the form $X = YC^t$ and obtain

$$[x_1, x_2] = \frac{1}{\sqrt{5}}[y_1, y_2]\begin{bmatrix} 2 & -1 \\ 1 & 2 \end{bmatrix}, \qquad x_1 = \frac{1}{\sqrt{5}}(2y_1 + y_2), \qquad x_2 = \frac{1}{\sqrt{5}}(-y_1 + 2y_2).$$

Therefore the linear part $4x_1 + 13x_2$ is transformed to

$$\frac{4}{\sqrt{5}}(2y_1 + y_2) + \frac{13}{\sqrt{5}}(-y_1 + 2y_2) = -\sqrt{5}\,y_1 + 6\sqrt{5}\,y_2.$$

The transformed Cartesian equation becomes

$$y_1^2 + 6y_2^2 - \sqrt{5}\,y_1 + 6\sqrt{5}\,y_2 - \tfrac{1}{4} = 0.$$

By completing the squares in y_1 and y_2 we rewrite this as follows:

$$(y_1 - \tfrac{1}{2}\sqrt{5})^2 + 6(y_2 + \tfrac{1}{2}\sqrt{5})^2 = 9.$$

This is the equation of an ellipse with its center at the point $(\frac{1}{2}\sqrt{5}, -\frac{1}{2}\sqrt{5})$ in the y_1y_2-system. The positive directions of the y_1 and y_2 axes are determined by the eigenvectors u_1 and u_2, as indicated in Figure 5.2.

We can simplify the equation further by writing

$$z_1 = y_1 - \tfrac{1}{2}\sqrt{5}, \qquad z_2 = y_2 + \tfrac{1}{2}\sqrt{5}.$$

Geometrically, this is the same as introducing a new system of coordinate axes parallel to the y_1y_2 axes but with the new origin at the center of the ellipse. In the z_1z_2-system the equation of the ellipse is simply

$$z_1^2 + 6z_2^2 = 9, \qquad \text{or} \qquad \frac{z_1^2}{9} + \frac{z_2^2}{3/2} = 1.$$

The ellipse and all three coordinate systems are shown in Figure 5.2.

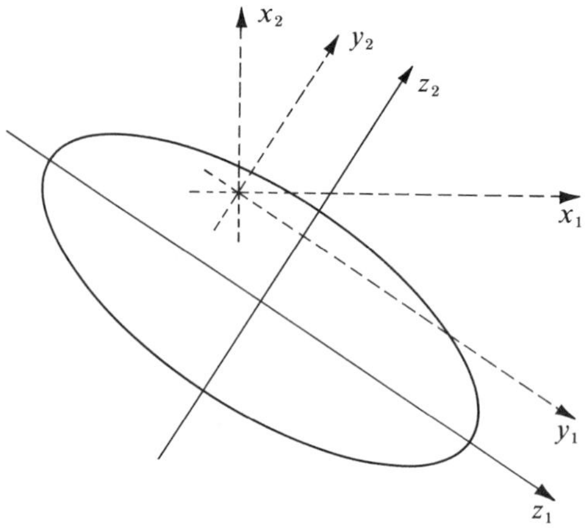

FIGURE 5.2 Rotation and translation of coordinate axes. The rotation $Y = XC$ is followed by the translation $z_1 = y_1 - \frac{1}{2}\sqrt{5}$, $z_2 = y_2 + \frac{1}{2}\sqrt{5}$.

EXAMPLE 2. $2x^2 - 4xy - y^2 - 4x + 10y - 13 = 0$. We rewrite this as

$$2x_1^2 - 4x_1x_2 - x_2^2 - 4x_1 + 10x_2 - 13 = 0.$$

The quadratic part is XAX^t, where $A = \begin{bmatrix} 2 & -2 \\ -2 & -1 \end{bmatrix}$. This matrix has the eigenvalues $\lambda_1 = 3$, $\lambda_2 = -2$. An orthonormal set of eigenvectors is $u_1 = t(2, -1)$, $u_2 = t(1, 2)$, where $t = 1/\sqrt{5}$. An orthogonal diagonalizing matrix is $C = t\begin{bmatrix} 2 & 1 \\ -1 & 2 \end{bmatrix}$. The equation of rotation $X = YC^t$ gves us

$$x_1 = \frac{1}{\sqrt{5}}(2y_1 + y_2), \qquad x_2 = \frac{1}{\sqrt{5}}(-y_1 + 2y_2).$$

Therefore the transformed equation becomes

$$3y_1^2 - 2y_2^2 - \frac{4}{\sqrt{5}}(2y_1 + y_2) + \frac{10}{\sqrt{5}}(-y_1 + 2y_2) - 13 = 0,$$

or

$$3y_1^2 - 2y_2^2 - \frac{18}{\sqrt{5}}y_1 + \frac{16}{\sqrt{5}}y_2 - 13 = 0.$$

By completing the squares in y_1 and y_2 we obtain the equation

$$3(y_1 - \tfrac{3}{5}\sqrt{5})^2 - 2(y_2 - \tfrac{4}{5}\sqrt{5})^2 = 12,$$

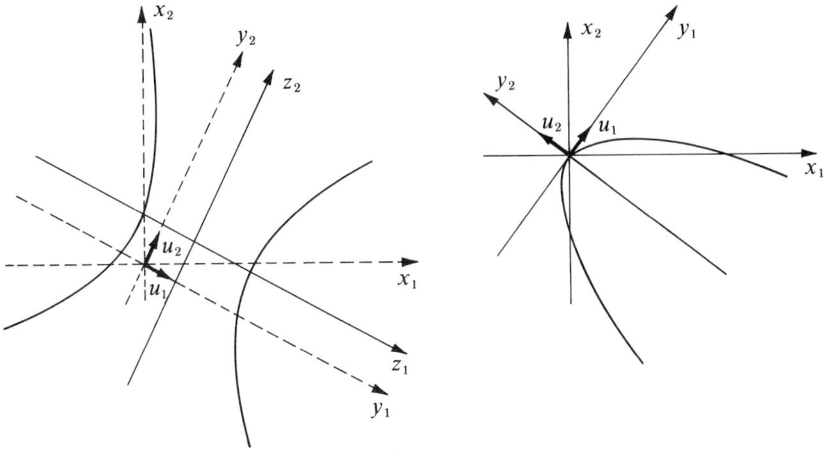

(a) Hyperbola: $3z_1^2 - 2z_2^2 = 12$ (b) Parabola: $y_1^2 + y_2 = 0$

FIGURE 5.3 The curves in Examples 2 and 3.

which represents a hyperbola with its center at $(\frac{3}{5}\sqrt{5}, \frac{4}{5}\sqrt{5})$ in the y_1y_2-system. The translation $z_1 = y_1 - \frac{3}{5}\sqrt{5}$, $z_2 = y_2 - \frac{4}{5}\sqrt{5}$ simplifies this equation further to

$$3z_1^2 - 2z_2^2 = 12, \qquad \text{or} \qquad \frac{z_1^2}{4} - \frac{z_2^2}{6} = 1.$$

The hyperbola is shown in Figure 5.3(a). The eigenvectors u_1 and u_2 determine the directions of the positive y_1 and y_2 axes.

EXAMPLE 3. $9x^2 + 24xy + 16y^2 - 20x + 15y = 0$. We rewrite this as

$$9x_1^2 + 24x_1x_2 + 16x_2^2 - 20x_1 + 15x_2 = 0.$$

The symmetric matrix for the quadratic part is $A = \begin{bmatrix} 9 & 12 \\ 12 & 16 \end{bmatrix}$. Its eigenvalues are $\lambda_1 = 25$, $\lambda_2 = 0$. An orthonormal set of eigenvectors is $u_1 = \frac{1}{5}(3, 4)$, $u_2 = \frac{1}{5}(-4, 3)$. An orthogonal diagonalizing matrix is $C = \frac{1}{5}\begin{bmatrix} 3 & -4 \\ 4 & 3 \end{bmatrix}$. The equation of rotation $X = YC^t$ gives us

$$x_1 = \tfrac{1}{5}(3y_1 - 4y_2), \qquad x_2 = \tfrac{1}{5}(4y_1 + 3y_2).$$

Therefore the transformed Cartesian equation becomes

$$25y_1^2 - \tfrac{20}{5}(3y_1 - 4y_2) + \tfrac{15}{5}(4y_1 + 3y_2) = 0.$$

This simplifies to $y_1^2 + y_2 = 0$, the equation of a parabola with its vertex at the origin. The parabola is shown in Figure 5.3(b).

EXAMPLE 4. *Degenerate cases.* A knowledge of the eigenvalues alone does not reveal whether the Cartesian equation represents a degenerate conic section. For example, the three equations $x^2 + 2y^2 = 1$, $x^2 + 2y^2 = 0$, and $x^2 + 2y^2 = -1$ all have the same eigenvalues; the first represents a nondegenerate ellipse, the second is satisfied only by $(x, y) = (0, 0)$, and the third represents the empty set. The last two can be regarded as degenerate cases of the ellipse.

The graph of the equation $y^2 = 0$ is the x-axis. The equation $y^2 - 1 = 0$ represents the two parallel lines $y = 1$ and $y = -1$. These can be regarded as degenerate cases of the parabola. The equation $x^2 - 4y^2 = 0$ represents two intersecting lines since it is satisfied if either $x - 2y = 0$ or $x + 2y = 0$. This can be regarded as a degenerate case of the hyperbola.

However, if the Cartesian equation $ax^2 + bxy + cy^2 + dx + ey + f = 0$ represents a nondegenerate conic section, then the *type* of conic can be determined quite easily. The characteristic polynomial of the matrix of the quadratic form $ax^2 + bxy + cy^2$ is

$$\det \begin{bmatrix} \lambda - a & -b/2 \\ -b/2 & \lambda - c \end{bmatrix} = \lambda^2 - (a + c)\lambda + (ac - \tfrac{1}{4}b^2) = (\lambda - \lambda_1)(\lambda - \lambda_2).$$

Therefore the product of the eigenvalues is

$$\lambda_1\lambda_2 = ac - \tfrac{1}{4}b^2 = \tfrac{1}{4}(4ac - b^2).$$

Since the type of conic is determined by the algebraic sign of the product $\lambda_1\lambda_2$, we see that the conic is an *ellipse*, *hyperbola*, or *parabola*, according as $4ac - b^2$ is *positive*, *negative*, or *zero*. The number $4ac - b^2$ is called the *discriminant* of the quadratic form $ax^2 + bxy + cy^2$. In Examples 1, 2 and 3 the discriminant has the values 34, -24, and 0, respectively.

5.15 Exercises

In each of Exercises 1 through 7, find (a) a symmetric matrix A for the quadratic form; (b) the eigenvalues of A; (c) an orthonormal set of eigenvectors; (d) an orthogonal diagonalizing matrix C.

1. $4x_1^2 + 4x_1x_2 + x_2^2$.
2. x_1x_2.
3. $x_1^2 + 2x_1x_2 - x_2^2$.
4. $34x_1^2 - 24x_1x_2 + 41x_2^2$.
5. $x_1^2 + x_1x_2 + x_1x_3 + x_2x_3$.
6. $2x_1^2 + 4x_1x_3 + x_2^2 - x_3^2$.
7. $3x_1^2 + 4x_1x_2 + 8x_1x_3 + 4x_2x_3 + 3x_3^2$.

In each of Exercises 8 through 18, identify and make a sketch of the conic section represented by the Cartesian equation.

8. $y^2 - 2xy + 2x^2 - 5 = 0$.
9. $y^2 - 2xy + 5x = 0$.
10. $y^2 - 2xy + x^2 - 5x = 0$.
11. $5x^2 - 4xy + 2y^2 - 6 = 0$.
12. $19x^2 + 4xy + 16y^2 - 212x + 104y = 356$.
13. $9x^2 + 24xy + 16y^2 - 52x + 14y = 6$.
14. $5x^2 + 6xy + 5y^2 - 2 = 0$.
15. $x^2 + 2xy + y^2 - 2x + 2y + 3 = 0$.
16. $2x^2 + 4xy + 5y^2 - 2x - y - 4 = 0$.
17. $x^2 + 4xy - 2y^2 - 12 = 0$.
18. $xy + y - 2x - 2 = 0$.
19. For what value (or values) of c will the graph of the Cartesian equation $2xy - 4x + 7y + c = 0$ be a pair of lines?
20. If the equation $ax^2 + bxy + cy^2 = 1$ represents an ellipse, prove that the area of the region it bounds is $2\pi/\sqrt{4ac - b^2}$. This gives a geometric meaning to the discriminant $4ac - b^2$.

★5.16† Eigenvalues of a symmetric transformation obtained as values of its quadratic form

Now we drop the requirement that V be finite-dimensional and we find a relation between the eigenvalues of a symmetric operator and its quadratic form.

Suppose x is an eigenvector with norm 1 belonging to an eigenvalue λ. Then $T(x) = \lambda x$ so we have

$$Q(x) = (T(x), x) = (\lambda x, x) = \lambda(x, x) = \lambda, \tag{5.13}$$

since $(x, x) = 1$. The set of all x in V satisfying $(x, x) = 1$ is called the *unit sphere* in V. Equation (5.13) proves the following theorem.

THEOREM 5.12. *Let $T\colon V \to V$ be a symmetric transformation on a real Euclidean space V, and let $Q(x) = (T(x), x)$. Then the eigenvalues of T (if any exist) are to be found among the values that Q takes on the unit sphere in V.*

EXAMPLE. Let $V = V_2(\mathbf{R})$ with the usual basis $(\boldsymbol{i}, \boldsymbol{j})$ and the usual dot product as inner product. Let T be the symmetric transformation with matrix $A = \begin{bmatrix} 4 & 0 \\ 0 & 8 \end{bmatrix}$. Then the quadratic form of T is given by

$$Q(x) = \sum_{i=1}^{2} \sum_{j=1}^{2} a_{ij} x_i x_j = 4x_1^2 + 8x_2^2 .$$

The eigenvalues of T are $\lambda_1 = 4$, $\lambda_2 = 8$. It is easy to see that these eigenvalues are, respectively, the minimum and maximum values which Q takes on the unit circle $x_1^2 + x_2^2 = 1$. In fact, on this circle we have

$$Q(x) = 4(x_1^2 + x_2^2) + 4x_2^2 = 4 + 4x_2^2, \qquad \text{where} \quad -1 \le x_2 \le 1 .$$

This has its smallest value, 4, when $x_2 = 0$ and its largest value, 8, when $x_2 = \pm 1$.

Figure 5.4 shows the unit circle and two ellipses. The inner ellipse has the Cartesian equation $4x_1^2 + 8x_2^2 = 4$. It consists of all points $x = (x_1, x_2)$ in the plane satisfying $Q(x) = 4$. The outer ellipse has Cartesian equation $4x_1^2 + 8x_2^2 = 8$ and consists of all points satisfying $Q(x) = 8$. The points $(\pm 1, 0)$ where the inner ellipse touches the unit circle are eigenvectors belonging to the eigenvalue 4. The points $(0, \pm 1)$ on the outer ellipse are eigenvectors belonging to the eigenvalue 8.

The foregoing example illustrates extremal properties of eigenvalues which hold more generally. In the next section we will prove that the smallest and largest eigenvalues (if they exist) are always the minimum and maximum values which Q takes on the unit sphere. Our discussion of these extremal properties will make use of the following theorem on quadratic forms. It should be noted that this theorem does not require that V be finite dimensional.

† Starred sections can be omitted or postponed without loss in continuity.

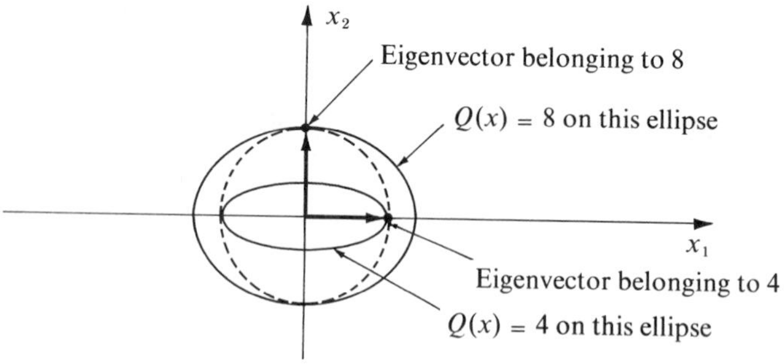

FIGURE 5.4 Geometric relation between the eigenvalues of T and the values of Q on the unit sphere, illustrated with a two-dimensional example.

THEOREM 5.13. *Let $T: V \to V$ be a symmetric transformation on a real Euclidean space V with quadratic form $Q(x) = (T(x), x)$. Assume that Q does not change sign on V. Then if $Q(x) = 0$ for some x in V we also have $T(x) = O$. In other words, if Q does not change sign, then Q vanishes only on the null space of T.*

Proof. Assume $Q(x) = 0$ for some x in V and let y be any element in V. Choose any real t and consider $Q(x + ty)$. Using linearity of T, linearity of the inner product, and symmetry of T, we have

$$\begin{aligned} Q(x + ty) &= (T(x + ty), x + ty) = (T(x) + tT(y), x + ty) \\ &= (T(x), x) + t(T(x), y) + t(T(y), x) + t^2(T(y), y) \\ &= Q(x) + 2t(T(x), y) + t^2Q(y) = at + bt^2, \end{aligned}$$

where $a = 2(T(x), y)$ and $b = Q(y)$. If Q is nonnegative on V we have the inequality

$$at + bt^2 \geq 0 \qquad \text{for all real } t.$$

In other words, the quadratic polynomial $p(t) = at + bt^2$ has its minimum at $t = 0$. Hence $p'(0) = 0$. But $p'(0) = a = 2(T(x), y)$, so $(T(x), y) = 0$. Since y was arbitrary, we can in particular take $y = T(x)$, getting $(T(x), T(x)) = 0$. This proves that $T(x) = O$.

If Q is nonpositive on V we get $p(t) = at + bt^2 \leq 0$ for all t, so p has its maximum at $t = 0$, and hence $p'(0) = 0$ as before.

★ 5.17 Extremal properties of eigenvalues of a symmetric transformation

Now we shall prove that the extreme values of a quadratic form on the unit sphere are eigenvalues.

THEOREM 5.14. *Let $T: V \to V$ be a symmetric linear transformation on a real Euclidean space V, and let $Q(x) = (T(x), x)$. Among all values that Q takes on the unit sphere, assume* *there is an extremum*† *(maximum or minimum) at a point u with $(u, u) = 1$. Then u is an eigenvector for T; the corresponding eigenvalue is $Q(u)$, the extreme value of Q on the unit sphere.*

Proof. Assume Q has a minimum at u. Then we have

$$Q(x) \geq Q(u) \qquad \text{for all } x \text{ with } (x, x) = 1. \tag{5.14}$$

Let $\lambda = Q(u)$. If $(x, x) = 1$ we have $Q(u) = \lambda(x, x) = (\lambda x, x)$ so inequality (5.14) can be written as

$$(T(x), x) \geq (\lambda x, x) \tag{5.15}$$

provided $(x, x) = 1$. Now we prove that (5.15) is valid for *all* x in V. Suppose $\|x\| = a$. Then $x = ay$, where $\|y\| = 1$. Hence

$$(T(x), x) = (T(ay), ay) = a^2(T(y), y) \qquad \text{and} \qquad (\lambda x, x) = a^2(\lambda y, y).$$

But $(T(y), y) \geq (\lambda y, y)$ since $(y, y) = 1$. Multiplying both members of this inequality by a^2 we get (5.15) for $x = ay$.

Since $(T(x), x) - (\lambda x, x) = (T(x) - \lambda x, x)$, we can rewrite inequality (5.15) in the form $(T(x) - \lambda x, x) \geq 0$, or

$$(S(x), x) \geq 0, \qquad \text{where} \quad S = T - \lambda I. \tag{5.16}$$

When $x = u$ we have equality in (5.14) and hence also in (5.16). The linear transformation S is symmetric. Inequality (5.16) states that the quadratic form Q_1 given by $Q_1(x) = (S(x), x)$ is nonnegative on V. When $x = u$ we have $Q_1(u) = 0$. Therefore, by Theorem 5.13 we must have $S(u) = O$. In other words, $T(u) = \lambda u$, so u is an eigenvector for T, and $\lambda = Q(u)$ is the corresponding eigenvalue. This completes the proof if Q has a minimum at u.

If there is a maximum at u all the inequalities in the foregoing proof are reversed and we apply Theorem 5.13 to the nonpositive quadratic form Q_1.

★ 5.18 The finite-dimensional case

Suppose now that $\dim V = n$. Then T has n real eigenvalues which can be arranged in increasing order, say

$$\lambda_1 \leq \lambda_2 \leq \cdots \leq \lambda_n.$$

According to Theorem 5.14, the smallest eigenvalue λ_1 is the minimum of Q on the unit sphere, and the largest eigenvalue is the maximum of Q on the unit sphere. Now we shall show that the intermediate eigenvalues also occur as extreme values of Q, restricted to certain subsets of the unit sphere.

† If V is infinite-dimensional, the quadratic form Q need not have an extremum on the unit sphere. This will be the case when T has no eigenvalues. In the finite-dimensional case, Q always has a maximum and a minimum somewhere on the unit sphere. This follows as a consequence of a more general theorem on extreme values of continuous functions. For a special case of this theorem see Section 9.16.

Let u_1 be an eigenvector on the unit sphere which minimizes Q. Then $\lambda_1 = Q(u_1)$. If λ is an eigenvalue different from λ_1 any eigenvector belonging to λ must be orthogonal to u_1. Therefore it is natural to search for such an eigenvector on the orthogonal complement of the subspace spanned by u_1.

Let S be the subspace spanned by u_1. The orthogonal complement $S^\perp$ consists of all elements in V orthogonal to u_1. In particular, $S^\perp$ contains all eigenvectors belonging to eigenvalues $\lambda \neq \lambda_1$. It is easily verified that $\dim S^\perp = n - 1$ and that T maps $S^\perp$ into itself.† Let S_{n-1} denote the unit sphere in the $(n-1)$-dimensional subspace $S^\perp$. (The unit sphere S_{n-1} is a subset of the unit sphere in V.) Applying Theorem 5.14 to the subspace $S^\perp$ we find that $\lambda_2 = Q(u_2)$, where u_2 is a point which minimizes Q on S_{n-1}.

The next eigenvector λ_3 can be obtained in a similar way as the minimum value of Q on the unit sphere S_{n-2} in the $(n-2)$-dimensional space consisting of those elements orthogonal to both u_1 and u_2. Continuing in this manner we find that each eigenvalue λ_k is the minimum value which Q takes on a unit sphere S_{n-k+1} in a subspace of dimension $n - k + 1$. The largest of these minima, λ_n, is also the *maximum* value which Q takes on each of the spheres S_{n-k+1}. The corresponding set of eigenvectors $u_1, \ldots, u_n$ form an orthonormal basis for V.

5.19 Unitary transformations

We conclude this chapter with a brief discussion of another important class of transformations known as unitary transformations. In the finite-dimensional case they are represented by unitary matrices.

DEFINITION. *Let E be a Euclidean space and V a subspace of E. A linear transformation $T: V \to E$ is called unitary on V if we have*

$$(T(x), T(y)) = (x, y) \qquad \text{for all } x \text{ and } y \text{ in } V. \tag{5.17}$$

When E is a real Euclidean space a unitary transformation is also called an orthogonal transformation.

Equation (5.17) is described by saying that T preserves inner products. Therefore it is natural to expect that T also preserves orthogonality and norms, since these are derived from the inner product.

THEOREM 5.15. *If $T: V \to E$ is a unitary transformation on V, then for all x and y in V we have:*

(a) $(x, y) = 0$ *implies* $(T(x), T(y)) = 0$ (*T preserves orthogonality*).
(b) $\|T(x)\| = \|x\|$ (*T preserves norms*).
(c) $\|T(x) - T(y)\| = \|x - y\|$ (*T preserves distances*).
(d) *T is invertible, and T^{-1} is unitary on $T(V)$.*

† This was done in the proof of Theorem 5.4, Section 5.6.

Proof. Part (a) follows at once from Equation (5.17). Part (b) follows by taking $x = y$ in (5.17). Part (c) follows from (b) because $T(x) - T(y) = T(x - y)$.

To prove (d) we use (b) which shows that $T(x) = O$ implies $x = O$, so T is invertible. If $x \in T(V)$ and $y \in T(V)$ we can write $x = T(u)$, $y = T(v)$, so we have

$$(T^{-1}(x), T^{-1}(y)) = (u, v) = (T(u), T(v)) = (x, y).$$

Therefore T^{-1} is unitary on $T(V)$.

Regarding eigenvalues and eigenvectors we have the following theorem.

THEOREM 5.16. *Let $T: V \to E$ be a unitary transformation on V.*

(a) *If T has an eigenvalue λ, then $|\lambda| = 1$.*

(b) *If x and y are eigenvectors belonging to distinct eigenvalues λ and μ, then x and y are orthogonal.*

(c) *If $V = E$ and* $\dim V = n$, *and if V is a complex space, then there exist eigenvectors $u_1, \ldots, u_n$ of T which form an orthonormal basis for V. The matrix of T relative to this basis is the diagonal matrix $\Lambda = \operatorname{diag}(\lambda_1, \ldots, \lambda_n)$, where λ_k is the eigenvalue belonging to u_k.*

Proof. To prove (a), let x be an eigenvector belonging to λ. Then $x \neq O$ and $T(x) = \lambda x$. Taking $y = x$ in Equation (5.17) we get

$$(\lambda x, \lambda x) = (x, x) \qquad \text{or} \qquad \lambda\bar{\lambda}(x, x) = (x, x).$$

Since $(x, x) > 0$ and $\lambda\bar{\lambda} = |\lambda|^2$, this implies $|\lambda| = 1$.

To prove (b), write $T(x) = \lambda x$, $T(y) = \mu y$ and compute the inner product $(T(x), T(y))$ in two ways. We have

$$(T(x), T(y)) = (x, y)$$

since T is unitary. We also have

$$(T(x), T(y)) = (\lambda x, \mu y) = \lambda\bar{\mu}(x, y)$$

since x and y are eigenvectors. Therefore $\lambda\bar{\mu}(x, y) = (x, y)$, so $(x, y) = 0$ unless $\lambda\bar{\mu} = 1$. But $\lambda\bar{\lambda} = 1$ by (a), so if we had $\lambda\bar{\mu} = 1$ we would also have $\lambda\bar{\lambda} = \lambda\bar{\mu}$, $\bar{\lambda} = \bar{\mu}$, $\lambda = \mu$, which contradicts the assumption that λ and μ are distinct. Therefore $\lambda\bar{\mu} \neq 1$ and $(x, y) = 0$.

Part (c) is proved by induction on n in much the same way that we proved Theorem 5.4, the corresponding result for Hermitian operators. The only change required is in that part of the proof which shows that T maps $S^{\perp}$ into itself, where

$$S^{\perp} = \{x \mid x \in V, \quad (x, u_1) = 0\}.$$

Here u_1 is an eigenvector of T with eigenvalue λ_1. From the equation $T(u_1) = \lambda_1 u_1$ we find

$$u_1 = \lambda_1^{-1} T(u_1) = \bar{\lambda}_1 T(u_1)$$

since $\lambda_1\bar{\lambda}_1 = |\lambda_1|^2 = 1$. Now choose any x in $S^\perp$ and note that

$$(T(x), u_1) = (T(x), \bar{\lambda}_1 T(u_1)) = \lambda_1(T(x), T(u_1)) = \lambda_1(x, u_1) = 0.$$

Hence $T(x) \in S^\perp$ if $x \in S^\perp$, so T maps $S^\perp$ into itself. The rest of the proof is identical with that of Theorem 5.4, so we shall not repeat the details.

The next two theorems describe properties of unitary transformations on a finite-dimensional space. We give only a brief outline of the proofs.

THEOREM 5.17. *Assume* $\dim V = n$ *and let* $E = (e_1, \ldots, e_n)$ *be a fixed basis for* V. *Then a linear transformation* $T: V \to V$ *is unitary if and only if*

(5.18) $$(T(e_i), T(e_j)) = (e_i, e_j) \quad \text{for all } i \text{ and } j.$$

In particular, if E is orthonormal then T is unitary if and only if T maps E onto an orthonormal basis.

Sketch of proof. Write $x = \sum x_i e_i$, $y = \sum y_j e_j$. Then we have

$$(x, y) = \left(\sum_{i=1}^n x_i e_i, \sum_{j=1}^n y_j e_j\right) = \sum_{i=1}^n \sum_{j=1}^n x_i \bar{y}_j (e_i, e_j),$$

and

$$(T(x), T(y)) = \left(\sum_{i=1}^n x_i T(e_i), \sum_{j=1}^n y_j T(e_j)\right) = \sum_{i=1}^n \sum_{j=1}^n x_i \bar{y}_j (T(e_i), T(e_j)).$$

Now compare (x, y) with $(T(x), T(y))$.

THEOREM 5.18. *Assume* $\dim V = n$ *and let* $(e_1, \ldots, e_n)$ *be an orthonormal basis for* V. *Let* $A = (a_{ij})$ *be the matrix representation of a linear transformation* $T: V \to V$ *relative to this basis. Then T is unitary if and only if A is unitary, that is, if and only if*

(5.19) $$A^*A = I.$$

Sketch of proof. Since (e_i, e_j) is the ij-entry of the identity matrix, Equation (5.19) implies

(5.20) $$(e_i, e_j) = \sum_{k=1}^n \bar{a}_{ki} a_{kj} = \sum_{k=1}^n a_{ki} \bar{a}_{kj}.$$

Since A is the matrix of T we have $T(e_i) = \sum_{k=1}^n a_{ki} e_k$, $T(e_j) = \sum_{r=1}^n a_{rj} e_r$, so

$$(T(e_i), T(e_j)) = \left(\sum_{k=1}^n a_{ki} e_k, \sum_{r=1}^n a_{rj} e_r\right) = \sum_{k=1}^n \sum_{r=1}^n a_{ki} \bar{a}_{rj} (e_k, e_r) = \sum_{k=1}^n a_{ki} \bar{a}_{kj}.$$

Now compare this with (5.20) and use Theorem 5.17.

THEOREM 5.19. *Every unitary matrix A has the following properties:*
(a) *A is nonsingular and* $A^{-1} = A^*$.
(b) *Each of A^t, $\bar{A}$, and A^* is a unitary matrix.*
(c) *The eigenvalues of A are complex numbers of absolute value* 1.
(d) $|\det A| = 1$; *if A is real, then* $\det A = \pm 1$.

The proof of Theorem 5.19 is left as an exercise for the reader.

5.20 Exercises

1. (a) Let $T: V \to V$ be the transformation given by $T(x) = cx$, where c is a fixed scalar. Prove that T is unitary if and only if $|c| = 1$.
(b) If V is one-dimensional, prove that the only unitary transformations on V are those described in (a). In particular, if V is a real one-dimensional space, there are only two orthogonal transformations, $T(x) = x$ and $T(x) = -x$.
2. Prove each of the following statements about a real orthogonal $n \times n$ matrix A.
(a) If λ is a real eigenvalue of A, then $\lambda = 1$ or $\lambda = -1$.
(b) If λ is a complex eigenvalue of A, then the complex conjugate $\bar{\lambda}$ is also an eigenvalue of A. In other words, the nonreal eigenvalues of A occur in conjugate pairs.
(c) If n is odd, then A has at least one real eigenvalue.
3. Let V be a real Euclidean space of dimension n. An orthogonal transformation $T: V \to V$ with determinant 1 is called a *rotation.* If n is odd, prove that 1 is an eigenvalue for T. This shows that every rotation of an odd-dimensional space has a fixed axis. [*Hint:* Use Exercise 2.]
4. Given a real orthogonal matrix A with -1 as an eigenvalue of multiplicity k. Prove that $\det A = (-1)^k$.
5. If T is linear and norm-preserving, prove that T is unitary.
6. If $T: V \to V$ is both unitary and Hermitian, prove that $T^2 = I$.
7. Let $(e_1, \ldots, e_n)$ and $(u_1, \ldots, u_n)$ be two orthonormal bases for a Euclidean space V. Prove that there is a unitary transformation T which maps one of these bases onto the other.
8. Find a real a such that the following matrix is unitary:

$$\begin{bmatrix} a & \frac{1}{2}i & \frac{1}{2}a(2i-1) \\ ia & \frac{1}{2}(1+i) & \frac{1}{2}a(1-i) \\ a & -\frac{1}{2} & \frac{1}{2}a(2-i) \end{bmatrix}.$$

9. If A is a skew-Hermitian matrix, prove that both $I - A$ and $I + A$ are nonsingular and that $(I - A)(I + A)^{-1}$ is unitary.
10. If A is a unitary matrix and if $I + A$ is nonsingular, prove that $(I - A)(I + A)^{-1}$ is skew-Hermitian.
11. If A is Hermitian, prove that $A - iI$ is nonsingular and that $(A - iI)^{-1}(A + iI)$ is unitary.
12. Prove that any unitary matrix can be diagonalized by a unitary matrix.
13. A square matrix is called *normal* if $AA^* = A^*A$. Determine which of the following types of matrices are normal.
(a) Hermitian matrices.
(b) Skew-Hermitian matrices.
(c) Symmetric matrices.
(d) Skew-symmetric matrices.
(e) Unitary matrices.
(f) Orthogonal matrices.
14. If A is a normal matrix ($AA^* = A^*A$) and if U is a unitary matrix, prove that U^*AU is normal.

6

LINEAR DIFFERENTIAL EQUATIONS

6.1 Historical introduction

The history of differential equations began in the 17th century when Newton, Leibniz, and the Bernoullis solved some simple differential equations of the first and second order arising from problems in geometry and mechanics. These early discoveries, beginning about 1690, seemed to suggest that the solutions of all differential equations based on geometric and physical problems could be expressed in terms of the familiar elementary functions of calculus. Therefore, much of the early work was aimed at developing ingenious techniques for solving differential equations by elementary means, that is to say, by addition, subtraction, multiplication, division, composition, and integration, applied only a finite number of times to the familiar functions of calculus.

Special methods such as separation of variables and the use of integrating factors were devised more or less haphazardly before the end of the 17th century. During the 18th century, more systematic procedures were developed, primarily by Euler, Lagrange, and Laplace. It soon became apparent that relatively few differential equations could be solved by elementary means. Little by little, mathematicians began to realize that it was hopeless to try to discover methods for solving all differential equations. Instead, they found it more fruitful to ask whether or not a given differential equation has any solution at all and, when it has, to try to deduce properties of the solution from the differential equation itself. Within this framework, mathematicians began to think of differential equations as new sources of functions.

An important phase in the theory developed early in the 19th century, paralleling the general trend toward a more rigorous approach to the calculus. In the 1820's, Cauchy obtained the first "existence theorem" for differential equations. He proved that every first-order equation of the form

$$y' = f(x, y)$$

has a solution whenever the right member, $f(x, y)$, satisfies certain general conditions. One important example is the Ricatti equation

$$y' = P(x)y^2 + Q(x)y + R(x),$$

where P, Q, and R are given functions. Cauchy's work implies the existence of a solution of the Ricatti equation in any open interval $(-r, r)$ about the origin, provided P, Q, and

R have power-series expansions in $(-r, r)$. In 1841 Joseph Liouville (1809–1882) showed that in some cases this solution cannot be obtained by elementary means.

Experience has shown that it is difficult to obtain results of much generality about solutions of differential equations, except for a few types. Among these are the so-called *linear* differential equations which occur in a great variety of scientific problems. Some simple types were discussed in Volume I—linear equations of first order and linear equations of second order with constant coefficients. The next section gives a review of the principal results concerning these equations.

6.2 Review of results concerning linear equations of first and second orders

A linear differential equation of first order is one of the form

$$y' + P(x)y = Q(x), \tag{6.1}$$

where P and Q are given functions. In Volume I we proved an existence-uniqueness theorem for this equation (Theorem 8.3) which we restate here.

THEOREM 6.1. *Assume P and Q are continuous on an open interval J. Choose any point a in J and let b be any real number. Then there is one and only one function $y = f(x)$ which satisfies the differential equation* (6.1) *and the initial condition $f(a) = b$. This function is given by the explicit formula*

$$f(x) = be^{-A(x)} + e^{-A(x)} \int_a^x Q(t)e^{A(t)}\, dt, \tag{6.2}$$

where $A(x) = \int_a^x P(t)\, dt$.

Linear equations of second order are those of the form

$$P_0(x)y'' + P_1(x)y' + P_2(x)y = R(x).$$

If the coefficients P_0, P_1, P_2 and the right-hand member R are continuous on some interval J, and if P_0 is never zero on J, an existence theorem (discussed in Section 6.5) guarantees that solutions always exist over the interval J. Nevertheless, there is no general formula analogous to (6.2) for expressing these solutions in terms of P_0, P_1, P_2, and R. Thus, even in this relatively simple generalization of (6.1), the theory is far from complete, except in special cases. If the coefficients are *constants* and if R is zero, all the solutions can be determined explicitly in terms of polynomials, exponential and trigonometric functions by the following theorem which was proved in Volume I (Theorem 8.7).

THEOREM 6.2. *Consider the differential equation*

$$y'' + ay' + by = 0, \tag{6.3}$$

where a and b are given real constants. Let $d = a^2 - 4b$. *Then every solution of* (6.3) *on the interval* $(-\infty, +\infty)$ *has the form*

$$y = e^{-ax/2}[c_1u_1(x) + c_2u_2(x)], \tag{6.4}$$

where c_1 *and* c_2 *are constants, and the functions* u_1 *and* u_2 *are determined according to the algebraic sign of d as follows:*

(a) *If* $d = 0$, *then* $u_1(x) = 1$ *and* $u_2(x) = x$.

(b) *If* $d > 0$, *then* $u_1(x) = e^{kx}$ *and* $u_2(x) = e^{-kx}$, *where* $k = \frac{1}{2}\sqrt{d}$.

(c) *If* $d < 0$, *then* $u_1(x) = \cos kx$ *and* $u_2(x) = \sin kx$, *where* $k = \frac{1}{2}\sqrt{-d}$.

The number $d = a^2 - 4b$ is the *discriminant* of the quadratic equation

$$r^2 + ar + b = 0. \tag{6.5}$$

This is called the *characteristic equation* of the differential equation (6.3). Its roots are given by

$$r_1 = \frac{-a + \sqrt{d}}{2}, \qquad r_2 = \frac{-a - \sqrt{d}}{2}.$$

The algebraic sign of d determines the nature of these roots. If $d > 0$ both roots are real and the solution in (6.4) can be expressed in the form

$$y = c_1e^{r_1x} + c_2e^{r_2x}.$$

If $d < 0$, the roots r_1 and r_2 are conjugate complex numbers. Each of the complex exponential functions $f_1(x) = e^{r_1x}$ and $f_2(x) = e^{r_2x}$ is a complex solution of the differential equation (6.3). We obtain real solutions by examining the real and imaginary parts of f_1 and f_2. Writing $r_1 = -\frac{1}{2}a + ik$, $r_2 = -\frac{1}{2}a - ik$, where $k = \frac{1}{2}\sqrt{-d}$, we have

$$f_1(x) = e^{r_1x} = e^{-ax/2}e^{ikx} = e^{-ax/2}\cos kx + ie^{-ax/2}\sin kx$$

and

$$f_2(x) = e^{r_2x} = e^{-ax/2}e^{-ikx} = e^{-ax/2}\cos kx - ie^{-ax/2}\sin kx.$$

The general solution appearing in Equation (6.4) is a linear combination of the real and imaginary parts of $f_1(x)$ and $f_2(x)$.

6.3 Exercises

These exercises have been selected from Chapter 8 in Volume I and are intended as a review of the introductory material on linear differential equations of first and second orders.

Linear equations of first order. In Exercises 1, 2, 3, solve the initial-value problem on the specified interval.

1. $y' - 3y = e^{2x}$ on $(-\infty, +\infty)$, with $y = 0$ when $x = 0$.
2. $xy' - 2y = x^5$ on $(0, +\infty)$, with $y = 1$ when $x = 1$.
3. $y' + y\tan x = \sin 2x$ on $(-\frac{1}{2}\pi, \frac{1}{2}\pi)$, with $y = 2$ when $x = 0$.

4. If a strain of bacteria grows at a rate proportional to the amount present and if the population doubles in one hour, by how much will it increase at the end of two hours?
5. A curve with Cartesian equation $y = f(x)$ passes through the origin. Lines drawn parallel to the coordinate axes through an arbitary point of the curve form a rectangle with two sides on the axes. The curve divides every such rectangle into two regions A and B, one of which has an area equal to n times the other. Find the function f.
6. (a) Let u be a nonzero solution of the second-order equation $y'' + P(x)y' + Q(x)y = 0$. Show that the substitution $y = uv$ converts the equation

$$y'' + P(x)y' + Q(x)y = R(x)$$

into a first-order linear equation for v'.
(b) Obtain a nonzero solution of the equation $y'' - 4y' + x^2(y' - 4y) = 0$ by inspection and use the method of part (a) to find a solution of

$$y'' - 4y' + x^2 (y' - 4y) = 2xe^{-x^3/3}$$

such that $y = 0$ and $y' = 4$ when $x = 0$.

Linear equations of second order with constant coefficients. In each of Exercises 7 through 10, find all solutions on $(-\infty, +\infty)$.

7. $y'' - 4y = 0$.
8. $y'' + 4y = 0$.
9. $y'' - 2y' + 5y = 0$.
10. $y'' + 2y' + y = 0$.

11. Find all values of the constant k such that the differential equation $y'' + ky = 0$ has a nontrivial solution $y = f_k(x)$ for which $f_k(0) = f_k(1) = 0$. For each permissible value of k, determine the corresponding solution $y = f_k(x)$. Consider both positive and negative values of k.
12. If (a, b) is a given point in the plane and if m is a given real number, prove that the differential equation $y'' + k^2y = 0$ has exactly one solution whose graph passes through (a, b) and has slope m there. Discuss separately the case $k = 0$.
13. In each case, find a linear differential equation of second order satisfied by u_1 and u_2.
(a) $u_1(x) = e^x, \quad u_2(x) = e^{-x}$.
(b) $u_1(x) = e^{2x}, \quad u_2(x) = xe^{2x}$.
(c) $u_1(x) = e^{-x/2}\cos x, \quad u_2(x) = e^{-x/2}\sin x$.
(d) $u_1(x) = \sin(2x + 1), \quad u_2(x) = \sin(2x + 2)$.
(e) $u_1(x) = \cosh x, \quad u_2(x) = \sinh x$.
14. A particle undergoes simple harmonic motion. Initially its displacement is 1, its velocity is 1 and its acceleration is -12. Compute its displacement and acceleration when the velocity is $\sqrt{8}$.

6.4 Linear differential equations of order *n*

A linear differential equation of order n is one of the form

$$P_0(x)y^{(n)} + P_1(x)y^{(n-1)} + \cdots + P_n(x)y = R(x). \tag{6.6}$$

The functions $P_0, P_1, \ldots, P_n$ multiplying the various derivatives of the unknown function y are called the *coefficients* of the equation. In our general discussion of the linear equation we shall assume that all the coefficients are continuous on some interval J. The word "interval" will refer either to a bounded or to an unbounded interval.

In the differential equation (6.6) the leading coefficient P_0 plays a special role, since it determines the order of the equation. Points at which $P_0(x) = 0$ are called *singular points* of the equation. The presence of singular points sometimes introduces complications that require special investigation. To avoid these difficulties we assume that the function P_0 is never zero on J. Then we can divide both sides of Equation (6.6) by P_0 and rewrite the differential equation in a form with leading coefficient 1. Therefore, in our general discussion we shall assume that the differential equation has the form

$$y^{(n)} + P_1(x)y^{(n-1)} + \cdots + P_n(x)y = R(x). \tag{6.7}$$

The discussion of linear equations can be simplified by the use of operator notation. Let $\mathscr{C}(J)$ denote the linear space consisting of all real-valued functions continuous on an interval J. Let $\mathscr{C}^n(J)$ denote the subspace consisting of all functions f whose first n derivatives $f', f'', \ldots, f^{(n)}$ exist and are continuous on J. Let $P_1, \ldots, P_n$ be n given functions in $\mathscr{C}(J)$ and consider the operator $L: \mathscr{C}^n(J) \to \mathscr{C}(J)$ given by

$$L(f) = f^{(n)} + P_1 f^{(n-1)} + \cdots + P_n f.$$

The operator L itself is sometimes written as

$$L = D^n + P_1 D^{n-1} + \cdots + P_n,$$

where D^k denotes the kth derivative operator. In operator notation the differential equation in (6.7) is written simply as

$$L(y) = R. \tag{6.8}$$

A solution of this equation is any function y in $\mathscr{C}^n(J)$ which satisfies (6.8) on the interval J.

It is easy to verify that $L(y_1 + y_2) = L(y_1) + L(y_2)$, and that $L(cy) = cL(y)$ for every constant c. Therefore L is a *linear* operator. This is why the equation $L(y) = R$ is referred to as a linear equation. The operator L is called a *linear differential operator of order n.*

With each linear equation $L(y) = R$ we may associate the equation

$$L(y) = 0,$$

in which the right-hand side has been replaced by zero. This is called the *homogeneous equation* corresponding to $L(y) = R$. When R is not identically zero, the equation $L(y) = R$ is called a *nonhomogeneous equation.* We shall find that we can always solve the nonhomogeneous equation whenever we can solve the corresponding homogeneous equation. Therefore, we begin our study with the homogeneous case.

The set of solutions of the homogeneous equation is the null space $N(L)$ of the operator L. This is also called the *solution space* of the equation. The solution space is a subspace of $\mathscr{C}^n(J)$. Although $\mathscr{C}^n(J)$ is infinite-dimensional, it turns out that the solution space $N(L)$ is always finite-dimensional. In fact, we shall prove that

$$\dim N(L) = n, \tag{6.9}$$

where n is the order of the operator L. Equation (6.9) is called the *dimensionality theorem* for linear differential operators. The dimensionality theorem will be deduced as a consequence of an existence-uniqueness theorem which we discuss next.

6.5 The existence-uniqueness theorem

THEOREM 6.3. EXISTENCE-UNIQUENESS THEOREM FOR LINEAR EQUATIONS OF ORDER n. *Let $P_1, P_2, \ldots, P_n$ be continuous functions on an open interval J, and let L be the linear differential operator*

$$L = D^n + P_1 D^{n-1} + \cdots + P_n .$$

If $x_0 \in J$ and if $k_0, k_1, \ldots, k_{n-1}$ are n given real numbers, then there exists one and only one function $y = f(x)$ which satisfies the homogeneous differential equation $L(y) = 0$ on J and which also satisfies the initial conditions

$$f(x_0) = k_0, f'(x_0) = k_1, \ldots, f^{(n-1)}(x_0) = k_{n-1} .$$

Note: The vector in n-space given by $(f(x_0), f'(x_0), \ldots, f^{(n-1)}(x_0))$ is called the *initial-value vector* of f at x_0, Theorem 6.3 tells us that if we choose a point x_0 in J and choose a vector in n-space, then the homogeneous equation $L(y) = 0$ has exactly one solution $y = f(x)$ on J with this vector as initial-value vector at x_0. For example, when $n = 2$ there is exactly one solution with prescribed value $f(x_0)$ and prescribed derivative $f'(x_0)$ at a prescribed point x_0.

The proof of the existence-uniqueness theorem will be obtained as a corollary of more general existence-uniqueness theorems discussed in Chapter 7. An alternate proof for the case of equations with constant coefficients is given in Section 7.9.

6.6 The dimension of the solution space of a homogeneous linear equation

THEOREM 6.4. DIMENSIONALITY THEOREM. *Let $L: \mathscr{C}^n(J) \to \mathscr{C}(J)$ be a linear differential operator of order n given by*

$$L = D^n + P_1 D^{n-1} + \cdots + P_n . \tag{6.10}$$

Then the solution space of the equation $L(y) = 0$ has dimension n.

Proof. Let V_n denote the n-dimensional linear space of n-tuples of scalars. Let T be the linear transformation that maps each function f in the solution space $N(L)$ onto the initial-value vector of f at x_0,

$$T(f) = (f(x_0), f'(x_0), \ldots, f^{(n-1)}(x_0)),$$

where x_0 is a fixed point in J. The uniqueness theorem tells us that $T(f) = 0$ implies $f = 0$. Therefore, by Theorem 2.10, T is one-to-one on $N(L)$. Hence T^{-1} is also one-to-one and maps V_n onto $N(L)$, and Theorem 2.11 shows that $\dim N(L) = \dim V_n = n$.

Now that we know that the solution space has dimension n, any set of n independent solutions will serve as a basis. Therefore, as a corollary of the dimensionality theorem we have:

THEOREM 6.5. *Let* $L: \mathscr{C}^n(J) \to \mathscr{C}(J)$ *be a linear differential operator of order* n. *If* $u_1, \ldots, u_n$ *are* n *independent solutions of the homogeneous differential equation* $L(y) = 0$ *on* J, *then every solution* $y = f(x)$ *on* J *can be expressed in the form*

$$f(x) = \sum_{k=1}^{n} c_k u_k(x), \tag{6.11}$$

where $c_1, \ldots, c_n$ *are constants.*

Note: Since all solutions of the differential equation $L(y) = 0$ are contained in formula (6.11), the linear combination on the right, with arbitrary constants $c_1, \ldots, c_n$, is sometimes called the *general solution* of the differential equation.

The dimensionality theorem tells us that the solution space of a homogeneous linear differential equation of order n always has a basis of n solutions, but it does not tell us how to determine such a basis. In fact, no simple method is known for determining a basis of solutions for every linear equation. However, special methods have been devised for special equations. Among these are differential equations with constant coefficients to which we turn now.

6.7 The algebra of constant-coefficient operators

A constant-coefficient operator A is a linear operator of the form

$$A = a_0 D^n + a_1 D^{n-1} + \cdots + a_{n-1} D + a_n, \tag{6.12}$$

where D is the derivative operator and $a_0, a_1, \ldots, a_n$ are real constants. If $a_0 \neq 0$ the operator is said to have order n. The operator A can be applied to any function y with n derivatives on some interval, the result being a function $A(y)$ given by

$$A(y) = a_0 y^{(n)} + a_1 y^{(n-1)} + \cdots + a_{n-1} y' + a_n y .$$

In this section, we restrict our attention to functions having derivatives of every order on $(-\infty, +\infty)$. The set of all such functions will be denoted by $\mathscr{C}^\infty$ and will be referred to as the class of *infinitely differentiable functions.* If $y \in \mathscr{C}^\infty$ then $A(y)$ is also in $\mathscr{C}^\infty$.

The usual algebraic operations on linear transformations (*addition*, *multiplication by scalars*, and *composition* or *multiplication*) can be applied, in particular, to constant-coefficient operators. Let A and B be two constant-coefficient operators (not necessarily of the same order). Since the sum $A + B$ and all scalar multiples λA are also constant-coefficient operators, the set of all constant-coefficient operators is a linear space. The product of A and B (in either order) is also a constant-coefficient operator. Therefore, sums, products, and scalar multiples of constant-coefficient operators satisfy the usual commutative, associative, and distributive laws satisfied by all linear transformations. Also, since we have $D^r D^s = D^s D^r$ for all positive integers r and s, any two constant-coefficient operators *commute;* $AB = BA$.

With each constant-coefficient operator A we associate a polynomial p_A called the characteristic polynomial of A. If A is given by (6.12), p_A is that polynomial which has the same coefficients as A. That is, for every real r we have

$$p_A(r) = a_0 r^n + a_1 r^{n-1} + \cdots + a_n .$$

Conversely, given any real polynomial p, there is a corresponding operator A whose coefficients are the same as those of p. The next theorem shows that this association between operators and polynomials is a one-to-one correspondence. Moreover, this correspondence associates with sums, products, and scalar multiples of operators the respective sums, products, and scalar multiples of their characteristic polynomials.

THEOREM 6.6. *Let A and B denote constant-coefficient operators with characteristic polynomials p_A and p_B, respectively, and let λ be a real number. Then we have:*

(a) $A = B$ *if and only if* $p_A = p_B$,

(b) $p_{A+B} = p_A + p_B$,

(c) $p_{AB} = p_A \cdot p_B$,

(d) $p_{\lambda A} = \lambda \cdot p_A$.

Proof. We consider part (a) first. Assume $p_A = p_B$. We wish to prove that $A(y) = B(y)$ for every y in $\mathscr{C}^\infty$. Since $p_A = p_B$, both polynomials have the same degree and the same coefficients. Therefore A and B have the same order and the same coefficients, so $A(y) = B(y)$ for every y in $\mathscr{C}^\infty$.

Next we prove that $A = B$ implies $p_A = p_B$. The relation $A = B$ means that $A(y) = B(y)$ for every y in $\mathscr{C}^\infty$. Take $y = e^{rx}$, where r is a constant. Since $y^{(k)} = r^k e^{rx}$ for every $k \geq 0$, we have

$$A(y) = p_A(r)e^{rx} \qquad \text{and} \qquad B(y) = p_B(r)e^{rx} .$$

The equation $A(y) = B(y)$ implies $p_A(r) = p_B(r)$. Since r is arbitrary we must have $p_A = p_B$. This completes the proof of part (a).

Parts (b), (c), and (d) follow at once from the definition of the characteristic polynomial.

From Theorem 6.6 it follows that every algebraic relation involving sums, products, and scalar multiples of polynomials p_A and p_B also holds for the operators A and B. In particular, if the characteristic polynomial p_A can be factored as a product of two or more polynomials, each factor must be the characteristic polynomial of some constant-coefficient operator, so, by Theorem 6.6, there is a corresponding factorization of the operator A. For example, if $p_A(r) = p_B(r)p_C(r)$, then $A = BC$. If $p_A(r)$ can be factored as a product of n linear factors, say

$$p_A(r) = a_0(r - r_1)(r - r_2) \cdots (r - r_n), \tag{6.13}$$

the corresponding factorization of A takes the form

$$A = a_0(D - r_1)(D - r_2) \cdots (D - r_n).$$

The fundamental theorem of algebra tells us that every polynomial $p_A(r)$ of degree $n \geq 1$ has a factorization of the form (6.13), where $r_1, r_2, \ldots, r_n$ are the roots of the equation,

$$p_A(r) = 0,$$

called the *characteristic equation* of A. Each root is written as often as its multiplicity indicates. The roots may be real or complex. Since $p_A(r)$ has real coefficients, the complex roots occur in conjugate pairs, $\alpha + i\beta$, $\alpha - i\beta$, if $\beta \neq 0$. The two linear factors corresponding to each such pair can be combined to give one quadratic factor $r^2 - 2\alpha r + \alpha^2 + \beta^2$ whose coefficients are real. Therefore, every polynomial $p_A(r)$ can be factored as a product of linear and quadratic polynomials *with real coefficients*. This gives a corresponding factorization of the operator A as a product of first-order and second-order constant-coefficient operators with real coefficients.

EXAMPLE 1. Let $A = D^2 - 5D + 6$. Since the characteristic polynomial $p_A(r)$ has the factorization $r^2 - 5r + 6 = (r - 2)(r - 3)$, the operator A has the factorization

$$D^2 - 5D + 6 = (D - 2)(D - 3).$$

EXAMPLE 2. Let $A = D^4 - 2D^3 + 2D^2 - 2D + 1$. The characteristic polynomial $p_A(r)$ has the factorization

$$r^4 - 2r^3 + 2r^2 - 2r + 1 = (r - 1)(r - 1)(r^2 + 1),$$

so A has the factorization $A = (D - 1)(D - 1)(D^2 + 1)$.

6.8 Determination of a basis of solutions for linear equations with constant coefficients by factorization of operators

The next theorem shows how factorization of constant-coefficient operators helps us to solve linear differential equations with constant coefficients.

THEOREM 6.7. *Let L be a constant-coefficient operator which can be factored as a product of constant-coefficient operators, say*

$$L = A_1 A_2 \cdots A_k.$$

Then the solution space of the linear differential equation $L(y) = 0$ contains the solution space of each differential equation $A_i(y) = 0$. In other words,

$$N(A_i) \subseteq N(L) \quad \text{for each } i = 1, 2, \ldots, k. \tag{6.14}$$

Proof. If u is in the null space of the last factor A_k we have $A_k(u) = 0$ so

$$L(u) = (A_1 A_2 \cdots A_k)(u) = (A_1 \cdots A_{k-1})A_k(u) = (A_1 \cdots A_{k-1})(0) = 0.$$

Therefore the null space of L contains the null space of the last factor A_k. But since constant-coefficient operators commute, we can rearrange the factors so that any one of them is the last factor. This proves (6.14).

If $L(u) = 0$, the operator L is said to *annihilate* u. Theorem 6.7 tells us that if a factor A_i of L annihilates u, then L also annihilates u.

We illustrate how the theorem can be used to solve homogeneous differential equations with constant coefficients. We have chosen examples to illustrate different features, depending on the nature of the roots of the characteristic equation.

CASE I. Real distinct roots.

EXAMPLE 1. Find a basis of solutions for the differential equation

$$(D^3 - 7D + 6)y = 0. \tag{6.15}$$

Solution. This has the form $L(y) = 0$ with

$$L = D^3 - 7D + 6 = (D - 1)(D - 2)(D + 3).$$

The null space of $D - 1$ contains $u_1(x) = e^x$, that of $D - 2$ contains $u_2(x) = e^{2x}$, and that of $D + 3$ contains $u_3(x) = e^{-3x}$. In Chapter 1 (p. 10) we proved that u_1, u_2, u_3 are independent. Since three independent solutions of a third order equation form a basis for the solution space, the general solution of (6.15) is given by

$$y = c_1 e^x + c_2 e^{2x} + c_3 e^{-3x}.$$

The method used to solve Example 1 enables us to find a basis for the solution space of any constant-coefficient operator that can be factored into distinct linear factors.

THEOREM 6.8. *Let L be a constant coefficient operator whose characteristic equation $p_L(r) = 0$ has n distinct real roots $r_1, r_2, \ldots, r_n$. Then the general solution of the differential equation $L(y) = 0$ on the interval $(-\infty, +\infty)$ is given by the formula*

$$y = \sum_{k=1}^{n} c_k e^{r_k x}. \tag{6.16}$$

Proof. We have the factorization

$$L = a_0(D - r_1)(D - r_2) \cdots (D - r_n).$$

Since the null space of $(D - r_k)$ contains $u_k(x) = e^{r_k x}$, the null space of L contains the n functions

$$u_1(x) = e^{r_1 x}, \qquad u_2(x) = e^{r_2 x}, \ldots, u_n(x) = e^{r_n x}. \tag{6.17}$$

In Chapter 1 (p. 10) we proved that these functions are independent. Therefore they form a basis for the solution space of the equation $L(y) = 0$, so the general solution is given by (6.16).

CASE II. Real roots, some of which are repeated.

If all the roots are real but not distinct, the functions in (6.17) are not independent and therefore do not form a basis for the solution space. If a root r occurs with multiplicity m, then $(D - r)^m$ is a factor of L. The next theorem shows how to obtain m independent solutions in the null space of this factor.

THEOREM 6.9. *The m functions*

$$u_1(x) = e^{rx}, \qquad u_2(x) = xe^{rx}, \ldots, u_m(x) = x^{m-1}e^{rx}$$

are m independent elements annihilated by the operator $(D - r)^m$.

Proof. The independence of these functions follows from the independence of the polynomials $1, x, x^2, \ldots, x^{m-1}$. To prove that $u_1, u_2, \ldots, u_m$ are annihilated by $(D - r)^m$ we use induction on m.

If $m = 1$ there is only one function, $u_1(x) = e^{rx}$, which is clearly annihilated by $(D - r)$. Suppose, then, that the theorem is true for $m - 1$. This means that the functions $u_1, \ldots, u_{m-1}$ are annihilated by $(D - r)^{m-1}$. Since

$$(D - r)^m = (D - r)(D - r)^{m-1}$$

the functions $u_1, \ldots, u_{m-1}$ are also annihilated by $(D - r)^m$. To complete the proof we must show that $(D - r)^m$ annihilates u_m. Therefore we consider

$$(D - r)^m u_m = (D - r)^{m-1}(D - r)(x^{m-1}e^{rx}).$$

We have

$$\begin{aligned}(D - r)(x^{m-1}e^{rx}) &= D(x^{m-1}e^{rx}) - rx^{m-1}e^{rx}\\ &= (m - 1)x^{m-2}e^{rx} + x^{m-1}re^{rx} - rx^{m-1}e^{rx}\\ &= (m - 1)x^{m-2}e^{rx} = (m - 1)u_{m-1}(x).\end{aligned}$$

When we apply $(D - r)^{m-1}$ to both members of this last equation we get 0 on the right since $(D - r)^{m-1}$ annihilates u_{m-1}. Hence $(D - r)^m u_m = 0$ so u_m is annihilated by $(D - r)^m$. This completes the proof.

EXAMPLE 2. Find the general solution of the differential equation $L(y) = 0$, where $L = D^3 - D^2 - 8D + 12$.

Solution. The operator L has the factorization

$$L = (D - 2)^2(D + 3).$$

By Theorem 6.9, the two functions

$$u_1(x) = e^{2x}, \qquad u_2(x) = xe^{2x}$$

are in the null space of $(D - 2)^2$. The function $u_3(x) = e^{-3x}$ is in the null space of $(D + 3)$. Since u_1, u_2, u_3 are independent (see Exercise 17 of Section 6.9) they form a basis for the null space of L, so the general solution of the differential equation is

$$y = c_1 e^{2x} + c_2 x e^{2x} + c_3 e^{-3x}.$$

Theorem 6.9 tells us how to find a basis of solutions for any nth order linear equation with constant coefficients whose characteristic equation has only real roots, some of which are repeated. If the *distinct* roots are $r_1, r_2, \ldots, r_k$ and if they occur with respective multiplicities $m_1, m_2, \ldots, m_k$, that part of the basis corresponding to r_p is given by the m_p functions

$$u_{q,p}(x) = x^{q-1}e^{r_p x}, \qquad \text{where} \quad q = 1, 2, \ldots, m_p.$$

As p takes the values $1, 2, \ldots, k$ we get $m_1 + \cdots + m_k$ functions altogether. In Exercise 17 of Section 6.9 we outline a proof showing that all these functions are independent. Since the sum of the multiplicities $m_1 + \cdots + m_k$ is equal to n, the order of the equation, the functions $u_{p,q}$ form a basis for the solution space of the equation.

EXAMPLE 3. Solve the equation $(D^6 + 2D^5 - 2D^3 - D^2)y = 0$.

Solution. We have $D^6 + 2D^5 - 2D^3 - D^2 = D^2(D - 1)(D + 1)^3$. The part of the basis corresponding to the factor D^2 is $u_1(x) = 1$, $u_2(x) = x$; the part corresponding to the factor $(D - 1)$ is $u_3(x) = e^x$; and the part corresponding to the factor $(D + 1)^3$ is $u_4(x) = e^{-x}$, $u_5(x) = xe^{-x}$, $u_6(x) = x^2e^{-x}$. The six functions $u_1, \ldots, u_6$ are independent so the general solution of the equation is

$$y = c_1 + c_2 x + c_3 e^x + (c_4 + c_5 x + c_6 x^2)e^{-x}.$$

CASE III. Complex roots.

If complex exponentials are used, there is no need to distinguish between real and complex roots of the characteristic equation of the differential equation $L(y) = 0$. If real-valued solutions are desired, we factor the operator L into linear and quadratic factors with real coefficients. Each pair of conjugate complex roots $\alpha + i\beta$, $\alpha - i\beta$ corresponds to a quadratic factor,

$$D^2 - 2\alpha D + \alpha^2 + \beta^2. \tag{6.18}$$

The null space of this second-order operator contains the two independent functions $u(x) = e^{\alpha x}\cos \beta x$ and $v(x) = e^{\alpha x}\sin \beta x$. If the pair of roots $\alpha \pm i\beta$ occurs with multiplicity m, the quadratic factor occurs to the mth power. The null space of the operator

$$[D^2 - 2\alpha D + \alpha^2 + \beta^2]^m$$

contains $2m$ independent functions,

$$u_q(x) = x^{q-1}e^{\alpha x}\cos\beta x, \qquad v_q(x) = x^{q-1}e^{\alpha x}\sin\beta x, \qquad q = 1, 2, \ldots, m.$$

These facts can be easily proved by induction on m. (Proofs are outlined in Exercise 20 of Section 6.9.) The following examples illustrate some of the possibilities.

EXAMPLE 4. $y''' - 4y'' + 13y' = 0$. The characteristic equation, $r^3 - 4r^2 + 13r = 0$, has the roots $0, 2 \pm 3i$; the general solution is

$$y = c_1 + e^{2x}(c_2\cos 3x + c_3\sin 3x).$$

EXAMPLE 5. $y''' - 2y'' + 4y' - 8y = 0$. The characteristic equation is

$$r^3 - 2r^2 + 4r - 8 = (r - 2)(r^2 + 4) = 0;$$

its roots are $2, 2i, -2i$, so the general solution of the differential equation is

$$y = c_1e^{2x} + c_2\cos 2x + c_3\sin 2x.$$

EXAMPLE 6. $y^{(5)} - 9y^{(4)} + 34y''' - 66y'' + 65y' - 25y = 0$. The characteristic equation can be written as

$$(r - 1)(r^2 - 4r + 5)^2 = 0;$$

its roots are $1, 2 \pm i, 2 \pm i$, so the general solution of the differential equation is

$$y = c_1e^x + e^{2x}[(c_2 + c_3x)\cos x + (c_4 + c_5x)\sin x].$$

6.9 Exercises

Find the general solution of each of the differential equations in Exercises 1 through 12.

1. $y''' - 2y'' - 3y' = 0$.
2. $y''' - y' = 0$.
3. $y''' + 4y'' + 4y' = 0$.
4. $y''' - 3y'' + 3y' - y = 0$.
5. $y^{(4)} + 4y''' + 6y'' + 4y' + y = 0$.
6. $y^{(4)} - 16y = 0$.
7. $y^{(4)} + 16y = 0$.
8. $y''' - y = 0$.
9. $y^{(4)} + 4y''' + 8y'' + 8y' + 4y = 0$.
10. $y^{(4)} + 2y'' + y = 0$.
11. $y^{(6)} + 4y^{(4)} + 4y'' = 0$.
12. $y^{(6)} + 8y^{(4)} + 16y'' = 0$.
13. If m is a positive constant, find that particular solution $y = f(x)$ of the differential equation

$$y''' - my'' + m^2y' - m^3y = 0$$

which satisfies the condition $f(0) = f'(0) = 0, f''(0) = 1$.

14. A linear differential equation with constant coefficients has characteristic equation $f(r) = 0$. If all the roots of the characteristic equation are negative, prove that every solution of the differential equation approaches zero as $x \to +\infty$. What can you conclude about the behavior of all solutions on the interval $[0, +\infty)$ if all the roots of the characteristic equation are nonpositive?

15. In each case, find a linear differential equation with constant coefficients satisfied by all the given functions.
(a) $u_1(x) = e^x$, $u_2(x) = e^{-x}$, $u_3(x) = e^{2x}$, $u_4(x) = e^{-2x}$.
(b) $u_1(x) = e^{-2x}$, $u_2(x) = xe^{-2x}$, $u_3(x) = x^2e^{-2x}$.
(c) $u_1(x) = 1$, $u_2(x) = x$, $u_3(x) = e^x$, $u_4(x) = xe^x$.
(d) $u_1(x) = x$, $u_2(x) = e^x$, $u_3(x) = xe^x$.
(e) $u_1(x) = x^2$, $u_2(x) = e^x$, $u_3(x) = xe^x$.
(f) $u_1(x) = e^{-2x}\cos 3x$, $u_2(x) = e^{-2x}\sin 3x$, $u_3(x) = e^{-2x}$, $u_4(x) = xe^{-2x}$.
(g) $u_1(x) = \cosh x$, $u_2(x) = \sinh x$, $u_3(x) = x\cosh x$, $u_4(x) = x\sinh x$.
(h) $u_1(x) = \cosh x \sin x$, $u_2(x) = \sinh x \cos x$, $u_3(x) = x$.
16. Let $r_1, \ldots, r_n$ be n distinct real numbers, and let $Q_1, \ldots, Q_n$ be n polynomials, none of which is the zero polynomial. Prove that the n functions

$$u_1(x) = Q_1(x)e^{r_1x}, \ldots, u_n(x) = Q_n(x)e^{r_nx}$$

are independent.

Outline of proof. Use induction on n. For $n = 1$ and $n = 2$ the result is easily verified. Assume the statement is true for $n = p$ and let $c_1, \ldots, c_{p+1}$ be $p + 1$ real scalars such that

$$\sum_{k=1}^{p+1} c_kQ_k(x)e^{r_kx} = 0.$$

Multiply both sides by $e^{-r_{p+1}x}$ and differentiate the resulting equation. Then use the induction hypothesis to show that all the scalars c_k are 0. An alternate proof can be given based on order of magnitude as $x \to +\infty$, as was done in Example 7 of Section 1.7 (p. 10).
17. Let $m_1, m_2, \ldots, m_k$ be k positive integers, let $r_1, r_2, \ldots, r_k$ be k distinct real numbers, and let $n = m_1 + \cdots + m_k$. For each pair of integers p, q satisfying $1 \le p \le k$, $1 \le q \le m_p$, let

$$u_{q,p}(x) = x^{q-1}e^{r_px}.$$

For example, when $p = 1$ the corresponding functions are

$$u_{1,1}(x) = e^{r_1x}, \qquad u_{2,1}(x) = xe^{r_1x}, \ldots, u_{m_1,1}(x) = x^{m_1-1}e^{r_1x}.$$

Prove that the n functions $u_{q,p}$ so defined are independent. [*Hint:* Use Exercise 16.]
18. Let L be a constant-coefficient linear differential operator of order n with characteristic polynomial $p(r)$. Let L' be the constant-coefficient operator whose characteristic polynomial is the derivative $p'(r)$. For example, if $L = 2D^2 - 3D + 1$ then $L' = 4D - 3$. More generally, define the mth derivative $L^{(m)}$ to be the operator whose characteristic polynomial is the mth derivative $p^{(m)}(r)$. (The operator $L^{(m)}$ should not be confused with the mth power L^m.)
(a) If u has n derivatives, prove that

$$L(u) = \sum_{k=0}^{n} \frac{p^{(k)}(0)}{k!} u^{(k)}.$$

(b) If u has $n - m$ derivatives, prove that

$$L^{(m)}(u) = \sum_{k=0}^{n-m} \frac{p^{(k+m)}(0)}{k!} u^{(k)} \qquad \text{for} \qquad m = 0, 1, 2, \ldots, n,$$

where $L^{(0)} = L$.

19. Refer to the notation of Exercise 18. If u and v have n derivatives, prove that

$$L(uv) = \sum_{k=0}^{n} \frac{L^{(k)}(u)}{k!} v^{(k)} .$$

[*Hint:* Use Exercise 18 along with Leibniz's formula for the kth derivative of a product:

$$(uv)^{(k)} = \sum_{r=0}^{k} \binom{k}{r} u^{(k-r)} v^{(r)} .]$$

20. (a) Let $p(t) = q(t)^m r(t)$, where q and r are polynomials and m is a positive integer. Prove that $p'(t) = q(t)^{m-1}s(t)$, where s is a polynomial.
(b) Let L be a constant-coefficient operator which annihilates u, where u is a given function of x. Let $M = L^m$, the mth power of L, where $m > 1$. Prove that each of the derivatives M', M'', . . . , $M^{(m-1)}$ also annihilates u.
(c) Use part (b) and Exercise 19 to prove that M annihilates each of the functions $u, xu, \ldots,$ $x^{m-1}u$.
(d) Use part (c) to show that the operator $(D^2 - 2\alpha D + \alpha^2 + \beta^2)^m$ annihilates each of the functions $x^q e^{\alpha x} \sin \beta x$ and $x^q e^{\alpha x} \cos \beta x$ for $q = 1, 2, \ldots, m - 1$.
21. Let L be a constant-coefficient operator of order n with characteristic polynomial $p(r)$. If α is constant and if u has n derivatives, prove that

$$L(e^{\alpha x}u(x)) = e^{\alpha x} \sum_{k=0}^{n} \frac{p^{(k)}(\alpha)}{k!} u^{(k)}(x) .$$

6.10 The relation between the homogeneous and nonhomogeneous equations

We return now to the general linear differential equation of order n with coefficients that are not necessarily constant. The next theorem describes the relation between solutions of a homogeneous equation $L(y) = 0$ and those of a nonhomogeneous equation $L(y) = R(x)$.

THEOREM 6.10. *Let $L : \mathscr{C}_n(J) \to \mathscr{C}(J)$ be a linear differential operator of order n. Let $u_1, \ldots,$ u_n be n independent solutions of the homogeneous equation $L(y) = 0$, and let y_1 be a particular solution of the nonhomogeneous equation $L(y) = R$, where $R \in \mathscr{C}(J)$. Then every solution $y = f(x)$ of the nonhomogeneous equation has the form*

$$f(x) = y_1(x) + \sum_{k=1}^{n} c_k u_k(x), \tag{6.19}$$

where $c_1, \ldots, c_n$ are constants.

Proof. By linearity we have $L(f - y_1) = L(f) - L(y_1) = R - R = 0$. Therefore $f - y_1$ is in the solution space of the homogeneous equation $L(y) = 0$, so $f - y_1$ is a linear combination of $u_1, \ldots, u_n$, say $f - y_1 = c_1u_1 + \cdots + c_nu_n$. This proves (6.19).

Since all solutions of $L(y) = R$ are found in (6.19), the sum on the right of (6.19) (with arbitrary constants $c_1, c_2, \ldots, c_n$) is called the general solution of the nonhomogeneous equation. Theorem 6.10 states that the general solution of the nonhomogeneous equation is obtained by adding to y_1 the general solution of the homogeneous equation.

Note: Theorem 6.10 has a simple geometric analogy which helps give an insight into its meaning. To determine all points on a plane we find a particular point on the plane and add to it all points on the parallel plane through the origin. To find all solutions of $L(y) = R$ we find a particular solution and add to it all solutions of the homogeneous equation $L(y) = 0$. The set of solutions of the nonhomogeneous equation is analogous to a plane through a particular point. The solution space of the homogeneous equation is analogous to a parallel plane through the origin.

To use Theorem 6.10 in practice we must solve two problems: (1) Find the general solution of the homogeneous equation $L(y) = 0$, and (2) find a particular solution of the nonhomogeneous equation $L(y) = R$. In the next section we show that we can always solve problem (2) if we can solve problem (1).

6.11 Determination of a particular solution of the nonhomogeneous equation. The method of variation of parameters

We turn now to the problem of determining one particular solution y_1 of the nonhomogeneous equation $L(y) = R$. We shall describe a method known as *variation of parameters* which tells us how to determine y_1 if we know n independent solutions $u_1, \ldots, u_n$ of the homogeneous equation $L(y) = 0$. The method provides a particular solution of the form

$$y_1 = v_1 u_1 + \cdots + v_n u_n, \tag{6.20}$$

where $v_1, \ldots, v_n$ are functions that can be calculated in terms of $u_1, \ldots, u_n$ and the right-hand member R. The method leads to a system of n linear algebraic equations satisfied by the derivatives $v_1', \ldots, v_n'$. This system can always be solved because it has a nonsingular coefficient matrix. Integration of the derivatives then gives the required functions $v_1, \ldots, v_n$. The method was first used by Johann Bernoulli to solve linear equations of first order, and then by Lagrange in 1774 to solve linear equations of second order.

For the nth order case the details can be simplified by using vector and matrix notation. The right-hand member of (6.20) can be written as an inner product,

$$y_1 = (v, u), \tag{6.21}$$

where v and u are n-dimensional vector functions given by

$$v = (v_1, \ldots, v_n), \qquad u = (u_1, \ldots, u_n).$$

We try to choose v so that the inner product defining y_1 will satisfy the nonhomogeneous equation $L(y) = R$, given that $L(u) = 0$, where $L(u) = (L(u_1), \ldots, L(u_n))$.

We begin by calculating the first derivative of y_1. We find

$$y_1' = (v, u') + (v', u). \tag{6.22}$$

We have n functions $v_1, \ldots, v_n$ to determine, so we should be able to put n conditions on them. If we impose the condition that the second term on the right of (6.22) should vanish, the formula for y_1' simplifies to

$$y_1' = (v, u'), \qquad \text{provided that } (v', u) = 0.$$

Differentiating the relation for y_1' we find

$$y_1'' = (v, u'') + (v', u').$$

If we can choose v so that $(v', u') = 0$ then the formula for y_1'' also simplifies and we get

$$y_1'' = (v, u''), \qquad \text{provided that also } (v', u') = 0.$$

If we continue in this manner for the first $n - 1$ derivatives of y_1 we find

$$y_1^{(n-1)} = (v, u^{(n-1)}), \qquad \text{provided that also } (v', u^{(n-2)}) = 0.$$

So far we have put $n - 1$ conditions on v. Differentiating once more we get

$$y_1^{(n)} = (v, u^{(n)}) + (v', u^{(n-1)}).$$

This time we impose the condition $(v', u^{(n-1)}) = R(x)$, and the last equation becomes

$$y_1^{(n)} = (v, u^{(n)}) + R(x), \qquad \text{provided that also } (v', u^{(n-1)}) = R(x).$$

Suppose, for the moment, that we can satisfy the n conditions imposed on v. Let $L = D^n + P_1(x)D^{n-1} + \cdots + P_n(x)$. When we apply L to y_1 we find

$$\begin{aligned} L(y_1) &= y_1^{(n)} + P_1(x)y_1^{(n-1)} + \cdots + P_n(x)y_1 \\ &= \{(v, u^{(n)}) + R(x)\} + P_1(x)(v, u^{(n-1)}) + \cdots + P_n(x)(v, u) \\ &= (v, L(u)) + R(x) = (v, 0) + R(x) = R(x). \end{aligned}$$

Thus $L(y_1) = R(x)$, so y_1 is a solution of the nonhomogeous equation.

The method will succeed if we can satisfy the n conditions we have imposed on v. These conditions state that $(v', u^{(k)}) = 0$ for $k = 0, 1, \ldots, n - 2$, and that $(v', u^{(n-1)}) = R(x)$. We can write these n equations as a single matrix equation,

$$W(x)v'(x) = R(x)\begin{bmatrix} 0 \\ \vdots \\ 0 \\ 1 \end{bmatrix}, \tag{6.23}$$

where $v'(x)$ is regarded as an $n \times 1$ column matrix, and where W is the $n \times n$ matrix function whose rows consist of the components of u and its successive derivatives:

$$W = \begin{bmatrix} u_1 & u_2 & \cdots & u_n \\ u_1' & u_2' & \cdots & u_n' \\ \cdot & & & \cdot \\ \cdot & & & \cdot \\ \cdot & & & \cdot \\ u_1^{(n-1)} & u_2^{(n-1)} & \cdots & u_n^{(n-1)} \end{bmatrix}.$$

The matrix W is called the *Wronskian matrix* of $u_1, \ldots, u_n$, after J. M. H. Wronski (1778–1853).

In the next section we shall prove that the Wronskian matrix is nonsingular. Therefore we can multiply both sides of (6.23) by $W(x)^{-1}$ to obtain

$$v'(x) = R(x)W(x)^{-1}\begin{bmatrix} 0 \\ \vdots \\ 0 \\ 1 \end{bmatrix}.$$

Choose two points c and x in the interval J under consideration and integrate this vector equation over the interval from c to x to obtain

$$v(x) = v(c) + \int_c^x R(t)W(t)^{-1}\begin{bmatrix} 0 \\ \vdots \\ 0 \\ 1 \end{bmatrix} dt = v(c) + z(x),$$

where

$$z(x) = \int_c^x R(t)W(t)^{-1}\begin{bmatrix} 0 \\ \vdots \\ 0 \\ 1 \end{bmatrix} dt.$$

The formula $y_1 = (u, v)$ for the particular solution now becomes

$$y_1 = (u, v) = (u, v(c) + z) = (u, v(c)) + (u, z).$$

The first term $(u, v(c))$ satisfies the homogeneous equation since it is a linear combination of $u_1, \ldots, u_n$. Therefore we can omit this term and use the second term (u, z) as a particular solution of the nonhomogeneous equation. In other words, a particular solution of

$L(y) = R$ is given by the inner product

$$(u(x), z(x)) = \left(u(x), \int_c^x R(t)W(t)^{-1}\begin{bmatrix}0\\ \vdots\\ 0\\ 1\end{bmatrix} dt\right).$$

Note that it is not necessary that the function R be continuous on the interval J. All that is required is that R be integrable on $[c, x]$.

We can summarize the results of this section by the following theorem.

THEOREM 6.11. *Let $u_1, \ldots, u_n$ be n independent solutions of the homogeneous nth order linear differential equation $L(y) = 0$ on an interval J. Then a particular solution y_1 of the nonhomogeneous equation $L(y) = R$ is given by the formula*

$$y_1(x) = \sum_{k=1}^{n} u_k(x)v_k(x),$$

where $v_1, \ldots, v_n$ are the entries in the $n \times 1$ column matrix v determined by the equation

$$v(x) = \int_c^x R(t)W(t)^{-1}\begin{bmatrix}0\\ \vdots\\ 0\\ 1\end{bmatrix} dt. \tag{6.24}$$

In this formula, W is the Wronskian matrix of $u_1, \ldots, u_n$, and c is any point in J.

Note: The definite integral in (6.24) can be replaced by any indefinite integral

$$\int R(x)W(x)^{-1}\begin{bmatrix}0\\ \vdots\\ 0\\ 1\end{bmatrix} dx.$$

EXAMPLE 1. Find the general solution of the differential equation

$$y'' - y = \frac{2}{1 + e^x}$$

on the interval $(-\infty, +\infty)$.

Solution. The homogeneous equation, $(D^2 - 1)y = 0$ has the two independent solutions $u_1(x) = e^x$, $u_2(x) = e^{-x}$. The Wronskian matrix of u_1 and u_2 is

$$W(x) = \begin{bmatrix} e^x & e^{-x} \\ e^x & -e^{-x}\end{bmatrix}.$$

Since $\det W(x) = -2$, the matrix is nonsingular and its inverse is given by

$$W(x)^{-1} = -\frac{1}{2}\begin{bmatrix} -e^{-x} & -e^{-x} \\ -e^{x} & e^{x} \end{bmatrix}.$$

Therefore

$$W(x)^{-1}\begin{bmatrix} 0 \\ 1 \end{bmatrix} = -\frac{1}{2}\begin{bmatrix} -e^{-x} \\ e^{x} \end{bmatrix}$$

and we have

$$R(x)W(x)^{-1}\begin{bmatrix} 0 \\ 1 \end{bmatrix} = -\frac{1}{2}\,\frac{2}{1+e^{x}}\begin{bmatrix} -e^{-x} \\ e^{x} \end{bmatrix} = \begin{bmatrix} \dfrac{e^{-x}}{1+e^{x}} \\[2ex] \dfrac{-e^{x}}{1+e^{x}} \end{bmatrix}.$$

Integrating each component of the vector on the right we find

$$v_1(x) = \int \frac{e^{-x}}{1+e^{x}}\,dx = \int \left(e^{-x} - 1 + \frac{e^{x}}{1+e^{x}}\right) dx = -e^{-x} - x + \log(1+e^{x})$$

and

$$v_2(x) = \int \frac{-e^{x}}{1+e^{x}}\,dx = -\log(1+e^{x}).$$

Therefore the general solution of the differential equation is

$$\begin{aligned} y &= c_1u_1(x) + c_2u_2(x) + v_1(x)u_1(x) + v_2(x)u_2(x) \\ &= c_1e^{x} + c_2e^{-x} - 1 - xe^{x} + (e^{x} - e^{-x})\log(1+e^{x}). \end{aligned}$$

6.12 Nonsingularity of the Wronskian matrix of *n* independent solutions of a homogeneous linear equation

In this section we prove that the Wronskian matrix W of n independent solutions $u_1, \ldots, u_n$ of a homogeneous equation $L(y) = 0$ is nonsingular. We do this by proving that the determinant of W is an exponential function which is never zero on the interval J under consideration.

Let $w(x) = \det W(x)$ for each x in J, and assume that the differential equation satisfied by $u_1, \ldots, u_n$ has the form

$$y^{(n)} + P_1(x)y^{(n-1)} + \cdots + P_n(x)y = 0. \tag{6.25}$$

Then we have:

THEOREM 6.12. *The Wronskian determinant satisfies the first-order differential equation*

$$w' + P_1(x)w = 0 \tag{6.26}$$

on J. Therefore, if $c \in J$ we have

(6.27) $$w(x) = w(c) \exp\left[-\int_c^x P_1(t)\,dt\right] \qquad \textit{(Abel's formula).}$$

Moreover, $w(x) \neq 0$ for all x in J.

Proof. Let u be the row-vector $u = (u_1, \ldots, u_n)$. Since each component of u satisfies the differential equation (6.25) the same is true of u. The rows of the Wronskian matrix W are the vectors $u, u', \ldots, u^{(n-1)}$. Hence we can write

$$w = \det W = \det (u, u', \ldots, u^{(n-1)}).$$

The derivative of w is the determinant of the matrix obtained by differentiating the last row of W (see Exercise 8 of Section 3.17). That is

$$w' = \det (u, u', \ldots, u^{(n-2)}, u^{(n)}).$$

Multiplying the last row of w by $P_1(x)$ we also have

$$P_1(x)w = \det (u, u', \ldots, u^{(n-2)}, P_1(x)u^{(n-1)}).$$

Adding these last two equations we find

$$w' + P_1(x)w = \det (u, u', \ldots, u^{(n-2)}, u^{(n)} + P_1(x)u^{(n-1)}).$$

But the rows of this last determinant are dependent since u satisfies the differential equation (6.25). Therefore the determinant is zero, which means that w satisfies (6.26). Solving (6.26) we obtain Abel's formula (6.27).

Next we prove that $w(c) \neq 0$ for some c in J. We do this by a contradiction argument. Suppose that $w(t) = 0$ for all t in J. Choose a fixed t in J, say $t = t_0$, and consider the linear system of algebraic equations

$$W(t_0)X = O,$$

where X is a column vector. Since $\det W(t_0) = 0$, the matrix $W(t_0)$ is singular so this system has a nonzero solution, say $X = (c_1, \ldots, c_n) \neq (0, \ldots, 0)$. Using the components of this nonzero vector, let f be the linear combination

$$f(t) = c_1u_1(t) + \cdots + c_nu_n(t).$$

The function f so defined satisfies $L(f) = 0$ on J since it is a linear combination of $u_1, \ldots, u_n$. The matrix equation $W(t_0)X = O$ implies that

$$f(t_0) = f'(t_0) = \cdots = f^{(n-1)}(t_0) = 0.$$

Therefore f has the initial-value vector O at $t = t_0$ so, by the uniqueness theorem, f is the zero solution. This means $c_1 = \cdots = c_n = 0$, which is a contradiction. Therefore $w(t) \neq 0$ for some t in J. Taking c to be this t in Abel's formula we see that $w(x) \neq 0$ for all x in J. This completes the proof of Theorem 6.12.

6.13 Special methods for determining a particular solution of the nonhomogeneous equation. Reduction to a system of first-order linear equations

Although variation of parameters provides a general method for determining a particular solution of $L(y) = R$, special methods are available that are often easier to apply when the equation has certain special forms. For example, if the equation has constant coefficients we can reduce the problem to that of solving a succession of linear equations of first order. The general method is best illustrated with a simple example.

EXAMPLE 1. Find a particular solution of the equation

$$(D - 1)(D - 2)y = xe^{x+x^2}. \tag{6.28}$$

Solution. Let $u = (D - 2)y$. Then the equation becomes

$$(D - 1)u = xe^{x+x^2}.$$

This is a first-order linear equation in u which can be solved using Theorem 6.1. A particular solution is

$$u = \tfrac{1}{2}e^{x+x^2}.$$

Substituting this in the equation $u = (D - 2)y$ we obtain

$$(D - 2)y = \tfrac{1}{2}e^{x+x^2},$$

a first-order linear equation for y. Solving this by Theorem 6.1 we find that a particular solution (with $y_1(0) = 0$) is given by

$$y_1(x) = \tfrac{1}{2}e^{2x} \int_0^x e^{t^2-t}\,dt.$$

Although the integral cannot be evaluated in terms of elementary functions we consider the equation as having been solved, since the solution is expressed in terms of integrals of familiar functions. The general solution of (6.28) is

$$y = c_1e^x + c_2e^{2x} + \tfrac{1}{2}e^{2x} \int_0^x e^{t^2-t}\,dt.$$

6.14 The annihilator method for determining a particular solution of the nonhomogeneous equation

We describe next a method which can be used if the equation $L(y) = R$ has constant coefficients and if the right-hand member R is itself annihilated by a constant-coefficient operator, say $A(R) = 0$. In principle, the method is very simple. We apply the operator A to both members of the differential equation $L(y) = R$ and obtain a new equation $AL(y) = 0$ which must be satisfied by all solutions of the original equation. Since AL is another constant-coefficient operator we can determine its null space by calculating the roots of the characteristic equation of AL. Then the problem remains of choosing from this null space a particular function y_1 that satisfies $L(y_1) = R$. The following examples illustrate the process.

EXAMPLE 1. Find a particular solution of the equation

$$(D^4 - 16)y = x^4 + x + 1.$$

Solution. The right-hand member, a polynomial of degree 4, is annihilated by the operator D^5. Therefore any solution of the given equation is also a solution of the equation

$$D^5(D^4 - 16)y = 0. \tag{6.29}$$

The roots of the characteristic equation are 0, 0, 0, 0, 0, 2, -2, $2i$, $-2i$, so all the solutions of (6.29) are to be found in the linear combination

$$y = c_1 + c_2x + c_3x^2 + c_4x^3 + c_5x^4 + c_6e^{2x} + c_7e^{-2x} + c_8\cos 2x + c_9\sin 2x.$$

We want to choose the c_i so that $L(y) = x^4 + x + 1$, where $L = D^4 - 16$. Since the last four terms are annihilated by L, we can take $c_6 = c_7 = c_8 = c_9 = 0$ and try to find $c_1, \ldots, c_5$ so that

$$L(c_1 + c_2x + c_3x^2 + c_4x^3 + c_5x^4) = x^4 + x + 1.$$

In other words, we seek a particular solution y_1 which is a polynomial of degree 4 satisfying $L(y_1) = x^4 + x + 1$. To simplify the algebra we write

$$16y_1 = ax^4 + bx^3 + cx^2 + dx + e.$$

This gives us $16y_1^{(4)} = 24a$, so $y_1^{(4)} = 3a/2$. Substituting in the differential equation $L(y_1) = x^4 + x + 1$, we must determine a, b, c, d, e to satisfy

$$\tfrac{3}{2}a - ax^4 - bx^3 - cx^2 - dx - e = x^4 + x + 1.$$

Equating coefficients of like powers of x we obtain

$$a = -1, \qquad b = c = 0, \qquad d = -1, \qquad e = -\tfrac{5}{2},$$

so the particular solution y_1 is given by

$$y_1 = -\tfrac{1}{16}x^4 - \tfrac{1}{16}x - \tfrac{5}{32}.$$

EXAMPLE 2. Solve the differential equation $y'' - 5y' + 6y = xe^x$.

Solution. The differential equation has the form

$$L(y) = R, \tag{6.30}$$

where $R(x) = xe^x$ and $L = D^2 - 5D + 6$. The corresponding homogeneous equation can be written as

$$(D - 2)(D - 3)y = 0\,;$$

it has the independent solutions $u_1(x) = e^{2x}$, $u_2(x) = e^{3x}$. Now we seek a particular solution y_1 of the nonhomogeneous equation. We recognize the function $R(x) = xe^x$ as a solution of the homogeneous equation

$$(D - 1)^2 y = 0\,.$$

Therefore, if we operate on both sides of (6.30) with the operator $(D - 1)^2$ we find that any function which satisfies (6.30) must also satisfy the equation

$$(D - 1)^2(D - 2)(D - 3)y = 0\,.$$

This differential equation has the characteristic roots $1, 1, 2, 3$, so all its solutions are to be found in the linear combination

$$y = ae^x + bxe^x + ce^{2x} + de^{3x}\,,$$

where a, b, c, d are constants. We want to choose a, b, c, d so that the resulting solution y_1 satisfies $L(y_1) = xe^x$. Since $L(ce^{2x} + de^{3x}) = 0$ for every choice of c and d, we need only choose a and b so that $L(ae^x + bxe^x) = xe^x$ and take $c = d = 0$. If we put

$$y_1 = ae^x + bxe^x\,,$$

we have

$$D(y_1) = (a + b)e^x + bxe^x\,, \qquad D^2(y_1) = (a + 2b)e^x + bxe^x\,,$$

so the equation $(D^2 - 5D + 6)y_1 = xe^x$ becomes

$$(2a - 3b)e^x + 2bxe^x = xe^x\,.$$

Canceling e^x and equating coefficients of like powers of x we find $a = \frac{3}{4}$, $b = \frac{1}{2}$. Therefore $y_1 = \frac{3}{4}e^x + \frac{1}{2}xe^x$ and the general solution of $L(y) = R$ is given by the formula

$$y = c_1e^{2x} + c_2e^{3x} + \tfrac{3}{4}e^x + \tfrac{1}{2}xe^x\,.$$

The method used in the foregoing examples is called the *annihilator method.* It will always work if we can find a constant coefficient operator A that annihilates R. From our knowledge of homogeneous linear differential equations with constant coefficients, we know that the only real-valued functions annihilated by constant-coefficient operators are linear combinations of terms of the form

$$x^{m-1}e^{\alpha x}\,, \qquad x^{m-1}e^{\alpha x}\cos\beta x\,, \qquad x^{m-1}e^{\alpha x}\sin\beta x\,,$$

where m is a positive integer and α and β are real constants. The function $y = x^{m-1}e^{\alpha x}$ is a solution of a differential equation with a characteristic root α having multiplicity m.

Therefore, this function has the annihilator $(D - \alpha)^m$. Each of the functions $y = x^{m-1}e^{\alpha x}\cos \beta x$ and $y = x^{m-1}e^{\alpha x}\sin \beta x$ is a solution of a differential equation with complex characteristic roots $\alpha \pm i\beta$, each occurring with multiplicity m, so they are annihilated by the operator $[D^2 - 2\alpha D + (\alpha^2 + \beta^2)]^m$. For ease of reference, we list these annihilators in Table 6.1, along with some of their special cases.

TABLE 6.1

Function	*Annihilator*
$y = x^{m-1}$	D^m
$y = e^{\alpha x}$	$D - \alpha$
$y = x^{m-1}e^{\alpha x}$	$(D - \alpha)^m$
$y = \cos \beta x$ or $y = \sin \beta x$	$D^2 + \beta^2$
$y = x^{m-1}\cos \beta x$ or $y = x^{m-1}\sin \beta x$	$(D^2 + \beta^2)^m$
$y = e^{\alpha x}\cos \beta x$ or $y = e^{\alpha x}\sin \beta x$	$D^2 - 2\alpha D + (\alpha^2 + \beta^2)$
$y = x^{m-1}e^{\alpha x}\cos \beta x$ or $y = x^{m-1}e^{\alpha x}\sin \beta x$	$[D^2 - 2\alpha D + (\alpha^2 + \beta^2)]^m$

Although the annihilator method is very efficient when applicable, it is limited to equations whose right members R have a constant-coefficient annihilator. If $R(x)$ has the form e^{x^2}, $\log x$, or $\tan x$, the method will not work; we must then use variation of parameters or some other method to find a particular solution.

6.15 Exercises

In each of Exercises 1 through 10, find the general solution on the interval $(-\infty, +\infty)$.

1. $y'' - y' = x^2$.
2. $y'' - 4y = e^{2x}$.
3. $y'' + 2y' = 3xe^x$.
4. $y'' + 4y = \sin x$.
5. $y'' - 2y' + y = e^x + e^{2x}$.
6. $y''' - y' = e^x$.
7. $y''' - y' = e^x + e^{-x}$.
8. $y''' + 3y'' + 3y' + y = xe^{-x}$.
9. $y'' + y = xe^x \sin 2x$.
10. $y^{(4)} - y = x^2e^{-x}$.
11. If a constant-coefficient operator A annihilates f and if a constant-coefficient operator B annihilates g, show that the product AB annihilates $f + g$.
12. Let A be a constant-coefficient operator with characteristic polynomial p_A.
(a) Use the annihilator method to prove that the differential equation $A(y) = e^{\alpha x}$ has a particular solution of the form

$$y_1 = \frac{e^{\alpha x}}{p_A(\alpha)}$$

if α is not a zero of the polynomial p_A.
(b) If α is a simple zero of p_A (multiplicity 1), prove that the equation $A(y) = e^{\alpha x}$ has the particular solution

$$y_1 = \frac{xe^{\alpha x}}{p'_A(\alpha)}.$$

(c) Generalize the results of (a) and (b) when α is a zero of p_A with multiplicity m.

13. Given two constant-coefficient operators A and B whose characteristic polynomials have no zeros in common. Let $C = AB$.
(a) Prove that every solution of the differential equation $C(y) = 0$ has the form $y = y_1 + y_2$, where $A(y_1) = 0$ and $B(y_2) = 0$.
(b) Prove that the functions y_1 and y_2 in part (a) are uniquely determined. That is, for a given y satisfying $C(y) = 0$ there is only one pair y_1, y_2 with the properties in part (a).
14. If $L(y) = y'' + ay' + by$, where a and b are constants, let f be that particular solution of $L(y) = 0$ satisfying the conditions $f(0) = 0$ and $f'(0) = 1$. Show that a particular solution of $L(y) = R$ is given by the formula

$$y_1(x) = \int_c^x f(x - t)R(t)\,dt$$

for any choice of c. In particular, if the roots of the characteristic equation are equal, say $r_1 = r_2 = m$, show that the formula for $y_1(x)$ becomes

$$y_1(x) = e^{mx}\int_c^x (x - t)e^{-mt}R(t)\,dt.$$

15. Let Q be the operator "multiplication by x." That is, $Q(y)(x) = x \cdot y(x)$ for each y in class $\mathscr{C}^\infty$ and each real x. Let I denote the identity operator, defined by $I(y) = y$ for each y in $\mathscr{C}^\infty$.
(a) Prove that $DQ - QD = I$.
(b) Show that $D^2Q - QD^2$ is a constant-coefficient operator of first order, and determine this operator explicitly as a linear polynomial in D.
(c) Show that $D^3Q - QD^3$ is a constant-coefficient operator of second order, and determine this operator explicitly as a quadratic polynomial in D.
(d) Guess the generalization suggested for the operator $D^nQ - QD^n$, and prove your result by induction.

In each of Exercises 16 through 20, find the general solution of the differential equation in the given interval.

16. $y'' - y = 1/x, \qquad (0, +\infty).$
17. $y'' + 4y = \sec 2x, \qquad \left(-\frac{\pi}{4}, \frac{\pi}{4}\right).$
18. $y'' - y = \sec^3 x - \sec x, \qquad \left(-\frac{\pi}{2}, \frac{\pi}{2}\right).$
19. $y'' - 2y' + y = e^{e^x}(e^x - 1)^2, \qquad (-\infty, +\infty).$
20. $y''' - 7y'' + 14y' - 8y = \log x, \qquad (0, +\infty).$

6.16 Miscellaneous exercises on linear differential equations

1. An integral curve $y = u(x)$ of the differential equation $y'' - 3y' - 4y = 0$ intersects an integral curve $y = v(x)$ of the differential equation $y'' + 4y' - 5y = 0$ at the origin. Determine the functions u and v if the two curves have equal slopes at the origin and if

$$\lim_{x\to\infty}\frac{[v(x)]^4}{u(x)} = \frac{5}{6}.$$

2. An integral curve $y = u(x)$ of the differential equation $y'' - 4y' + 29y = 0$ intersects an integral curve $y = v(x)$ of the differential equation $y'' + 4y' + 13y = 0$ at the origin. The two curves have equal slopes at the origin. Determine u and v if $u'(\pi/2) = 1$.

3. Given that the differential equation $y'' + 4xy' + Q(x)y = 0$ has two solutions of the form $y_1 = u(x)$ and $y_2 = xu(x)$, where $u(0) = 1$. Determine both $u(x)$ and $Q(x)$ explicitly in terms of x.
4. Let $L(y) = y'' + P_1y' + P_2y$. To solve the nonhomogeneous equation $L(y) = R$ by variation of parameters, we need to know two linearly independent solutions of the homogeneous equation. This exercise shows that if *one* solution u_1 of $L(y) = 0$ is known, and if u_1 is never zero on an interval J, a second solution u_2 of the homogeneous equation is given by the formula

$$u_2(x) = u_1(x) \int_c^x \frac{Q(t)}{[u_1(t)]^2}\,dt,$$

where $Q(x) = e^{-\int P_1(x)\,dx}$, and c is any point in J. These two solutions are independent on J.
(a) Prove that the function u_2 does, indeed, satisfy $L(y) = 0$.
(b) Prove that u_1 and u_2 are independent on J.
5. Find the general solution of the equation

$$xy'' - 2(x + 1)y' + (x + 2)y = x^3e^{2x}$$

for $x > 0$, given that the homogeneous equation has a solution of the form $y = e^{mx}$.
6. Obtain one nonzero solution by inspection and then find the general solution of the differential equation

$$(y'' - 4y') + x^2(y' - 4y) = 0.$$

7. Find the general solution of the differential equation

$$4x^2y'' + 4xy' - y = 0,$$

given that there is a particular solution of the form $y = x^m$ for $x > 0$.
8. Find a solution of the homogeneous equation by trial, and then find the general solution of the equation

$$x(1 - x)y'' - (1 - 2x)y' + (x^2 - 3x + 1)y = (1 - x)^3.$$

9. Find the general solution of the equation

$$(2x - 3x^3)y'' + 4y' + 6xy = 0,$$

given that it has a solution that is a polynomial in x.
10. Find the general solution of the equation

$$x^2(1 - x)y'' + 2x(2 - x)y' + 2(1 + x)y = x^2,$$

given that the homogeneous equation has a solution of the form $y = x^c$.
11. Let $g(x) = \int_1^x e^t/t\,dt$ if $x > 0$. (Do not attempt to evaluate this integral.) Find all values of the constant a such that the function f defined by

$$f(x) = \frac{1}{x}e^{ag(x)}$$

satisfies the linear differential equation

$$x^2y'' + (3x - x^2)y' + (1 - x - e^{2x})y = 0.$$

Use this information to determine the general solution of the equation on the interval $(0, +\infty)$.

6.17 Linear equations of second order with analytic coefficients

A function f is said to be analytic on an interval $(x_0 - r, x_0 + r)$ if f has a power-series expansion in this interval,

$$f(x) = \sum_{n=0}^{\infty} a_n(x - x_0)^n ,$$

convergent for $|x - x_0| < r$. If the coefficients of a homogeneous linear differential equation

$$y^{(n)} + P_1(x)y^{(n-1)} + \cdots + P_n(x)y = 0$$

are analytic in an interval $(x_0 - r, x_0 + r)$, then it can be shown that there exist n independent solutions $u_1, \ldots, u_n$, each of which is analytic on the same interval. We shall prove this theorem for equations of second order and then discuss an important example that occurs in many applications.

THEOREM 6.13. *Let P_1 and P_2 be analytic on an open interval $(x_0 - r, x_0 + r)$, say*

$$P_1(x) = \sum_{n=0}^{\infty} b_n(x - x_0)^n , \qquad P_2(x) = \sum_{n=0}^{\infty} c_n(x - x_0)^n .$$

Then the differential equation

$$y'' + P_1(x)y' + P_2(x)y = 0 \tag{6.31}$$

has two independent solutions u_1 and u_2 which are analytic on the same interval.

Proof. We try to find a power-series solution of the form

$$y = \sum_{n=0}^{\infty} a_n(x - x_0)^n , \tag{6.32}$$

convergent in the given interval. To do this, we substitute the given series for P_1 and P_2 in the differential equation and then determine relations which the coefficients a_n must satisfy so that the function y given by (6.32) will satisfy the equation.

The derivatives y' and y'' can be obtained by differentiating the power series for y term by term (see Theorem 11.9 in Volume I). This gives us

$$y' = \sum_{n=1}^{\infty} na_n(x - x_0)^{n-1} = \sum_{n=0}^{\infty} (n + 1)a_{n+1}(x - x_0)^n ,$$

$$y'' = \sum_{n=2}^{\infty} n(n - 1)a_n(x - x_0)^{n-2} = \sum_{n=0}^{\infty} (n + 2)(n + 1)a_{n+2}(x - x_0)^n .$$

The products $P_1(x)y'$ and $P_2(x)y$ are given by the power series†

$$P_1(x)y' = \sum_{n=0}^{\infty}\left(\sum_{k=0}^{n}(k+1)a_{k+1}b_{n-k}\right)(x-x_0)^n$$

and

$$P_2(x)y = \sum_{n=0}^{\infty}\left(\sum_{k=0}^{n}a_k c_{n-k}\right)(x-x_0)^n .$$

When these series are substituted in the differential equation (6.31) we find

$$\sum_{n=0}^{\infty}\left\{(n+2)(n+1)a_{n+2} + \sum_{k=0}^{n}[(k+1)a_{k+1}b_{n-k} + a_k c_{n-k}]\right\}(x-x_0)^n = 0 .$$

Therefore the differential equation will be satisfied if we choose the coefficients a_n so that they satisfy the recursion formula

$$(6.33) \qquad (n+2)(n+1)a_{n+2} = -\sum_{k=0}^{n}[(k+1)a_{k+1}b_{n-k} + a_k c_{n-k}]$$

for $n = 0, 1, 2, \ldots$. This formula expresses a_{n+2} in terms of the earlier coefficients $a_0, a_1, \ldots, a_{n+1}$ and the coefficients of the given functions P_1 and P_2. We choose arbitrary values of the first two coefficients a_0 and a_1 and use the recursion formula to define the remaining coefficients $a_2, a_3, \ldots$, in terms of a_0 and a_1. This guarantees that the power series in (6.32) will satisfy the differential equation (6.31). The next step in the proof is to show that the series so defined actually converges for every x in the interval $(x_0 - r, x_0 + r)$. This is done by dominating the series in (6.32) by another power series known to converge. Finally, we show that we can choose a_0 and a_1 to obtain two independent solutions.

We prove now that the series (6.32) whose coefficients are defined by (6.33) converges in the required interval.

Choose a fixed point $x_1 \neq x_0$ in the interval $(x_0 - r, x_0 + r)$ and let $t = |x_1 - x_0|$. Since the series for P_1 and P_2 converge absolutely for $x = x_1$ the terms of these series are bounded, say

$$|b_k|\, t^k \le M_1 \qquad \text{and} \qquad |c_k|\, t^k \le M_2 ,$$

for some $M_1 > 0$, $M_2 > 0$. Let M be the larger of M_1 and tM_2. Then we have

$$|b_k| \le \frac{M}{t^k} \qquad \text{and} \qquad |c_k| \le \frac{M}{t^{k+1}} .$$

The recursion formula implies the inequality

$$\begin{aligned}
(n+2)(n+1)\,|a_{n+2}| &\le \sum_{k=0}^{n}\left\{(k+1)\,|a_{k+1}|\,\frac{M}{t^{n-k}} + |a_k|\,\frac{M}{t^{n-k+1}}\right\} \\
&= \frac{M}{t^{n+1}}\left\{\sum_{k=0}^{n}(k+1)\,|a_{k+1}|\,t^{k+1} + \sum_{k=0}^{n}|a_{k+1}|\,t^{k+1} + |a_0| - |a_{n+1}|\,t^{n+1}\right\} \\
&\le \frac{M}{t^{n+1}}\left\{\sum_{k=0}^{n}(k+2)\,|a_{k+1}|\,t^{k+1} + |a_0|\right\} = \frac{M}{t^{n+1}}\sum_{k=0}^{n+1}(k+1)\,|a_k|\,t^k .
\end{aligned}$$

† Those readers not familiar with multiplication of power series may consult Exercise 7 of Section 6.21.

Now let $A_0 = |a_0|$, $A_1 = |a_1|$, and define $A_2, A_3, \ldots$ successively by the recursion formula

$$(6.34) \qquad (n + 2)(n + 1)A_{n+2} = \frac{M}{t^{n+1}} \sum_{k=0}^{n+1} (k + 1)A_k t^k$$

for $n \geq 0$. Then $|a_n| \leq A_n$ for all $n \geq 0$, so the series $\sum a_n(x - x_0)^n$ is dominated by the series $\sum A_n |x - x_0|^n$. Now we use the ratio test to show that $\sum A_n |x - x_0|^n$ converges if $|x - x_0| < t$.

Replacing n by $n - 1$ in (6.34) and subtracting t^{-1} times the resulting equation from (6.34) we find that $(n + 2)(n + 1)A_{n+2} - t^{-1}(n + 1)nA_{n+1} = M(n + 2)A_{n+1}$. Therefore

$$A_{n+2} = A_{n+1} \frac{(n + 1)n + (n + 2)Mt}{(n + 2)(n + 1)t},$$

and we find

$$\frac{A_{n+2} |x - x_0|^{n+2}}{A_{n+1} |x - x_0|^{n+1}} = \frac{(n + 1)n + (n + 2)Mt}{(n + 2)(n + 1)t} |x - x_0| \to \frac{|x - x_0|}{t}$$

as $n \to \infty$. This limit is less than 1 if $|x - x_0| < t$. Hence $\sum a_n(x - x_0)^n$ converges if $|x - x_0| < t$. But since $t = |x_1 - x_0|$ and since x_1 was an arbitrary point in the interval $(x_0 - r, x_0 + r)$, the series $\sum a_n(x - x_0)^n$ converges for all x in $(x_0 - r, x_0 + r)$.

The first two coefficients a_0 and a_1 represent the initial values of y and its derivative at the point x_0. If we let u_1 be the power-series solution with $a_0 = 1$ and $a_1 = 0$, so that

$$u_1(x_0) = 1 \qquad \text{and} \qquad u_1'(x_0) = 0,$$

and let u_2 be the solution with $a_0 = 0$ and $a_1 = 1$, so that

$$u_2(x_0) = 0 \qquad \text{and} \qquad u_2'(x_0) = 1,$$

then the solutions u_1 and u_2 will be independent. This completes the proof.

6.18 The Legendre equation

In this section we find power-series solutions for the Legendre equation,

$$(6.35) \qquad (1 - x^2)y'' - 2xy' + \alpha(\alpha + 1)y = 0,$$

where α is any real constant. This equation occurs in problems of attraction and in heat-flow problems with spherical symmetry. When α is a positive integer we shall find that the equation has polynomial solutions called *Legendre polynomials.* These are the same polynomials we encountered earlier in connection with the Gram-Schmidt process (Chapter 1, page 26).

The Legendre equation can be written as

$$[(x^2 - 1)y']' = \alpha(\alpha + 1)y,$$

which has the form

$$T(y) = \lambda y,$$

where T is a Sturm-Liouville operator, $T(f) = (pf')'$, with $p(x) = x^2 - 1$ and $\lambda = \alpha(\alpha + 1)$. Therefore the nonzero solutions of the Legendre equation are eigenfunctions of T belonging to the eigenvalue $\alpha(\alpha + 1)$. Since $p(x)$ satisfies the boundary conditions

$$p(1) = p(-1) = 0,$$

the operator T is symmetric with respect to the inner product

$$(f, g) = \int_{-1}^{1} f(x)g(x)\, dx.$$

The general theory of symmetric operators tells us that eigenfunctions belonging to distinct eigenvalues are orthogonal (Theorem 5.3).

In the differential equation treated in Theorem 6.13 the coefficient of y'' is 1. The Legendre equation can be put in this form if we divide through by $1 - x^2$. From (6.35) we obtain

$$y'' + P_1(x)y' + P_2(x)y = 0,$$

where

$$P_1(x) = -\frac{2x}{1 - x^2} \quad \text{and} \quad P_2(x) = \frac{\alpha(\alpha + 1)}{1 - x^2},$$

if $x^2 \neq 1$. Since $1/(1 - x^2) = \sum_{n=0}^{\infty} x^{2n}$ for $|x| < 1$, both P_1 and P_2 have power-series expansions in the open interval $(-1, 1)$ so Theorem 6.13 is applicable. To find the recursion formula for the coefficients it is simpler to leave the equation in the form (6.35) and try to find a power-series solution of the form

$$y = \sum_{n=0}^{\infty} a_n x^n$$

valid in the open interval $(-1, 1)$. Differentiating this series term by term we obtain

$$y' = \sum_{n=1}^{\infty} n a_n x^{n-1} \quad \text{and} \quad y'' = \sum_{n=2}^{\infty} n(n-1)a_n x^{n-2}.$$

Therefore we have

$$2xy' = \sum_{n=1}^{\infty} 2na_n x^n = \sum_{n=0}^{\infty} 2na_n x^n,$$

and

$$\begin{aligned}(1 - x^2)y'' &= \sum_{n=2}^{\infty} n(n-1)a_n x^{n-2} - \sum_{n=2}^{\infty} n(n-1)a_n x^n \\ &= \sum_{n=0}^{\infty} (n+2)(n+1)a_{n+2} x^n - \sum_{n=0}^{\infty} n(n-1)a_n x^n \\ &= \sum_{n=0}^{\infty} [(n+2)(n+1)a_{n+2} - n(n-1)a_n]x^n.\end{aligned}$$

If we substitute these series in the differential equation (6.35), we see that the equation will be satisfied if, and only if, the coefficients satisfy the relation

$$(n+2)(n+1)a_{n+2} - n(n-1)a_n - 2na_n + \alpha(\alpha+1)a_n = 0$$

for all $n \geq 0$. This equation is the same as

$$(n+2)(n+1)a_{n+2} - (n-\alpha)(n+1+\alpha)a_n = 0,$$

or

$$a_{n+2} = -\frac{(\alpha - n)(\alpha + n + 1)}{(n+1)(n+2)}\, a_n . \tag{6.36}$$

This relation enables us to determine $a_2, a_4, a_6, \ldots$, successively in terms of a_0. Similarly, we can compute $a_3, a_5, a_7, \ldots$, in terms of a_1. For the coefficients with even subscripts we have

$$a_2 = -\frac{\alpha(\alpha+1)}{1\cdot 2}\, a_0,$$

$$a_4 = -\frac{(\alpha-2)(\alpha+3)}{3\cdot 4}\, a_2 = (-1)^2 \frac{\alpha(\alpha-2)(\alpha+1)(\alpha+3)}{4!}\, a_0,$$

and, in general,

$$a_{2n} = (-1)^n \frac{\alpha(\alpha-2)\cdots(\alpha-2n+2)\cdot(\alpha+1)(\alpha+3)\cdots(\alpha+2n-1)}{(2n)!}\, a_0 .$$

This can be proved by induction. For the coefficients with odd subscripts we find

$$a_{2n+1} = (-1)^n \frac{(\alpha-1)(\alpha-3)\cdots(\alpha-2n+1)\cdot(\alpha+2)(\alpha+4)\cdots(\alpha+2n)}{(2n+1)!}\, a_1 .$$

Therefore the series for y can be written as

$$y = a_0 u_1(x) + a_1 u_2(x), \tag{6.37}$$

where

(6.38)

$$u_1(x) = 1 + \sum_{n=1}^{\infty} (-1)^n \frac{\alpha(\alpha-2)\cdots(\alpha-2n+2)\cdot(\alpha+1)(\alpha+3)\cdots(\alpha+2n-1)}{(2n)!}\, x^{2n}$$

and

(6.39)

$$u_2(x) = x + \sum_{n=1}^{\infty} (-1)^n \frac{(\alpha-1)(\alpha-3)\cdots(\alpha-2n+1)\cdot(\alpha+2)(\alpha+4)\cdots(\alpha+2n)}{(2n+1)!}\, x^{2n+1} .$$

The ratio test shows that each of these series converges for $|x| < 1$. Also, since the relation (6.36) is satisfied separately by the even and odd coefficients, each of u_1 and u_2 is a solution of the differential equation (6.35). These solutions satisfy the initial conditions

$$u_1(0) = 1, \qquad u_1'(0) = 0, \qquad u_2(0) = 0, \qquad u_2'(0) = 1 .$$

Since u_1 and u_2 are independent, the general solution of the Legendre equation (6.35) over the open interval $(-1, 1)$ is given by the linear combination (6.37) with arbitrary constants a_0 and a_1.

When α is 0 or a positive even integer, say $\alpha = 2m$, the series for $u_1(x)$ becomes a polynomial of degree $2m$ containing only even powers of x. Since we have

$$\alpha(\alpha - 2)\cdots(\alpha - 2n + 2) = 2m(2m - 2)\cdots(2m - 2n + 2) = \frac{2^n m!}{(m - n)!}$$

and

$$(\alpha + 1)(\alpha + 3)\cdots(\alpha + 2n - 1) = (2m + 1)(2m + 3)\cdots(2m + 2n - 1)$$
$$= \frac{(2m + 2n)!\, m!}{2^n (2m)!\, (m + n)!}$$

the formula for $u_1(x)$ in this case becomes

$$u_1(x) = 1 + \frac{(m!)^2}{(2m!)} \sum_{k=1}^{m} (-1)^k \frac{(2m + 2k)!}{(m - k)!\,(m + k)!\,(2k)!} x^{2k}. \tag{6.40}$$

For example, when $\alpha = 0, 2, 4, 6$ $(m = 0, 1, 2, 3)$ the corresponding polynomials are

$$u_1(x) = 1, \qquad 1 - 3x^3, \qquad 1 - 10x^2 + \tfrac{35}{3}x^4, \qquad 1 - 21x^2 + 63x^4 - \tfrac{231}{5}x^6.$$

The series for $u_2(x)$ is not a polynomial when α is even because the coefficient of x^{2n+1} is never zero.

When α is an *odd* positive integer, the roles of u_1 and u_2 are reversed; the series for $u_2(x)$ becomes a polynomial and the series for $u_1(x)$ is not a polynomial. Specifically if, $\alpha = 2m + 1$ we have

$$u_2(x) = x + \frac{(m!)^2}{(2m + 1)!} \sum_{k=1}^{m} (-1)^k \frac{(2m + 2k + 1)!}{(m - k)!\,(m + k)!\,(2k + 1)!} x^{2k+1}. \tag{6.41}$$

For example, when $\alpha = 1, 3, 5$ $(m = 0, 1, 2)$, the corresponding polynomials are

$$u_2(x) = x, \qquad x - \tfrac{5}{3}x^3, \qquad x - \tfrac{14}{3}x^3 + \tfrac{21}{5}x^5.$$

6.19 The Legendre polynomials

Some of the properties of the polynomial solutions of the Legendre equation can be deduced directly from the differential equation or from the formulas in (6.40) and (6.41). Others are more easily deduced from an alternative formula for these polynomials which we shall now derive.

First we shall obtain a single formula which contains (aside from constant factors) both the polynomials in (6.40) and (6.41). Let

$$P_n(x) = \frac{1}{2^n} \sum_{r=0}^{[n/2]} \frac{(-1)^r (2n - 2r)!}{r!\,(n - r)!\,(n - 2r)!} x^{n-2r}, \tag{6.42}$$

where $[n/2]$ denotes the greatest integer $\leq n/2$. We will show presently that this is the *Legendre polynomial* of degree n introduced in Chapter 1. When n is even, it is a constant multiple of the polynomial $u_1(x)$ in Equation (6.40); when n is odd, it is a constant multiple of the polynomial $u_2(x)$ in (6.41).† The first seven Legendre polynomials are given by the formulas

$$P_0(x) = 1, \quad P_1(x) = x, \quad P_2(x) = \tfrac{1}{2}(3x^2 - 1), \quad P_3(x) = \tfrac{1}{2}(5x^3 - 3x),$$

$$P_4(x) = \tfrac{1}{8}(35x^4 - 30x^2 + 3), \quad P_5(x) = \tfrac{1}{8}(63x^5 - 70x^3 + 15x),$$

$$P_6(x) = \tfrac{1}{16}(231x^6 - 315x^4 + 105x^2 - 5).$$

Figure 6.1 shows the graphs of the first five of these functions over the interval $[-1, 1]$.

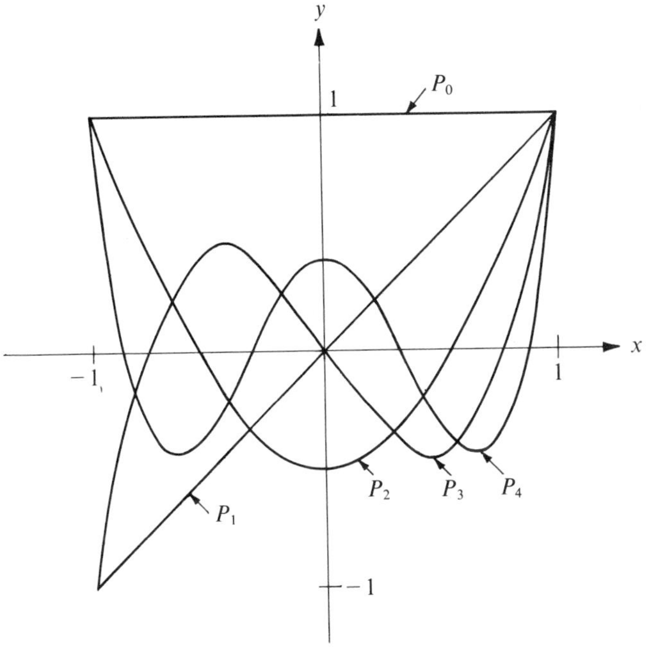

FIGURE 6.1 Graphs of Legendre polynomials over the interval $[-1, 1]$.

Now we can show that, except for scalar factors, the Legendre polynomials are those obtained by applying the Gram-Schmidt orthogonalization process to the sequence of polynomials $1, x, x^2, \ldots$, with the inner product

$$(f, g) = \int_{-1}^{1} f(x)g(x)\,dx.$$

† When n is even, say $n = 2m$, we may replace the index of summation k in Equation (6.40) by a new index r, where $r = m - k$; we find that the sum in (6.40) is a constant multiple of $P_n(x)$. Similarly, when n is odd, a change of index transforms the sum in (6.41) to a constant multiple of $P_n(x)$.

First we note that if $m \neq n$ the polynomials P_n and P_m are orthogonal because they are eigenfunctions of a symmetric operator belonging to distinct eigenvalues. Also, since P_n has degree n and $P_0 = 1$, the polynomials $P_0(x), P_1(x), \ldots, P_n(x)$ span the same subspace as $1, x, \ldots, x^n$. In Section 1.14, Example 2, we constructed another orthogonal set of polynomials $y_0, y_1, y_2, \ldots$, such that $y_0(x), y_1(x), \ldots, y_n(x)$ spans the same subspace as $1, x, \ldots, x^n$ for each n. The orthogonalization theorem (Theorem 1.13) tells us that, except for scalar factors, there is only one set of orthogonal functions with this property. Hence we must have

$$P_n(x) = c_n y_n(x)$$

for some scalars c_n. The coefficient of x^n in $y_n(x)$ is 1, so c_n is the coefficient of x^n in $P_n(x)$. From (6.42) we see that

$$c_n = \frac{(2n)!}{2^n(n!)^2}.$$

6.20 Rodrigues' formula for the Legendre polynomials

In the sum (6.42) defining $P_n(x)$ we note that

$$\frac{(2n-2r)!}{(n-2r)!}\, x^{n-2r} = \frac{d^n}{dx^n}\, x^{2n-2r} \quad \text{and} \quad \frac{1}{r!\,(n-r)!} = \frac{1}{n!}\binom{n}{r},$$

where $\binom{n}{r}$ is the binomial coefficient, and we write the sum in the form

$$P_n(x) = \frac{1}{2^n n!}\frac{d^n}{dx^n}\sum_{r=0}^{[n/2]} (-1)^r \binom{n}{r} x^{2n-2r}.$$

When $[n/2] < r \leq n$, the term x^{2n-2r} has degree less than n, so its nth derivative is zero. Therefore we do not alter the sum if we allow r to run from 0 to n. This gives us

$$P_n(x) = \frac{1}{2^n n!}\frac{d^n}{dx^n}\sum_{r=0}^{n} (-1)^r \binom{n}{r} x^{2n-2r}.$$

Now we recognize the sum on the right as the binomial expansion of $(x^2 - 1)^n$. Therefore we have

$$P_n(x) = \frac{1}{2^n n!}\frac{d^n}{dx^n}(x^2-1)^n.$$

This is known as *Rodrigues' formula*, in honor of Olinde Rodrigues (1794–1851), a French economist and reformer.

Using Rodrigues' formula and the differential equation, we can derive a number of important properties of the Legendre polynomials. Some of these properties are listed below. Their proofs are outlined in the next set of exercises.

For each $n \geq 0$ we have

$$P_n(1) = 1.$$

Moreover, $P_n(x)$ is the only polynomial which satisfies the Legendre equation

$$(1 - x^2)y'' - 2xy' + n(n + 1)y = 0$$

and has the value 1 when $x = 1$.

For each $n \geq 0$ we have

$$P_n(-x) = (-1)^n P_n(x).$$

This shows that P_n is an even function when n is even, and an odd function when n is odd.

We have already mentioned the orthogonality relation,

$$\int_{-1}^{1} P_n(x)P_m(x)\,dx = 0 \qquad \text{if} \quad m \neq n.$$

When $m = n$ we have the norm relation

$$\|P_n\|^2 = \int_{-1}^{1} [P_n(x)]^2\,dx = \frac{2}{2n + 1}.$$

Every polynomial of degree n can be expressed as a linear combination of the Legendre polynomials $P_0, P_1, \ldots, P_n$. In fact, if f is a polynomial of degree n we have

$$f(x) = \sum_{k=0}^{n} c_k P_k(x),$$

where

$$c_k = \frac{2k + 1}{2} \int_{-1}^{1} f(x)P_k(x)\,dx.$$

From the orthogonality relation it follows that

$$\int_{-1}^{1} g(x)P_n(x)\,dx = 0$$

for every polynomial g of degree less than n. This property can be used to prove that the Legendre polynomial P_n has n distinct real zeros and that they all lie in the open interval $(-1, 1)$.

6.21 Exercises

1. The Legendre equation (6.35) with $\alpha = 0$ has the polynomial solution $u_1(x) = 1$ and a solution u_2, not a polynomial, given by the series in Equation (6.41).

   (a) Show that the sum of the series for u_2 is given by

$$u_2(x) = \frac{1}{2} \log \frac{1 + x}{1 - x} \qquad \text{for} \quad |x| < 1.$$

   (b) Verify directly that the function u_2 in part (a) is a solution of the Legendre equation when $\alpha = 0$.

2. Show that the function f defined by the equation

$$f(x) = 1 - \frac{x}{2}\log\frac{1+x}{1-x}$$

for $|x| < 1$ satisfies the Legendre equation (6.35) with $\alpha = 1$. Express this function as a linear combination of the solutions u_1 and u_2 given in Equations (6.38) and (6.39).

3. The Legendre equation (6.35) can be written in the form

$$[(x^2 - 1)y']' - \alpha(\alpha + 1)y = 0 .$$

(a) If a, b, c are constants with $a > b$ and $4c + 1 > 0$, show that a differential equation of the type

$$[(x - a)(x - b)y']' - cy = 0$$

can be transformed to a Legendre equation by a change of variable of the form $x = At + B$, with $A > 0$. Determine A and B in terms of a and b.

(b) Use the method suggested in part (a) to transform the equation

$$(x^2 - x)y'' + (2x - 1)y' - 2y = 0$$

to a Legendre equation.

4. Find two independent power-series solutions of the *Hermite equation*

$$y'' - 2xy' + 2\alpha y = 0$$

on an interval of the form $(-r, r)$. Show that one of these solutions is a polynomial when α is a nonnegative integer.

5. Find a power-series solution of the differential equation

$$xy'' + (3 + x^3)y' + 3x^2y = 0$$

valid for all x. Find a second solution of the form $y = x^{-2}\sum a_n x^n$ valid for all $x \neq 0$.

6. Find a power-series solution of the differential equation

$$x^2y'' + x^2y' - (\alpha x + 2)y = 0$$

valid on an interval of the form $(-r, r)$.

7. Given two functions A and B analytic on an interval $(x_0 - r, x_0 + r)$, say

$$A(x) = \sum_{n=0}^{\infty} a_n(x - x_0)^n, \qquad B(x) = \sum_{n=0}^{\infty} b_n(x - x_0)^n .$$

It can be shown that the product $C(x) = A(x)B(x)$ is also analytic on $(x_0 - r, x_0 + r)$. This exercise shows that C has the power-series expansion

$$C(x) = \sum_{n=0}^{\infty} c_n(x - x_0)^n, \qquad \text{where } c_n = \sum_{k=0}^{n} a_k b_{n-k} .$$

(a) Use Leibniz's rule for the nth derivative of a product to show that the nth derivative of C is given by

$$C^{(n)}(x) = \sum_{k=0}^{n} \binom{n}{k} A^{(k)}(x) B^{(n-k)}(x).$$

(b) Now use the fact that $A^{(k)}(x_0) = k!\, a_k$ and $B^{(n-k)}(x_0) = (n-k)!\, b_{n-k}$ to obtain

$$C^{(n)}(x_0) = n! \sum_{k=0}^{n} a_k b_{n-k}.$$

Since $C^{(n)}(x_0) = n!\, c_n$, this proves the required formula for c_n.

In Exercises 8 through 14, $P_n(x)$ denotes the Legendre polynomial of degree n. These exercises outline proofs of the properties of the Legendre polynomials described in Section 6.20.

8. (a) Use Rodrigues' formula to show that

$$P_n(x) = \frac{1}{2^n}(x+1)^n + (x-1)Q_n(x),$$

where $Q_n(x)$ is a polynomial.
(b) Prove that $P_n(1) = 1$ and that $P_n(-1) = (-1)^n$.
(c) Prove that $P_n(x)$ is the only polynomial solution of Legendre's equation (with $\alpha = n$) having the value 1 when $x = 1$.

9. (a) Use the differential equations satisfied by P_n and P_m to show that

$$[(1-x^2)(P_n P_m' - P_n' P_m)]' = [n(n+1) - m(m+1)]P_n P_m.$$

(b) If $m \neq n$, integrate the equation in (a) from -1 to 1 to give an alternate proof of the orthogonality relation

$$\int_{-1}^{1} P_n(x) P_m(x)\, dx = 0.$$

10. (a) Let $f(x) = (x^2-1)^n$. Use integration by parts to show that

$$\int_{-1}^{1} f^{(n)}(x) f^{(n)}(x)\, dx = -\int_{-1}^{1} f^{(n+1)}(x) f^{(n-1)}(x)\, dx.$$

Apply this formula repeatedly to deduce that the integral on the left is equal to

$$2(2n)! \int_0^1 (1-x^2)^n\, dx.$$

(b) The substitution $x = \cos t$ transforms the integral $\int_0^1 (1-x^2)^n\, dx$ to $\int_0^{\pi/2} \sin^{2n+1} t\, dt$. Use the relation

$$\int_0^{\pi/2} \sin^{2n+1} t\, dt = \frac{2n(2n-2)\cdots 2}{(2n+1)(2n-1)\cdots 3\cdot 1}$$

and Rodrigues' formula to obtain

$$\int_{-1}^{1} [P_n(x)]^2\, dx = \frac{2}{2n+1}.$$

11. (a) Show that

$$P_n(x) = \frac{(2n)!}{2^n(n!)^2} x^n + Q_n(x),$$

where $Q_n(x)$ is a polynomial of degree less than n.
(b) Express the polynomial $f(x) = x^4$ as a linear combination of P_0, P_1, P_2, P_3, and P_4.
(c) Show that every polynomial f of degree n can be expressed as a linear combination of the Legendre polynomials $P_0, P_1, \ldots, P_n$.

12. (a) If f is a polynomial of degree n, write

$$f(x) = \sum_{k=0}^{n} c_k P_k(x).$$

[This is possible because of Exercise 11(c).] For a fixed m, $0 \le m \le n$, multiply both sides of this equation by $P_m(x)$ and integrate from -1 to 1. Use Exercises 9(b) and 10(b) to deduce the relation

$$c_m = \frac{2m+1}{2} \int_{-1}^{1} f(x) P_m(x)\, dx.$$

13. Use Exercises 9 and 11 to show that $\int_{-1}^{1} g(x)P_n(x)\, dx = 0$ for every polynomial g of degree less than n.

14. (a) Use Rolle's theorem to show that P_n cannot have any multiple zeros in the open interval $(-1, 1)$. In other words, any zeros of P_n which lie in $(-1, 1)$ must be simple zeros.
(b) Assume P_n has m zeros in the interval $(-1, 1)$. If $m = 0$, let $Q_0(x) = 1$. If $m \ge 1$, let

$$Q_m(x) = (x - x_1)(x - x_2) \cdots (x - x_m),$$

where $x_1, x_2, \ldots, x_m$ are the m zeros of P_n in $(-1, 1)$. Show that, at each point x in $(-1, 1)$, $Q_m(x)$ has the same sign as $P_n(x)$.
(c) Use part (b), along with Exercise 13, to show that the inequality $m < n$ leads to a contradiction. This shows that P_n has n distinct real zeros, all of which lie in the open interval $(-1, 1)$.

15. (a) Show that the value of the integral $\int_{-1}^{1} P_n(x)P'_{n+1}(x)\, dx$ is independent of n.
(b) Evaluate the integral $\int_{-1}^{1} x\, P_n(x)P_{n-1}(x)\, dx$.

6.22 The method of Frobenius

In Section 6.17 we learned how to find power-series solutions of the differential equation

$$y'' + P_1(x)y' + P_2(x)y = 0 \tag{6.43}$$

in an interval about a point x_0 where the coefficients P_1 and P_2 are analytic. If either P_1 or P_2 is not analytic near x_0, power-series solutions valid near x_0 may or may not exist. For example, suppose we try to find a power-series solution of the differential equation

$$x^2y'' - y' - y = 0 \tag{6.44}$$

near $x_0 = 0$. If we assume that a solution $y = \sum a_k x^k$ exists and substitute this series in the differential equation we are led to the recursion formula

$$a_{n+1} = \frac{n^2 - n - 1}{n + 1} a_n.$$

Although this gives us a power series $y = \sum a_k x^k$ which formally satisfies (6.44), the ratio test shows that this power series converges *only* for $x = 0$. Thus, there is no power-series solution of (6.44) valid in any open interval about $x_0 = 0$. This example does not violate Theorem 6.13 because when we put Equation (6.44) in the form (6.43) we find that the coefficients P_1 and P_2 are given by

$$P_1(x) = -\frac{1}{x^2} \quad \text{and} \quad P_2(x) = -\frac{1}{x^2}.$$

These functions do not have power-series expansions about the origin. The difficulty here is that the coefficient of y'' in (6.44) has the value 0 when $x = 0$; in other words, the differential equation has a singular point at $x = 0$.

A knowledge of the theory of functions of a complex variable is needed to appreciate the difficulties encountered in the investigation of differential equations near a singular point. However, some important special cases of equations with singular points can be treated by elementary methods. For example, suppose the differential equation in (6.43) is equivalent to an equation of the form

$$(6.45) \qquad (x - x_0)^2 y'' + (x - x_0)P(x)y' + Q(x)y = 0,$$

where P and Q have power-series expansions in some open interval $(x_0 - r, x_0 + r)$. In this case we say that x_0 is a *regular* singular point of the equation. If we divide both sides of (6.45) by $(x - x_0)^2$ the equation becomes

$$y'' + \frac{P(x)}{x - x_0} y' + \frac{Q(x)}{(x - x_0)^2} y = 0$$

for $x \neq x_0$. If $P(x_0) \neq 0$ or $Q(x_0) \neq 0$, or if $Q(x_0) = 0$ and $Q'(x_0) \neq 0$, either the coefficient of y' or the coefficient of y will not have a power-series expansion about the point x_0, so Theorem 6.13 will not be applicable. In 1873 the German mathematician Georg Frobenius (1849–1917) developed a useful method for treating such equations. We shall describe the theorem of Frobenius but we shall not present its proof.† In the next section we give the details of the proof for an important special case, the Bessel equation.

Frobenius' theorem splits into two parts, depending on the nature of the roots of the quadratic equation

$$(6.46) \qquad t(t - 1) + P(x_0)t + Q(x_0) = 0.$$

This quadratic equation is called the *indicial equation* of the given differential equation (6.45). The coefficients $P(x_0)$ and $Q(x_0)$ are the constant terms in the power-series expansions of P and Q. Let α_1 and α_2 denote the roots of the indicial equation. These roots may be real or complex, equal or distinct. The type of solution obtained by the Frobenius method depends on whether or not these roots differ by an integer.

† For a proof see E. Hille, *Analysis*, Vol. II, Blaisdell Publishing Co., 1966, or E. A. Coddington, *An Introduction to Ordinary Differential Equations*, Prentice-Hall, 1961.

THEOREM 6.14. FIRST CASE OF FROBENIUS' THEOREM. *Let α_1 and α_2 be the roots of the indicial equation and assume that $\alpha_1 - \alpha_2$ is not an integer. Then the differential equation* (6.45) *has two independent solutions u_1 and u_2 of the form*

$$u_1(x) = |x - x_0|^{\alpha_1} \sum_{n=0}^{\infty} a_n(x - x_0)^n, \quad \text{with} \quad a_0 = 1, \tag{6.47}$$

and

$$u_2(x) = |x - x_0|^{\alpha_2} \sum_{n=0}^{\infty} b_n(x - x_0)^n, \quad \text{with} \quad b_0 = 1. \tag{6.48}$$

Both series converge in the interval $|x - x_0| < r$, and the differential equation is satisfied for $0 < |x - x_0| < r$.

THEOREM 6.15. SECOND CASE OF FROBENIUS' THEOREM. *Let α_1, α_2 be the roots of the indicial equation and assume that $\alpha_1 - \alpha_2 = N$, a nonnegative integer. Then the differential equation* (6.45) *has a solution u_1 of the form* (6.47) *and another independent solution u_2 of the form*

$$u_2(x) = |x - x_0|^{\alpha_2} \sum_{n=0}^{\infty} b_n(x - x_0)^n + C\, u_1(x) \log |x - x_0|, \tag{6.49}$$

where $b_0 = 1$. The constant C is nonzero if $N = 0$. If $N > 0$, the constant C may or may not be zero. As in Case 1, both series converge in the interval $|x - x_0| < r$, and the solutions are valid for $0 < |x - x_0| < r$.

6.23 The Bessel equation

In this section we use the method suggested by Frobenius to solve the Bessel equation

$$x^2y'' + xy' + (x^2 - \alpha^2)y = 0,$$

where α is a nonnegative constant. This equation is used in problems concerning vibrations of membranes, heat flow in cylinders, and propagation of electric currents in cylindrical conductors. Some of its solutions are known as *Bessel functions.* Bessel functions also arise in certain problems in Analytic Number Theory. The equation is named after the German astronomer F. W. Bessel (1784–1846), although it appeared earlier in the researches of Daniel Bernoulli (1732) and Euler (1764).

The Bessel equation has the form (6.45) with $x_0 = 0$, $P(x) = 1$, and $Q(x) = x^2 - \alpha^2$, so the point x_0 is a regular singular point. Since P and Q are analytic on the entire real line, we try to find solutions of the form

$$y = |x|^t \sum_{n=0}^{\infty} a_n x^n, \tag{6.50}$$

with $a_0 \neq 0$, valid for all real x with the possible exception of $x = 0$.

First we keep $x > 0$, so that $|x|^t = x^t$. Differentiation of (6.50) gives us

$$y' = tx^{t-1} \sum_{n=0}^{\infty} a_n x^n + x^t \sum_{n=0}^{\infty} n a_n x^{n-1} = x^{t-1} \sum_{n=0}^{\infty} (n + t) a_n x^n.$$

Similarly, we obtain

$$y'' = x^{t-2}\sum_{n=0}^{\infty}(n+t)(n+t-1)a_n x^n .$$

If $L(y) = x^2y'' + xy' + (x^2 - \alpha^2)y$, we find

$$L(y) = x^t\sum_{n=0}^{\infty}(n+t)(n+t-1)a_n x^n + x^t\sum_{n=0}^{\infty}(n+t)a_n x^n$$
$$+ x^t\sum_{n=0}^{\infty}a_n x^{n+2} - x^t\sum_{n=0}^{\infty}\alpha^2 a_n x^n = x^t\sum_{n=0}^{\infty}[(n+t)^2 - \alpha^2]a_n x^n + x^t\sum_{n=0}^{\infty}a_n x^{n+2}.$$

Now we put $L(y) = 0$, cancel x^t, and try to determine the a_n so that the coefficient of each power of x will vanish. For the constant term we need $(t^2 - \alpha^2)a_0 = 0$. Since we seek a solution with $a_0 \neq 0$, this requires that

(6.51) $$t^2 - \alpha^2 = 0 .$$

This is the *indicial equation.* Its roots α and $-\alpha$ are the only possible values of t that can give us a solution of the desired type.

Consider first the choice $t = \alpha$. For this t the remaining equations for determining the coefficients become

(6.52) $$[(1+\alpha)^2 - \alpha^2]a_1 = 0 \qquad \text{and} \qquad [(n+\alpha)^2 - \alpha^2]a_n + a_{n-2} = 0$$

for $n \geq 2$. Since $\alpha \geq 0$, the first of these implies that $a_1 = 0$. The second formula can be written as

(6.53) $$a_n = -\frac{a_{n-2}}{(n+\alpha)^2 - \alpha^2} = -\frac{a_{n-2}}{n(n+2\alpha)},$$

so $a_3 = a_5 = a_7 = \cdots = 0$. For the coefficients with even subscripts we have

$$a_2 = \frac{-a_0}{2(2+2\alpha)} = \frac{-a_0}{2^2(1+\alpha)}, \qquad a_4 = \frac{-a_2}{4(4+2\alpha)} = \frac{(-1)^2 a_0}{2^4 2!\,(1+\alpha)(2+\alpha)},$$

$$a_6 = \frac{-a_4}{6(6+2\alpha)} = \frac{(-1)^3 a_0}{2^6 3!\,(1+\alpha)(2+\alpha)(3+\alpha)},$$

and, in general,

$$a_{2n} = \frac{(-1)^n a_0}{2^{2n} n!\,(1+\alpha)(2+\alpha)\cdots(n+\alpha)}.$$

Therefore the choice $t = \alpha$ gives us the solution

$$y = a_0 x^\alpha\left(1 + \sum_{n=1}^{\infty}\frac{(-1)^n x^{2n}}{2^{2n} n!\,(1+\alpha)(2+\alpha)\cdots(n+\alpha)}\right).$$

The ratio test shows that the power series appearing in this formula converges for all real x.

In this discussion we assumed that $x > 0$. If $x < 0$ we can repeat the discussion with x^t replaced by $(-x)^t$. We again find that t must satisfy the equation $t^2 - \alpha^2 = 0$. Taking $t = \alpha$ we then obtain the same solution, except that the outside factor x^α is replaced by $(-x)^\alpha$. Therefore the function f_α given by the equation

$$(6.54) \qquad f_\alpha(x) = a_0 |x|^\alpha \left(1 + \sum_{n=1}^{\infty} \frac{(-1)^n x^{2n}}{2^{2n} n!\,(1+\alpha)(2+\alpha)\cdots(n+\alpha)}\right)$$

is a solution of the Bessel equation valid for all real $x \neq 0$. For those values of α for which $f_\alpha'(0)$ and $f_\alpha''(0)$ exist the solution is also valid for $x = 0$.

Now consider the root $t = -\alpha$ of the indicial equation. We obtain, in place of (6.52), the equations

$$[(1-\alpha)^2 - \alpha^2]a_1 = 0 \qquad \text{and} \qquad [(n-\alpha)^2 - \alpha^2]a_n + a_{n-2} = 0,$$

which become

$$(1 - 2\alpha)a_1 = 0 \qquad \text{and} \qquad n(n-2\alpha)a_n + a_{n-2} = 0.$$

If 2α is not an integer these equations give us $a_1 = 0$ and

$$a_n = -\frac{a_{n-2}}{n(n-2\alpha)}$$

for $n \geq 2$. Since this recursion formula is the same as (6.53), with α replaced by $-\alpha$, we are led to the solution

$$(6.55) \qquad f_{-\alpha}(x) = a_0 |x|^{-\alpha} \left(1 + \sum_{n=1}^{\infty} \frac{(-1)^n x^{2n}}{2^{2n} n!\,(1-\alpha)(2-\alpha)\cdots(n-\alpha)}\right)$$

valid for all real $x \neq 0$.

The solution $f_{-\alpha}$ was obtained under the hypothesis that 2α is not a positive integer. However, the series for $f_{-\alpha}$ is meaningful even if 2α is a positive integer, so long as α is not a positive integer. It can be verified that $f_{-\alpha}$ satisfies the Bessel equation for such α. Therefore, for each $\alpha \geq 0$ we have the series solution f_α, given by Equation (6.54); and if α is not a nonnegative integer we have found another solution $f_{-\alpha}$ given by Equation (6.55). The two solutions f_α and $f_{-\alpha}$ are independent, since one of them $\to \infty$ as $x \to 0$, and the other does not. Next we shall simplify the form of the solutions. To do this we need some properties of Euler's gamma function, and we digress briefly to recall these properties.

For each real $s > 0$ we define $\Gamma(s)$ by the improper integral

$$\Gamma(s) = \int_{0+}^{\infty} t^{s-1} e^{-t}\, dt.$$

This integral converges if $s > 0$ and diverges if $s \leq 0$. Integration by parts leads to the functional equation

$$\Gamma(s+1) = s\,\Gamma(s). \tag{6.56}$$

This implies that

$$\Gamma(s+2) = (s+1)\Gamma(s+1) = (s+1)s\,\Gamma(s),$$

$$\Gamma(s+3) = (s+2)\Gamma(s+2) = (s+2)\,(s+1)s\,\Gamma(s),$$

and, in general,

$$\Gamma(s+n) = (s+n-1)\cdots(s+1)s\,\Gamma(s) \tag{6.57}$$

for every positive integer n. Since $\Gamma(1) = \int_0^\infty e^{-t}\,dt = 1$, when we put $s = 1$ in (6.57) we find

$$\Gamma(n+1) = n!.$$

Thus, the gamma function is an extension of the factorial function from integers to positive real numbers.

The functional equation (6.56) can be used to extend the definition of $\Gamma(s)$ to negative values of s that are not integers. We write (6.56) in the form

$$\Gamma(s) = \frac{\Gamma(s+1)}{s}. \tag{6.58}$$

The right-hand member is meaningful if $s + 1 > 0$ and $s \neq 0$. Therefore, we can use this equation to *define* $\Gamma(s)$ if $-1 < s < 0$. The right-hand member of (6.58) is now meaningful if $s + 2 > 0$, $s \neq -1$, $s \neq 0$, and we can use this equation to define $\Gamma(s)$ for $-2 < s < -1$. Continuing in this manner, we can extend the definition of $\Gamma(s)$ by induction to every open interval of the form $-n < s < -n + 1$, where n is a positive integer. The functional equation (6.56) and its extension in (6.57) are now valid for all real s for which both sides are meaningful.

We return now to the discussion of the Bessel equation. The series for f_α in Equation (6.54) contains the product $(1+\alpha)\,(2+\alpha)\cdots(n+\alpha)$. We can express this product in terms of the gamma function by taking $s = 1 + \alpha$ in (6.57). This gives us

$$(1+\alpha)(2+\alpha)\cdots(n+\alpha) = \frac{\Gamma(n+1+\alpha)}{\Gamma(1+\alpha)}.$$

Therefore, if we choose $a_0 = 2^{-\alpha}/\Gamma(1+\alpha)$ in Equation (6.54) and denote the resulting function $f_\alpha(x)$ by $J_\alpha(x)$ when $x > 0$, the solution for $x > 0$ can be written as

$$J_\alpha(x) = \left(\frac{x}{2}\right)^{\alpha} \sum_{n=0}^{\infty} \frac{(-1)^n}{n!\,\Gamma(n+1+\alpha)} \left(\frac{x}{2}\right)^{2n}. \tag{6.59}$$

The function J_α defined by this equation for $x > 0$ and $\alpha \geq 0$ is called the *Bessel function of the first kind of order* α. When α is a nonnegative integer, say $\alpha = p$, the Bessel function J_p is given by the power series

$$J_p(x) = \sum_{n=0}^{\infty} \frac{(-1)^n}{n!\,(n+p)!}\left(\frac{x}{2}\right)^{2n+p} \qquad (p = 0, 1, 2, \ldots).$$

This is also a solution of the Bessel equation for $x < 0$. Extensive tables of Bessel functions have been constructed. The graphs of the two functions J_0 and J_1 are shown in Figure 6.2.

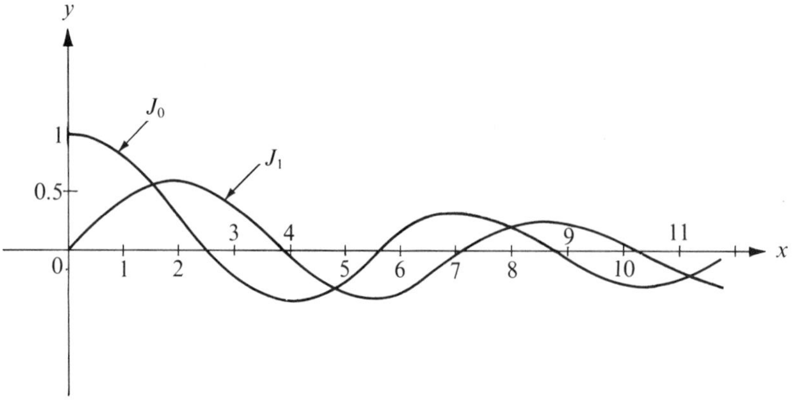

FIGURE 6.2 Graphs of the Bessel functions J_0 and J_1.

We can define a new function $J_{-\alpha}$ by replacing α by $-\alpha$ in Equation (6.59), if α is such that $\Gamma(n + 1 - \alpha)$ is meaningful; that is, if α is not a positive integer. Therefore, if $x > 0$ and $\alpha > 0$, $\alpha \neq 1, 2, 3, \ldots$, we define

$$J_{-\alpha}(x) = \left(\frac{x}{2}\right)^{-\alpha} \sum_{n=0}^{\infty} \frac{(-1)^n}{n!\,\Gamma(n+1-\alpha)}\left(\frac{x}{2}\right)^{2n}.$$

Taking $s = 1 - \alpha$ in (6.57) we obtain

$$\Gamma(n + 1 - \alpha) = (1 - \alpha)\,(2 - \alpha)\cdots(n - \alpha)\,\Gamma(1 - \alpha)$$

and we see that the series for $J_{-\alpha}(x)$ is the same as that for $f_{-\alpha}(x)$ in Equation (6.55) with $a_0 = 2^\alpha/\Gamma(1 - \alpha)$, $x > 0$. Therefore, if α is not a positive integer, $J_{-\alpha}$ is a solution of the Bessel equation for $x > 0$.

If α is not an integer, the two solutions $J_\alpha(x)$ and $J_{-\alpha}(x)$ are linearly independent on the positive real axis (since their ratio is not constant) and the general solution of the Bessel equation for $x > 0$ is

$$y = c_1 J_\alpha(x) + c_2 J_{-\alpha}(x).$$

If α is a nonnegative integer, say $\alpha = p$, we have found only the solution J_p and its constant multiples valid for $x > 0$. Another solution, independent of this one, can be found by the method described in Exercise 4 of Section 6.16. This states that if u_1 is a solution of $y'' + P_1y' + P_2y = 0$ that never vanishes on an interval I, a second solution u_2 independent of u_1 is given by the integral

$$u_2(x) = u_1(x) \int_c^x \frac{Q(t)}{[u_1(t)]^2}\, dt,$$

where $Q(x) = e^{-\int P_1(x)\,dx}$. For the Bessel equation we have $P_1(x) = 1/x$, so $Q(x) = 1/x$ and a second solution u_2 is given by the formula

$$u_2(x) = J_p(x) \int_c^x \frac{1}{t[J_p(t)]^2}\, dt, \tag{6.60}$$

if c and x lie in an interval I in which J_p does not vanish.

This second solution can be put in other forms. For example, from Equation (6.59) we may write

$$\frac{1}{[J_p(t)]^2} = \frac{1}{t^{2p}}\, g_p(t),$$

where $g_p(0) \neq 0$. In the interval I the function g_p has a power-series expansion

$$g_p(t) = \sum_{n=0}^{\infty} A_n t^n$$

which could be determined by equating coefficients in the identity $g_p(t)\,[J_p(t)]^2 = t^{2p}$. If we assume the existence of such an expansion, the integrand in (6.60) takes the form

$$\frac{1}{t[J_p(t)]^2} = \frac{1}{t^{2p+1}} \sum_{n=0}^{\infty} A_n t^n .$$

Integrating this formula term by term from c to x we obtain a logarithmic term $A_{2p} \log x$ (from the power t^{-1}) plus a series of the form $x^{-2p} \sum B_n x^n$. Therefore Equation (6.60) takes the form

$$u_2(x) = A_{2p} J_p(x) \log x + J_p(x) x^{-2p} \sum_{n=0}^{\infty} B_n x^n .$$

It can be shown that the coefficient $A_{2p} \neq 0$. If we multiply $u_2(x)$ by $1/A_{2p}$ the resulting solution is denoted by $K_p(x)$ and has the form

$$K_p(x) = J_p(x) \log x + x^{-p} \sum_{n=0}^{\infty} C_n x^n .$$

This is the form of the solution promised by the second case of Frobenius' theorem.

Having arrived at this formula, we can verify that a solution of this form actually exists by substituting the right-hand member in the Bessel equation and determining the coefficients C_n so as to satisfy the equation. The details of this calculation are lengthy and will be omitted. The final result can be expressed as

$$K_p(x) = J_p(x)\log x - \frac{1}{2}\left(\frac{x}{2}\right)^{-p}\sum_{n=0}^{p-1}\frac{(p-n-1)!}{n!}\left(\frac{x}{2}\right)^{2n} - \frac{1}{2}\left(\frac{x}{2}\right)^{p}\sum_{n=0}^{\infty}(-1)^n\frac{h_n + h_{n+p}}{n!\,(n+p)!}\left(\frac{x}{2}\right)^{2n},$$

where $h_0 = 0$ and $h_n = 1 + \frac{1}{2} + \cdots + 1/n$ for $n \geq 1$. The series on the right converges for all real x. The function K_p defined for $x > 0$ by this formula is called the *Bessel function of the second kind of order p*. Since K_p is not a constant multiple of J_p, the general solution of the Bessel equation in this case for $x > 0$ is

$$y = c_1 J_p(x) + c_2 K_p(x).$$

Further properties of the Bessel functions are discussed in the next set of exercises.

6.24 Exercises

1. (a) Let f be any solution of the Bessel equation of order α and let $g(x) = x^{1/2}f(x)$ for $x > 0$. Show that g satisfies the differential equation

$$y'' + \left(1 + \frac{1 - 4\alpha^2}{4x^2}\right)y = 0.$$

(b) When $4\alpha^2 = 1$ the differential equation in (a) becomes $y'' + y = 0$; its general solution is $y = A\cos x + B\sin x$. Use this information and the equation† $\Gamma(\frac{1}{2}) = \sqrt{\pi}$ to show that, for $x > 0$,

$$J_{1/2}(x) = \left(\frac{2}{\pi x}\right)^{1/2}\sin x \quad \text{and} \quad J_{-1/2}(x) = \left(\frac{2}{\pi x}\right)^{1/2}\cos x.$$

(c) Deduce the formulas in part (b) directly from the series for $J_{1/2}(x)$ and $J_{-1/2}(x)$.

2. Use the series representation for Bessel functions to show that

(a) $\dfrac{d}{dx}(x^\alpha J_\alpha(x)) = x^\alpha J_{\alpha-1}(x)$,

(b) $\dfrac{d}{dx}(x^{-\alpha}J_\alpha(x)) = -x^{-\alpha}J_{\alpha+1}(x)$.

3. Let $F_\alpha(x) = x^\alpha J_\alpha(x)$ and $G_\alpha(x) = x^{-\alpha}J_\alpha(x)$ for $x > 0$. Note that each positive zero of J_α is a zero of F_α and is also a zero of G_α. Use Rolle's theorem and Exercise 2 to prove that the positive zeros of J_α and $J_{\alpha+1}$ interlace. That is, there is a zero of J_α between each pair of positive zeros of $J_{\alpha+1}$, and a zero of $J_{\alpha+1}$ between each pair of positive zeros of J_α. (See Figure 6.2.)

† The change of variable $t = u^2$ gives us

$$\Gamma(\tfrac{1}{2}) = \int_{0+}^{\infty} t^{-1/2}e^{-t}\,dt = 2\int_0^\infty e^{-u^2}\,du = \sqrt{\pi}.$$

(See Exercise 16 of Section 11.28 for a proof that $2\int_0^\infty e^{-u^2}\,du = \sqrt{\pi}$.)

4. (a) From the relations in Exercise 2 deduce the recurrence relations

$$\frac{\alpha}{x}J_\alpha(x) + J_\alpha'(x) = J_{\alpha-1}(x) \quad \text{and} \quad \frac{\alpha}{x}J_\alpha(x) - J_\alpha'(x) = J_{\alpha+1}(x).$$

(b) Use the relations in part (a) to deduce the formulas

$$J_{\alpha-1}(x) + J_{\alpha+1}(x) = \frac{2\alpha}{x}J_\alpha(x) \quad \text{and} \quad J_{\alpha-1}(x) - J_{\alpha+1}(x) = 2J_\alpha'(x).$$

5. Use Exercise 1(b) and a suitable recurrence formula to show that

$$J_{3/2}(x) = \left(\frac{2}{\pi x}\right)^{1/2}\left(\frac{\sin x}{x} - \cos x\right).$$

Find a similar formula for $J_{-3/2}(x)$. *Note:* $J_\alpha(x)$ is an elementary function for every α which is half an odd integer.

6. Prove that

$$\frac{1}{2}\frac{d}{dx}(J_\alpha^2(x) + J_{\alpha+1}^2(x)) = \frac{\alpha}{x}J_\alpha^2(x) - \frac{\alpha+1}{x}J_{\alpha+1}^2(x)$$

and

$$\frac{d}{dx}(xJ_\alpha(x)J_{\alpha+1}(x)) = x(J_\alpha^2(x) - J_{\alpha+1}^2(x)).$$

7. (a) Use the identities in Exercise 6 to show that

$$J_0^2(x) + 2\sum_{n=1}^{\infty} J_n^2(x) = 1 \quad \text{and} \quad \sum_{n=0}^{\infty}(2n+1)J_n(x)J_{n+1}(x) = \tfrac{1}{2}x.$$

(b) From part (a), deduce that $|J_0(x)| \le 1$ and $|J_n(x)| \le \frac{1}{2}\sqrt{2}$ for $n = 1, 2, 3, \ldots$, and all $x \ge 0$.

8. Let $g_\alpha(x) = x^{1/2}f_\alpha(ax^b)$ for $x > 0$, where a and b are nonzero constants. Show that g_α satisfies the differential equation

$$x^2y'' + (a^2b^2x^{2b} + \tfrac{1}{4} - \alpha^2b^2)y = 0$$

if, and only if, f_α is a solution of the Bessel equation of order α.

9. Use Exercise 8 to express the general solution of each of the following differential equations in terms of Bessel functions for $x > 0$.
(a) $y'' + xy = 0$.
(b) $y'' + x^2y = 0$.
(c) $y'' + x^my = 0$.
(d) $x^2y'' + (x^4 + \frac{1}{8})y = 0$.

10. Generalize Exercise 8 when f_α and g_α are related by the equation $g_\alpha(x) = x^cf_\alpha(ax^b)$ for $x > 0$. Then find the general solution of each of the following equations in terms of Bessel functions for $x > 0$.
(a) $xy'' + 6y' + y = 0$.
(b) $xy'' + 6y' + xy = 0$.
(c) $xy'' + 6y' + x^4y = 0$.
(d) $x^2y'' - xy' + (x + 1)y = 0$.

11. A Bessel function identity exists of the form

$$J_2(x) - J_0(x) = aJ_c''(x),$$

where a and c are constants. Determine a and c.

12. Find a power series solution of the differential equation $xy'' + y' + y = 0$ convergent for $-\infty < x < +\infty$. Show that for $x > 0$ it can be expressed in terms of a Bessel function.
13. Consider a linear second-order differential equation of the form

$$x^2A(x)y'' + xP(x)y' + Q(x)y = 0,$$

where $A(x)$, $P(x)$, and $Q(x)$ have power series expansions,

$$A(x) = \sum_{k=0}^{\infty} a_k x^k, \qquad P(x) = \sum_{k=0}^{\infty} p_k x^k, \qquad Q(x) = \sum_{k=0}^{\infty} q_k x^k,$$

with $a_0 \neq 0$, each convergent in an open interval $(-r, r)$. If the differential equation has a series solution of the form

$$y = x^t \sum_{n=0}^{\infty} c_n x^n,$$

valid for $0 < x < r$, show that t satisfies a quadratic equation of the form $t^2 + bt + c = 0$, and determine b and c in terms of coefficients of the series for $A(x)$, $P(x)$, and $Q(x)$.
14. Consider a special case of Exercise 13 in which $A(x) = 1 - x$, $P(x) = \frac{1}{2}$, and $Q(x) = -\frac{1}{4}x$. Find a series solution with t not an integer.
15. The differential equation $2x^2y'' + (x^2 - x)y' + y = 0$ has two independent solutions of the form

$$y = x^t \sum_{n=0}^{\infty} c_n x^n,$$

valid for $x > 0$. Determine these solutions.
16. The nonlinear differential equation $y'' + y + \alpha y^2 = 0$ is only "mildly" nonlinear if α is a small nonzero constant. Assume there is a solution which can be expressed as a power series in α of the form

$$y = \sum_{n=0}^{\infty} u_n(x)\alpha^n \qquad \text{(valid in some interval } 0 < \alpha < r)$$

and that this solution satisfies the initial conditions $y = 1$ and $y' = 0$ when $x = 0$. To conform with these initial conditions, we try to choose the coefficients $u_n(x)$ so that $u_0(0) = 1$, $u_0'(0) = 0$ and $u_n(0) = u_n'(0) = 0$ for $n \geq 1$. Substitute this series in the differential equation, equate suitable powers of α and thereby determine $u_0(x)$ and $u_1(x)$.

7

SYSTEMS OF DIFFERENTIAL EQUATIONS

7.1 Introduction

Although the study of differential equations began in the 17th century, it was not until the 19th century that mathematicians realized that relatively few differential equations could be solved by elementary means. The work of Cauchy, Liouville, and others showed the importance of establishing general theorems to guarantee the existence of solutions to certain specific classes of differential equations. Chapter 6 illustrated the use of an existence-uniqueness theorem in the study of linear differential equations. This chapter is concerned with a proof of this theorem and related topics.

Existence theory for differential equations of higher order can be reduced to the first-order case by the introduction of *systems* of equations. For example, the second-order equation

$$y'' + 2ty' - y = e^t \tag{7.1}$$

can be transformed to a system of two first-order equations by introducing two unknown functions y_1 and y_2, where

$$y_1 = y\,, \qquad y_2 = y_1'\,.$$

Then we have $y_2' = y_1'' = y''$, so (7.1) can be written as a system of two first-order equations:

$$\begin{aligned} y_1' &= y_2 \\ y_2' &= y_1 - 2ty_2 + e^t\,. \end{aligned} \tag{7.2}$$

We cannot solve the equations separately by the methods of Chapter 6 because each of them involves two unknown functions.

In this chapter we consider systems consisting of n linear differential equations of first order involving n unknown functions $y_1, \ldots, y_n$. These systems have the form

$$\begin{aligned} y_1' &= p_{11}(t)y_1 + p_{12}(t)y_2 + \cdots + p_{1n}(t)y_n + q_1(t) \\ &\;\vdots \\ y_n' &= p_{n1}(t)y_1 + p_{n2}(t)y_2 + \cdots + p_{nn}(t)y_n + q_n(t)\,. \end{aligned} \tag{7.3}$$

The functions p_{ik} and q_i which appear in (7.3) are considered as given functions defined on a given interval J. The functions $y_1, \ldots, y_n$ are unknown functions to be determined. Systems of this type are called *first-order linear systems.* In general, each equation in the system involves more than one unknown function so the equations cannot be solved separately.

A linear differential equation of order n can always be transformed to a linear system. Suppose the given nth order equation is

$$y^{(n)} + a_1 y^{(n-1)} + \cdots + a_n y = R(t), \tag{7.4}$$

where the coefficients a_i are given functions. To transform this to a system we write $y_1 = y$ and introduce a new unknown function for each of the successive derivatives of y. That is, we put

$$y_1 = y, \qquad y_2 = y_1', \qquad y_3 = y_2', \ldots, y_n = y_{n-1}',$$

and rewrite (7.4) as the system

$$\begin{aligned} y_1' &= y_2 \\ y_2' &= y_3 \\ &\cdot \\ &\cdot \\ &\cdot \\ y_{n-1}' &= y_n \\ y_n' &= -a_n y_1 - a_{n-1} y_2 - \cdots - a_1 y_n + R(t). \end{aligned} \tag{7.5}$$

The discussion of systems may be simplified considerably by the use of vector and matrix notation. Consider the general system (7.3) and introduce vector-valued functions $Y = (y_1, \ldots, y_n)$, $Q = (q_1, \ldots, q_n)$, and a matrix-valued function $P = [p_{ij}]$, defined by the equations

$$Y(t) = (y_1(t), \ldots, y_n(t)), \qquad Q(t) = (q_1(t), \ldots, q_n(t)), \qquad P(t) = [p_{ji}(t)]$$

for each t in J. We regard the vectors as $n \times 1$ column matrices and write the system (7.3) in the simpler form

$$Y' = P(t)Y + Q(t). \tag{7.6}$$

For example, in system (7.2) we have

$$Y = \begin{bmatrix} y_1 \\ y_2 \end{bmatrix}, \qquad P(t) = \begin{bmatrix} 0 & 1 \\ 1 & -2t \end{bmatrix}, \qquad Q(t) = \begin{bmatrix} 0 \\ e^t \end{bmatrix}.$$

In system (7.5) we have

$$Y = \begin{bmatrix} y_1 \\ y_2 \\ \cdot \\ \cdot \\ \cdot \\ y_n \end{bmatrix}, \quad P(t) = \begin{bmatrix} 0 & 1 & 0 & \cdots & 0 \\ 0 & 0 & 1 & \cdots & 0 \\ \cdot & \cdot & \cdot & & \cdot \\ \cdot & \cdot & \cdot & & \cdot \\ \cdot & \cdot & \cdot & & \cdot \\ 0 & 0 & 0 & \cdots & 1 \\ -a_n & -a_{n-1} & -a_{n-2} & \cdots & -a_1 \end{bmatrix}, \quad Q(t) = \begin{bmatrix} 0 \\ 0 \\ \cdot \\ \cdot \\ \cdot \\ 0 \\ R(t) \end{bmatrix}.$$

An initial-value problem for system (7.6) is to find a vector-valued function Y which satisfies (7.6) and which also satisfies an initial condition of the form $Y(a) = B$, where $a \in J$ and $B = (b_1, \ldots, b_n)$ is a given n-dimensional vector.

In the case $n = 1$ (the scalar case) we know from Theorem 6.1 that, if P and Q are continuous on J, all solutions of (7.6) are given by the explicit formula

$$\text{(7.7)} \qquad Y(x) = e^{A(x)}Y(a) + e^{A(x)} \int_a^x e^{-A(t)}Q(t)\,dt,$$

where $A(x) = \int_a^x P(t)\,dt$, and a is any point in J. We will show that this formula can be suitably generalized for systems, that is, when $P(t)$ is an $n \times n$ matrix function and $Q(t)$ is an n-dimensional vector function. To do this we must assign a meaning to integrals of matrices and to exponentials of matrices. Therefore, we digress briefly to discuss the calculus of matrix functions.

7.2 Calculus of matrix functions

The generalization of the concepts of integral and derivative for matrix functions is straightforward. If $P(t) = [p_{ij}(t)]$, we define the integral $\int_a^b P(t)\,dt$ by the equation

$$\int_a^b P(t)\,dt = \left[\int_a^b p_{ij}(t)\,dt\right].$$

That is, the integral of matrix $P(t)$ is the matrix obtained by integrating each entry of $P(t)$, assuming of course, that each entry is integrable on $[a, b]$. The reader can verify that the linearity property for integrals generalizes to matrix functions.

Continuity and differentiability of matrix functions are also defined in terms of the entries. We say that a matrix function $P = [p_{ij}]$ is continuous at t if each entry p_{ij} is continuous at t. The derivative P' is defined by differentiating each entry,

$$P'(t) = [p'_{ij}(t)],$$

whenever all derivatives $p'_{ij}(t)$ exist. It is easy to verify the basic differentiation rules for sums and products. For example, if P and Q are differentiable matrix functions, we have

$$(P + Q)' = P' + Q'$$

if P and Q are of the same size, and we also have

$$(PQ)' = PQ' + P'Q$$

if the product PQ is defined. The chain rule also holds. That is, if $F(t) = P[g(t)]$, where P is a differentiable matrix function and g is a differentiable scalar function, then $F'(t) = g'(t)P'[g(t)]$. The zero-derivative theorem, and the first and second fundamental theorems of calculus are also valid for matrix functions. Proofs of these properties are requested in the next set of exercises.

The definition of the exponential of a matrix is not so simple and requires further preparation. This is discussed in the next section.

7.3 Infinite series of matrices. Norms of matrices

Let $A = [a_{ij}]$ be an $n \times n$ matrix of real or complex entries. We wish to define the exponential e^A in such a way that it possesses some of the fundamental properties of the ordinary real or complex-valued exponential. In particular, we shall require the law of exponents in the form

$$e^{tA}e^{sA} = e^{(t+s)A} \qquad \text{for all real } s \text{ and } t, \tag{7.8}$$

and the relation

$$e^{O} = I, \tag{7.9}$$

where O and I are the $n \times n$ zero and identity matrices, respectively. It might seem natural to define e^A to be the matrix $[e^{a_{ij}}]$. However, this is unacceptable since it satisfies neither of properties (7.8) or (7.9). Instead, we shall define e^A by means of a power series expansion,

$$e^A = \sum_{k=0}^{\infty} \frac{A^k}{k!}.$$

We know that this formula holds if A is a real or complex number, and we will prove that it implies properties (7.8) and (7.9) if A is a matrix. Before we can do this we need to explain what is meant by a convergent series of matrices.

DEFINITION OF CONVERGENT SERIES OF MATRICES. *Given an infinite sequence of $m \times n$ matrices $\{C_k\}$ whose entries are real or complex numbers, denote the ij-entry of C_k by $c_{ij}^{(k)}$. If all mn series*

$$\sum_{k=1}^{\infty} c_{ij}^{(k)} \qquad (i = 1, \ldots, m; j = 1, \ldots, n) \tag{7.10}$$

are convergent, then we say the series of matrices $\sum_{k=1}^{\infty} C_k$ is convergent, and its sum is defined to be the $m \times n$ matrix whose ij-entry is the series in (7.10).

A simple and useful test for convergence of a series of matrices can be given in terms of the *norm* of a matrix, a generalization of the absolute value of a number.

DEFINITION OF NORM OF A MATRIX. *If $A = [a_{ij}]$ is an $m \times n$ matrix of real or complex entries, the norm of A, denoted by $\|A\|$, is defined to be the nonnegative number given by the formula*

$$\|A\| = \sum_{i=1}^{m} \sum_{j=1}^{n} |a_{ij}| . \tag{7.11}$$

In other words, the norm of A is the sum of the absolute values of all its entries. There are other definitions of norms that are sometimes used, but we have chosen this one because of the ease with which we can prove the following properties.

THEOREM 7.1. FUNDAMENTAL PROPERTIES OF NORMS. *For rectangular matrices A and B, and all real or complex scalars c we have*

$$\|A + B\| \le \|A\| + \|B\| , \; \|AB\| \le \|A\| \, \|B\| , \; \|cA\| = |c| \, \|A\| .$$

Proof. We prove only the result for $\|AB\|$, assuming that A is $m \times n$ and B is $n \times p$. The proofs of the others are simpler and are left as exercises.

Writing $A = [a_{ik}]$, $B = [b_{kj}]$, we have $AB = [\sum_{k=1}^{n} a_{ik} b_{kj}]$, so from (7.11) we obtain

$$\|AB\| = \sum_{i=1}^{m} \sum_{j=1}^{p} \left| \sum_{k=1}^{n} a_{ik} b_{kj} \right| \le \sum_{i=1}^{m} \sum_{k=1}^{n} |a_{ik}| \sum_{j=1}^{p} |b_{kj}| \le \sum_{i=1}^{m} \sum_{k=1}^{n} |a_{ik}| \, \|B\| = \|A\| \, \|B\| .$$

Note that in the special case $B = A$ the inequality for $\|AB\|$ becomes $\|A^2\| \le \|A\|^2$. By induction we also have

$$\|A^k\| \le \|A\|^k \qquad \text{for} \quad k = 1, 2, 3, \ldots .$$

These inequalities will be useful in the discussion of the exponential matrix.

The next theorem gives a useful sufficient condition for convergence of a series of matrices.

THEOREM 7.2. TEST FOR CONVERGENCE OF A MATRIX SERIES. *If $\{C_k\}$ is a sequence of $m \times n$ matrices such that $\sum_{k=1}^{\infty} \|C_k\|$ converges, then the matrix series $\sum_{k=1}^{\infty} C_k$ also converges.*

Proof. Let the ij-entry of C_k be denoted by $c_{ij}^{(k)}$. Since $|c_{ij}^{(k)}| \le \|C_k\|$, convergence of $\sum_{k=1}^{\infty} \|C_k\|$ implies absolute convergence of each series $\sum_{k=1}^{\infty} c_{ij}^{(k)}$. Hence each series $\sum_{k=1}^{\infty} c_{ij}^{(k)}$ is convergent, so the matrix series $\sum_{k=1}^{\infty} C_k$ is convergent.

7.4 Exercises

1. Verify that the linearity property of integrals also holds for integrals of matrix functions.
2. Verify each of the following differentiation rules for matrix functions, assuming P and Q are differentiable. In (a), P and Q must be of the same size so that $P + Q$ is meaningful. In

(b) and (d) they need not be of the same size provided the products are meaningful. In (c) and (d), Q is assumed to be nonsingular.

(a) $(P + Q)' = P' + Q'$.
(b) $(PQ)' = PQ' + P'Q$.
(c) $(Q^{-1})' = -Q^{-1}Q'Q^{-1}$.
(d) $(PQ^{-1})' = -PQ^{-1}Q'Q^{-1} + P'Q^{-1}$.

3. (a) Let P be a differentiable matrix function. Prove that the derivatives of P^2 and P^3 are given by the formulas

$$(P^2)' = PP' + P'P, \qquad (P^3)' = P^2P' + PP'P + P'P^2.$$

(b) Guess a general formula for the derivative of P^k and prove it by induction.

4. Let P be a differentiable matrix function and let g be a differentiable scalar function whose range is a subset of the domain of P. Define the composite function $F(t) = P[g(t)]$ and prove the chain rule, $F'(t) = g'(t)P'[g(t)]$.

5. Prove the zero-derivative theorem for matrix functions: *If $P'(t) = O$ for every t in an open interval (a, b), then the matrix function P is constant on (a, b).*

6. State and prove generalizations of the first and second fundamental theorems of calculus for matrix functions.

7. State and prove a formula for integration by parts in which the integrands are matrix functions.

8. Prove the following properties of matrix norms:

$$\|A + B\| \le \|A\| + \|B\|, \qquad \|cA\| = |c| \, \|A\|.$$

9. If a matrix function P is integrable on an interval $[a, b]$ prove that

$$\left\| \int_a^b P(t)\, dt \right\| \le \int_a^b \|P(t)\| \, dt.$$

10. Let D be an $n \times n$ diagonal matrix, say $D = \text{diag}\,(\lambda_1, \ldots, \lambda_n)$. Prove that the matrix series $\sum_{k=0}^{\infty} D^k/k!$ converges and is also a diagonal matrix,

$$\sum_{k=0}^{\infty} \frac{D^k}{k!} = \text{diag}\,(e^{\lambda_1}, \ldots, e^{\lambda_n}).$$

(The term corresponding to $k = 0$ is understood to be the identity matrix I.)

11. Let D be an $n \times n$ diagonal matrix, $D = \text{diag}\,(\lambda_1, \ldots, \lambda_n)$. If the matrix series $\sum_{k=0}^{\infty} c_k D^k$ converges, prove that

$$\sum_{k=0}^{\infty} c_k D^k = \text{diag}\left(\sum_{k=0}^{\infty} c_k \lambda_1^k, \ldots, \sum_{k=0}^{\infty} c_k \lambda_n^k\right).$$

12. Assume that the matrix series $\sum_{k=1}^{\infty} C_k$ converges, where each C_k is an $n \times n$ matrix. Prove that the matrix series $\sum_{k=1}^{\infty} (AC_kB)$ also converges and that its sum is the matrix

$$A\left(\sum_{k=1}^{\infty} C_k\right)B.$$

Here A and B are matrices such that the products AC_kB are meaningful.

7.5 The exponential matrix

Using Theorem 7.2 it is easy to prove that the matrix series

(7.12) $$\sum_{k=0}^{\infty} \frac{A^k}{k!}$$

converges for every square matrix A with real or complex entries. (The term corresponding to $k = 0$ is understood to be the identity matrix I.) The norm of each term satisfies the inequality

$$\left\| \frac{A^k}{k!} \right\| \le \frac{\|A\|^k}{k!}.$$

Since the series $\sum a^k/k!$ converges for every real a, Theorem 7.2 implies that the series in (7.12) converges for every square matrix A.

DEFINITION OF THE EXPONENTIAL MATRIX. *For any $n \times n$ matrix A with real or complex entries we define the exponential e^A to be the $n \times n$ matrix given by the convergent series in* (7.12). *That is,*

$$e^A = \sum_{k=0}^{\infty} \frac{A^k}{k!}.$$

Note that this definition implies $e^O = I$, where O is the zero matrix. Further properties of the exponential will be developed with the help of differential equations.

7.6 The differential equation satisfied by e^{tA}

Let t be a real number, let A be an $n \times n$ matrix, and let $E(t)$ be the $n \times n$ matrix given by

$$E(t) = e^{tA}.$$

We shall keep A fixed and study this matrix as a function of t. First we obtain a differential equation satisfied by E.

THEOREM 7.3. *For every real t the matrix function E defined by $E(t) = e^{tA}$ satisfies the matrix differential equation*

$$E'(t) = E(t)A = AE(t).$$

Proof. From the definition of the exponential matrix we have

$$E(t) = \sum_{k=0}^{\infty} \frac{(tA)^k}{k!} = \sum_{k=0}^{\infty} \frac{t^k A^k}{k!}.$$

Let $c_{ij}^{(k)}$ denote the ij-entry of A^k. Then the ij-entry of $t^kA^k/k!$ is $t^kc_{ij}^{(k)}/k!$. Hence, from the definition of a matrix series, we have

$$(7.13) \qquad \sum_{k=0}^{\infty} \frac{t^kA^k}{k!} = \left[\sum_{k=0}^{\infty} \frac{t^k}{k!} c_{ij}^{(k)}\right].$$

Each entry on the right of (7.13) is a power series in t, convergent for all t. Therefore its derivative exists for all t and is given by the differentiated series

$$\sum_{k=1}^{\infty} \frac{kt^{k-1}}{k!} c_{ij}^{(k)} = \sum_{k=0}^{\infty} \frac{t^k}{k!} c_{ij}^{(k+1)}.$$

This shows that the derivative $E'(t)$ exists and is given by the matrix series

$$E'(t) = \sum_{k=0}^{\infty} \frac{t^kA^{k+1}}{k!} = \left(\sum_{k=0}^{\infty} \frac{t^kA^k}{k!}\right)A = E(t)A.$$

In the last equation we used the property $A^{k+1} = A^kA$. Since A commutes with A^k we could also have written $A^{k+1} = AA^k$ to obtain the relation $E'(t) = AE(t)$. This completes the proof.

Note: The foregoing proof also shows that A commutes with e^{tA}.

7.7 Uniqueness theorem for the matrix differential equation $F'(t) = AF(t)$

In this section we prove a uniqueness theorem which characterizes all solutions of the matrix differential equation $F'(t) = AF(t)$. The proof makes use of the following theorem.

THEOREM 7.4. NONSINGULARITY OF e^{tA}. *For any $n \times n$ matrix A and any scalar t we have*

$$(7.14) \qquad e^{tA}e^{-tA} = I.$$

Hence e^{tA} is nonsingular, and its inverse is e^{-tA}.

Proof. Let F be the matrix function defined for all real t by the equation

$$F(t) = e^{tA}e^{-tA}.$$

We shall prove that $F(t)$ is the identity matrix I by showing that the derivative $F'(t)$ is the zero matrix. Differentiating F as a product, using the result of Theorem 7.3, we find

$$\begin{aligned} F'(t) &= e^{tA}(e^{-tA})' + (e^{tA})'e^{-tA} = e^{tA}(-Ae^{-tA}) + Ae^{tA}e^{-tA} \\ &= -Ae^{tA}e^{-tA} + Ae^{tA}e^{-tA} = O, \end{aligned}$$

since A commutes with e^{tA}. Therefore, by the zero-derivative theorem, F is a constant matrix. But $F(0) = e^{OA}e^{OA} = I$, so $F(t) = I$ for all t. This proves (7.14).

THEOREM 7.5. UNIQUENESS THEOREM. *Let A and B be given $n \times n$ constant matrices. Then the only $n \times n$ matrix function F satisfying the initial-value problem*

$$F'(t) = AF(t), \qquad F(0) = B$$

for $-\infty < t < +\infty$ is

$$F(t) = e^{tA}B. \tag{7.15}$$

Proof. First we note that $e^{tA}B$ is a solution. Now let F be any solution and consider the matrix function

$$G(t) = e^{-tA}F(t).$$

Differentiating this product we obtain

$$G'(t) = e^{-tA}F'(t) - Ae^{-tA}F(t) = e^{-tA}AF(t) - e^{-tA}AF(t) = O.$$

Therefore $G(t)$ is a constant matrix,

$$G(t) = G(0) = F(0) = B.$$

In other words, $e^{-tA}F(t) = B$. Multiplying by e^{tA} and using (7.14) we obtain (7.15).

Note: The same type of proof shows that $F(t) = Be^{tA}$ is the only solution of the initial-value problem

$$F'(t) = F(t)A, \qquad F(0) = B.$$

7.8 The law of exponents for exponential matrices

The law of exponents $e^Ae^B = e^{A+B}$ is not always true for matrix exponentials. A counter example is given in Exercise 13 of Section 7.12. However, it is not difficult to prove that the formula is true for matrices A and B which commute.

THEOREM 7.6. *Let A and B be two $n \times n$ matrices which commute, $AB = BA$. Then we have*

$$e^{A+B} = e^Ae^B. \tag{7.16}$$

Proof. From the equation $AB = BA$ we find that

$$A^2B = A(BA) = (AB)A = (BA)A = BA^2,$$

so B commutes with A^2. By induction, B commutes with every power of A. By writing e^{tA} as a power series we find that B also commutes with e^{tA} for every real t.

Now let F be the matrix function defined by the equation

$$F(t) = e^{t(A+B)} - e^{tA}e^{tB}.$$

Differentiating $F(t)$ and using the fact that B commutes with e^{tA} we find

$$F'(t) = (A + B)e^{t(A+B)} - Ae^{tA}e^{tB} - e^{tA}Be^{tB}$$
$$= (A + B)e^{t(A+B)} - (A + B)e^{tA}e^{tB} = (A + B)F(t).$$

By the uniqueness theorem we have

$$F(t) = e^{t(A+B)}F(0).$$

But $F(0) = O$, so $F(t) = O$ for all t. Hence

$$e^{t(A+B)} = e^{tA}e^{tB}.$$

When $t = 1$ we obtain (7.16).

EXAMPLE. The matrices sA and tA commute for all scalars s and t. Hence we have

$$e^{sA}e^{tA} = e^{(s+t)A}.$$

7.9 Existence and uniqueness theorems for homogeneous linear systems with constant coefficients

The vector differential equation $Y'(t) = AY(t)$, where A is an $n \times n$ constant matrix and Y is an n-dimensional vector function (regarded as an $n \times 1$ column matrix) is called a *homogeneous linear system with constant coefficients.* We shall use the exponential matrix to give an explicit formula for the solution of such a system.

THEOREM 7.7. *Let A be a given $n \times n$ constant matrix and let B be a given n-dimensional vector. Then the initial-value problem*

$$Y'(t) = AY(t), \qquad Y(0) = B, \tag{7.17}$$

has a unique solution on the interval $-\infty < t < +\infty$. This solution is given by the formula

$$Y(t) = e^{tA}B. \tag{7.18}$$

More generally, the unique solution of the initial value problem

$$Y'(t) = AY(t), \qquad Y(a) = B,$$

is $Y(t) = e^{(t-a)A}B$.

Proof. Differentiation of (7.18) gives us $Y'(t) = Ae^{tA}B = AY(t)$. Since $Y(0) = B$, this is a solution of the initial-value problem (7.17).

To prove that it is the only solution we argue as in the proof of Theorem 7.5. Let $Z(t)$ be another vector function satisfying $Z'(t) = AZ(t)$ with $Z(0) = B$, and let $G(t) = e^{-tA}Z(t)$. Then we easily verify that $G'(t) = O$, so $G(t) = G(0) = Z(0) = B$. In other words,

$e^{-tA}Z(t) = B$, so $Z(t) = e^{tA}B = Y(t)$. The more general case with initial value $Y(a) = B$ is treated in exactly the same way.

7.10 The problem of calculating e^{tA}

Although Theorem 7.7 gives an explicit formula for the solution of a homogeneous system with constant coefficients, there still remains the problem of actually computing the exponential matrix e^{tA}. If we were to calculate e^{tA} directly from the series definition we would have to compute all the powers A^k for $k = 0, 1, 2, \ldots$, and then compute the sum of each series $\sum_{k=0}^{\infty} t^k c_{ij}^{(k)}/k!$, where $c_{ij}^{(k)}$ is the ij-entry of A^k. In general this is a hopeless task unless A is a matrix whose powers may be readily calculated. For example, if A is a diagonal matrix, say

$$A = \text{diag}\,(\lambda_1, \ldots, \lambda_n),$$

then every power of A is also a diagonal matrix, in fact,

$$A^k = \text{diag}\,(\lambda_1^k, \ldots, \lambda_n^k).$$

Therefore in this case e^{tA} is a diagonal matrix given by

$$e^{tA} = \text{diag}\left(\sum_{k=0}^{\infty} \frac{t^k}{k!}\lambda_1^k, \ldots, \sum_{k=0}^{\infty} \frac{t^k}{k!}\lambda_n^k\right) = \text{diag}\,(e^{t\lambda_1}, \ldots, e^{t\lambda_n}).$$

Another easy case to handle is when A is a matrix which can be diagonalized. For example, if there is a nonsingular matrix C such that $C^{-1}AC$ is a diagonal matrix, say $C^{-1}AC = D$, then we have $A = CDC^{-1}$, from which we find

$$A^2 = (CDC^{-1})(CDC^{-1}) = CD^2C^{-1},$$

and, more generally,

$$A^k = CD^kC^{-1}.$$

Therefore in this case we have

$$e^{tA} = \sum_{k=0}^{\infty} \frac{t^kA^k}{k!} = \sum_{k=0}^{\infty} \frac{t^k}{k!} CD^kC^{-1} = C\left(\sum_{k=0}^{\infty} \frac{t^kD^k}{k!}\right)C^{-1} = Ce^{tD}C^{-1}.$$

Here the difficulty lies in determining C and its inverse. Once these are known, e^{tA} is easily calculated. Of course, not every matrix can be diagonalized so the usefulness of the foregoing remarks is limited.

EXAMPLE 1. Calculate e^{tA} for the 2×2 matrix $A = \begin{bmatrix} 5 & 4 \\ 1 & 2 \end{bmatrix}$.

Solution. This matrix has distinct eigenvalues $\lambda_1 = 6$, $\lambda_2 = 1$, so there is a nonsingular matrix $C = \begin{bmatrix} a & b \\ c & d \end{bmatrix}$ such that $C^{-1}AC = D$, where $D = \text{diag}\,(\lambda_1, \lambda_2) = \begin{bmatrix} 6 & 0 \\ 0 & 1 \end{bmatrix}$. To determine C we can write $AC = CD$, or

$$\begin{bmatrix} 5 & 4 \\ 1 & 2 \end{bmatrix}\begin{bmatrix} a & b \\ c & d \end{bmatrix} = \begin{bmatrix} a & b \\ c & d \end{bmatrix}\begin{bmatrix} 6 & 0 \\ 0 & 1 \end{bmatrix}.$$

Multiplying the matrices, we find that this equation is satisfied for any scalars a, b, c, d with $a = 4c$, $b = -d$. Taking $c = d = 1$ we choose

$$C = \begin{bmatrix} 4 & -1 \\ 1 & 1 \end{bmatrix}, \qquad C^{-1} = \frac{1}{5}\begin{bmatrix} 1 & 1 \\ -1 & 4 \end{bmatrix}.$$

Therefore

$$e^{tA} = Ce^{tD}C^{-1} = \frac{1}{5}\begin{bmatrix} 4 & -1 \\ 1 & 1 \end{bmatrix}\begin{bmatrix} e^{6t} & 0 \\ 0 & e^{t} \end{bmatrix}\begin{bmatrix} 1 & 1 \\ -1 & 4 \end{bmatrix}$$

$$= \frac{1}{5}\begin{bmatrix} 4 & -1 \\ 1 & 1 \end{bmatrix}\begin{bmatrix} e^{6t} & e^{6t} \\ -e^{t} & 4e^{t} \end{bmatrix} = \frac{1}{5}\begin{bmatrix} 4e^{6t} + e^{t} & 4e^{6t} - 4e^{t} \\ e^{6t} - e^{t} & e^{6t} + 4e^{t} \end{bmatrix}.$$

EXAMPLE 2. Solve the linear system

$$y_1' = 5y_1 + 4y_2$$
$$y_2' = y_1 + 2y_2$$

subject to the initial conditions $y_1(0) = 2$, $y_2(0) = 3$.

Solution. In matrix form the system can be written as

$$Y'(t) = AY(t), \qquad Y(0) = \begin{bmatrix} 2 \\ 3 \end{bmatrix}, \qquad \text{where} \quad A = \begin{bmatrix} 5 & 4 \\ 1 & 2 \end{bmatrix}.$$

By Theorem 7.7 the solution is $Y(t) = e^{tA}Y(0)$. Using the matrix e^{tA} calculated in Example 1 we find

$$\begin{bmatrix} y_1 \\ y_2 \end{bmatrix} = \frac{1}{5}\begin{bmatrix} 4e^{6t} + e^{t} & 4e^{6t} - 4e^{t} \\ e^{6t} - e^{t} & e^{6t} + 4e^{t} \end{bmatrix}\begin{bmatrix} 2 \\ 3 \end{bmatrix}$$

from which we obtain

$$y_1 = 4e^{6t} - 2e^{t}, \qquad y_2 = e^{6t} + 2e^{t}.$$

There are many methods known for calculating e^{tA} when A cannot be diagonalized. Most of these methods are rather complicated and require preliminary matrix transformations, the nature of which depends on the multiplicities of the eigenvalues of A. In a later section we shall discuss a practical and straightforward method for calculating e^{tA} which can be used whether or not A can be diagonalized. It is valid for *all* matrices A and requires no preliminary transformations of any kind. This method was developed by E. J. Putzer in a paper in the *American Mathematical Monthly*, Vol. 73 (1966), pp. 2–7. It is based on a famous theorem attributed to Arthur Cayley (1821–1895) and William Rowan Hamilton

(1805–1865) which states that every square matrix satisfies its characteristic equation. First we shall prove the Cayley-Hamilton theorem and then we shall use it to obtain Putzer's formulas for calculating e^{tA}.

7.11 The Cayley-Hamilton theorem

THEOREM 7.8. CAYLEY-HAMILTON THEOREM. *Let A be an $n \times n$ matrix and let*

$$f(\lambda) = \det(\lambda I - A) = \lambda^n + c_{n-1}\lambda^{n-1} + \cdots + c_1\lambda + c_0 \tag{7.19}$$

be its characteristic polynomial. Then $f(A) = O$. In other words, A satisfies the equation

$$A^n + c_{n-1}A^{n-1} + \cdots + c_1A + c_0I = O. \tag{7.20}$$

Proof. The proof is based on Theorem 3.12 which states that for any square matrix A we have

$$A\,(\text{cof}\,A)^t = (\det A)I. \tag{7.21}$$

We apply this formula with A replaced by $\lambda I - A$. Since $\det(\lambda I - A) = f(\lambda)$, Equation (7.21) becomes

$$(\lambda I - A)\{\text{cof}\,(\lambda I - A)\}^t = f(\lambda)I. \tag{7.22}$$

This equation is valid for all real λ. The idea of the proof is to show that it is also valid when λ is replaced by A.

The entries of the matrix $\text{cof}\,(\lambda I - A)$ are the cofactors of $\lambda I - A$. Except for a factor ± 1, each such cofactor is the determinant of a minor of $\lambda I - A$ of order $n - 1$. Therefore each entry of $\text{cof}\,(\lambda I - A)$, and hence of $\{\text{cof}\,(\lambda I - A)\}^t$, is a polynomial in λ of degree $\le n - 1$. Therefore

$$\{\text{cof}\,(\lambda I - A)\}^t = \sum_{k=0}^{n-1} \lambda^k B_k,$$

where each coefficient B_k is an $n \times n$ matrix with scalar entries. Using this in (7.22) we obtain the relation

$$(\lambda I - A)\sum_{k=0}^{n-1} \lambda^k B_k = f(\lambda)I \tag{7.23}$$

which can be rewritten in the form

$$\lambda^n B_{n-1} + \sum_{k=1}^{n-1} \lambda^k(B_{k-1} - AB_k) - AB_0 = \lambda^n I + \sum_{k=1}^{n-1} \lambda^k c_k I + c_0 I. \tag{7.24}$$

At this stage we equate coefficients of like powers of λ in (7.24) to obtain the equations

$$
\begin{aligned}
B_{n-1} &= I \\
B_{n-2} - AB_{n-1} &= c_{n-1}I \\
&\;\;\vdots \\
B_0 - AB_1 &= c_1 I \\
-AB_0 &= c_0 I.
\end{aligned}
\tag{7.25}
$$

Equating coefficients is permissible because (7.24) is equivalent to n^2 scalar equations, in each of which we may equate coefficients of like powers of λ. Now we multiply the equations in (7.25) in succession by $A^n, A^{n-1}, \ldots, A, I$ and add the results. The terms on the left cancel and we obtain

$$O = A^n + c_{n-1}A^{n-1} + \cdots + c_1 A + c_0 I.$$

This proves the Cayley-Hamilton theorem.

Note: Hamilton proved the theorem in 1853 for a special class of matrices. A few years later, Cayley announced that the theorem is true for all matrices, but gave no proof.

EXAMPLE. The matrix $A = \begin{bmatrix} 5 & 4 & 0 \\ 1 & 2 & 0 \\ 1 & 2 & 2 \end{bmatrix}$ has characteristic polynomial

$$f(\lambda) = (\lambda - 1)(\lambda - 2)(\lambda - 6) = \lambda^3 - 9\lambda^2 + 20\lambda - 12.$$

The Cayley-Hamilton theorem states that A satisfies the equation

$$A^3 - 9A^2 + 20A - 12I = O. \tag{7.26}$$

This equation can be used to express A^3 and all higher powers of A in terms of I, A, and A^2. For example, we have

$$
\begin{aligned}
A^3 &= 9A^2 - 20A + 12I, \\
A^4 &= 9A^3 - 20A^2 + 12A = 9(9A^2 - 20A + 12I) - 20A^2 + 12A \\
&= 61A^2 - 168A + 108I.
\end{aligned}
$$

It can also be used to express A^{-1} as a polynomial in A. From (7.26) we write $A(A^2 - 9A + 20I) = 12I$, and we obtain

$$A^{-1} = \tfrac{1}{12}(A^2 - 9A + 20I).$$

7.12 Exercises

In each of Exercises 1 through 4, (a) express A^{-1}, A^2 and all higher powers of A as a linear combination of I and A. (The Cayley-Hamilton theorem can be of help.) (b) Calculate e^{tA}.

1. $A = \begin{bmatrix} 1 & 0 \\ 1 & 1 \end{bmatrix}$.
2. $A = \begin{bmatrix} 1 & 0 \\ 1 & 2 \end{bmatrix}$.
3. $A = \begin{bmatrix} 0 & 1 \\ 1 & 0 \end{bmatrix}$.
4. $A = \begin{bmatrix} -1 & 0 \\ 0 & 1 \end{bmatrix}$.

5. (a) If $A = \begin{bmatrix} 0 & 1 \\ -1 & 0 \end{bmatrix}$, prove that $e^{tA} = \begin{bmatrix} \cos t & \sin t \\ -\sin t & \cos t \end{bmatrix}$.

   (b) Find a corresponding formula for e^{tA} when $A = \begin{bmatrix} a & b \\ -b & a \end{bmatrix}$, a, b real.

6. If $F(t) = \begin{bmatrix} t & t-1 \\ 0 & 1 \end{bmatrix}$, prove that $e^{F(t)} = eF(et^{-1})$.

7. If $A(t)$ is a scalar function of t, the derivative of $e^{A(t)}$ is $e^{A(t)}A'(t)$. Compute the derivative of $e^{A(t)}$ when $A(t) = \begin{bmatrix} 1 & t \\ 0 & 0 \end{bmatrix}$ and show that the result is not equal to either of the two products $e^{A(t)}A'(t)$ or $A'(t)e^{A(t)}$.

In each of Exercises 8, 9, 10, (a) calculate A^n, and express A^3 in terms of I, A, A^2. (b) Calculate e^{tA}.

8. $A = \begin{bmatrix} 0 & 1 & 1 \\ 0 & 0 & 1 \\ 0 & 0 & 0 \end{bmatrix}$.
9. $A = \begin{bmatrix} 0 & 1 & 1 \\ 0 & 1 & 1 \\ 0 & 0 & 0 \end{bmatrix}$.
10. $A = \begin{bmatrix} 2 & 0 & 0 \\ 0 & 1 & 0 \\ 0 & 1 & 1 \end{bmatrix}$.

11. If $A = \begin{bmatrix} 0 & -1 & 0 \\ 1 & 0 & 1 \\ 0 & 1 & 0 \end{bmatrix}$, express e^{tA} as a linear combination of I, A, A^2.

12. If $A = \begin{bmatrix} 0 & 1 & 0 \\ 2 & 0 & 2 \\ 0 & 1 & 0 \end{bmatrix}$, prove that $e^{A} = \begin{bmatrix} x^2 & xy & y^2 \\ 2xy & x^2+y^2 & 2xy \\ y^2 & xy & x^2 \end{bmatrix}$, where $x = \cosh 1$ and $y = \sinh 1$.

13. This example shows that the equation $e^{A+B} = e^A e^B$ is not always true for matrix exponentials. Compute each of the matrices $e^A e^B$, $e^B e^A$, e^{A+B} when $A = \begin{bmatrix} 1 & 1 \\ 0 & 0 \end{bmatrix}$ and $B = \begin{bmatrix} 1 & -1 \\ 0 & 0 \end{bmatrix}$, and note that the three results are distinct.

7.13 Putzer's method for calculating e^{tA}

The Cayley-Hamilton theorem shows that the nth power of any $n \times n$ matrix A can be expressed as a linear combination of the lower powers $I, A, A^2, \ldots, A^{n-1}$. It follows that each of the higher powers $A^{n+1}, A^{n+2}, \ldots$, can also be expressed as a linear combination of

$I, A, A^2, \ldots, A^{n-1}$. Therefore, in the infinite series defining e^{tA}, each term $t^k A^k/k!$ with $k \geq n$ is a linear combination of $t^k I, t^k A, t^k A^2, \ldots, t^k A^{n-1}$. Hence we can expect that e^{tA} should be expressible as a polynomial in A of the form

$$e^{tA} = \sum_{k=0}^{n-1} q_k(t) A^k, \tag{7.27}$$

where the scalar coefficients $q_k(t)$ depend on t. Putzer developed two useful methods for expressing e^{tA} as a polynomial in A. The next theorem describes the simpler of the two methods.

THEOREM 7.9. *Let $\lambda_1, \ldots, \lambda_n$ be the eigenvalues of an $n \times n$ matrix A, and define a sequence of polynomials in A as follows:*

$$P_0(A) = I, \qquad P_k(A) = \prod_{m=1}^{k} (A - \lambda_m I), \qquad \textit{for} \quad k = 1, 2, \ldots, n. \tag{7.28}$$

Then we have

$$e^{tA} = \sum_{k=0}^{n-1} r_{k+1}(t) P_k(A), \tag{7.29}$$

where the scalar coefficients $r_1(t), \ldots, r_n(t)$ are determined recursively from the system of linear differential equations

$$\begin{aligned} r_1'(t) &= \lambda_1 r_1(t), \qquad r_1(0) = 1, \\ r_{k+1}'(t) &= \lambda_{k+1} r_{k+1}(t) + r_k(t), \qquad r_{k+1}(0) = 0, \qquad (k = 1, 2, \ldots, n-1). \end{aligned} \tag{7.30}$$

Note: Equation (7.29) does not express e^{tA} directly in powers of A as indicated in (7.27), but as a linear combination of the polynomials $P_0(A), P_1(A), \ldots, P_{n-1}(A)$. These polynomials are easily calculated once the eigenvalues of A are determined. Also the multipliers $r_1(t), \ldots, r_n(t)$ in (7.30) are easily calculated. Although this requires solving a system of linear differential equations, this particular system has a triangular matrix and the solutions can be determined in succession.

Proof. Let $r_1(t), \ldots, r_n(t)$ be the scalar functions determined by (7.30) and define a matrix function F by the equation

$$F(t) = \sum_{k=0}^{n-1} r_{k+1}(t) P_k(A). \tag{7.31}$$

Note that $F(0) = r_1(0)P_0(A) = I$. We will prove that $F(t) = e^{tA}$ by showing that F satisfies the same differential equation as e^{tA}, namely, $F'(t) = AF(t)$.

Differentiating (7.31) and using the recursion formulas (7.30) we obtain

$$F'(t) = \sum_{k=0}^{n-1} r_{k+1}'(t) P_k(A) = \sum_{k=0}^{n-1} \{r_k(t) + \lambda_{k+1} r_{k+1}(t)\} P_k(A),$$

where $r_0(t)$ is defined to be 0. We rewrite this in the form

$$F'(t) = \sum_{k=0}^{n-2} r_{k+1}(t)P_{k+1}(A) + \sum_{k=0}^{n-1} \lambda_{k+1}r_{k+1}(t)P_k(A),$$

then subtract $\lambda_n F(t) = \sum_{k=0}^{n-1} \lambda_n r_{k+1}(t)P_k(A)$ to obtain the relation

$$F'(t) - \lambda_n F(t) = \sum_{k=0}^{n-2} r_{k+1}(t)\{P_{k+1}(A) + (\lambda_{k+1} - \lambda_n)P_k(A)\}. \tag{7.32}$$

But from (7.28) we see that $P_{k+1}(A) = (A - \lambda_{k+1}I)P_k(A)$, so

$$\begin{aligned} P_{k+1}(A) + (\lambda_{k+1} - \lambda_n)P_k(A) &= (A - \lambda_{k+1}I)P_k(A) + (\lambda_{k+1} - \lambda_n)P_k(A) \\ &= (A - \lambda_n I)P_k(A). \end{aligned}$$

Therefore Equation (7.32) becomes

$$\begin{aligned} F'(t) - \lambda_n F(t) = (A - \lambda_n I)\sum_{k=0}^{n-2} r_{k+1}(t)P_k(A) &= (A - \lambda_n I)\{F(t) - r_n(t)P_{n-1}(A)\} \\ &= (A - \lambda_n I)F(t) - r_n(t)P_n(A). \end{aligned}$$

The Cayley-Hamilton theorem implies that $P_n(A) = O$, so the last equation becomes

$$F'(t) - \lambda_n F(t) = (A - \lambda_n I)F(t) = AF(t) - \lambda_n F(t),$$

from which we find $F'(t) = AF(t)$. Since $F(0) = I$, the uniqueness theorem (Theorem 7.7) shows that $F(t) = e^{tA}$.

EXAMPLE 1. Express e^{tA} as a linear combination of I and A if A is a 2×2 matrix with both its eigenvalues equal to λ.

Solution. Writing $\lambda_1 = \lambda_2 = \lambda$, we are to solve the system of differential equations

$$\begin{aligned} r_1'(t) &= \lambda r_1(t), & r_1(0) &= 1, \\ r_2'(t) &= \lambda r_2(t) + r_1(t), & r_2(0) &= 0. \end{aligned}$$

Solving these first-order equations in succession we find

$$r_1(t) = e^{\lambda t}, \qquad r_2(t) = te^{\lambda t}.$$

Since $P_0(A) = I$ and $P_1(A) = A - \lambda I$, the required formula for e^{tA} is

$$e^{tA} = e^{\lambda t}I + te^{\lambda t}(A - \lambda I) = e^{\lambda t}(1 - \lambda t)I + te^{\lambda t}A. \tag{7.33}$$

EXAMPLE 2. Solve Example 1 if the eigenvalues of A are λ and μ, where $\lambda \neq \mu$.

Solution. In this case the system of differential equations is

$$r_1'(t) = \lambda r_1(t), \qquad r_1(0) = 1,$$
$$r_2'(t) = \mu r_2(t) + r_1(t), \qquad r_2(0) = 0.$$

Its solutions are given by

$$r_1(t) = e^{\lambda t}, \qquad r_2(t) = \frac{e^{\lambda t} - e^{\mu t}}{\lambda - \mu}.$$

Since $P_0(A) = I$ and $P_1(A) = A - \lambda I$ the required formula for e^{tA} is

$$e^{tA} = e^{\lambda t} I + \frac{e^{\lambda t} - e^{\mu t}}{\lambda - \mu}(A - \lambda I) = \frac{\lambda e^{\mu t} - \mu e^{\lambda t}}{\lambda - \mu} I + \frac{e^{\lambda t} - e^{\mu t}}{\lambda - \mu} A. \tag{7.34}$$

If the eigenvalues λ, μ are complex numbers, the exponentials $e^{\lambda t}$ and $e^{\mu t}$ will also be complex numbers. But if λ and μ are complex conjugates, the scalars multiplying I and A in (7.34) will be real. For example, suppose

$$\lambda = \alpha + i\beta, \qquad \mu = \alpha - i\beta, \qquad \beta \neq 0.$$

Then $\lambda - \mu = 2i\beta$ so Equation (7.34) becomes

$$\begin{aligned} e^{tA} &= e^{(\alpha + i\beta)t} I + \frac{e^{(\alpha+i\beta)t} - e^{(\alpha-i\beta)t}}{2i\beta} [A - (\alpha + i\beta)I] \\ &= e^{\alpha t}\left\{ e^{i\beta t} I + \frac{e^{i\beta t} - e^{-i\beta t}}{2i\beta}(A - \alpha I - i\beta I) \right\} \\ &= e^{\alpha t}\left\{ (\cos \beta t + i \sin \beta t) I + \frac{\sin \beta t}{\beta}(A - \alpha I - i\beta I) \right\}. \end{aligned}$$

The terms involving i cancel and we get

$$e^{tA} = \frac{e^{\alpha t}}{\beta}\{(\beta \cos \beta t - \alpha \sin \beta t) I + \sin \beta t \, A\}. \tag{7.35}$$

7.14 Alternate methods for calculating e^{tA} in special cases

Putzer's method for expressing e^{tA} as a polynomial in A is completely general because it is valid for all square matrices A. A general method is not always the simplest method to use in certain special cases. In this section we give simpler methods for computing e^{tA} in three special cases: (a) When all the eigenvalues of A are equal, (b) when all the eigenvalues of A are distinct, and (c) when A has two distinct eigenvalues, exactly one of which has multiplicity 1.

THEOREM 7.10. *If A is an $n \times n$ matrix with all its eigenvalues equal to λ, then we have*

$$(7.36) \qquad e^{tA} = e^{\lambda t}\sum_{k=0}^{n-1}\frac{t^k}{k!}(A - \lambda I)^k .$$

Proof. Since the matrices $\lambda t I$ and $t(A - \lambda I)$ commute we have

$$e^{tA} = e^{\lambda t I}e^{t(A-\lambda I)} = (e^{\lambda t}I)\sum_{k=0}^{\infty}\frac{t^k}{k!}(A - \lambda I)^k .$$

The Cayley-Hamilton theorem implies that $(A - \lambda I)^k = O$ for every $k \geq n$, so the theorem is proved.

THEOREM 7.11. *If A is an $n \times n$ matrix with n distinct eigenvalues $\lambda_1, \lambda_2, \ldots, \lambda_n$, then we have*

$$e^{tA} = \sum_{k=1}^{n} e^{t\lambda_k}L_k(A),$$

where $L_k(A)$ is a polynomial in A of degree $n - 1$ given by the formula

$$L_k(A) = \prod_{\substack{j=1\\ j\neq k}}^{n} \frac{A - \lambda_j I}{\lambda_k - \lambda_j} \qquad \textit{for} \quad k = 1, 2, \ldots, n .$$

Note: The polynomials $L_k(A)$ are called *Lagrange interpolation coefficients.*

Proof. We define a matrix function F by the equation

$$(7.37) \qquad F(t) = \sum_{k=1}^{n} e^{t\lambda_k}L_k(A)$$

and verify that F satisfies the differential equation $F'(t) = AF(t)$ and the initial condition $F(0) = I$. From (7.37) we see that

$$AF(t) - F'(t) = \sum_{k=1}^{n} e^{t\lambda_k}(A - \lambda_k I)L_k(A) .$$

By the Cayley-Hamilton theorem we have $(A - \lambda_k I)L_k(A) = O$ for each k, so F satisfies the differential equation $F'(t) = AF(t)$.

To complete the proof we need to show that F satisfies the initial condition $F(0) = I$, which becomes

$$(7.38) \qquad \sum_{k=1}^{n} L_k(A) = I .$$

A proof of (7.38) is outlined in Exercise 16 of Section 7.15.

The next theorem treats the case when A has two distinct eigenvalues, exactly one of which has multiplicity 1.

THEOREM 7.12. *Let A be an $n \times n$ matrix ($n \geq 3$) with two distinct eigenvalues λ and μ, where λ has multiplicity $n - 1$ and μ has multiplicity 1. Then we have*

$$e^{tA} = e^{\lambda t} \sum_{k=0}^{n-2} \frac{t^k}{k!} (A - \lambda I)^k + \left\{ \frac{e^{\mu t}}{(\mu - \lambda)^{n-1}} - \frac{e^{\lambda t}}{(\mu - \lambda)^{n-1}} \sum_{k=0}^{n-2} \frac{t^k}{k!} (\mu - \lambda)^k \right\} (A - \lambda I)^{n-1}.$$

Proof. As in the proof of Theorem 7.10 we begin by writing

$$e^{tA} = e^{\lambda t} \sum_{k=0}^{\infty} \frac{t^k}{k!} (A - \lambda I)^k = e^{\lambda t} \sum_{k=0}^{n-2} \frac{t^k}{k!} (A - \lambda I)^k + e^{\lambda t} \sum_{k=n-1}^{\infty} \frac{t^k}{k!} (A - \lambda I)^k$$

$$= e^{\lambda t} \sum_{k=0}^{n-2} \frac{t^k}{k!} (A - \lambda I)^k + e^{\lambda t} \sum_{r=0}^{\infty} \frac{t^{n-1+r}}{(n - 1 + r)!} (A - \lambda I)^{n-1+r}.$$

Now we evaluate the series over r in closed form by using the Cayley-Hamilton theorem. Since

$$A - \mu I = A - \lambda I - (\mu - \lambda) I$$

we find

$$(A - \lambda I)^{n-1}(A - \mu I) = (A - \lambda I)^n - (\mu - \lambda)(A - \lambda I)^{n-1}.$$

The left member is O by the Cayley-Hamilton theorem so

$$(A - \lambda I)^n = (\mu - \lambda)(A - \lambda I)^{n-1}.$$

Using this relation repeatedly we find

$$(A - \lambda I)^{n-1+r} = (\mu - \lambda)^r (A - \lambda I)^{n-1}.$$

Therefore the series over r becomes

$$\sum_{r=0}^{\infty} \frac{t^{n-1+r}}{(n - 1 + r)!} (\mu - \lambda)^r (A - \lambda I)^{n-1} = \frac{1}{(\mu - \lambda)^{n-1}} \sum_{k=n-1}^{\infty} \frac{t^k}{k!} (\mu - \lambda)^k (A - \lambda I)^{n-1}$$

$$= \frac{1}{(\mu - \lambda)^{n-1}} \left\{ e^{t(\mu - \lambda)} - \sum_{k=0}^{n-2} \frac{t^k}{k!} (\mu - \lambda)^k \right\} (A - \lambda I)^{n-1}.$$

This completes the proof.

The explicit formula in Theorem 7.12 can also be deduced by applying Putzer's method, but the details are more complicated.

The explicit formulas in Theorems 7.10, 7.11 and 7.12 cover all matrices of order $n \leq 3$. Since the 3×3 case often arises in practice, the formulas in this case are listed below for easy reference.

CASE 1. If a 3×3 matrix A has eigenvalues $\lambda, \lambda, \lambda$, then

$$e^{tA} = e^{\lambda t} \{ I + t(A - \lambda I) + \tfrac{1}{2} t^2 (A - \lambda I)^2 \}.$$

CASE 2. If a 3×3 matrix A has distinct eigenvalues λ, μ, ν, then

$$e^{tA} = e^{\lambda t}\frac{(A-\mu I)(A-\nu I)}{(\lambda-\mu)(\lambda-\nu)} + e^{\mu t}\frac{(A-\lambda I)(A-\nu I)}{(\mu-\lambda)(\mu-\nu)} + e^{\nu t}\frac{(A-\lambda I)(A-\mu I)}{(\nu-\lambda)(\nu-\mu)}.$$

CASE 3. If a 3×3 matrix A has eigenvalues λ, λ, μ, with $\lambda \neq \mu$, then

$$e^{tA} = e^{\lambda t}\{I + t(A-\lambda I)\} + \frac{e^{\mu t}-e^{\lambda t}}{(\mu-\lambda)^2}(A-\lambda I)^2 - \frac{te^{\lambda t}}{\mu-\lambda}(A-\lambda I)^2.$$

EXAMPLE. Compute e^{tA} when $A = \begin{bmatrix} 0 & 1 & 0 \\ 0 & 0 & 1 \\ 2 & -5 & 4 \end{bmatrix}$.

Solution. The eigenvalues of A are 1, 1, 2, so the formula of Case 3 gives us

$$e^{tA} = e^t\{I + t(A-I)\} + (e^{2t}-e^t)(A-I)^2 - te^t(A-I)^2. \tag{7.39}$$

By collecting powers of A we can also write this as follows,

$$e^{tA} = (-2te^t + e^{2t})I + \{(3t+2)e^t - 2e^{2t}\}A - \{(t+1)e^t - e^{2t}\}A^2. \tag{7.40}$$

At this stage we can calculate $(A-I)^2$ or A^2 and perform the indicated operations in (7.39) or (7.40) to write the result as a 3×3 matrix,

$$e^{tA} = \begin{bmatrix} -2te^t + e^{2t} & (3t+2)e^t - 2e^{2t} & -(t+1)e^t + e^{2t} \\ -2(t+1)e^t + 2e^{2t} & (3t+5)e^t - 4e^{2t} & -(t+2)e^t + 2e^{2t} \\ -2(t+2)e^t + 4e^{2t} & (3t+8)e^t - 8e^{2t} & -(t+4)e^t + 4e^{2t} \end{bmatrix}.$$

7.15 Exercises

For each of the matrices in Exercises 1 through 6, express e^{tA} as a polynomial in A.

1. $A = \begin{bmatrix} 5 & -2 \\ 4 & -1 \end{bmatrix}$.
2. $A = \begin{bmatrix} 1 & 2 \\ 2 & -1 \end{bmatrix}$.
3. $A = \begin{bmatrix} 1 & 0 & 2 \\ 0 & 1 & 3 \\ 0 & 0 & 1 \end{bmatrix}$.
4. $A = \begin{bmatrix} 0 & 1 & 0 \\ 0 & 0 & 1 \\ -6 & -11 & -6 \end{bmatrix}$.
5. $A = \begin{bmatrix} 3 & -1 & 1 \\ 2 & 0 & 1 \\ 1 & -1 & 2 \end{bmatrix}$.
6. $A = \begin{bmatrix} 1 & 1 & 0 & 0 \\ 0 & 2 & 1 & 0 \\ 0 & 0 & 3 & 0 \\ 0 & 0 & 0 & 4 \end{bmatrix}$.
7. (a) A 3×3 matrix A is known to have all its eigenvalues equal to λ. Prove that

$$e^{tA} = \tfrac{1}{2}e^{\lambda t}\{(\lambda^2t^2 - 2\lambda t + 2)I + (-2\lambda t^2 + 2t)A + t^2A^2\}.$$

(b) Find a corresponding formula if A is a 4×4 matrix with all its eigenvalues equal to λ.

In each of Exercises 8 through 15, solve the system $Y' = AY$ subject to the given initial condition.

8. $A = \begin{bmatrix} 1 & 2 \\ 2 & -1 \end{bmatrix}, \quad Y(0) = \begin{bmatrix} c_1 \\ c_2 \end{bmatrix}.$ 9. $A = \begin{bmatrix} -5 & 3 \\ -15 & 7 \end{bmatrix}, \quad Y(0) = \begin{bmatrix} 1 \\ 1 \end{bmatrix}.$

10. $A = \begin{bmatrix} 3 & -1 & 1 \\ 2 & 0 & 1 \\ 1 & -1 & 2 \end{bmatrix}, \quad Y(0) = \begin{bmatrix} 1 \\ -1 \\ 2 \end{bmatrix}.$ 11. $A = \begin{bmatrix} 2 & 0 & 0 \\ 0 & 1 & 0 \\ 0 & 1 & 1 \end{bmatrix}, \quad Y(0) = \begin{bmatrix} c_1 \\ c_2 \\ c_3 \end{bmatrix}.$

12. $A = \begin{bmatrix} 0 & 1 & 0 \\ 0 & 0 & 1 \\ -6 & -11 & -6 \end{bmatrix}, \quad Y(0) = \begin{bmatrix} 1 \\ 0 \\ 0 \end{bmatrix}.$ 13. $A = \begin{bmatrix} -2 & 2 & -3 \\ 2 & 1 & -6 \\ -1 & -2 & 0 \end{bmatrix}, \quad Y(0) = \begin{bmatrix} 8 \\ 0 \\ 0 \end{bmatrix}.$

14. $A = \begin{bmatrix} 1 & 1 & 0 & 0 \\ 0 & 2 & 1 & 0 \\ 0 & 0 & 3 & 0 \\ 0 & 0 & 0 & 4 \end{bmatrix}, \quad Y(0) = \begin{bmatrix} 0 \\ 0 \\ 1 \\ 1 \end{bmatrix}.$ 15. $A = \begin{bmatrix} 0 & 0 & 2 & 0 \\ 1 & 0 & 0 & 2 \\ 0 & 0 & 0 & 4 \\ 0 & 0 & 1 & 0 \end{bmatrix}, \quad Y(0) = \begin{bmatrix} 1 \\ 0 \\ 2 \\ 1 \end{bmatrix}.$

16. This exercise outlines a proof of Equation (7.38) used in the proof of Theorem 7.11. Let $L_k(\lambda)$ be the polynomial in λ of degree $n - 1$ defined by the equation

$$L_k(\lambda) = \prod_{\substack{j=1 \\ j \neq k}}^{n} \frac{\lambda - \lambda_j}{\lambda_k - \lambda_j},$$

where $\lambda_1, \ldots, \lambda_n$ are n distinct scalars.

(a) Prove that

$$L_k(\lambda_i) = \begin{cases} 0 & \text{if } \lambda_i \neq \lambda_k, \\ 1 & \text{if } \lambda_i = \lambda_k. \end{cases}$$

(b) Let $y_1, \ldots, y_n$ be n arbitrary scalars, and let

$$p(\lambda) = \sum_{k=1}^{n} y_k L_k(\lambda).$$

Prove that $p(\lambda)$ is the only polynomial of degree $\leq n - 1$ which satisfies the n equations

$$p(\lambda_k) = y_k \qquad \text{for } k = 1, 2, \ldots, n.$$

(c) Prove that $\sum_{k=1}^{n} L_k(\lambda) = 1$ for every λ, and deduce that for every square matrix A we have

$$\sum_{k=1}^{n} L_k(A) = I,$$

where I is the identity matrix.

7.16 Nonhomogeneous linear systems with constant coefficients

We consider next the nonhomogeneous initial-value problem

$$Y'(t) = AY(t) + Q(t), \qquad Y(a) = B, \tag{7.41}$$

on an interval J. Here A is an $n \times n$ constant matrix, Q is an n-dimensional vector function (regarded as an $n \times 1$ column matrix) continuous on J, and a is a given point in J. We can obtain an explicit formula for the solution of this problem by the same process used to treat the scalar case.

First we multiply both members of (7.41) by the exponential matrix e^{-tA} and rewrite the differential equation in the form

$$e^{-tA}\{Y'(t) - AY(t)\} = e^{-tA}Q(t). \tag{7.42}$$

The left member of (7.42) is the derivative of the product $e^{-tA}Y(t)$. Therefore, if we integrate both members of (7.42) from a to x, where $x \in J$, we obtain

$$e^{-xA}Y(x) - e^{-aA}Y(a) = \int_a^x e^{-tA}Q(t)\,dt.$$

Multiplying by e^{xA} we obtain the explicit formula (7.43) which appears in the following theorem.

THEOREM 7.13. *Let A be an $n \times n$ constant matrix and let Q be an n-dimensional vector function continuous on an interval J. Then the initial-value problem*

$$Y'(t) = AY(t) + Q(t), \qquad Y(a) = B,$$

has a unique solution on J given by the explicit formula

$$Y(x) = e^{(x-a)A}B + e^{xA}\int_a^x e^{-tA}Q(t)\,dt. \tag{7.43}$$

As in the homogeneous case, the difficulty in applying this formula in practice lies in the calculation of the exponential matrices.

Note that the first term, $e^{(x-a)A}B$, is the solution of the homogeneous problem $Y'(t) = AY(t)$, $Y(a) = B$. The second term is the solution of the nonhomogeneous problem

$$Y'(t) = AY(t) + Q(t), \qquad Y(a) = O.$$

We illustrate Theorem 7.13 with an example.

EXAMPLE. Solve the initial-value problem

$$Y'(t) = AY(t) + Q(t), \qquad Y(0) = B,$$

on the interval $(-\infty, +\infty)$, where

$$A = \begin{bmatrix} 2 & -1 & 1 \\ 0 & 3 & -1 \\ 2 & 1 & 3 \end{bmatrix}, \qquad Q(t) = \begin{bmatrix} e^{2t} \\ 0 \\ te^{2t} \end{bmatrix}, \qquad B = \begin{bmatrix} 0 \\ 0 \\ 0 \end{bmatrix}.$$

Solution. According to Theorem 7.13, the solution is given by

$$Y(x) = e^{xA} \int_0^x e^{-tA} Q(t)\, dt = \int_0^x e^{(x-t)A} Q(t)\, dt. \tag{7.44}$$

The eigenvalues of A are 2, 2, and 4. To calculate e^{xA} we use the formula of Case 3, Section 7.14, to obtain

$$\begin{aligned} e^{xA} &= e^{2x}\{I + x(A - 2I)\} + \tfrac{1}{4}(e^{4x} - e^{2x})(A - 2I)^2 - \tfrac{1}{2}xe^{2x}(A - 2I)^2 \\ &= e^{2x}\{I + x(A - 2I) + \tfrac{1}{4}(e^{2x} - 2x - 1)(A - 2I)^2\}. \end{aligned}$$

We can replace x by $x - t$ in this formula to obtain $e^{(x-t)A}$. Therefore the integrand in (7.44) is

$$\begin{aligned} e^{(x-t)A}Q(t) &= e^{2(x-t)}\{I + (x - t)(A - 2I) + \tfrac{1}{4}[e^{2(x-t)} - 2(x - t) - 1](A - 2I)^2\}Q(t) \\ &= e^{2x}\begin{bmatrix} 1 \\ 0 \\ t \end{bmatrix} + (A - 2I)e^{2x}\begin{bmatrix} x - t \\ 0 \\ t(x - t) \end{bmatrix} + \frac{1}{4}(A - 2I)^2 e^{2x}\begin{bmatrix} e^{2x}e^{-2t} - 2(x - t) - 1 \\ 0 \\ e^{2x}te^{-2t} - 2t(x - t) - t \end{bmatrix}. \end{aligned}$$

Integrating, we find

$$\begin{aligned} Y(x) = \int_0^x e^{(x-t)A}Q(t)\, dt &= e^{2x}\begin{bmatrix} x \\ 0 \\ \tfrac{1}{2}x^2 \end{bmatrix} + (A - 2I)e^{2x}\begin{bmatrix} \tfrac{1}{2}x^2 \\ 0 \\ \tfrac{1}{6}x^3 \end{bmatrix} \\ &\quad + \frac{1}{4}(A - 2I)^2 e^{2x}\begin{bmatrix} \tfrac{1}{2}e^{2x} - \tfrac{1}{2} - x - x^2 \\ 0 \\ \tfrac{1}{4}e^{2x} - \tfrac{1}{4} - \tfrac{1}{2}x - \tfrac{1}{2}x^2 - \tfrac{1}{3}x^3 \end{bmatrix}. \end{aligned}$$

Since we have

$$A - 2I = \begin{bmatrix} 0 & -1 & 1 \\ 0 & 1 & -1 \\ 2 & 1 & 1 \end{bmatrix} \quad \text{and} \quad (A - 2I)^2 = \begin{bmatrix} 2 & 0 & 2 \\ -2 & 0 & -2 \\ 2 & 0 & 2 \end{bmatrix},$$

we find

$$\begin{aligned} Y(x) &= e^{2x}\begin{bmatrix} x \\ 0 \\ \tfrac{1}{2}x^2 \end{bmatrix} + e^{2x}\begin{bmatrix} \tfrac{1}{6}x^3 \\ -\tfrac{1}{6}x^3 \\ x^2 + \tfrac{1}{6}x^3 \end{bmatrix} + e^{2x}\begin{bmatrix} \tfrac{3}{8}e^{2x} - \tfrac{3}{8} - \tfrac{3}{4}x - \tfrac{3}{4}x^2 - \tfrac{1}{6}x^3 \\ -\tfrac{3}{8}e^{2x} + \tfrac{3}{8} + \tfrac{3}{4}x + \tfrac{3}{4}x^2 + \tfrac{1}{6}x^3 \\ \tfrac{3}{8}e^{2x} - \tfrac{3}{8} - \tfrac{3}{4}x - \tfrac{3}{4}x^2 - \tfrac{1}{6}x^3 \end{bmatrix} \\ &= e^{2x}\begin{bmatrix} \tfrac{3}{8}e^{2x} - \tfrac{3}{8} + \tfrac{1}{4}x - \tfrac{3}{4}x^2 \\ -\tfrac{3}{8}e^{2x} + \tfrac{3}{8} + \tfrac{3}{4}x + \tfrac{3}{4}x^2 \\ \tfrac{3}{8}e^{2x} - \tfrac{3}{8} - \tfrac{3}{4}x + \tfrac{3}{4}x^2 \end{bmatrix}. \end{aligned}$$

The rows of this matrix are the required functions y_1, y_2, y_3.

7.17 Exercises

1. Let Z be a solution of the nonhomogeneous system

$$Z'(t) = AZ(t) + Q(t),$$

on an interval J with initial value $Z(a)$. Prove that there is only one solution of the nonhomogeneous system

$$Y'(t) = AY(t) + Q(t)$$

on J with initial value $Y(a)$ and that it is given by the formula

$$Y(t) = Z(t) + e^{(t-a)A}\{Y(a) - Z(a)\}.$$

Special methods are often available for determining a particular solution $Z(t)$ which resembles the given function $Q(t)$. Exercises 2, 3, 5, and 7 indicate such methods for $Q(t) = C$, $Q(t) = e^{\alpha t}C$, $Q(t) = t^m C$, and $Q(t) = (\cos \alpha t)C + (\sin \alpha t)D$, where C and D are constant vectors. If the particular solution $Z(t)$ so obtained does not have the required initial value, we modify $Z(t)$ as indicated in Exercise 1 to obtain another solution $Y(t)$ with the required initial value.

2. (a) Let A be a constant $n \times n$ matrix, B and C constant n-dimensional vectors. Prove that the solution of the system

$$Y'(t) = AY(t) + C, \qquad Y(a) = B,$$

on $(-\infty, +\infty)$ is given by the formula

$$Y(x) = e^{(x-a)A}B + \left(\int_0^{x-a} e^{uA}\, du\right)C.$$

(b) If A is nonsingular, show that the integral in part (a) has the value $\{e^{(x-a)A} - I\}A^{-1}$.
(c) Compute $Y(x)$ explicitly when

$$A = \begin{bmatrix} -1 & 2 \\ -2 & 3 \end{bmatrix}, \quad C = \begin{bmatrix} 1 \\ 2 \end{bmatrix}, \quad B = \begin{bmatrix} b \\ c \end{bmatrix}, \quad a = 0.$$

3. Let A be an $n \times n$ constant matrix, let B and C be n-dimensional constant vectors, and let α be a given scalar.
(a) Prove that the nonhomogeneous system $Z'(t) = AZ(t) + e^{\alpha t}C$ has a solution of the form $Z(t) = e^{\alpha t}B$ if, and only if, $(\alpha I - A)B = C$.
(b) If α is not an eigenvalue of A, prove that the vector B can always be chosen so that the system in (a) has a solution of the form $Z(t) = e^{\alpha t}B$.
(c) If α is not an eigenvalue of A, prove that every solution of the system $Y'(t) = AY(t) + e^{\alpha t}C$ has the form $Y(t) = e^{tA}(Y(0) - B) + e^{\alpha t}B$, where $B = (\alpha I - A)^{-1}C$.

4. Use the method suggested by Exercise 3 to find a solution of the nonhomogeneous system $Y'(t) = AY(t) + e^{2t}C$, with

$$A = \begin{bmatrix} 3 & 1 \\ 2 & 2 \end{bmatrix}, \quad C = \begin{bmatrix} -1 \\ -1 \end{bmatrix}, \quad Y(0) = \begin{bmatrix} 0 \\ 1 \end{bmatrix}.$$

5. Let A be an $n \times n$ constant matrix, let B and C be n-dimensional constant vectors, and let m be a positive integer.
(a) Prove that the nonhomogeneous system $Y'(t) = AY(t) + t^m C$, $Y(0) = B$, has a particular solution of the form

$$Y(t) = B_0 + tB_1 + t^2 B_2 + \cdots + t^m B_m,$$

where $B_0, B_1, \ldots, B_m$ are constant vectors, if and only if

$$C = -\frac{1}{m!} A^{m+1} B.$$

Determine the coefficients $B_0, B_1, \ldots, B_m$ for such a solution.
(b) If A is nonsingular, prove that the initial vector B can always be chosen so that the system in (a) has a solution of the specified form.

6. Consider the nonhomogeneous system

$$y_1' = 3y_1 + y_2 + t^3$$

$$y_2' = 2y_1 + 2y_2 + t^3.$$

(a) Find a particular solution of the form $Y(t) = B_0 + tB_1 + t^2 B_2 + t^3 B_3$.
(b) Find a solution of the system with $y_1(0) = y_2(0) = 1$.

7. Let A be an $n \times n$ constant matrix, let B, C, D be n-dimensional constant vectors, and let α be a given nonzero real number. Prove that the nonhomogeneous system

$$Y'(t) = AY(t) + (\cos \alpha t)C + (\sin \alpha t)D, \qquad Y(0) = B,$$

has a particular solution of the form

$$Y(t) = (\cos \alpha t)E + (\sin \alpha t)F,$$

where E and F are constant vectors, if and only if

$$(A^2 + \alpha^2 I)B = -(AC + \alpha D).$$

Determine E and F in terms of A, B, C for such a solution. Note that if $A^2 + \alpha^2 I$ is nonsingular, the initial vector B can always be chosen so that the system has a solution of the specified form.

8. (a) Find a particular solution of the nonhomogeneous system

$$y_1' = y_1 + 3y_2 + 4 \sin 2t$$

$$y_2' = y_1 - y_2.$$

(b) Find a solution of the system with $y_1(0) = y_2(0) = 1$.

In each of Exercises 9 through 12, solve the nonhomogeneous system $Y'(t) = AY(t) + Q(t)$ subject to the given initial condition.

9. $A = \begin{bmatrix} 4 & 1 \\ -2 & 1 \end{bmatrix}, \quad Q(t) = \begin{bmatrix} 0 \\ -2e^t \end{bmatrix}, \quad Y(0) = \begin{bmatrix} 1 \\ 0 \end{bmatrix}.$

10. $A = \begin{bmatrix} -5 & -1 \\ 1 & -3 \end{bmatrix}, \quad Q(t) = \begin{bmatrix} e \\ e^{2t} \end{bmatrix}, \quad Y(0) = \begin{bmatrix} c_1 \\ c_2 \end{bmatrix}.$

11. $A = \begin{bmatrix} -5 & -1 \\ 2 & -3 \end{bmatrix}, \quad Q(t) = \begin{bmatrix} 7e^t - 27 \\ -3e^t + 12 \end{bmatrix}, \quad Y(0) = \begin{bmatrix} -\frac{1007}{442} \\ \frac{707}{221} \end{bmatrix}.$

12. $A = \begin{bmatrix} -1 & -1 & 0 \\ 0 & -1 & -1 \\ 0 & 0 & -1 \end{bmatrix}, \quad Q(t) = \begin{bmatrix} t^2 \\ 2t \\ t \end{bmatrix}, \quad Y(0) = \begin{bmatrix} 6 \\ -2 \\ 1 \end{bmatrix}.$

7.18 The general linear system $Y'(t) = P(t)Y(t) + Q(t)$

Theorem 7.13 gives an explicit formula for the solution of the linear system

$$Y'(t) = AY(t) + Q(t), \qquad Y(a) = B,$$

where A is a constant $n \times n$ matrix and $Q(t)$, $Y(t)$ are $n \times 1$ column matrices. We turn now to the more general case

$$(7.45) \qquad Y'(t) = P(t)Y(t) + Q(t), \qquad Y(a) = B,$$

where the $n \times n$ matrix $P(t)$ is not necessarily constant.

If P and Q are continuous on an open interval J, a general existence-uniqueness theorem which we shall prove in a later section tells us that for each a in J and each initial vector B there is exactly one solution to the initial-value problem (7.45). In this section we use this result to obtain a formula for the solution, generalizing Theorem 7.13.

In the scalar case ($n = 1$) the differential equation (7.45) can be solved as follows. We let $A(x) = \int_a^x P(t)\,dt$, then multiply both members of (7.45) by $e^{-A(t)}$ to rewrite the differential equation in the form

$$(7.46) \qquad e^{-A(t)}\{Y'(t) - P(t)Y(t)\} = e^{-A(t)}Q(t).$$

Now the left member is the derivative of the product $e^{-A(t)}\,Y(t)$. Therefore, we can integrate both members from a to x, where a and x are points in J, to obtain

$$e^{-A(x)}Y(x) - e^{-A(a)}Y(a) = \int_a^x e^{-A(t)}Q(t)\,dt.$$

Multiplying by $e^{A(x)}$ we obtain the explicit formula

$$(7.47) \qquad Y(x) = e^{A(x)}e^{-A(a)}Y(a) + e^{A(x)}\int_a^x e^{-A(t)}Q(t)\,dt.$$

The only part of this argument that does not apply immediately to matrix functions is the statement that the left-hand member of (7.46) is the derivative of the product $e^{-A(t)}Y(t)$. At this stage we used the fact that the derivative of $e^{-A(t)}$ is $-P(t)e^{-A(t)}$. In the scalar case this is a consequence of the following formula for differentiating exponential functions:

$$\text{If} \quad E(t) = e^{A(t)}, \quad \text{then} \quad E'(t) = A'(t)e^{A(t)}.$$

Unfortunately, this differentiation formula is not always true when A is a matrix function. For example, it is false for the 2×2 matrix function $A(t) = \begin{bmatrix} 1 & t \\ 0 & 0 \end{bmatrix}$. (See Exercise 7 of Section 7.12.) Therefore a modified argument is needed to extend Equation (7.47) to the matrix case.

Suppose we multiply each member of (7.45) by an unspecified $n \times n$ matrix $F(t)$. This gives us the relation

$$F(t)Y'(t) = F(t)P(t)Y(t) + F(t)Q(t).$$

Now we add $F'(t)Y(t)$ to both members in order to transform the left member to the derivative of the product $F(t)Y(t)$. If we do this, the last equation gives us

$$\{F(t)Y(t)\}' = \{F'(t) + F(t)P(t)\}Y(t) + F(t)Q(t).$$

If we can choose the matrix $F(t)$ so that the sum $\{F'(t) + F(t)P(t)\}$ on the right is the zero matrix, the last equation simplifies to

$$\{F(t)Y(t)\}' = F(t)Q(t).$$

Integrating this from a to x we obtain

$$F(x)Y(x) - F(a)Y(a) = \int_a^x F(t)Q(t)\,dt.$$

If, in addition, the matrix $F(x)$ is nonsingular, we obtain the explicit formula

$$(7.48) \qquad Y(x) = F(x)^{-1}F(a)Y(a) + F(x)^{-1}\int_a^x F(t)Q(t)\,dt.$$

This is a generalization of the scalar formula (7.47). The process will work if we can find a $n \times n$ matrix function $F(t)$ which satisfies the matrix differential equation

$$F'(t) = -F(t)P(t)$$

and which is nonsingular.

Note that this differential equation is very much like the original differential equation (7.45) with $Q(t) = 0$, except that the unknown function $F(t)$ is a square matrix instead of a column matrix. Also, the unknown function is multiplied on the right by $-P(t)$ instead of on the left by $P(t)$.

We shall prove next that the differential equation for F always has a nonsingular solution. The proof will depend on the following existence theorem for homogeneous linear systems.

THEOREM 7.14. *Assume $A(t)$ is an $n \times n$ matrix function continuous on an open interval J. If $a \in J$ and if B is a given n-dimensional vector, the homogeneous linear system*

$$Y'(t) = A(t)Y(t), \qquad Y(a) = B,$$

has an n-dimensional vector solution Y on J.

A proof of Theorem 7.14 appears in Section 7.21. With the help of this theorem we can prove the following.

THEOREM 7.15. *Given an $n \times n$ matrix function P, continuous on an open interval J, and given any point a in J, there exists an $n \times n$ matrix function F which satisfies the matrix differential equation*

$$F'(x) = -F(x)P(x) \tag{7.49}$$

on J with initial value $F(a) = I$. Moreover, $F(x)$ is nonsingular for each x in J.

Proof. Let $Y_k(x)$ be a vector solution of the differential equation

$$Y_k'(x) = -P(x)^t Y_k(x)$$

on J with initial vector $Y_k(a) = I_k$, where I_k is the kth column of the $n \times n$ identity matrix I. Here $P(x)^t$ denotes the transpose of $P(x)$. Let $G(x)$ be the $n \times n$ matrix whose kth column is $Y_k(x)$. Then G satisfies the matrix differential equation

$$G'(x) = -P(x)^t G(x) \tag{7.50}$$

on J with initial value $G(a) = I$. Now take the transpose of each member of (7.50). Since the transpose of a product is the product of transposes in reverse order, we obtain

$$\{G'(x)\}^t = -G(x)^t P(x).$$

Also, the transpose of the derivative G' is the derivative of the transpose G^t. Therefore the matrix $F(x) = G(x)^t$ satisfies the differential equation (7.49) with initial value $F(a) = I$.

Now we prove that $F(x)$ is nonsingular by exhibiting its inverse. Let H be the $n \times n$ matrix function whose kth column is the solution of the differential equation

$$Y'(x) = P(x)Y(x)$$

with initial vector $Y(a) = I_k$, the kth column of I. Then H satisfies the initial-value problem

$$H'(x) = P(x)H(x), \qquad H(a) = I,$$

on J. The product $F(x)H(x)$ has derivative

$$F(x)H'(x) + F'(x)H(x) = F(x)P(x)H(x) - F(x)P(x)H(x) = O$$

for each x in J. Therefore the product $F(x)H(x)$ is constant, $F(x)H(x) = F(a)H(a) = I$, so $H(x)$ is the inverse of $F(x)$. This completes the proof.

The results of this section are summarized in the following theorem.

THEOREM 7.16. *Given an $n \times n$ matrix function P and an n-dimensional vector function Q, both continuous on an open interval J, the solution of the initial-value problem*

$$Y'(x) = P(x)Y(x) + Q(x), \qquad Y(a) = B, \tag{7.51}$$

on J is given by the formula

$$Y(x) = F(x)^{-1}Y(a) + F(x)^{-1}\int_a^x F(t)Q(t)\,dt. \tag{7.52}$$

The $n \times n$ matrix $F(x)$ is the transpose of the matrix whose kth column is the solution of the initial-value problem

$$Y'(x) = -P(x)^t Y(x), \qquad Y(a) = I_k, \tag{7.53}$$

where I_k is the kth column of the identity matrix I.

Although Theorem 7.16 provides an explicit formula for the solution of the general linear system (7.51), the formula is not always a useful one for calculating the solution because of the difficulty involved in determining the matrix function F. The determination of F requires the solution of n homogeneous linear systems (7.53). The next section describes a power-series method that is sometimes used to solve homogeneous linear systems.

We remind the reader once more that the proof of Theorem 7.16 was based on Theorem 7.14, the existence theorem for homogeneous linear systems, which we have not yet proved.

7.19 A power-series method for solving homogeneous linear systems

Consider a homogeneous linear system

$$Y'(x) = A(x)Y(x), \qquad Y(0) = B, \tag{7.54}$$

in which the given $n \times n$ matrix $A(x)$ has a power-series expansion in x convergent in some open interval containing the origin, say

$$A(x) = A_0 + xA_1 + x^2A_2 + \cdots + x^kA_k + \ldots, \qquad \text{for } |x| < r_1,$$

where the coefficients $A_0, A_1, A_2, \ldots$ are given $n \times n$ matrices. Let us try to find a power-series solution of the form

$$Y(x) = B_0 + xB_1 + x^2B_2 + \cdots + x^kB_k + \cdots,$$

with vector coefficients $B_0, B_1, B_2, \ldots$. Since $Y(0) = B_0$, the initial condition will be satisfied by taking $B_0 = B$, the prescribed initial vector. To determine the remaining coefficients we substitute the power series for $Y(x)$ in the differential equation and equate coefficients of like powers of x to obtain the following system of equations:

$$(7.55) \qquad B_1 = A_0 B, \qquad (k+1)B_{k+1} = \sum_{r=0}^{k} A_r B_{k-r} \quad \text{for } k = 1, 2, \ldots.$$

These equations can be solved in succession for the vectors $B_1, B_2, \ldots$. If the resulting power series for $Y(x)$ converges in some interval $|x| < r_2$, then $Y(x)$ will be a solution of the initial-value problem (7.54) in the interval $|x| < r$, where $r = \min \{r_1, r_2\}$.

For example, if $A(x)$ is a constant matrix A, then $A_0 = A$ and $A_k = O$ for $k \geq 1$, so the system of equations in (7.55) becomes

$$B_1 = AB, \qquad (k+1)B_{k+1} = AB_k \quad \text{for } k \geq 1.$$

Solving these equations in succession we find

$$B_k = \frac{1}{k!} A^k B \quad \text{for } k \geq 1.$$

Therefore the series solution in this case becomes

$$Y(x) = B + \sum_{k=1}^{\infty} \frac{x^k}{k!} A^k B = e^{xA} B.$$

This agrees with the result obtained earlier for homogeneous linear systems with constant coefficients.

7.20 Exercises

1. Let p be a real-valued function and Q an $n \times 1$ matrix function, both continuous on an interval J. Let A be an $n \times n$ constant matrix. Prove that the initial-value problem

$$Y'(x) = p(x)AY(x) + Q(x), \qquad Y(a) = B,$$

has the solution

$$Y(x) = e^{q(x)A}B + e^{q(x)A}\int_a^x e^{-q(t)A}Q(t)\,dt$$

on J, where $q(x) = \int_a^x p(t)\,dt$.

2. Consider the special case of Exercise 1 in which A is nonsingular, $a = 0$, $p(x) = 2x$, and $Q(x) = xC$, where C is a constant vector. Show that the solution becomes

$$Y(x) = e^{x^2 A}(B + \tfrac{1}{2}A^{-1}C) - \tfrac{1}{2}A^{-1}C.$$

3. Let $A(t)$ be an $n \times n$ matrix function and let $E(t) = e^{A(t)}$. Let $Q(t)$, $Y(t)$, and B be $n \times 1$ column matrices. Assume that

$$E'(t) = A'(t)E(t)$$

on an open interval J. If $a \in J$ and if A' and Q are continuous on J, prove that the initial-value problem

$$Y'(t) = A'(t)Y(t) + Q(t), \qquad Y(a) = B,$$

has the following solution on J:

$$Y(x) = e^{A(x)}e^{-A(a)}B + e^{A(x)}\int_a^x e^{-A(t)}Q(t)\,dt.$$

4. Let $E(t) = e^{A(t)}$. This exercise describes examples of matrix functions $A(t)$ for which $E'(t) = A'(t)E(t)$.
(a) Let $A(t) = t^r A$, where A is an $n \times n$ constant matrix and r is a positive integer. Prove that $E'(t) = A'(t)E(t)$ on $(-\infty, \infty)$.
(b) Let $A(t)$ be a polynomial in t with matrix coefficients, say

$$A(t) = \sum_{r=0}^{m} t^r A_r,$$

where the coefficients commute, $A_r A_s = A_s A_r$ for all r and s. Prove that $E'(t) = A'(t)E(t)$ on $(-\infty, \infty)$.
(c) Solve the homogeneous linear system

$$Y'(t) = (I + tA)Y(t), \qquad Y(0) = B$$

on the interval $(-\infty, \infty)$, where A is an $n \times n$ constant matrix.

5. Assume that the $n \times n$ matrix function $A(x)$ has a power-series expansion convergent for $|x| < r$. Develop a power-series procedure for solving the following homogeneous linear system of second order:

$$Y''(x) = A(x)Y(x), \qquad \text{with} \quad Y(0) = B, \qquad Y'(0) = C.$$

6. Consider the second-order system $Y''(x) + AY(x) = 0$, with $Y(0) = B$, $Y'(0) = C$, where A is a constant $n \times n$ matrix. Prove that the system has the power-series solution

$$Y(x) = \left(I + \sum_{k=1}^{\infty} \frac{(-1)^k x^{2k} A^k}{(2k)!}\right)B + \left(xI + \sum_{k=1}^{\infty} \frac{(-1)^k x^{2k+1} A^k}{(2k+1)!}\right)C$$

convergent for $-\infty < x < +\infty$.

7.21 Proof of the existence theorem by the method of successive approximations

In this section we prove the existence and uniqueness of a solution for any homogeneous linear system

$$Y'(t) = A(t)Y(t), \tag{7.56}$$

where $A(t)$ is an $n \times n$ matrix function, continuous on an open interval J. We shall prove that for any point a in J and any given initial-vector B there exists exactly one solution $Y(t)$ on J satisfying the initial condition $Y(a) = B$.

We shall use the *method of successive approximations*, an iterative method which also has applications in many other problems. The method was first published by Liouville in 1838 in connection with the study of linear differential equations of second order. It was later extended by J. Caqué in 1864, L. Fuchs in 1870, and G. Peano in 1888 to the study of linear equations of order n. In 1890 Émile Picard (1856–1941) extended the method to encompass nonlinear differential equations as well. In recognition of his fundamental contributions, some writers refer to the method as *Picard's method.* The method is not only of theoretical interest but can also be used to obtain numerical approximations to solutions in some cases.

The method begins with an initial guess at a solution of the equation (7.56). We take as initial guess the given initial vector B, although this is not essential. We then substitute this guess in the right-hand member of the equation and obtain a new differential equation,

$$Y'(t) = A(t)B.$$

In this equation the right-hand member no longer contains the unknown function, so the equation can be solved immediately by integrating both members from a to x, where x is an arbitrary point in J. This equation has exactly one solution Y_1 on J satisfying the initial condition $Y_1(a) = B$, namely

$$Y_1(x) = B + \int_a^x A(t)B\,dt.$$

Now we replace $Y(t)$ by $Y_1(t)$ in the right-hand member of the original differential equation (7.56) to obtain a new differential equation

$$Y'(t) = A(t)Y_1(t).$$

This equation has a unique solution Y_2 on J with $Y_2(a) = B$,

$$(7.57) \qquad Y_2(x) = B + \int_a^x A(t)Y_1(t)\,dt.$$

We then substitute Y_2 in the right-hand member of (7.56) and solve the resulting equation to determine Y_3 with $Y_3(a) = B$, and so on. This process generates a sequence of functions $Y_0, Y_1, Y_2, \ldots$, where $Y_0 = B$ and where Y_{k+1} is determined from Y_k by the recursion formula

$$(7.58) \qquad Y_{k+1}(x) = B + \int_a^x A(t)Y_k(t)\,dt \qquad \text{for} \quad k = 0, 1, 2, \ldots.$$

Our goal is to prove that the sequence of functions so defined converges to a limit function Y which is a solution of the differential equation (7.56) on J and which also satisfies the initial condition $Y(a) = B$. The functions $Y_0, Y_1, Y_2, \ldots$ are called *successive approximations* to Y. Before we investigate the convergence of the process we illustrate the method with an example.

EXAMPLE. Consider the initial-value problem $Y'(t) = AY(t)$, $Y(0) = B$, where A is a constant $n \times n$ matrix. We know that the solution is given by the formula $Y(x) = e^{xA}B$ for all real x. We will show how this solution can be obtained by the method of successive approximations.

The initial guess is $Y_0(x) = B$. The recursion formula (7.58) gives us

$$Y_1(x) = B + \int_0^x AB\,dt = B + xAB,$$

$$Y_2(x) = B + \int_0^x AY_1(t)\,dt = B + \int_0^x (AB + tA^2B)\,dt = B + xAB + \tfrac{1}{2}x^2A^2B.$$

By induction we find

$$Y_k(x) = B + xAB + \tfrac{1}{2}x^2A^2B + \cdots + \frac{1}{k!}x^kA^kB = \left(\sum_{r=0}^{k}\frac{(xA)^r}{r!}\right)B.$$

The sum on the right is a partial sum of the series for e^{xA}. Therefore when $k \to \infty$ we find

$$\lim_{k\to\infty} Y_k(x) = e^{xA}B$$

for all x. Thus, in this example we can show directly that the successive approximations converge to a solution of the initial-value problem on $(-\infty, +\infty)$.

Proof of convergence of the sequence of successive approximations. We return now to the general sequence defined by the recursion formula (7.58). To prove that the sequence converges we write each term $Y_k(x)$ as a telescoping sum,

$$Y_k(x) = Y_0(x) + \sum_{m=0}^{k-1}\{Y_{m+1}(x) - Y_m(x)\}. \tag{7.59}$$

To prove that $Y_k(x)$ tends to a limit as $k \to \infty$ we shall prove that the infinite series

$$\sum_{m=0}^{\infty}\{Y_{m+1}(x) - Y_m(x)\} \tag{7.60}$$

eonverges for each x in J. For this purpose it suffices to prove that the series

$$\sum_{m=0}^{\infty}\|Y_{m+1}(x) - Y_m(x)\| \tag{7.61}$$

converges. In this series we use the matrix norm introduced in Section 7.3; the norm of a matrix is the sum of the absolute values of all its entries.

Consider a closed and bounded subinterval J_1 of J containing a. We shall prove that for every x in J_1 the series in (7.61) is dominated by a convergent series of constants independent of x. This implies that the series converges *uniformly* on J_1.

To estimate the size of the terms in (7.61) we use the recursion formula repeatedly. Initially, we have

$$Y_1(x) - Y_0(x) = \int_a^x A(t)B\,dt.$$

For simplicity, we assume that $a < x$. Then we can write

$$\|Y_1(x) - Y_0(x)\| = \left\| \int_a^x A(t)B\, dt \right\| \le \int_a^x \|A(t)\|\, \|B\|\, dt. \tag{7.62}$$

Since each entry of $A(t)$ is continuous on J, each entry is bounded on the closed bounded interval J_1. Therefore $\|A(t)\| \le M$, where M is the sum of the bounds of all the entries of $A(t)$ on the interval J_1. (The number M depends on J_1.) Therefore the integrand in (7.62) is bounded by $\|B\|\, M$, so we have

$$\|Y_1(x) - Y_0(x)\| \le \int_a^x \|B\|\, M\, dt = \|B\|\, M(x - a)$$

for all $x > a$ in J_1.

Now we use the recursion formula once more to express the difference $Y_2 - Y_1$ in terms of $Y_1 - Y_0$, and then use the estimate just obtained for $Y_1 - Y_0$ to obtain

$$\|Y_2(x) - Y_1(x)\| = \left\| \int_a^x A(t)\{Y_1(t) - Y_0(t)\}\, dt \right\| \le \int_a^x \|A(t)\|\, \|B\|\, M(t - a)\, dt$$

$$\le \|B\|\, M^2 \int_a^x (t - a)\, dt = \|B\| \frac{M^2(x - a)^2}{2!}$$

for all $x > a$ in J_1. By induction we find

$$\|Y_{m+1}(x) - Y_m(x)\| \le \|B\| \frac{M^{m+1}(x - a)^{m+1}}{(m + 1)!} \qquad \text{for} \quad m = 0, 1, 2, \ldots,$$

and for all $x > a$ in J_1. If $x < a$ a similar argument gives the same inequality with $|x - a|$ appearing instead of $(x - a)$. If we denote by L the length of the interval J_1, then we have $|x - a| \le L$ for all x in J_1 so we obtain the estimate

$$\|Y_{m+1}(x) - Y_m(x)\| \le \|B\| \frac{M^{m+1}L^{m+1}}{(m + 1)!} \qquad \text{for} \quad m = 0, 1, 2, \ldots,$$

and for all x in J_1. Therefore the series in (7.61) is dominated by the convergent series

$$\|B\| \sum_{m=0}^{\infty} \frac{(ML)^{m+1}}{(m + 1)!} = \|B\|\, (e^{ML} - 1).$$

This proves that the series in (7.61) converges uniformly on J_1.

The foregoing argument shows that the sequence of successive approximations always converges and the convergence is uniform on J_1. Let Y denote the limit function. That is, define $Y(x)$ for each x in J_1 by the equation

$$Y(x) = \lim_{k \to \infty} Y_k(x).$$

We shall prove that Y has the following properties:

(a) Y is continuous on J_1.

(b) $Y(x) = B + \int_a^x A(t)Y(t)\,dt$ for all x in J_1.

(c) $Y(a) = B$ and $Y'(x) = A(x)Y(x)$ for all x in J_1.

Part (c) shows that Y is a solution of the initial-value problem on J_1.

Proof of (a). Each function Y_k is a column matrix whose entries are scalar functions, continuous on J_1. Each entry of the limit function Y is the limit of a uniformly convergent sequence of continuous functions so, by Theorem 11.1 of Volume I, each entry of Y is also continuous on J_1. Therefore Y itself is continuous on J_1.

Proof of (b). The recursion formula (7.58) states that

$$Y_{k+1}(x) = B + \int_a^x A(t)Y_k(t)\,dt.$$

Therefore

$$Y(x) = \lim_{k\to\infty} Y_{k+1}(x) = B + \lim_{k\to\infty}\int_a^x A(t)Y_k(t)\,dt = B + \int_a^x A(t)\lim_{k\to\infty} Y_k(t)\,dt$$
$$= B + \int_a^x A(t)Y(t)\,dt.$$

The interchange of the limit symbol with the integral sign is valid because of the uniform convergence of the sequence $\{Y_k\}$ on J_1.

Proof of (c). The equation $Y(a) = B$ follows at once from (b). Because of (a), the integrand in (b) is continuous on J_1 so, by the first fundamental theorem of calculus, $Y'(x)$ exists and equals $A(x)Y(x)$ on J_1.

The interval J_1 was any closed and bounded subinterval of J containing a. If J_1 is enlarged, the process for obtaining $Y(x)$ doesn't change because it only involves integration from a to x. Since for every x in J there is a closed bounded subinterval of J containing a and x, a solution exists over the full interval J.

THEOREM 7.17. UNIQUENESS THEOREM FOR HOMOGENEOUS LINEAR SYSTEMS. *If $A(t)$ is continuous on an open interval J, the differential equation*

$$Y'(t) = A(t)Y(t)$$

has at most one solution on J satisfying a given initial condition $Y(a) = B$.

Proof. Let Y and Z be two solutions on J. Let J_1 be any closed and bounded subinterval of J containing a. We will prove that $Z(x) = Y(x)$ for every x in J_1. This implies that $Z = Y$ on the full interval J.

Since both Y and Z are solutions we have

$$Z'(t) - Y'(t) = A(t)\{Z(t) - Y(t)\}.$$

Choose x in J_1 and integrate this equation from a to x to obtain

$$Z(x) - Y(x) = \int_a^x A(t)\{Z(t) - Y(t)\}\, dt\,.$$

This implies the inequality

$$\|Z(x) - Y(x)\| \le M \left| \int_a^x \|Z(t) - Y(t)\|\, dt \right|, \tag{7.63}$$

where M is an upper bound for $\|A(t)\|$ on J_1. Let M_1 be an upper bound for the continuous function $\|Z(t) - Y(t)\|$ on J_1. Then the inequality (7.63) gives us

$$\|Z(x) - Y(x)\| \le MM_1 \,|x - a|\,. \tag{7.64}$$

Using (7.64) in the right-hand member of (7.63) we obtain

$$\|Z(x) - Y(x)\| \le M^2 M_1 \left| \int_a^x |t - a|\, dt \right| = M^2 M_1 \frac{|x - a|^2}{2}\,.$$

By induction we find

$$\|Z(x) - Y(x)\| \le M^m M_1 \frac{|x - a|^m}{m!}\,. \tag{7.65}$$

When $m \to \infty$ the right-hand member approaches 0, so $Z(x) = Y(x)$. This completes the proof.

The results of this section can be summarized in the following existence-uniqueness theorem.

THEOREM 7.18. *Let A be an $n \times n$ matrix function continuous on an open interval J. If $a \in J$ and if B is any n-dimensional vector, the homogeneous linear system*

$$Y'(t) = A(t)\,Y(t)\,, \qquad Y(a) = B\,,$$

has one and only one n-dimensional vector solution on J.

7.22 The method of successive approximations applied to first-order nonlinear systems

The method of successive approximations can also be applied to some nonlinear systems. Consider a first-order system of the form

$$Y' = F(t, Y)\,, \tag{7.66}$$

where F is a given n-dimensional vector-valued function, and Y is an unknown n-dimensional vector-valued function to be determined. We seek a solution Y which satisfies the equation

$$Y'(t) = F[t, Y(t)]$$

for each t in some interval J and which also satisfies a given initial condition, say $Y(a) = B$, where $a \in J$ and B is a given n-dimensional vector.

In a manner parallel to the linear case, we construct a sequence of successive approximations $Y_0, Y_1, Y_2, \ldots,$ by taking $Y_0 = B$ and defining Y_{k+1} in terms of Y_k by the recursion formula

$$(7.67) \qquad Y_{k+1}(x) = B + \int_a^x F[t, Y_k(t)]\, dt \qquad \text{for} \quad k = 0, 1, 2, \ldots.$$

Under certain conditions on F, this sequence will converge to a limit function Y which will satisfy the given differential equation and the given initial condition.

Before we investigate the convergence of the process we discuss some one-dimensional examples chosen to illustrate some of the difficulties that can arise in practice.

EXAMPLE 1. Consider the nonlinear initial-value problem $y' = x^2 + y^2$ with $y = 0$ when $x = 0$. We shall compute a few approximations to the solution. We choose $Y_0(x) = 0$ and determine the next three approximations as follows:

$$Y_1(x) = \int_0^x t^2\, dt = \frac{x^3}{3},$$

$$Y_2(x) = \int_0^x [t^2 + Y_1^2(t)]\, dt = \int_0^x \left(t^2 + \frac{t^6}{9}\right) dt = \frac{x^3}{3} + \frac{x^7}{63},$$

$$Y_3(x) = \int_0^x \left[t^2 + \left(\frac{t^3}{3} + \frac{t^7}{63}\right)^2\right] dt = \frac{x^3}{3} + \frac{x^7}{63} + \frac{2x^{11}}{2079} + \frac{x^{15}}{59535}.$$

It is now apparent that a great deal of labor will be needed to compute further approximations. For example, the next two approximations Y_4 and Y_5 will be polynomials of degrees 31 and 63, respectively.

The next example exhibits a further difficulty that can arise in the computation of the successive approximations.

EXAMPLE 2. Consider the nonlinear initial-value problem $y' = 2x + e^y$, with $y = 0$ when $x = 0$. We begin with the initial guess $Y_0(x) = 0$ and we find

$$Y_1(x) = \int_0^x (2t + 1)\, dt = x^2 + x,$$

$$Y_2(x) = \int_0^x (2t + e^{t^2+t})\, dt = x^2 + \int_0^x e^{t^2+t}\, dt.$$

Here further progress is impeded by the fact that the last integral cannot be evaluated in terms of elementary functions. However, for a given x it is possible to calculate a numerical approximation to the integral and thereby obtain an approximation to $Y_2(x)$.

Because of the difficulties displayed in the last two examples, the method of successive approximations is sometimes not very useful for the explicit determination of solutions in practice. The real value of the method is its use in establishing existence theorems.

7.23 Proof of an existence-uniqueness theorem for first-order nonlinear systems

We turn now to an existence-uniqueness theorem for first-order nonlinear systems. By placing suitable restrictions on the function which appears in the right-hand member of the differential equation

$$Y' = F(x, Y),$$

we can extend the method of proof used for the linear case in Section 7.21.

Let J denote the open interval over which we seek a solution. Assume $a \in J$ and let B be a given n-dimensional vector. Let S denote a set in $(n + 1)$-space given by

$$S = \{(x,Y) \mid \ |x - a| \le h, \|Y - B\| \le k\},$$

where $h > 0$ and $k > 0$. [If $n = 1$ this is a rectangle with center at (a, B) and with base $2h$ and altitude $2k$.] We assume that the domain of F includes a set S of this type and that F is bounded on S, say

$$\|F(x, Y)\| \le M \tag{7.68}$$

for all (x, Y) in S, where M is a positive constant.

Next, we assume that the composite function $G(x) = F(x, Y(x))$ is continuous on the interval $(a - h, a + h)$ for every function Y which is continuous on $(a - h, a + h)$ and which has the property that $(x, Y(x)) \in S$ for all x in $(a - h, a + h)$. This assumption guarantees the existence of the integrals that occur in the method of successive approximations, and it also implies continuity of the functions so constructed.

Finally, we assume that F satisfies a condition of the form

$$\|F(x, Y) - F(x, Z)\| \le A \|Y - Z\|$$

for every pair of points (x, Y) and (x, Z) in S, where A is a positive constant. This is called a *Lipschitz condition* in honor of Rudolph Lipschitz who first introduced it in 1876. A Lipschitz condition does not restrict a function very seriously and it enables us to extend the proof of existence and uniqueness from the linear to the nonlinear case.

THEOREM 7.19. EXISTENCE AND UNIQUENESS OF SOLUTIONS TO FIRST-ORDER NONLINEAR SYSTEMS. *Assume F satisfies the boundedness, continuity, and Lipschitz conditions specified above on a set S. Let I denote the open interval $(a - c, a + c)$, where $c = \min\{h, k/M\}$. Then there is one and only one function Y defined on I with $Y(a) = B$ such that $(x, Y(x)) \in S$ and*

$$Y'(x) = F(x, Y(x)) \qquad \textit{for each } x \textit{ in } I.$$

Proof. Since the proof is analogous to that for the linear case we sketch only the principal steps. We let $Y_0(x) = B$ and define vector-valued functions $Y_1, Y_2, \ldots$ on I by the recursion formula

$$Y_{m+1}(x) = B + \int_a^x F[t, Y_m(t)]\,dt \qquad \text{for} \quad m = 0, 1, 2, \ldots. \tag{7.69}$$

For the recursion formula to be meaningful we need to know that $(x, Y_m(x)) \in S$ for each x in I. This is easily proved by induction on m. When $m = 0$ we have $(x, Y_0(x)) = (x, B)$, which is in S. Assume then that $(x, Y_m(x)) \in S$ for some m and each x in I. Using (7.69) and (7.68) we obtain

$$\|Y_{m+1}(x) - B\| \le \left|\int_a^x \|F[t, Y_m(t)]\| \, dt\right| \le M \left|\int_a^x dt\right| = M\,|x - a|.$$

Since $|x - a| \le c$ for x in I, this implies that

$$\|Y_{m+1}(x) - B\| \le Mc \le k,$$

which shows that $(x, Y_{m+1}(x)) \in S$ for each x in I. Therefore the recursion formula is meaningful for every $m \ge 0$ and every x in I.

The convergence of the sequence $\{Y_m(x)\}$ is now established exactly as in Section 7.21. We write

$$Y_k(x) = Y_0(x) + \sum_{m=0}^{k-1} \{Y_{m+1}(x) - Y_m(x)\}$$

and prove that $Y_k(x)$ tends to a limit as $k \to \infty$ by proving that the infinite series

$$\sum_{m=0}^{\infty} \|Y_{m+1}(x) - Y_m(x)\|$$

converges on I. This is deduced from the inequality

$$\|Y_{m+1}(x) - Y_m(x)\| \le \frac{MA^m\,|x - a|^{m+1}}{(m+1)!} \le \frac{MA^m c^{m+1}}{(m+1)!}$$

which is proved by induction, using the recursion formula and the Lipschitz condition. We then define the limit function Y by the equation

$$Y(x) = \lim_{m \to \infty} Y_m(x)$$

for each x in I and verify that it satisfies the integral equation

$$Y(x) = B + \int_a^x F[t, Y(t)]\, dt,$$

exactly as in the linear case. This proves the existence of a solution. The uniqueness may then be proved by the same method used to prove Theorem 7.17.

7.24 Exercises

1. Consider the linear initial-value problem

$$y' + y = 2e^x, \qquad \text{with} \quad y = 1 \text{ when } x = 0.$$

(a) Find the exact solution Y of this problem.
(b) Apply the method of successive approximations, starting with the initial guess $Y_0(x) = 1$. Determine $Y_n(x)$ explicitly and show that

$$\lim_{n\to\infty} Y_n(x) = Y(x)$$

for all real x.

2. Apply the method of successive approximations to the nonlinear initial-value problem

$$y' = x + y^2, \qquad \text{with} \quad y = 0 \text{ when } x = 0.$$

Take $Y_0(x) = 0$ as the initial guess and compute $Y_3(x)$.

3. Apply the method of successive approximations to the nonlinear initial-value problem

$$y' = 1 + xy^2, \qquad \text{with} \quad y = 0 \text{ when } x = 0.$$

Take $Y_0(x) = 0$ as the initial guess and compute $Y_3(x)$.

4. Apply the method of successive approximations to the nonlinear initial-value problem

$$y' = x^2 + y^2, \qquad \text{with} \quad y = 0 \text{ when } x = 0.$$

Start with the "bad" initial guess $Y_0(x) = 1$, compute $Y_3(x)$, and compare with the results of Example 1 in Section 7.22.

5. Consider the nonlinear initial-value problem

$$y' = x^2 + y^2, \qquad \text{with} \quad y = 1 \text{ when } x = 0.$$

(a) Apply the method of successive approximations, starting with the initial guess $Y_0(x) = 1$, and compute $Y_2(x)$.
(b) Let $R = [-1, 1] \times [-1, 1]$. Find the smallest M such that $|f(x, y)| \le M$ on R. Find an interval $I = (-c, c)$ such that the graph of every approximating function Y_n over I will lie in R.
(c) Assume the solution $y = Y(x)$ has a power-series expansion in a neighborhood of the origin. Determine the first six nonzero terms of this expansion and compare with the result of part (a).

6. Consider the initial-value problem

$$y' = 1 + y^2, \qquad \text{with} \quad y = 0 \text{ when } x = 0.$$

(a) Apply the method of successive approximations, starting with the initial guess $Y_0(x) = 0$, and compute $Y_4(x)$.
(b) Prove that every approximating function Y_n is defined on the entire real axis.
(c) Use Theorem 7.19 to show that the initial-value problem has at most one solution in any interval of the form $(-h, h)$.
(d) Solve the differential equation by separation of variables and thereby show that there is exactly one solution Y of the initial-value problem on the interval $(-\pi/2, \pi/2)$ and no solution on any larger interval. In this example, the successive approximations are defined on the entire real axis, but they converge to a limit function only on the interval $(-\pi/2, \pi/2)$.

7. We seek two functions $y = Y(x)$ and $z = Z(x)$ that simultaneously satisfy the system of equations

$$y' = z, \qquad z' = x^3(y + z)$$

with initial conditions $y = 1$ and $z = 1/2$ when $x = 0$. Start with the initial guesses $Y_0(x) = 1$, $Z_0(x) = 1/2$, and use the method of successive approximations to obtain the approximating functions

$$Y_3(x) = 1 + \frac{x}{2} + \frac{3x^5}{40} + \frac{x^6}{60} + \frac{x^9}{192},$$

$$Z_3(x) = \frac{1}{2} + \frac{3x^4}{8} + \frac{x^5}{10} + \frac{3x^8}{64} + \frac{7x^9}{360} + \frac{x^{12}}{256}.$$

8. Consider the system of equations

$$y' = 2x + z, \qquad z' = 3xy + x^2 z,$$

with initial conditions $y = 2$ and $z = 0$ when $x = 0$. Start with the initial guesses $Y_0(x) = 2$, $Z_0(x) = 0$, use the method of successive approximations, and determine $Y_3(x)$ and $Z_3(x)$.

9. Consider the initial-value problem

$$y'' = x^2 y' + x^4 y, \quad \text{with} \quad y = 5 \quad \text{and} \quad y' = 1 \text{ when } x = 0.$$

Change this problem to an equivalent problem involving a system of two equations for two unknown functions $y = Y(x)$ and $z = Z(x)$, where $z = y'$. Then use the method of successive approximations, starting with initial guesses $Y_0(x) = 5$ and $Z_0(x) = 1$, and determine $Y_3(x)$ and $Z_3(x)$.

10. Let f be defined on the rectangle $R = [-1, 1] \times [-1, 1]$ as follows:

$$f(x, y) = \begin{cases} 0 & \text{if } x = 0, \\ 2y/x & \text{if } x \neq 0 \text{ and } |y| \leq x^2, \\ 2x & \text{if } x \neq 0 \text{ and } y > x^2, \\ -2x & \text{if } x \neq 0 \text{ and } y < -x^2. \end{cases}$$

(a) Prove that $|f(x, y)| \leq 2$ for all (x, y) in R.
(b) Show that f does not satisfy a Lipschitz condition on R.
(c) For each constant C satisfying $|C| \leq 1$, show that $y = Cx^2$ is a solution of the initial-value problem $y' = f(x, y)$, with $y = 0$ when $x = 0$. Show also that the graph of each of these solutions over $(-1, 1)$ lies in R.
(d) Apply the method of successive approximations to this initial-value problem, starting with initial guess $Y_0(x) = 0$. Determine $Y_n(x)$ and show that the approximations converge to a solution of the problem on the interval $(-1, 1)$.
(e) Repeat part (d), starting with initial guess $Y_0(x) = x$. Determine $Y_n(x)$ and show that the approximating functions converge to a solution different from any of those in part (c).
(f) Repeat part (d), starting with the initial guess $Y_0(x) = x^3$.
(g) Repeat part (d), starting with the initial guess $Y_0(x) = x^{1/3}$.

★ 7.25 Successive approximations and fixed points of operators

The basic idea underlying the method of successive approximations can be used not only to establish existence theorems for differential equations but also for many other important problems in analysis. The rest of this chapter reformulates the method of successive approximations in a setting that greatly increases the scope of its applications.

In the proof of Theorem 7.18 we constructed a sequence of functions $\{Y_k\}$ according to the recursion formula

$$Y_{k+1}(x) = B + \int_a^x AY_k(t)\,dt\,.$$

The right-hand member of this formula can be regarded as an operator T which converts certain functions Y into new functions $T(Y)$ according to the equation

$$T(Y) = B + \int_a^x AY(t)\,dt\,.$$

In the proof of Theorem 7.18 we found that the solution Y of the initial-value problem $Y'(t) = AY(t)$, $Y(a) = B$, satisfies the integral equation

$$Y = B + \int_a^x AY(t)\,dt\,.$$

In operator notation this states that $Y = T(Y)$. In other words, the solution Y remains unaltered by the operator T. Such a function Y is called a *fixed point* of the operator T.

Many important problems in analysis can be formulated so their solution depends on the existence of a fixed point for some operator. Therefore it is worthwhile to try to discover properties of operators that guarantee the existence of a fixed point. We turn now to a systematic treatment of this problem.

★ 7.26 Normed linear spaces

To formulate the method of successive approximations in a general form it is convenient to work within the framework of linear spaces. Let S be an arbitrary linear space. When we speak of approximating one element x in S by another element y in S, we consider the difference $x - y$, which we call the *error* of the approximation. To measure the size of this error we introduce a norm in the space.

DEFINITION OF A NORM. *Let S be any linear space. A real-valued function N defined on S is called a norm if it has the following properties:*

(a) $N(x) \geq 0$ *for each x in S.*
(b) $N(cx) = |c|\,N(x)$ *for each x in S and each scalar c.*
(c) $N(x + y) \leq N(x) + N(y)$ *for all x and y in S.*
(d) $N(x) = 0$ *implies* $x = O$.

A linear space with a norm assigned to it is called a *normed linear space.*

The norm of x is sometimes written $\|x\|$ instead of $N(x)$. In this notation the fundamental properties become:

(a) $\|x\| \geq 0$ for all x in S.
(b) $\|cx\| = |c|\,\|x\|$ for all x in S and all scalars c.
(c) $\|x + y\| \leq \|x\| + \|y\|$ for all x and y in S.
(d) $\|x\| = 0$ implies $x = O$.

If x and y are in S, we refer to $\|x - y\|$ as the *distance* from x to y.

If the space S is Euclidean, then it always has a norm which it inherits from the inner product, namely, $\|x\| = (x, x)^{1/2}$. However, we shall be interested in a particular norm which does not arise from an inner product.

EXAMPLE. *The* max *norm.* Let $C(J)$ denote the linear space of real-valued functions continuous on a closed and bounded interval J. If $\varphi \in C(J)$, define

$$\|\varphi\| = \max_{x \in J} |\varphi(x)|,$$

where the symbol on the right stands for the maximum absolute value of φ on J. The reader may verify that this norm has the four fundamental properties.

The max norm is not derived from an inner product. To prove this we show that the max norm violates some property possessed by all inner-product norms. For example, if a norm is derived from an inner product, then the "parallelogram law"

$$\|x + y\|^2 + \|x - y\|^2 = 2\,\|x\|^2 + 2\,\|y\|^2$$

holds for all x and y in S. (See Exercise 16 in Section 1.13.) The parallelogram law is not always satisfied by the max norm. For example, let x and y be the functions on the interval $[0, 1]$ given by

$$x(t) = t, \qquad y(t) = 1 - t.$$

Then we have $\|x\| = \|y\| = \|x + y\| = \|x - y\| = 1$, so the parallelogram law is violated.

★ 7.27 Contraction operators

In this section we consider the normed linear space $C(J)$ of all real functions continuous on a closed bounded interval J in which $\|\varphi\|$ is the max norm. Consider an operator

$$T\colon C(J) \to C(J)$$

whose domain is $C(J)$ and whose range is a subset of $C(J)$. That is, if φ is continuous on J, then $T(\varphi)$ is also continuous on J. The following formulas illustrate a few simple examples of such operators. In each case φ is an arbitrary function in $C(J)$ and $T(\varphi)(x)$ is defined for each x in J by the formula given:

$$T(\varphi)(x) = \lambda\varphi(x), \qquad \text{where } \lambda \text{ is a fixed real number},$$

$$T(\varphi)(x) = \int_c^x \varphi(t)\,dt, \qquad \text{where } c \text{ is a given point in } J,$$

$$T(\varphi)(x) = b + \int_c^x f[t, \varphi(t)]\,dt,$$

where b is a constant and the composition $f[t, \varphi(t)]$ is continuous on J.

We are interested in those operators T for which the distance $\|T(\varphi) - T(\psi)\|$ is less than a fixed constant multiple $\alpha < 1$ of $\|\varphi - \psi\|$. These are called *contraction operators;* they are defined as follows.

DEFINITION OF A CONTRACTION OPERATOR. *An operator $T: C(J) \to C(J)$ is called a contraction operator if there is a constant α satisfying $0 \le \alpha < 1$ such that for every pair of functions φ and ψ in $C(J)$ we have*

$$\|T(\varphi) - T(\psi)\| \le \alpha \|\varphi - \psi\| . \tag{7.70}$$

The constant α is called a contraction constant for T.

Note: Inequality (7.70) holds if and only if we have

$$|T(\varphi)(x) - T(\psi)(x)| \le \alpha \|\varphi - \psi\| \qquad \text{for every } x \text{ in } J.$$

EXAMPLE 1. Let T be the operator defined by $T(\varphi)(x) = \lambda\varphi(x)$, where λ is constant. Since

$$|T(\varphi)(x) - T(\psi)(x)| = |\lambda| \, |\varphi(x) - \psi(x)| ,$$

we have $\|T(\varphi) - T(\psi)\| = |\lambda| \, \|\varphi - \psi\|$. Therefore this operator is a contraction operator if and only if $|\lambda| < 1$, in which case $|\lambda|$ may be used as a contraction constant.

EXAMPLE 2. Let $T(\varphi)(x) = b + \int_c^x f[t, \varphi(t)]\, dt$, where f satisfies a Lipschitz condition of the form

$$|f(x, y) - f(x, z)| \le K|y - z|$$

for all x in J and all real y and z; here K is a positive constant. Let $L(J)$ denote the length of the interval J. If $KL(J) < 1$ we can easily show that T is a contraction operator with contraction constant $KL(J)$. In fact, for every x in J we have

$$\begin{aligned} |T(\varphi)(x) - T(\psi)(x)| &= \left| \int_c^x \{f[t, \varphi(t)] - f[t, \psi(t)]\}\, dt \right| \le K \left| \int_c^x |\varphi(t) - \psi(t)|\, dt \right| \\ &\le K\, \|\varphi - \psi\| \left| \int_c^x dt \right| \le KL(J)\, \|\varphi - \psi\| . \end{aligned}$$

If $KL(J) < 1$, then T is a contraction operator with contraction constant $\alpha = KL(J)$.

★ 7.28 Fixed-point theorem for contraction operators

The next theorem shows that every contraction operator has a unique fixed point.

THEOREM 7.20. *Let $T: C(J) \to C(J)$ be a contraction operator. Then there exists one and only one function φ in $C(J)$ such that*

$$T(\varphi) = \varphi . \tag{7.71}$$

Proof. Let φ_0 be any function in $C(J)$ and define a sequence of functions $\{\varphi_n\}$ by the recursion formula

$$\varphi_{n+1} = T(\varphi_n) \qquad \text{for} \quad n = 0, 1, 2, \ldots .$$

Note that $\varphi_{n+1} \in C(J)$ for each n. We shall prove that the sequence $\{\varphi_n\}$ converges to a limit function φ in $C(J)$. The method is similar to that used in the proof of Theorem 7.18. We write each φ_n as a telescoping sum,

$$\varphi_n(x) = \varphi_0(x) + \sum_{k=0}^{n-1} \{\varphi_{k+1}(x) - \varphi_k(x)\} \tag{7.72}$$

and prove convergence of $\{\varphi_n\}$ by showing that the infinite series

$$\varphi_0(x) + \sum_{k=0}^{\infty} \{\varphi_{k+1}(x) - \varphi_k(x)\} \tag{7.73}$$

converges uniformly on J. Then we show that the sum of this series is the required fixed point.

The uniform convergence of the series will be established by comparing it with the convergent geometric series

$$M \sum_{k=0}^{\infty} \alpha^k ,$$

where $M = \|\varphi_0\| + \|\varphi_1\|$, and α is a contraction constant for T. The comparison is provided by the inequality

$$|\varphi_{k+1}(x) - \varphi_k(x)| \leq M\alpha^k \tag{7.74}$$

which holds for every x in J and every $k \geq 1$. To prove (7.74) we note that

$$|\varphi_{k+1}(x) - \varphi_k(x)| = |T(\varphi_k)(x) - T(\varphi_{k-1})(x)| \leq \alpha \, \|\varphi_k - \varphi_{k-1}\| .$$

Therefore the inequality in (7.74) will be proved if we show that

$$\|\varphi_k - \varphi_{k-1}\| \leq M\alpha^{k-1} \tag{7.75}$$

for every $k \geq 1$. We now prove (7.75) by induction. For $k = 1$ we have

$$\|\varphi_1 - \varphi_0\| \leq \|\varphi_1\| + \|\varphi_0\| = M ,$$

which is the same as (7.75). To prove that (7.75) holds for $k + 1$ if it holds for k we note that

$$|\varphi_{k+1}(x) - \varphi_k(x)| = |T(\varphi_k)(x) - T(\varphi_{k-1})(x)| \leq \alpha \, \|\varphi_k - \varphi_{k-1}\| \leq M\alpha^k .$$

Since this is valid for each x in J we must also have

$$\|\varphi_{k+1} - \varphi_k\| \leq M\alpha^k .$$

This proves (7.75) by induction. Therefore the series in (7.73) converges uniformly on J. If we let $\varphi(x)$ denote its sum we have

$$\varphi(x) = \lim_{n \to \infty} \varphi_n(x) = \varphi_0(x) + \sum_{k=0}^{\infty} \{\varphi_{k+1}(x) - \varphi_k(x)\} . \tag{7.76}$$

The function φ is continuous on J because it is the sum of a uniformly convergent series of continuous functions. To prove that φ is a fixed point of T we compare $T(\varphi)$ with $\varphi_{n+1} = T(\varphi_n)$. Using the contraction property of T we have

$$(7.77) \qquad |T(\varphi)(x) - \varphi_{n+1}(x)| = |T(\varphi)(x) - T(\varphi_n)(x)| \leq \alpha\, |\varphi(x) - \varphi_n(x)|\,.$$

But from (7.72) and (7.76) we find

$$|\varphi(x) - \varphi_n(x)| = \left| \sum_{k=n}^{\infty} \{\varphi_{k+1}(x) - \varphi_k(x)\} \right| \leq \sum_{k=n}^{\infty} |\varphi_{k+1}(x) - \varphi_k(x)| \leq M \sum_{k=n}^{\infty} \alpha^k,$$

where in the last step we used (7.74). Therefore (7.77) implies

$$|T(\varphi)(x) - \varphi_{n+1}(x)| \leq M \sum_{k=n}^{\infty} \alpha^{k+1}.$$

When $n \to \infty$ the series on the right tends to 0, so $\varphi_{n+1}(x) \to T(\varphi)(x)$. But since $\varphi_{n+1}(x) \to \varphi(x)$ as $n \to \infty$, this proves that $\varphi(x) = T(\varphi)(x)$ for each x in J. Therefore $\varphi = T(\varphi)$, so φ is a fixed point.

Finally we prove that the fixed point φ is unique. Let ψ be another function in $C(J)$ such that $T(\psi) = \psi$. Then we have

$$\|\varphi - \psi\| = \|T(\varphi) - T(\psi)\| \leq \alpha\, \|\varphi - \psi\|\,.$$

This gives us $(1 - \alpha)\, \|\varphi - \psi\| \leq 0$. Since $\alpha < 1$ we may divide by $1 - \alpha$ to obtain the inequality $\|\varphi - \psi\| \leq 0$. But since we also have $\|\varphi - \psi\| \geq 0$ this means that $\|\varphi - \psi\| = 0$, and hence $\varphi - \psi = 0$. The proof of the fixed-point theorem is now complete.

★ 7.29 Applications of the fixed-point theorem

To indicate the broad scope of applications of the fixed point theorem we use it to prove two important theorems. The first gives a sufficient condition for an equation of the form $f(x, y) = 0$ to define y as a function of x.

THEOREM 7.21. AN IMPLICIT-FUNCTION THEOREM. *Let f be defined on a rectangular strip R of the form*

$$R = \{(x, y) \mid a \leq x \leq b,\ -\infty < y < +\infty\}.$$

Assume that the partial derivative $D_2 f(x, y)$ exists† and satisfies an inequality of the form

$$(7.78) \qquad 0 < m \leq D_2 f(x, y) \leq M$$

† $D_2 f(x, y)$ is the derivative of $f(x, y)$ with respect to y, holding x fixed.

for all (x, y) *in* R, *where* m *and* M *are constants with* $m \le M$. *Assume also that for each function* φ *continuous on* $[a, b]$ *the composite function* $g(x) = f[x, \varphi(x)]$ *is continuous on* $[a, b]$. *Then there exists one and only one function* $y = Y(x)$, *continuous on* $[a, b]$, *such that*

$$f[x, Y(x)] = 0 \tag{7.79}$$

for all x *in* $[a, b]$.

Note: We describe this result by saying that the equation $f(x, y) = 0$ serves to define y implicitly as a function of x in $[a, b]$.

Proof. Let C denote the linear space of continuous functions on $[a, b]$, and define an operator $T: C \to C$ by the equation

$$T(\varphi)(x) = \varphi(x) - \frac{1}{M} f[x, \varphi(x)]$$

for each x in $[a, b]$. Here M is the positive constant in (7.78). The function $T(\varphi) \in C$ whenever $\varphi \in C$. We shall prove that T is a contraction operator. Once we know this it follows that T has a unique fixed point Y in C. For this function Y we have $Y = T(Y)$ which means

$$Y(x) = Y(x) - \frac{1}{M} f[x, Y(x)]$$

for each x in $[a, b]$. This gives us (7.79), as required.

To show that T is a contraction operator we consider the difference

$$T(\varphi)(x) - T(\psi)(x) = \varphi(x) - \psi(x) - \frac{f[x, \varphi(x)] - f[x, \psi(x)]}{M}. \tag{7.80}$$

By the mean-value theorem for derivatives we have

$$f[x, \varphi(x)] - f[x, \psi(x)] = D_2 f[x, z(x)][\varphi(x) - \psi(x)],$$

where $z(x)$ lies between $\varphi(x)$ and $\psi(x)$. Therefore (7.80) gives us

$$T(\varphi)(x) - T(\psi)(x) = [\varphi(x) - \psi(x)]\left(1 - \frac{D_2 f[x, z(x)]}{M}\right). \tag{7.81}$$

The hypothesis (7.78) implies that

$$0 \le 1 - \frac{D_2 f[x, z(x)]}{M} \le 1 - \frac{m}{M}.$$

Therefore (7.81) gives us the inequality

$$|T(\varphi)(x) - T(\psi)(x)| \le |\varphi(x) - \psi(x)| \left(1 - \frac{m}{M}\right) \le \alpha \, \|\varphi - \psi\|, \tag{7.82}$$

where $\alpha = 1 - m/M$. Since $0 < m \le M$, we have $0 \le \alpha < 1$. Inequality (7.82) is valid for every x in $[a, b]$. Hence T is a contraction operator. This completes the proof.

The next application of the fixed-point theorem establishes an existence theorem for the integral equation

$$\varphi(x) = \psi(x) + \lambda \int_a^b K(x, t)\varphi(t)\, dt\,. \tag{7.83}$$

Here ψ is a given function, continuous on $[a, b]$, λ is a given constant, and K is a given function defined and bounded on the square

$$S = \{(x, y) \mid a \le x \le b, a \le y \le b\}\,.$$

The function K is called the *kernel* of the integral equation. Let C be the linear space of continuous functions on $[a, b]$. We assume that the kernel K is such that the operator T given by

$$T(\varphi)(x) = \psi(x) + \lambda \int_a^b K(x, t)\varphi(t)\, dt$$

maps C into C. In other words, we assume that $T(\varphi) \in C$ whenever $\varphi \in C$. A solution of the integral equation is any function φ in C that satisfies (7.83).

THEOREM 7.22. AN EXISTENCE THEOREM FOR INTEGRAL EQUATIONS. *If, under the foregoing conditions, we have*

$$|K(x, y)| \le M \tag{7.84}$$

for all (x, y) *in* S, *where* $M > 0$, *then for each* λ *such that*

$$|\lambda| < \frac{1}{M(b - a)} \tag{7.85}$$

there is one and only one function φ *in* C *that satisfies the integral equation* (7.83).

Proof. We shall prove that T is a contraction operator. Take any two functions φ_1 and φ_2 in C and consider the difference

$$T(\varphi_1)(x) - T(\varphi_2)(x) = \lambda \int_a^b K(x, t)[\varphi_1(t) - \varphi_2(t)]\, dt\,.$$

Using the inequality (7.84) we may write

$$|T(\varphi_1)(x) - T(\varphi_2)(x)| \le |\lambda|\, M(b - a)\, \|\varphi_1 - \varphi_2\| = \alpha\, \|\varphi_1 - \varphi_2\|\,,$$

where $\alpha = |\lambda|\, M(b - a)$. Because of (7.85) we have $0 \le \alpha < 1$, so T is a contraction operator with contraction constant α. Therefore T has a unique fixed point φ in C. This function φ satisfies (7.83).

PART 2

NONLINEAR ANALYSIS

8

DIFFERENTIAL CALCULUS OF SCALAR AND VECTOR FIELDS

8.1 Functions from $\mathbf{R}^n$ to $\mathbf{R}^m$. Scalar and vector fields

Part 1 of this volume dealt primarily with linear transformations

$$T: V \rightarrow W$$

from one linear space V into another linear space W. In Part 2 we drop the requirement that T be linear but restrict the spaces V and W to be finite-dimensional. Specifically, we shall consider functions with domain in n-space $\mathbf{R}^n$ and with range in m-space $\mathbf{R}^m$.

When both n and m are equal to 1, such a function is called a real-valued function of a real variable. When $n = 1$ and $m > 1$ it is called a vector-valued function of a real variable. Examples of such functions were studied extensively in Volume I.

In this chapter we assume that $n > 1$ and $m \geq 1$. When $m = 1$ the function is called a real-valued function of a vector variable or, more briefly, a *scalar field.* When $m > 1$ it is called a vector-valued function of a vector variable, or simply a *vector field.*

This chapter extends the concepts of limit, continuity, and derivative to scalar and vector fields. Chapters 10 and 11 extend the concept of the integral.

Notation: Scalars will be denoted by light-faced type, and vectors by bold-faced type. If f is a scalar field defined at a point $\boldsymbol{x} = (x_1, \ldots, x_n)$ in $\mathbf{R}^n$, the notations $f(\boldsymbol{x})$ and $f(x_1, \ldots, x_n)$ are both used to denote the value of f at that particular point. If $\boldsymbol{f}$ is a vector field we write $\boldsymbol{f}(\boldsymbol{x})$ or $\boldsymbol{f}(x_1, \ldots, x_n)$ for the function value at $\boldsymbol{x}$. We shall use the inner product

$$\boldsymbol{x} \cdot \boldsymbol{y} = \sum_{k=1}^{n} x_k y_k$$

and the corresponding norm $\|\boldsymbol{x}\| = (\boldsymbol{x} \cdot \boldsymbol{x})^{1/2}$, where $\boldsymbol{x} = (x_1, \ldots, x_n)$ and $\boldsymbol{y} = (y_1, \ldots, y_n)$. Points in $\mathbf{R}^2$ are usually denoted by (x, y) instead of (x_1, x_2); points in $\mathbf{R}^3$ by (x, y, z) instead of (x_1, x_2, x_3).

Scalar and vector fields defined on subsets of $\mathbf{R}^2$ and $\mathbf{R}^3$ occur frequently in the applications of mathematics to science and engineering. For example, if at each point $\boldsymbol{x}$ of the atmosphere we assign a real number $f(\boldsymbol{x})$ which represents the temperature at $\boldsymbol{x}$, the function

f so defined is a scalar field. If we assign a vector which represents the wind velocity at that point, we obtain an example of a vector field.

In physical problems dealing with either scalar or vector fields it is important to know how the field changes as we move from one point to another. In the one-dimensional case the derivative is the mathematical tool used to study such changes. Derivative theory in the one-dimensional case deals with functions defined on open intervals. To extend the theory to $\mathbf{R}^n$ we consider generalizations of open intervals called *open sets*.

8.2 Open balls and open sets

Let $\boldsymbol{a}$ be a given point in $\mathbf{R}^n$ and let r be a given positive number. The set of all points $\boldsymbol{x}$ in $\mathbf{R}^n$ such that

$$\|\boldsymbol{x} - \boldsymbol{a}\| < r$$

is called an *open n-ball* of radius r and center $\boldsymbol{a}$. We denote this set by $B(\boldsymbol{a})$ or by $B(\boldsymbol{a}; r)$.

The ball $B(\boldsymbol{a}; r)$ consists of all points whose distance from $\boldsymbol{a}$ is less than r. In $\mathbf{R}^1$ this is simply an open interval with center at $\boldsymbol{a}$. In $\mathbf{R}^2$ it is a circular disk, and in $\mathbf{R}^3$ it is a spherical solid with center at $\boldsymbol{a}$ and radius r.

DEFINITION OF AN INTERIOR POINT. *Let S be a subset of $\mathbf{R}^n$, and assume that $\boldsymbol{a} \in S$. Then $\boldsymbol{a}$ is called an interior point of S if there is an open n-ball with center at $\boldsymbol{a}$, all of whose points belong to S.*

In other words, every interior point $\boldsymbol{a}$ of S can be surrounded by an n-ball $B(\boldsymbol{a})$ such that $B(\boldsymbol{a}) \subseteq S$. The set of all interior points of S is called the *interior* of S and is denoted by int S. An open set containing a point $\boldsymbol{a}$ is sometimes called a *neighborhood of* $\boldsymbol{a}$.

DEFINITION OF AN OPEN SET. *A set S in $\mathbf{R}^n$ is called open if all its points are interior points. In other words, S is open if and only if $S = \text{int } S$.*

EXAMPLES. In $\mathbf{R}^1$ the simplest type of open set is an open interval. The union of two or more open intervals is also open. A closed interval $[a, b]$ is not an open set because neither endpoint of the interval can be enclosed in a 1-ball lying within the given interval.

The 2-ball $S = B(O; 1)$ shown in Figure 8.1 is an example of an open set in $\mathbf{R}^2$. Every point $\boldsymbol{a}$ of S is the center of a disk lying entirely in S. For some points the radius of this disk is very small.

Some open sets in $\mathbf{R}^2$ can be constructed by taking the Cartesian product of open sets in $\mathbf{R}^1$. If A_1 and A_2 are subsets of $\mathbf{R}^1$, their Cartesian product $A_1 \times A_2$ is the set in $\mathbf{R}^2$ defined by

$$A_1 \times A_2 = \{(a_1, a_2) \mid a_1 \in A_1 \quad \text{and} \quad a_2 \in A_2\}.$$

An example is shown in Figure 8.2. The sets A_1 and A_2 are intervals, and $A_1 \times A_2$ is a rectangle.

If A_1 and A_2 are open subsets of $\mathbf{R}^1$, then $A_1 \times A_2$ will be an open subset of $\mathbf{R}^2$. To prove this, choose any point $\boldsymbol{a} = (a_1, a_2)$ in $A_1 \times A_2$. We must show that $\boldsymbol{a}$ is an interior point

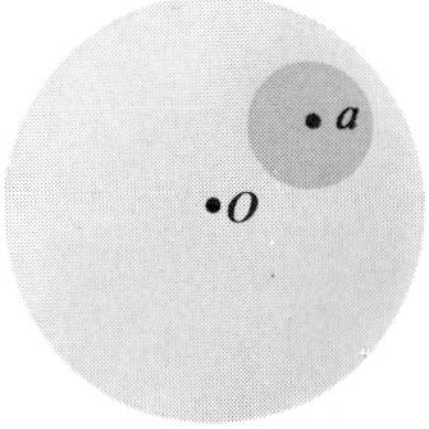

FIGURE 8.1 The disk $B(O; 1)$ is an open set in $\mathbf{R}^2$.

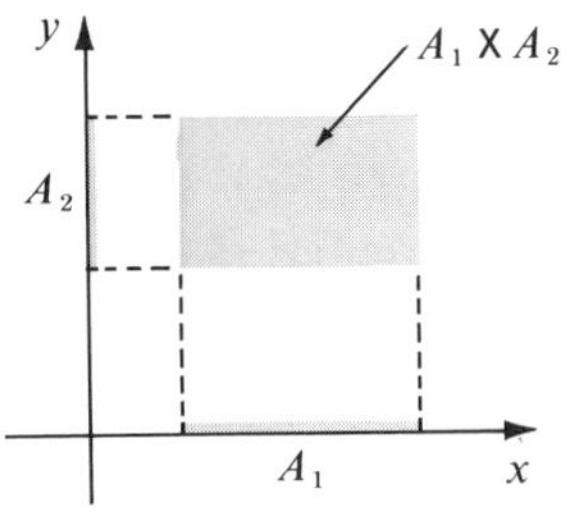

FIGURE 8.2 The Cartesian product of two open intervals is an open rectangle.

of $A_1 \times A_2$. Since A_1 and A_2 are open in $\mathbf{R}^1$ there is a 1-ball $B(a_1; r_1)$ in A_1 and a 1-ball $B(a_2; r_2)$ in A_2. Let $r = \min\{r_1, r_2\}$. We can easily show that the 2-ball $B(\boldsymbol{a}; r) \subseteq A_1 \times A_2$. In fact, if $\boldsymbol{x} = (x_1, x_2)$ is any point of $B(\boldsymbol{a}; r)$ then $\|\boldsymbol{x} - \boldsymbol{a}\| < r$, so $|x_1 - a_1| < r_1$ and $|x_2 - a_2| < r_2$. Hence $x_1 \in B(a_1; r_1)$ and $x_2 \in B(a_2; r_2)$. Therefore $x_1 \in A_1$ and $x_2 \in A_2$, so $(x_1, x_2) \in A_1 \times A_2$. This proves that every point of $B(\boldsymbol{a}; r)$ is in $A_1 \times A_2$. Therefore every point of $A_1 \times A_2$ is an interior point, so $A_1 \times A_2$ is open.

The reader should realize that an open subset of $\mathbf{R}^1$ is no longer an open set when it is considered as a subset of $\mathbf{R}^2$, because a subset of $\mathbf{R}^1$ cannot contain a 2-ball.

DEFINITIONS OF EXTERIOR AND BOUNDARY. *A point $\boldsymbol{x}$ is said to be exterior to a set S in $\mathbf{R}^n$ if there is an n-ball $B(\boldsymbol{x})$ containing no points of S. The set of all points in $\mathbf{R}^n$ exterior to S is called the exterior of S and is denoted by* ext S. *A point which is neither exterior to S nor an interior point of S is called a boundary point of S. The set of all boundary points of S is called the boundary of S and is denoted by ∂S.*

These concepts are illustrated in Figure 8.1. The exterior of S is the set of all $\boldsymbol{x}$ with $\|\boldsymbol{x}\| > 1$. The boundary of S consists of all $\boldsymbol{x}$ with $\|\boldsymbol{x}\| = 1$.

8.3 Exercises

1. Let f be a scalar field defined on a set S and let c be a given real number. The set of all points $\boldsymbol{x}$ in S such that $f(\boldsymbol{x}) = c$ is called a *level set* of f. (Geometric and physical problems dealing with level sets will be discussed later in this chapter.) For each of the following scalar fields, S is the whole space $\mathbf{R}^n$. Make a sketch to describe the level sets corresponding to the given values of c.
(a) $f(x, y) = x^2 + y^2$, $\quad c = 0, 1, 4, 9$.
(b) $f(x, y) = e^{xy}$, $\quad c = e^{-2}, e^{-1}, 1, e, e^2, e^3$.
(c) $f(x, y) = \cos(x + y)$, $\quad c = -1, 0, \frac{1}{2}, \frac{1}{2}\sqrt{2}, 1$.
(d) $f(x, y, z) = x + y + z$, $\quad c = -1, 0, 1$.
(e) $f(x, y, z) = x^2 + 2y^2 + 3z^2$, $\quad c = 0, 6, 12$.
(f) $f(x, y, z) = \sin(x^2 + y^2 + z^2)$, $\quad c = -1, -\frac{1}{2}, 0, \frac{1}{2}\sqrt{2}, 1$.

2. In each of the following cases, let S be the set of all points (x, y) in the plane satisfying the given inequalities. Make a sketch showing the set S and explain, by a geometric argument, whether or not S is open. Indicate the boundary of S on your sketch.
   (a) $x^2 + y^2 < 1$.
   (b) $3x^2 + 2y^2 < 6$.
   (c) $|x| < 1$ and $|y| < 1$.
   (d) $x \geq 0$ and $y > 0$.
   (e) $|x| \leq 1$ and $|y| \leq 1$.
   (f) $x > 0$ and $y < 0$.
   (g) $xy < 1$.
   (h) $1 \leq x \leq 2$ and $3 < y < 4$.
   (i) $1 < x < 2$ and $y > 0$.
   (j) $x \geq y$.
   (k) $x > y$.
   (l) $y > x^2$ and $|x| < 2$.
   (m) $(x^2 + y^2 - 1)(4 - x^2 - y^2) > 0$.
   (n) $(2x - x^2 - y^2)(x^2 + y^2 - x) > 0$.
3. In each of the following, let S be the set of all points (x, y, z) in 3-space satisfying the given inequalities and determine whether or not S is open.
   (a) $z^2 - x^2 - y^2 - 1 > 0$.
   (b) $|x| < 1$, $|y| < 1$, and $|z| < 1$.
   (c) $x + y + z < 1$.
   (d) $|x| \leq 1$, $|y| < 1$, and $|z| < 1$.
   (e) $x + y + z < 1$ and $x > 0$, $y > 0$, $z > 0$.
   (f) $x^2 + 4y^2 + 4z^2 - 2x + 16y + 40z + 113 < 0$.
4. (a) If A is an open set in n-space and if $\boldsymbol{x} \in A$, show that the set $A - \{\boldsymbol{x}\}$, obtained by removing the point $\boldsymbol{x}$ from A, is open.
   (b) If A is an open interval on the real line and B is a closed subinterval of A, show that $A - B$ is open.†
   (c) If A and B are open intervals on the real line, show that $A \cup B$ and $A \cap B$ are open.
   (d) If A is a closed interval on the real line, show that its complement (relative to the whole real line) is open.
5. Prove the following properties of open sets in $\mathbf{R}^n$:
   (a) The empty set $\varnothing$ is open.
   (b) $\mathbf{R}^n$ is open.
   (c) The union of any collection of open sets is open.
   (d) The intersection of a finite collection of open sets is open.
   (e) Give an example to show that the intersection of an infinite collection of open sets is not necessarily open.

Closed sets. A set S in $\mathbf{R}^n$ is called *closed* if its complement $\mathbf{R}^n - S$ is open. The next three exercises discuss properties of closed sets.

6. In each of the following cases, let S be the set of all points (x, y) in $\mathbf{R}^2$ satisfying the given conditions. Make a sketch showing the set S and give a geometric argument to explain whether S is open, closed, both open and closed, or neither open nor closed.
   (a) $x^2 + y^2 \geq 0$.
   (b) $x^2 + y^2 < 0$.
   (c) $x^2 + y^2 \leq 1$.
   (d) $1 < x^2 + y^2 < 2$.
   (e) $1 \leq x^2 + y^2 \leq 2$.
   (f) $1 < x^2 + y^2 \leq 2$.
   (g) $1 \leq x \leq 2$, $3 \leq y \leq 4$.
   (h) $1 \leq x \leq 2$, $3 \leq y < 4$.
   (i) $y = x^2$.
   (j) $y \geq x^2$.
   (k) $y \geq x^2$ and $|x| < 2$.
   (l) $y \geq x^2$ and $|x| \leq 2$.
7. (a) If A is a closed set in n-space and $\boldsymbol{x}$ is a point not in A, prove that $A \cup \{\boldsymbol{x}\}$ is also closed.
   (b) Prove that a closed interval $[a, b]$ on the real line is a closed set.
   (c) If A and B are closed intervals on the real line, show that $A \cup B$ and $A \cap B$ are closed.

---

† If A and B are sets, the difference $A - B$ (called the *complement of B relative to A*) is the set of all elements of A which are not in B.

8. Prove the following properties of closed sets in $\mathbf{R}^n$. You may use the results of Exercise 5.
   (a) The empty set $\varnothing$ is closed.
   (b) $\mathbf{R}^n$ is closed.
   (c) The intersection of any collection of closed sets is closed.
   (d) The union of a finite number of closed sets is closed.
   (e) Give an example to show that the union of an infinite collection of closed sets is not necessarily closed.
9. Let S be a subset of $\mathbf{R}^n$.
   (a) Prove that both int S and ext S are open sets.
   (b) Prove that $\mathbf{R}^n = (\text{int } S) \cup (\text{ext } S) \cup \partial S$, a union of disjoint sets, and use this to deduce that the boundary ∂S is always a closed set.
10. Given a set S in $\mathbf{R}^n$ and a point $\boldsymbol{x}$ with the property that every ball $B(\boldsymbol{x})$ contains both interior points of S and points exterior to S. Prove that $\boldsymbol{x}$ is a boundary point of S. Is the converse statement true? That is, does every boundary point of S necessarily have this property?
11. Let S be a subset of $\mathbf{R}^n$. Prove that ext $S = \text{int}(\mathbf{R}^n - S)$.
12. Prove that a set S in $\mathbf{R}^n$ is closed if and only if $S = (\text{int } S) \cup \partial S$.

8.4 Limits and continuity

The concepts of limit and continuity are easily extended to scalar and vector fields. We shall formulate the definitions for vector fields; they apply also to scalar fields.

We consider a function $\boldsymbol{f}: S \to \mathbf{R}^m$, where S is a subset of $\mathbf{R}^n$. If $\boldsymbol{a} \in \mathbf{R}^n$ and $\boldsymbol{b} \in \mathbf{R}^m$ we write

$$\lim_{\boldsymbol{x}\to\boldsymbol{a}} \boldsymbol{f}(\boldsymbol{x}) = \boldsymbol{b} \qquad (\text{or}, \boldsymbol{f}(\boldsymbol{x}) \to \boldsymbol{b} \text{ as } \boldsymbol{x} \to \boldsymbol{a}) \tag{8.1}$$

to mean that

$$\lim_{\|\boldsymbol{x}-\boldsymbol{a}\|\to 0} \|\boldsymbol{f}(\boldsymbol{x}) - \boldsymbol{b}\| = 0. \tag{8.2}$$

The limit symbol in equation (8.2) is the usual limit of elementary calculus. In this definition it is not required that $\boldsymbol{f}$ be defined at the point $\boldsymbol{a}$ itself.

If we write $\boldsymbol{h} = \boldsymbol{x} - \boldsymbol{a}$, Equation (8.2) becomes

$$\lim_{\|\boldsymbol{h}\|\to 0} \|\boldsymbol{f}(\boldsymbol{a} + \boldsymbol{h}) - \boldsymbol{b}\| = 0.$$

For points in $\mathbf{R}^2$ we write (x, y) for $\boldsymbol{x}$ and (a, b) for $\boldsymbol{a}$ and express the limit relation (8.1) as follows:

$$\lim_{(x,y)\to(a,b)} \boldsymbol{f}(x, y) = \boldsymbol{b}.$$

For points in $\mathbf{R}^3$ we put $\boldsymbol{x} = (x, y, z)$ and $\boldsymbol{a} = (a, b, c)$ and write

$$\lim_{(x,y,z)\to(a,b,c)} \boldsymbol{f}(x, y, z) = \boldsymbol{b}.$$

A function $\boldsymbol{f}$ is said to be *continuous* at $\boldsymbol{a}$ if $\boldsymbol{f}$ is defined at $\boldsymbol{a}$ and if

$$\lim_{\boldsymbol{x}\to\boldsymbol{a}} \boldsymbol{f}(\boldsymbol{x}) = \boldsymbol{f}(\boldsymbol{a}).$$

We say $\boldsymbol{f}$ is *continuous on a set* S if $\boldsymbol{f}$ is continuous at each point of S.

Since these definitions are straightforward extensions of those in the one-dimensional case, it is not surprising to learn that many familiar properties of limits and continuity can also be extended. For example, the usual theorems concerning limits and continuity of sums, products, and quotients also hold for scalar fields. For vector fields, quotients are not defined but we have the following theorem concerning sums, multiplication by scalars, inner products, and norms.

THEOREM 8.1. *If* $\lim_{x \to a} f(x) = b$ *and* $\lim_{x \to a} g(x) = c$, *then we also have:*

(a) $\lim_{x \to a} [f(x) + g(x)] = b + c$.

(b) $\lim_{x \to a} \lambda f(x) = \lambda b$ *for every scalar* λ.

(c) $\lim_{x \to a} f(x) \cdot g(x) = b \cdot c$.

(d) $\lim_{x \to a} \|f(x)\| = \|b\|$.

Proof. We prove only parts (c) and (d); proofs of (a) and (b) are left as exercises for the reader.

To prove (c) we write

$$f(x) \cdot g(x) - b \cdot c = [f(x) - b] \cdot [g(x) - c] + b \cdot [g(x) - c] + c \cdot [f(x) - b].$$

Now we use the triangle inequality and the Cauchy-Schwarz inequality to obtain

$$0 \le \|f(x) \cdot g(x) - b \cdot c\| \le \|f(x) - b\| \, \|g(x) - c\| + \|b\| \, \|g(x) - c\| + \|c\| \, \|f(x) - b\|.$$

Since $\|f(x) - b\| \to 0$ and $\|g(x) - c\| \to 0$ as $x \to a$, this shows that $\|f(x) \cdot g(x) - b \cdot c\| \to 0$ as $x \to a$, which proves (c).

Taking $f(x) = g(x)$ in part (c) we find

$$\lim_{x \to a} \|f(x)\|^2 = \|b\|^2,$$

from which we obtain (d).

EXAMPLE 1. *Continuity and components of a vector field.* If a vector field f has values in $\mathbf{R}^m$, each function value $f(x)$ has m components and we can write

$$f(x) = (f_1(x), \ldots, f_m(x)).$$

The m scalar fields $f_1, \ldots, f_m$ are called *components* of the vector field f. We shall prove that f is continuous at a point if, and only if, each component f_k is continuous at that point.

Let e_k denote the kth unit coordinate vector (all the components of e_k are 0 except the kth, which is equal to 1). Then $f_k(x)$ is given by the dot product

$$f_k(x) = f(x) \cdot e_k.$$

Therefore, part (c) of Theorem 8.1 shows that each point of continuity of f is also a point of continuity of f_k. Moreover, since we have

$$f(x) = \sum_{k=1}^{m} f_k(x)e_k,$$

repeated application of parts (a) and (b) of Theorem 8.1 shows that a point of continuity of all m components $f_1, \ldots, f_m$ is also a point of continuity of f.

EXAMPLE 2. *Continuity of the identity function.* The identity function, $f(x) = x$, is continuous everywhere in $\mathbf{R}^n$. Therefore its components are also continuous everywhere in $\mathbf{R}^n$. These are the n scalar fields given by

$$f_1(x) = x_1, f_2(x) = x_2, \ldots, f_n(x) = x_n.$$

EXAMPLE 3. *Continuity of linear transformations.* Let $f: \mathbf{R}^n \to \mathbf{R}^m$ be a linear transformation. We will prove that f is continuous at each point a in $\mathbf{R}^n$. By linearity we have

$$f(a + h) = f(a) + f(h).$$

Therefore, it suffices to prove that $f(h) \to O$ as $h \to O$. Writing h in terms of its components we have $h = h_1e_1 + \cdots + h_ne_n$. Using linearity again we find $f(h) = h_1f(e_1) + \cdots + h_nf(e_n)$. This shows that $f(h) \to O$ as $h \to O$.

EXAMPLE 4. *Continuity of polynomials in n variables.* A scalar field P defined on $\mathbf{R}^n$ by a formula of the form

$$P(x) = \sum_{k_1=0}^{p_1} \cdots \sum_{k_n=0}^{p_n} c_{k_1 \cdots k_n} x_1^{k_1} \cdots x_n^{k_n}$$

is called a polynomial in n variables $x_1, \ldots, x_n$. A polynomial is continuous everywhere in $\mathbf{R}^n$ because it is a finite sum of products of scalar fields continuous everywhere in $\mathbf{R}^n$. For example, a polynomial in two variables x and y, given by

$$P(x, y) = \sum_{i=0}^{p} \sum_{j=0}^{q} c_{ij} x^i y^j$$

is continuous at every point (x, y) in $\mathbf{R}^2$.

EXAMPLE 5. *Continuity of rational functions.* A scalar field f given by $f(x) = P(x)/Q(x)$, where P and Q are polynomials in the components of x, is called a rational function. A rational function is continuous at each point where $Q(x) \neq 0$.

Further examples of continuous function can be constructed with the help of the next theorem, which is concerned with continuity of composite functions.

THEOREM 8.2. *Let f and g be functions such that the composite function $f \circ g$ is defined at a, where*

$$(f \circ g)(x) = f[g(x)].$$

If g is continuous at a and if f is continuous at $g(a)$, then the composition $f \circ g$ is continuous at a.

Proof. Let $y = f(x)$ and $b = g(a)$. Then we have

$$f[g(x)] - f[g(a)] = f(y) - f(b).$$

By hypothesis, $y \to b$ as $x \to a$, so we have

$$\lim_{\|x-a\|\to 0} \|f[g(x)] - f[g(a)]\| = \lim_{\|y-b\|\to 0} \|f(y) - f(b)\| = 0.$$

Therefore $\lim_{x\to a} f[g(x)] = f[g(a)]$, so $f \circ g$ is continuous at a.

EXAMPLE 6. The foregoing theorem implies continuity of scalar fields h, where $h(x, y)$ is given by formulas such as

$$\sin (x^2y), \qquad \log (x^2 + y^2), \qquad \frac{e^{x+y}}{x + y}, \qquad \log [\cos (x^2 + y^2)].$$

These examples are continuous at all points at which the functions are defined. The first is continuous at all points in the plane, and the second at all points except the origin. The third is continuous at all points (x, y) at which $x + y \neq 0$, and the fourth at all points at which $x^2 + y^2$ is not an odd multiple of $\pi/2$. [The set of (x, y) such that $x^2 + y^2 = n\pi/2$, $n = 1, 3, 5, \ldots$, is a family of circles centered at the origin.] These examples show that the discontinuities of a function of two variables may consist of isolated points, entire curves, or families of curves.

EXAMPLE 7. A function of two variables may be continuous in each variable separately and yet be discontinuous as a function of the two variables together. This is illustrated by the following example:

$$f(x, y) = \frac{xy}{x^2 + y^2} \quad \text{if} \quad (x, y) \neq (0, 0), \quad f(0, 0) = 0.$$

For points (x, y) on the x-axis we have $y = 0$ and $f(x, y) = f(x, 0) = 0$, so the function has the constant value 0 everywhere on the x-axis. Therefore, if we put $y = 0$ and think of f as a function of x alone, f is continuous at $x = 0$. Similarly, f has the constant value 0 at all points on the y-axis, so if we put $x = 0$ and think of f as a function of y alone, f is continuous at $y = 0$. However, as a function of two variables, f is not continuous at the origin. In fact, at each point of the line $y = x$ (except at the origin) the function has the constant value $\frac{1}{2}$ because $f(x, x) = x^2/(2x^2) = \frac{1}{2}$; since there are points on this line arbitrarily close to the origin and since $f(0, 0) \neq \frac{1}{2}$, the function is not continuous at (0, 0).

8.5 Exercises

The exercises in this section are concerned with limits and continuity of scalar fields defined on subsets of the plane.

1. In each of the following examples a scalar field f is defined by the given equation for all points (x, y) in the plane for which the expression on the right is defined. In each example determine the set of points (x, y) at which f is continuous.

(a) $f(x, y) = x^4 + y^4 - 4x^2y^2$.

(b) $f(x, y) = \log (x^2 + y^2)$.

(c) $f(x, y) = \dfrac{1}{y} \cos x^2$.

(d) $f(x, y) = \tan \dfrac{x^2}{y}$.

(e) $f(x, y) = \arctan \dfrac{y}{x}$.

(f) $f(x, y) = \arcsin \dfrac{x}{\sqrt{x^2 + y^2}}$.

(g) $f(x, y) = \arctan \dfrac{x + y}{1 - xy}$.

(h) $f(x, y) = \dfrac{x}{\sqrt{x^2 + y^2}}$.

(i) $f(x, y) = x^{(y^2)}$.

(j) $f(x, y) = \arccos \sqrt{x/y}$.

2. If $\lim_{(x,y)\to(a,b)} f(x, y) = L$, and if the one-dimensional limits

$$\lim_{x\to a} f(x, y) \quad \text{and} \quad \lim_{y\to b} f(x, y)$$

both exist, prove that

$$\lim_{x\to a} [\lim_{y\to b} f(x, y)] = \lim_{y\to b} [\lim_{x\to a} f(x, y)] = L.$$

The two limits in this equation are called *iterated* limits; the exercise shows that the existence of the two-dimensional limit and of the two one-dimensional limits implies the existence and equality of the two iterated limits. (The converse is not always true. A counter example is given in Exercise 4.)

3. Let $f(x, y) = (x - y)/(x + y)$ if $x + y \neq 0$. Show that

$$\lim_{x\to 0} [\lim_{y\to 0} f(x, y)] = 1 \quad \text{but that} \quad \lim_{y\to 0} [\lim_{x\to 0} f(x, y)] = -1.$$

Use this result with Exercise 2 to deduce that $f(x, y)$ does not tend to a limit as $(x, y) \to (0, 0)$.

4. Let

$$f(x, y) = \frac{x^2y^2}{x^2y^2 + (x - y)^2} \quad \text{whenever} \quad x^2y^2 + (x - y)^2 \neq 0.$$

Show that

$$\lim_{x\to 0} [\lim_{y\to 0} f(x, y)] = \lim_{y\to 0} [\lim_{x\to 0} f(x, y)] = 0$$

but that $f(x, y)$ does not tend to a limit as $(x, y) \to (0, 0)$. [*Hint:* Examine f on the line $y = x$.] This example shows that the converse of Exercise 2 is not always true.

5. Let

$$f(x, y) = \begin{cases} x \sin \dfrac{1}{y} & \text{if } y \neq 0, \\ 0 & \text{if } y = 0. \end{cases}$$

Show that $f(x, y) \to 0$ as $(x, y) \to (0, 0)$ but that

$$\lim_{y \to 0} [\lim_{x \to 0} f(x, y)] \neq \lim_{x \to 0} [\lim_{y \to 0} f(x, y)].$$

Explain why this does not contradict Exercise 2.

6. If $(x, y) \neq (0, 0)$, let $f(x, y) = (x^2 - y^2)/(x^2 + y^2)$. Find the limit of $f(x, y)$ as $(x, y) \to (0, 0)$ along the line $y = mx$. Is it possible to define $f(0, 0)$ so as to make f continuous at $(0, 0)$?
7. Let $f(x, y) = 0$ if $y \leq 0$ or if $y \geq x^2$ and let $f(x, y) = 1$ if $0 < y < x^2$. Show that $f(x, y) \to 0$ as $(x, y) \to (0, 0)$ along any straight line through the origin. Find a curve through the origin along which (except at the origin) $f(x, y)$ has the constant-value 1. Is f continuous at the origin?
8. If $f(x, y) = [\sin (x^2 + y^2)]/(x^2 + y^2)$ when $(x, y) \neq (0, 0)$ how must $f(0, 0)$ be defined so as to make f continuous at the origin?
9. Let f be a scalar field continuous at an interior point $\boldsymbol{a}$ of a set S in $\mathbf{R}^n$. If $f(\boldsymbol{a}) \neq 0$, prove that there is an n-ball $B(\boldsymbol{a})$ in which f has the same sign as $f(\boldsymbol{a})$.

8.6 The derivative of a scalar field with respect to a vector

This section introduces derivatives of scalar fields. Derivatives of vector fields are discussed in Section 8.18.

Let f be a scalar field defined on a set S in $\mathbf{R}^n$, and let $\boldsymbol{a}$ be an interior point of S. We wish to study how the field changes as we move from $\boldsymbol{a}$ to a nearby point. For example, suppose $f(\boldsymbol{a})$ is the temperature at a point $\boldsymbol{a}$ in a heated room with an open window. If we move toward the window the temperature will decrease, but if we move toward the heater it will increase. In general, the manner in which a field changes will depend on the direction in which we move away from $\boldsymbol{a}$.

Suppose we specify this direction by another vector $\boldsymbol{y}$. That is, suppose we move from $\boldsymbol{a}$ toward another point $\boldsymbol{a} + \boldsymbol{y}$ along the line segment joining $\boldsymbol{a}$ and $\boldsymbol{a} + \boldsymbol{y}$. Each point on this segment has the form $\boldsymbol{a} + h\boldsymbol{y}$, where h is a real number. An example is shown in Figure 8.3. The distance from $\boldsymbol{a}$ to $\boldsymbol{a} + h\boldsymbol{y}$ is $\|h\boldsymbol{y}\| = |h| \, \|\boldsymbol{y}\|$.

Since $\boldsymbol{a}$ is an interior point of S, there is an n-ball $B(\boldsymbol{a}; r)$ lying entirely in S. If h is chosen so that $|h| \, \|\boldsymbol{y}\| < r$, the segment from $\boldsymbol{a}$ to $\boldsymbol{a} + h\boldsymbol{y}$ will lie in S. (See Figure 8.4.) We keep

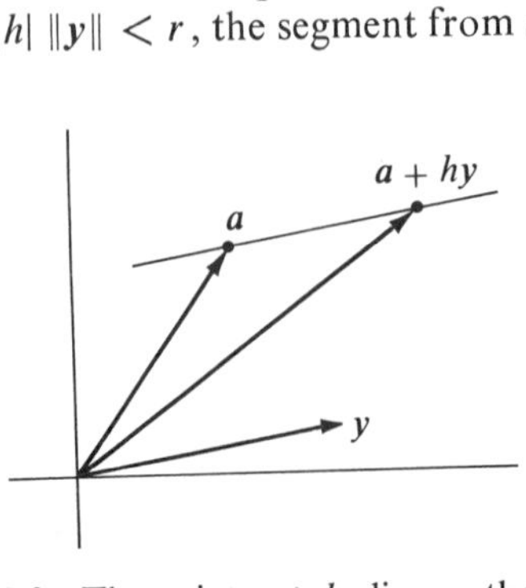

FIGURE 8.3 The point $\boldsymbol{a} + h\boldsymbol{y}$ lies on the line through $\boldsymbol{a}$ parallel to $\boldsymbol{y}$.

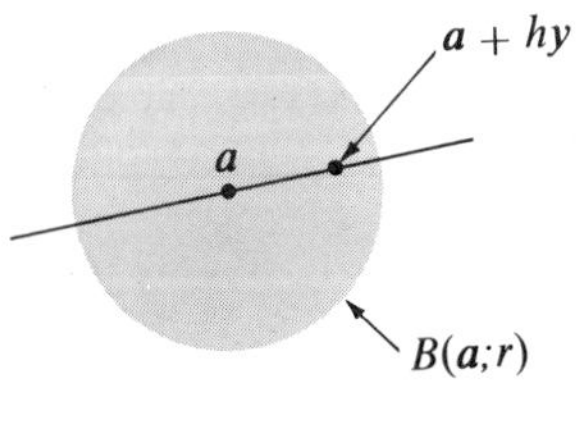

FIGURE 8.4 The point $\boldsymbol{a} + h\boldsymbol{y}$ lies in the n-ball $B(\boldsymbol{a}; r)$ if $\|h\boldsymbol{y}\| < r$.

$h \neq 0$ but small enough to guarantee that $\boldsymbol{a} + h\boldsymbol{y} \in S$ and we form the difference quotient

$$\frac{f(\boldsymbol{a} + h\boldsymbol{y}) - f(\boldsymbol{a})}{h}. \tag{8.3}$$

The numerator of this quotient tells us how much the function changes when we move from $\boldsymbol{a}$ to $\boldsymbol{a} + h\boldsymbol{y}$. The quotient itself is called the *average rate of change* of f over the line segment joining $\boldsymbol{a}$ to $\boldsymbol{a} + h\boldsymbol{y}$. We are interested in the behavior of this quotient as $h \to 0$. This leads us to the following definition.

DEFINITION OF THE DERIVATIVE OF A SCALAR FIELD WITH RESPECT TO A VECTOR. *Given a scalar field $f\colon S \to \mathbf{R}$, where $S \subseteq \mathbf{R}^n$. Let $\boldsymbol{a}$ be an interior point of S and let $\boldsymbol{y}$ be an arbitrary point in $\mathbf{R}^n$. The derivative of f at $\boldsymbol{a}$ with respect to $\boldsymbol{y}$ is denoted by the symbol $f'(\boldsymbol{a}; \boldsymbol{y})$ and is defined by the equation*

$$f'(\boldsymbol{a}; \boldsymbol{y}) = \lim_{h \to 0} \frac{f(\boldsymbol{a} + h\boldsymbol{y}) - f(\boldsymbol{a})}{h} \tag{8.4}$$

when the limit on the right exists.

EXAMPLE 1. If $\boldsymbol{y} = \boldsymbol{O}$, the difference quotient (8.3) is 0 for every $h \neq 0$, so $f'(\boldsymbol{a}; \boldsymbol{O})$ always exists and equals 0.

EXAMPLE 2. *Derivative of a linear transformation.* If $f\colon S \to \mathbf{R}$ is linear, then $f(\boldsymbol{a} + h\boldsymbol{y}) = f(\boldsymbol{a}) + hf(\boldsymbol{y})$ and the difference quotient (8.3) is equal to $f(\boldsymbol{y})$ for every $h \neq 0$. In this case, $f'(\boldsymbol{a}; \boldsymbol{y})$ always exists and is given by

$$f'(\boldsymbol{a}; \boldsymbol{y}) = f(\boldsymbol{y})$$

for every $\boldsymbol{a}$ in S and every $\boldsymbol{y}$ in $\mathbf{R}^n$. In other words, the derivative of a linear transformation with respect to $\boldsymbol{y}$ is equal to the value of the function at $\boldsymbol{y}$.

To study how f behaves on the line passing through $\boldsymbol{a}$ and $\boldsymbol{a} + \boldsymbol{y}$ for $\boldsymbol{y} \neq \boldsymbol{O}$ we introduce the function

$$g(t) = f(\boldsymbol{a} + t\boldsymbol{y}).$$

The next theorem relates the derivatives $g'(t)$ and $f'(\boldsymbol{a} + t\boldsymbol{y}; \boldsymbol{y})$.

THEOREM 8.3. *Let $g(t) = f(\boldsymbol{a} + t\boldsymbol{y})$. If one of the derivatives $g'(t)$ or $f'(\boldsymbol{a} + t\boldsymbol{y}; \boldsymbol{y})$ exists then the other also exists and the two are equal,*

$$g'(t) = f'(\boldsymbol{a} + t\boldsymbol{y}; \boldsymbol{y}). \tag{8.5}$$

In particular, when $t = 0$ we have $g'(0) = f'(\boldsymbol{a}; \boldsymbol{y})$.

Proof. Forming the difference quotient for g, we have

$$\frac{g(t+h)-g(t)}{h}=\frac{f(\boldsymbol{a}+t\boldsymbol{y}+h\boldsymbol{y})-f(\boldsymbol{a}+t\boldsymbol{y})}{h}.$$

Letting $h \to 0$ we obtain (8.5).

EXAMPLE 3. Compute $f'(\boldsymbol{a};\boldsymbol{y})$ if $f(\boldsymbol{x}) = \|\boldsymbol{x}\|^2$ for all $\boldsymbol{x}$ in $\mathbf{R}^n$.

Solution. We let $g(t) = f(\boldsymbol{a}+t\boldsymbol{y}) = (\boldsymbol{a}+t\boldsymbol{y})\cdot(\boldsymbol{a}+t\boldsymbol{y}) = \boldsymbol{a}\cdot\boldsymbol{a}+2t\boldsymbol{a}\cdot\boldsymbol{y}+t^2\boldsymbol{y}\cdot\boldsymbol{y}$. Therefore $g'(t) = 2\boldsymbol{a}\cdot\boldsymbol{y}+2t\boldsymbol{y}\cdot\boldsymbol{y}$, so $g'(0) = f'(\boldsymbol{a};\boldsymbol{y}) = 2\boldsymbol{a}\cdot\boldsymbol{y}$.

A simple corollary of Theorem 8.3 is the mean-value theorem for scalar fields.

THEOREM 8.4. MEAN-VALUE THEOREM FOR DERIVATIVES OF SCALAR FIELDS. *Assume the derivative $f'(\boldsymbol{a}+t\boldsymbol{y};\boldsymbol{y})$ exists for each t in the interval $0 \le t \le 1$. Then for some real θ in the open interval $0 < \theta < 1$ we have*

$$f(\boldsymbol{a}+\boldsymbol{y})-f(\boldsymbol{a}) = f'(\boldsymbol{z};\boldsymbol{y}), \qquad \textit{where} \quad \boldsymbol{z} = \boldsymbol{a}+\theta\boldsymbol{y}.$$

Proof. Let $g(t) = f(\boldsymbol{a}+t\boldsymbol{y})$. Applying the one-dimensional mean-value theorem to g on the interval $[0, 1]$ we have

$$g(1)-g(0) = g'(\theta), \qquad \text{where} \quad 0<\theta<1.$$

Since $g(1)-g(0) = f(\boldsymbol{a}+\boldsymbol{y})-f(\boldsymbol{a})$ and $g'(\theta) = f'(\boldsymbol{a}+\theta\boldsymbol{y};\boldsymbol{y})$, this completes the proof.

8.7 Directional derivatives and partial derivatives

In the special case when $\boldsymbol{y}$ is a *unit* vector, that is, when $\|\boldsymbol{y}\| = 1$, the distance between $\boldsymbol{a}$ and $\boldsymbol{a}+h\boldsymbol{y}$ is $|h|$. In this case the difference quotient (8.3) represents the average rate of change of f *per unit distance* along the segment joining $\boldsymbol{a}$ to $\boldsymbol{a}+h\boldsymbol{y}$; the derivative $f'(\boldsymbol{a};\boldsymbol{y})$ is called a *directional derivative.*

DEFINITION OF DIRECTIONAL AND PARTIAL DERIVATIVES. *If $\boldsymbol{y}$ is a unit vector, the derivative $f'(\boldsymbol{a};\boldsymbol{y})$ is called the directional derivative of f at $\boldsymbol{a}$ in the direction of $\boldsymbol{y}$. In particular, if $\boldsymbol{y} = \boldsymbol{e}_k$ (the kth unit coordinate vector) the directional derivative $f'(\boldsymbol{a};\boldsymbol{e}_k)$ is called the partial derivative with respect to $\boldsymbol{e}_k$ and is also denoted by the symbol $D_kf(\boldsymbol{a})$. Thus,*

$$D_kf(\boldsymbol{a}) = f'(\boldsymbol{a};\boldsymbol{e}_k).$$

The following notations are also used for the partial derivative $D_kf(\boldsymbol{a})$:

$$D_kf(a_1,\ldots,a_n), \qquad \frac{\partial f}{\partial x_k}(a_1,\ldots,a_n), \qquad \text{and} \qquad f'_{x_k}(a_1,\ldots,a_n).$$

Sometimes the derivative f'_{x_k} is written without the prime as f_{x_k} or even more simply as f_k.

In $\mathbf{R}^2$ the unit coordinate vectors are denoted by $\boldsymbol{i}$ and $\boldsymbol{j}$. If $\boldsymbol{a} = (a, b)$ the partial derivatives $f'(\boldsymbol{a}; \boldsymbol{i})$ and $f'(\boldsymbol{a}; \boldsymbol{j})$ are also written as

$$\frac{\partial f}{\partial x}(a, b) \quad \text{and} \quad \frac{\partial f}{\partial y}(a, b),$$

respectively. In $\mathbf{R}^3$, if $\boldsymbol{a} = (a, b, c)$ the partial derivatives $D_1 f(\boldsymbol{a})$, $D_2 f(\boldsymbol{a})$, and $D_3 f(\boldsymbol{a})$ are also denoted by

$$\frac{\partial f}{\partial x}(a, b, c), \quad \frac{\partial f}{\partial y}(a, b, c), \quad \text{and} \quad \frac{\partial f}{\partial z}(a, b, c).$$

8.8 Partial derivatives of higher order

Partial differentiation produces new scalar fields $D_1 f, \ldots, D_n f$ from a given scalar field f. The partial derivatives of $D_1 f, \ldots, D_n f$ are called *second-order partial derivatives* of f. For functions of two variables there are four second-order partial derivatives, which are written as follows:

$$D_1(D_1 f) = \frac{\partial^2 f}{\partial x^2}, \quad D_1(D_2 f) = \frac{\partial^2 f}{\partial x\, \partial y}, \quad D_2(D_1 f) = \frac{\partial^2 f}{\partial y\, \partial x}, \quad D_2(D_2 f) = \frac{\partial^2 f}{\partial y^2}.$$

Note that $D_1(D_2 f)$ means the partial derivative of $D_2 f$ with respect to the first variable. We sometimes use the notation $D_{i,j} f$ for the second-order partial derivative $D_i(D_j f)$. For example, $D_{1,2} f = D_1(D_2 f)$. In the ∂-notation we indicate the order of derivatives by writing

$$\frac{\partial^2 f}{\partial x\, \partial y} = \frac{\partial}{\partial x}\left(\frac{\partial f}{\partial y}\right).$$

This may or may not be equal to the other mixed partial derivative,

$$\frac{\partial^2 f}{\partial y\, \partial x} = \frac{\partial}{\partial y}\left(\frac{\partial f}{\partial x}\right).$$

In Section 8.23 we shall prove that the two mixed partials $D_1(D_2 f)$ and $D_2(D_1 f)$ are equal at a point if one of them is continuous in a neighborhood of the point. Section 8.23 also contains an example in which $D_1(D_2 f) \neq D_2(D_1 f)$ at a point.

8.9 Exercises

1. A scalar field f is defined on $\mathbf{R}^n$ by the equation $f(\boldsymbol{x}) = \boldsymbol{a} \cdot \boldsymbol{x}$, where $\boldsymbol{a}$ is a constant vector. Compute $f'(\boldsymbol{x}; \boldsymbol{y})$ for arbitrary $\boldsymbol{x}$ and $\boldsymbol{y}$.

2. (a) Solve Exercise 1 when $f(\boldsymbol{x}) = \|\boldsymbol{x}\|^4$.
   (b) Take $n = 2$ in (a) and find all points (x, y) for which $f'(2\boldsymbol{i} + 3\boldsymbol{j}; x\boldsymbol{i} + y\boldsymbol{j}) = 6$.
   (c) Take $n = 3$ in (a) and find all points (x, y, z) for which $f'(\boldsymbol{i} + 2\boldsymbol{j} + 3\boldsymbol{k}; x\boldsymbol{i} + y\boldsymbol{j} + z\boldsymbol{k}) = 0$.
3. Let $T\colon \mathbf{R}^n \to \mathbf{R}^n$ be a given linear transformation. Compute the derivative $f'(\boldsymbol{x}; \boldsymbol{y})$ for the scalar field defined on $\mathbf{R}^n$ by the equation $f(\boldsymbol{x}) = \boldsymbol{x} \cdot T(\boldsymbol{x})$.

In each of Exercises 4 through 9, compute all first-order partial derivatives of the given scalar field. The fields in Exercises 8 and 9 are defined on $\mathbf{R}^n$.

4. $f(x, y) = x^2 + y^2 \sin(xy)$.
5. $f(x, y) = \sqrt{x^2 + y^2}$.
6. $f(x, y) = \dfrac{x}{\sqrt{x^2 + y^2}}, \quad (x, y) \neq (0, 0)$.
7. $f(x, y) = \dfrac{x + y}{x - y}, \quad x \neq y$.
8. $f(\boldsymbol{x}) = \boldsymbol{a} \cdot \boldsymbol{x}, \quad \boldsymbol{a}$ fixed.
9. $f(\boldsymbol{x}) = \sum_{i=1}^{n} \sum_{j=1}^{n} a_{ij} x_i x_j, \quad a_{ij} = a_{ji}$.

In each of Exercises 10 through 17, compute all first-order partial derivatives. In each of Exercises 10, 11, and 12 verify that the mixed partials $D_1(D_2 f)$ and $D_2(D_1 f)$ are equal.

10. $f(x, y) = x^4 + y^4 - 4x^2y^2$.
11. $f(x, y) = \log(x^2 + y^2), \quad (x, y) \neq (0, 0)$.
12. $f(x, y) = \dfrac{1}{y} \cos x^2, \quad y \neq 0$.
13. $f(x, y) = \tan(x^2/y), \quad y \neq 0$.
14. $f(x, y) = \arctan(y/x), \quad x \neq 0$.
15. $f(x, y) = \arctan \dfrac{x + y}{1 - xy}, \quad xy \neq 1$.
16. $f(x, y) = x^{(y^2)}, \quad x > 0$.
17. $f(x, y) = \arccos \sqrt{x/y}, \quad y \neq 0$.

18. Let $v(r, t) = t^n e^{-r^2/(4t)}$. Find a value of the constant n such that v satisfies the following equation:

$$\frac{\partial v}{\partial t} = \frac{1}{r^2} \frac{\partial}{\partial r}\left(r^2 \frac{\partial v}{\partial r}\right).$$

19. Given $z = u(x, y)e^{ax+by}$ and $\partial^2 u/(\partial x\, \partial y) = 0$. Find values of the constants a and b such that

$$\frac{\partial^2 z}{\partial x\, \partial y} - \frac{\partial z}{\partial x} - \frac{\partial z}{\partial y} + z = 0.$$

20. (a) Assume that $f'(\boldsymbol{x}; \boldsymbol{y}) = 0$ for every $\boldsymbol{x}$ in some n-ball $B(\boldsymbol{a})$ and for every vector $\boldsymbol{y}$. Use the mean-value theorem to prove that f is constant on $B(\boldsymbol{a})$.
    (b) Suppose that $f'(\boldsymbol{x}; \boldsymbol{y}) = 0$ for a *fixed* vector $\boldsymbol{y}$ and for every $\boldsymbol{x}$ in $B(\boldsymbol{a})$. What can you conclude about f in this case?
21. A set S in $\mathbf{R}^n$ is called *convex* if for every pair of points $\boldsymbol{a}$ and $\boldsymbol{b}$ in S the line segment from $\boldsymbol{a}$ to $\boldsymbol{b}$ is also in S; in other words, $t\boldsymbol{a} + (1 - t)\boldsymbol{b} \in S$ for each t in the interval $0 \leq t \leq 1$.
    (a) Prove that every n-ball is convex.
    (b) If $f'(\boldsymbol{x}; \boldsymbol{y}) = 0$ for every $\boldsymbol{x}$ in an open convex set S and for every $\boldsymbol{y}$ in $\mathbf{R}^n$, prove that f is constant on S.
22. (a) Prove that there is no scalar field f such that $f'(\boldsymbol{a}; \boldsymbol{y}) > 0$ for a fixed vector $\boldsymbol{a}$ and every nonzero vector $\boldsymbol{y}$.
    (b) Give an example of a scalar field f such that $f'(\boldsymbol{x}; \boldsymbol{y}) > 0$ for a fixed vector $\boldsymbol{y}$ and every vector $\boldsymbol{x}$.

8.10 Directional derivatives and continuity

In the one-dimensional theory, existence of the derivative of a function f at a point implies continuity at that point. This is easily proved by choosing an $h \neq 0$ and writing

$$f(a + h) - f(a) = \frac{f(a + h) - f(a)}{h} \cdot h .$$

As $h \to 0$ the right side tends to the limit $f'(a) \cdot 0 = 0$ and hence $f(a + h) \to f(a)$. This shows that the existence of $f'(a)$ implies continuity of f at a.

Suppose we apply the same argument to a general scalar field. Assume the derivative $f'(\boldsymbol{a}; \boldsymbol{y})$ exists for some $\boldsymbol{y}$. Then if $h \neq 0$ we can write

$$f(\boldsymbol{a} + h\boldsymbol{y}) - f(\boldsymbol{a}) = \frac{f(\boldsymbol{a} + h\boldsymbol{y}) - f(\boldsymbol{a})}{h} \cdot h .$$

As $h \to 0$ the right side tends to the limit $f'(\boldsymbol{a}; \boldsymbol{y}) \cdot 0 = 0$; hence the existence of $f'(\boldsymbol{a}; \boldsymbol{y})$ for a given $\boldsymbol{y}$ implies that

$$\lim_{h \to 0} f(\boldsymbol{a} + h\boldsymbol{y}) = f(\boldsymbol{a})$$

for the same $\boldsymbol{y}$. This means that $f(\boldsymbol{x}) \to f(\boldsymbol{a})$ as $\boldsymbol{x} \to \boldsymbol{a}$ *along a straight line through* $\boldsymbol{a}$ *having the direction* $\boldsymbol{y}$. If $f'(\boldsymbol{a}; \boldsymbol{y})$ exists for *every* vector $\boldsymbol{y}$, then $f(\boldsymbol{x}) \to f(\boldsymbol{a})$ as $\boldsymbol{x} \to \boldsymbol{a}$ along every line through $\boldsymbol{a}$. This seems to suggest that f is continuous at $\boldsymbol{a}$. Surprisingly enough, this conclusion need not be true. The next example describes a scalar field which has a directional derivative in every direction at $\boldsymbol{O}$ but which is not continuous at $\boldsymbol{O}$.

EXAMPLE. Let f be the scalar field defined on $\mathbf{R}^2$ as follows:

$$f(x, y) = \frac{xy^2}{x^2 + y^4} \quad \text{if} \quad x \neq 0, \qquad f(0, y) = 0 .$$

Let $\boldsymbol{a} = (0, 0)$ and let $\boldsymbol{y} = (a, b)$ be any vector. If $a \neq 0$ and $h \neq 0$ we have

$$\frac{f(\boldsymbol{a} + h\boldsymbol{y}) - f(\boldsymbol{a})}{h} = \frac{f(h\boldsymbol{y}) - f(\boldsymbol{O})}{h} = \frac{f(ha, hb)}{h} = \frac{ab^2}{a^2 + h^2b^4} .$$

Letting $h \to 0$ we find $f'(\boldsymbol{O}; \boldsymbol{y}) = b^2/a$. If $\boldsymbol{y} = (0, b)$ we find, in a similar way, that $f'(\boldsymbol{O}; \boldsymbol{y}) = 0$. Therefore $f'(\boldsymbol{O}; \boldsymbol{y})$ exists for all directions $\boldsymbol{y}$. Also, $f(\boldsymbol{x}) \to 0$ as $\boldsymbol{x} \to \boldsymbol{O}$ along any straight line through the origin. However, at each point of the parabola $x = y^2$ (except at the origin) the function f has the value $\frac{1}{2}$. Since such points exist arbitrarily close to the origin and since $f(\boldsymbol{O}) = 0$, the function f is not continuous at $\boldsymbol{O}$.

The foregoing example shows that the existence of *all* directional derivatives at a point fails to imply continuity at that point. For this reason, directional derivatives are a somewhat unsatisfactory extension of the one-dimensional concept of derivative. A more suitable generalization exists which implies continuity and, at the same time, permits us to extend the principal theorems of one-dimensional derivative theory to the higher dimensional case. This is called the *total derivative.*

8.11 The total derivative

We recall that in the one-dimensional case a function f with a derivative at a can be approximated near a by a linear Taylor polynomial. If $f'(a)$ exists we let $E(a, h)$ denote the difference

$$E(a, h) = \frac{f(a + h) - f(a)}{h} - f'(a) \quad \text{if} \quad h \neq 0, \tag{8.6}$$

and we define $E(a, 0) = 0$. From (8.6) we obtain the formula

$$f(a + h) = f(a) + f'(a)h + hE(a, h),$$

an equation which holds also for $h = 0$. This is the first-order Taylor formula for approximating $f(a + h) - f(a)$ by $f'(a)h$. The error committed is $hE(a, h)$. From (8.6) we see that $E(a, h) \to 0$ as $h \to 0$. Therefore the error $hE(a, h)$ is of smaller order than h for small h.

This property of approximating a differentiable function by a linear function suggests a way of extending the concept of differentiability to the higher-dimensional case.

Let $f\colon S \to \mathbf{R}$ be a scalar field defined on a set S in $\mathbf{R}^n$. Let $\boldsymbol{a}$ be an interior point of S, and let $B(\boldsymbol{a}; r)$ be an n-ball lying in S. Let $\boldsymbol{v}$ be a vector with $\|\boldsymbol{v}\| < r$, so that $\boldsymbol{a} + \boldsymbol{v} \in B(\boldsymbol{a}; r)$.

DEFINITION OF A DIFFERENTIABLE SCALAR FIELD. *We say that f is differentiable at $\boldsymbol{a}$ if there exists a linear transformation*

$$T_{\boldsymbol{a}}\colon \mathbf{R}^n \to \mathbf{R}$$

from $\mathbf{R}^n$ to $\mathbf{R}$, and a scalar function $E(\boldsymbol{a}, \boldsymbol{v})$ such that

$$f(\boldsymbol{a} + \boldsymbol{v}) = f(\boldsymbol{a}) + T_{\boldsymbol{a}}(\boldsymbol{v}) + \|\boldsymbol{v}\|\, E(\boldsymbol{a}, \boldsymbol{v}), \tag{8.7}$$

for $\|\boldsymbol{v}\| < r$, where $E(\boldsymbol{a}, \boldsymbol{v}) \to 0$ as $\|\boldsymbol{v}\| \to 0$. The linear transformation $T_{\boldsymbol{a}}$ is called the total derivative of f at $\boldsymbol{a}$.

Note: The total derivative $T_{\boldsymbol{a}}$ is a linear transformation, not a number. The function value $T_{\boldsymbol{a}}(\boldsymbol{v})$ is a real number; it is defined for every point $\boldsymbol{v}$ in $\mathbf{R}^n$. The total derivative was introduced by W. H. Young in 1908 and by M. Fréchet in 1911 in a more general context.

Equation (8.7), which holds for $\|\boldsymbol{v}\| < r$, is called a *first-order Taylor formula* for $f(\boldsymbol{a} + \boldsymbol{v})$. It gives a linear approximation, $T_{\boldsymbol{a}}(\boldsymbol{v})$, to the difference $f(\boldsymbol{a} + \boldsymbol{v}) - f(\boldsymbol{a})$. The error in the approximation is $\|\boldsymbol{v}\|\, E(\boldsymbol{a}, \boldsymbol{v})$, a term which is of smaller order than $\|\boldsymbol{v}\|$ as $\|\boldsymbol{v}\| \to 0$; that is, $E(\boldsymbol{a}, \boldsymbol{v}) = o(\|\boldsymbol{v}\|)$ as $\|\boldsymbol{v}\| \to 0$.

The next theorem shows that if the total derivative exists it is unique. It also tells us how to compute $T_{\boldsymbol{a}}(\boldsymbol{y})$ for every $\boldsymbol{y}$ in $\mathbf{R}^n$.

THEOREM 8.5. *Assume f is differentiable at $\boldsymbol{a}$ with total derivative $T_{\boldsymbol{a}}$. Then the derivative $f'(\boldsymbol{a}; \boldsymbol{y})$ exists for every $\boldsymbol{y}$ in $\mathbf{R}^n$ and we have*

$$T_{\boldsymbol{a}}(\boldsymbol{y}) = f'(\boldsymbol{a}; \boldsymbol{y}). \tag{8.8}$$

Moreover, $f'(\boldsymbol{a}; \boldsymbol{y})$ is a linear combination of the components of $\boldsymbol{y}$. In fact, if $\boldsymbol{y} = (y_1, \ldots, y_n)$, we have

$$f'(\boldsymbol{a}; \boldsymbol{y}) = \sum_{k=1}^{n} D_k f(\boldsymbol{a}) y_k. \tag{8.9}$$

Proof. Equation (8.8) holds trivially if $\boldsymbol{y} = \boldsymbol{O}$ since both $T_{\boldsymbol{a}}(\boldsymbol{O}) = 0$ and $f'(\boldsymbol{a}; \boldsymbol{O}) = 0$. Therefore we can assume that $\boldsymbol{y} \neq \boldsymbol{O}$.

Since f is differentiable at $\boldsymbol{a}$ we have a Taylor formula,

$$f(\boldsymbol{a} + \boldsymbol{v}) = f(\boldsymbol{a}) + T_{\boldsymbol{a}}(\boldsymbol{v}) + \|\boldsymbol{v}\| \, E(\boldsymbol{a}, \boldsymbol{v}) \tag{8.10}$$

for $\|\boldsymbol{v}\| < r$ for some $r > 0$, where $E(\boldsymbol{a}, \boldsymbol{v}) \to 0$ as $\|\boldsymbol{v}\| \to 0$. In this formula we take $\boldsymbol{v} = h\boldsymbol{y}$, where $h \neq 0$ and $|h| \, \|\boldsymbol{y}\| < r$. Then $\|\boldsymbol{v}\| < r$. Since $T_{\boldsymbol{a}}$ is linear we have $T_{\boldsymbol{a}}(\boldsymbol{v}) = T_{\boldsymbol{a}}(h\boldsymbol{y}) = hT_{\boldsymbol{a}}(\boldsymbol{y})$. Therefore (8.10) gives us

$$\frac{f(\boldsymbol{a} + h\boldsymbol{y}) - f(\boldsymbol{a})}{h} = T_{\boldsymbol{a}}(\boldsymbol{y}) + \frac{|h| \, \|\boldsymbol{y}\|}{h} E(\boldsymbol{a}, \boldsymbol{v}). \tag{8.11}$$

Since $\|\boldsymbol{v}\| \to 0$ as $h \to 0$ and since $|h|/h = \pm 1$, the right-hand member of (8.11) tends to the limit $T_{\boldsymbol{a}}(\boldsymbol{y})$ as $h \to 0$. Therefore the left-hand member tends to the same limit. This proves (8.8).

Now we use the linearity of $T_{\boldsymbol{a}}$ to deduce (8.9). If $\boldsymbol{y} = (y_1, \ldots, y_n)$ we have $\boldsymbol{y} = \sum_{k=1}^{n} y_k \boldsymbol{e}_k$, hence

$$T_{\boldsymbol{a}}(\boldsymbol{y}) = T_{\boldsymbol{a}}\left(\sum_{k=1}^{n} y_k \boldsymbol{e}_k\right) = \sum_{k=1}^{n} y_k T_{\boldsymbol{a}}(\boldsymbol{e}_k) = \sum_{k=1}^{n} y_k f'(\boldsymbol{a}; \boldsymbol{e}_k) = \sum_{k=1}^{n} y_k D_k f(\boldsymbol{a}).$$

8.12 The gradient of a scalar field

The formula in Theorem 8.5, which expresses $f'(\boldsymbol{a}; \boldsymbol{y})$ as a linear combination of the components of $\boldsymbol{y}$, can be written as a dot product,

$$f'(\boldsymbol{a}; \boldsymbol{y}) = \sum_{k=1}^{n} D_k f(\boldsymbol{a}) y_k = \nabla f(\boldsymbol{a}) \cdot \boldsymbol{y},$$

where $\nabla f(\boldsymbol{a})$ is the vector whose components are the partial derivatives of f at $\boldsymbol{a}$,

$$\nabla f(\boldsymbol{a}) = (D_1 f(\boldsymbol{a}), \ldots, D_n f(\boldsymbol{a})).$$

This is called the *gradient* of f. The gradient ∇f is a vector field defined at each point $\boldsymbol{a}$ where the partial derivatives $D_1 f(\boldsymbol{a}), \ldots, D_n f(\boldsymbol{a})$ exist. We also write grad f for ∇f. The symbol ∇ is pronounced "del."

The first-order Taylor formula (8.10) can now be written in the form

$$f(\boldsymbol{a} + \boldsymbol{v}) = f(\boldsymbol{a}) + \nabla f(\boldsymbol{a}) \cdot \boldsymbol{v} + \|\boldsymbol{v}\|\, E(\boldsymbol{a}, \boldsymbol{v}), \tag{8.12}$$

where $E(\boldsymbol{a}, \boldsymbol{v}) \to 0$ as $\|\boldsymbol{v}\| \to 0$. In this form it resembles the one-dimensional Taylor formula, with the gradient vector $\nabla f(\boldsymbol{a})$ playing the role of the derivative $f'(\boldsymbol{a})$.

From the Taylor formula we can easily prove that differentiability implies continuity.

THEOREM 8.6. *If a scalar field f is differentiable at $\boldsymbol{a}$, then f is continuous at $\boldsymbol{a}$.*

Proof. From (8.12) we have

$$|f(\boldsymbol{a} + \boldsymbol{v}) - f(\boldsymbol{a})| = |\nabla f(\boldsymbol{a}) \cdot \boldsymbol{v} + \|\boldsymbol{v}\|\, E(\boldsymbol{a}, \boldsymbol{v})| .$$

Applying the triangle inequality and the Cauchy-Schwarz inequality we find

$$0 \le |f(\boldsymbol{a} + \boldsymbol{v}) - f(\boldsymbol{a})| \le \|\nabla f(\boldsymbol{a})\|\, \|\boldsymbol{v}\| + \|\boldsymbol{v}\|\, |E(\boldsymbol{a}, \boldsymbol{v})| .$$

This shows that $f(\boldsymbol{a} + \boldsymbol{v}) \to f(\boldsymbol{a})$ as $\|\boldsymbol{v}\| \to 0$, so f is continuous at $\boldsymbol{a}$.

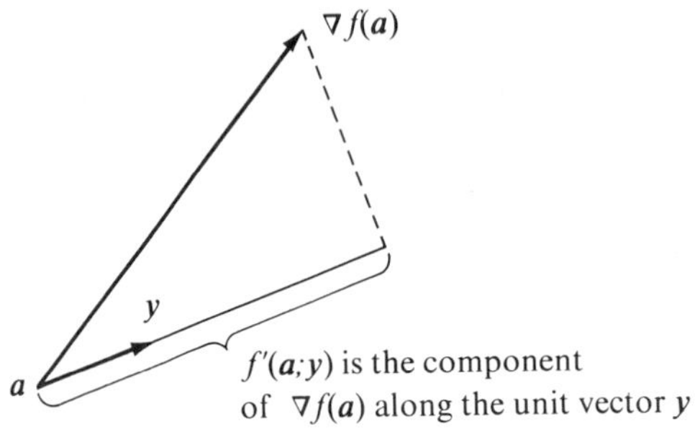

FIGURE 8.5 Geometric relation of the directional derivative to the gradient vector.

When $\boldsymbol{y}$ is a *unit* vector the directional derivative $f'(\boldsymbol{a}; \boldsymbol{y})$ has a simple geometric relation to the gradient vector. Assume that $\nabla f(\boldsymbol{a}) \ne \boldsymbol{O}$ and let θ denote the angle between $\boldsymbol{y}$ and $\nabla f(\boldsymbol{a})$. Then we have

$$f'(\boldsymbol{a}; \boldsymbol{y}) = \nabla f(\boldsymbol{a}) \cdot \boldsymbol{y} = \|\nabla f(\boldsymbol{a})\|\, \|\boldsymbol{y}\| \cos\theta = \|\nabla f(\boldsymbol{a})\| \cos\theta .$$

This shows that the directional derivative is simply the component of the gradient vector in the direction of $\boldsymbol{y}$. Figure 8.5 shows the vectors $\nabla f(\boldsymbol{a})$ and $\boldsymbol{y}$ attached to the point $\boldsymbol{a}$. The derivative is largest when $\cos\theta = 1$, that is, when $\boldsymbol{y}$ has the same direction as $\nabla f(\boldsymbol{a})$. In other words, at a given point $\boldsymbol{a}$, the scalar field undergoes its maximum rate of change in the direction of the gradient vector; moreover, this maximum is equal to the length of the gradient vector. When $\nabla f(\boldsymbol{a})$ is orthogonal to $\boldsymbol{y}$, the directional derivative $f'(\boldsymbol{a}; \boldsymbol{y})$ is 0.

In 2-space the gradient vector is often written as

$$\nabla f(x, y) = \frac{\partial f(x, y)}{\partial x} \boldsymbol{i} + \frac{\partial f(x, y)}{\partial y} \boldsymbol{j}.$$

In 3-space the corresponding formula is

$$\nabla f(x, y, z) = \frac{\partial f(x, y, z)}{\partial x} \boldsymbol{i} + \frac{\partial f(x, y, z)}{\partial y} \boldsymbol{j} + \frac{\partial f(x, y, z)}{\partial z} \boldsymbol{k}.$$

8.13 A sufficient condition for differentiability

If f is differentiable at $\boldsymbol{a}$, then all partial derivatives $D_1 f(\boldsymbol{a}), \ldots, D_n f(\boldsymbol{a})$ exist. However, the existence of all these partials does not necessarily imply that f is differentiable at $\boldsymbol{a}$. A counter example is provided by the function

$$f(x, y) = \frac{xy^2}{x^2 + y^4} \quad \text{if} \quad x \neq 0, \qquad f(0, y) = 0,$$

discussed in Section 8.10. For this function, both partial derivatives $D_1 f(\boldsymbol{O})$ and $D_2 f(\boldsymbol{O})$ exist but f is not continuous at $\boldsymbol{O}$, hence f cannot be differentiable at $\boldsymbol{O}$.

The next theorem shows that the existence of *continuous* partial derivatives at a point implies differentiability at that point.

THEOREM 8.7. A SUFFICIENT CONDITION FOR DIFFERENTIABILITY. *Assume that the partial derivatives $D_1 f, \ldots, D_n f$ exist in some n-ball $B(\boldsymbol{a})$ and are continuous at $\boldsymbol{a}$. Then f is differentiable at $\boldsymbol{a}$.*

Note: A scalar field satisfying the hypothesis of Theorem 8.7 is said to be *continuously differentiable* at $\boldsymbol{a}$.

Proof. The only candidate for $T_{\boldsymbol{a}}(\boldsymbol{v})$ is $\nabla f(\boldsymbol{a}) \cdot \boldsymbol{v}$. We will show that

$$f(\boldsymbol{a} + \boldsymbol{v}) - f(\boldsymbol{a}) = \nabla f(\boldsymbol{a}) \cdot \boldsymbol{v} + \|\boldsymbol{v}\| \, E(\boldsymbol{a}, \boldsymbol{v}),$$

where $E(\boldsymbol{a}, \boldsymbol{v}) \to 0$ as $\|\boldsymbol{v}\| \to 0$. This will prove the theorem.

Let $\lambda = \|\boldsymbol{v}\|$. Then $\boldsymbol{v} = \lambda \boldsymbol{u}$, where $\|\boldsymbol{u}\| = 1$. We keep λ small enough so that $\boldsymbol{a} + \boldsymbol{v}$ lies in the ball $B(\boldsymbol{a})$ in which the partial derivatives $D_1 f, \ldots, D_n f$ exist. Expressing $\boldsymbol{u}$ in terms of its components we have

$$\boldsymbol{u} = u_1 \boldsymbol{e}_1 + \cdots + u_n \boldsymbol{e}_n,$$

where $\boldsymbol{e}_1, \ldots, \boldsymbol{e}_n$ are the unit coordinate vectors. Now we write the difference $f(\boldsymbol{a} + \boldsymbol{v}) - f(\boldsymbol{a})$ as a telescoping sum,

$$f(\boldsymbol{a} + \boldsymbol{v}) - f(\boldsymbol{a}) = f(\boldsymbol{a} + \lambda \boldsymbol{u}) - f(\boldsymbol{a}) = \sum_{k=1}^{n} \{f(\boldsymbol{a} + \lambda \boldsymbol{v}_k) - f(\boldsymbol{a} + \lambda \boldsymbol{v}_{k-1})\}, \tag{8.13}$$

where $v_0, v_1, \ldots, v_n$ are any vectors in $\mathbf{R}^n$ such that $v_0 = O$ and $v_n = u$. We choose these vectors so they satisfy the recurrence relation $v_k = v_{k-1} + u_k e_k$. That is, we take

$$v_0 = O, \qquad v_1 = u_1 e_1, \qquad v_2 = u_1 e_1 + u_2 e_2, \quad \ldots, \quad v_n = u_1 e_1 + \cdots + u_n e_n .$$

Then the kth term of the sum in (8.13) becomes

$$f(a + \lambda v_{k-1} + \lambda u_k e_k) - f(a + \lambda v_{k-1}) = f(b_k + \lambda u_k e_k) - f(b_k),$$

where $b_k = a + \lambda v_{k-1}$. The two points b_k and $b_k + \lambda u_k e_k$ differ only in their kth component. Therefore we can apply the mean-value theorem of differential calculus to write

$$(8.14) \qquad f(b_k + \lambda u_k e_k) - f(b_k) = \lambda u_k \, D_k f(c_k),$$

where c_k lies on the line segment joining b_k to $b_k + \lambda u_k e_k$. Note that $b_k \to a$ and hence $c_k \to a$ as $\lambda \to 0$.

Using (8.14) in (8.13) we obtain

$$f(a + v) - f(a) = \lambda \sum_{k=1}^{n} D_k f(c_k) u_k .$$

But $\nabla f(a) \cdot v = \lambda \, \nabla f(a) \cdot u = \lambda \sum_{k=1}^{n} D_k f(a) u_k$, so

$$f(a + v) - f(a) - \nabla f(a) \cdot v = \lambda \sum_{k=1}^{n} \{D_k f(c_k) - D_k f(a)\} u_k = \|v\| \, E(a, v),$$

where

$$E(a, v) = \sum_{k=1}^{n} \{D_k f(c_k) - D_k f(a)\} u_k .$$

Since $c_k \to a$ as $\|v\| \to 0$, and since each partial derivative $D_k f$ is continuous at a, we see that $E(a, v) \to 0$ as $\|v\| \to 0$. This completes the proof.

8.14 Exercises

1. Find the gradient vector at each point at which it exists for the scalar fields defined by the following equations:
   (a) $f(x, y) = x^2 + y^2 \sin(xy)$.
   (b) $f(x, y) = e^x \cos y$.
   (c) $f(x, y, z) = x^2 y^3 z^4$.
   (d) $f(x, y, z) = x^2 - y^2 + 2z^2$.
   (e) $f(x, y, z) = \log(x^2 + 2y^2 - 3z^2)$.
   (f) $f(x, y, z) = x^{y^z}$.
2. Evaluate the directional derivatives of the following scalar fields for the points and directions given:
   (a) $f(x, y, z) = x^2 + 2y^2 + 3z^2$ at $(1, 1, 0)$ in the direction of $i - j + 2k$.
   (b) $f(x, y, z) = (x/y)^z$ at $(1, 1, 1)$ in the direction of $2i + j - k$.
3. Find the points (x, y) and the directions for which the directional derivative of $f(x, y) = 3x^2 + y^2$ has its largest value, if (x, y) is restricted to be on the circle $x^2 + y^2 = 1$.
4. A differentiable scalar field f has, at the point $(1, 2)$, directional derivatives $+2$ in the direction toward $(2, 2)$ and -2 in the direction toward $(1, 1)$. Determine the gradient vector at $(1, 2)$ and compute the directional derivative in the direction toward $(4, 6)$.
5. Find values of the constants a, b, and c such that the directional derivative of $f(x, y, z) = axy^2 + byz + cz^2x^3$ at the point $(1, 2, -1)$ has a maximum value of 64 in a direction parallel to the z-axis.

6. Given a scalar field differentiable at a point $\boldsymbol{a}$ in $\mathbf{R}^2$. Suppose that $f'(\boldsymbol{a}; \boldsymbol{y}) = 1$ and $f'(\boldsymbol{a}; \boldsymbol{z}) = 2$, where $\boldsymbol{y} = 2\boldsymbol{i} + 3\boldsymbol{j}$ and $\boldsymbol{z} = \boldsymbol{i} + \boldsymbol{j}$. Make a sketch showing the set of all points (x, y) for which $f'(\boldsymbol{a}; x\boldsymbol{i} + y\boldsymbol{j}) = 6$. Also, calculate the gradient $\nabla f(\boldsymbol{a})$.
7. Let f and g denote scalar fields that are differentiable on an open set S. Derive the following properties of the gradient:
(a) $\text{grad}\, f = \boldsymbol{O}$ if f is constant on S.
(b) $\text{grad}\,(f + g) = \text{grad}\, f + \text{grad}\, g$.
(c) $\text{grad}\,(cf) = c\, \text{grad}\, f$ if c is a constant.
(d) $\text{grad}\,(fg) = f\, \text{grad}\, g + g\, \text{grad}\, f$.
(e) $\text{grad}\left(\dfrac{f}{g}\right) = \dfrac{g\, \text{grad}\, f - f\, \text{grad}\, g}{g^2}$ at points at which $g \neq 0$.
8. In $\mathbf{R}^3$ let $\boldsymbol{r}(x, y, z) = x\boldsymbol{i} + y\boldsymbol{j} + z\boldsymbol{k}$, and let $r(x, y, z) = \|\boldsymbol{r}(x, y, z)\|$.
(a) Show that $\nabla r(x, y, z)$ is a unit vector in the direction of $\boldsymbol{r}(x, y, z)$.
(b) Show that $\nabla(r^n) = nr^{n-2}\boldsymbol{r}$ if n is a positive integer. [*Hint:* Use Exercise 7(d).]
(c) Is the formula of part (b) valid when n is a negative integer or zero?
(d) Find a scalar field f such that $\nabla f = \boldsymbol{r}$.
9. Assume f is differentiable at each point of an n-ball $B(\boldsymbol{a})$. If $f'(\boldsymbol{x}; \boldsymbol{y}) = 0$ for n independent vectors $\boldsymbol{y}_1, \ldots, \boldsymbol{y}_n$ and for every $\boldsymbol{x}$ in $B(\boldsymbol{a})$, prove that f is constant on $B(\boldsymbol{a})$.
10. Assume f is differentiable at each point of an n-ball $B(\boldsymbol{a})$.
(a) If $\nabla f(\boldsymbol{x}) = \boldsymbol{O}$ for every $\boldsymbol{x}$ in $B(\boldsymbol{a})$, prove that f is constant on $B(\boldsymbol{a})$.
(b) If $f(\boldsymbol{x}) \leq f(\boldsymbol{a})$ for all $\boldsymbol{x}$ in $B(\boldsymbol{a})$, prove that $\nabla f(\boldsymbol{a}) = \boldsymbol{O}$.
11. Consider the following six statements about a scalar field $f: S \to \mathbf{R}$, where $S \subseteq \mathbf{R}^n$ and $\boldsymbol{a}$ is an interior point of S.
(a) f is continuous at $\boldsymbol{a}$.
(b) f is differentiable at $\boldsymbol{a}$.
(c) $f'(\boldsymbol{a}; \boldsymbol{y})$ exists for every $\boldsymbol{y}$ in $\mathbf{R}^n$.
(d) All the first-order partial derivatives of f exist in a neighborhood of $\boldsymbol{a}$ and are continuous at $\boldsymbol{a}$.
(e) $\nabla f(\boldsymbol{a}) = \boldsymbol{O}$.
(f) $f(\boldsymbol{x}) = \|\boldsymbol{x} - \boldsymbol{a}\|$ for all $\boldsymbol{x}$ in $\mathbf{R}^n$.

In a table like the one shown here, mark T in the appropriate square if the statement in row (x) always implies the statement in column (y). For example, if (a) always implies (b), mark T in the second square of the first row. The main diagonal has already been filled in for you.

	a	b	c	d	e	f
a	T					
b		T				
c			T			
d				T		
e					T	
f						T

8.15 A chain rule for derivatives of scalar fields

In one-dimensional derivative theory, the chain rule enables us to compute the derivative of a composite function $g(t) = f[r(t)]$ by the formula

$$g'(t) = f'[r(t)] \cdot r'(t).$$

This section provides an extension of the formula when f is replaced by a scalar field defined on a set in n-space and r is replaced by a vector-valued function of a real variable with values in the domain of f.

In a later section we further extend the formula to cover the case in which both f and r are vector fields.

It is easy to conceive of examples in which the composition of a scalar field and a vector field might arise. For instance, suppose $f(\boldsymbol{x})$ measures the temperature at a point $\boldsymbol{x}$ of a solid in 3-space, and suppose we wish to know how the temperature changes as the point $\boldsymbol{x}$ varies along a curve C lying in the solid. If the curve is described by a vector-valued function $\boldsymbol{r}$ defined on an interval $[a, b]$, we can introduce a new function g by means of the formula

$$g(t) = f[\boldsymbol{r}(t)] \quad \text{if} \quad a \le t \le b .$$

This composite function g expresses the temperature as a function of the parameter t, and its derivative $g'(t)$ measures the rate of change of the temperature along the curve. The following extension of the chain rule enables us to compute the derivative $g'(t)$ without determining $g(t)$ explicitly.

THEOREM 8.8. CHAIN RULE. *Let f be a scalar field defined on an open set S in $\mathbf{R}^n$, and let $\boldsymbol{r}$ be a vector-valued function which maps an interval J from $\mathbf{R}^1$ into S. Define the composite function $g = f \circ \boldsymbol{r}$ on J by the equation*

$$g(t) = f[\boldsymbol{r}(t)] \quad \textit{if} \quad t \in J .$$

Let t be a point in J at which $\boldsymbol{r}'(t)$ exists and assume that f is differentiable at $\boldsymbol{r}(t)$. Then $g'(t)$ exists and is equal to the dot product

$$g'(t) = \nabla f(\boldsymbol{a}) \cdot \boldsymbol{r}'(t), \qquad \textit{where} \quad \boldsymbol{a} = \boldsymbol{r}(t). \tag{8.15}$$

Proof. Let $\boldsymbol{a} = \boldsymbol{r}(t)$, where t is a point in J at which $\boldsymbol{r}'(t)$ exists. Since S is open there is an n-ball $B(\boldsymbol{a})$ lying in S. We take $h \neq 0$ but small enough so that $\boldsymbol{r}(t+h)$ lies in $B(\boldsymbol{a})$, and we let $\boldsymbol{y} = \boldsymbol{r}(t+h) - \boldsymbol{r}(t)$. Note that $\boldsymbol{y} \to \boldsymbol{O}$ as $h \to 0$. Now we have

$$g(t+h) - g(t) = f[\boldsymbol{r}(t+h)] - f[\boldsymbol{r}(t)] = f(\boldsymbol{a} + \boldsymbol{y}) - f(\boldsymbol{a}) .$$

Applying the first-order Taylor formula for f we have

$$f(\boldsymbol{a} + \boldsymbol{y}) - f(\boldsymbol{a}) = \nabla f(\boldsymbol{a}) \cdot \boldsymbol{y} + \|\boldsymbol{y}\| \, E(\boldsymbol{a}, \boldsymbol{y}),$$

where $E(\boldsymbol{a}, \boldsymbol{y}) \to 0$ as $\|\boldsymbol{y}\| \to 0$. Since $\boldsymbol{y} = \boldsymbol{r}(t+h) - \boldsymbol{r}(t)$ this gives us

$$\frac{g(t+h) - g(t)}{h} = \nabla f(\boldsymbol{a}) \cdot \frac{\boldsymbol{r}(t+h) - \boldsymbol{r}(t)}{h} + \frac{\|\boldsymbol{r}(t+h) - \boldsymbol{r}(t)\|}{h} E(\boldsymbol{a}, \boldsymbol{y}) .$$

Letting $h \to 0$ we obtain (8.15).

EXAMPLE 1. *Directional derivative along a curve.* When the function $\boldsymbol{r}$ describes a curve C, the derivative $\boldsymbol{r}'$ is the velocity vector (tangent to the curve) and the derivative g' in Equation (8.15) is the derivative of f with respect to the velocity vector, assuming that $\boldsymbol{r}' \neq \boldsymbol{O}$. If $\boldsymbol{T}(t)$ is a unit vector in the direction of $\boldsymbol{r}'(t)$ ($\boldsymbol{T}$ is the unit tangent vector), the dot product $\nabla f[\boldsymbol{r}(t)] \cdot \boldsymbol{T}(t)$ is called the *directional derivative of f along the curve C* or *in the direction of C*. For a plane curve we can write

$$\boldsymbol{T}(t) = \cos \alpha(t)\, \boldsymbol{i} + \cos \beta(t)\, \boldsymbol{j},$$

where $\alpha(t)$ and $\beta(t)$ are the angles made by the vector $\boldsymbol{T}(t)$ and the positive x- and y-axes; the directional derivative of f along C becomes

$$\nabla f[\boldsymbol{r}(t)] \cdot \boldsymbol{T}(t) = D_1 f[\boldsymbol{r}(t)] \cos \alpha(t) + D_2 f[\boldsymbol{r}(t)] \cos \beta(t).$$

This formula is often written more briefly as

$$\nabla f \cdot \boldsymbol{T} = \frac{\partial f}{\partial x} \cos \alpha + \frac{\partial f}{\partial y} \cos \beta.$$

Some authors write df/ds for the directional derivative $\nabla f \cdot \boldsymbol{T}$. Since the directional derivative along C is defined in terms of $\boldsymbol{T}$, its value depends on the parametric representation chosen for C. A change of the representation could reverse the direction of $\boldsymbol{T}$; this, in turn, would reverse the sign of the directional derivative.

EXAMPLE 2. Find the directional derivative of the scalar field $f(x, y) = x^2 - 3xy$ along the parabola $y = x^2 - x + 2$ at the point (1,2).

Solution. At an arbitrary point (x, y) the gradient vector is

$$\nabla f(x, y) = \frac{\partial f}{\partial x}\boldsymbol{i} + \frac{\partial f}{\partial y}\boldsymbol{j} = (2x - 3y)\boldsymbol{i} - 3x\boldsymbol{j}.$$

At the point $(1, 2)$ we have $\nabla f(1, 2) = -4\boldsymbol{i} - 3\boldsymbol{j}$. The parabola can be represented parametrically by the vector equation $\boldsymbol{r}(t) = t\boldsymbol{i} + (t^2 - t + 2)\boldsymbol{j}$. Therefore, $\boldsymbol{r}(1) = \boldsymbol{i} + 2\boldsymbol{j}$, $\boldsymbol{r}'(t) = \boldsymbol{i} + (2t - 1)\boldsymbol{j}$, and $\boldsymbol{r}'(1) = \boldsymbol{i} + \boldsymbol{j}$. For this representation of C the unit tangent vector $\boldsymbol{T}(1)$ is $(\boldsymbol{i} + \boldsymbol{j})/\sqrt{2}$ and the required directional derivative is $\nabla f(1, 2) \cdot \boldsymbol{T}(1) = -7/\sqrt{2}$.

EXAMPLE 3. Let f be a nonconstant scalar field, differentiable everywhere in the plane, and let c be a constant. Assume the Cartesian equation $f(x, y) = c$ describes a curve C having a tangent at each of its points. Prove that f has the following properties at each point of C:

(a) The gradient vector ∇f is normal to C.
(b) The directional derivative of f is zero along C.
(c) The directional derivative of f has its largest value in a direction normal to C.

Solution. If $\boldsymbol{T}$ is a unit tangent vector to C, the directional derivative of f along C is the dot product $\nabla f \cdot \boldsymbol{T}$. This product is zero if ∇f is perpendicular to $\boldsymbol{T}$, and it has its largest value if ∇f is parallel to $\boldsymbol{T}$. Therefore both statements (b) and (c) are consequences of (a). To prove (a), consider any plane curve Γ with a vector equation of the form $\boldsymbol{r}(t) = X(t)\boldsymbol{i} + Y(t)\boldsymbol{j}$ and introduce the function $g(t) = f[\boldsymbol{r}(t)]$. By the chain rule we have $g'(t) = \nabla f[\boldsymbol{r}(t)] \cdot \boldsymbol{r}'(t)$. When $\Gamma = C$, the function g has the constant value c so $g'(t) = 0$ if $\boldsymbol{r}(t) \in C$. Since $g' = \nabla f \cdot \boldsymbol{r}'$, this shows that ∇f is perpendicular to $\boldsymbol{r}'$ on C; hence ∇f is normal to C.

8.16 Applications to geometry. Level sets. Tangent planes

The chain rule can be used to deduce geometric properties of the gradient vector. Let f be a scalar field defined on a set S in $\mathbf{R}^n$ and consider those points $\boldsymbol{x}$ in S for which $f(\boldsymbol{x})$ has a constant value, say $f(\boldsymbol{x}) = c$. Denote this set by $L(c)$, so that

$$L(c) = \{\boldsymbol{x} \mid \boldsymbol{x} \in S \text{ and } f(\boldsymbol{x}) = c\}.$$

The set $L(c)$ is called a *level set* of f. In $\mathbf{R}^2$, $L(c)$ is called a *level curve*; in $\mathbf{R}^3$ it is called a *level surface*.

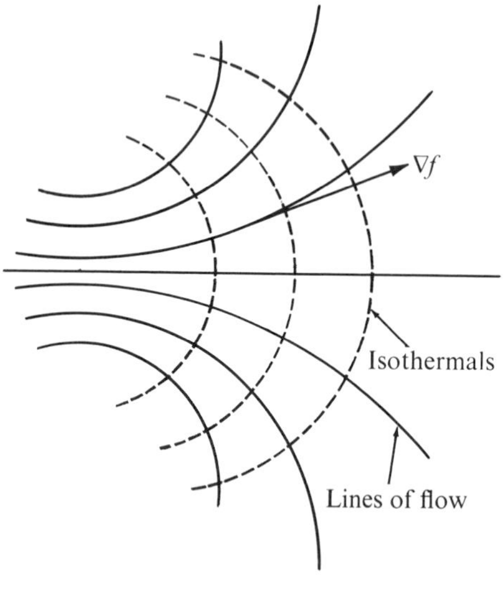

FIGURE 8.6 The dotted curves are isothermals: $f(x, y) = c$. The gradient vector ∇f points in the direction of the lines of flow.

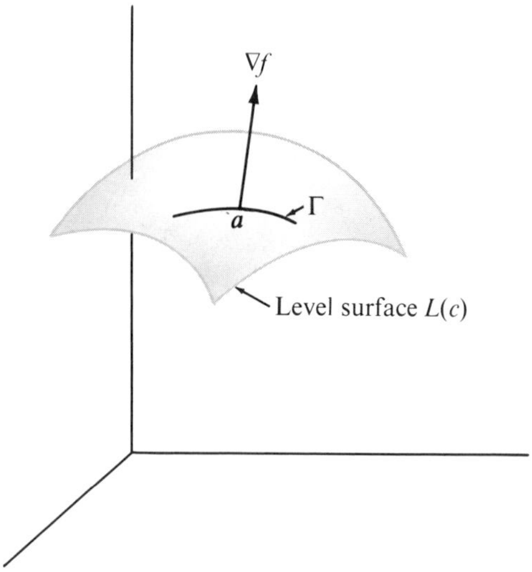

FIGURE 8.7 The gradient vector ∇f is normal to each curve Γ on the level surface $f(x, y, z) = c$.

Families of level sets occur in many physical applications. For example, if $f(x, y)$ represents temperature at (x, y), the level curves of f (curves of constant temperature) are called *isothermals*. The flow of heat takes place in the direction of most rapid change in temperature. As was shown in Example 3 of the foregoing section, this direction is normal to the isothermals. Hence, in a thin flat sheet the flow of heat is along a family of curves orthogonal to the isothermals. These are called the *lines of flow*; they are the orthogonal trajectories of the isothermals. Examples are shown in Figure 8.6.

Now consider a scalar field f differentiable on an open set S in $\mathbf{R}^3$, and examine one of its level surfaces, $L(c)$. Let $\boldsymbol{a}$ be a point on this surface, and consider a curve Γ which lies on the surface and passes through $\boldsymbol{a}$, as suggested by Figure 8.7. We shall prove that the gradient vector $\nabla f(\boldsymbol{a})$ is normal to this curve at $\boldsymbol{a}$. That is, we shall prove that $\nabla f(\boldsymbol{a})$ is perpendicular to the tangent vector of Γ at $\boldsymbol{a}$. For this purpose we assume that Γ is

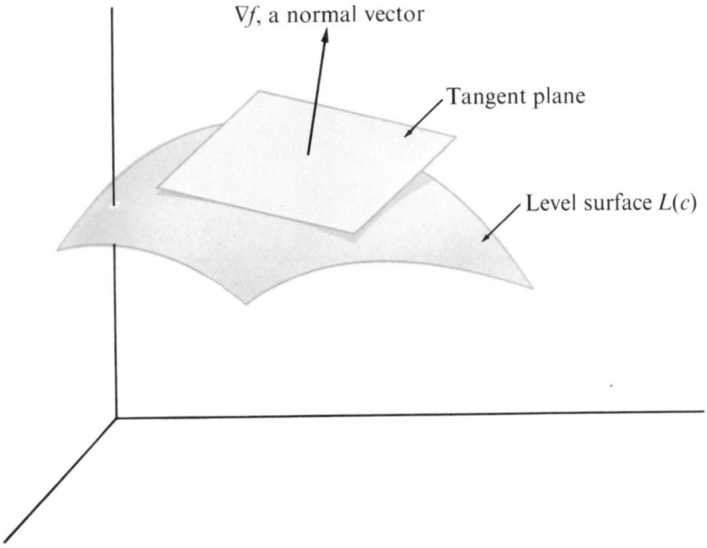

FIGURE 8.8 The gradient vector ∇f is normal to the tangent plane of a level surface $f(x, y, z) = c$.

described parametrically by a differentiable vector-valued function $\boldsymbol{r}$ defined on some interval J in $\mathbf{R}^1$. Since Γ lies on the level surface $L(c)$, the function $\boldsymbol{r}$ satisfies the equation

$$f[\boldsymbol{r}(t)] = c \qquad \text{for all } t \text{ in } J.$$

If $g(t) = f[\boldsymbol{r}(t)]$ for t in J, the chain rule states that

$$g'(t) = \nabla f[\boldsymbol{r}(t)] \cdot \boldsymbol{r}'(t).$$

Since g is constant on J, we have $g'(t) = 0$ on J. In particular, choosing t_1 so that $g(t_1) = \boldsymbol{a}$ we find that

$$\nabla f(\boldsymbol{a}) \cdot \boldsymbol{r}'(t_1) = 0.$$

In other words, the gradient of f at $\boldsymbol{a}$ is perpendicular to the tangent vector $\boldsymbol{r}'(t_1)$, as asserted.

Now we take a family of curves on the level surface $L(c)$, all passing through the point $\boldsymbol{a}$. According to the foregoing discussion, the tangent vectors of all these curves are perpendicular to the gradient vector $\nabla f(\boldsymbol{a})$. If $\nabla f(\boldsymbol{a})$ is not the zero vector, these tangent vectors determine a plane, and the gradient $\nabla f(\boldsymbol{a})$ is normal to this plane. (See Figure 8.8.). This particular plane is called the *tangent plane* of the surface $L(c)$ at $\boldsymbol{a}$.

We know from Volume I that a plane through $\boldsymbol{a}$ with normal vector $\boldsymbol{N}$ consists of all points $\boldsymbol{x}$ in $\mathbf{R}^3$ satisfying $\boldsymbol{N} \cdot (\boldsymbol{x} - \boldsymbol{a}) = 0$. Therefore the tangent plane to the level surface $L(c)$ at $\boldsymbol{a}$ consists of all $\boldsymbol{x}$ in $\mathbf{R}^3$ satisfying

$$\nabla f(\boldsymbol{a}) \cdot (\boldsymbol{x} - \boldsymbol{a}) = 0 .$$

To obtain a Cartesian equation for this plane we express $\boldsymbol{x}$, $\boldsymbol{a}$, and $\nabla f(\boldsymbol{a})$ in terms of their components. Writing $\boldsymbol{x} = (x, y, z)$, $\boldsymbol{a} = (x_1, y_1, z_1)$, and

$$\nabla f(\boldsymbol{a}) = D_1 f(\boldsymbol{a})\boldsymbol{i} + D_2 f(\boldsymbol{a})\boldsymbol{j} + D_3 f(\boldsymbol{a})\boldsymbol{k} ,$$

we obtain the Cartesian equation

$$D_1 f(\boldsymbol{a})(x - x_1) + D_2 f(\boldsymbol{a})(y - y_1) + D_3 f(\boldsymbol{a})(z - z_1) = 0 .$$

A similar discussion applies to scalar fields defined in $\mathbf{R}^2$. In Example 3 of the foregoing section we proved that the gradient vector $\nabla f(\boldsymbol{a})$ at a point $\boldsymbol{a}$ of a level curve is perpendicular to the tangent vector of the curve at $\boldsymbol{a}$. Therefore the tangent line of the level curve $L(c)$ at a point $\boldsymbol{a} = (x_1, y_1)$ has the Cartesian equation

$$D_1 f(\boldsymbol{a})(x - x_1) + D_2 f(\boldsymbol{a})(y - y_1) = 0 .$$

8.17 Exercises

1. In this exercise you may assume the existence and continuity of all derivatives under consideration. The equations $u = f(x, y)$, $x = X(t)$, $y = Y(t)$ define u as a function of t, say $u = F(t)$.
   (a) Use the chain rule to show that
   $$F'(t) = \frac{\partial f}{\partial x} X'(t) + \frac{\partial f}{\partial y} Y'(t),$$
   where $\partial f / \partial x$ and $\partial f / \partial y$ are to be evaluated at $[X(t), Y(t)]$.
   (b) In a similar way, express the second derivative $F''(t)$ in terms of derivatives of f, X, and Y. Remember that the partial derivatives in the formula of part (a) are composite functions given by
   $$\frac{\partial f}{\partial x} = D_1 f[X(t), Y(t)], \qquad \frac{\partial f}{\partial y} = D_2 f[X(t), Y(t)].$$

2. Refer to Exercise 1 and compute $F'(t)$ and $F''(t)$ in terms of t for each of the following special cases:
   (a) $f(x, y) = x^2 + y^2$, $X(t) = t$, $Y(t) = t^2$.
   (b) $f(x, y) = e^{xy} \cos (xy^2)$, $X(t) = \cos t$, $Y(t) = \sin t$.
   (c) $f(x, y) = \log [(1 + e^{x^2})/(1 + e^{y^2})]$, $X(t) = e$, $Y(t)^t = e^{-t}$.

3. In each case, evaluate the directional derivative of f for the points and directions specified:
(a) $f(x, y, z) = 3x - 5y + 2z$ at $(2, 2, 1)$ in the direction of the outward normal to the sphere $x^2 + y^2 + z^2 = 9$.
(b) $f(x, y, z) = x^2 - y^2$ at a general point of the surface $x^2 + y^2 + z^2 = 4$ in the direction of the outward normal at that point.
(c) $f(x, y, z) = x^2 + y^2 - z^2$ at $(3, 4, 5)$ along the curve of intersection of the two surfaces $2x^2 + 2y^2 - z^2 = 25$ and $x^2 + y^2 = z^2$.
4. (a) Find a vector $V(x, y, z)$ normal to the surface

$$z = \sqrt{x^2 + y^2} + (x^2 + y^2)^{3/2}$$

at a general point (x, y, z) of the surface, $(x, y, z) \neq (0, 0, 0)$.
(b) Find the cosine of the angle θ between $V(x, y, z)$ and the z-axis and determine the limit of $\cos \theta$ as $(x, y, z) \to (0, 0, 0)$.
5. The two equations $e^u \cos v = x$ and $e^u \sin v = y$ define u and v as functions of x and y, say $u = U(x, y)$ and $v = V(x, y)$. Find explicit formulas for $U(x, y)$ and $V(x, y)$, valid for $x > 0$, and show that the gradient vectors $\nabla U(x, y)$ and $\nabla V(x, y)$ are perpendicular at each point (x, y).
6. Let $f(x, y) = \sqrt{|xy|}$.
(a) Verify that $\partial f/\partial x$ and $\partial f/\partial y$ are both zero at the origin.
(b) Does the surface $z = f(x, y)$ have a tangent plane at the origin? [*Hint:* Consider the section of the surface made by the plane $x = y$.]
7. If (x_0, y_0, z_0) is a point on the surface $z = xy$, then the two lines $z = y_0 x, y = y_0$ and $z = x_0 y, x = x_0$ intersect at (x_0, y_0, z_0) and lie on the surface. Verify that the tangent plane to this surface at the point (x_0, y_0, z_0) contains these two lines.
8. Find a Cartesian equation for the tangent plane to the surface $xyz = a^3$ at a general point (x_0, y_0, z_0). Show that the volume of the tetrahedron bounded by this plane and the three coordinate plane is $9a^3/2$.
9. Find a pair of linear Cartesian equations for the line which is tangent to both the surfaces $x^2 + y^2 + 2z^2 = 4$ and $z = e^{x-y}$ at the point $(1, 1, 1)$.
10. Find a constant c such that at any point of intersection of the two spheres

$$(x - c)^2 + y^2 + z^2 = 3 \quad \text{and} \quad x^2 + (y - 1)^2 + z^2 = 1$$

the corresponding tangent planes will be perpendicular to each other.
11. If r_1 and r_2 denote the distances from a point (x, y) on an ellipse to its foci, show that the equation $r_1 + r_2 =$ constant (satisfied by these distances) implies the relation

$$\boldsymbol{T} \cdot \nabla(r_1 + r_2) = 0,$$

where $\boldsymbol{T}$ is the unit tangent to the curve. Interpret this result geometrically, thereby showing that the tangent makes equal angles with the lines joining (x, y) to the foci.
12. If $\nabla f(x, y, z)$ is always parallel to $x\boldsymbol{i} + y\boldsymbol{j} + z\boldsymbol{k}$, show that f must assume equal values at the points $(0, 0, a)$ and $(0, 0, -a)$.

8.18 Derivatives of vector fields

Derivative theory for vector fields is a straightforward extension of that for scalar fields. Let $\boldsymbol{f}: S \to \mathbf{R}^m$ be a vector field defined on a subset S of $\mathbf{R}^n$. If $\boldsymbol{a}$ is an interior point of S

and if $\boldsymbol{y}$ is any vector in $\mathbf{R}^n$ we define the derivative $\boldsymbol{f}'(\boldsymbol{a}; \boldsymbol{y})$ by the formula

$$\boldsymbol{f}'(\boldsymbol{a}; \boldsymbol{y}) = \lim_{h \to 0} \frac{\boldsymbol{f}(\boldsymbol{a} + h\boldsymbol{y}) - \boldsymbol{f}(\boldsymbol{a})}{h},$$

whenever the limit exists. The derivative $\boldsymbol{f}'(\boldsymbol{a}; \boldsymbol{y})$ is a vector in $\mathbf{R}^m$.

Let f_k denote the kth component of $\boldsymbol{f}$. We note that the derivative $\boldsymbol{f}'(\boldsymbol{a}; \boldsymbol{y})$ exists if and only if $f_k'(\boldsymbol{a}; \boldsymbol{y})$ exists for each $k = 1, 2, \ldots, m$, in which case we have

$$\boldsymbol{f}'(\boldsymbol{a}; \boldsymbol{y}) = (f_1'(\boldsymbol{a}; \boldsymbol{y}), \ldots, f_m'(\boldsymbol{a}; \boldsymbol{y})) = \sum_{k=1}^{m} f_k'(\boldsymbol{a}; \boldsymbol{y})\boldsymbol{e}_k,$$

where $\boldsymbol{e}_k$ is the kth unit coordinate vector.

We say that $\boldsymbol{f}$ is *differentiable* at an interior point $\boldsymbol{a}$ if there is a linear transformation

$$\boldsymbol{T}_a : \mathbf{R}^n \to \mathbf{R}^m$$

such that

$$\boldsymbol{f}(\boldsymbol{a} + \boldsymbol{v}) = \boldsymbol{f}(\boldsymbol{a}) + \boldsymbol{T}_a(\boldsymbol{v}) + \|\boldsymbol{v}\|\, \boldsymbol{E}(\boldsymbol{a}, \boldsymbol{v}), \tag{8.16}$$

where $\boldsymbol{E}(\boldsymbol{a}, \boldsymbol{v}) \to \boldsymbol{O}$ as $\boldsymbol{v} \to \boldsymbol{O}$. The first-order Taylor formula (8.16) is to hold for all $\boldsymbol{v}$ with $\|\boldsymbol{v}\| < r$ for some $r > 0$. The term $\boldsymbol{E}(\boldsymbol{a}, \boldsymbol{v})$ is a vector in $\mathbf{R}^m$. The linear transformation $\boldsymbol{T}_a$ is called the *total derivative* of $\boldsymbol{f}$ at $\boldsymbol{a}$.

For scalar fields we proved that $T_a(\boldsymbol{y})$ is the dot product of the gradient vector $\nabla f(\boldsymbol{a})$ with $\boldsymbol{y}$. For vector fields we will prove that $\boldsymbol{T}_a(\boldsymbol{y})$ is a vector whose kth component is the dot product $\nabla f_k(\boldsymbol{a}) \cdot \boldsymbol{y}$.

THEOREM 8.9. *Assume $\boldsymbol{f}$ is differentiable at $\boldsymbol{a}$ with total derivative $\boldsymbol{T}_a$. Then the derivative $\boldsymbol{f}'(\boldsymbol{a}; \boldsymbol{y})$ exists for every $\boldsymbol{a}$ in $\mathbf{R}^n$, and we have*

$$\boldsymbol{T}_a(\boldsymbol{y}) = \boldsymbol{f}'(\boldsymbol{a}; \boldsymbol{y}). \tag{8.17}$$

Moreover, if $\boldsymbol{f} = (f_1, \ldots, f_m)$ and if $\boldsymbol{y} = (y_1, \ldots, y_n)$, we have

$$\boldsymbol{T}_a(\boldsymbol{y}) = \sum_{k=1}^{m} \nabla f_k(\boldsymbol{a}) \cdot \boldsymbol{y}\, \boldsymbol{e}_k = (\nabla f_1(\boldsymbol{a}) \cdot \boldsymbol{y}, \ldots, \nabla f_m(\boldsymbol{a}) \cdot \boldsymbol{y}). \tag{8.18}$$

Proof. We argue as in the scalar case. If $\boldsymbol{y} = \boldsymbol{O}$, then $\boldsymbol{f}'(\boldsymbol{a}; \boldsymbol{y}) = \boldsymbol{O}$ and $\boldsymbol{T}_a(\boldsymbol{O}) = \boldsymbol{O}$. Therefore we can assume that $\boldsymbol{y} \neq \boldsymbol{O}$. Taking $\boldsymbol{v} = h\boldsymbol{y}$ in the Taylor formula (8.16) we have

$$\boldsymbol{f}(\boldsymbol{a} + h\boldsymbol{y}) - \boldsymbol{f}(\boldsymbol{a}) = \boldsymbol{T}_a(h\boldsymbol{y}) + \|h\boldsymbol{y}\|\, \boldsymbol{E}(\boldsymbol{a}, \boldsymbol{v}) = h\boldsymbol{T}_a(\boldsymbol{y}) + |h|\, \|\boldsymbol{y}\|\, \boldsymbol{E}(\boldsymbol{a}, \boldsymbol{v}).$$

Dividing by h and letting $h \to 0$ we obtain (8.17).

To prove (8.18) we simply note that

$$\boldsymbol{f}'(\boldsymbol{a}; \boldsymbol{y}) = \sum_{k=1}^{m} f_k'(\boldsymbol{a}; \boldsymbol{y})\, \boldsymbol{e}_k = \sum_{k=1}^{m} \nabla f_k(\boldsymbol{a}) \cdot \boldsymbol{y}\, \boldsymbol{e}_k.$$

Equation (8.18) can also be written more simply as a matrix product,

$$T_a(y) = Df(a)y,$$

where $Df(a)$ is the $m \times n$ matrix whose kth row is $\nabla f_k(a)$, and where y is regarded as an $n \times 1$ column matrix. The matrix $Df(a)$ is called the *Jacobian matrix* of f at a. Its kj entry is the partial derivative $D_j f_k(a)$. Thus, we have

$$Df(a) = \begin{bmatrix} D_1 f_1(a) & D_2 f_1(a) & \cdots & D_n f_1(a) \\ D_1 f_2(a) & D_2 f_2(a) & \cdots & D_n f_2(a) \\ \cdot & & & \cdot \\ \cdot & & & \cdot \\ \cdot & & & \cdot \\ D_1 f_m(a) & D_2 f_m(a) & \cdots & D_n f_m(a) \end{bmatrix}.$$

The Jacobian matrix $Df(a)$ is defined at each point where the mn partial derivatives $D_j f_k(a)$ exist.

The total derivative T_a is also written as $f'(a)$. The derivative $f'(a)$ is a linear transformation; the Jacobian $Df(a)$ is a matrix representation for this transformation.

The first-order Taylor formula takes the form

$$f(a + v) = f(a) + f'(a)(v) + \|v\| \, E(a, v), \tag{8.19}$$

where $E(a, v) \to O$ as $v \to O$. This resembles the one-dimensional Taylor formula. To compute the components of the vector $f'(a)(v)$ we can use the matrix product $Df(a)v$ or formula (8.18) of Theorem 8.9.

8.19 Differentiability implies continuity

THEOREM 8.10. *If a vector field f is differentiable at a, then f is continuous at a.*

Proof. As in the scalar case, we use the Taylor formula to prove this theorem. If we let $v \to O$ in (8.19) the error term $\|v\| \, E(a, v) \to O$. The linear part $f'(a)(v)$ also tends to O because linear transformations are continuous at O. This completes the proof.

At this point it is convenient to derive an inequality which will be used in the proof of the chain rule in the next section. The inequality concerns a vector field f differentiable at a; it states that

$$\|f'(a)(v)\| \le M_f(a) \, \|v\|, \qquad \text{where} \quad M_f(a) = \sum_{k=1}^{m} \|\nabla f_k(a)\|. \tag{8.20}$$

To prove this we use Equation (8.18) along with the triangle inequality and the Cauchy-Schwarz inequality to obtain

$$\|f'(a)(v)\| = \left\| \sum_{k=1}^{m} \nabla f_k(a) \cdot v \, e_k \right\| \le \sum_{k=1}^{m} |\nabla f_k(a) \cdot v| \le \sum_{k=1}^{m} \|\nabla f_k(a)\| \, \|v\|.$$

8.20 The chain rule for derivatives of vector fields

THEOREM 8.11. CHAIN RULE. *Let f and g be vector fields such that the composition $h = f \circ g$ is defined in a neighborhood of a point a. Assume that g is differentiable at a, with total derivative $g'(a)$. Let $b = g(a)$ and assume that f is differentiable at b, with total derivative $f'(b)$. Then h is differentiable at a, and the total derivative $h'(a)$ is given by*

$$h'(a) = f'(b) \circ g'(a),$$

the composition of the linear transformations $f'(b)$ and $g'(a)$.

Proof. We consider the difference $h(a + y) - h(a)$ for small $\|y\|$, and show that we have a first-order Taylor formula. From the definition of h we have

(8.21) $$h(a + y) - h(a) = f[g(a + y)] - f[g(a)] = f(b + v) - f(b),$$

where $v = g(a + y) - g(a)$. The Taylor formula for $g(a + y)$ gives us

(8.22) $$v = g'(a)(y) + \|y\| \, E_g(a, y), \qquad \text{where} \quad E_g(a, y) \to O \quad \text{as} \quad y \to O.$$

The Taylor formula for $f(b + v)$ gives us

(8.23) $$f(b + v) - f(b) = f'(b)(v) + \|v\| \, E_f(b, v),$$

where $E_f(b, v) \to O$ as $v \to O$. Using (8.22) in (8.23) we obtain

(8.24) $$\begin{aligned} f(b + v) - f(b) &= f'(b)g'(a)(y) + f'(b)(\|y\| \, E_g(a, y)) + \|v\| \, E_f(b, v) \\ &= f'(b)g'(a)(y) + \|y\| \, E(a, y), \end{aligned}$$

where $E(a, O) = O$ and

(8.25) $$E(a, y) = f'(b)(E_g(a, y)) + \frac{\|v\|}{\|y\|} E_f(b, v) \qquad \text{if} \quad y \neq O.$$

To complete the proof we need to show that $E(a, y) \to O$ as $y \to O$.

The first term on the right of (8.25) tends to O as $y \to O$ because $E_g(a, y) \to O$ as $y \to O$ and linear transformations are continuous at O.

In the second term on the right of (8.25) the factor $E_f(b, v) \to O$ because $v \to O$ as $y \to O$. The quotient $\|v\|/\|y\|$ remains bounded because, by (8.22) and (8.20) we have

$$\|v\| \le M_g(a) \, \|y\| + \|y\| \, \|E_g(a, y)\|.$$

Therefore both terms on the right of (8.25) tend to O as $y \to O$, so $E(a, y) \to O$.

Thus, from (8.24) and (8.21) we obtain the Taylor formula

$$h(a + y) - h(a) = f'(b)g'(a)(y) + \|y\| \, E(a, y),$$

where $E(a, y) \to O$ as $y \to O$. This proves that h is differentiable at a and that the total derivative $h'(a)$ is equal to the composition $f'(b) \circ g'(a)$.

8.21 Matrix form of the chain rule

Let $h = f \circ g$, where g is differentiable at a and f is differentiable at $b = g(a)$. The chain rule states that

$$h'(a) = f'(b) \circ g'(a).$$

We can express the chain rule in terms of the Jacobian matrices $Dh(a)$, $Df(b)$, and $Dg(a)$ which represent the linear transformations $h'(a)$, $f'(b)$, and $g'(a)$, respectively. Since composition of linear transformations corresponds to multiplication of their matrices, we obtain

$$Dh(a) = Df(b)\, Dg(a), \qquad \text{where} \quad b = g(a). \tag{8.26}$$

This is called the *matrix form of the chain rule.* It can also be written as a set of scalar equations by expressing each matrix in terms of its entries.

Suppose that $a \in \mathbf{R}^p$, $b = g(a) \in \mathbf{R}^n$, and $f(b) \in \mathbf{R}^m$. Then $h(a) \in \mathbf{R}^m$ and we can write

$$g = (g_1, \ldots, g_n), \qquad f = (f_1, \ldots, f_m), \qquad h = (h_1, \ldots, h_m).$$

Then $Dh(a)$ is an $m \times p$ matrix, $Df(b)$ is an $m \times n$ matrix, and $Dg(a)$ is an $n \times p$ matrix, given by

$$Dh(a) = [D_j h_i(a)]_{i,j=1}^{m,p}, \qquad Df(b) = [D_k f_i(b)]_{i,k=1}^{m,n}, \qquad Dg(a) = [D_j g_k(a)]_{k,j=1}^{n,p}.$$

The matrix equation (8.26) is equivalent to mp scalar equations,

$$D_j h_i(a) = \sum_{k=1}^{n} D_k f_i(b) D_j g_k(a), \qquad \text{for} \quad i = 1, 2, \ldots, m \qquad \text{and} \qquad j = 1, 2, \ldots, p.$$

These equations express the partial derivatives of the components of h in terms of the partial derivatives of the components of f and g.

EXAMPLE 1. *Extended chain rule for scalar fields.* Suppose f is a scalar field ($m = 1$). Then h is also a scalar field and there are p equations in the chain rule, one for each of the partial derivatives of h:

$$D_j h(a) = \sum_{k=1}^{n} D_k f(b) D_j g_k(a), \qquad \text{for} \quad j = 1, 2, \ldots, p.$$

The special case $p = 1$ was already considered in Section 8.15. In this case we get only one equation,

$$h'(a) = \sum_{k=1}^{n} D_k f(b) g_k'(a).$$

Now take $p = 2$ and $n = 2$. Write $\boldsymbol{a} = (s, t)$ and $\boldsymbol{b} = (x, y)$. Then the components x and y are related to s and t by the equations

$$x = g_1(s, t), \qquad y = g_2(s, t).$$

The chain rule gives a pair of equations for the partial derivatives of h:

$$D_1h(s, t) = D_1f(x, y)\, D_1g_1(s, t) + D_2f(x, y)\, D_1g_2(s, t),$$
$$D_2h(s, t) = D_1f(x, y)\, D_2g_1(s, t) + D_2f(x, y)\, D_2g_2(s, t).$$

In the ∂-notation, this pair of equations is usually written as

(8.27) $$\frac{\partial h}{\partial s} = \frac{\partial f}{\partial x}\frac{\partial x}{\partial s} + \frac{\partial f}{\partial y}\frac{\partial y}{\partial s}, \qquad \frac{\partial h}{\partial t} = \frac{\partial f}{\partial x}\frac{\partial x}{\partial t} + \frac{\partial f}{\partial y}\frac{\partial y}{\partial t}.$$

EXAMPLE 2. *Polar coordinates.* The temperature of a thin plate is described by a scalar field f, the temperature at (x, y) being $f(x, y)$. Polar coordinates $x = r\cos\theta$, $y = r\sin\theta$ are introduced, and the temperature becomes a function of r and θ determined by the equation

$$\varphi(r, \theta) = f(r\cos\theta, r\sin\theta).$$

Express the partial derivatives $\partial\varphi/\partial r$ and $\partial\varphi/\partial\theta$ in terms of the partial derivatives $\partial f/\partial x$ and $\partial f/\partial y$.

Solution. We use the chain rule as expressed in Equation (8.27), writing (r, θ) instead of (s, t), and φ instead of h. The equations

$$x = r\cos\theta, \qquad y = r\sin\theta$$

give us

$$\frac{\partial x}{\partial r} = \cos\theta, \qquad \frac{\partial y}{\partial r} = \sin\theta, \qquad \frac{\partial x}{\partial\theta} = -r\sin\theta, \qquad \frac{\partial y}{\partial\theta} = r\cos\theta.$$

Substituting these formulas in (8.27) we obtain

(8.28) $$\frac{\partial\varphi}{\partial r} = \frac{\partial f}{\partial x}\cos\theta + \frac{\partial f}{\partial y}\sin\theta, \qquad \frac{\partial\varphi}{\partial\theta} = -r\frac{\partial f}{\partial x}\sin\theta + r\frac{\partial f}{\partial y}\cos\theta.$$

These are the required formulas for $\partial\varphi/\partial r$ and $\partial\varphi/\partial\theta$.

EXAMPLE 3. *Second-order partial derivatives.* Refer to Example 2 and express the second-order partial derivative $\partial^2\varphi/\partial\theta^2$ in terms of partial derivatives of f.

Solution. We begin with the formula for $\partial\varphi/\partial\theta$ in (8.28) and differentiate with respect to θ, treating r as a constant. There are two terms on the right, each of which must be differentiated as a product. Thus we have

$$(8.29)\qquad \begin{aligned}\frac{\partial^2\varphi}{\partial\theta^2} &= -r\frac{\partial f}{\partial x}\frac{\partial(\sin\theta)}{\partial\theta} - r\sin\theta\frac{\partial}{\partial\theta}\left(\frac{\partial f}{\partial x}\right) + r\frac{\partial f}{\partial y}\frac{\partial(\cos\theta)}{\partial\theta} + r\cos\theta\frac{\partial}{\partial\theta}\left(\frac{\partial f}{\partial y}\right)\\ &= -r\cos\theta\frac{\partial f}{\partial x} - r\sin\theta\frac{\partial}{\partial\theta}\left(\frac{\partial f}{\partial x}\right) - r\sin\theta\frac{\partial f}{\partial y} + r\cos\theta\frac{\partial}{\partial\theta}\left(\frac{\partial f}{\partial y}\right).\end{aligned}$$

To compute the derivatives of $\partial f/\partial x$ and $\partial f/\partial y$ with respect to θ we must keep in mind that, as functions of r and θ, $\partial f/\partial x$ and $\partial f/\partial y$ are *composite functions* given by

$$\frac{\partial f}{\partial x} = D_1 f(r\cos\theta, r\sin\theta) \qquad\text{and}\qquad \frac{\partial f}{\partial y} = D_2 f(r\cos\theta, r\sin\theta).$$

Therefore, their derivatives with respect to θ must be determined by use of the chain rule. We again use (8.27), with f replaced by $D_1 f$, to obtain

$$\frac{\partial}{\partial\theta}\left(\frac{\partial f}{\partial x}\right) = \frac{\partial(D_1 f)}{\partial x}\frac{\partial x}{\partial\theta} + \frac{\partial(D_1 f)}{\partial y}\frac{\partial y}{\partial\theta} = \frac{\partial^2 f}{\partial x^2}(-r\sin\theta) + \frac{\partial^2 f}{\partial y\,\partial x}(r\cos\theta).$$

Similarly, using (8.27) with f replaced by $D_2 f$, we find

$$\frac{\partial}{\partial\theta}\left(\frac{\partial f}{\partial y}\right) = \frac{\partial(D_2 f)}{\partial x}\frac{\partial x}{\partial\theta} + \frac{\partial(D_2 f)}{\partial y}\frac{\partial y}{\partial\theta} = \frac{\partial^2 f}{\partial x\,\partial y}(-r\sin\theta) + \frac{\partial^2 f}{\partial y^2}(r\cos\theta).$$

When these formulas are used in (8.29) we obtain

$$\begin{aligned}\frac{\partial^2\varphi}{\partial\theta^2} &= -r\cos\theta\frac{\partial f}{\partial x} + r^2\sin^2\theta\frac{\partial^2 f}{\partial x^2} - r^2\sin\theta\cos\theta\frac{\partial^2 f}{\partial y\,\partial x}\\ &\quad - r\sin\theta\frac{\partial f}{\partial y} - r^2\sin\theta\cos\theta\frac{\partial^2 f}{\partial x\,\partial y} + r^2\cos^2\theta\frac{\partial^2 f}{\partial y^2}.\end{aligned}$$

This is the required formula for $\partial^2\varphi/\partial\theta^2$. Analogous formulas for the second-order partial derivatives $\partial^2\varphi/\partial r^2$, $\partial^2\varphi/(\partial r\,\partial\theta)$, and $\partial^2\varphi/(\partial\theta\,\partial r)$ are requested in Exercise 5 of the next section.

8.22 Exercises

In these exercises you may assume differentiability of all functions under consideration.

1. The substitution $t = g(x, y)$ converts $F(t)$ into $f(x, y)$, where $f(x, y) = F[g(x, y)]$.
   (a) Show that
   $$\frac{\partial f}{\partial x} = F'[g(x, y)]\frac{\partial g}{\partial x} \qquad\text{and}\qquad \frac{\partial f}{\partial y} = F'[g(x, y)]\frac{\partial g}{\partial y}.$$
   (b) Consider the special case $F(t) = e^{\sin t}$, $g(x, y) = \cos(x^2 + y^2)$. Compute $\partial f/\partial x$ and $\partial f/\partial y$ by use of the formulas in part (a). To check your result, determine $f(x, y)$ explicitly in terms of x and y and compute $\partial f/\partial x$ and $\partial f/\partial y$ directly from f.

2. The substitution $u = (x - y)/2$, $v = (x + y)/2$ changes $f(u, v)$ into $F(x, y)$. Use an appropriate form of the chain rule to express the partial derivatives $\partial F/\partial x$ and $\partial F/\partial y$ in terms of the partial derivatives $\partial f/\partial u$ and $\partial f/\partial v$.
3. The equations $u = f(x, y)$, $x = X(s, t)$, $y = Y(s, t)$ define u as a function of s and t, say $u = F(s, t)$.
   (a) Use an appropriate form of the chain rule to express the partial derivatives $\partial F/\partial s$ and $\partial F/\partial t$ in terms of $\partial f/\partial x$, $\partial f/\partial y$, $\partial X/\partial s$, $\partial X/\partial t$, $\partial Y/\partial s$, $\partial Y/\partial t$.
   (b) If $\partial^2 f/(\partial x\, \partial y) = \partial^2 f/(\partial y\, \partial x)$, show that

$$\frac{\partial^2 F}{\partial s^2} = \frac{\partial f}{\partial x}\frac{\partial^2 X}{\partial s^2} + \frac{\partial^2 f}{\partial x^2}\left(\frac{\partial X}{\partial s}\right)^2 + 2\frac{\partial X}{\partial s}\frac{\partial Y}{\partial s}\frac{\partial^2 f}{\partial x\, \partial y} + \frac{\partial f}{\partial y}\frac{\partial^2 Y}{\partial s^2} + \frac{\partial^2 f}{\partial y^2}\left(\frac{\partial Y}{\partial s}\right)^2.$$

   (c) Find similar formulas for the partial derivatives $\partial^2 F/(\partial s\, \partial t)$ and $\partial^2 F/\partial t^2$.
4. Solve Exercise 3 in each of the following special cases:
   (a) $X(s, t) = s + t$, $Y(s, t) = st$.
   (b) $X(s, t) = st$, $Y(s, t) = s/t$.
   (c) $X(s, t) = (s - t)/2$, $Y(s, t) = (s + t)/2$.
5. The introduction of polar coordinates changes $f(x, y)$ into $\varphi(r, \theta)$, where $x = r\cos\theta$ and $y = r\sin\theta$. Express the second-order partial derivatives $\partial^2\varphi/\partial r^2$, $\partial^2\varphi/(\partial r\, \partial\theta)$, and $\partial^2\varphi/(\partial\theta\, \partial r)$ in terms of the partial derivatives of f. You may use the formulas derived in Example 2 of Section 8.21.
6. The equations $u = f(x, y, z)$, $x = X(r, s, t)$, $y = Y(r, s, t)$, and $z = Z(r, s, t)$ define u as a function of r, s, and t, say $u = F(r, s, t)$. Use an appropriate form of the chain rule to express the partial derivatives $\partial F/\partial r$, $\partial F/\partial s$, and $\partial F/\partial t$ in terms of partial derivatives of f, X, Y, and Z.
7. Solve Exercise 6 in each of the following special cases:
   (a) $X(r, s, t) = r + s + t$, $Y(r, s, t) = r - 2s + 3t$, $Z(r, s, t) = 2r + s - t$.
   (b) $X(r, s, t) = r^2 + s^2 + t^2$, $Y(r, s, t) = r^2 - s^2 - t^2$, $Z(r, s, t) = r^2 - s^2 + t^2$.
8. The equations $u = f(x, y, z)$, $x = X(s, t)$, $y = Y(s, t)$, $z = Z(s, t)$ define u as a function of s and t, say $u = F(s, t)$. Use an appropriate form of the chain rule to express the partial derivatives $\partial F/\partial s$ and $\partial F/\partial t$ in terms of partial derivatives of f, X, Y, and Z.
9. Solve Exercise 8 in each of the following special cases:
   (a) $X(s, t) = s^2 + t^2$, $Y(s, t) = s^2 - t^2$, $Z(s, t) = 2st$.
   (b) $X(s, t) = s + t$, $Y(s, t) = s - t$, $Z(s, t) = st$.
10. The equations $u = f(x, y)$, $x = X(r, s, t)$, $y = Y(r, s, t)$ define u as a function of r, s, and t, say $u = F(r, s, t)$. Use an appropriate form of the chain rule to express the partial derivatives $\partial F/\partial r$, $\partial F/\partial s$, and $\partial F/\partial t$ in terms of partial derivatives of f, X, and Y.
11. Solve Exercise 10 in each of the following special cases:
   (a) $X(r, s, t) = r + s$, $Y(r, s, t) = t$.
   (b) $X(r, s, t) = r + s + t$, $Y(r, s, t) = r^2 + s^2 + t^2$.
   (c) $X(r, s, t) = r/s$, $Y(r, s, t) = s/t$.
12. Let $h(\boldsymbol{x}) = f[\boldsymbol{g}(\boldsymbol{x})]$, where $\boldsymbol{g} = (g_1, \ldots, g_n)$ is a vector field differentiable at $\boldsymbol{a}$, and f is a scalar field differentiable at $\boldsymbol{b} = \boldsymbol{g}(\boldsymbol{a})$. Use the chain rule to show that the gradient of h can be expressed as a linear combination of the gradient vectors of the components of $\boldsymbol{g}$, as follows:

$$\nabla h(\boldsymbol{a}) = \sum_{k=1}^{n} D_k f(\boldsymbol{b}) \nabla g_k(\boldsymbol{a}).$$

13. (a) If $\boldsymbol{f}(x, y, z) = x\boldsymbol{i} + y\boldsymbol{j} + z\boldsymbol{k}$, prove that the Jacobian matrix $D\boldsymbol{f}(x, y, z)$ is the identity matrix of order 3.

(b) Find all differentiable vector fields $f: \mathbf{R}^3 \to \mathbf{R}^3$ for which the Jacobian matrix $Df(x, y, z)$ is the identity matrix of order 3.
(c) Find all differentiable vector fields $f: \mathbf{R}^3 \to \mathbf{R}^3$ for which the Jacobian matrix is a diagonal matrix of the form diag $(p(x), q(y), r(z))$, where p, q, and r are given continuous functions.

14. Let $f: \mathbf{R}^2 \to \mathbf{R}^2$ and $g: \mathbf{R}^3 \to \mathbf{R}^2$ be two vector fields defined as follows:

$$f(x, y) = e^{x+2y}\boldsymbol{i} + \sin(y + 2x)\boldsymbol{j},$$

$$g(u, v, w) = (u + 2v^2 + 3w^3)\boldsymbol{i} + (2v - u^2)\boldsymbol{j}.$$

(a) Compute each of the Jacobian matrices $Df(x, y)$ and $Dg(u, v, w)$.
(b) Compute the composition $h(u, v, w) = f[g(u, v, w)]$.
(c) Compute the Jacobian matrix $Dh(1, -1, 1)$.

15. Let $f: \mathbf{R}^3 \to \mathbf{R}^2$ and $g: \mathbf{R}^3 \to \mathbf{R}^3$ be two vector fields defined as follows:

$$f(x, y, z) = (x^2 + y + z)\boldsymbol{i} + (2x + y + z^2)\boldsymbol{j},$$

$$g(u, v, w) = uv^2w^2\boldsymbol{i} + w^2 \sin v\,\boldsymbol{j} + u^2e^v\boldsymbol{k}.$$

(a) Compute each of the Jacobian matrices $Df(x, y, z)$ and $Dg(u, v, w)$.
(b) Compute the composition $h(u, v, w) = f[g(u, v, w)]$.
(c) Compute the Jacobian matrix $Dh(u, 0, w)$.

★8.23 Sufficient conditions for the equality of mixed partial derivatives

If f is a real-valued function of two variables, the two mixed partial derivatives $D_{1,2}f$ and $D_{2,1}f$ are not necessarily equal. By $D_{1,2}f$ we mean $D_1(D_2f) = \partial^2 f/(\partial x\, \partial y)$, and by $D_{2,1}f$ we mean $D_2(D_1f) = \partial^2 f/(\partial y\, \partial x)$. For example, if f is defined by the equations

$$f(x, y) = xy\frac{x^2 - y^2}{x^2 + y^2} \quad \text{for} \quad (x, y) \neq (0, 0), \qquad f(0, 0) = 0,$$

it is easy to prove that $D_{2,1}f(0, 0) = -1$ and $D_{1,2}f(0, 0) = 1$. This may be seen as follows: The definition of $D_{2,1}f(0, 0)$ states that

$$D_{2,1}f(0, 0) = \lim_{k\to 0}\frac{D_1f(0, k) - D_1f(0, 0)}{k}. \tag{8.30}$$

Now we have

$$D_1f(0, 0) = \lim_{h\to 0}\frac{f(h, 0) - f(0, 0)}{h} = 0$$

and, if $(x, y) \neq (0, 0)$, we find

$$D_1f(x, y) = \frac{y(x^4 + 4x^2y^2 - y^4)}{(x^2 + y^2)^2}.$$

Therefore, if $k \neq 0$ we have $D_1f(0, k) = -k^5/k^4 = -k$ and hence

$$\frac{D_1f(0, k) - D_1f(0, 0)}{k} = -1.$$

Using this in (8.30) we find that $D_{2,1}f(0, 0) = -1$. A similar argument shows that $D_{1,2}f(0, 0) = 1$, and hence $D_{2,1}f(0, 0) \neq D_{1,2}f(0, 0)$.

In the example just treated the two mixed partials $D_{1,2}f$ and $D_{2,1}f$ are not both continuous at the origin. It can be shown that the two mixed partials *are* equal at a point (a, b) if at least one of them is continuous in a neighborhood of the point. We shall prove first that they are equal if *both* are continuous. More precisely, we have the following theorem.

THEOREM 8.12. A SUFFICIENT CONDITION FOR EQUALITY OF MIXED PARTIAL DERIVATIVES. *Assume f is a scalar field such that the partial derivatives D_1f, D_2f, $D_{1,2}f$, and $D_{2,1}f$ exist on an open set S. If (a, b) is a point in S at which both $D_{1,2}f$ and $D_{2,1}f$ are continuous, we have*

$$D_{1,2}f(a, b) = D_{2,1}f(a, b). \tag{8.31}$$

Proof. Choose nonzero h and k such that the rectangle $R(h, k)$ with vertices (a, b), $(a + h, b)$, $(a + h, b + k)$, and $(a, b + k)$ lies in S. (An example is shown in Figure 8.9.)

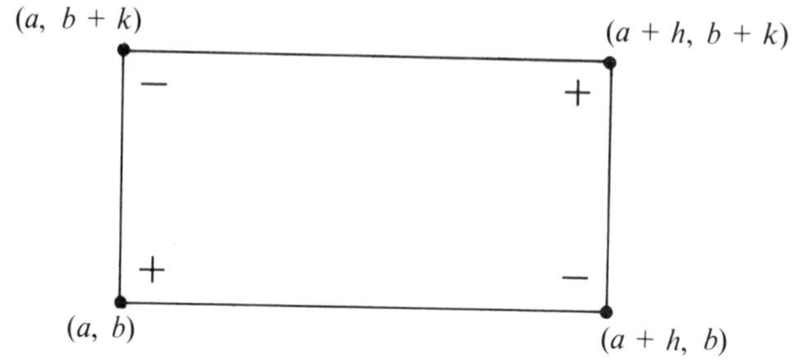

FIGURE 8.9 $\Delta(h, k)$ is a combination of values of f at the vertices.

Consider the expression

$$\Delta(h, k) = f(a + h, b + k) - f(a + h, b) - f(a, b + k) + f(a, b).$$

This is a combination of the values of f at the vertices of $R(h, k)$, taken with the algebraic signs indicated in Figure 8.9. We shall express $\Delta(h, k)$ in terms of $D_{2,1}f$ and also in terms of $D_{1,2}f$.

We consider a new function G of one variable defined by the equation

$$G(x) = f(x, b + k) - f(x, b)$$

for all x between a and $a + h$. (Geometrically, we are considering the values of f at those points at which an arbitrary vertical line cuts the horizontal edges of $R(h, k)$.) Then we have

$$\Delta(h, k) = G(a + h) - G(a). \tag{8.32}$$

Applying the one-dimensional mean-value theorem to the right-hand member of (8.32) we obtain $G(a + h) - G(a) = hG'(x_1)$, where x_1 lies between a and $a + h$. Since $G'(x) = D_1f(x, b + k) - D_1f(x, b)$, Equation (8.32) becomes

$$\Delta(h, k) = h[D_1f(x_1, b + k) - D_1f(x_1, b)]. \tag{8.33}$$

Applying the mean-value theorem to the right-hand member of (8.33) we obtain

$$\Delta(h, k) = hkD_{2,1}f(x_1, y_1), \tag{8.34}$$

where y_1 lies between b and $b + k$. The point (x_1, y_1) lies somewhere in the rectangle $R(h, k)$.

Applying the same procedure to the function $H(y) = f(a + h, y) - f(a, y)$ we find a second expression for $\Delta(h, k)$, namely,

$$\Delta(h, k) = hkD_{1,2}f(x_2, y_2), \tag{8.35}$$

where (x_2, y_2) also lies in $R(h, k)$. Equating the two expressions for $\Delta(h, k)$ and cancelling hk we obtain

$$D_{1,2}f(x_1, y_1) = D_{2,1}f(x_2, y_2).$$

Now we let $(h, k) \to (0, 0)$ and use the continuity of $D_{1,2}f$ and $D_{2,1}f$ to obtain (8.31).

The foregoing argument can be modified to prove a stronger version of Theorem 8.12.

THEOREM 8.13. *Let f be a scalar field such that the partial derivatives D_1f, D_2f, and $D_{2,1}f$ exist on an open set S containing (a, b). Assume further that $D_{2,1}f$ is continuous on S. Then the derivative $D_{1,2}f(a, b)$ exists and we have*

$$D_{1,2}f(a, b) = D_{2,1}f(a, b).$$

Proof. We define $\Delta(h, k)$ as in the proof of Theorem 8.12. The part of the proof leading to Equation (8.34) is still valid, giving us

$$\frac{\Delta(h, k)}{hk} = D_{2,1}f(x_1, y_1) \tag{8.36}$$

for some (x_1, y_1) in the rectangle $R(h, k)$. The rest of the proof is not applicable since it requires the existence of the derivative $D_{1,2}f(a, b)$, which we now wish to prove.

The definition of $D_{1,2}f(a, b)$ states that

$$D_{1,2}f(a, b) = \lim_{h \to 0} \frac{D_2f(a + h, b) - D_2f(a, b)}{h}. \tag{8.37}$$

We are to prove that this limit exists and has the value $D_{2,1}f(a, b)$. From the definition of $D_2 f$ we have

$$D_2 f(a, b) = \lim_{k \to 0} \frac{f(a, b + k) - f(a, b)}{k}$$

and

$$D_2 f(a + h, b) = \lim_{k \to 0} \frac{f(a + h, b + k) - f(a + h, b)}{k}.$$

Therefore the difference quotient in (8.37) can be written as

$$\frac{D_2 f(a + h, b) - D_2 f(a, b)}{h} = \lim_{k \to 0} \frac{\Delta(h, k)}{hk}.$$

Using (8.36) we can rewrite this as

(8.38)
$$\frac{D_2 f(a + h, b) - D_2 f(a, b)}{h} = \lim_{k \to 0} D_{2,1} f(x_1, y_1).$$

To complete the proof we must show that

(8.39)
$$\lim_{h \to 0} \left[\lim_{k \to 0} D_{2,1} f(x_1, y_1) \right] = D_{2,1} f(a, b).$$

When $k \to 0$, the point $y_1 \to b$, but the behavior of x_1 as a function of k is unknown. If we knew that x_1 approached some limit, say $\bar{x}$, as $k \to 0$, we could use the continuity of $D_{2,1}f$ to deduce that

$$\lim_{k \to 0} D_{2,1} f(x_1, y_1) = D_{2,1} f(\bar{x}, b).$$

Since the limit $\bar{x}$ would have to lie in the interval $a \leq \bar{x} \leq a + h$, we could then let $h \to 0$ and deduce (8.39). However, the fact that $\bar{x}$ depends on k in an unknown fashion makes a slightly more involved argument necessary.

Because of Equation (8.38) we know that the following limit exists:

$$\lim_{k \to 0} D_{2,1} f(x_1, y_1).$$

Let us denote this limit by $F(h)$. To complete the proof we must show that

$$\lim_{h \to 0} F(h) = D_{2,1} f(a, b).$$

For this purpose we appeal to the definition of continuity of $D_{2,1}f$ at (a, b).

Let ϵ be a given positive number. Continuity of $D_{2,1}f$ at (a, b) means that there is an open disk N with center (a, b) and radius δ, say, such that

(8.40)
$$|D_{2,1} f(x, y) - D_{2,1} f(a, b)| < \frac{\epsilon}{2} \quad \text{whenever} \quad (x, y) \in N.$$

If we choose h and k so that $|h| < \delta/2$ and $|k| < \delta/2$, the entire rectangle shown in Figure 8.9 will lie in the neighborhood N and, specifically, the point (x_1, y_1) will be in N. Therefore (8.40) is valid when $(x, y) = (x_1, y_1)$ and we can write

$$0 \le |D_{2,1}f(x_1, y_1) - D_{2,1}f(a, b)| < \frac{\epsilon}{2}. \tag{8.41}$$

Now keep h fixed and let $k \to 0$. The term $D_{2,1}f(x_1, y_1)$ approaches $F(h)$ and the other terms in (8.41) are independent of k. Therefore we have

$$0 \le |F(h) - D_{2,1}f(a, b)| \le \frac{\epsilon}{2} < \epsilon,$$

provided that $0 < |h| < \delta/2$. But this is precisely the meaning of the statement

$$\lim_{h\to 0} F(h) = D_{2,1}f(a, b)$$

and, as we have already remarked, this completes the proof.

Note: It should be observed that the theorem is also valid if the roles of the two derivatives $D_{1,2}f$ and $D_{2,1}f$ are interchanged.

8.24 Miscellaneous exercises

1. Find a scalar field f satisfying both the following conditions:
   (a) The partial derivatives $D_1f(0, 0)$ and $D_2f(0, 0)$ exist and are zero.
   (b) The directional derivative at the origin in the direction of the vector $\boldsymbol{i} + \boldsymbol{j}$ exists and has the value 3. Explain why such an f cannot be differentiable at (0, 0).
2. Let f be defined as follows:

$$f(x, y) = y\frac{x^2 - y^2}{x^2 + y^2} \quad \text{if} \quad (x, y) \ne (0, 0), \qquad f(0, 0) = 0.$$

   Compute the following partial derivatives, when they exist: $D_1f(0, 0)$, $D_2f(0, 0)$, $D_{2,1}f(0, 0)$, $D_{1,2}f(0, 0)$.
3. Let $f(x, y) = \dfrac{xy^3}{x^3 + y^6}$ if $(x, y) \ne (0, 0)$, and define $f(0, 0) = 0$.
   (a) Prove that the derivative $f'(\boldsymbol{O}; \boldsymbol{a})$ exists for every vector $\boldsymbol{a}$ and compute its value in terms of the components of $\boldsymbol{a}$.
   (b) Determine whether or not f is continuous at the origin.
4. Define $f(x, y) = \int_0^{\sqrt{xy}} e^{-t^2}\, dt$ for $x > 0, y > 0$. Compute $\partial f/\partial x$ in terms of x and y.
5. Assume that the equations $u = f(x, y)$, $x = X(t)$, $y = Y(t)$ define u as a function of t, say $u = F(t)$. Compute the third derivative $F'''(t)$ in terms of derivatives of f, X, and Y.
6. The change of variables $x = u + v$, $y = uv^2$ transforms $f(x, y)$ into $g(u, v)$. Compute the value of $\partial^2 g/(\partial v\, \partial u)$ at the point at which $u = 1, v = 1$, given that

$$\frac{\partial f}{\partial y} = \frac{\partial^2 f}{\partial x^2} = \frac{\partial^2 f}{\partial y^2} = \frac{\partial^2 f}{\partial x\, \partial y} = \frac{\partial^2 f}{\partial y\, \partial x} = 1$$

   at that point.

7. The change of variables $x = uv$, $y = \frac{1}{2}(u^2 - v^2)$ transforms $f(x, y)$ to $g(u, v)$.
(a) Calculate $\partial g/\partial u$, $\partial g/\partial v$, and $\partial^2 g/(\partial u\, \partial v)$ in terms of partial derivatives of f. (You may assume equality of mixed partials.)
(b) If $\|\nabla f(x, y)\|^2 = 2$ for all x and y, determine constants a and b such that

$$a\left(\frac{\partial g}{\partial u}\right)^2 - b\left(\frac{\partial g}{\partial v}\right)^2 = u^2 + v^2.$$

8. Two functions F and G of one variable and a function z of two variables are related by the equation

$$[F(x) + G(y)]^2 e^{z(x,y)} = 2F'(x)G'(y)$$

whenever $F(x) + G(y) \neq 0$. Show that the mixed partial derivative $D_{2,1}z(x, y)$ is never zero. (You may assume the existence and continuity of all derivatives encountered.)
9. A scalar field f is bounded and continuous on a rectangle $R = [a, b] \times [c, d]$. A new scalar field g is defined on R as follows:

$$g(u, v) = \int_c^v \left[\int_a^u f(x, y)\, dx\right] dy.$$

(a) It can be shown that for each fixed u in $[a, b]$ the function A defined on $[c, d]$ by the equation $A(y) = \int_a^u f(x, y)\, dx$ is continuous on $[c, d]$. Use this fact to prove that $\partial g/\partial v$ exists and is continuous on the open rectangle $S = (a, b) \times (c, d)$ (the interior of R).
(b) Assume that

$$\int_c^v \left[\int_a^u f(x, y)\, dx\right] dy = \int_a^u \left[\int_c^v f(x, y)\, dy\right] dx$$

for all (u, v) in R. Prove that g is differentiable on S and that the mixed partial derivatives $D_{1,2}g(u, v)$ and $D_{2,1}g(u, v)$ exist and are equal to $f(u, v)$ at each point of S.
10. Refer to Exercise 9. Suppose u and v are expressed parametrically as follows: $u = A(t)$, $v = B(t)$; and let $\varphi(t) = g[A(t), B(t)]$.
(a) Determine $\varphi'(t)$ in terms of f, A', and B'.
(b) Compute $\varphi'(t)$ in terms of t when $f(x, y) = e^{x+y}$ and $A(t) = B(t) = t^2$. (Assume R lies in the first quadrant.)
11. If $f(x, y, z) = (\boldsymbol{r} \times \boldsymbol{A}) \cdot (\boldsymbol{r} \times \boldsymbol{B})$, where $\boldsymbol{r} = x\boldsymbol{i} + y\boldsymbol{j} + z\boldsymbol{k}$ and $\boldsymbol{A}$ and $\boldsymbol{B}$ are constant vectors, show that $\nabla f(x, y, z) = \boldsymbol{B} \times (\boldsymbol{r} \times \boldsymbol{A}) + \boldsymbol{A} \times (\boldsymbol{r} \times \boldsymbol{B})$.
12. Let $\boldsymbol{r} = x\boldsymbol{i} + y\boldsymbol{j} + z\boldsymbol{k}$ and let $r = \|\boldsymbol{r}\|$. If $\boldsymbol{A}$ and $\boldsymbol{B}$ are constant vectors, show that:

(a) $\boldsymbol{A} \cdot \nabla\left(\dfrac{1}{r}\right) = -\dfrac{\boldsymbol{A} \cdot \boldsymbol{r}}{r^3}$.

(b) $\boldsymbol{B} \cdot \nabla\left(\boldsymbol{A} \cdot \nabla\left(\dfrac{1}{r}\right)\right) = \dfrac{3\boldsymbol{A} \cdot \boldsymbol{r}\,\boldsymbol{B} \cdot \boldsymbol{r}}{r^5} - \dfrac{\boldsymbol{A} \cdot \boldsymbol{B}}{r^3}$.

13. Find the set of all points (a, b, c) in 3-space for which the two spheres $(x - a)^2 + (y - b)^2 + (z - c)^2 = 1$ and $x^2 + y^2 + z^2 = 1$ intersect orthogonally. (Their tangent planes should be perpendicular at each point of intersection.)
14. A cylinder whose equation is $y = f(x)$ is tangent to the surface $z^2 + 2xz + y = 0$ at all points common to the two surfaces. Find $f(x)$.

9

APPLICATIONS OF DIFFERENTIAL CALCULUS

9.1 Partial differential equations

The theorems of differential calculus developed in Chapter 8 have a wide variety of applications. This chapter illustrates their use in some examples related to partial differential equations, implicit functions, and extremum problems. We begin with some elementary remarks concerning partial differential equations.

An equation involving a scalar field f and its partial derivatives is called a *partial differential equation*. Two simple examples in which f is a function of two variables are the first-order equation

$$\frac{\partial f(x, y)}{\partial x} = 0, \tag{9.1}$$

and the second-order equation

$$\frac{\partial^2 f(x, y)}{\partial x^2} + \frac{\partial^2 f(x, y)}{\partial y^2} = 0. \tag{9.2}$$

Each of these is a homogeneous *linear* partial differential equation. That is, each has the form $L(f) = 0$, where L is a linear differential operator involving one or more partial derivatives. Equation (9.2) is called the two-dimensional *Laplace equation*.

Some of the theory of linear ordinary differential equations can be extended to partial differential equations. For example, it is easy to verify that for each of Equations (9.1) and (9.2) the set of solutions is a linear space. However, there is an important difference between ordinary and partial linear differential equations that should be realized at the outset. We illustrate this difference by comparing the partial differential equation (9.1) with the ordinary differential equation

$$f'(x) = 0. \tag{9.3}$$

The most general function satisfying (9.3) is $f(x) = C$, where C is an arbitrary constant. In other words, the solution-space of (9.3) is one-dimensional. But the most general function satisfying (9.1) is

$$f(x, y) = g(y),$$

where g is any function of y. Since g is arbitrary we can easily obtain an infinite set of independent solutions. For example, we can take $g(y) = e^{cy}$ and let c vary over all real numbers. Thus, the solution-space of (9.1) is *infinite*-dimensional.

In some respects this example is typical of what happens in general. Somewhere in the process of solving a first-order partial differential equation, an integration is required to remove each partial derivative. At this step an arbitrary function is introduced in the solution. This results in an infinite-dimensional solution space.

In many problems involving partial differential equations it is necessary to select from the wealth of solutions a particular solution satisfying one or more auxiliary conditions. As might be expected, the nature of these conditions has a profound effect on the existence or uniqueness of solutions. A systematic study of such problems will not be attempted in this book. Instead, we will treat some special cases to illustrate the ideas introduced in Chapter 8.

9.2 A first-order partial differential equation with constant coefficients

Consider the first-order partial differential equation

$$3\frac{\partial f(x, y)}{\partial x} + 2\frac{\partial f(x, y)}{\partial y} = 0. \tag{9.4}$$

All the solutions of this equation can be found by geometric considerations. We express the left member as a dot product, and write the equation in the form

$$(3\boldsymbol{i} + 2\boldsymbol{j}) \cdot \nabla f(x, y) = 0.$$

This tells us that the gradient vector $\nabla f(x, y)$ is orthogonal to the vector $3\boldsymbol{i} + 2\boldsymbol{j}$ at each point (x, y). But we also know that $\nabla f(x, y)$ is orthogonal to the level curves of f. Hence these level curves must be straight lines parallel to $3\boldsymbol{i} + 2\boldsymbol{j}$. In other words, the level curves of f are the lines

$$2x - 3y = c.$$

Therefore $f(x, y)$ is constant when $2x - 3y$ is constant. This suggests that

$$f(x, y) = g(2x - 3y) \tag{9.5}$$

for some function g.

Now we verify that, for each differentiable function g, the scalar field f defined by (9.5) does, indeed, satisfy (9.4). Using the chain rule to compute the partial derivatives of f we find

$$\frac{\partial f}{\partial x} = 2g'(2x - 3y), \qquad \frac{\partial f}{\partial y} = -3g'(2x - 3y),$$

$$3\frac{\partial f}{\partial x} + 2\frac{\partial f}{\partial y} = 6g'(2x - 3y) - 6g'(2x - 3y) = 0.$$

Therefore, f satisfies (9.4).

Conversely, we can show that every differentiable f which satisfies (9.4) must necessarily have the form (9.5) for some g. To do this, we introduce a linear change of variables,

(9.6) $$x = Au + Bv, \qquad y = Cu + Dv.$$

This transforms $f(x, y)$ into a function of u and v, say

$$h(u, v) = f(Au + Bv, Cu + Dv).$$

We shall choose the constants A, B, C, D so that h satisfies the simpler equation

(9.7) $$\frac{\partial h(u, v)}{\partial u} = 0.$$

Then we shall solve this equation and show that f has the required form.

Using the chain rule we find

$$\frac{\partial h}{\partial u} = \frac{\partial f}{\partial x}\frac{\partial x}{\partial u} + \frac{\partial f}{\partial y}\frac{\partial y}{\partial u} = \frac{\partial f}{\partial x} A + \frac{\partial f}{\partial y} C.$$

Since f satisfies (9.4) we have $\partial f/\partial y = -(3/2)(\partial f/\partial x)$, so the equation for $\partial h/\partial u$ becomes

$$\frac{\partial h}{\partial u} = \frac{\partial f}{\partial x}\left(A - \frac{3}{2}C\right).$$

Therefore, h will satisfy (9.7) if we choose $A = \frac{3}{2}C$. Taking $A = 3$ and $C = 2$ we find

(9.8) $$x = 3u + Bv, \qquad y = 2u + Dv.$$

For this choice of A and C, the function h satisfies (9.7), so $h(u, v)$ is a function of v alone, say

$$h(u, v) = g(v)$$

for some function g. To express v in terms of x and y we eliminate u from (9.8) and obtain $2x - 3y = (2B - 3D)v$. Now we choose B and D to make $2B - 3D = 1$, say $B = 2$, $D = 1$. For this choice the transformation (9.6) is nonsingular; we have $v = 2x - 3y$, and hence

$$f(x, y) = h(u, v) = g(v) = g(2x - 3y).$$

This shows that every differentiable solution f of (9.4) has the form (9.5).

Exactly the same type of argument proves the following theorem for first-order equations with constant coefficients.

THEOREM 9.1. *Let g be differentiable on $\mathbf{R}^1$, and let f be the scalar field defined on $\mathbf{R}^2$ by the equation*

(9.9) $$f(x, y) = g(bx - ay),$$

where a and b are constants, not both zero. Then f satisfies the first-order partial differential equation

$$a\frac{\partial f(x, y)}{\partial x} + b\frac{\partial f(x, y)}{\partial y} = 0 \tag{9.10}$$

everywhere in $\mathbf{R}^2$. *Conversely, every differentiable solution of* (9.10) *necessarily has the form* (9.9) *for some g.*

9.3 Exercises

In this set of exercises you may assume differentiability of all functions under consideration.

1. Determine that solution of the partial differential equation

$$4\frac{\partial f(x, y)}{\partial x} + 3\frac{\partial f(x, y)}{\partial y} = 0$$

which satisfies the condition $f(x, 0) = \sin x$ for all x.

2. Determine that solution of the partial differential equation

$$5\frac{\partial f(x, y)}{\partial x} - 2\frac{\partial f(x, y)}{\partial y} = 0$$

which satisfies the conditions $f(0, 0) = 0$ and $D_1 f(x, 0) = e^x$ for all x.

3. (a) If $u(x, y) = f(xy)$, prove that u satisfies the partial differential equation

$$x\frac{\partial u}{\partial x} - y\frac{\partial u}{\partial y} = 0.$$

Find a solution such that $u(x, x) = x^4 e^{x^2}$ for all x.

(b) If $v(x, y) = f(x/y)$ for $y \neq 0$, prove that v satisfies the partial-differential equation

$$x\frac{\partial v}{\partial x} + y\frac{\partial v}{\partial y} = 0.$$

Find a solution such that $v(1, 1) = 2$ and $D_1 v(x, 1/x) = 1/x$ for all $x \neq 0$.

4. If $g(u, v)$ satisfies the partial differential equation

$$\frac{\partial^2 g(u, v)}{\partial u\, \partial v} = 0,$$

prove that $g(u, v) = \varphi_1(u) + \varphi_2(v)$, where $\varphi_1(u)$ is a function of u alone and $\varphi_2(v)$ is a function of v alone.

5. Assume f satisfies the partial differential equation

$$\frac{\partial^2 f}{\partial x^2} - 2\frac{\partial^2 f}{\partial x\, \partial y} - 3\frac{\partial^2 f}{\partial y^2} = 0.$$

Introduce the linear change of variables, $x = Au + Bv$, $y = Cu + Dv$, where A, B, C, D are constant, and let $g(u, v) = f(Au + Bv, Cu + Dv)$. Compute nonzero integer values of

A, B, C, D such that g satisfies $\partial^2 g/(\partial u\, \partial v) = 0$. Solve this equation for g and thereby determine f. (Assume equality of the mixed partials.)

6. A function u is defined by an equation of the form

$$u(x, y) = xy f\left(\frac{x + y}{xy}\right).$$

Show that u satisfies a partial differential equation of the form

$$x^2 \frac{\partial u}{\partial x} - y^2 \frac{\partial u}{\partial y} = G(x, y)u,$$

and find $G(x, y)$.

7. The substitution $x = e^s$, $y = e^t$ converts $f(x, y)$ into $g(s, t)$, where $g(s, t) = f(e^s, e^t)$. If f is known to satisfy the partial differential equation

$$x^2 \frac{\partial^2 f}{\partial x^2} + y^2 \frac{\partial^2 f}{\partial y^2} + x \frac{\partial f}{\partial x} + y \frac{\partial f}{\partial y} = 0,$$

show that g satisfies the partial-differential equation

$$\frac{\partial^2 g}{\partial s^2} + \frac{\partial^2 g}{\partial t^2} = 0.$$

8. Let f be a scalar field that is differentiable on an open set S in $\mathbf{R}^n$. We say that f is *homogeneous of degree p* over S if

$$f(t\boldsymbol{x}) = t^p f(\boldsymbol{x})$$

for every $t > 0$ and every $\boldsymbol{x}$ in S for which $t\boldsymbol{x} \in S$. For a homogeneous scalar field of degree p show that we have

$$\boldsymbol{x} \cdot \nabla f(\boldsymbol{x}) = p f(\boldsymbol{x}) \qquad \text{for each } \boldsymbol{x} \text{ in } S.$$

This is known as *Euler's theorem for homogeneous functions.* If $\boldsymbol{x} = (x_1, \ldots, x_n)$ it can be expressed as

$$x_1 \frac{\partial f}{\partial x_1} + \cdots + x_n \frac{\partial f}{\partial x_n} = p f(x_1, \ldots, x_n).$$

[*Hint:* For fixed $\boldsymbol{x}$, define $g(t) = f(t\boldsymbol{x})$ and compute $g'(1)$.]

9. Prove the converse of Euler's theorem. That is, if f satisfies $\boldsymbol{x} \cdot \nabla f(\boldsymbol{x}) = p f(\boldsymbol{x})$ for all $\boldsymbol{x}$ in an open set S, then f must be homogeneous of degree p over S. [*Hint:* For fixed $\boldsymbol{x}$, define $g(t) = f(t\boldsymbol{x}) - t^p f(\boldsymbol{x})$ and compute $g'(t)$.]
10. Prove the following extension of Euler's theorem for homogeneous functions of degree p in the 2-dimensional case. (Assume equality of the mixed partials.)

$$x^2 \frac{\partial^2 f}{\partial x^2} + 2xy \frac{\partial^2 f}{\partial x\, \partial y} + y^2 \frac{\partial^2 f}{\partial y^2} = p(p - 1)f.$$

9.4 The one-dimensional wave equation

Imagine a string of infinite length stretched along the x-axis and allowed to vibrate in the xy-plane. We denote by $y = f(x, t)$ the vertical displacement of the string at the point x at time t. We assume that, at time $t = 0$, the string is displaced along a prescribed curve, $y = F(x)$. An example is shown in Figure 9.1(a). Figures 9.1(b) and (c) show possible displacement curves for later values of t. We regard the displacement $f(x, t)$ as an unknown function of x and t to be determined. A mathematical model for this problem (suggested by physical considerations which we shall not discuss here) is the partial differential equation

$$\frac{\partial^2 f}{\partial t^2} = c^2 \frac{\partial^2 f}{\partial x^2},$$

where c is a positive constant depending on the physical characteristics of the string. This equation is called the *one-dimensional wave equation.* We will solve this equation subject to certain auxiliary conditions.

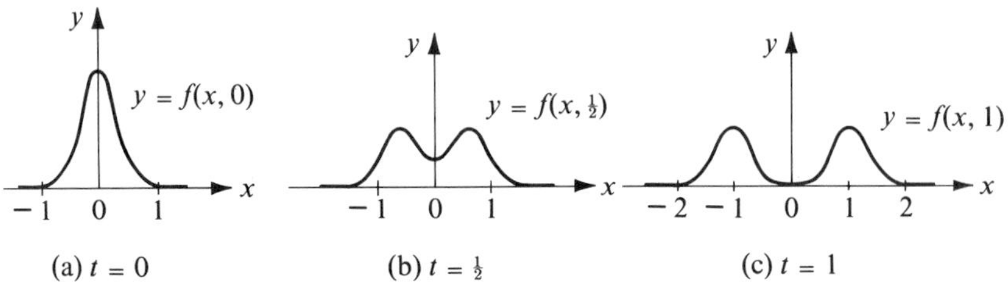

FIGURE 9.1 The displacement curve $y = f(x, t)$ shown for various values of t.

Since the initial displacement is the prescribed curve $y = F(x)$, we seek a solution satisfying the condition

$$f(x, 0) = F(x).$$

We also assume that $\partial y/\partial t$, the velocity of the vertical displacement, is prescribed at time $t = 0$, say

$$D_2 f(x, 0) = G(x),$$

where G is a given function. It seems reasonable to expect that this information should suffice to determine the subsequent motion of the string. We will show that, indeed, this is true by determining the function f in terms of F and G. The solution is expressed in a form given by Jean d'Alembert (1717–1783), a French mathematician and philosopher.

THEOREM 9.2. D'ALEMBERT'S SOLUTION OF THE WAVE EQUATION. *Let F and G be given functions such that G is differentiable and F is twice differentiable on $\mathbf{R}^1$. Then the function f given by the formula*

$$f(x, t) = \frac{F(x + ct) + F(x - ct)}{2} + \frac{1}{2c}\int_{x-ct}^{x+ct} G(s)\, ds \tag{9.11}$$

satisfies the one-dimensional wave equation

$$\frac{\partial^2 f}{\partial t^2} = c^2 \frac{\partial^2 f}{\partial x^2} \tag{9.12}$$

and the initial conditions

$$f(x, 0) = F(x), \qquad D_2 f(x, 0) = G(x). \tag{9.13}$$

Conversely, any function f with equal mixed partials which satisfies (9.12) *and* (9.13) *necessarily has the form* (9.11).

Proof. It is a straightforward exercise to verify that the function f given by (9.11) satisfies the wave equation and the given initial conditions. This verification is left to the reader. We shall prove the converse.

One way to proceed is to assume that f is a solution of the wave equation, introduce a linear change of variables,

$$x = Au + Bv, \qquad t = Cu + Dv,$$

which transforms $f(x, t)$ into a function of u and v, say

$$g(u, v) = f(Au + Bv, Cu + Dv),$$

and choose the constants A, B, C, D so that g satisfies the simpler equation

$$\frac{\partial^2 g}{\partial u\, \partial v} = 0.$$

Solving this equation for g we find that $g(u, v) = \varphi_1(u) + \varphi_2(v)$, where $\varphi_1(u)$ is a function of u alone and $\varphi_2(v)$ is a function of v alone. The constants A, B, C, D can be chosen so that $u = x + ct$, $v = x - ct$, from which we obtain

$$f(x, t) = \varphi_1(x + ct) + \varphi_2(x - ct). \tag{9.14}$$

Then we use the initial conditions (9.13) to determine the functions φ_1 and φ_2 in terms of the given functions F and G.

We will obtain (9.14) by another method which makes use of Theorem 9.1 and avoids the change of variables. First we rewrite the wave equation in the form

$$L_1(L_2 f) = 0, \tag{9.15}$$

where L_1 and L_2 are the first-order linear differential operators given by

$$L_1 = \frac{\partial}{\partial t} - c\frac{\partial}{\partial x}, \qquad L_2 = \frac{\partial}{\partial t} + c\frac{\partial}{\partial x}.$$

Let f be a solution of (9.15) and let

$$u(x, t) = L_2 f(x, t).$$

Equation (9.15) states that u satisfies the first-order equation $L_1(u) = 0$. Hence, by Theorem 9.1 we have

$$u(x, t) = \varphi(x + ct)$$

for some function φ. Let Φ be any primitive of φ, say $\Phi(y) = \int_0^y \varphi(s)\, ds$, and let

$$v(x, t) = \frac{1}{2c}\Phi(x + ct).$$

We will show that $L_2(v) = L_2(f)$. We have

$$\frac{\partial v}{\partial x} = \frac{1}{2c}\Phi'(x + ct) \qquad \text{and} \qquad \frac{\partial v}{\partial t} = \frac{1}{2}\Phi'(x + ct),$$

so

$$L_2 v = \frac{\partial v}{\partial t} + c\frac{\partial v}{\partial x} = \Phi'(x + ct) = \varphi(x + ct) = u(x, t) = L_2 f.$$

In other words, the difference $f - v$ satisfies the first-order equation

$$L_2(f - v) = 0.$$

By Theorem 9.1 we must have $f(x, t) - v(x, t) = \psi(x - ct)$ for some function ψ. Therefore

$$f(x, t) = v(x, t) + \psi(x - ct) = \frac{1}{2c}\Phi(x + ct) + \psi(x - ct).$$

This proves (9.14) with $\varphi_1 = \dfrac{1}{2c}\Phi$ and $\varphi_2 = \psi$.

Now we use the initial conditions (9.13) to determine the functions φ_1 and φ_2 in terms of the given functions F and G. The relation $f(x, 0) = F(x)$ implies

(9.16) $$\varphi_1(x) + \varphi_2(x) = F(x).$$

The other initial condition, $D_2 f(x, 0) = G(x)$, implies

(9.17) $$c\varphi_1'(x) - c\varphi_2'(x) = G(x).$$

Differentiating (9.16) we obtain

(9.18) $$\varphi_1'(x) + \varphi_2'(x) = F'(x).$$

Solving (9.17) and (9.18) for $\varphi_1'(x)$ and $\varphi_2'(x)$ we find

$$\varphi_1'(x) = \frac{1}{2}F'(x) + \frac{1}{2c}G(x), \qquad \varphi_2'(x) = \frac{1}{2}F'(x) - \frac{1}{2c}G(x).$$

Integrating these relations we get

$$\varphi_1(x) - \varphi_1(0) = \frac{F(x) - F(0)}{2} + \frac{1}{2c}\int_0^x G(s)\,ds\,,$$

$$\varphi_2(x) - \varphi_2(0) = \frac{F(x) - F(0)}{2} - \frac{1}{2c}\int_0^x G(s)\,ds\,.$$

In the first equation we replace x by $x + ct$; in the second equation we replace x by $x - ct$. Then we add the two resulting equations and use the fact that $\varphi_1(0) + \varphi_2(0) = F(0)$ to obtain

$$f(x, t) = \varphi_1(x + ct) + \varphi_2(x - ct) = \frac{F(x + ct) + F(x - ct)}{2} + \frac{1}{2c}\int_{x-ct}^{x+ct} G(s)\,ds\,.$$

This completes the proof.

EXAMPLE. Assume the initial displacement is given by the formula

$$F(x) = \begin{cases} 1 + \cos \pi x & \text{for } -1 \le x \le 1\,, \\ 0 & \text{for } |x| \ge 1\,. \end{cases}$$

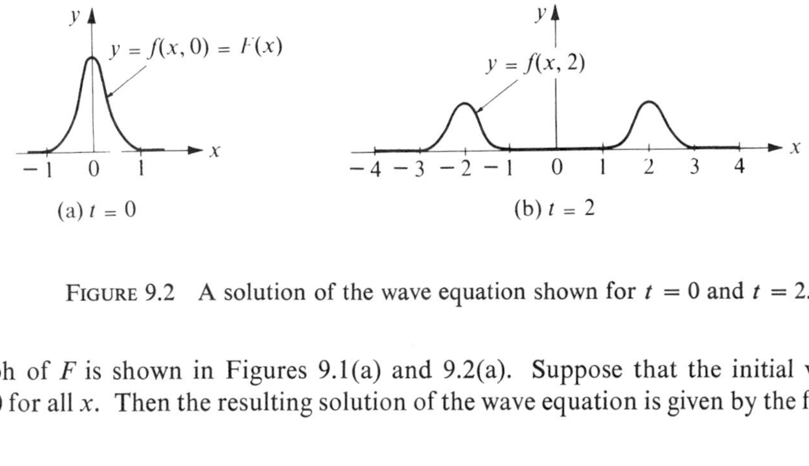

FIGURE 9.2 A solution of the wave equation shown for $t = 0$ and $t = 2$.

The graph of F is shown in Figures 9.1(a) and 9.2(a). Suppose that the initial velocity $G(x) = 0$ for all x. Then the resulting solution of the wave equation is given by the formula

$$f(x, t) = \frac{F(x + ct) + F(x - ct)}{2}\,.$$

Figures 9.1 and 9.2 show the curve $y = f(x, t)$ for various values of t. The figures illustrate that the solution of the wave equation is a combination of two standing waves, one traveling to the right, the other to the left, each with speed c.

Further examples illustrating the use of the chain rule in the study of partial differential equations are given in the next set of exercises.

9.5 Exercises

In this set of exercises you may assume differentiability of all functions under consideration.

1. If k is a positive constant and $g(x, t) = \frac{1}{2}x/\sqrt{kt}$, let

$$f(x, t) = \int_0^{g(x,t)} e^{-u^2}\, du .$$

(a) Show that $\dfrac{\partial f}{\partial x} = e^{-g^2}\dfrac{\partial g}{\partial x}$ and $\dfrac{\partial f}{\partial t} = e^{-g^2}\dfrac{\partial g}{\partial t}$.

(b) Show that f satisfies the partial differential equation

$$k\frac{\partial^2 f}{\partial x^2} = \frac{\partial f}{\partial t} \quad \text{(the \textit{heat equation})}.$$

2. Consider a scalar field f defined in $\mathbf{R}^2$ such that $f(x, y)$ depends only on the distance r of (x, y) from the origin, say $f(x, y) = g(r)$, where $r = (x^2 + y^2)^{1/2}$.

(a) Prove that for $(x, y) \neq (0, 0)$ we have

$$\frac{\partial^2 f}{\partial x^2} + \frac{\partial^2 f}{\partial y^2} = \frac{1}{r}g'(r) + g''(r).$$

(b) Now assume further that f satisfies *Laplace's equation*,

$$\frac{\partial^2 f}{\partial x^2} + \frac{\partial^2 f}{\partial y^2} = 0,$$

for all $(x, y) \neq (0, 0)$. Use part (a) to prove that $f(x, y) = a \log (x^2 + y^2) + b$ for $(x, y) \neq (0, 0)$, where a and b are constants.

3. Repeat Exercise 2 for the n-dimensional case, where $n \geq 3$. That is, assume that $f(\boldsymbol{x}) = f(x_1, \ldots, x_n) = g(r)$, where $r = \|\boldsymbol{x}\|$. Show that

$$\frac{\partial^2 f}{\partial x_1^2} + \cdots + \frac{\partial^2 f}{\partial x_n^2} = \frac{n-1}{r}g'(r) + g''(r)$$

for $\boldsymbol{x} \neq \boldsymbol{O}$. If f satisfies the *n-dimensional Laplace equation*,

$$\frac{\partial^2 f}{\partial x_1^2} + \cdots + \frac{\partial^2 f}{\partial x_n^2} = 0,$$

for all $\boldsymbol{x} \neq \boldsymbol{O}$, deduce that $f(\boldsymbol{x}) = a\,\|\boldsymbol{x}\|^{2-n} + b$ for $\boldsymbol{x} \neq \boldsymbol{O}$, where a, b are constants.

Note: The linear operator ∇^2 defined by the equation

$$\nabla^2 f = \frac{\partial^2 f}{\partial x_1^2} + \cdots + \frac{\partial^2 f}{\partial x_n^2}$$

is called the *n-dimensional Laplacian.*

4. *Two-dimensional Laplacian in polar coordinates.* The introduction of polar coordinates $x = r\cos\theta$, $y = r\sin\theta$, converts $f(x, y)$ into $g(r, \theta)$. Verify the following formulas:

(a) $$\|\nabla f(r\cos\theta, r\sin\theta)\|^2 = \left(\frac{\partial g}{\partial r}\right)^2 + \frac{1}{r^2}\left(\frac{\partial g}{\partial \theta}\right)^2.$$

(b) $$\frac{\partial^2 f}{\partial x^2} + \frac{\partial^2 f}{\partial y^2} = \frac{\partial^2 g}{\partial r^2} + \frac{1}{r^2}\frac{\partial^2 g}{\partial \theta^2} + \frac{1}{r}\frac{\partial g}{\partial r}.$$

5. *Three-dimensional Laplacian in spherical coordinates.* The introduction of spherical coordinates

$$x = \rho\cos\theta\sin\varphi, \qquad y = \rho\sin\theta\sin\varphi, \qquad z = \rho\cos\varphi,$$

transforms $f(x, y, z)$ to $F(\rho, \theta, \varphi)$. This exercise shows how to express the Laplacian $\nabla^2 f$ in terms of partial derivatives of F.

(a) First introduce polar coordinates $x = r\cos\theta$, $y = r\sin\theta$ to transform $f(x, y, z)$ to $g(r, \theta, z)$. Use Exercise 4 to show that

$$\nabla^2 f = \frac{\partial^2 g}{\partial r^2} + \frac{1}{r^2}\frac{\partial^2 g}{\partial \theta^2} + \frac{1}{r}\frac{\partial g}{\partial r} + \frac{\partial^2 g}{\partial z^2}.$$

(b) Now transform $g(r, \theta, z)$ to $F(\rho, \theta, \varphi)$ by taking $z = \rho\cos\varphi$, $r = \rho\sin\varphi$. Note that, except for a change in notation, this transformation is the same as that used in part (a). Deduce that

$$\nabla^2 f = \frac{\partial^2 F}{\partial \rho^2} + \frac{2}{\rho}\frac{\partial F}{\partial \rho} + \frac{1}{\rho^2}\frac{\partial^2 F}{\partial \varphi^2} + \frac{\cos\varphi}{\rho^2\sin\varphi}\frac{\partial F}{\partial \varphi} + \frac{1}{\rho^2\sin\varphi}\frac{\partial^2 F}{\partial \theta^2}.$$

6. This exercise shows how Legendre's differential equation arises when we seek solutions of Laplace's equation having a special form. Let f be a scalar field satisfying the three-dimensional Laplace equation, $\nabla^2 f = 0$. Introduce spherical coordinates as in Exercise 5 and let $F(\rho, \theta, \varphi) = f(x, y, z)$.

(a) Suppose we seek solutions f of Laplace's equation such that $F(\rho, \theta, \varphi)$ is independent of θ and has the special form $F(\rho, \theta, \varphi) = \rho^n G(\varphi)$. Show that f satisfies Laplace's equation if G satisfies the second-order equation

$$\frac{d^2 G}{d\varphi^2} + \cot\varphi\frac{dG}{d\varphi} + n(n + 1)G = 0.$$

(b) The change of variable $x = \cos\varphi$ ($\varphi = \arccos x$, $-1 \le x \le 1$) transforms $G(\varphi)$ to $g(x)$. Show that g satisfies the Legendre equation

$$(1 - x^2)\frac{d^2 g}{dx^2} - 2x\frac{dg}{dx} + n(n + 1)g = 0.$$

7. *Two-dimensional wave equation.* A thin flexible membrane is stretched over the xy-plane and allowed to vibrate. Let $z = f(x, y, t)$ denote the vertical displacement of the membrane at the point (x, y) at time t. Physical considerations suggest that f satisfies the two-dimensional wave equation,

$$\frac{\partial^2 f}{\partial t^2} = c^2\left(\frac{\partial^2 f}{\partial x^2} + \frac{\partial^2 f}{\partial y^2}\right),$$

where c is a positive constant depending on the physical characteristics of the membrane. This exercise reveals a connection between this equation and Bessel's differential equation.
(a) Introduce polar coordinates $x = r\cos\theta$, $y = r\sin\theta$, and let $F(r, \theta, t) = f(r\cos\theta, r\sin\theta, t)$. If f satisfies the wave equation show that F satisfies the equation

$$\frac{\partial^2 F}{\partial t^2} = c^2\left(\frac{\partial^2 F}{\partial r^2} + \frac{1}{r^2}\frac{\partial^2 F}{\partial \theta^2} + \frac{1}{r}\frac{\partial F}{\partial r}\right).$$

(b) If $F(r, \theta, t)$ is independent of θ, say $F(r, \theta, t) = \varphi(r, t)$ the equation in (a) simplifies to

$$\frac{\partial^2 \varphi}{\partial t^2} = c^2\left(\frac{\partial^2 \varphi}{\partial r^2} + \frac{1}{r}\frac{\partial \varphi}{\partial r}\right).$$

Now let φ be a solution such that $\varphi(r, t)$ factors into a function of r times a function of t, say $\varphi(r, t) = R(r)T(t)$. Show that each of the functions R and T satisfies an ordinary linear differential equation of second order.
(c) If the function T in part (b) is periodic with period $2\pi/c$, show that R satisfies the Bessel equation $r^2R'' + rR' + r^2R = 0$.

9.6 Derivatives of functions defined implicitly

Some surfaces in 3-space are described by Cartesian equations of the form

$$F(x, y, z) = 0.$$

An equation like this is said to provide an *implicit representation* of the surface. For example, the equation $x^2 + y^2 + z^2 - 1 = 0$ represents the surface of a unit sphere with center at the origin. Sometimes it is possible to solve the equation $F(x, y, z) = 0$ for one of the variables in terms of the other two, say for z in terms of x and y. This leads to one or more equations of the form

$$z = f(x, y).$$

For the sphere we have two solutions,

$$z = \sqrt{1 - x^2 - y^2} \quad \text{and} \quad z = -\sqrt{1 - x^2 - y^2},$$

one representing the upper hemisphere, the other the lower hemisphere.

In the general case it may not be an easy matter to obtain an explicit formula for z in terms of x and y. For example, there is no easy method for solving for z in the equation $y^2 + xz + z^2 - e^z - 4 = 0$. Nevertheless, a judicious use of the chain rule makes it possible to deduce various properties of the partial derivatives $\partial f/\partial x$ and $\partial f/\partial y$ without an explicit knowledge of $f(x, y)$. The procedure is described in this section.

We assume that there is a function $f(x, y)$ such that

$$F[x, y, f(x, y)] = 0 \tag{9.19}$$

for all (x, y) in some open set S, although we may not have explicit formulas for calculating $f(x, y)$. We describe this by saying that the equation $F(x, y, z) = 0$ defines z *implicitly* as a function of x and y, and we write

$$z = f(x, y).$$

Now we introduce an auxiliary function g defined on S as follows:

$$g(x, y) = F[x, y, f(x, y)].$$

Equation (9.19) states that $g(x, y) = 0$ on S; hence the partial derivatives $\partial g/\partial x$ and $\partial g/\partial y$ are also 0 on S. But we can also compute these partial derivatives by the chain rule. To do this we write

$$g(x, y) = F[u_1(x, y), u_2(x, y), u_3(x, y)],$$

where $u_1(x, y) = x$, $u_2(x, y) = y$, and $u_3(x, y) = f(x, y)$. The chain rule gives us the formulas

$$\frac{\partial g}{\partial x} = D_1F\,\frac{\partial u_1}{\partial x} + D_2F\,\frac{\partial u_2}{\partial x} + D_3F\,\frac{\partial u_3}{\partial x} \quad \text{and} \quad \frac{\partial g}{\partial y} = D_1F\,\frac{\partial u_1}{\partial y} + D_2F\,\frac{\partial u_2}{\partial y} + D_3F\,\frac{\partial u_3}{\partial y},$$

where each partial derivative D_kF is to be evaluated at $(x, y, f(x, y))$. Since we have

$$\frac{\partial u_1}{\partial x} = 1, \quad \frac{\partial u_2}{\partial x} = 0, \quad \frac{\partial u_3}{\partial x} = \frac{\partial f}{\partial x}, \quad \text{and} \quad \frac{\partial g}{\partial x} = 0,$$

the first of the foregoing equations becomes

$$D_1F + D_3F\,\frac{\partial f}{\partial x} = 0.$$

Solving this for $\partial f/\partial x$ we obtain

$$(9.20) \qquad \frac{\partial f}{\partial x} = -\frac{D_1F[x, y, f(x, y)]}{D_3F[x, y, f(x, y)]}$$

at those points at which $D_3F[x, y, f(x, y)] \neq 0$. By a similar argument we obtain a corresponding formula for $\partial f/\partial y$:

$$(9.21) \qquad \frac{\partial f}{\partial y} = -\frac{D_2F[x, y, f(x, y)]}{D_3F[x, y, f(x, y)]}$$

at those points at which $D_3F[x, y, f(x, y)] \neq 0$. These formulas are usually written more briefly as follows:

$$\frac{\partial f}{\partial x} = -\frac{\partial F/\partial x}{\partial F/\partial z}, \qquad \frac{\partial f}{\partial y} = -\frac{\partial F/\partial y}{\partial F/\partial z}.$$

EXAMPLE. Assume that the equation $y^2 + xz + z^2 - e^z - c = 0$ defines z as a function of x and y, say $z = f(x, y)$. Find a value of the constant c such that $f(0, e) = 2$, and compute the partial derivatives $\partial f/\partial x$ and $\partial f/\partial y$ at the point $(x, y) = (0, e)$.

Solution. When $x = 0$, $y = e$, and $z = 2$, the equation becomes $e^2 + 4 - e^2 - c = 0$, and this is satisfied by $c = 4$. Let $F(x, y, z) = y^2 + xz + z^2 - e^z - 4$. From (9.20) and (9.21) we have

$$\frac{\partial f}{\partial x} = -\frac{z}{x + 2z - e^z}, \qquad \frac{\partial f}{\partial y} = -\frac{2y}{x + 2z - e^z}.$$

When $x = 0$, $y = e$, and $z = 2$ we find $\partial f/\partial x = 2/(e^2 - 4)$ and $\partial f/\partial y = 2e/(e^2 - 4)$. Note that we were able to compute the partial derivatives $\partial f/\partial x$ and $\partial f/\partial y$ using only the value of $f(x, y)$ at the single point $(0, e)$.

The foregoing discussion can be extended to functions of more than two variables.

THEOREM 9.3. *Let F be a scalar field differentiable on an open set T in $\mathbf{R}^n$. Assume that the equation*

$$F(x_1, \ldots, x_n) = 0$$

defines x_n implicitly as a differentiable function of $x_1, \ldots, x_{n-1}$, say

$$x_n = f(x_1, \ldots, x_{n-1}),$$

for all points $(x_1, \ldots, x_{n-1})$ in some open set S in $\mathbf{R}^{n-1}$. Then for each $k = 1, 2, \ldots, n - 1$, the partial derivative $D_k f$ is given by the formula

$$D_k f = -\frac{D_k F}{D_n F} \tag{9.22}$$

at those points at which $D_n F \neq 0$. The partial derivatives $D_k F$ and $D_n F$ which appear in (9.22) are to be evaluated at the point $(x_1, x_2, \ldots, x_{n-1}, f(x_1, \ldots, x_{n-1}))$.

The proof is a direct extension of the argument used to derive Equations (9.20) and (9.21) and is left to the reader.

The discussion can be generalized in another way. Suppose we have two surfaces with the following implicit representations:

$$F(x, y, z) = 0, \qquad G(x, y, z) = 0. \tag{9.23}$$

If these surfaces intersect along a curve C, it may be possible to obtain a parametric representation of C by solving the two equations in (9.23) simultaneously for two of the variables in terms of the third, say for x and y in terms of z. Let us suppose that it is possible to solve for x and y and that solutions are given by the equations

$$x = X(z), \qquad y = Y(z)$$

for all z in some open interval (a, b). Then when x and y are replaced by $X(z)$ and $Y(z)$, respectively, the two equations in (9.23) are identically satisfied. That is, we can write

$F[X(z), Y(z), z] = 0$ and $G[X(z), Y(z), z] = 0$ for all z in (a, b). Again, by using the chain rule, we can compute the derivatives $X'(z)$ and $Y'(z)$ without an explicit knowledge of $X(z)$ and $Y(z)$. To do this we introduce new functions f and g by means of the equations

$$f(z) = F[X(z), Y(z), z] \quad \text{and} \quad g(z) = G[X(z), Y(z), z].$$

Then $f(z) = g(z) = 0$ for every z in (a, b) and hence the derivatives $f'(z)$ and $g'(z)$ are also zero on (a, b). By the chain rule these derivatives are given by the formula

$$f'(z) = \frac{\partial F}{\partial x} X'(z) + \frac{\partial F}{\partial y} Y'(z) + \frac{\partial F}{\partial z}, \qquad g'(z) = \frac{\partial G}{\partial x} X'(z) + \frac{\partial G}{\partial y} Y'(z) + \frac{\partial G}{\partial z}.$$

Since $f'(z)$ and $g'(z)$ are both zero we can determine $X'(z)$ and $Y'(z)$ by solving the following pair of simultaneous *linear* equations:

$$\frac{\partial F}{\partial x} X'(z) + \frac{\partial F}{\partial y} Y'(z) = -\frac{\partial F}{\partial z},$$

$$\frac{\partial G}{\partial x} X'(z) + \frac{\partial G}{\partial y} Y'(z) = -\frac{\partial G}{\partial z}.$$

At those points at which the determinant of the system is not zero, these equations have a unique solution which can be expressed as follows, using Cramer's rule:

$$(9.24) \qquad X'(z) = -\frac{\begin{vmatrix} \dfrac{\partial F}{\partial z} & \dfrac{\partial F}{\partial y} \\ \dfrac{\partial G}{\partial z} & \dfrac{\partial G}{\partial y} \end{vmatrix}}{\begin{vmatrix} \dfrac{\partial F}{\partial x} & \dfrac{\partial F}{\partial y} \\ \dfrac{\partial G}{\partial x} & \dfrac{\partial G}{\partial y} \end{vmatrix}}, \qquad Y'(z) = -\frac{\begin{vmatrix} \dfrac{\partial F}{\partial x} & \dfrac{\partial F}{\partial z} \\ \dfrac{\partial G}{\partial x} & \dfrac{\partial G}{\partial z} \end{vmatrix}}{\begin{vmatrix} \dfrac{\partial F}{\partial x} & \dfrac{\partial F}{\partial y} \\ \dfrac{\partial G}{\partial x} & \dfrac{\partial G}{\partial y} \end{vmatrix}}.$$

The determinants which appear in (9.24) are determinants of Jacobian matrices and are called *Jacobian determinants*. A special notation is often used to denote Jacobian determinants. We write

$$\frac{\partial(f_1, \ldots, f_n)}{\partial(x_1, \ldots, x_n)} = \det \begin{bmatrix} \dfrac{\partial f_1}{\partial x_1} & \dfrac{\partial f_1}{\partial x_2} & \cdots & \dfrac{\partial f_1}{\partial x_n} \\ \cdot & & & \cdot \\ \cdot & & & \cdot \\ \cdot & & & \cdot \\ \dfrac{\partial f_n}{\partial x_1} & \dfrac{\partial f_n}{\partial x_2} & \cdots & \dfrac{\partial f_n}{\partial x_n} \end{bmatrix}.$$

In this notation, the formulas in (9.24) can be expressed more briefly in the form

$$X'(z) = \frac{\partial(F, G)/\partial(y, z)}{\partial(F, G)/\partial(x, y)}, \qquad Y'(z) = \frac{\partial(F, G)/\partial(z, x)}{\partial(F, G)/\partial(x, y)}. \tag{9.25}$$

(The minus sign has been incorporated into the numerators by interchanging the columns.)

The method can be extended to treat more general situations in which m equations in n variables are given, where $n > m$ and we solve for m of the variables in terms of the remaining $n - m$ variables. The partial derivatives of the new functions so defined can be expressed as quotients of Jacobian determinants, generalizing (9.25). An example with $m = 2$ and $n = 4$ is described in Exercise 3 of Section 9.8.

9.7 Worked examples

In this section we illustrate some of the concepts of the foregoing section by solving various types of problems dealing with functions defined implicitly.

EXAMPLE 1. Assume that the equation $g(x, y) = 0$ determines y as a differentiable function of x, say $y = Y(x)$ for all x in some open interval (a, b). Express the derivative $Y'(x)$ in terms of the partial derivatives of g.

Solution. Let $G(x) = g[x, Y(x)]$ for x in (a, b). Then the equation $g(x, y) = 0$ implies $G(x) = 0$ in (a, b). By the chain rule we have

$$G'(x) = \frac{\partial g}{\partial x} \cdot 1 + \frac{\partial g}{\partial y} Y'(x),$$

from which we obtain

$$Y'(x) = -\frac{\partial g/\partial x}{\partial g/\partial y} \tag{9.26}$$

at those points x in (a, b) at which $\partial g/\partial y \neq 0$. The partial derivatives $\partial g/\partial x$ and $\partial g/\partial y$ are given by the formulas $\partial g/\partial x = D_1 g[x, Y(x)]$ and $\partial g/\partial y = D_2 g[x, Y(x)]$.

EXAMPLE 2. When y is eliminated from the two equations $z = f(x, y)$ and $g(x, y) = 0$, the result can be expressed in the form $z = h(x)$. Express the derivative $h'(x)$ in terms of the partial derivatives of f and g.

Solution. Let us assume that the equation $g(x, y) = 0$ may be solved for y in terms of x and that a solution is given by $y = Y(x)$ for all x in some open interval (a, b). Then the function h is given by the formula

$$h(x) = f[x, Y(x)] \quad \text{if} \quad x \in (a, b).$$

Applying the chain rule we have

$$h'(x) = \frac{\partial f}{\partial x} + \frac{\partial f}{\partial y} Y'(x).$$

Using Equation (9.26) of Example 1 we obtain the formula

$$h'(x) = \frac{\dfrac{\partial g}{\partial y}\dfrac{\partial f}{\partial x} - \dfrac{\partial f}{\partial y}\dfrac{\partial g}{\partial x}}{\dfrac{\partial g}{\partial y}}.$$

The partial derivatives on the right are to be evaluated at the point $(x, Y(x))$. Note that the numerator can also be expressed as a Jacobian determinant, giving us

$$h'(x) = \frac{\partial(f, g)/\partial(x, y)}{\partial g/\partial y}.$$

EXAMPLE 3. The two equations $2x = v^2 - u^2$ and $y = uv$ define u and v as functions of x and y. Find formulas for $\partial u/\partial x, \partial u/\partial y, \partial v/\partial x, \partial v/\partial y$.

Solution. If we hold y fixed and differentiate the two equations in question with respect to x, remembering that u and v are functions of x and y, we obtain

$$2 = 2v\frac{\partial v}{\partial x} - 2u\frac{\partial u}{\partial x} \quad \text{and} \quad 0 = u\frac{\partial v}{\partial x} + v\frac{\partial u}{\partial x}.$$

Solving these simultaneously for $\partial u/\partial x$ and $\partial v/\partial x$ we find

$$\frac{\partial u}{\partial x} = -\frac{u}{u^2 + v^2} \quad \text{and} \quad \frac{\partial v}{\partial x} = \frac{v}{u^2 + v^2}.$$

On the other hand, if we hold x fixed and differentiate the two given equations with respect to y we obtain the equations

$$0 = 2v\frac{\partial v}{\partial y} - 2u\frac{\partial u}{\partial y} \quad \text{and} \quad 1 = u\frac{\partial v}{\partial y} + v\frac{\partial u}{\partial y}.$$

Solving these simultaneously we find

$$\frac{\partial u}{\partial y} = \frac{v}{u^2 + v^2} \quad \text{and} \quad \frac{\partial v}{\partial y} = \frac{u}{u^2 + v^2}.$$

EXAMPLE 4. Let u be defined as a function of x and y by means of the equation

$$u = F(x + u, yu).$$

Find $\partial u/\partial x$ and $\partial u/\partial y$ in terms of the partial derivatives of F.

Solution. Suppose that $u = g(x, y)$ for all (x, y) in some open set S. Substituting $g(x, y)$ for u in the original equation we must have

$$g(x, y) = F[u_1(x, y), u_2(x, y)], \tag{9.27}$$

where $u_1(x, y) = x + g(x, y)$ and $u_2(x, y) = y\,g(x, y)$. Now we hold y fixed and differentiate both sides of (9.27) with respect to x, using the chain rule on the right, to obtain

$$\frac{\partial g}{\partial x} = D_1F\,\frac{\partial u_1}{\partial x} + D_2F\,\frac{\partial u_2}{\partial x}. \tag{9.28}$$

But $\partial u_1/\partial x = 1 + \partial g/\partial x$, and $\partial u_2/\partial x = y\,\partial g/\partial x$. Hence (9.28) becomes

$$\frac{\partial g}{\partial x} = D_1F \cdot \left(1 + \frac{\partial g}{\partial x}\right) + D_2F \cdot \left(y\,\frac{\partial g}{\partial x}\right).$$

Solving this equation for $\partial g/\partial x$ (and writing $\partial u/\partial x$ for $\partial g/\partial x$) we obtain

$$\frac{\partial u}{\partial x} = \frac{-D_1F}{D_1F + y\,D_2F - 1}.$$

In a similar way we find

$$\frac{\partial g}{\partial y} = D_1F\,\frac{\partial u_1}{\partial y} + D_2F\,\frac{\partial u_2}{\partial y} = D_1F\,\frac{\partial g}{\partial y} + D_2F\left(y\,\frac{\partial g}{\partial y} + g(x, y)\right).$$

This leads to the equation

$$\frac{\partial u}{\partial y} = \frac{-g(x, y)\,D_2F}{D_1F + y\,D_2F - 1}.$$

The partial derivatives D_1F and D_2F are to be evaluated at the point $(x + g(x, y), y\,g(x, y))$.

EXAMPLE 5. When u is eliminated from the two equations $x = u + v$ and $y = uv^2$, we get an equation of the form $F(x, y, v) = 0$ which defines v implicitly as a function of x and y, say $v = h(x, y)$. Prove that

$$\frac{\partial h}{\partial x} = \frac{h(x, y)}{3h(x, y) - 2x}$$

and find a similar formula for $\partial h/\partial y$.

Solution. Eliminating u from the two given equations, we obtain the relation

$$xv^2 - v^3 - y = 0.$$

Let F be the function defined by the equation

$$F(x, y, v) = xv^2 - v^3 - y.$$

The discussion in Section 9.6 is now applicable and we can write

$$\frac{\partial h}{\partial x} = -\frac{\partial F/\partial x}{\partial F/\partial v} \quad \text{and} \quad \frac{\partial h}{\partial y} = -\frac{\partial F/\partial y}{\partial F/\partial v}. \tag{9.29}$$

But $\partial F/\partial x = v^2$, $\partial F/\partial v = 2xv - 3v^2$, and $\partial F/\partial y = -1$. Hence the equations in (9.29) become

$$\frac{\partial h}{\partial x} = -\frac{v^2}{2xv - 3v^2} = -\frac{v}{2x - 3v} = \frac{h(x, y)}{3h(x, y) - 2x}$$

and

$$\frac{\partial h}{\partial y} = -\frac{-1}{2xv - 3v^2} = \frac{1}{2xh(x, y) - 3h^2(x, y)}.$$

EXAMPLE 6. The equation $F(x, y, z) = 0$ defines z implicitly as a function of x and y, say $z = f(x, y)$. Assuming that $\partial^2 F/(\partial x\, \partial z) = \partial^2 F/(\partial z\, \partial x)$, show that

$$\text{(9.30)} \qquad \frac{\partial^2 f}{\partial x^2} = -\frac{\left(\dfrac{\partial^2 F}{\partial z^2}\right)\left(\dfrac{\partial F}{\partial x}\right)^2 - 2\left(\dfrac{\partial^2 F}{\partial x\, \partial z}\right)\left(\dfrac{\partial F}{\partial z}\right)\left(\dfrac{\partial F}{\partial x}\right) + \left(\dfrac{\partial F}{\partial z}\right)^2\left(\dfrac{\partial^2 F}{\partial x^2}\right)}{\left(\dfrac{\partial F}{\partial z}\right)^3},$$

where the partial derivatives on the right are to be evaluated at $(x, y, f(x, y))$.

Solution. By Equation (9.20) of Section 9.6 we have

$$\text{(9.31)} \qquad \frac{\partial f}{\partial x} = -\frac{\partial F/\partial x}{\partial F/\partial z}.$$

We must remember that this quotient really means

$$-\frac{D_1 F[x, y, f(x, y)]}{D_3 F[x, y, f(x, y)]}.$$

Let us introduce $G(x, y) = D_1 F[x, y, f(x, y)]$ and $H(x, y) = D_3 F[x, y, f(x, y)]$. Our object is to evaluate the partial derivative with respect to x of the quotient

$$\frac{\partial f}{\partial x} = -\frac{G(x, y)}{H(x, y)},$$

holding y fixed. The rule for differentiating quotients gives us

$$\text{(9.32)} \qquad \frac{\partial^2 f}{\partial x^2} = -\frac{H\dfrac{\partial G}{\partial x} - G\dfrac{\partial H}{\partial x}}{H^2}.$$

Since G and H are composite functions, we use the chain rule to compute the partial derivatives $\partial G/\partial x$ and $\partial H/\partial x$. For $\partial G/\partial x$ we have

$$\begin{aligned}\frac{\partial G}{\partial x} &= D_1(D_1 F)\cdot 1 + D_2(D_1 F)\cdot 0 + D_3(D_1 F)\cdot \frac{\partial f}{\partial x}\\ &= \frac{\partial^2 F}{\partial x^2} + \frac{\partial^2 F}{\partial z\, \partial x}\frac{\partial f}{\partial x}.\end{aligned}$$

Similarly, we find

$$\frac{\partial H}{\partial x} = D_1(D_3F)\cdot 1 + D_2(D_3F)\cdot 0 + D_3(D_3F)\cdot\frac{\partial f}{\partial x}$$
$$= \frac{\partial^2 F}{\partial x\,\partial z} + \frac{\partial^2 F}{\partial z^2}\frac{\partial f}{\partial x}.$$

Substituting these in (9.32) and replacing $\partial f/\partial x$ by the quotient in (9.31) we obtain the formula in (9.30).

9.8 Exercises

In the exercises in this section you may assume the existence and continuity of all derivatives under consideration.

1. The two equations $x + y = uv$ and $xy = u - v$ determine x and y implicitly as functions of u and v, say $x = X(u, v)$ and $y = Y(u, v)$. Show that $\partial X/\partial u = (xv - 1)/(x - y)$ if $x \neq y$, and find similar formulas for $\partial X/\partial v$, $\partial Y/\partial u$, $\partial Y/\partial v$.
2. The two equations $x + y = uv$ and $xy = u - v$ determine x and v as functions of u and y, say $x = X(u, y)$ and $v = V(u, y)$. Show that $\partial X/\partial u = (u + v)/(1 + yu)$ if $1 + yu \neq 0$, and find similar formulas for $\partial X/\partial y$, $\partial V/\partial u$, $\partial V/\partial y$.
3. The two equations $F(x, y, u, v) = 0$ and $G(x, y, u, v) = 0$ determine x and y implicitly as functions of u and v, say $x = X(u, v)$ and $y = Y(u, v)$. Show that

$$\frac{\partial X}{\partial u} = \frac{\partial(F, G)/\partial(y, u)}{\partial(F, G)/\partial(x, y)}$$

at points at which the Jacobian $\partial(F, G)/\partial(x, y) \neq 0$, and find similar formulas for the partial derivatives $\partial X/\partial v$, $\partial Y/\partial u$, and $\partial Y/\partial v$.
4. The intersection of the two surfaces given by the Cartesian equations $2x^2 + 3y^2 - z^2 = 25$ and $x^2 + y^2 = z^2$ contains a curve C passing through the point $P = (\sqrt{7}, 3, 4)$. These equations may be solved for x and y in terms of z to give a parametric representation of C with z as parameter.
(a) Find a unit tangent vector T to C at the point P without using an explicit knowledge of the parametric representation.
(b) Check the result in part (a) by determining a parametric representation of C with z as parameter.
5. The three equations $F(u, v) = 0$, $u = xy$, and $v = \sqrt{x^2 + z^2}$ define a surface in xyz-space. Find a normal vector to this surface at the point $x = 1$, $y = 1$, $z = \sqrt{3}$ if it is known that $D_1F(1, 2) = 1$ and $D_2F(1, 2) = 2$.
6. The three equations

$$x^2 - y\cos(uv) + z^2 = 0,$$
$$x^2 + y^2 - \sin(uv) + 2z^2 = 2,$$
$$xy - \sin u \cos v + z = 0,$$

define x, y, and z as functions of u and v. Compute the partial derivatives $\partial x/\partial u$ and $\partial x/\partial v$ at the point $x = y = 1$, $u = \pi/2$, $v = 0$, $z = 0$.

7. The equation $f(y/x, z/x) = 0$ defines z implicitly as a function of x and y, say $z = g(x, y)$. Show that

$$x\frac{\partial g}{\partial x} + y\frac{\partial g}{\partial y} = g(x, y)$$

at those points at which $D_2 f[y/x, g(x, y)/x]$ is not zero.

8. Let F be a real-valued function of two real variables and assume that the partial derivatives $D_1 F$ and $D_2 F$ are never zero. Let u be another real-valued function of two real variables such that the partial derivatives $\partial u/\partial x$ and $\partial u/\partial y$ are related by the equation $F(\partial u/\partial x, \partial u/\partial y) = 0$. Prove that a constant n exists such that

$$\frac{\partial^2 u}{\partial x^2}\frac{\partial^2 u}{\partial y^2} = \left(\frac{\partial^2 u}{\partial x\, \partial y}\right)^n,$$

and find n. Assume that $\partial^2 u/(\partial x\, \partial y) = \partial^2 u/(\partial y\, \partial x)$.

9. The equation $x + z + (y + z)^2 = 6$ defines z implicitly as a function of x and y, say $z = f(x, y)$. Compute the partial derivatives $\partial f/\partial x$, $\partial f/\partial y$, and $\partial^2 f/(\partial x\, \partial y)$ in terms of x, y, and z.

10. The equation $\sin(x + y) + \sin(y + z) = 1$ defines z implicitly as a function of x and y, say $z = f(x, y)$. Compute the second derivative $D_{1,2} f$ in terms of x, y, and z.

11. The equation $F(x + y + z, x^2 + y^2 + z^2) = 0$ defines z implicitly as a function of x and y, say $z = f(x, y)$. Determine the partial derivatives $\partial f/\partial x$ and $\partial f/\partial y$ in terms of x, y, z and the partial derivatives $D_1 F$ and $D_2 F$.

12. Let f and g be functions of one real variable and define $F(x, y) = f[x + g(y)]$. Find formulas for all the partial derivatives of F of first and second order, expressed in terms of the derivatives of f and g. Verify the relation

$$\frac{\partial F}{\partial x}\frac{\partial^2 F}{\partial x\, \partial y} = \frac{\partial F}{\partial y}\frac{\partial^2 F}{\partial x^2}.$$

9.9 Maxima, minima, and saddle points

A surface that is described explicitly by an equation of the form $z = f(x, y)$ can be thought of as a level surface of the scalar field F defined by the equation

$$F(x, y, z) = f(x, y) - z.$$

If f is differentiable, the gradient of this field is given by the vector

$$\nabla F = \frac{\partial f}{\partial x}\boldsymbol{i} + \frac{\partial f}{\partial y}\boldsymbol{j} - \boldsymbol{k}.$$

A linear equation for the tangent plane at a point $P_1 = (x_1, y_1, z_1)$ can be written in the form

$$z - z_1 = A(x - x_1) + B(y - y_1),$$

where

$$A = D_1 f(x_1, y_1) \quad \text{and} \quad B = D_2 f(x_1, y_1).$$

When both coefficients A and B are zero, the point P_1 is called a *stationary point* of the surface and the point (x_1, y_1) is called a *stationary point* or a *critical point* of the function f.

The tangent plane is horizontal at a stationary point. The stationary points of a surface are usually classified into three categories: maxima, minima, and saddle points. If the surface is thought of as a mountain landscape, these categories correspond, respectively, to mountain tops, bottoms of valleys, and mountain passes.

The concepts of maxima, minima, and saddle points can be introduced for arbitrary scalar fields defined on subsets of $\mathbf{R}^n$.

DEFINITION. *A scalar field f is said to have an absolute maximum at a point $\boldsymbol{a}$ of a set S in $\mathbf{R}^n$ if*

$$f(\boldsymbol{x}) \leq f(\boldsymbol{a}) \tag{9.33}$$

for all $\boldsymbol{x}$ in S. The number $f(\boldsymbol{a})$ is called the absolute maximum value of f on S. The function f is said to have a relative maximum at $\boldsymbol{a}$ if the inequality in (9.33) *is satisfied for every $\boldsymbol{x}$ in some n-ball $B(\boldsymbol{a})$ lying in S.*

In other words, a relative maximum at $\boldsymbol{a}$ is the absolute maximum in some neighborhood of $\boldsymbol{a}$. The terms *absolute minimum* and *relative minimum* are defined in an analogous fashion, using the inequality opposite to that in (9.33). The adjectives *global* and *local* are sometimes used in place of *absolute* and *relative*, respectively.

DEFINITION. *A number which is either a relative maximum or a relative minimum of f is called an extremum of f.*

If f has an extremum at an interior point $\boldsymbol{a}$ and is differentiable there, then all first-order partial derivatives $D_1f(\boldsymbol{a}), \ldots, D_nf(\boldsymbol{a})$ must be zero. In other words, $\nabla f(\boldsymbol{a}) = \boldsymbol{O}$. (This is easily proved by holding each component fixed and reducing the problem to the one-dimensional case.) In the case $n = 2$, this means that there is a horizontal tangent plane to the surface $z = f(x, y)$ at the point $(\boldsymbol{a}, f(\boldsymbol{a}))$. On the other hand, it is easy to find examples in which the vanishing of all partial derivatives at $\boldsymbol{a}$ does not necessarily imply an extremum at $\boldsymbol{a}$. This occurs at the so-called *saddle points* which are defined as follows.

DEFINITION. *Assume f is differentiable at $\boldsymbol{a}$. If $\nabla f(\boldsymbol{a}) = \boldsymbol{O}$ the point $\boldsymbol{a}$ is called a stationary point of f. A stationary point is called a saddle point if every n-ball $B(\boldsymbol{a})$ contains points $\boldsymbol{x}$ such that $f(\boldsymbol{x}) < f(\boldsymbol{a})$ and other points such that $f(\boldsymbol{x}) > f(\boldsymbol{a})$.*

The situation is somewhat analogous to the one-dimensional case in which stationary points of a function are classified as maxima, minima, and points of inflection. The following examples illustrate several types of stationary points. In each case the stationary point in question is at the origin.

EXAMPLE 1. *Relative maximum.* $z = f(x, y) = 2 - x^2 - y^2$. This surface is a paraboloid of revolution. In the vicinity of the origin it has the shape shown in Figure 9.3(a). Its

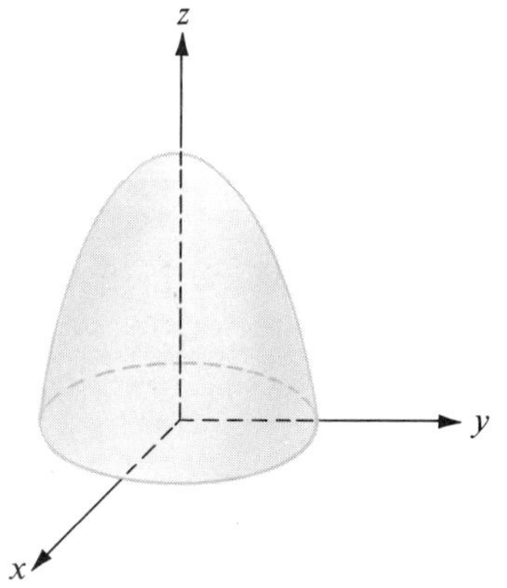

(a) $z = 2 - x^2 - y^2$

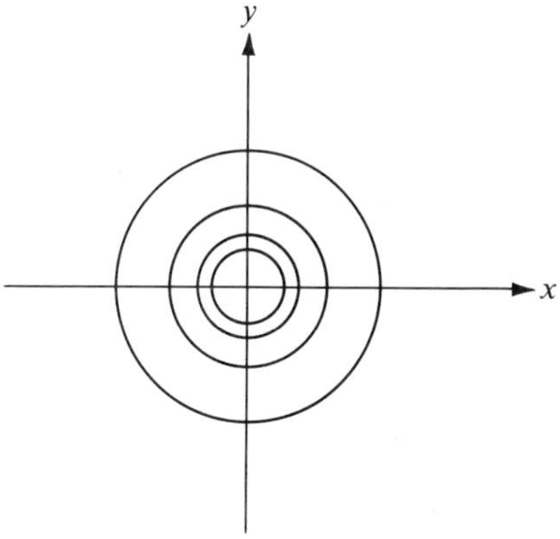

(b) Level curves: $x^2 + y^2 = c$

Example 1. Relative maximum at the origin.

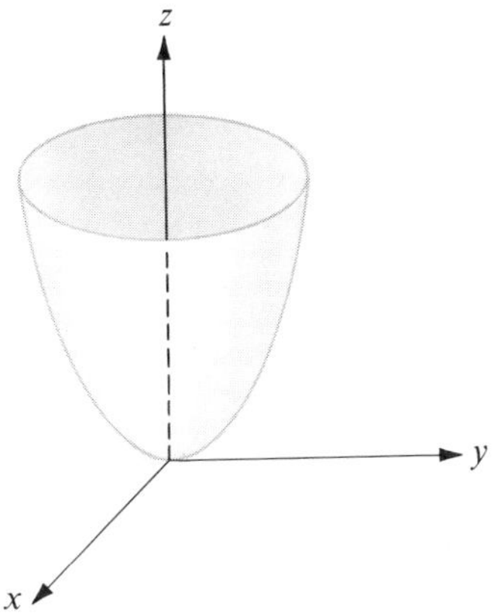

(c) $z = x^2 + y^2$

Example 2. Relative minimum at the origin.

FIGURE 9.3 Examples 1 and 2.

level curves are circles, some of which are shown in Figure 9.3(b). Since $f(x, y) = 2 - (x^2 + y^2) \leq 2 = f(0, 0)$ for all (x, y), it follows that f not only has a relative maximum at $(0, 0)$ but also an *absolute* maximum there. Both partial derivatives $\partial f/\partial x$ and $\partial f/\partial y$ vanish at the origin.

EXAMPLE 2. *Relative minimum.* $z = f(x, y) = x^2 + y^2$. This example, another paraboloid of revolution, is essentially the same as Example 1, except that there is a minimum at the origin rather than a maximum. The appearance of the surface near the origin is illustrated in Figure 9.3(c) and some of the level curves are shown in Figure 9.3(b).

EXAMPLE 3. *Saddle point.* $z = f(x, y) = xy$. This surface is a hyperbolic paraboloid. Near the origin the surface is saddle shaped, as shown in Figure 9.4(a). Both partial derivatives $\partial f/\partial x$ and $\partial f/\partial y$ are zero at the origin but there is neither a relative maximum nor a relative minimum there. In fact, for points (x, y) in the first or third quadrants, x and y have the same sign, giving us $f(x, y) > 0 = f(0, 0)$, whereas for points in the second and fourth quadrants x and y have opposite signs, giving us $f(x, y) < 0 = f(0, 0)$. Therefore, in every neighborhood of the origin there are points at which the function is less than $f(0, 0)$ and points at which the function exceeds $f(0, 0)$, so the origin is a saddle

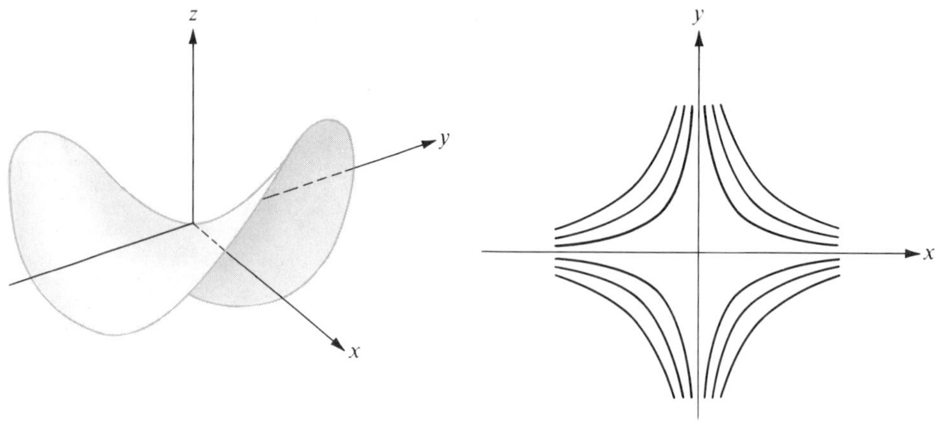

(a) $z = xy$ (b) Level curves: $xy = c$

FIGURE 9.4 Example 3. Saddle point at the origin.

point. The presence of the saddle point is also revealed by Figure 9.4(b), which shows some of the level curves near (0, 0). These are hyperbolas having the x- and y-axes as asymptotes.

EXAMPLE 4. *Saddle point.* $z = f(x, y) = x^3 - 3xy^2$. Near the origin, this surface has the appearance of a mountain pass in the vicinity of three peaks. This surface, sometimes referred to as a "monkey saddle," is shown in Figure 9.5(a). Some of the level curves are illustrated in Figure 9.5(b). It is clear that there is a saddle point at the origin.

EXAMPLE 5. *Relative minimum.* $z = f(x, y) = x^2y^2$. This surface has the appearance of a valley surrounded by four mountains, as suggested by Figure 9.6(a). There is an absolute minimum at the origin, since $f(x, y) \geq f(0, 0)$ for all (x, y). The level curves [shown in Figure 9.6(b)] are hyperbolas having the x- and y-axes as asymptotes. Note that these level curves are similar to those in Example 3. In this case, however, the function assumes only nonnegative values on all its level curves.

EXAMPLE 6. *Relative maximum.* $z = f(x, y) = 1 - x^2$. In this case the surface is a cylinder with generators parallel to the y-axis, as shown in Figure 9.7(a). Cross sections cut by planes parallel to the x-axis are parabolas. There is obviously an absolute maximum at the origin because $f(x, y) = 1 - x^2 \leq 1 = f(0, 0)$ for all (x, y). The level curves form a family of parallel straight lines as shown in Figure 9.7(b).

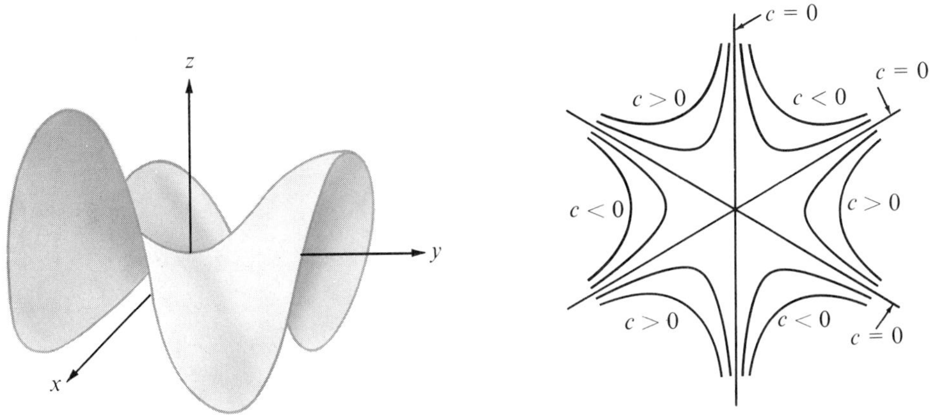

(a) $z = x^3 - 3xy^2$. (b) Level curves: $x^3 - 3xy^2 = c$.

FIGURE 9.5 Example 4. Saddle point at the origin.

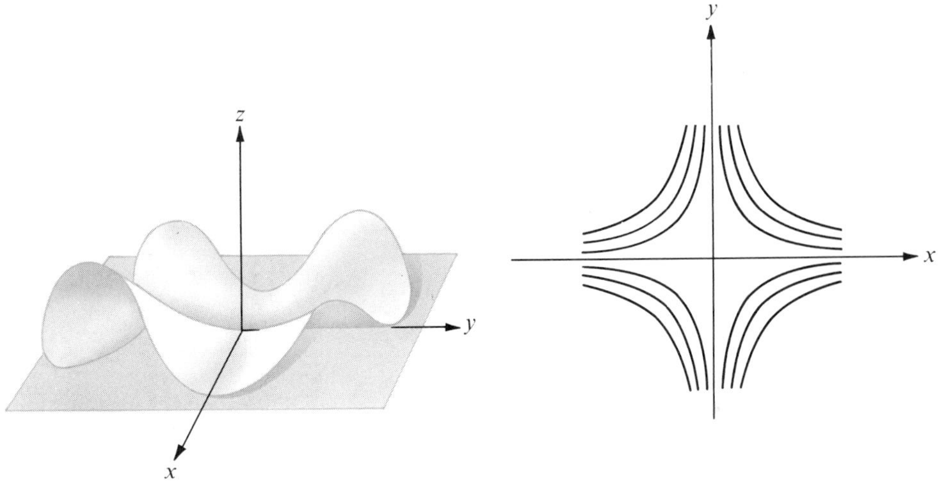

(a) $z = x^2y^2$ (b) Level curves: $x^2y^2 = c$

FIGURE 9.6 Example 5. Relative minimum at the origin.

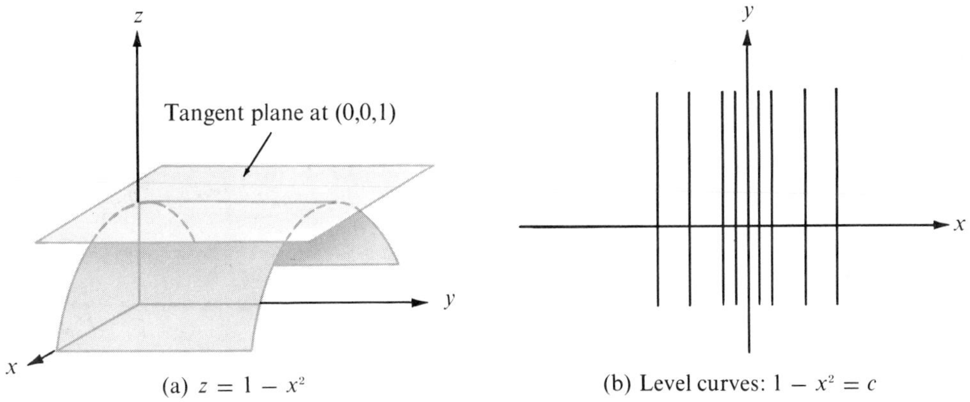

(a) $z = 1 - x^2$ (b) Level curves: $1 - x^2 = c$

FIGURE 9.7 Example 6. Relative maximum at the origin.

9.10 Second-order Taylor formula for scalar fields

If a differentiable scalar field f has a stationary point at $\boldsymbol{a}$, the nature of the stationary point is determined by the algebraic sign of the difference $f(\boldsymbol{x}) - f(\boldsymbol{a})$ for $\boldsymbol{x}$ near $\boldsymbol{a}$. If $\boldsymbol{x} = \boldsymbol{a} + \boldsymbol{y}$, we have the first-order Taylor formula

$$f(\boldsymbol{a} + \boldsymbol{y}) - f(\boldsymbol{a}) = \nabla f(\boldsymbol{a}) \cdot \boldsymbol{y} + \|\boldsymbol{y}\|\, E(\boldsymbol{a}, \boldsymbol{y}), \qquad \text{where} \quad E(\boldsymbol{a}, \boldsymbol{y}) \to 0 \quad \text{as } \boldsymbol{y} \to \boldsymbol{O}.$$

At a stationary point, $\nabla f(\boldsymbol{a}) = \boldsymbol{O}$ and the Taylor formula becomes

$$f(\boldsymbol{a} + \boldsymbol{y}) - f(\boldsymbol{a}) = \|\boldsymbol{y}\|\, E(\boldsymbol{a}, \boldsymbol{y}).$$

To determine the algebraic sign of $f(\boldsymbol{a} + \boldsymbol{y}) - f(\boldsymbol{a})$ we need more information about the error term $\|\boldsymbol{y}\|\, E(\boldsymbol{a}, \boldsymbol{y})$. The next theorem shows that if f has continuous second-order partial derivatives at $\boldsymbol{a}$, the error term is equal to a quadratic form,

$$\frac{1}{2} \sum_{i=1}^{n} \sum_{j=1}^{n} D_{ij} f(\boldsymbol{a}) y_i y_j$$

plus a term of smaller order than $\|\boldsymbol{y}\|^2$. The coefficients of the quadratic form are the second-order partial derivatives $D_{ij} f = D_i(D_j f)$, evaluated at $\boldsymbol{a}$. The $n \times n$ matrix of second-order derivatives $D_{ij} f(\boldsymbol{x})$ is called the *Hessian matrix*† and is denoted by $H(\boldsymbol{x})$. Thus, we have

$$H(\boldsymbol{x}) = [D_{ij} f(\boldsymbol{x})]_{i,j=1}^{n}$$

whenever the derivatives exist. The quadratic form can be written more simply in matrix notation as follows:

$$\sum_{i=1}^{n} \sum_{j=1}^{n} D_{ij} f(\boldsymbol{a}) y_i y_j = \boldsymbol{y} H(\boldsymbol{a}) \boldsymbol{y}^t,$$

where $\boldsymbol{y} = (y_1, \ldots, y_n)$ is considered as a $1 \times n$ row matrix, and $\boldsymbol{y}^t$ is its transpose, an $n \times 1$ column matrix. When the partial derivatives $D_{ij} f$ are continuous we have $D_{ij} f = D_{ji} f$ and the matrix $H(\boldsymbol{a})$ is symmetric.

Taylor's formula, giving a quadratic approximation to $f(\boldsymbol{a} + \boldsymbol{y}) - f(\boldsymbol{a})$, now takes the following form.

THEOREM 9.4. SECOND-ORDER TAYLOR FORMULA FOR SCALAR FIELDS. *Let f be a scalar field with continuous second-order partial derivatives $D_{ij} f$ in an n-ball $B(\boldsymbol{a})$. Then for all $\boldsymbol{y}$ in $\mathbf{R}^n$ such that $\boldsymbol{a} + \boldsymbol{y} \in B(\boldsymbol{a})$ we have*

$$(9.34) \quad f(\boldsymbol{a} + \boldsymbol{y}) - f(\boldsymbol{a}) = \nabla f(\boldsymbol{a}) \cdot \boldsymbol{y} + \frac{1}{2!} \boldsymbol{y} H(\boldsymbol{a} + c\boldsymbol{y}) \boldsymbol{y}^t, \qquad \textit{where} \quad 0 < c < 1.$$

† Named for Ludwig Otto Hesse (1811–1874), a German mathematician who made many contributions to the theory of surfaces.

This can also be written in the form

$$(9.35) \qquad f(\boldsymbol{a}+\boldsymbol{y}) - f(\boldsymbol{a}) = \nabla f(\boldsymbol{a}) \cdot \boldsymbol{y} + \frac{1}{2!}\boldsymbol{y}H(\boldsymbol{a})\boldsymbol{y}^t + \|\boldsymbol{y}\|^2 E_2(\boldsymbol{a}, \boldsymbol{y}),$$

where $E_2(\boldsymbol{a}, \boldsymbol{y}) \to 0$ *as* $\boldsymbol{y} \to \boldsymbol{O}$.

Proof. Keep $\boldsymbol{y}$ fixed and define $g(u)$ for real u by the equation

$$g(u) = f(\boldsymbol{a} + u\boldsymbol{y}) \qquad \text{for} \quad -1 \le u \le 1 .$$

Then $f(\boldsymbol{a}+\boldsymbol{y}) - f(\boldsymbol{a}) = g(1) - g(0)$. We will prove the theorem by applying the second-order Taylor formula to g on the interval [0, 1]. We obtain

$$(9.36) \qquad g(1) - g(0) = g'(0) + \frac{1}{2!} g''(c), \qquad \text{where} \quad 0 < c < 1 .$$

Here we have used Lagrange's form of the remainder (see Section 7.7 of Volume I).

Since g is a composite function given by $g(u) = f[\boldsymbol{r}(u)]$, where $\boldsymbol{r}(u) = \boldsymbol{a} + u\boldsymbol{y}$, we can compute its derivative by the chain rule. We have $\boldsymbol{r}'(u) = \boldsymbol{y}$ so the chain rule gives us

$$g'(u) = \nabla f[\boldsymbol{r}(u)] \cdot \boldsymbol{r}'(u) = \nabla f[\boldsymbol{r}(u)] \cdot \boldsymbol{y} = \sum_{j=1}^{n} D_j f[\boldsymbol{r}(u)] y_j ,$$

provided $\boldsymbol{r}(u) \in B(\boldsymbol{a})$. In particular, $g'(0) = \nabla f(\boldsymbol{a}) \cdot \boldsymbol{y}$. Using the chain rule once more we find

$$g''(u) = \sum_{i=1}^{n} D_i \left(\sum_{j=1}^{n} D_j f[\boldsymbol{r}(u)] y_j \right) y_i = \sum_{i=1}^{n} \sum_{j=1}^{n} D_{ij} f[\boldsymbol{r}(u)] y_i y_j = \boldsymbol{y} H[\boldsymbol{r}(u)] \boldsymbol{y}^t .$$

Hence $g''(c) = \boldsymbol{y}H(\boldsymbol{a} + c\boldsymbol{y})\boldsymbol{y}^t$, so Equation (9.36) becomes (9.34).

To prove (9.35) we define $E_2(\boldsymbol{a}, \boldsymbol{y})$ by the equation

$$(9.37) \qquad \|\boldsymbol{y}\|^2 E_2(\boldsymbol{a}, \boldsymbol{y}) = \frac{1}{2!} \boldsymbol{y}\{H(\boldsymbol{a} + c\boldsymbol{y}) - H(\boldsymbol{a})\}\boldsymbol{y}^t \qquad \text{if} \quad \boldsymbol{y} \ne \boldsymbol{O},$$

and let $E_2(\boldsymbol{a}, \boldsymbol{O}) = 0$. Then Equation (9.34) takes the form

$$f(\boldsymbol{a}+\boldsymbol{y}) - f(\boldsymbol{a}) = \nabla f(\boldsymbol{a}) \cdot \boldsymbol{y} + \frac{1}{2!}\boldsymbol{y}H(\boldsymbol{a})\boldsymbol{y}^t + \|\boldsymbol{y}\|^2 E_2(\boldsymbol{a}, \boldsymbol{y}).$$

To complete the proof we need to show that $E_2(\boldsymbol{a}, \boldsymbol{y}) \to 0$ as $\boldsymbol{y} \to \boldsymbol{O}$.

From (9.37) we find that

$$\|\boldsymbol{y}\|^2 |E_2(\boldsymbol{a}, \boldsymbol{y})| = \frac{1}{2} \left| \sum_{i=1}^{n} \sum_{j=1}^{n} \{D_{ij} f(\boldsymbol{a} + c\boldsymbol{y}) - D_{ij} f(\boldsymbol{a})\} y_i y_j \right|$$

$$\le \frac{1}{2} \sum_{i=1}^{n} \sum_{j=1}^{n} |D_{ij} f(\boldsymbol{a} + c\boldsymbol{y}) - D_{ij} f(\boldsymbol{a})| \, \|\boldsymbol{y}\|^2 .$$

Dividing by $\|y\|^2$ we obtain the inequality

$$|E_2(a, y)| \le \frac{1}{2} \sum_{i=1}^{n} \sum_{j=1}^{n} |D_{ij}f(a + cy) - D_{ij}f(a)|$$

for $y \neq O$. Since each second-order partial derivative $D_{ij}f$ is continuous at a, we have $D_{ij}f(a + cy) \to D_{ij}f(a)$ as $y \to O$, so $E_2(a, y) \to 0$ as $y \to O$. This completes the proof.

9.11 The nature of a stationary point determined by the eigenvalues of the Hessian matrix

At a stationary point we have $\nabla f(a) = O$, so the Taylor formula in Equation (9.35) becomes

$$f(a + y) - f(a) = \tfrac{1}{2} yH(a)y^t + \|y\|^2 E_2(a, y).$$

Since the error term $\|y\|^2 E_2(a, y)$ tends to zero faster than $\|y\|^2$, it seems reasonable to expect that for small y the algebraic sign of $f(a + y) - f(a)$ is the same as that of the quadratic form $yH(a)y^t$; hence the nature of the stationary point should be determined by the algebraic sign of the quadratic form. This section is devoted to a proof of this fact.

First we give a connection between the algebraic sign of a quadratic form and its eigenvalues.

THEOREM 9.5. *Let $A = [a_{ij}]$ be an $n \times n$ real symmetric matrix, and let*

$$Q(y) = yAy^t = \sum_{i=1}^{n} \sum_{j=1}^{n} a_{ij} y_i y_j .$$

Then we have:

(a) *$Q(y) > 0$ for all $y \neq O$ if and only if all the eigenvalues of A are positive.*
(b) *$Q(y) < 0$ for all $y \neq O$ if and only if all the eigenvalues of A are negative.*

Note: In case (a), the quadratic form is called *positive definite*; in case (b) it is called *negative definite.*

Proof. According to Theorem 5.11 there is an orthogonal matrix C that reduces the quadratic form yAy^t to a diagonal form. That is

$$Q(y) = yAy^t = \sum_{i=1}^{n} \lambda_i x_i^2 \tag{9.38}$$

where $x = (x_1, \ldots, x_n)$ is the row matrix $x = yC$, and $\lambda_1, \ldots, \lambda_n$ are the eigenvalues of A. The eigenvalues are real since A is symmetric.

If all the eigenvalues are positive, Equation (9.38) shows that $Q(y) > 0$ whenever $x \neq O$. But since $x = yC$ we have $y = xC^{-1}$, so $x \neq O$ if and only if $y \neq O$. Therefore $Q(y) > 0$ for all $y \neq O$.

Conversely, if $Q(y) > 0$ for all $y \neq O$ we can choose y so that $x = yC$ is the kth coordinate vector e_k. For this y, Equation (9.38) gives us $Q(y) = \lambda_k$, so each $\lambda_k > 0$. This proves part (a). The proof of (b) is entirely analogous.

The next theorem describes the nature of a stationary point in terms of the algebraic sign of the quadratic form $yH(a)y^t$.

THEOREM 9.6. *Let f be a scalar field with continuous second-order partial derivatives $D_{ij}f$ in an n-ball $B(a)$, and let $H(a)$ denote the Hessian matrix at a stationary point a. Then we have:*

(a) *If all the eigenvalues of $H(a)$ are positive, f has a relative minimum at a.*
(b) *If all the eigenvalues of $H(a)$ are negative, f has a relative maximum at a.*
(c) *If $H(a)$ has both positive and negative eigenvalues, then f has a saddle point at a.*

Proof. Let $Q(y) = yH(a)y^t$. The Taylor formula gives us

$$f(a + y) - f(a) = \tfrac{1}{2}Q(y) + \|y\|^2 E_2(a, y), \tag{9.39}$$

where $E_2(a, y) \to 0$ as $y \to O$. We will prove that there is a positive number r such that, if $0 < \|y\| < r$, the algebraic sign of $f(a + y) - f(a)$ is the same as that of $Q(y)$.

Assume first that all the eigenvalues $\lambda_1, \ldots, \lambda_n$ of $H(a)$ are positive. Let h be the smallest eigenvalue. If $u < h$, the n numbers

$$\lambda_1 - u, \ldots, \lambda_n - u$$

are also positive. These numbers are the eigenvalues of the real symmetric matrix $H(a) - uI$, where I is the $n \times n$ identity matrix. By Theorem 9.5, the quadratic form $y[H(a) - uI]y^t$ is positive definite, and hence $y[H(a) - uI]y^t > 0$ for all $y \neq O$. Therefore

$$yH(a)y^t > y(uI)y^t = u\,\|y\|^2$$

for all real $u < h$. Taking $u = \tfrac{1}{2}h$ we obtain the inequality

$$Q(y) > \tfrac{1}{2}h\,\|y\|^2$$

for all $y \neq O$. Since $E_2(a, y) \to O$ as $y \to O$, there is a positive number r such that $|E_2(a, y)| < \tfrac{1}{4}h$ whenever $0 < \|y\| < r$. For such y we have

$$0 \leq \|y\|^2\,|E_2(a, y)| < \tfrac{1}{4}h\,\|y\|^2 < \tfrac{1}{2}Q(y),$$

and Taylor's formula (9.39) shows that

$$f(a + y) - f(a) \geq \tfrac{1}{2}Q(y) - \|y\|^2\,|E_2(a, y)| > 0.$$

Therefore f has a relative minimum at a, which proves part (a). To prove (b) we can use a similar argument, or simply apply part (a) to $-f$.

To prove (c), let λ_1 and λ_2 be two eigenvalues of $H(a)$ of opposite signs. Let $h = \min\{|\lambda_1|, |\lambda_2|\}$. Then for each real u satisfying $-h < u < h$ the numbers

$$\lambda_1 - u \quad \text{and} \quad \lambda_2 - u$$

are eigenvalues of opposite sign for the matrix $H(\boldsymbol{a}) - uI$. Therefore, if $u \in (-h, h)$, the quadratic form $\boldsymbol{y}[H(\boldsymbol{a}) - uI]\boldsymbol{y}^t$ takes both positive and negative values in every neighborhood of $\boldsymbol{y} = \boldsymbol{O}$. Choose $r > 0$ as above so that $|E_2(\boldsymbol{a}, \boldsymbol{y})| < \frac{1}{4}h$ whenever $0 < \|\boldsymbol{y}\| < r$. Then, arguing as above, we see that for such $\boldsymbol{y}$ the algebraic sign of $f(\boldsymbol{a} + \boldsymbol{y}) - f(\boldsymbol{a})$ is the same as that of $Q(\boldsymbol{y})$. Since both positive and negative values occur as $\boldsymbol{y} \to \boldsymbol{O}$, f has a saddle point at $\boldsymbol{a}$. This completes the proof.

Note: If all the eigenvalues of $H(\boldsymbol{a})$ are zero, Theorem 9.6 gives no information concerning the stationary point. Tests involving higher order derivatives can be used to treat such examples, but we shall not discuss them here.

9.12 Second-derivative test for extrema of functions of two variables

In the case $n = 2$ the nature of the stationary point can also be determined by the algebraic sign of the second derivative $D_{1,1}f(\boldsymbol{a})$ and the determinant of the Hessian matrix.

THEOREM 9.7. *Let $\boldsymbol{a}$ be a stationary point of a scalar field $f(x_1, x_2)$ with continuous second-order partial derivatives in a 2-ball $B(\boldsymbol{a})$. Let*

$$A = D_{1,1}f(\boldsymbol{a}), \qquad B = D_{1,2}f(\boldsymbol{a}), \qquad C = D_{2,2}f(\boldsymbol{a}),$$

and let

$$\Delta = \det H(\boldsymbol{a}) = \det \begin{bmatrix} A & B \\ B & C \end{bmatrix} = AC - B^2 .$$

Then we have:

(a) *If $\Delta < 0$, f has a saddle point at $\boldsymbol{a}$.*
(b) *If $\Delta > 0$ and $A > 0$, f has a relative minimum at $\boldsymbol{a}$.*
(c) *If $\Delta > 0$ and $A < 0$, f has a relative maximum at $\boldsymbol{a}$.*
(d) *If $\Delta = 0$, the test is inconclusive.*

Proof. In this case the characteristic equation $\det[\lambda I - H(\boldsymbol{a})] = 0$ is a quadratic equation,

$$\lambda^2 - (A + C)\lambda + \Delta = 0 .$$

The eigenvalues λ_1, λ_2 are related to the coefficients by the equations

$$\lambda_1 + \lambda_2 = A + C, \qquad \lambda_1\lambda_2 = \Delta .$$

If $\Delta < 0$ the eigenvalues have opposite signs, so f has a saddle point at $\boldsymbol{a}$, which proves (a). If $\Delta > 0$, the eigenvalues have the same sign. In this case $AC > B^2 \geq 0$, so A and C have the same sign. This sign must be that of λ_1 and λ_2 since $A + C = \lambda_1 + \lambda_2$. This proves (b) and (c).

To prove (d) we refer to Examples 4 and 5 of Section 9.9. In both these examples we have $\Delta = 0$ at the origin. In Example 4 the origin is a saddle point, and in Example 5 it is a relative minimum.

Even when Theorem 9.7 is applicable it may not be the simplest way to determine the nature of a stationary point. For example, when $f(x, y) = e^{1/g(x,y)}$, where $g(x, y) = x^2 + 2 + \cos^2 y - 2\cos y$, the test is applicable, but the computations are lengthy. In this case we may express $g(x, y)$ as a sum of squares by writing $g(x, y) = 1 + x^2 + (1 - \cos y)^2$. We see at once that f has relative maxima at the points at which $x^2 = 0$ and $(1 - \cos y)^2 = 0$. These are the points $(0, 2n\pi)$, when n is any integer.

9.13 Exercises

In Exercises 1 through 15, locate and classify the stationary points (if any) of the surfaces having the Cartesian equations given.

1. $z = x^2 + (y - 1)^2$.
2. $z = x^2 - (y - 1)^2$.
3. $z = 1 + x^2 - y^2$.
4. $z = (x - y + 1)^2$.
5. $z = 2x^2 - xy - 3y^2 - 3x + 7y$.
6. $z = x^2 - xy + y^2 - 2x + y$.
7. $z = x^3 - 3xy^2 + y^3$.
8. $z = x^2y^3(6 - x - y)$.
9. $z = x^3 + y^3 - 3xy$.
10. $z = \sin x \cosh y$.
11. $z = e^{2x+3y}(8x^2 - 6xy + 3y^2)$.
12. $z = (5x + 7y - 25)e^{-(x^2+xy+y^2)}$.
13. $z = \sin x \sin y \sin (x + y)$, $\quad 0 \le x \le \pi, 0 \le y \le \pi$.
14. $z = x - 2y + \log \sqrt{x^2 + y^2} + 3 \arctan \dfrac{y}{x}$, $\quad x > 0$.
15. $z = (x^2 + y^2)e^{-(x^2+y^2)}$.
16. Let $f(x, y) = 3x^4 - 4x^2y + y^2$. Show that on every line $y = mx$ the function has a minimum at $(0, 0)$, but that there is no relative minimum in any two-dimensional neighborhood of the origin. Make a sketch indicating the set of points (x, y) at which $f(x, y) > 0$ and the set at which $f(x, y) < 0$.
17. Let $f(x, y) = (3 - x)(3 - y)(x + y - 3)$.
    (a) Make a sketch indicating the set of points (x, y) at which $f(x, y) \ge 0$.
    (b) Find all points (x, y) in the plane at which $D_1 f(x, y) = D_2 f(x, y) = 0$. [*Hint:* $D_1 f(x, y)$ has $(3 - y)$ as a factor.]
    (c) Which of the stationary points are relative maxima? Which are relative minima? Which are neither? Give reasons for your answers.
    (d) Does f have an absolute minimum or an absolute maximum on the whole plane? Give reasons for your answers.
18. Determine all the relative and absolute extreme values and the saddle points for the function $f(x, y) = xy(1 - x^2 - y^2)$ on the square $0 \le x \le 1, 0 \le y \le 1$.
19. Determine constants a and b such that the integral

$$\int_0^1 \{ax + b - f(x)\}^2\, dx$$

will be as small as possible if (a) $f(x) = x^2$; (b) $f(x) = (x^2 + 1)^{-1}$.

20. Let $f(x, y) = Ax^2 + 2Bxy + Cy^2 + 2Dx + 2Ey + F$, where $A > 0$ and $B^2 < AC$.
    (a) Prove that a point (x_1, y_1) exists at which f has a minimum. [*Hint:* Transform the quadratic part to a sum of squares.]
    (b) Prove that $f(x_1, y_1) = Dx_1 + Ey_1 + F$ at this minimum.
    (c) Show that

$$f(x_1, y_1) = \frac{1}{AC - B^2} \begin{vmatrix} A & B & D \\ B & C & E \\ D & E & F \end{vmatrix}.$$

21. *Method of least squares.* Given n distinct numbers $x_1, \ldots, x_n$ and n further numbers $y_1, \ldots, y_n$ (not necessarily distinct), it is generally impossible to find a straight line $f(x) = ax + b$ which passes through all the points (x_i, y_i), that is, such that $f(x_i) = y_i$ for each i. However, we can try a linear function which makes the "total square error"

$$E(a, b) = \sum_{i=1}^{n} [f(x_i) - y_i]^2$$

a minimum. Determine values of a and b which do this.

22. Extend the method of least squares to 3-space. That is, find a linear function $f(x, y) = ax + by + c$ which minimizes the total square error

$$E(a, b, c) = \sum_{i=1}^{n} [f(x_i, y_i) - z_i]^2,$$

where (x_i, y_i) are n given distinct points and $z_1, \ldots, z_n$ are n given real numbers.

23. Let $z_1, \ldots, z_n$ be n distinct points in m-space. If $\boldsymbol{x} \in \mathbf{R}^m$, define

$$f(\boldsymbol{x}) = \sum_{k=1}^{n} \|\boldsymbol{x} - \boldsymbol{z}_k\|^2.$$

Prove that f has a minimum at the point $\boldsymbol{a} = \dfrac{1}{n} \displaystyle\sum_{k=1}^{n} \boldsymbol{z}_k$ (the centroid).

24. Let $\boldsymbol{a}$ be a stationary point of a scalar field f with continuous second-order partial derivatives in an n-ball $B(\boldsymbol{a})$. Prove that f has a saddle point at $\boldsymbol{a}$ if at least two of the diagonal entries of the Hessian matrix $H(\boldsymbol{a})$ have opposite signs.

25. Verify that the scalar field $f(x, y, z) = x^4 + y^4 + z^4 - 4xyz$ has a stationary point at $(1, 1, 1)$, and determine the nature of this stationary point by computing the eigenvalues of its Hessian matrix.

9.14 Extrema with constraints. Lagrange's multipliers

We begin this section with two examples of extremum problems with constraints.

EXAMPLE 1. Given a surface S not passing through the origin, determine those points of S which are nearest to the origin.

EXAMPLE 2. If $f(x, y, z)$ denotes the temperature at (x, y, z), determine the maximum and minimum values of the temperature on a given curve C in 3-space.

Both these examples are special cases of the following general problem: *Determine the extreme values of a scalar field $f(\boldsymbol{x})$ when $\boldsymbol{x}$ is restricted to lie in a given subset of the domain of f.*

In Example 1 the scalar field to be minimized is the distance function,

$$f(x, y, z) = (x^2 + y^2 + z^2)^{1/2};$$

the constraining subset is the given surface S. In Example 2 the constraining subset is the given curve C.

Constrained extremum problems are often very difficult; no general method is known for attacking them in their fullest generality. Special methods are available when the constraining subset has a fairly simple structure, for instance, if it is a surface as in Example 1, or a curve as in Example 2. This section discusses the method of Lagrange's multipliers for solving such problems. First we describe the method in its general form, and then we give geometric arguments to show why it works in the two examples mentioned above.

The method of Lagrange's multipliers. If a scalar field $f(x_1, \ldots, x_n)$ has a relative extremum when it is subject to m constraints, say

$$(9.40) \qquad g_1(x_1, \ldots, x_n) = 0, \qquad \ldots, \qquad g_m(x_1, \ldots, x_n) = 0,$$

where $m < n$, then there exist m scalars $\lambda_1, \ldots, \lambda_m$ such that

$$(9.41) \qquad \nabla f = \lambda_1 \nabla g_1 + \cdots + \lambda_m \nabla g_m$$

at each extremum point.

To determine the extremum points in practice we consider the system of $n + m$ equations obtained by taking the m constraint equations in (9.40) along with the n scalar equations determined by the vector relation (9.41). These equations are to be solved (if possible) for the $n + m$ unknowns $x_1, \ldots, x_n$ and $\lambda_1, \ldots, \lambda_m$. The points $(x_1, \ldots, x_n)$ at which relative extrema occur are found among the solutions to these equations.

The scalars $\lambda_1, \ldots, \lambda_m$ which are introduced to help us solve this type of problem are called *Lagrange's multipliers.* One multiplier is introduced for each constraint. The scalar field f and the constraint functions $g_1, \ldots, g_m$ are assumed to be differentiable. The method is valid if the number of constraints, m, is less than the number of variables, n, and if not all the Jacobian determinants of the constraint functions with respect to m of the variables $x_1, \ldots, x_n$ are zero at the extreme value in question. The proof of the validity of the method is an important result in advanced calculus and will not be discussed here. (See Chapter 7 of the author's *Mathematical Analysis*, Addison-Wesley, Reading, Mass., 1957.) Instead we give geometric arguments to show why the method works in the two examples described at the beginning of this section.

Geometric solution of Example 1. We wish to determine those points on a given surface S which are nearest to the origin. A point (x, y, z) in 3-space lies at a distance r from the origin if and only if it lies on the sphere

$$x^2 + y^2 + z^2 = r^2 .$$

This sphere is a level surface of the function $f(x, y, z) = (x^2 + y^2 + z^2)^{1/2}$ which is being minimized. If we start with $r = 0$ and let r increase until the corresponding level surface first touches the given surface S, each point of contact will be a point of S nearest to the origin.

To determine the coordinates of the contact points we assume that S is described by a Cartesian equation $g(x, y, z) = 0$. If S has a tangent plane at a point of contact, this plane must also be tangent to the contacting level surface. Therefore the gradient vector of the surface $g(x, y, z) = 0$ must be parallel to the gradient vector of the contacting level surface $f(x, y, z) = r$. Hence there is a constant λ such that

$$\nabla f(x, y, z) = \lambda \nabla g(x, y, z)$$

at each contact point. This is the vector equation (9.41) provided by Lagrange's method when there is one constraint.

Geometric solution to Example 2. We seek the extreme values of a temperature function $f(x, y, z)$ on a given curve C. If we regard the curve C as the intersection of two surfaces, say

$$g_1(x, y, z) = 0 \quad \text{and} \quad g_2(x, y, z) = 0,$$

we have an extremum problem with two constraints. The two gradient vectors ∇g_1 and ∇g_2 are normals to these surfaces, hence they are also normal to C, the curve of intersection. (See Figure 9.8.) We show next that the gradient vector ∇f of the temperature

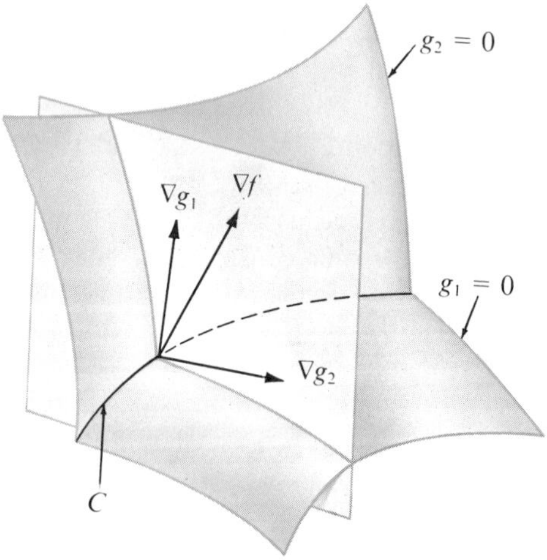

FIGURE 9.8 The vectors ∇g_1, ∇g_2, and ∇f shown lying in the same plane.

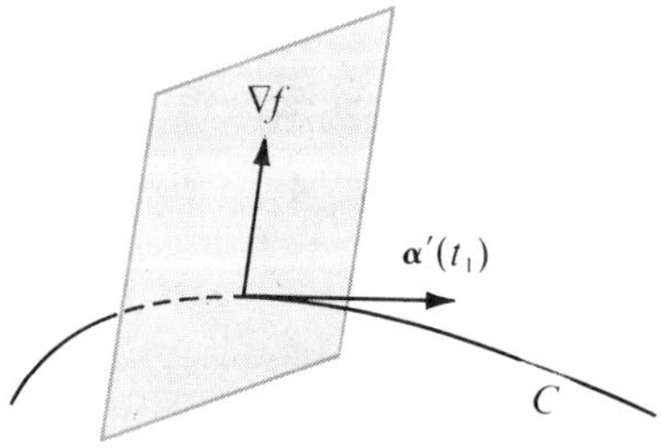

FIGURE 9.9 The gradient vector ∇f lies in a plane normal to C.

function is also normal to C at each relative extremum on C. This implies that ∇f lies in the same plane as ∇g_1 and ∇g_2; hence if ∇g_1 and ∇g_2 are independent we can express ∇f as a linear combination of ∇g_1 and ∇g_2, say

$$\nabla f = \lambda_1 \nabla g_1 + \lambda_2 \nabla g_2 .$$

This is the vector equation (9.41) provided by Lagrange's method when there are two constraints.

To show that ∇f is normal to C at an extremum we imagine C as being described by a vector-valued function $\boldsymbol{\alpha}(t)$, where t varies over an interval $[a, b]$. On the curve C the temperature becomes a function of t, say $\varphi(t) = f[\boldsymbol{\alpha}(t)]$. If φ has a relative extremum at an interior point t_1 of $[a, b]$ we must have $\varphi'(t_1) = 0$. On the other hand, the chain rule tells us that $\varphi'(t)$ is given by the dot product

$$\varphi'(t) = \nabla f[\boldsymbol{\alpha}(t)] \cdot \boldsymbol{\alpha}'(t).$$

This dot product is zero at t_1, hence ∇f is perpendicular to $\boldsymbol{\alpha}'(t_1)$. But $\boldsymbol{\alpha}'(t_1)$ is tangent to C, so $\nabla f[\boldsymbol{\alpha}(t_1)]$ lies in the plane normal to C, as shown in Figure 9.9.

The two gradient vectors ∇g_1 and ∇g_2 are independent if and only if their cross product is nonzero. The cross product is given by

$$\nabla g_1 \times \nabla g_2 = \begin{vmatrix} \boldsymbol{i} & \boldsymbol{j} & \boldsymbol{k} \\ \dfrac{\partial g_1}{\partial x} & \dfrac{\partial g_1}{\partial y} & \dfrac{\partial g_1}{\partial z} \\ \dfrac{\partial g_2}{\partial x} & \dfrac{\partial g_2}{\partial y} & \dfrac{\partial g_2}{\partial z} \end{vmatrix} = \frac{\partial(g_1, g_2)}{\partial(y, z)}\boldsymbol{i} + \frac{\partial(g_1, g_2)}{\partial(z, x)}\boldsymbol{j} + \frac{\partial(g_1, g_2)}{\partial(x, y)}\boldsymbol{k}.$$

Therefore, independence of ∇g_1 and ∇g_2 means that not all three of the Jacobian determinants on the right are zero. As remarked earlier, Lagrange's method is applicable whenever this condition is satisfied.

If ∇g_1 and ∇g_2 are dependent the method may fail. For example, suppose we try to apply Lagrange's method to find the extreme values of $f(x, y, z) = x^2 + y^2$ on the curve of intersection of the two surfaces $g_1(x, y, z) = 0$ and $g_2(x, y, z) = 0$, where $g_1(x, y, z) = z$ and $g_2(x, y, z) = z^2 - (y - 1)^3$. The two surfaces, a plane and a cylinder, intersect along the straight line C shown in Figure 9.10. The problem obviously has a solution, because

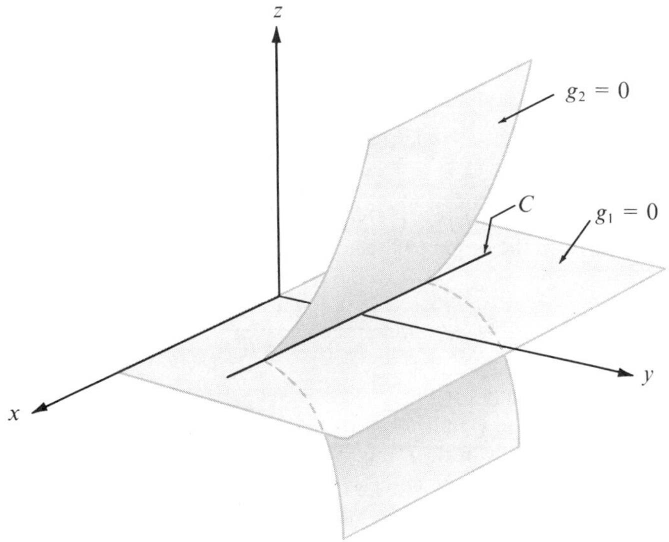

FIGURE 9.10 An example where Lagrange's method is not applicable.

$f(x, y, z)$ represents the distance of the point (x, y, z) from the z-axis and this distance is a minimum on C when the point is at $(0, 1, 0)$. However, at this point the gradient vectors are $\nabla g_1 = \boldsymbol{k}$, $\nabla g_2 = \boldsymbol{O}$, and $\nabla f = 2\boldsymbol{j}$, and it is clear that there are no scalars λ_1 and λ_2 that satisfy Equation (9.41).

9.15 Exercises

1. Find the extreme values of $z = xy$ subject to the condition $x + y = 1$.
2. Find the maximum and minimum distances from the origin to the curve $5x^2 + 6xy + 5y^2 = 8$.
3. Assume a and b are fixed positive numbers.
   (a) Find the extreme values of $z = x/a + y/b$ subject to the condition $x^2 + y^2 = 1$.
   (b) Find the extreme values of $z = x^2 + y^2$ subject to the condition $x/a + y/b = 1$.
   In each case, interpret the problem geometrically.
4. Find the extreme values of $z = \cos^2 x + \cos^2 y$ subject to the side condition $x - y = \pi/4$.
5. Find the extreme values of the scalar field $f(x, y, z) = x - 2y + 2z$ on the sphere $x^2 + y^2 + z^2 = 1$.
6. Find the points of the surface $z^2 - xy = 1$ nearest to the origin.
7. Find the shortest distance from the point $(1, 0)$ to the parabola $y^2 = 4x$.
8. Find the points on the curve of intersection of the two surfaces

$$x^2 - xy + y^2 - z^2 = 1 \qquad \text{and} \qquad x^2 + y^2 = 1$$

   which are nearest to the origin.
9. If a, b, and c are positive numbers, find the maximum value of $f(x, y, z) = x^a y^b z^c$ subject to the side condition $x + y + z = 1$.
10. Find the minimum volume bounded by the planes $x = 0$, $y = 0$, $z = 0$, and a plane which is tangent to the ellipsoid

$$\frac{x^2}{a^2} + \frac{y^2}{b^2} + \frac{z^2}{c^2} = 1$$

   at a point in the octant $x > 0$, $y > 0$, $z > 0$.
11. Find the maximum of $\log x + \log y + 3 \log z$ on that portion of the sphere $x^2 + y^2 + z^2 = 5r^2$ where $x > 0$, $y > 0$, $z > 0$. Use the result to prove that for real positive numbers a, b, c we have

$$abc^3 \le 27\left(\frac{a + b + c}{5}\right)^5 .$$

12. Given the conic section $Ax^2 + 2Bxy + Cy^2 = 1$, where $A > 0$ and $B^2 < AC$. Let m and M denote the distances from the origin to the nearest and furthest points of the conic. Show that

$$M^2 = \frac{A + C + \sqrt{(A - C)^2 + 4B^2}}{2(AC - B^2)}$$

   and find a companion formula for m^2.
13. Use the method of Lagrange's multipliers to find the greatest and least distances of a point on the ellipse $x^2 + 4y^2 = 4$ from the straight line $x + y = 4$.
14. The cross section of a trough is an isosceles trapezoid. If the trough is made by bending up the sides of a strip of metal c inches wide, what should be the angle of inclination of the sides and the width across the bottom if the cross-sectional area is to be a maximum?

9.16 The extreme-value theorem for continuous scalar fields

The extreme-value theorem for real-valued functions continuous on a closed and bounded interval can be extended to scalar fields. We consider scalar fields continuous on a closed n-dimensional interval. Such an interval is defined as the Cartesian product of n one-dimensional closed intervals. If $\boldsymbol{a} = (a_1, \ldots, a_n)$ and $\boldsymbol{b} = (b_1, \ldots, b_n)$ we write

$$[\boldsymbol{a}, \boldsymbol{b}] = [a_1, b_1] \times \cdots \times [a_n, b_n] = \{(x_1, \ldots, x_n) \mid x_1 \in [a_1, b_1], \ldots, x_n \in [a_n, b_n]\}.$$

For example, when $n = 2$ the Cartesian product $[\boldsymbol{a}, \boldsymbol{b}]$ is a rectangle.

The proof of the extreme-value theorem parallels the proof given in Volume I for the 1-dimensional case. First we prove that continuity of f implies boundedness, then we prove that f actually attains its maximum and minimum values somewhere in $[\boldsymbol{a}, \boldsymbol{b}]$.

THEOREM 9.8. BOUNDEDNESS THEOREM FOR CONTINUOUS SCALAR FIELDS. *If f is a scalar field continuous at each point of a closed interval $[\boldsymbol{a}, \boldsymbol{b}]$ in $\mathbf{R}^n$, then f is bounded on $[\boldsymbol{a}, \boldsymbol{b}]$. That is, there is a number $C \geq 0$ such that $|f(\boldsymbol{x})| \leq C$ for all $\boldsymbol{x}$ in $[\boldsymbol{a}, \boldsymbol{b}]$.*

Proof. We argue by contradiction, using the method of successive bisection. Figure 9.11 illustrates the method for the case $n = 2$.

Assume f is unbounded on $[\boldsymbol{a}, \boldsymbol{b}]$. Let $I^{(1)} = [\boldsymbol{a}, \boldsymbol{b}]$ and let $I_k^{(1)} = [a_k, b_k]$, so that

$$I^{(1)} = I_1^{(1)} \times \cdots \times I_n^{(1)}.$$

Bisect each one-dimensional interval $I_k^{(1)}$ to form two subintervals, a left half $I_{k,1}^{(1)}$ and a right half $I_{k,2}^{(1)}$. Now consider all possible Cartesian products of the form

$$I_{1,j_1}^{(1)} \times \cdots \times I_{n,j_n}^{(1)},$$

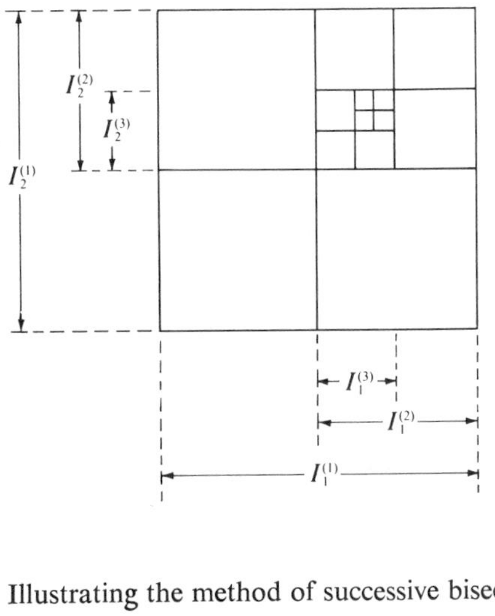

FIGURE 9.11 Illustrating the method of successive bisection in the plane.

where each $j_i = 1$ or 2. There are exactly 2^n such products. Each product is an n-dimensional subinterval of $[\boldsymbol{a}, \boldsymbol{b}]$, and their union is equal to $[\boldsymbol{a}, \boldsymbol{b}]$. The function f is unbounded in *at least one* of these subintervals (if it were bounded in each of them it would also be bounded on $[\boldsymbol{a}, \boldsymbol{b}]$). One of these we denote by $I^{(2)}$ which we express as

$$I^{(2)} = I_1^{(2)} \times \cdots \times I_n^{(2)},$$

where each $I_k^{(2)}$ is one of the one-dimensional subintervals of $I_k^{(1)}$, of length $\frac{1}{2}(b_k - a_k)$.

We now proceed with $I^{(2)}$ as we did with $I^{(1)}$, bisecting each one-dimensional component interval $I_k^{(2)}$ and arriving at an n-dimensional interval $I^{(3)}$ in which f is unbounded. We continue the process, obtaining an infinite set of n-dimensional intervals

$$I^{(1)}, I^{(2)}, \ldots, \qquad \text{with} \quad I^{(m+1)} \subseteq I^{(m)},$$

in each of which f is unbounded. The mth interval $I^{(m)}$ can be expressed in the form

$$I^{(m)} = I_1^{(m)} \times \cdots \times I_n^{(m)}.$$

Since each one-dimensional interval $I_k^{(m)}$ is obtained by $m - 1$ successive bisections of $[a_k, b_k]$, if we write $I_k^{(m)} = [a_k^{(m)}, b_k^{(m)}]$ we have

$$\text{(9.42)} \qquad b_k^{(m)} - a_k^{(m)} = \frac{b_k - a_k}{2^{m-1}}, \qquad \text{for} \quad k = 1, 2, \ldots, n.$$

For each fixed k, the supremum of all left endpoints $a_k^{(m)}$ $(m = 1, 2, \ldots)$ must therefore be equal to the infimum of all right endpoints $b_k^{(m)}$ $(m = 1, 2, \ldots)$, and their common value we denote by t_k. The point $\boldsymbol{t} = (t_1, \ldots, t_n)$ lies in $[\boldsymbol{a}, \boldsymbol{b}]$. By continuity of f at $\boldsymbol{t}$ there is an n-ball $B(\boldsymbol{t}; r)$ in which we have

$$|f(\boldsymbol{x}) - f(\boldsymbol{t})| \leq 1 \qquad \text{for all } \boldsymbol{x} \text{ in } B(\boldsymbol{t}; r) \cap [\boldsymbol{a}, \boldsymbol{b}].$$

This inequality implies

$$|f(\boldsymbol{x})| < 1 + |f(\boldsymbol{t})| \qquad \text{for all } \boldsymbol{x} \text{ in } B(\boldsymbol{t}; r) \cap [\boldsymbol{a}, \boldsymbol{b}],$$

so f is bounded on the set $B(\boldsymbol{t}; r) \cap [\boldsymbol{a}, \boldsymbol{b}]$. But this set contains the entire interval $I^{(m)}$ when m is large enough so that each of the n numbers in (9.42) is less than $r/\sqrt{n}$. Therefore for such m the function f is bounded on $I^{(m)}$, contradicting the fact that f is unbounded on $I^{(m)}$. This contradiction completes the proof.

If f is bounded on $[\boldsymbol{a}, \boldsymbol{b}]$, the set of all function values $f(\boldsymbol{x})$ is a set of real numbers bounded above and below. Therefore this set has a supremum and an infimum which we denote by $\sup f$ and $\inf f$, respectively. That is, we write

$$\sup f = \sup \{f(\boldsymbol{x}) \mid \boldsymbol{x} \in [\boldsymbol{a}, \boldsymbol{b}]\}, \qquad \inf f = \inf \{f(\boldsymbol{x}) \mid \boldsymbol{x} \in [\boldsymbol{a}, \boldsymbol{b}]\}.$$

Now we prove that a continuous function takes on both values $\inf f$ and $\sup f$ somewhere in $[\boldsymbol{a}, \boldsymbol{b}]$.

THEOREM 9.9 EXTREME-VALUE THEOREM FOR CONTINUOUS SCALAR FIELDS. *If f is continuous on a closed interval $[\boldsymbol{a}, \boldsymbol{b}]$ in $\mathbf{R}^n$, then there exist points $\boldsymbol{c}$ and $\boldsymbol{d}$ in $[\boldsymbol{a}, \boldsymbol{b}]$ such that*

$$f(\boldsymbol{c}) = \sup f \qquad \textit{and} \qquad f(\boldsymbol{d}) = \inf f.$$

Proof. It suffices to prove that f attains its supremum in $[\boldsymbol{a}, \boldsymbol{b}]$. The result for the infimum then follows as a consequence because the infimum of f is the supremum of $-f$.

Let $M = \sup f$. We shall assume that there is no $\boldsymbol{x}$ in $[\boldsymbol{a}, \boldsymbol{b}]$ for which $f(\boldsymbol{x}) = M$ and obtain a contradiction. Let $g(\boldsymbol{x}) = M - f(\boldsymbol{x})$. Then $g(\boldsymbol{x}) > 0$ for all $\boldsymbol{x}$ in $[\boldsymbol{a}, \boldsymbol{b}]$ so the reciprocal $1/g$ is continuous on $[\boldsymbol{a}, \boldsymbol{b}]$. By the boundedness theorem, $1/g$ is bounded on $[\boldsymbol{a}, \boldsymbol{b}]$, say $1/g(\boldsymbol{x}) < C$ for all $\boldsymbol{x}$ in $[\boldsymbol{a}, \boldsymbol{b}]$, where $C > 0$. This implies $M - f(\boldsymbol{x}) > 1/C$, so that $f(\boldsymbol{x}) < M - 1/C$ for all $\boldsymbol{x}$ in $[\boldsymbol{a}, \boldsymbol{b}]$. This contradicts the fact that M is the least upper bound of f on $[\boldsymbol{a}, \boldsymbol{b}]$. Hence $f(\boldsymbol{x}) = M$ for at least one $\boldsymbol{x}$ in $[\boldsymbol{a}, \boldsymbol{b}]$.

9.17 The small-span theorem for continuous scalar fields (uniform continuity)

Let f be continuous on a bounded closed interval $[\boldsymbol{a}, \boldsymbol{b}]$ in $\mathbf{R}^n$, and let $M(f)$ and $m(f)$ denote, respectively, the maximum and minimum values of f on $[\boldsymbol{a}, \boldsymbol{b}]$. The difference

$$M(f) - m(f)$$

is called the *span* of f on $[\boldsymbol{a}, \boldsymbol{b}]$. As in the one-dimensional case we have a small-span theorem for continuous functions which tells us that the interval $[\boldsymbol{a}, \boldsymbol{b}]$ can be partitioned so that the span of f in each subinterval is arbitrarily small.

Write $[\boldsymbol{a}, \boldsymbol{b}] = [a_1, b_1] \times \cdots \times [a_n, b_n]$, and let P_k be a partition of the interval $[a_k, b_k]$. That is, P_k is a set of points

$$P_k = \{x_0, x_1, \ldots, x_{r-1}, x_r\}$$

such that $a_k = x_0 \le x_1 \le \cdots \le x_{r-1} \le x_r = b_k$. The Cartesian product

$$P = P_1 \times \cdots \times P_n$$

is called a partition of the interval $[\boldsymbol{a}, \boldsymbol{b}]$. The small-span theorem, also called the theorem on uniform continuity, now takes the following form.

THEOREM 9.10. *Let f be a scalar field continuous on a closed interval $[\boldsymbol{a}, \boldsymbol{b}]$ in $\mathbf{R}^n$. Then for every $\epsilon > 0$ there is a partition of $[\boldsymbol{a}, \boldsymbol{b}]$ into a finite number of subintervals such that the span of f in every subinterval is less than ϵ.*

Proof. The proof is entirely analogous to the one-dimensional case so we only outline the principal steps. We argue by contradiction, using the method of successive bisection. We assume the theorem is false; that is, we assume that for some ϵ_0 the interval $[\boldsymbol{a}, \boldsymbol{b}]$

cannot be partitioned into a finite number of subintervals in each of which the span of f is less than ϵ_0. By successive bisection we obtain an infinite set of subintervals $I^{(1)}, I^{(2)}, \ldots,$ in each of which the span of f is at least ϵ_0. By considering the least upper bound of the leftmost endpoints of the component intervals of $I^{(1)}, I^{(2)}, \ldots$ we obtain a point $\boldsymbol{t}$ in $[\boldsymbol{a}, \boldsymbol{b}]$ lying in all these intervals. By continuity of f at $\boldsymbol{t}$ there is an n-ball $B(\boldsymbol{t}; r)$ such that the span of f is less than $\frac{1}{2}\epsilon_0$ in $B(\boldsymbol{t}; r) \cap [\boldsymbol{a}, \boldsymbol{b}]$. But, when m is sufficiently large, the interval $I^{(m)}$ lies in the set $B(\boldsymbol{t}; r) \cap [\boldsymbol{a}, \boldsymbol{b}]$, so the span of f is no larger than $\frac{1}{2}\epsilon_0$ in $I^{(m)}$, contradicting the fact that the span of f is at least ϵ_0 in $I^{(m)}$.

10

LINE INTEGRALS

10.1 Introduction

In Volume I we discussed the integral $\int_a^b f(x)\,dx$, first for real-valued functions defined and bounded on finite intervals, and then for unbounded functions and infinite intervals. The concept was later extended to vector-valued functions and, in Chapter 7 of Volume II, to matrix functions.

This chapter extends the notion of integral in another direction. The interval $[a, b]$ is replaced by a curve in n-space described by a vector-valued function $\boldsymbol{\alpha}$, and the integrand is a vector field $\boldsymbol{f}$ defined and bounded on this curve. The resulting integral is called a *line integral*, a *curvilinear integral*, or a *contour integral*, and is denoted by $\int \boldsymbol{f} \cdot d\boldsymbol{\alpha}$ or by some similar symbol. The dot is used purposely to suggest an inner product of two vectors. The curve is called a *path of integration.*

Line integrals are of fundamental importance in both pure and applied mathematics. They occur in connection with work, potential energy, heat flow, change in entropy, circulation of a fluid, and other physical situations in which the behavior of a vector or scalar field is studied along a curve.

10.2 Paths and line integrals

Before defining line integrals we recall the definition of curve given in Volume I. Let $\boldsymbol{\alpha}$ be a vector-valued function defined on a finite closed interval $J = [a, b]$. As t runs through J, the function values $\boldsymbol{\alpha}(t)$ trace out a set of points in n-space called the *graph* of the function. If $\boldsymbol{\alpha}$ is continuous on J the graph is called a *curve*; more specifically, the curve described by $\boldsymbol{\alpha}$.

In our study of curves in Volume I we found that different functions can trace out the same curve in different ways, for example, in different directions or with different velocities. In the study of line integrals we are concerned not only with the set of points on a curve but with the actual manner in which the curve is traced out, that is, with the function $\boldsymbol{\alpha}$ itself. Such a function will be called a *continuous path.*

DEFINITION. *Let $J = [a, b]$ be a finite closed interval in $\mathbf{R}^1$. A function $\boldsymbol{\alpha}: J \to \mathbf{R}^n$ which is continuous on J is called a continuous path in n-space. The path is called smooth if the derivative $\boldsymbol{\alpha}'$ exists and is continuous in the open interval (a, b). The path is called piecewise* *smooth if the interval* $[a, b]$ *can be partitioned into a finite number of subintervals in each of which the path is smooth.*

Figure 10.1 shows the graph of a piecewise smooth path. In this example the curve has a tangent line at all but a finite number of its points. These exceptional points subdivide the curve into arcs, along each of which the tangent line turns continuously.

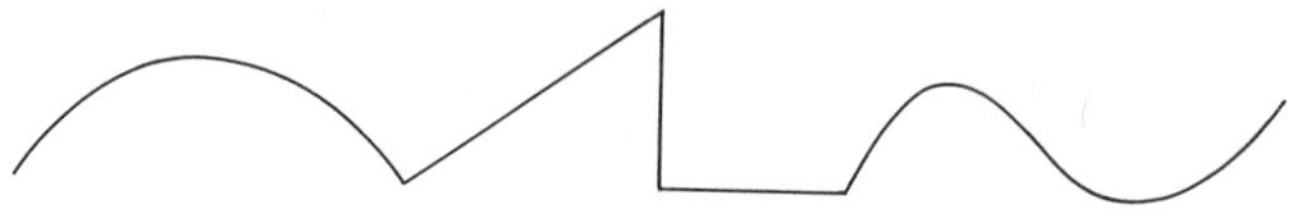

FIGURE 10.1 The graph of a piecewise smooth path in the plane.

DEFINITION OF LINE INTEGRAL. *Let* $\boldsymbol{\alpha}$ *be a piecewise smooth path in n-space defined on an interval* $[a, b]$, *and let* $\boldsymbol{f}$ *be a vector field defined and bounded on the graph of* $\boldsymbol{\alpha}$. *The line integral of* $\boldsymbol{f}$ *along* $\boldsymbol{\alpha}$ *is denoted by the symbol* $\int \boldsymbol{f} \cdot d\boldsymbol{\alpha}$ *and is defined by the equation*

$$\int \boldsymbol{f} \cdot d\boldsymbol{\alpha} = \int_a^b \boldsymbol{f}[\boldsymbol{\alpha}(t)] \cdot \boldsymbol{\alpha}'(t)\, dt\,, \tag{10.1}$$

whenever the integral on the right exists, either as a proper or improper integral.

Note: In most examples that occur in practice the dot product $\boldsymbol{f}[\boldsymbol{\alpha}(t)] \cdot \boldsymbol{\alpha}'(t)$ is bounded on $[a, b]$ and continuous except possibly at a finite number of points, in which case the integral exists as a proper integral.

10.3 Other notations for line integrals

If C denotes the graph of $\boldsymbol{\alpha}$, the line integral $\int \boldsymbol{f} \cdot d\boldsymbol{\alpha}$ is also written as $\int_C \boldsymbol{f} \cdot d\boldsymbol{\alpha}$ and is called *the integral of* $\boldsymbol{f}$ *along* C.

If $\boldsymbol{a} = \boldsymbol{\alpha}(a)$ and $\boldsymbol{b} = \boldsymbol{\alpha}(b)$ denote the end points of C, the line integral is sometimes written as $\int_{\boldsymbol{a}}^{\boldsymbol{b}} \boldsymbol{f}$ or as $\int_{\boldsymbol{a}}^{\boldsymbol{b}} \boldsymbol{f} \cdot d\boldsymbol{\alpha}$ and is called the line integral of $\boldsymbol{f}$ from $\boldsymbol{a}$ to $\boldsymbol{b}$ along $\boldsymbol{\alpha}$. When the notation $\int_{\boldsymbol{a}}^{\boldsymbol{b}} \boldsymbol{f}$ is used it should be kept in mind that the integral depends not only on the end points $\boldsymbol{a}$ and $\boldsymbol{b}$ but also on the path $\boldsymbol{\alpha}$ joining them.

When $\boldsymbol{a} = \boldsymbol{b}$ the path is said to be *closed.* The symbol $\oint$ is often used to indicate integration along a closed path.

When $\boldsymbol{f}$ and $\boldsymbol{\alpha}$ are expressed in terms of their components, say

$$\boldsymbol{f} = (f_1, \ldots, f_n) \qquad \text{and} \qquad \boldsymbol{\alpha} = (\alpha_1, \ldots, \alpha_n),$$

the integral on the right of (10.1) becomes a sum of integrals,

$$\sum_{k=1}^{n} \int_a^b f_k[\boldsymbol{\alpha}(t)]\alpha_k'(t)\, dt\,.$$

In this case the line integral is also written as $\int f_1\, d\alpha_1 + \cdots + f_n\, d\alpha_n$.

In the two-dimensional case the path $\boldsymbol{\alpha}$ is usually described by a pair of parametric equations,

$$x = \alpha_1(t), \qquad y = \alpha_2(t),$$

and the line integral $\int_C \boldsymbol{f} \cdot d\boldsymbol{\alpha}$ is written as $\int_C f_1\, dx + f_2\, dy$, or as $\int_C f_1(x, y)\, dx + f_2(x, y)\, dy$.

In the three-dimensional case we use three parametric equations,

$$x = \alpha_1(t), \qquad y = \alpha_2(t), \qquad z = \alpha_3(t)$$

and write the line integral $\int_C \boldsymbol{f} \cdot d\boldsymbol{\alpha}$ as $\int_C f_1\, dx + f_2\, dy + f_3\, dz$, or as

$$\int_C f_1(x, y, z)\, dx + f_2(x, y, z)\, dy + f_3(x, y, z)\, dz.$$

EXAMPLE. Let $\boldsymbol{f}$ be a two-dimensional vector field given by

$$\boldsymbol{f}(x, y) = \sqrt{y}\, \boldsymbol{i} + (x^3 + y)\boldsymbol{j}$$

for all (x, y) with $y \geq 0$. Calculate the line integral of $\boldsymbol{f}$ from $(0, 0)$ to $(1, 1)$ along each of the following paths:

(a) the line with parametric equations $x = t$, $y = t$, $0 \leq t \leq 1$;

(b) the path with parametric equations $x = t^2$, $y = t^3$, $0 \leq t \leq 1$.

Solution. For the path in part (a) we take $\boldsymbol{\alpha}(t) = t\boldsymbol{i} + t\boldsymbol{j}$. Then $\boldsymbol{\alpha}'(t) = \boldsymbol{i} + \boldsymbol{j}$ and $\boldsymbol{f}[\boldsymbol{\alpha}(t)] = \sqrt{t}\, \boldsymbol{i} + (t^3 + t)\boldsymbol{j}$. Therefore the dot product of $\boldsymbol{f}[\boldsymbol{\alpha}(t)]$ and $\boldsymbol{\alpha}'(t)$ is equal to $\sqrt{t} + t^3 + t$ and we find

$$\int_{(0,0)}^{(1,1)} \boldsymbol{f} \cdot d\boldsymbol{\alpha} = \int_0^1 (\sqrt{t} + t^3 + t)\, dt = \frac{17}{12}.$$

For the path in part (b) we take $\boldsymbol{\alpha}(t) = t^2\boldsymbol{i} + t^3\boldsymbol{j}$. Then $\boldsymbol{\alpha}'(t) = 2t\boldsymbol{i} + 3t^2\boldsymbol{j}$ and $\boldsymbol{f}[\boldsymbol{\alpha}(t)] = t^{3/2}\boldsymbol{i} + (t^6 + t^3)\boldsymbol{j}$. Therefore

$$\boldsymbol{f}[\boldsymbol{\alpha}(t)] \cdot \boldsymbol{\alpha}'(t) = 2t^{5/2} + 3t^8 + 3t^5,$$

so

$$\int_{(0,0)}^{(1,1)} \boldsymbol{f} \cdot d\boldsymbol{\alpha} = \int_0^1 (2t^{5/2} + 3t^8 + 3t^5)\, dt = \frac{59}{42}.$$

This example shows that the integral from one point to another may depend on the path joining the two points.

Now let us carry out the calculation for part (b) once more, using the same curve but with a different parametric representation. The same curve can be described by the function

$$\boldsymbol{\beta}(t) = t\boldsymbol{i} + t^{3/2}\boldsymbol{j}, \qquad \text{where} \quad 0 \leq t \leq 1.$$

This leads to the relation

$$\boldsymbol{f}[\boldsymbol{\beta}(t)] \cdot \boldsymbol{\beta}'(t) = (t^{3/4}\boldsymbol{i} + (t^3 + t^{3/2})\boldsymbol{j}) \cdot (\boldsymbol{i} + \tfrac{3}{2}t^{1/2}\boldsymbol{j}) = t^{3/4} + \tfrac{3}{2}t^{7/2} + \tfrac{3}{2}t^2,$$

the integral of which from 0 to 1 is 59/42, as before. This calculation illustrates that the value of the integral is independent of the parametric representation used to describe the curve. This is a general property of line integrals which is proved in the next section.

10.4 Basic properties of line integrals

Since line integrals are defined in terms of ordinary integrals, it is not surprising to find that they share many of the properties of ordinary integrals. For example, they have a *linearity property* with respect to the integrand,

$$\int (a\boldsymbol{f} + b\boldsymbol{g}) \cdot d\boldsymbol{\alpha} = a \int \boldsymbol{f} \cdot d\boldsymbol{\alpha} + b \int \boldsymbol{g} \cdot d\boldsymbol{\alpha},$$

and an *additive property* with respect to the path of integration:

$$\int_C \boldsymbol{f} \cdot d\boldsymbol{\alpha} = \int_{C_1} \boldsymbol{f} \cdot d\boldsymbol{\alpha} + \int_{C_2} \boldsymbol{f} \cdot d\boldsymbol{\alpha},$$

where the two curves C_1 and C_2 make up the curve C. That is, C is described by a function $\boldsymbol{\alpha}$ defined on an interval $[a, b]$, and the curves C_1 and C_2 are those traced out by $\boldsymbol{\alpha}(t)$ as t varies over subintervals $[a, c]$ and $[c, b]$, respectively, for some c satisfying $a < c < b$. The proofs of these properties follow immediately from the definition of the line integral; they are left as exercises for the reader.

Next we examine the behavior of line integrals under a change of parameter. Let $\boldsymbol{\alpha}$ be a continuous path defined on an interval $[a, b]$, let u be a real-valued function that is differentiable, with u' never zero on an interval $[c, d]$, and such that the range of u is $[a, b]$. Then the function $\boldsymbol{\beta}$ defined on $[c, d]$ by the equation

$$\boldsymbol{\beta}(t) = \boldsymbol{\alpha}[u(t)]$$

is a continuous path having the same graph as $\boldsymbol{\alpha}$. Two paths $\boldsymbol{\alpha}$ and $\boldsymbol{\beta}$ so related are called *equivalent*. They are said to provide different parametric representations of the same curve. The function u is said to define a change of parameter.

Let C denote the common graph of two equivalent paths $\boldsymbol{\alpha}$ and $\boldsymbol{\beta}$. If the derivative of u is always positive on $[c, d]$ the function u is increasing and we say that the two paths $\boldsymbol{\alpha}$ and $\boldsymbol{\beta}$ trace out C in the *same direction*. If the derivative of u is always negative we say that $\boldsymbol{\alpha}$ and $\boldsymbol{\beta}$ trace out C in *opposite directions*. In the first case the function u is said to be *orientation-preserving*; in the second case u is said to be *orientation-reversing*. An example is shown in Figure 10.2.

The next theorem shows that a line integral remains unchanged under a change of parameter that preserves orientation; it reverses its sign if the change of parameter reverses orientation. We assume both intergals $\int \boldsymbol{f} \cdot d\boldsymbol{\alpha}$ and $\int \boldsymbol{f} \cdot d\boldsymbol{\beta}$ exist.

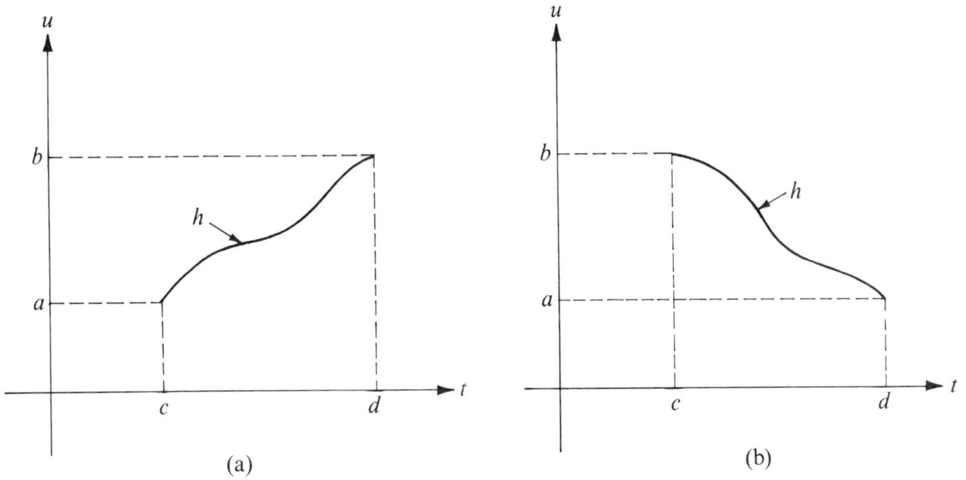

FIGURE 10.2 A change of parameter defined by $u = h(t)$. In (a), the function h preserves orientation. In (b), the function h reverses the orientation.

THEOREM 10.1. BEHAVIOR OF A LINE INTEGRAL UNDER A CHANGE OF PARAMETER. *Let $\boldsymbol{\alpha}$ and $\boldsymbol{\beta}$ be equivalent piecewise smooth paths. Then we have*

$$\int_C \boldsymbol{f} \cdot d\boldsymbol{\alpha} = \int_C \boldsymbol{f} \cdot d\boldsymbol{\beta}$$

if $\boldsymbol{\alpha}$ and $\boldsymbol{\beta}$ trace out C in the same direction; and

$$\int_C \boldsymbol{f} \cdot d\boldsymbol{\alpha} = -\int_C \boldsymbol{f} \cdot d\boldsymbol{\beta}$$

if $\boldsymbol{\alpha}$ and $\boldsymbol{\beta}$ trace out C in opposite directions.

Proof. It suffices to prove the theorem for smooth paths; then we invoke the additive property with respect to the path of integration to deduce the result for piecewise smooth paths.

The proof is a simple application of the chain rule. The paths $\boldsymbol{\alpha}$ and $\boldsymbol{\beta}$ are related by an equation of the form $\boldsymbol{\beta}(t) = \boldsymbol{\alpha}[u(t)]$, where u is defined on an interval $[c, d]$ and $\boldsymbol{\alpha}$ is defined on an interval $[a, b]$. From the chain rule we have

$$\boldsymbol{\beta}'(t) = \boldsymbol{\alpha}'[u(t)]u'(t).$$

Therefore we find

$$\int_C \boldsymbol{f} \cdot d\boldsymbol{\beta} = \int_c^d \boldsymbol{f}[\boldsymbol{\beta}(t)] \cdot \boldsymbol{\beta}'(t)\, dt = \int_c^d \boldsymbol{f}(\boldsymbol{\alpha}[u(t)]) \cdot \boldsymbol{\alpha}'[u(t)]u'(t)\, dt.$$

In the last integral we introduce the substitution $v = u(t)$, $dv = u'(t)\, dt$ to obtain

$$\int_C \boldsymbol{f} \cdot d\boldsymbol{\beta} = \int_{u(c)}^{u(d)} \boldsymbol{f}(\boldsymbol{\alpha}(v)) \cdot \boldsymbol{\alpha}'(v)\, dv = \pm \int_a^b \boldsymbol{f}(\boldsymbol{\alpha}(v)) \cdot \boldsymbol{\alpha}'(v)\, dv = \pm \int_C \boldsymbol{f} \cdot d\boldsymbol{\alpha},$$

where the + sign is used if $a = u(c)$ and $b = u(d)$, and the − sign is used if $a = u(d)$ and $b = u(c)$. The first case occurs if $\boldsymbol{\alpha}$ and $\boldsymbol{\beta}$ trace out C in the same direction, the second if they trace out C in opposite directions.

10.5 Exercises

In each of Exercises 1 through 8 calculate the line integral of the vector field f along the path described.

1. $f(x, y) = (x^2 - 2xy)\boldsymbol{i} + (y^2 - 2xy)\boldsymbol{j}$, from $(-1, 1)$ to $(1, 1)$ along the parabola $y = x^2$.
2. $f(x, y) = (2a - y)\boldsymbol{i} + x\boldsymbol{j}$, along the path described by $\boldsymbol{\alpha}(t) = a(t - \sin t)\boldsymbol{i} + a(1 - \cos t)\boldsymbol{j}$, $0 \le t \le 2\pi$.
3. $f(x, y, z) = (y^2 - z^2)\boldsymbol{i} + 2yz\boldsymbol{j} - x^2\boldsymbol{k}$, along the path described by $\boldsymbol{\alpha}(t) = t\boldsymbol{i} + t^2\boldsymbol{j} + t^3\boldsymbol{k}$, $0 \le t \le 1$.
4. $f(x, y) = (x^2 + y^2)\boldsymbol{i} + (x^2 - y^2)\boldsymbol{j}$, from $(0, 0)$ to $(2, 0)$ along the curve $y = 1 - |1 - x|$.
5. $f(x, y) = (x + y)\boldsymbol{i} + (x - y)\boldsymbol{j}$, once around the ellipse $b^2x^2 + a^2y^2 = a^2b^2$ in a counterclockwise direction.
6. $f(x, y, z) = 2xy\boldsymbol{i} + (x^2 + z)\boldsymbol{j} + y\boldsymbol{k}$, from $(1, 0, 2)$ to $(3, 4, 1)$ along a line segment.
7. $f(x, y, z) = x\boldsymbol{i} + y\boldsymbol{j} + (xz - y)\boldsymbol{k}$, from $(0, 0, 0)$ to $(1, 2, 4)$ along a line segment.
8. $f(x, y, z) = x\boldsymbol{i} + y\boldsymbol{j} + (xz - y)\boldsymbol{k}$, along the path described by $\boldsymbol{\alpha}(t) = t^2\boldsymbol{i} + 2t\boldsymbol{j} + 4t^3\boldsymbol{k}$, $0 \le t \le 1$.

In each of Exercises 9 through 12, compute the value of the given line integral.

9. $\int_C (x^2 - 2xy)\,dx + (y^2 - 2xy)\,dy$, where C is a path from $(-2, 4)$ to $(1, 1)$ along the parabola $y = x^2$.
10. $\displaystyle\int_C \frac{(x + y)\,dx - (x - y)\,dy}{x^2 + y^2}$, where C is the circle $x^2 + y^2 = a^2$, traversed once in a counterclockwise direction.
11. $\displaystyle\int_C \frac{dx + dy}{|x| + |y|}$, where C is the square with vertices $(1, 0)$, $(0, 1)$, $(-1, 0)$, and $(0, -1)$, traversed once in a counterclockwise direction.
12. $\int_C y\,dx + z\,dy + x\,dz$, where
    (a) C is the curve of intersection of the two surfaces $x + y = 2$ and $x^2 + y^2 + z^2 = 2(x + y)$. The curve is to be traversed once in a direction that appears clockwise when viewed from the origin.
    (b) C is the intersection of the two surfaces $z = xy$ and $x^2 + y^2 = 1$, traversed once in a direction that appears counterclockwise when viewed from high above the xy-plane.

10.6 The concept of work as a line intregal

Consider a particle which moves along a curve under the action of a force field f. If the curve is the graph of a piecewise smooth path $\boldsymbol{\alpha}$, the *work* done by f is defined to be the line integral $\int f \cdot d\boldsymbol{\alpha}$. The following examples illustrate some fundamental properties of work.

EXAMPLE 1. *Work done by a constant force.* If f is a constant force, say $f = \boldsymbol{c}$, it can be shown that the work done by f in moving a particle from a point $\boldsymbol{a}$ to a point $\boldsymbol{b}$ along any piecewise smooth path joining $\boldsymbol{a}$ and $\boldsymbol{b}$ is $\boldsymbol{c} \cdot (\boldsymbol{b} - \boldsymbol{a})$, the dot product of the force and the displacement vector $\boldsymbol{b} - \boldsymbol{a}$. We shall prove this in a special case.

We let $\boldsymbol{\alpha} = (\alpha_1, \ldots, \alpha_n)$ be a path joining $\boldsymbol{a}$ and $\boldsymbol{b}$, say $\boldsymbol{\alpha}(a) = \boldsymbol{a}$ and $\boldsymbol{\alpha}(b) = \boldsymbol{b}$, and we write $\boldsymbol{c} = (c_1, \ldots, c_n)$. Assume $\boldsymbol{\alpha}'$ is continuous on $[a, b]$. Then the work done by $\boldsymbol{f}$ is equal to

$$\int \boldsymbol{f} \cdot d\boldsymbol{\alpha} = \sum_{k=1}^{n} c_k \int_a^b \alpha_k'(t)\, dt = \sum_{k=1}^{n} c_k[\alpha_k(b) - \alpha_k(a)] = \boldsymbol{c} \cdot [\boldsymbol{\alpha}(b) - \boldsymbol{\alpha}(a)] = \boldsymbol{c} \cdot (\boldsymbol{b} - \boldsymbol{a}).$$

For this force field the work depends only on the end points $\boldsymbol{a}$ and $\boldsymbol{b}$ and not on the curve joining them. Not all force fields have this property. Those which do are called *conservative*. The example on p. 325 is a nonconservative force field. In a later section we shall determine all conservative force fields.

EXAMPLE 2. *The principle of work and energy.* A particle of mass m moves along a curve under the action of a force field $\boldsymbol{f}$. If the speed of the particle at time t is $v(t)$, its kinetic energy is defined to be $\frac{1}{2}mv^2(t)$. Prove that *the change in kinetic energy in any time interval is equal to the work done by* $\boldsymbol{f}$ *during this time interval.*

Solution. Let $\boldsymbol{r}(t)$ denote the position of the particle at time t. The work done by $\boldsymbol{f}$ during a time interval $[a, b]$ is $\int_{\boldsymbol{r}(a)}^{\boldsymbol{r}(b)} \boldsymbol{f} \cdot d\boldsymbol{r}$. We wish to prove that

$$\int_{\boldsymbol{r}(a)}^{\boldsymbol{r}(b)} \boldsymbol{f} \cdot d\boldsymbol{r} = \tfrac{1}{2}mv^2(b) - \tfrac{1}{2}mv^2(a).$$

From Newton's second law of motion we have

$$\boldsymbol{f}[\boldsymbol{r}(t)] = m\boldsymbol{r}''(t) = m\boldsymbol{v}'(t),$$

where $\boldsymbol{v}(t)$ denotes the velocity vector at time t. The speed is the length of the velocity vector, $v(t) = \|\boldsymbol{v}(t)\|$. Therefore

$$\boldsymbol{f}[\boldsymbol{r}(t)] \cdot \boldsymbol{r}'(t) = \boldsymbol{f}[\boldsymbol{r}(t)] \cdot \boldsymbol{v}(t) = m\boldsymbol{v}'(t) \cdot \boldsymbol{v}(t) = \tfrac{1}{2}m \frac{d}{dt}(\boldsymbol{v}(t) \cdot \boldsymbol{v}(t)) = \tfrac{1}{2}m \frac{d}{dt}(v^2(t)).$$

Integrating from a to b we obtain

$$\int_{\boldsymbol{r}(a)}^{\boldsymbol{r}(b)} \boldsymbol{f} \cdot d\boldsymbol{r} = \int_a^b \boldsymbol{f}[\boldsymbol{r}(t)] \cdot \boldsymbol{r}'(t)\, dt = \tfrac{1}{2}mv^2(t)\Big|_a^b = \tfrac{1}{2}mv^2(b) - \tfrac{1}{2}mv^2(a),$$

as required.

10.7 Line integrals with respect to arc length

Let $\boldsymbol{\alpha}$ be a path with $\boldsymbol{\alpha}'$ continuous on an interval $[a, b]$. The graph of $\boldsymbol{\alpha}$ is a rectifiable curve. In Volume I we proved that the corresponding arc-length function s is given by the integral

$$s(t) = \int_a^t \|\boldsymbol{\alpha}'(u)\|\, du.$$

The derivative of arc length is given by

$$s'(t) = \|\boldsymbol{\alpha}'(t)\|.$$

Let φ be a scalar field defined and bounded on C, the graph of $\boldsymbol{\alpha}$. The *line integral of* φ *with respect to arc length along* C is denoted by the symbol $\int_C \varphi\, ds$ and is defined by the equation

$$\int_C \varphi\, ds = \int_a^b \varphi[\boldsymbol{\alpha}(t)]s'(t)\, dt\,,$$

whenever the integral on the right exists.

Now consider a scalar field φ given by $\varphi[\boldsymbol{\alpha}(t)] = \boldsymbol{f}[\boldsymbol{\alpha}(t)] \cdot \boldsymbol{T}(t)$, the dot product of a vector field $\boldsymbol{f}$ defined on C and the unit tangent vector $\boldsymbol{T}(t) = (d\boldsymbol{\alpha}/ds)$. In this case the line integral $\int_C \varphi\, ds$ is the same as the line integral $\int_C \boldsymbol{f} \cdot d\boldsymbol{\alpha}$ because

$$\boldsymbol{f}[\boldsymbol{\alpha}(t)] \cdot \boldsymbol{\alpha}'(t) = \boldsymbol{f}[\boldsymbol{\alpha}(t)] \cdot \frac{d\boldsymbol{\alpha}}{ds}\frac{ds}{dt} = \boldsymbol{f}[\boldsymbol{\alpha}(t)] \cdot \boldsymbol{T}(t)s'(t) = \varphi[\boldsymbol{\alpha}(t)]s'(t)\,.$$

When $\boldsymbol{f}$ denotes a velocity field, the dot product $\boldsymbol{f} \cdot \boldsymbol{T}$ is the tangential component of velocity, and the line integral $\int_C \boldsymbol{f} \cdot \boldsymbol{T}\, ds$ is called the *flow integral* of $\boldsymbol{f}$ along C. When C is a closed curve the flow integral is called the *circulation* of $\boldsymbol{f}$ along C. These terms are commonly used in the theory of fluid flow.

10.8 Further applications of line integrals

Line integrals with respect to arc length also occur in problems concerned with mass distribution along a curve. For example, think of a curve C in 3-space as a wire made of a thin material of varying density. Assume the density is described by a scalar field φ, where $\varphi(x, y, z)$ is the *mass per unit length* at the point (x, y, z) of C. Then the total mass M of the wire is defined to be the line integral of φ with respect to arc length:

$$M = \int_C \varphi(x, y, z)\, ds\,.$$

The *center of mass* of the wire is defined to be the point $(\bar{x}, \bar{y}, \bar{z})$ whose coordinates are determined by the equations

$$\bar{x}M = \int_C x\varphi(x, y, z)\, ds\,, \qquad \bar{y}M = \int_C y\varphi(x, y, z)\, ds\,, \qquad \bar{z}M = \int_C z\varphi(x, y, z)\, ds\,.$$

A wire of constant density is called *uniform*. In this case the center of mass is also called the *centroid*.

EXAMPLE 1. Compute the mass M of one coil of a spring having the shape of the helix whose vector equation is

$$\boldsymbol{\alpha}(t) = a\cos t\,\boldsymbol{i} + a\sin t\,\boldsymbol{j} + bt\boldsymbol{k}$$

if the density at (x, y, z) is $x^2 + y^2 + z^2$.

Solution. The integral for M is

$$M = \int_C (x^2 + y^2 + z^2)\, ds = \int_0^{2\pi} (a^2\cos^2 t + a^2\sin^2 t + b^2t^2)s'(t)\, dt\,.$$

Since $s'(t) = \|\boldsymbol{\alpha}'(t)\|$ and $\boldsymbol{\alpha}'(t) = -a \sin t\, \boldsymbol{i} + a \cos t\, \boldsymbol{j} + b\boldsymbol{k}$, we have $s'(t) = \sqrt{a^2 + b^2}$ and hence

$$M = \sqrt{a^2 + b^2} \int_0^{2\pi} (a^2 + b^2t^2)\, dt = \sqrt{a^2 + b^2}\left(2\pi a^2 + \frac{8}{3}\pi^3 b^2\right).$$

In this example the z-coordinate $\bar{z}$ of the center of mass is given by

$$\begin{aligned} \bar{z}M &= \int_C z(x^2 + y^2 + z^2)\, ds = \sqrt{a^2 + b^2} \int_0^{2\pi} bt(a^2 + b^2t^2)\, dt \\ &= b\sqrt{a^2 + b^2}\,(2\pi^2 a^2 + 4\pi^4 b^2). \end{aligned}$$

The coordinates $\bar{x}$ and $\bar{y}$ are requested in Exercise 15 of Section 10.9.

Line integrals can be used to define the moment of inertia of a wire with respect to an axis. If $\delta(x, y, z)$ represents the perpendicular distance from a point (x, y, z) of C to an axis L, the moment of inertia I_L is defined to be the line integral

$$I_L = \int_C \delta^2(x, y, z)\varphi(x, y, z)\, ds,$$

where $\varphi(x, y, z)$ is the density at (x, y, z). The moments of inertia about the coordinate axes are denoted by I_x, I_y, and I_z.

EXAMPLE 2. Compute the moment of inertia I_z of the spring coil in Example 1.

Solution. Here $\delta^2(x, y, z) = x^2 + y^2 = a^2$ and $\varphi(x, y, z) = x^2 + y^2 + z^2$, so we have

$$I_z = \int_C (x^2 + y^2)(x^2 + y^2 + z^2)\, ds = a^2 \int_C (x^2 + y^2 + z^2)\, ds = Ma^2,$$

where M is the mass, as computed in Example 1.

10.9 Exercises

1. A force field f in 3-space is given by $f(x, y, z) = x\boldsymbol{i} + y\boldsymbol{j} + (xz - y)\boldsymbol{k}$. Compute the work done by this force in moving a particle from $(0, 0, 0)$ to $(1, 2, 4)$ along the line segment joining these two points.
2. Find the amount of work done by the force $f(x, y) = (x^2 - y^2)\boldsymbol{i} + 2xy\boldsymbol{j}$ in moving a particle (in a counterclockwise direction) once around the square bounded by the coordinate axes and the lines $x = a$ and $y = a$, $a > 0$.
3. A two-dimensional force field f is given by the equation $f(x, y) = cxy\boldsymbol{i} + x^6y^2\boldsymbol{j}$, where c is a positive constant. This force acts on a particle which must move from $(0, 0)$ to the line $x = 1$ along a curve of the form

$$y = ax^b, \qquad \text{where} \quad a > 0 \quad \text{and} \quad b > 0.$$

Find a value of a (in terms of c) such that the work done by this force is independent of b.
4. A force field f in 3-space is given by the formula $f(x, y, z) = yz\boldsymbol{i} + xz\boldsymbol{j} + x(y + 1)\boldsymbol{k}$. Calculate the work done by f in moving a particle once around the triangle with vertices $(0, 0, 0)$, $(1, 1, 1)$, $(-1, 1, -1)$ in that order.

5. Calculate the work done by the force field $f(x, y, z) = (y - z)\boldsymbol{i} + (z - x)\boldsymbol{j} + (x - y)\boldsymbol{k}$ along the curve of intersection of the sphere $x^2 + y^2 + z^2 = 4$ and the plane $z = y \tan \theta$, where $0 < \theta < \pi/2$. The path is transversed in a direction that appears counterclockwise when viewed from high above the xy-plane.
6. Calculate the work done by the force field $f(x, y, z) = y^2\boldsymbol{i} + z^2\boldsymbol{j} + x^2\boldsymbol{k}$ along the curve of intersection of the sphere $x^2 + y^2 + z^2 = a^2$ and the cylinder $x^2 + y^2 = ax$, where $z \geq 0$ and $a > 0$. The path is traversed in a direction that appears clockwise when viewed from high above the xy-plane.

Calculate the line integral with respect to arc length in each of Exercises 7 through 10.

7. $\int_C (x + y)\, ds$, where C is the triangle with vertices (0, 0), (1, 0), and (0, 1), traversed in a counterclockwise direction.
8. $\int_C y^2\, ds$, where C has the vector equation

$$\boldsymbol{\alpha}(t) = a(t - \sin t)\boldsymbol{i} + a(1 - \cos t)\boldsymbol{j}, \qquad 0 \leq t \leq 2\pi.$$

9. $\int_C (x^2 + y^2)\, ds$, where C has the vector equation

$$\boldsymbol{\alpha}(t) = a(\cos t + t \sin t)\boldsymbol{i} + a(\sin t - t \cos t)\boldsymbol{j}, \qquad 0 \leq t \leq 2\pi.$$

10. $\int_C z\, ds$, where C has the vector equation

$$\boldsymbol{\alpha}(t) = t \cos t\, \boldsymbol{i} + t \sin t\, \boldsymbol{j} + t\boldsymbol{k}, \qquad 0 \leq t \leq t_0.$$

11. Consider a uniform semicircular wire of radius a.
(a) Show that the centroid lies on the axis of symmetry at a distance $2a/\pi$ from the center.
(b) Show that the moment of inertia about the diameter through the end points of the wire is $\frac{1}{2}Ma^2$, where M is the mass of the wire.
12. A wire has the shape of the circle $x^2 + y^2 = a^2$. Determine its mass and moment of inertia about a diameter if the density at (x, y) is $|x| + |y|$.
13. Find the mass of a wire whose shape is that of the curve of intersection of the sphere $x^2 + y^2 + z^2 = 1$ and the plane $x + y + z = 0$ if the density of the wire at (x, y, z) is x^2.
14. A uniform wire has the shape of that portion of the curve of intersection of the two surfaces $x^2 + y^2 = z^2$ and $y^2 = x$ connecting the points (0, 0, 0) and $(1, 1, \sqrt{2})$. Find the z-coordinate of its centroid.
15. Determine the coordinates $\bar{x}$ and $\bar{y}$ of the center of mass of the spring coil described in Example 1 of Section 10.8.
16. For the spring coil described in Example 1 of Section 10.8, compute the moments of inertia I_x and I_y.

10.10 Open connected sets. Independence of the path

Let S be an open set in $\mathbf{R}^n$. The set S is called *connected* if every pair of points in S can be joined by a piecewise smooth path whose graph lies in S. That is, for every pair of points $\boldsymbol{a}$ and $\boldsymbol{b}$ in S there is a piecewise smooth path $\boldsymbol{\alpha}$ defined on an interval $[a, b]$ such that $\boldsymbol{\alpha}(t) \in S$ for each t in $[a, b]$, with $\boldsymbol{\alpha}(a) = \boldsymbol{a}$ and $\boldsymbol{\alpha}(b) = \boldsymbol{b}$.

Three examples of open connected sets in the plane are shown in Figure 10.3. Examples in 3-space analogous to these would be (a) a solid ellipsoid, (b) a solid polyhedron, and (c) a solid torus; in each case only the interior points are considered.

An open set S is said to be *disconnected* if S is the union of two or more disjoint nonempty open sets. An example is shown in Figure 10.4. It can be shown that the class of open connected sets is identical with the class of open sets that are not disconnected.†

Now let f be a vector field that is continuous on an open connected set S. Choose two points $\boldsymbol{a}$ and $\boldsymbol{b}$ in S and consider the line integral of $\boldsymbol{f}$ from $\boldsymbol{a}$ to $\boldsymbol{b}$ along some piecewise smooth path in S. The value of the integral depends, in general, on the path joining $\boldsymbol{a}$ to $\boldsymbol{b}$. For some vector fields, the integral depends only on the end points $\boldsymbol{a}$ and $\boldsymbol{b}$ and not on the path which joins them. In this case we say the integral is *independent of the path from* $\boldsymbol{a}$ *to* $\boldsymbol{b}$. We say the line integral of f is *independent of the path in* S if it is independent of the path from $\boldsymbol{a}$ to $\boldsymbol{b}$ for every pair of points $\boldsymbol{a}$ and $\boldsymbol{b}$ in S.

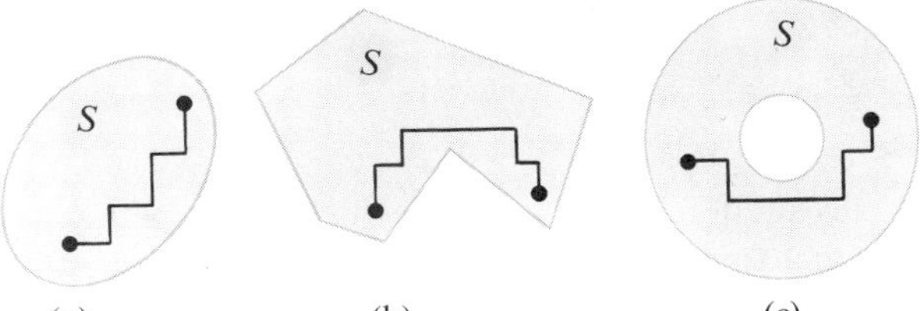

FIGURE 10.3 Examples of open connected sets.

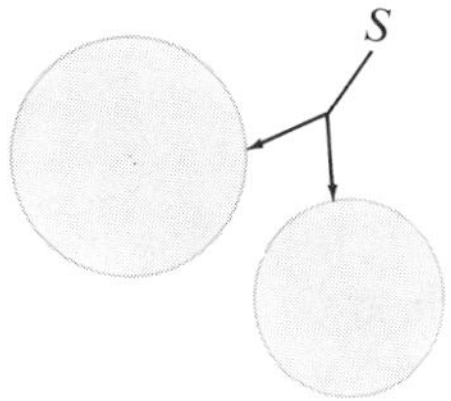

FIGURE 10.4 A disconnected set S, the union of two disjoint circular disks.

Which vector fields have line integrals independent of the path? To answer this question we extend the first and second fundamental theorems of calculus to line integrals.

10.11 The second fundamental theorem of calculus for line integrals

The second fundamental theorem for real functions, as proved in Volume I (Theorem 5.3), states that

$$\int_a^b \varphi'(t)\,dt = \varphi(b) - \varphi(a),$$

provided that φ' is continuous on some open interval containing both a and b. To extend this result to line integrals we need a slightly stronger version of the theorem in which continuity of φ' is assumed only in the open interval (a, b).

THEOREM 10.2. *Let φ be a real function that is continuous on a closed interval $[a, b]$ and assume that the integral $\int_a^b \varphi'(t)\,dt$ exists. If φ' is continuous on the open interval (a, b), we have*

$$\int_a^b \varphi'(t)\,dt = \varphi(b) - \varphi(a).$$

Proof. For each x in $[a, b]$ define $f(x) = \int_a^x \varphi'(t)\,dt$. We wish to prove that

$$f(b) = \varphi(b) - \varphi(a). \tag{10.2}$$

† For a further discussion of connectedness, see Chapter 8 of the author's *Mathematical Analysis*, Addison-Wesley, Reading, Mass., 1957.

By Theorem 3.4 of Volume I, f is continuous on the closed interval $[a, b]$. By Theorem 5.1 of Volume I, f is differentiable on the open interval (a, b), with $f'(x) = \varphi'(x)$ for each x in (a, b). Therefore, by the zero-derivative theorem (Theorem 5.2 of Volume I), the difference $f - \varphi$ is constant on the open interval (a, b). By continuity, $f - \varphi$ is also constant on the closed interval $[a, b]$. In particular, $f(b) - \varphi(b) = f(a) - \varphi(a)$. But since $f(a) = 0$, this proves (10.2).

THEOREM 10.3. SECOND FUNDAMENTAL THEOREM OF CALCULUS FOR LINE INTEGRALS. *Let φ be a differentiable scalar field with a continuous gradient $\nabla\varphi$ on an open connected set S in $\mathbf{R}^n$. Then for any two points $\boldsymbol{a}$ and $\boldsymbol{b}$ joined by a piecewise smooth path $\boldsymbol{\alpha}$ in S we have*

$$\int_{\boldsymbol{a}}^{\boldsymbol{b}} \nabla\varphi \cdot d\boldsymbol{\alpha} = \varphi(\boldsymbol{b}) - \varphi(\boldsymbol{a}).$$

Proof. Choose any two points $\boldsymbol{a}$ and $\boldsymbol{b}$ in S and join them by a piecewise smooth path $\boldsymbol{\alpha}$ in S defined on an interval $[a, b]$. Assume first that $\boldsymbol{\alpha}$ is *smooth* on $[a, b]$. Then the line integral of $\nabla\varphi$ from $\boldsymbol{a}$ to $\boldsymbol{b}$ along $\boldsymbol{\alpha}$ is given by

$$\int_{\boldsymbol{a}}^{\boldsymbol{b}} \nabla\varphi \cdot d\boldsymbol{\alpha} = \int_a^b \nabla\varphi[\boldsymbol{\alpha}(t)] \cdot \boldsymbol{\alpha}'(t)\, dt.$$

By the chain rule we have

$$\nabla\varphi[\boldsymbol{\alpha}(t)] \cdot \boldsymbol{\alpha}'(t) = g'(t),$$

where g is the composite function defined on $[a, b]$ by the formula

$$g(t) = \varphi[\boldsymbol{\alpha}(t)].$$

The derivative g' is continuous on the open interval (a, b) because $\nabla\varphi$ is continuous on S and $\boldsymbol{\alpha}$ is smooth. Therefore we can apply Theorem 10.2 to g to obtain

$$\int_{\boldsymbol{a}}^{\boldsymbol{b}} \nabla\varphi \cdot d\boldsymbol{\alpha} = \int_a^b g'(t)\, dt = g(b) - g(a) = \varphi[\boldsymbol{\alpha}(b)] - \varphi[\boldsymbol{\alpha}(a)] = \varphi(\boldsymbol{b}) - \varphi(\boldsymbol{a}).$$

This proves the theorem if $\boldsymbol{\alpha}$ is smooth.

When $\boldsymbol{\alpha}$ is piecewise smooth we partition the interval $[a, b]$ into a finite number (say r) of subintervals $[t_{k-1}, t_k]$, in each of which $\boldsymbol{\alpha}$ is smooth, and we apply the result just proved to each subinterval. This gives us

$$\int_{\boldsymbol{a}}^{\boldsymbol{b}} \nabla\varphi = \sum_{k=1}^{r} \int_{\boldsymbol{\alpha}(t_{k-1})}^{\boldsymbol{\alpha}(t_k)} \nabla\varphi = \sum_{k=1}^{r} \{\varphi[\boldsymbol{\alpha}(t_k)] - \varphi[\boldsymbol{\alpha}(t_{k-1})]\} = \varphi(\boldsymbol{b}) - \varphi(\boldsymbol{a}),$$

as required.

As a consequence of Theorem 10.3 we see that the line integral of a gradient is independent of the path in any open connected set S in which the gradient is continuous. For a closed path we have $\boldsymbol{b} = \boldsymbol{a}$, so $\varphi(\boldsymbol{b}) - \varphi(\boldsymbol{a}) = 0$. In other words, *the line integral of a continuous gradient is zero around every piecewise smooth closed path in S.* In Section 10.14 we shall prove (in Theorem 10.4) that gradients are the *only* continuous vector fields with this property.

10.12 Applications to mechanics

If a vector field f is the gradient of a scalar field φ, then φ is called a *potential function* for f. In 3-space, the level sets of φ are called *equipotential surfaces*; in 2-space they are called *equipotential lines.* (If φ denotes temperature, the word "equipotential" is replaced by "isothermal"; if φ denotes pressure the word "isobaric" is used.)

EXAMPLE 1. In 3-space, let $\varphi(x, y, z) = r^n$, where $r = (x^2 + y^2 + z^2)^{1/2}$. For every integer n we have

$$\nabla(r^n) = nr^{n-2}\boldsymbol{r},$$

where $\boldsymbol{r} = x\boldsymbol{i} + y\boldsymbol{j} + z\boldsymbol{k}$. (See Exercise 8 of Section 8.14.) Therefore φ is a potential of the vector field

$$\boldsymbol{f}(x, y, z) = nr^{n-2}\boldsymbol{r}.$$

The equipotential surfaces of φ are concentric spheres centered at the origin.

EXAMPLE 2. *The Newtonian potential.* Newton's gravitation law states that the force $\boldsymbol{f}$ which a particle of mass M exerts on another particle of mass m is a vector of length GmM/r^2, where G is a constant and r is the distance between the two particles. Place the origin at the particle of mass M, and let $\boldsymbol{r} = x\boldsymbol{i} + y\boldsymbol{j} + z\boldsymbol{k}$ be the position vector from the origin to the particle of mass m. Then $r = \|\boldsymbol{r}\|$ and $-\boldsymbol{r}/r$ is a unit vector with the same direction as $\boldsymbol{f}$, so Newton's law becomes

$$\boldsymbol{f} = -GmMr^{-3}\boldsymbol{r}.$$

Taking $n = -1$ in Example 1 we see that the gravitational force $\boldsymbol{f}$ is the gradient of the scalar field given by

$$\varphi(x, y, z) = GmMr^{-1}.$$

This is called the *Newtonian potential.*

The work done by the gravitational force in moving the particle of mass m from (x_1, y_1, z_1) to (x_2, y_2, z_2) is

$$\varphi(x_1, y_1, z_1) - \varphi(x_2, y_2, z_2) = GmM\left(\frac{1}{r_1} - \frac{1}{r_2}\right),$$

where $r_1 = (x_1^2 + y_1^2 + z_1^2)^{1/2}$ and $r_2 = (x_2^2 + y_2^2 + z_2^2)^{1/2}$. If the two points lie on the same equipotential surface then $r_1 = r_2$ and no work is done.

EXAMPLE 3. *The principle of conservation of mechanical energy.* Let $\boldsymbol{f}$ be a continuous force field having a potential φ in an open connected set S. Theorem 10.3 tells us that the work done by $\boldsymbol{f}$ in moving a particle from $\boldsymbol{a}$ to $\boldsymbol{x}$ along any piecewise smooth path in S is

$\varphi(\boldsymbol{x}) - \varphi(\boldsymbol{a})$, the change in the potential function. In Example 2 of Section 10.6 we proved that this work is also equal to the change in kinetic energy of the particle, $k(\boldsymbol{x}) - k(\boldsymbol{a})$ where $k(\boldsymbol{x})$ denotes the kinetic energy of the particle when it is located at $\boldsymbol{x}$. Thus, we have

$$k(\boldsymbol{x}) - k(\boldsymbol{a}) = \varphi(\boldsymbol{x}) - \varphi(\boldsymbol{a}),$$

or

$$k(\boldsymbol{x}) - \varphi(\boldsymbol{x}) = k(\boldsymbol{a}) - \varphi(\boldsymbol{a}). \tag{10.3}$$

The scalar $-\varphi(\boldsymbol{x})$ is called the *potential energy*† of the particle.

If $\boldsymbol{a}$ is kept fixed and $\boldsymbol{x}$ is allowed to vary over the set S, Equation (10.3) tells us that the sum of $k(\boldsymbol{x})$ and $-\varphi(\boldsymbol{x})$ is constant. In other words, *if a force field is a gradient, the sum of the kinetic and potential energies of a particle moving in this field is constant.* In mechanics this is called the principle of conservation of (mechanical) energy. A force field with a potential function is said to be *conservative* because the total energy, kinetic plus potential, is conserved. In a conservative field, no net work is done in moving a particle around a closed curve back to its starting point. A force field will not be conservative if friction or viscosity exists in the system, since these tend to convert mechanical energy into heat energy.

10.13 Exercises

1. Determine which of the following open sets S in $\mathbf{R}^2$ are connected. For each connected set, choose two arbitrary distinct points in S and explain how you would find a piecewise smooth curve in S connecting the two points.
   (a) $S = \{(x, y) \mid x^2 + y^2 \geq 0\}$. (c) $S = \{(x, y) \mid x^2 + y^2 < 1\}$.
   (b) $S = \{(x, y) \mid x^2 + y^2 > 0\}$. (d) $S = \{(x, y) \mid 1 < x^2 + y^2 < 2\}$.
   (e) $S = \{(x, y) \mid x^2 + y^2 > 1 \text{ and } (x - 3)^2 + y^2 > 1\}$.
   (f) $S = \{(x, y) \mid x^2 + y^2 < 1 \text{ or } (x - 3)^2 + y^2 < 1\}$.
2. Given a two-dimensional vector field

$$\boldsymbol{f}(x, y) = P(x, y)\boldsymbol{i} + Q(x, y)\boldsymbol{j},$$

   where the partial derivatives $\partial P/\partial y$ and $\partial Q/\partial x$ are continuous on an open set S. If $\boldsymbol{f}$ is the gradient of some potential φ, prove that

$$\frac{\partial P}{\partial y} = \frac{\partial Q}{\partial x}$$

   at each point of S.
3. For each of the following vector fields, use the result of Exercise 2 to prove that $\boldsymbol{f}$ is *not* a gradient. Then find a closed path C such that $\oint_C \boldsymbol{f} \neq 0$.
   (a) $\boldsymbol{f}(x, y) = y\boldsymbol{i} - x\boldsymbol{j}$.
   (b) $\boldsymbol{f}(x, y) = y\boldsymbol{i} + (xy - x)\boldsymbol{j}$.
4. Given a three-dimensional vector field

$$\boldsymbol{f}(x, y, z) = P(x, y, z)\boldsymbol{i} + Q(x, y, z)\boldsymbol{j} + R(x, y, z)\boldsymbol{k},$$

---

† Some authors refer to $-\varphi$ as the potential function of $\boldsymbol{f}$ so that the potential energy at $\boldsymbol{x}$ will be equal to the value of the potential function φ at $\boldsymbol{x}$.

where the partial derivatives

$$\frac{\partial P}{\partial y}, \frac{\partial P}{\partial z}, \frac{\partial Q}{\partial x}, \frac{\partial Q}{\partial z}, \frac{\partial R}{\partial x}, \frac{\partial R}{\partial y},$$

are continuous on an open set S. If f is the gradient of some potential function φ, prove that

$$\frac{\partial P}{\partial y} = \frac{\partial Q}{\partial x}, \qquad \frac{\partial P}{\partial z} = \frac{\partial R}{\partial x}, \qquad \frac{\partial Q}{\partial z} = \frac{\partial R}{\partial y}$$

at each point of S.

5. For each of the following vector fields, use the result of Exercise 4 to prove that f is *not* a gradient. Then find a closed path C such that $\oint_C f \neq 0$.
(a) $f(x, y, z) = y\boldsymbol{i} + x\boldsymbol{j} + x\boldsymbol{k}$.
(b) $f(x, y, z) = xy\boldsymbol{i} + (x^2 + 1)\boldsymbol{j} + z^2\boldsymbol{k}$.
6. A force field f is defined in 3-space by the equation

$$f(x, y, z) = y\boldsymbol{i} + z\boldsymbol{j} + yz\boldsymbol{k}.$$

(a) Determine whether or not f is conservative.
(b) Calculate the work done in moving a particle along the curve described by

$$\boldsymbol{\alpha}(t) = \cos t\,\boldsymbol{i} + \sin t\,\boldsymbol{j} + e^t\boldsymbol{k}$$

as t runs from 0 to π.
7. A two-dimensional force field $\boldsymbol{F}$ is described by the equation

$$\boldsymbol{F}(x, y) = (x + y)\boldsymbol{i} + (x - y)\boldsymbol{j}.$$

(a) Show that the work done by this force in moving a particle along a curve

$$\boldsymbol{\alpha}(t) = f(t)\boldsymbol{i} + g(t)\boldsymbol{j}, \qquad a \le t \le b,$$

depends only on $f(a), f(b), g(a), g(b)$.
(b) Find the amount of work done when $f(a) = 1, f(b) = 2, g(a) = 3, g(b) = 4$.
8. A force field is given in polar coordinates by the equation

$$\boldsymbol{F}(r,\theta) = -4 \sin \theta\,\boldsymbol{i} + 4 \sin \theta\,\boldsymbol{j}.$$

Compute the work done in moving a particle from the point (1, 0) to the origin along the spiral whose polar equation is $r = e^{-\theta}$.
9. A radial or "central" force field $\boldsymbol{F}$ in the plane can be written in the form $\boldsymbol{F}(x, y) = f(r)\boldsymbol{r}$, where $\boldsymbol{r} = x\boldsymbol{i} + y\boldsymbol{j}$ and $r = \|\boldsymbol{r}\|$. Show that such a force field is conservative.
10. Find the work done by force $\boldsymbol{F}(x, y) = (3y^2 + 2)\boldsymbol{i} + 16x\boldsymbol{j}$. in moving a particle from $(-1, 0)$ to $(1, 0)$ along the upper half of the ellipse $b^2x^2 + y^2 = b^2$. Which ellipse (that is, which value of b) makes the work a minimum?

10.14 The first fundamental theorem of calculus for line integrals

Section 10.11 extended the second fundamental theorem of calculus to line integrals. This section extends the first fundamental theorem. We recall that the first fundamental theorem states that every indefinite integral of a continuous function f has a derivative equal to f. That is, if

$$\varphi(x) = \int_a^x f(t)\,dt,$$

then at the points of continuity of f we have

$$\varphi'(x) = f(x).$$

To extend this theorem to line integrals we begin with a vector field $\boldsymbol{f}$, continuous on an open connected set S, and integrate it along a piecewise smooth curve C from a fixed point $\boldsymbol{a}$ in S to an arbitrary point $\boldsymbol{x}$. Then we let φ denote the scalar field defined by the line integral

$$\varphi(\boldsymbol{x}) = \int_{\boldsymbol{a}}^{\boldsymbol{x}} \boldsymbol{f} \cdot d\boldsymbol{\alpha},$$

where $\boldsymbol{\alpha}$ describes C. Since S is connected, each point $\boldsymbol{x}$ in S can be reached by such a curve. For this definition of $\varphi(\boldsymbol{x})$ to be unambiguous, we need to know that the integral depends only on $\boldsymbol{x}$ and not on the particular path used to join $\boldsymbol{a}$ to $\boldsymbol{x}$. Therefore, it is natural to require the line integral of $\boldsymbol{f}$ to be independent of the path in S. Under these conditions, the extension of the first fundamental theorem takes the following form:

THEOREM 10.4. FIRST FUNDAMENTAL THEOREM FOR LINE INTEGRALS. *Let $\boldsymbol{f}$ be a vector field that is continuous on an open connected set S in $\mathbf{R}^n$, and assume that the line integral of $\boldsymbol{f}$ is independent of the path in S. Let $\boldsymbol{a}$ be a fixed point of S and define a scalar field φ on S by the equation*

$$\varphi(\boldsymbol{x}) = \int_{\boldsymbol{a}}^{\boldsymbol{x}} \boldsymbol{f} \cdot d\boldsymbol{\alpha},$$

where $\boldsymbol{\alpha}$ is any piecewise smooth path in S joining $\boldsymbol{a}$ to $\boldsymbol{x}$. Then the gradient of φ exists and is equal to $\boldsymbol{f}$; that is,

$$\nabla\varphi(\boldsymbol{x}) = \boldsymbol{f}(\boldsymbol{x}) \qquad \textit{for every } \boldsymbol{x} \textit{ in } S.$$

Proof. We shall prove that the partial derivative $D_k\varphi(\boldsymbol{x})$ exists and is equal to $f_k(\boldsymbol{x})$, the kth component of $\boldsymbol{f}(\boldsymbol{x})$, for each $k = 1, 2, \ldots, n$ and each $\boldsymbol{x}$ in S.

Let $B(\boldsymbol{x}; r)$ be an n-ball with center at $\boldsymbol{x}$ and radius r lying in S. If $\boldsymbol{y}$ is a unit vector, the point $\boldsymbol{x} + h\boldsymbol{y}$ also lies in S for every real h satisfying $0 < |h| < r$, and we can form the difference quotient

$$\frac{\varphi(\boldsymbol{x} + h\boldsymbol{y}) - \varphi(\boldsymbol{x})}{h}.$$

Because of the additive property of line integrals, the numerator of this quotient can be written as

$$\varphi(\boldsymbol{x} + h\boldsymbol{y}) - \varphi(\boldsymbol{x}) = \int_{\boldsymbol{x}}^{\boldsymbol{x}+h\boldsymbol{y}} \boldsymbol{f} \cdot d\boldsymbol{\alpha},$$

and the path joining $\boldsymbol{x}$ to $\boldsymbol{x} + h\boldsymbol{y}$ can be any piecewise smooth path lying in S. In particular, we can use the line segment described by

$$\boldsymbol{\alpha}(t) = \boldsymbol{x} + th\boldsymbol{y}, \qquad \text{where} \quad 0 \le t \le 1.$$

Since $\boldsymbol{\alpha}'(t) = h\boldsymbol{y}$, the difference quotient becomes

$$\frac{\varphi(\boldsymbol{x} + h\boldsymbol{y}) - \varphi(\boldsymbol{x})}{h} = \int_0^1 \boldsymbol{f}(\boldsymbol{x} + th\boldsymbol{y}) \cdot \boldsymbol{y}\, dt. \tag{10.4}$$

Now we take $\boldsymbol{y} = \boldsymbol{e}_k$, the kth unit coordinate vector, and note that the integrand becomes $\boldsymbol{f}(\boldsymbol{x} + th\boldsymbol{y}) \cdot \boldsymbol{y} = f_k(\boldsymbol{x} + th\boldsymbol{e}_k)$. Then we make the change of variable $u = ht$, $du = h\,dt$, and we write (10.4) in the form

$$\frac{\varphi(\boldsymbol{x} + h\boldsymbol{e}_k) - \varphi(\boldsymbol{x})}{h} = \frac{1}{h}\int_0^h f_k(\boldsymbol{x} + u\boldsymbol{e}_k)\, du = \frac{g(h) - g(0)}{h}, \tag{10.5}$$

where g is the function defined on the open interval $(-r, r)$ by the equation

$$g(t) = \int_0^t f_k(\boldsymbol{x} + u\boldsymbol{e}_k)\, du.$$

Since each component f_k is continuous on S, the first fundamental theorem for ordinary integrals tells us that $g'(t)$ exists for each t in $(-r, r)$ and that

$$g'(t) = f_k(\boldsymbol{x} + t\boldsymbol{e}_k).$$

In particular, $g'(0) = f_k(\boldsymbol{x})$. Therefore, if we let $h \to 0$ in (10.5) we find that

$$\lim_{h \to 0} \frac{\varphi(\boldsymbol{x} + h\boldsymbol{e}_k) - \varphi(\boldsymbol{x})}{h} = \lim_{h \to 0} \frac{g(h) - g(0)}{h} = g'(0) = f_k(\boldsymbol{x}).$$

This proves that the partial derivative $D_k\varphi(\boldsymbol{x})$ exists and equals $f_k(\boldsymbol{x})$, as asserted.

10.15 Necessary and sufficient conditions for a vector field to be a gradient

The first and second fundamental theorems for line integrals together tell us that a necessary and sufficient condition for a continuous vector field to be a gradient on an open connected set is for its line integral between any two points to be independent of the path. We shall prove now that this condition is equivalent to the statement that the line integral is zero around every piecewise smooth *closed* path. All these conditions are summarized in the following theorem.

THEOREM 10.5. NECESSARY AND SUFFICIENT CONDITIONS FOR A VECTOR FIELD TO BE A GRADIENT. *Let $\boldsymbol{f}$ be a vector field continuous on an open connected set S in $\mathbf{R}^n$. Then the following three statements are equivalent.*

(a) *$\boldsymbol{f}$ is the gradient of some potential function in S.*
(b) *The line integral of $\boldsymbol{f}$ is independent of the path in S.*
(c) *The line integral of $\boldsymbol{f}$ is zero around every piecewise smooth closed path in S.*

Proof. We shall prove that (b) implies (a), (a) implies (c), and (c) implies (b). Statement (b) implies (a) because of the first fundamental theorem. The second fundamental theorem shows that (a) implies (c).

To complete the proof we show that (c) implies (b). Assume (c) holds and let C_1 and C_2 be any two piecewise smooth curves in S with the same end points. Let C_1 be the graph of a function $\boldsymbol{\alpha}$ defined on an interval $[a, b]$, and let C_2 be the graph of a function $\boldsymbol{\beta}$ defined on $[c, d]$.

Define a new function $\boldsymbol{\gamma}$ as follows:

$$\boldsymbol{\gamma}(t) = \begin{cases} \boldsymbol{\alpha}(t) & \text{if } a \le t \le b, \\ \boldsymbol{\beta}(b + d - t) & \text{if } b \le t \le b + d - c. \end{cases}$$

Then $\boldsymbol{\gamma}$ describes a closed curve C such that

$$\oint_C \boldsymbol{f} \cdot d\boldsymbol{\gamma} = \int_{C_1} \boldsymbol{f} \cdot d\boldsymbol{\alpha} - \int_{C_2} \boldsymbol{f} \cdot d\boldsymbol{\beta}.$$

Since $\oint_C \boldsymbol{f} \cdot d\boldsymbol{\gamma} = 0$ because of (c), we have $\int_{C_1} \boldsymbol{f} \cdot d\boldsymbol{\alpha} = \int_{C_2} \boldsymbol{f} \cdot d\boldsymbol{\beta}$, so the integral of $\boldsymbol{f}$ is independent of the path. This proves (b). Thus, (a), (b), and (c) are equivalent.

Note: If $\oint_C \boldsymbol{f} \neq 0$ for a particular closed curve C, then $\boldsymbol{f}$ is *not* a gradient. However, if a line integral $\oint_C \boldsymbol{f}$ is zero for a particular closed curve C or even for infinitely many closed curves, it does not necessarily follow that $\boldsymbol{f}$ *is* a gradient. For example, the reader can easily verify that the line integral of the vector field $\boldsymbol{f}(x, y) = x\boldsymbol{i} + xy\boldsymbol{j}$ is zero for every circle C with center at the origin. Nevertheless, this particular vector field is not a gradient.

10.16 Necessary conditions for a vector field to be a gradient

The first fundamental theorem can be used to determine whether or not a given vector field is a gradient on an open connected set S. If the line integral of $\boldsymbol{f}$ is independent of the path in S, we simply define a scalar field φ by integrating $\boldsymbol{f}$ from some fixed point to an arbitrary point $\boldsymbol{x}$ in S along a convenient path in S. Then we compute the partial derivatives of φ and compare $D_k\varphi$ with f_k, the kth component of $\boldsymbol{f}$. If $D_k\varphi(\boldsymbol{x}) = f_k(\boldsymbol{x})$ for every $\boldsymbol{x}$ in S and every k, then $\boldsymbol{f}$ is a gradient on S and φ is a potential. If $D_k\varphi(\boldsymbol{x}) \neq f_k(\boldsymbol{x})$ for some k and some $\boldsymbol{x}$, then $\boldsymbol{f}$ is *not* a gradient on S.

The next theorem gives another test for determining when a vector field $\boldsymbol{f}$ is *not* a gradient. This test is especially useful in practice because it does not require any integration.

THEOREM 10.6. NECESSARY CONDITIONS FOR A VECTOR FIELD TO BE A GRADIENT. *Let $\boldsymbol{f} = (f_1, \ldots, f_n)$ be a continuously differentiable vector field on an open set S in $\mathbf{R}^n$. If $\boldsymbol{f}$ is a gradient on S, then the partial derivatives of the components of $\boldsymbol{f}$ are related by the equations*

$$D_i f_j(\boldsymbol{x}) = D_j f_i(\boldsymbol{x}) \tag{10.6}$$

for $i, j = 1, 2, \ldots, n$ and every $\boldsymbol{x}$ in S.

Proof. If f is a gradient, then $f = \nabla \varphi$ for some potential function φ. This means that

$$f_j = D_j \varphi$$

for each $j = 1, 2, \ldots, n$. Differentiating both members of this equation with respect to x_i we find

$$D_i f_j = D_i D_j \varphi .$$

Similarly, we have

$$D_j f_i = D_j D_i \varphi .$$

Since the partial derivatives $D_i f_j$ and $D_j f_i$ are continuous on S, the two mixed partial derivatives $D_i D_j \varphi$ and $D_j D_i \varphi$ must be equal on S. This proves (10.6).

EXAMPLE 1. Determine whether or not the vector field

$$\boldsymbol{f}(x, y) = 3x^2 y \boldsymbol{i} + x^3 y \boldsymbol{j}$$

is a gradient on any open subset of $\mathbf{R}^2$.

Solution. Here we have

$$f_1(x, y) = 3x^2 y, \qquad f_2(x, y) = x^3 y.$$

The partial derivatives $D_2 f_1$ and $D_1 f_2$ are given by

$$D_2 f_1(x, y) = 3x^2, \qquad D_1 f_2(x, y) = 3x^2 y .$$

Since $D_2 f_1(x, y) \neq D_1 f_2(x, y)$ except when $x = 0$ or $y = 1$, this vector field is not a gradient on any open subset of $\mathbf{R}^2$.

The next example shows that the conditions of Theorem 10.6 are not always sufficient for a vector field to be a gradient.

EXAMPLE 2. Let S be the set of all $(x, y) \neq (0, 0)$ in $\mathbf{R}^2$, and let $\boldsymbol{f}$ be the vector field defined on S by the equation

$$\boldsymbol{f}(x, y) = \frac{-y}{x^2 + y^2} \boldsymbol{i} + \frac{x}{x^2 + y^2} \boldsymbol{j} .$$

Show that $D_1 f_2 = D_2 f_1$ everywhere on S but that, nevertheless, $\boldsymbol{f}$ is not a gradient on S.

Solution. The reader can easily verify that $D_1 f_2(x, y) = D_2 f_1(x, y)$ for all (x, y) in S. (See Exercise 17 in Section 10.18.)

To prove that $\boldsymbol{f}$ is not a gradient on S we compute the line integral of $\boldsymbol{f}$ around the unit circle given by

$$\boldsymbol{\alpha}(t) = \cos t \, \boldsymbol{i} + \sin t \, \boldsymbol{j}, \qquad 0 \le t \le 2\pi .$$

We obtain

$$\oint f \cdot d\alpha = \int_0^{2\pi} f[\alpha(t)] \cdot \alpha'(t)\, dt = \int_0^{2\pi} (\sin^2 t + \cos^2 t)\, dt = 2\pi .$$

Since the line integral around this closed path is not zero, f is not a gradient on S. Further properties of this vector field are discussed in Exercise 18 of Section 10.18.

At the end of this chapter we shall prove that the necessary conditions of Theorem 10.6 are also sufficient if they are satisfied on an open *convex* set. (See Theorem 10.9.)

10.17 Special methods for constructing potential functions

The first fundamental theorem for line integrals also gives us a method for constructing potential functions. If f is a continuous gradient on an open connected set S, the line integral of f is independent of the path in S. Therefore we can find a potential φ simply by integrating f from some fixed point $\boldsymbol{a}$ to an arbitrary point $\boldsymbol{x}$ in S, using any piecewise smooth path lying in S. The scalar field so obtained depends on the choice of the initial point $\boldsymbol{a}$. If we start from another initial point, say $\boldsymbol{b}$, we obtain a new potential ψ. But, because of the additive property of line integrals, φ and ψ can differ only by a constant, this constant being the integral of f from $\boldsymbol{a}$ to $\boldsymbol{b}$.

The following examples illustrate the use of different choices of the path of integration.

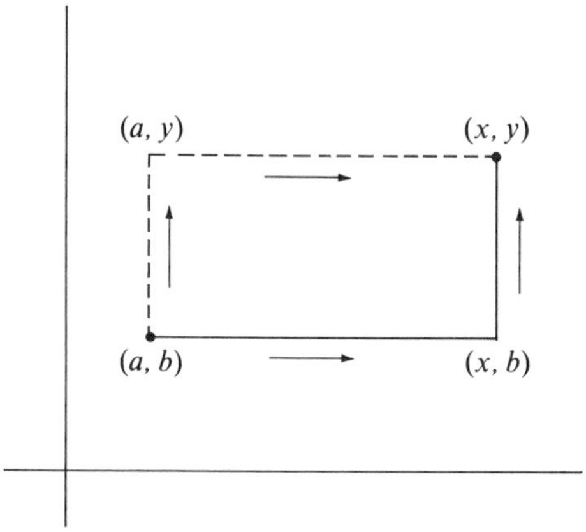

FIGURE 10.5 Two polygonal paths from (a, b) to (x, y).

EXAMPLE 1. *Construction of a potential on an open rectangle.* If f is a continuous gradient on an open rectangle in $\mathbf{R}^n$, a potential φ can be constructed by integrating from a fixed point to an arbitrary point along a set of line segments parallel to the coordinate axes. A two-dimensional example is shown in Figure 10.5. We can integrate first from (a, b) to (x, b) along a horizontal segment, then from (x, b) to (x, y) along a vertical segment. Along the horizontal segment we use the parametric representation

$$\alpha(t) = t\boldsymbol{i} + b\boldsymbol{j}, \qquad a \le t \le x,$$

and along the vertical segment we use the representation

$$\boldsymbol{\alpha}(t) = x\boldsymbol{i} + t\boldsymbol{j}, \qquad b \le t \le y.$$

If $\boldsymbol{f}(x, y) = f_1(x, y)\boldsymbol{i} + f_2(x, y)\boldsymbol{j}$, the resulting formula for a potential $\varphi(x, y)$ is

$$\varphi(x, y) = \int_a^x f_1(t, b)\, dt + \int_b^y f_2(x, t)\, dt. \tag{10.7}$$

We could also integrate first from (a, b) to (a, y) along a vertical segment and then from (a, y) to (x, y) along a horizontal segment as indicated by the dotted lines in Figure 10.5. This gives us another formula for $\varphi(x, y)$,

$$\varphi(x, y) = \int_b^y f_2(a, t)\, dt + \int_a^x f_1(t, y)\, dt. \tag{10.8}$$

Both formulas (10.7) and (10.8) give the same value for $\varphi(x, y)$ because the line integral of a gradient is independent of the path.

EXAMPLE 2. *Construction of a potential function by the use of indefinite integrals.* The use of indefinite integrals often helps to simplify the calculation of potential functions. For example, suppose a three-dimensional vector field $\boldsymbol{f} = (f_1, f_2, f_3)$ is the gradient of a potential function φ on an open set S in $\mathbf{R}^3$. Then we have

$$\frac{\partial \varphi}{\partial x} = f_1, \qquad \frac{\partial \varphi}{\partial y} = f_2, \qquad \frac{\partial \varphi}{\partial z} = f_3$$

everywhere on S. Using indefinite integrals and integrating the first of these equations with respect to x (holding y and z constant) we find

$$\varphi(x, y, z) = \int f_1(x, y, z)\, dx + A(y, z),$$

where $A(y, z)$ is a "constant of integration" to be determined. Similarly, if we integrate the equation $\partial\varphi/\partial y = f_2$ with respect to y and the equation $\partial\varphi/\partial z = f_3$ with respect to z we obtain the further relations

$$\varphi(x, y, z) = \int f_2(x, y, z)\, dy + B(x, z)$$

and

$$\varphi(x, y, z) = \int f_3(x, y, z)\, dz + C(x, y),$$

where $B(x, z)$ and $C(x, y)$ are functions to be determined. Finding φ means finding three functions $A(y, z)$, $B(x, z)$, and $C(x, y)$ such that all three equations for $\varphi(x, y, z)$ agree in their right-hand members. In many cases this can be done by inspection, as illustrated by the following example.

EXAMPLE 3. Find a potential function φ for the vector field defined on $\mathbf{R}^3$ by the equation

$$f(x, y, z) = (2xyz + z^2 - 2y^2 + 1)\boldsymbol{i} + (x^2z - 4xy)\boldsymbol{j} + (x^2y + 2xz - 2)\boldsymbol{k}.$$

Solution. Without knowing in advance whether or not f has a potential function φ, we try to construct a potential as outlined in Example 2, assuming that a potential φ exists. Integrating the component f_1 with respect to x we find

$$\varphi(x, y, z) = \int (2xyz + z^2 - 2y^2 + 1)\, dx + A(y, z) = x^2yz + xz^2 - 2xy^2 + x + A(y, z).$$

Integrating f_2 with respect to y, and then f_3 with respect to z, we find

$$\varphi(x, y, z) = \int (x^2z - 4xy)\, dy + B(x, z) = x^2yz - 2xy^2 + B(x, z),$$

$$\varphi(x, y, z) = \int (x^2y + 2xz - 2)\, dz + C(x, y) = x^2yz + xz^2 - 2z + C(x, y).$$

By inspection we see that the choices $A(y, z) = -2z$, $B(x, z) = xz^2 + x - 2z$, and $C(x, y) = x - 2xy^2$ will make all three equations agree; hence the function φ given by the equation

$$\varphi(x, y, z) = x^2yz + xz^2 - 2xy^2 + x - 2z$$

is a potential for f on $\mathbf{R}^3$.

EXAMPLE 4. *Construction of a potential on a convex set.* A set S in $\mathbf{R}^n$ is called *convex* if every pair of points in S can be joined by a line segment, all of whose points lie in S. An example is shown in Figure 10.6. Every open convex set is connected.

If f is a continuous gradient on an open convex set, then a potential φ can be constructed by integrating f from a fixed point $\boldsymbol{a}$ in S to an arbitrary point $\boldsymbol{x}$ along the line segment joining $\boldsymbol{a}$ to $\boldsymbol{x}$. The line segment can be parametrized by the function

$$\boldsymbol{\alpha}(t) = \boldsymbol{a} + t(\boldsymbol{x} - \boldsymbol{a}), \qquad \text{where} \quad 0 \le t \le 1.$$

This gives us $\boldsymbol{\alpha}'(t) = \boldsymbol{x} - \boldsymbol{a}$, so the corresponding potential is given by the integral

$$\varphi(\boldsymbol{x}) = \int_0^1 f(\boldsymbol{a} + t(\boldsymbol{x} - \boldsymbol{a})) \cdot (\boldsymbol{x} - \boldsymbol{a})\, dt. \tag{10.9}$$

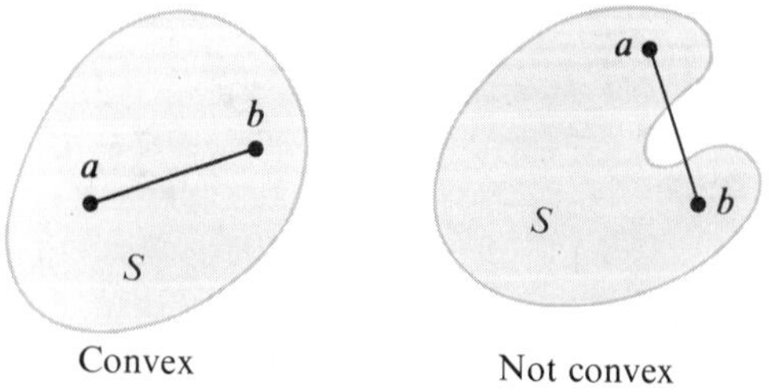

FIGURE 10.6 In a convex set S, the segment joining $\boldsymbol{a}$ and $\boldsymbol{b}$ is in S for all points $\boldsymbol{a}$ and $\boldsymbol{b}$ in S.

If S contains the origin we can take $\boldsymbol{a} = \boldsymbol{O}$ and write (10.9) more simply as

(10.10)
$$\varphi(\boldsymbol{x}) = \int_0^1 \boldsymbol{f}(t\boldsymbol{x}) \cdot \boldsymbol{x}\, dt .$$

10.18 Exercises

In each of Exercises 1 through 12, a vector field $\boldsymbol{f}$ is defined by the formulas given. In each case determine whether or not $\boldsymbol{f}$ is the gradient of a scalar field. When $\boldsymbol{f}$ is a gradient, find a corresponding potential function φ.

1. $\boldsymbol{f}(x, y) = x\boldsymbol{i} + y\boldsymbol{j}$.
2. $\boldsymbol{f}(x, y) = 3x^2y\boldsymbol{i} + x^3\boldsymbol{j}$.
3. $\boldsymbol{f}(x, y) = (2xe^y + y)\boldsymbol{i} + (x^2e^y + x - 2y)\boldsymbol{j}$.
4. $\boldsymbol{f}(x, y) = (\sin y - y \sin x + x)\boldsymbol{i} + (\cos x + x \cos y + y)\boldsymbol{j}$.
5. $\boldsymbol{f}(x, y) = [\sin (xy) + xy \cos (xy)]\boldsymbol{i} + x^2 \cos (xy)\boldsymbol{j}$.
6. $\boldsymbol{f}(x, y, z) = x\boldsymbol{i} + y\boldsymbol{j} + z\boldsymbol{k}$.
7. $\boldsymbol{f}(x, y, z) = (x + z)\boldsymbol{i} - (y + z)\boldsymbol{j} + (x - y)\boldsymbol{k}$.
8. $\boldsymbol{f}(x, y, z) = 2xy^3\boldsymbol{i} + x^2z^3\boldsymbol{j} + 3x^2yz^2\boldsymbol{k}$.
9. $\boldsymbol{f}(x, y, z) = 3y^4z^2\boldsymbol{i} + 4x^3z^2\boldsymbol{j} - 3x^2y^2\boldsymbol{k}$.
10. $\boldsymbol{f}(x, y, z) = (2x^2 + 8xy^2)\boldsymbol{i} + (3x^3y - 3xy)\boldsymbol{j} - (4y^2z^2 + 2x^3z)\boldsymbol{k}$.
11. $\boldsymbol{f}(x, y, z) = (y^2 \cos x + z^3)\boldsymbol{i} - (4 - 2y \sin x)\boldsymbol{j} + (3xz^2 + 2)\boldsymbol{k}$.
12. $\boldsymbol{f}(x, y, z) = (4xy - 3x^2z^2 + 1)\boldsymbol{i} + 2(x^2 + 1)\boldsymbol{j} - (2x^3z + 3z^2)\boldsymbol{k}$.

13. A fluid flows in the xy-plane, each particle moving directly away from the origin. If a particle is at a distance r from the origin its speed is ar^n, where a and n are constants.
(a) Determine those values of a and n for which the velocity vector field is the gradient of some scalar field.
(b) Find a potential function of the velocity whenever the velocity is a gradient. The case $n = -1$ should be treated separately.
14. If both φ and ψ are potential functions for a continuous vector field $\boldsymbol{f}$ on an open connected set S in $\mathbf{R}^n$, prove that $\varphi - \psi$ is constant on S.
15. Let S be the set of all $\boldsymbol{x} \neq \boldsymbol{O}$ in $\mathbf{R}^n$. Let $r = \|\boldsymbol{x}\|$, and let $\boldsymbol{f}$ be the vector field defined on S by the equation
$$\boldsymbol{f}(\boldsymbol{x}) = r^p\boldsymbol{x},$$
where p is a real constant. Find a potential function for $\boldsymbol{f}$ on S. The case $p = -2$ should be treated separately.
16. Solve Exercise 15 for the vector field defined by the equation
$$\boldsymbol{f}(\boldsymbol{x}) = \frac{g'(r)}{r}\boldsymbol{x},$$
where g is a real function with a continuous derivative everywhere on $\mathbf{R}^1$.

The following exercises deal with the vector field $\boldsymbol{f}$ defined on the set S of all points $(x, y) \neq (0, 0)$ in $\mathbf{R}^2$ by the equation
$$\boldsymbol{f}(x, y) = -\frac{y}{x^2 + y^2}\boldsymbol{i} + \frac{x}{x^2 + y^2}\boldsymbol{j}.$$

In Example 2 of Section 10.16 we showed that f is not a gradient on S, even though $D_1 f_2 = D_2 f_1$ everywhere on S.

17. Verify that for every point (x, y) in S we have

$$D_1 f_2(x, y) = D_2 f_1(x, y) = \frac{y^2 - x^2}{(x^2 + y^2)^2}.$$

18. This exercise shows that f *is* a gradient on the set

$$T = \mathbf{R}^2 - \{(x, y) \mid y = 0, \quad x \leq 0\},$$

consisting of all points-in the xy-plane except those on the nonpositive x-axis.

(a) If $(x, y) \in T$, express x and y in polar coordinates,

$$x = r \cos \theta, \qquad y = r \sin \theta,$$

where $r > 0$ and $-\pi < \theta < \pi$. Prove that θ is given by the formulas

$$\theta = \begin{cases} \arctan \dfrac{y}{x} & \text{if } \; x > 0, \\[2ex] \dfrac{\pi}{2} & \text{if } \; x = 0, \\[2ex] \arctan \dfrac{y}{x} + \pi & \text{if } \; x < 0. \end{cases}$$

[Recall the definition of the arc tangent function: For any real t, $\arctan t$ is the unique real number φ which satisfies the two conditions $\tan \varphi = t$ and $-\pi/2 < \varphi < \pi/2$.]

(b) Deduce that

$$\frac{\partial \theta}{\partial x} = -\frac{y}{x^2 + y^2}, \qquad \frac{\partial \theta}{\partial y} = \frac{x}{x^2 + y^2}$$

for all (x, y) in T. This proves that θ is a potential function for f on the set T.

This exercise illustrates that the existence of a potential function depends not only on the vector field f but also on the set in which the vector field is defined.

10.19 Applications to exact differential equations of first order

Some differential equations of the first order can be solved with the aid of potential functions. Suppose we have a first-order differential equation of the form

$$y' = f(x, y).$$

If we multiply both sides by a nonvanishing factor $Q(x, y)$ we transform this equation to the form $Q(x, y)y' - f(x, y)Q(x, y) = 0$. If we then write $P(x, y)$ for $-f(x, y)Q(x, y)$ and use the Leibniz notation for derivatives, writing dy/dx for y', the differential equation takes the form

$$P(x, y)\, dx + Q(x, y)\, dy = 0. \tag{10.11}$$

We assume that P and Q are continuous on some open connected set S in the plane. With each such differential equation we can associate a vector field $\boldsymbol{V}$, where

$$\boldsymbol{V}(x, y) = P(x, y)\boldsymbol{i} + Q(x, y)\boldsymbol{j}.$$

The components P and Q are the coefficients of dx and dy in Equation (10.11). The differential equation in (10.11) is said to be *exact* in S if the vector field $\boldsymbol{V}$ is the gradient of a potential; that is, if $\boldsymbol{V}(x, y) = \nabla\varphi(x, y)$ for each point (x, y) in S, where φ is some scalar field. When such a φ exists we have $\partial\varphi/\partial x = P$ and $\partial\varphi/\partial y = Q$, and the differential equation in (10.11) becomes

$$\frac{\partial\varphi}{\partial x}\,dx + \frac{\partial\varphi}{\partial y}\,dy = 0.$$

We shall prove now that each solution of this differential equation satisfies the relation $\varphi(x, y) = C$, where C is a constant. More precisely, assume there is a solution Y of the differential equation (10.11) defined on an open interval (a, b) such that the point $(x, Y(x))$ is in S for each x in (a, b). We shall prove that

$$\varphi[x, Y(x)] = C$$

for some constant C. For this purpose we introduce the composite function g defined on (a, b) by the equation

$$g(x) = \varphi[x, Y(x)].$$

By the chain rule, the derivative of g is given by

$$g'(x) = D_1\varphi[x, Y(x)] + D_2\varphi[x, Y(x)]Y'(x) = P(x, y) + Q(x, y)y', \tag{10.12}$$

where $y = Y(x)$ and $y' = Y'(x)$. If y satisfies (10.11), then $P(x, y) + Q(x, y)y' = 0$, so $g'(x) = 0$ for each x in (a, b) and, therefore, g is constant on (a, b). This proves that every solution y satisfies the equation $\varphi(x, y) = C$.

Now we may turn this argument around to find a solution of the differential equation. Suppose the equation

$$\varphi(x, y) = C \tag{10.13}$$

defines y as a differentiable function of x, say $y = Y(x)$ for x in an interval (a, b), and let $g(x) = \varphi[x, Y(x)]$. Equation (10.13) implies that g is constant on (a, b). Hence, by (10.12), $P(x, y) + Q(x, y)y' = 0$, so y is a solution. Therefore, we have proved the following theorem:

THEOREM 10.7. *Assume that the differential equation*

$$P(x, y)\,dx + Q(x, y)\,dy = 0 \tag{10.14}$$

is exact in an open connected set S, and let φ be a scalar field satisfying

$$\frac{\partial \varphi}{\partial x} = P \qquad \text{and} \qquad \frac{\partial \varphi}{\partial y} = Q$$

everywhere in S. Then every solution $y = Y(x)$ of (10.14) *whose graph lies in S satisfies the equation $\varphi[x, Y(x)] = C$ for some constant C. Conversely, if the equation*

$$\varphi(x, y) = C$$

defines y implicitly as a differentiable function of x, then this function is a solution of the differential equation (10.14).

The foregoing theorem provides a straightforward method for solving exact differential equations of the first order. We simply construct a potential function φ and then write the equation $\varphi(x, y) = C$, where C is a constant. Whenever this equation defines y implicitly as a function of x, the corresponding y satisfies (10.14). Therefore we can use Equation (10.13) as a representation of a one-parameter family of integral curves. Of course, the only admissible values of C are those for which $\varphi(x_0, y_0) = C$ for some (x_0, y_0) in S.

EXAMPLE 1. Consider the differential equation

$$\frac{dy}{dx} = -\frac{3x^2 + 6xy^2}{6x^2y + 4y^3}.$$

Clearing the fractions we may write the equation as

$$(3x^2 + 6xy^2)\,dx + (6x^2y + 4y^3)\,dy = 0.$$

This is now a special case of (10.14) with $P(x, y) = 3x^2 + 6xy^2$ and $Q(x, y) = 6x^2y + 4y^3$. Integrating P with respect to x and Q with respect to y and comparing results, we find that a potential function φ is given by the formula

$$\varphi(x, y) = x^3 + 3x^2y^2 + y^4.$$

By Theorem 10.7, each solution of the differential equation satisfies

$$x^3 + 3x^2y^2 + y^4 = C$$

for some C. This provides an implicit representation of a family of integral curves. In this particular case the equation is quadratic in y^2 and can be solved to give an explicit formula for y in terms of x and C.

EXAMPLE 2. Consider the first-order differential equation

$$y\,dx + 2x\,dy = 0. \tag{10.15}$$

Here $P(x, y) = y$ and $Q(x, y) = 2x$. Since $\partial P/\partial y = 1$ and $\partial Q/\partial x = 2$, this differential equation is not exact. However, if we multiply both sides by y we obtain an equation that *is* exact:

$$y^2\,dx + 2xy\,dy = 0. \tag{10.16}$$

A potential of the vector field $y^2\boldsymbol{i} + 2xy\boldsymbol{j}$ is $\varphi(x, y) = xy^2$, and every solution of (10.16) satisfies the relation $xy^2 = C$. This relation also represents a family of integral curves for Equation (10.15).

The multiplier y which converted (10.15) into an exact equation is called an *integrating factor.* In general, if multiplication of a first-order linear equation by a nonzero factor $\mu(x, y)$ results in an exact equation, the multiplier $\mu(x, y)$ is called an integrating factor of the original equation. A differential equation may have more than one integrating factor. For example, $\mu(x, y) = 2xy^3$ is another integrating factor of (10.15). Some special differential equations for which integrating factors can easily be found are discussed in the following set of exercises.

10.20 Exercises

Show that the differential equations in Exercises 1 through 5 are exact, and in each case find a one-parameter family of integral curves.

1. $(x + 2y)\,dx + (2x + y)\,dy = 0.$
2. $2xy\,dx + x^2\,dy = 0.$
3. $(x^2 - y)\,dx - (x + \sin^2 y)\,dy = 0.$
4. $4\sin x \sin 3y \cos x\,dx - 3\cos 3y \cos 2x\,dy = 0.$
5. $(3x^2y + 8xy^2)\,dx + (x^3 + 8x^2y + 12ye^y)\,dy = 0.$

6. Show that a linear first-order equation, $y' + P(x)y = Q(x)$, has the integrating factor $\mu(x) = e^{\int P(x)\,dx}$. Use this to solve the equation.
7. Let $\mu(x, y)$ be an integrating factor of the differential equation $P(x, y)\,dx + Q(x, y)\,dy = 0$. Show that
$$\frac{\partial P}{\partial y} - \frac{\partial Q}{\partial x} = Q\frac{\partial}{\partial x}\log|\mu| - P\frac{\partial}{\partial y}\log|\mu|.$$
Use this equation to deduce the following rules for finding integrating factors:
(a) If $(\partial P/\partial y - \partial Q/\partial x)/Q$ is a function of x alone, say $f(x)$, then $e^{\int f(x)\,dx}$ is an integrating factor.
(b) If $(\partial Q/\partial x - \partial P/\partial y)/P$ is a function of y alone, say $g(y)$, then $e^{\int g(y)\,dy}$ is an integrating factor.
8. Use Exercise 7 to find integrating factors for the following equations, and determine a one-parameter family of integral curves.
(a) $y\,dx - (2x + y)\,dy = 0.$
(b) $(x^3 + y^3)\,dx - xy^2\,dy = 0.$
9. If $\partial P/\partial y - \partial Q/\partial x = f(x)Q(x, y) - g(y)P(x, y)$, show that $\exp\{\int f(x)\,dx + \int g(y)\,dy\}$ is an integrating factor of the differential equation $P(x, y)\,dx + Q(x, y)\,dy = 0$. Find such an integrating factor for each of the following equations and obtain a one-parameter family of integral curves.
(a) $(2x^2y + y^2)\,dx + (2x^3 - xy)\,dy = 0.$
(b) $(e^x \sec y - \tan y)\,dx + dy = 0.$

10. The following differential equations have an integrating factor in common. Find such an integrating factor and obtain a one-parameter family of integral curves for each equation.

$$(3y + 4xy^2)\,dx + (4x + 5x^2y)\,dy = 0,$$
$$(6y + x^2y^2)\,dx + (8x + x^3y)\,dy = 0.$$

10.21 Potential functions on convex sets

In Theorem 10.6 we proved that the conditions

$$D_i f_j(\boldsymbol{x}) = D_j f_i(\boldsymbol{x})$$

are *necessary* for a continuously differentiable vector field $\boldsymbol{f} = (f_1, \ldots, f_n)$ to be a gradient on an open set S in $\mathbf{R}^n$. We then showed, by an example, that these conditions are not always sufficient. In this section we prove that the conditions are sufficient whenever the set S is convex. The proof will make use of the following theorem concerning differentiation under the integral sign.

THEOREM 10.8. *Let S be a closed interval in $\mathbf{R}^n$ with nonempty interior and let $J = [a, b]$ be a closed interval in $\mathbf{R}^1$. Let J_{n+1} be the closed interval $S \times J$ in $\mathbf{R}^{n+1}$. Write each point in J_{n+1} as $(\boldsymbol{x}, t)$, where $\boldsymbol{x} \in S$ and $t \in J$,*

$$(\boldsymbol{x}, t) = (x_1, \ldots, x_n, t).$$

Assume that ψ is a scalar field defined on J_{n+1} such that the partial derivative $D_k\psi$ is continuous on J_{n+1}, where k is one of $1, 2, \ldots, n$. Define a scalar field φ on S by the equation

$$\varphi(\boldsymbol{x}) = \int_a^b \psi(\boldsymbol{x}, t)\,dt.$$

Then the partial derivative $D_k\varphi$ exists at each interior point of S and is given by the formula

$$D_k\varphi(\boldsymbol{x}) = \int_a^b D_k\psi(\boldsymbol{x}, t)\,dt.$$

In other words, we have

$$D_k \int_a^b \psi(\boldsymbol{x}, t)\,dt = \int_a^b D_k\psi(\boldsymbol{x}, t)\,dt.$$

Note: This theorem is usually described by saying that we can differentiate under the integral sign.

Proof. Choose any $\boldsymbol{x}$ in the interior of S. Since int S is open, there is an $r > 0$ such that $\boldsymbol{x} + h\boldsymbol{e}_k \in \text{int } S$ for all h satisfying $0 < |h| < r$. Here $\boldsymbol{e}_k$ is the kth unit coordinate vector in $\mathbf{R}^n$. For such h we have

$$\varphi(\boldsymbol{x} + h\boldsymbol{e}_k) - \varphi(\boldsymbol{x}) = \int_a^b \{\psi(\boldsymbol{x} + h\boldsymbol{e}_k, t) - \psi(\boldsymbol{x}, t)\}\,dt. \tag{10.17}$$

Applying the mean value theorem to the integrand we have

$$\psi(x + he_k, t) - \psi(x, t) = h\, D_k\psi(z, t),$$

where z lies on the line segment joining x and $x + he_k$. Hence (10.17) becomes

$$\frac{\varphi(x + he_k) - \varphi(x)}{h} = \int_a^b D_k\psi(z, t)\, dt .$$

Therefore

$$\frac{\varphi(x + he_k) - \varphi(x)}{h} - \int_a^b D_k\psi(x, t)\, dt = \int_a^b \{D_k\psi(z, t) - D_k\psi(x, t)\}\, dt .$$

The last integral has absolute value not exceeding

$$\int_a^b |D_k\psi(z, t) - D_k\psi(x, t)|\, dt \le (b - a) \max |D_k\psi(z, t) - D_k\psi(x, t)|$$

where the maximum is taken for all z on the segment joining x to $x + he_k$, and all t in $[a, b]$. Now we invoke the uniform continuity of D_k on $S \times J$ (Theorem 9.10) to conclude that for every $\epsilon > 0$ there is a $\delta > 0$ such that this maximum is $<\epsilon/(b - a)$, whenever $0 < |h| < \delta$. Therefore

$$\left| \frac{\varphi(x + he_k) - \varphi(x)}{h} - \int_a^b D_k\psi(x, t)\, dt \right| < \epsilon \qquad \text{whenever} \quad 0 < |h| < \delta .$$

This proves that $D_k\varphi(x)$ exists and equals $\int_a^b D_k\psi(x, t)\, dt$, as required.

Now we shall use this theorem to give the following necessary and sufficient condition for a vector field to be a gradient on a convex set.

THEOREM 10.9. *Let $f = (f_1, \ldots, f_n)$ be a continuously differentiable vector field on an open convex set S in $\mathbf{R}^n$. Then f is a gradient on S if and only if we have*

(10.18) $$D_k f_j(x) = D_j f_k(x)$$

for each x in S and all $k, j = 1, 2, \ldots, n$.

Proof. We know, from Theorem 10.6, that the conditions are necessary. To prove sufficiency, we assume (10.18) and construct a potential φ on S.

For simplicity, we assume that S contains the origin. Let $\varphi(x)$ be the integral of f along the line segment from O to an arbitrary point x in S. As shown earlier in Equation (10.10) we have

$$\varphi(x) = \int_0^1 f(tx) \cdot x\, dt = \int_0^1 \psi(x, t)\, dt,$$

where $\psi(x, t) = f(tx) \cdot x$. There is a closed n-dimensional subinterval T of S with nonempty interior such that ψ satisfies the hypotheses of Theorem 10.8 in $T \times J$, where $J = [0, 1]$. Therefore the partial derivative $D_k\varphi(x)$ exists for each $k = 1, 2, \ldots, n$ and can be computed by differentiating under the integral sign,

$$D_k\varphi(x) = \int_0^1 D_k\psi(x, t)\, dt.$$

To compute $D_k\psi(x, t)$, we differentiate the dot product $f(tx) \cdot x$ and obtain

$$\begin{aligned} D_k\psi(x, t) &= f(tx) \cdot D_k x + D_k\{f(tx)\} \cdot x \\ &= f(tx) \cdot e_k + t(D_k f_1(tx), \ldots, D_k f_n(tx)) \cdot x \\ &= f_k(tx) + t(D_1 f_k(tx), \ldots, D_n f_k(tx)) \cdot x, \end{aligned}$$

where in the last step we used the relation (10.18). Therefore we have

$$D_k\psi(x, t) = f_k(tx) + t\nabla f_k(tx) \cdot x.$$

Now let $g(t) = f_k(tx)$. By the chain rule we have

$$g'(t) = \nabla f_k(tx) \cdot x$$

so the last formula for $D_k\psi(x, t)$ becomes

$$D_k\psi(x, t) = g(t) + tg'(t).$$

Integrating this from 0 to 1 we find

$$D_k\varphi(x) = \int_0^1 D_k\psi(x, t) = \int_0^1 g(t)\, dt + \int_0^1 tg'(t)\, dt. \tag{10.19}$$

We integrate by parts in the last integral to obtain

$$\int_0^1 tg'(t)\, dt = tg(t)\Big|_0^1 - \int_0^1 g(t)\, dt = g(1) - \int_0^1 g(t)\, dt.$$

Therefore (10.19) becomes

$$D_k\varphi(x) = g(1) = f_k(x).$$

This shows that $\nabla\varphi = f$ on S, which completes the proof.

11

MULTIPLE INTEGRALS

11.1 Introduction

Volume I discussed integrals $\int_a^b f(x)\,dx$, first for functions defined and bounded on finite intervals, and later for unbounded functions and infinite intervals. Chapter 10 of Volume II generalized the concept by introducing line integrals. This chapter extends the concept in yet another direction. The one-dimensional interval $[a, b]$ is replaced by a two-dimensional set Q, called the *region of integration.* First we consider rectangular regions; later we consider more general regions with curvilinear boundaries. The integrand is a scalar field f defined and bounded on Q. The resulting integral is called a *double integral* and is denoted by the symbol

$$\iint_Q f, \qquad \text{or by} \qquad \iint_Q f(x, y)\,dx\,dy\,.$$

As in the one-dimensional case, the symbols dx and dy play no role in the definition of the double integral; however, they are useful in computations and transformations of integrals.

The program in this chapter consists of several stages. First we discuss the definition of the double integral. The approach here is analogous to the one-dimensional case treated in Volume I. The integral is defined first for step functions and then for more general functions. As in the one-dimensional case, the definition does not provide a useful procedure for actual computation of integrals. We shall find that most double integrals occurring in practice can be computed by repeated one-dimensional integration. We shall also find a connection between double integrals and line integrals. Applications of double integrals to problems involving area, volume, mass, center of mass, and related concepts are also given. Finally, we indicate how the concepts can be extended to n-space.

11.2 Partitions of rectangles. Step functions

Let Q be a rectangle, the Cartesian product of two closed intervals $[a, b]$ and $[c, d]$,

$$Q = [a, b] \times [c, d] = \{(x, y) \mid x \in [a, b] \quad \text{and} \quad y \in [c, d]\}\,.$$

An example is shown in Figure 11.1. Consider two partitions P_1 and P_2 of $[a, b]$ and $[c, d]$,

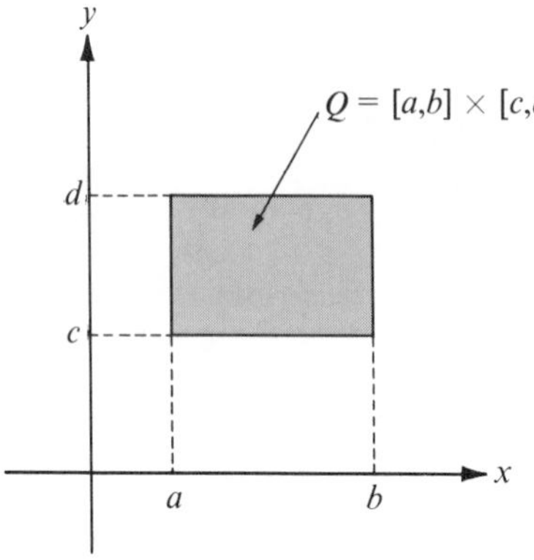

FIGURE 11.1 A rectangle Q, the Cartesian product of two intervals.

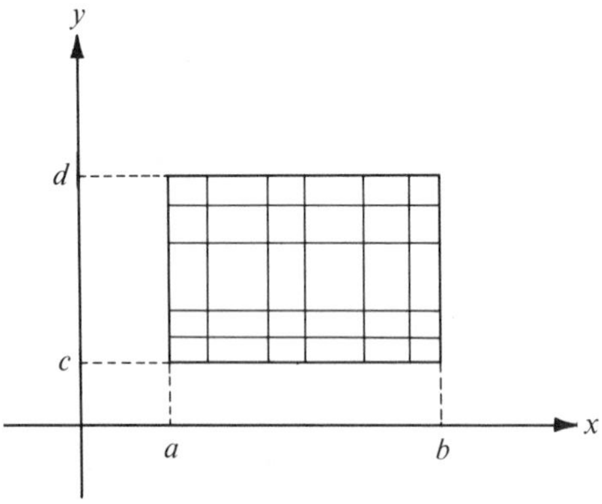

FIGURE 11.2 A partition of a rectangle Q.

respectively, say

$$P_1 = \{x_0, x_1, \ldots, x_{n-1}, x_n\} \quad \text{and} \quad P_2 = \{y_0, y_1, \ldots, y_{m-1}, y_m\},$$

where $x_0 = a$, $x_n = b$, $y_0 = c$, $y_m = d$. The Cartesian product $P_1 \times P_2$ is said to be a partition of Q. Since P_1 decomposes $[a, b]$ into n subintervals and P_2 decomposes $[c, d]$ into m subintervals, the partition $P = P_1 \times P_2$ decomposes Q into mn subrectangles. Figure 11.2 illustrates an example of a partition of Q into 30 subrectangles. A partition P' of Q is said to be finer than P if $P \subseteq P'$, that is, if every point in P is also in P'.

The Cartesian product of two open subintervals of P_1 and P_2 is a subrectangle with its edges missing. This is called an open subrectangle of P or of Q.

DEFINITION OF STEP FUNCTION. *A function f defined on a rectangle Q is said to be a step function if a partition P of Q exists such that f is constant on each of the open subrectangles of P.*

The graph of an example is shown in Figure 11.3. Most of the graph consists of horizontal rectangular patches. A step function also has well-defined values at each of the boundary

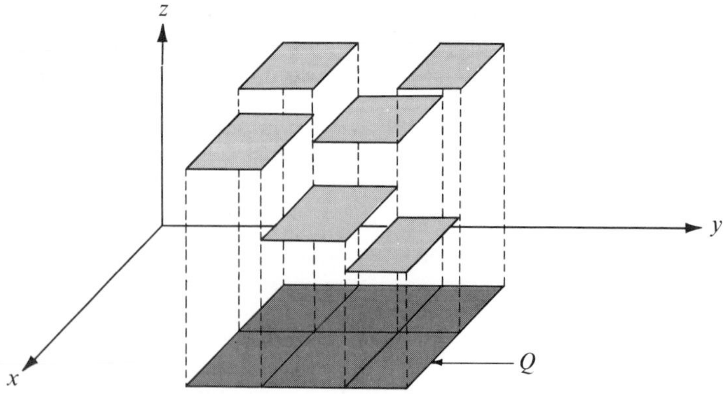

FIGURE 11.3 The graph of a step function defined on a rectangle Q.

points of the subrectangles, but the actual values at these points are not relevant to integration theory.

If f and g are two step functions defined on a given rectangle Q, then the linear combination $c_1 f + c_2 g$ is also a step function. In fact, if P and P' are partitions of Q such that f is constant on the open subrectangles of P and g is constant on the open subrectangles of P', then $c_1 f + c_2 g$ is constant on the open subrectangles of the union $P \cup P'$ (which we may call a common refinement of P and P'). Thus, the set of step functions defined on Q forms a linear space.

11.3 The double integral of a step function

Let $P = P_1 \times P_2$ be a partition of a rectangle Q into mn subrectangles and let f be a step function that is constant on the open subrectangles of Q. Let the subrectangle determined by $[x_{i-1}, x_i]$ and $[y_{j-1}, y_j]$ be denoted by Q_{ij} and let c_{ij} denote the constant value that f takes at the interior points of Q_{ij}. If f is positive, the volume of the rectangular box with base Q_{ij} and altitude c_{ij} is the product

$$c_{ij} \cdot (x_i - x_{i-1})(y_j - y_{j-1}).$$

For any step function f, positive or not, the sum of all these products is defined to be the double integral of f over Q. Thus, we have the following definition.

DEFINITION OF THE DOUBLE INTEGRAL OF A STEP FUNCTION. *Let f be a step function which takes the constant value c_{ij} on the open subrectangle $(x_{i-1}, x_i) \times (y_{j-1}, y_j)$ of a rectangle Q. The double integral of f over Q is defined by the formula*

$$\iint_Q f = \sum_{i=1}^{n} \sum_{j=1}^{m} c_{ij} \cdot (x_i - x_{i-1})(y_j - y_{j-1}). \tag{11.1}$$

As in the one-dimensional case, the value of the integral does not change if the partition P is replaced by any finer partition P'. Thus, the value of the integral is independent of the choice of P so long as f is constant on the open subrectangles of Q.

For brevity, we sometimes write Δx_i instead of $(x_i - x_{i-1})$ and Δy_j instead of $(y_j - y_{j-1})$, and the sum in (11.1) becomes

$$\sum_{i=1}^{n} \sum_{j=1}^{m} c_{ij}\, \Delta x_i\, \Delta y_j .$$

To remind ourselves how this sum is formed, we write the symbol for the integral as

$$\iint_Q f(x, y)\, dx\, dy .$$

This symbol is merely an alternative notation for $\iint_Q f$.

Note that if f is constant on the interior of Q, say $f(x, y) = k$ when $a < x < b$ and $c < y < d$, we have

$$\iint_Q f = k(b - a)(d - c), \tag{11.2}$$

regardless of the values of f on the edges of Q. Since we have

$$b - a = \int_a^b dx \qquad \text{and} \qquad d - c = \int_c^d dy,$$

formula (11.2) can also be written as

$$(11.3) \qquad \iint_Q f = \int_c^d \left[\int_a^b f(x, y)\, dx\right] dy = \int_a^b \left[\int_c^d f(x, y)\, dy\right] dx.$$

The integrals which appear on the right are one-dimensional integrals, and the formula is said to provide an evaluation of the double integral by *repeated* or *iterated* integration. In particular, when f is a step function of the type described above, we can write

$$\iint_{Q_{ij}} f = \int_{y_{j-1}}^{y_j} \left[\int_{x_{i-1}}^{x_i} f(x, y)\, dx\right] dy = \int_{x_{i-1}}^{x_i} \left[\int_{y_{j-1}}^{y_j} f(x, y)\, dy\right] dx.$$

Summing on i and j and using (11.1), we find that (11.3) holds for step functions.

The following further properties of the double integral of a step function are generalizations of the corresponding one-dimensional theorems. They may be proved as direct consequences of the definition in (11.1) or by use of formula (11.3) and the companion theorems for one-dimensional integrals. In the following theorems the symbols s and t denote step functions defined on a rectangle Q. To avoid trivial special cases we assume that Q is a nondegenerate rectangle; in other words, that Q is not merely a single point or a line segment.

THEOREM 11.1. LINEARITY PROPERTY. *For every real c_1 and c_2 we have*

$$\iint_Q [c_1 s(x, y) + c_2 t(x, y)]\, dx\, dy = c_1 \iint_Q s(x, y)\, dx\, dy + c_2 \iint_Q t(x, y)\, dx\, dy.$$

THEOREM 11.2. ADDITIVE PROPERTY. *If Q is subdivided into two rectangles Q_1 and Q_2, then*

$$\iint_Q s(x, y)\, dx\, dy = \iint_{Q_1} s(x, y)\, dx\, dy + \iint_{Q_2} s(x, y)\, dx\, dy.$$

THEOREM 11.3 COMPARISON THEOREM. *If $s(x, y) \le t(x, y)$ for every (x, y) in Q, we have*

$$\iint_Q s(x, y)\, dx\, dy \le \iint_Q t(x, y)\, dx\, dy.$$

In particular, if $t(x, y) \ge 0$ for every (x, y) in Q, then

$$\iint_Q t(x, y)\, dx\, dy \ge 0.$$

The proofs of these theorems are left as exercises.

11.4 The definition of the double integral of a function defined and bounded on a rectangle

Let f be a function that is defined and bounded on a rectangle Q; specifically, suppose that

$$|f(x, y)| \leq M \quad \text{if} \quad (x, y) \in Q.$$

Then f may be surrounded from above and from below by two constant step functions s and t, where $s(x, y) = -M$ and $t(x, y) = M$ for all (x, y) in Q. Now consider *any* two step functions s and t, defined on Q, such that

$$s(x, y) \leq f(x, y) \leq t(x, y) \quad \text{for every point } (x, y) \text{ in } Q. \tag{11.4}$$

DEFINITION OF THE INTEGRAL OF A BOUNDED FUNCTION OVER A RECTANGLE. *If there is one and only one number I such that*

$$\iint_Q s \leq I \leq \iint_Q t \tag{11.5}$$

for every pair of step functions satisfying the inequalities in (11.4), *this number I is called the double integral of f over Q and is denoted by the symbol*

$$\iint_Q f \quad or \quad \iint_Q f(x, y)\, dx\, dy.$$

When such an I exists the function f is said to be integrable on Q.

11.5 Upper and lower double integrals

The definition of the double integral is entirely analogous to the one-dimensional case. Upper and lower double integrals can also be defined as was done in the one-dimensional case.

Assume f is bounded on a rectangle Q and let s and t be step functions satisfying (11.4). We say that s is *below* f, and t is *above* f, and we write $s \leq f \leq t$. Let S denote the set of all numbers $\iint_Q s$ obtained as s runs through all step functions below f, and let T be the set of all numbers $\iint_Q t$ obtained as t runs through all step functions above f. Both sets S and T are nonempty since f is bounded. Also, $\iint_Q s \leq \iint_Q t$ if $s \leq f \leq t$, so every number in S is less than every number in T. Therefore S has a supremum, and T has an infimum, and they satisfy the inequalities

$$\iint_Q s \leq \sup S \leq \inf T \leq \iint_Q t$$

for all s and t satisfying $s \leq f \leq t$. This shows that both numbers $\sup S$ and $\inf T$ satisfy (11.5). Therefore, f is integrable on Q if and only if $\sup S = \inf T$, in which case we have

$$\iint_Q f = \sup S = \inf T.$$

The number sup S is called the *lower integral* of f and is denoted by $\underline{I}(f)$. The number inf T is called the *upper integral* of f and is denoted by $\bar{I}(f)$. Thus, we have

$$\underline{I}(f) = \sup\left\{\iint_Q s \mid s \leq f\right\}, \qquad \bar{I}(f) = \inf\left\{\iint_Q t \mid f \leq t\right\}.$$

The foregoing argument proves the following theorem.

THEOREM 11.4. *Every function f which is bounded on a rectangle Q has a lower integral $\underline{I}(f)$ and an upper integral $\bar{I}(f)$ satisfying the inequalities*

$$\iint_Q s \leq \underline{I}(f) \leq \bar{I}(f) \leq \iint_Q t$$

for all step functions s and t with $s \leq f \leq t$. The function f is integrable on Q if and only if its upper and lower integrals are equal, in which case we have

$$\iint_Q f = \underline{I}(f) = \bar{I}(f).$$

Since the foregoing definitions are entirely analogous to the one-dimensional case, it is not surprising to learn that the linearity property, the additive property, and the comparison theorem as stated for step functions in Section 11.3, also hold for double integrals in general. The proofs of these statements are analogous to those in the one-dimensional case and will be omitted.

11.6 Evaluation of a double integral by repeated one-dimensional integration

In one-dimensional integration theory, the second fundamental theorem of calculus provides a practical method for calculating integrals. The next theorem accomplishes the same result in the two-dimensional theory; it enables us to evaluate certain double integrals by means of two successive one-dimensional integrations. The result is an extension of formula (11.3), which we have already proved for step functions.

THEOREM 11.5. *Let f be defined and bounded on a rectangle $Q = [a, b] \times [c, d]$, and assume that f is integrable on Q. For each fixed y in $[c, d]$ assume that the one-dimensional integral $\int_a^b f(x, y)\, dx$ exists, and denote its value by $A(y)$. If the integral $\int_c^d A(y)\, dy$ exists it is equal to the double integral $\iint_Q f$. In other words, we have the formula*

$$\iint_Q f(x, y)\, dx\, dy = \int_c^d \left[\int_a^b f(x, y)\, dx\right] dy. \tag{11.6}$$

Proof. Choose any two step functions s and t satisfying $s \leq f \leq t$ on Q. Integrating with respect to x over the interval $[a, b]$ we have

$$\int_a^b s(x, y)\, dx \leq A(y) \leq \int_a^b t(x, y)\, dx.$$

Since the integral $\int_c^d A(y)\,dy$ exists, we can integrate both these inequalities with respect to y over $[c, d]$ and use Equation (11.3) to obtain

$$\iint_Q s \le \int_c^d A(y)\,dy \le \iint_Q t .$$

Therefore $\int_c^d A(y)\,dy$ is a number which lies between $\iint_Q s$ and $\iint_Q t$ for all step functions s and t approximating f from below and from above, respectively. Since f is integrable on Q, the only number with this property is the double integral of f over Q. Therefore $\int_c^d A(y)\,dy = \iint_Q f$, which proves Equation (11.6).

Formula (11.6) is said to provide an evaluation of the double integral by repeated or iterated integration. The process is described by saying that first we integrate f with respect to x from a to b (holding y fixed), and then we integrate the result with respect to y from c to d. If we interchange the order of integration, we have a similar formula, namely,

$$\text{(11.7)} \qquad \iint_Q f(x, y)\,dx\,dy = \int_a^b \left[\int_c^d f(x, y)\,dy\right] dx ,$$

which holds if we assume that $\int_c^d f(x, y)\,dy$ exists for each fixed x in $[a, b]$ and is integrable on $[a, b]$.

11.7 Geometric interpretation of the double integral as a volume

Theorem 11.5 has a simple geometric interpretation, illustrated in Figure 11.4. If f is nonnegative, the set S of points (x, y, z) in 3-space with (x, y) in Q and $0 \le z \le f(x, y)$ is

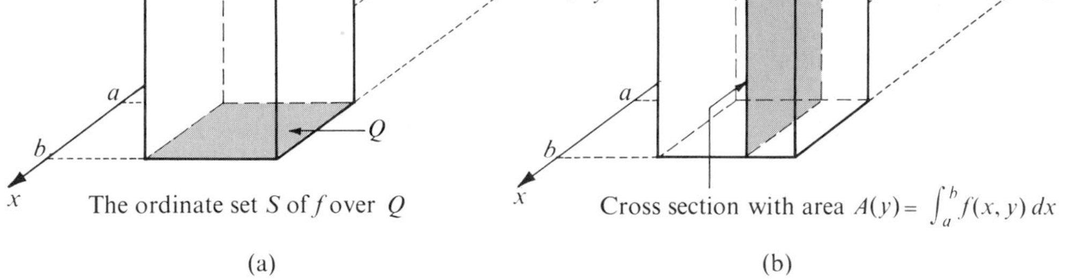

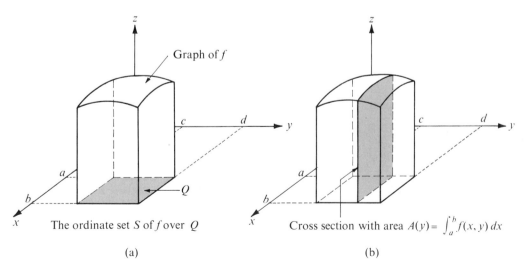

FIGURE 11.4 The volume of S is the integral of the cross-sectional area:
$$v(S) = \int_c^d A(y)\,dy.$$

called the *ordinate set of f over Q*. It consists of those points between the rectangle Q and the surface $z = f(x, y)$. (See Figure 11.4(a).) For each y in the interval $[c, d]$, the integral

$$A(y) = \int_a^b f(x, y)\, dx$$

is the area of the cross section of S cut by a plane parallel to the xz-plane (the shaded region in Figure 11.4(b)). Since the cross-sectional area $A(y)$ is integrable on $[c, d]$, Theorem 2.7 of Volume I tells us that the integral $\int_c^d A(y)\, dy$ is equal to $v(S)$, the volume of S. Thus, for nonnegative integrands, Theorem 11.5 shows that the volume of the ordinate set of f over Q is equal to the double integral $\iint_Q f$.

Equation (11.7) gives another way of computing the volume of the ordinate set. This time we integrate the area of the cross sections cut by planes parallel to the yz-plane.

11.8 Worked examples

In this section we illustrate Theorem 11.5 with two numerical examples.

EXAMPLE 1. If $Q = [-1, 1] \times [0, \pi/2]$, evaluate $\iint_Q (x \sin y - ye^x)\, dx\, dy$, given that the integral exists. The region of integration is shown in Figure 11.5.

Solution. Integrating first with respect to x and calling the result $A(y)$, we have

$$A(y) = \int_{-1}^{1} (x \sin y - ye^x)\, dx = \left(\frac{x^2}{2} \sin y - ye^x\right)\Bigg|_{x=-1}^{x=1} = -ey + y/e.$$

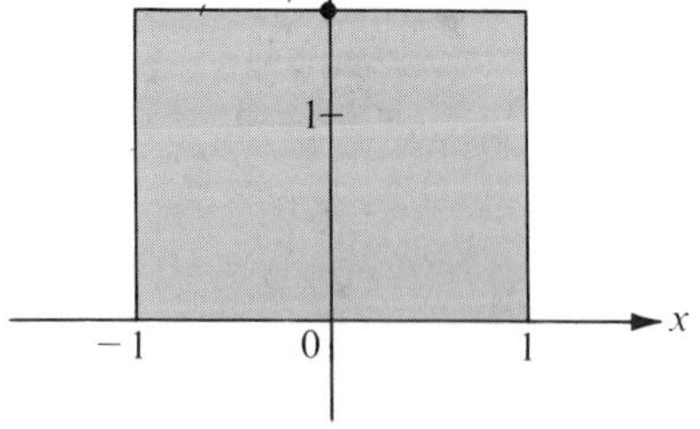

FIGURE 11.5 The region of integration for Example 1.

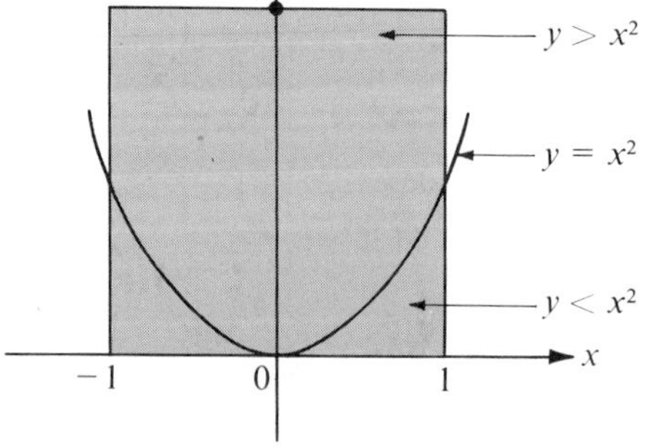

FIGURE 11.6 The region of integration for Example 2.

Applying Theorem 11.5 we find

$$\iint_Q (x \sin y - ye^x)\,dx\,dy = \int_0^{\pi/2} A(y)\,dy = \int_0^{\pi/2} (-ey + y/e)\,dy$$

$$= (1/e - e)\int_0^{\pi/2} y\,dy = (1/e - e)\pi^2/8\,.$$

As a check on the calculations we may integrate first with respect to y:

$$\iint_Q (x \sin y - ye^x)\,dx\,dy = \int_{-1}^{1}\left[\int_0^{\pi/2} (x \sin y - ye^x)\,dy\right]dx$$

$$= \int_{-1}^{1} (-x \cos y - \tfrac{1}{2}y^2e^x)\Big|_{y=0}^{y=\pi/2}\,dx$$

$$= \int_{-1}^{1} (-\pi^2e^x/8 + x)\,dx = (1/e - e)\pi^2/8\,.$$

EXAMPLE 2. If $Q = [-1, 1] \times [0, 2]$, evaluate the double integral $\iint_Q \sqrt{|y - x^2|}\,dx\,dy$, given that it exists.

Solution. If we integrate first with respect to y and call the result $H(x)$, we have $H(x) = \int_0^2 \sqrt{|y - x^2|}\,dy$. The region of integration is the rectangle shown in Figure 11.6. The parabola $y = x^2$ is also shown because of the presence of $|y - x^2|$ in the integrand. Above this parabola we have $y > x^2$ and below it we have $y < x^2$. This suggests that we split the integral for $H(x)$ as follows:

$$H(x) = \int_0^2 \sqrt{|y - x^2|}\,dy = \int_0^{x^2} \sqrt{x^2 - y}\,dy + \int_{x^2}^2 \sqrt{y - x^2}\,dy\,.$$

We remember that x is treated as a constant in each of these integrals. In the first integral we make the change of variable $t = x^2 - y$ and in the second we put $t = y - x^2$. This gives us

$$H(x) = \int_0^2 \sqrt{|y - x^2|}\,dy = -\int_{x^2}^0 \sqrt{t}\,dt + \int_0^{2-x^2} \sqrt{t}\,dt = \tfrac{2}{3}x^3 + \tfrac{2}{3}(2 - x^2)^{3/2}.$$

Applying Theorem 11.5 we find

$$\iint_Q \sqrt{|y - x^2|}\,dx\,dy = \int_{-1}^{1} [\tfrac{2}{3}x^3 + \tfrac{2}{3}(2 - x^2)^{3/2}]\,dx = \tfrac{4}{3}\int_0^1 (2 - x^2)^{3/2}\,dx$$

$$= \frac{1}{3}\left[x(2 - x^2)^{3/2} + 3x\sqrt{2 - x^2} + 6 \arcsin\left(\frac{x}{\sqrt{2}}\right)\right]\Big|_0^1 = \frac{4}{3} + \frac{\pi}{2}\,.$$

The same result may be obtained by integrating first with respect to x, but the calculations are more complicated.

11.9 Exercises

Evaluate the double integrals in Exercises 1 through 8 by repeated integration, given that each integral exists.

1. $\iint_Q xy(x + y)\,dx\,dy,\quad$ where $Q = [0, 1] \times [0, 1]$.

2. $\iint_Q (x^3 + 3x^2y + y^3)\,dx\,dy,\quad$ where $Q = [0, 1] \times [0, 1]$.

3. $\iint_Q (\sqrt{y} + x - 3xy^2)\,dx\,dy,\quad$ where $Q = [0, 1] \times [1, 3]$.

4. $\iint_Q \sin^2 x \sin^2 y\,dx\,dy,\quad$ where $Q = [0, \pi] \times [0, \pi]$.

5. $\iint_Q \sin(x + y)\,dx\,dy,\quad$ where $Q = [0, \pi/2] \times [0, \pi/2]$.

6. $\iint_Q |\cos(x + y)|\,dx\,dy,\quad$ where $Q = [0, \pi] \times [0, \pi]$.

7. $\iint_Q f(x + y)\,dx\,dy,\quad$ where $Q = [0, 2] \times [0, 2]$, and $f(t)$ denotes the greatest integer $\le t$.

8. $\iint_Q y^{-3} e^{tx/y}\,dx\,dy,\quad$ where $Q = [0, t] \times [1, t]$, $t > 0$.

9. If Q is a rectangle, show that a double integral of the form $\iint_Q f(x)g(y)\,dx\,dy$ is equal to the product of two one-dimensional integrals. State the assumptions about existence that you are using.

10. Let f be defined on the rectangle $Q = [0, 1] \times [0, 1]$ as follows:

$$f(x, y) = \begin{cases} 1 - x - y & \text{if } x + y \le 1, \\ 0 & \text{otherwise.} \end{cases}$$

Make a sketch of the ordinate set of f over Q and compute the volume of this ordinate set by double integration. (Assume the integral exists.)

11. Solve Exercise 10 when

$$f(x, y) = \begin{cases} x + y & \text{if } x^2 \le y \le 2x^2, \\ 0 & \text{otherwise.} \end{cases}$$

12. Solve Exercise 10 when $Q = [-1, 1] \times [-1, 1]$ and

$$f(x, y) = \begin{cases} x^2 + y^2 & \text{if } x^2 + y^2 \le 1, \\ 0 & \text{otherwise.} \end{cases}$$

13. Let f be defined on the rectangle $Q = [1, 2] \times [1, 4]$ as follows:

$$f(x, y) = \begin{cases} (x + y)^{-2} & \text{if } x \le y \le 2x, \\ 0 & \text{otherwise}. \end{cases}$$

Indicate, by means of a sketch, the portion of Q in which f is nonzero and compute the value of the double integral $\iint_Q f$, given that the integral exists.

14. Let f be defined on the rectangle $Q = [0, 1] \times [0, 1]$ as follows:

$$f(x, y) = \begin{cases} 1 & \text{if } x = y, \\ 0 & \text{if } x \ne y. \end{cases}$$

Prove that the double integral $\iint_Q f$ exists and equals 0.

11.10 Integrability of continuous functions

The small-span theorem (Theorem 9.10) can be used to prove integrability of a function which is continuous on a rectangle.

THEOREM 11.6. INTEGRABILITY OF CONTINUOUS FUNCTIONS. *If a function f is continuous on a rectangle $Q = [a, b] \times [c, d]$, then f is integrable on Q. Moreover, the value of the integral can be obtained by iterated integration,*

$$\iint_Q f = \int_c^d \left[\int_a^b f(x, y)\, dx\right] dy = \int_a^b \left[\int_c^d f(x, y)\, dy\right] dx. \tag{11.8}$$

Proof. Theorem 9.8 shows that f is bounded on Q, so f has an upper integral and a lower integral. We shall prove that $\underline{I}(f) = \bar{I}(f)$. Choose $\epsilon > 0$. By the small-span theorem, for this choice of ϵ there is a partition P of Q into a finite number (say n) of subrectangles $Q_1, \ldots, Q_n$ such that the span of f in every subrectangle is less than ϵ. Denote by $M_k(f)$ and $m_k(f)$, respectively, the absolute maximum and minimum values of f in Q_k. Then we have

$$M_k(f) - m_k(f) < \epsilon$$

for each $k = 1, 2, \ldots, n$. Now let s and t be two step functions defined on the interior of each Q_k as follows:

$$s(\boldsymbol{x}) = m_k(f), \qquad t(\boldsymbol{x}) = M_k(f) \qquad \text{if } \boldsymbol{x} \in \operatorname{int} Q_k.$$

At the boundary points we define $s(\boldsymbol{x}) = m$ and $t(\boldsymbol{x}) = M$, where m and M are, respectively, the absolute minimum and maximum values of f on Q. Then we have $s \le f \le t$ for all $\boldsymbol{x}$ in Q. Also, we have

$$\iint_Q s = \sum_{k=1}^n m_k(f)a(Q_k) \qquad \text{and} \qquad \iint_Q t = \sum_{k=1}^n M_k(f)a(Q_k),$$

where $a(Q_k)$ is the area of rectangle Q_k. The difference of these two integrals is

$$\iint_Q t - \iint_Q s = \sum_{k=1}^{n} \{M_k(f) - m_k(f)\}a(Q_k) < \epsilon \sum_{k=1}^{n} a(Q_k) = \epsilon a(Q),$$

where $a(Q)$ is the area of Q. Since $\iint_Q s \le \underline{I}(f) \le \bar{I}(f) \le \iint_Q t$, we obtain the inequality

$$0 \le \bar{I}(f) - \underline{I}(f) \le \epsilon a(Q).$$

Letting $\epsilon \to 0$ we see that $\underline{I}(f) = \bar{I}(f)$, so f is integrable on Q.

Next we prove that the double integral is equal to the first iterated integral in (11.8). For each fixed y in $[c, d]$ the one-dimensional integral $\int_a^b f(x, y)\,dx$ exists since the integrand is continuous on Q. Let $A(y) = \int_a^b f(x, y)\,dx$. We shall prove that A is continuous on $[c, d]$. If y and y_1 are any two points in $[c, d]$ we have

$$A(y) - A(y_1) = \int_a^b \{f(x, y) - f(x, y_1)\}\,dx$$

from which we find

$$|A(y) - A(y_1)| \le (b - a) \max_{a \le x \le b} |f(x, y) - f(x, y_1)| = (b - a)\,|f(x_1, y) - f(x_1, y_1)|$$

where x_1 is a point in $[a, b]$ where $|f(x, y) - f(x, y_1)|$ attains its maximum. This inequality shows that $A(y) \to A(y_1)$ as $y \to y_1$, so A is continuous at y_1. Therefore the integral $\int_c^d A(y)\,dy$ exists and, by Theorem 11.5, it is equal to $\iint_Q f$. A similar argument works when the iteration is taken in the reverse order.

11.11 Integrability of bounded functions with discontinuities

Let f be defined and bounded on a rectangle Q. In Theorem 11.6 we proved that the double integral of f over Q exists if f is continuous everywhere on Q. In this section we prove that the integral also exists if f has discontinuities in Q, provided the set of discontinuities is not too large. To measure the size of the set of discontinuities we introduce the following concept.

DEFINITION OF A BOUNDED SET OF CONTENT ZERO. *Let A be a bounded subset of the plane. The set A is said to have content zero if for every $\epsilon > 0$ there is a finite set of rectangles whose union contains A and the sum of whose areas does not exceed ϵ.*

In other words, a bounded plane set of content zero can be enclosed in a union of rectangles whose total area is arbitrarily small.

The following statements about bounded sets of content zero are easy consequences of this definition. Proofs are left as exercises for the reader.

(a) Any finite set of points in the plane has content zero.
(b) The union of a finite number of bounded sets of content zero is also of content zero.
(c) Every subset of a set of content zero has content zero.
(d) Every line segment has content zero.

THEOREM 11.7. *Let f be defined and bounded on a rectangle $Q = [a, b] \times [c, d]$. If the set of discontinuities of f in Q is a set of content zero then the double integral $\iint_Q f$ exists.*

Proof. Let $M > 0$ be such that $|f| \leq M$ on Q. Let D denote the set of discontinuities of f in Q. Given $\delta > 0$, let P be a partition of Q such that the sum of the areas of all the subrectangles of P which contain points of D is less than δ. (This is possible since D has content zero.) On these subrectangles define step functions s and t as follows:

$$s(x) = -M, \qquad t(x) = M.$$

On the remaining subrectangles of P define s and t as was done in the proof of Theorem 11.6. Then we have $s \leq f \leq t$ throughout Q. By arguing as in the proof of Theorem 11.6 we obtain the inequality

$$\iint_Q t - \iint_Q s \leq \epsilon a(Q) + 2M\delta. \tag{11.9}$$

The first term, $\epsilon a(Q)$, comes from estimating the integral of $t - s$ over the subrectangles containing only points of continuity of f; the second term, $2M\delta$, comes from estimating the integral of $t - s$ over the subrectangles which contain points of D. From (11.9) we obtain the inequality

$$0 \leq \bar{I}(f) - \underline{I}(f) \leq \epsilon a(Q) + 2M\delta.$$

Letting $\epsilon \to 0$ we find $0 \leq \bar{I}(f) - \underline{I}(f) \leq 2M\delta$. Since δ is arbitrary this implies $\bar{I}(f) = \underline{I}(f)$, so f is integrable on Q.

11.12 Double integrals extended over more general regions

Up to this point the double integral has been defined only for rectangular regions of integration. However, it is not difficult to extend the concept to more general regions.

Let S be a bounded region, and enclose S in a rectangle Q. Let f be defined and bounded on S. Define a new function $\tilde{f}$ on Q as follows:

$$\tilde{f}(x, y) = \begin{cases} f(x, y) & \text{if } (x, y) \in S, \\ 0 & \text{if } (x, y) \in Q - S. \end{cases} \tag{11.10}$$

In other words, extend the definition of f to the whole rectangle Q by making the function values equal to 0 outside S. Now ask whether or not the extended function $\tilde{f}$ is integrable on Q. If so, we say that f is integrable on S and that, *by definition*,

$$\iint_S f = \iint_Q \tilde{f}.$$

First we consider sets of points S in the xy-plane described as follows:

$$S = \{(x, y) \mid a \leq x \leq b \quad \text{and} \quad \varphi_1(x) \leq y \leq \varphi_2(x)\},$$

where φ_1 and φ_2 are functions continuous on a closed interval $[a, b]$ and satisfying $\varphi_1 \leq \varphi_2$. An example of such a region, which we call a region of Type I, is shown in Figure 11.7. In a region of Type I, for each point t in $[a, b]$ the vertical line $x = t$ intersects S in a line segment joining the curve $y = \varphi_1(x)$ to $y = \varphi_2(x)$. Such a region is bounded because φ_1 and φ_2 are continuous and hence bounded on $[a, b]$.

Another type of region T (Type II) can be described as follows:

$$T = \{(x, y) \mid c \leq y \leq d \quad \text{and} \quad \psi_1(y) \leq x \leq \psi_2(y)\},$$

where ψ_1 and ψ_2 are continuous on an interval $[c, d]$ with $\psi_1 \leq \psi_2$. An example is shown in Figure 11.8. In this case horizontal lines intersect T in line segments. Regions of Type II are also bounded. All the regions we shall consider are either of one of these two types or can be split into a finite number of pieces, each of which is of one of these two types.

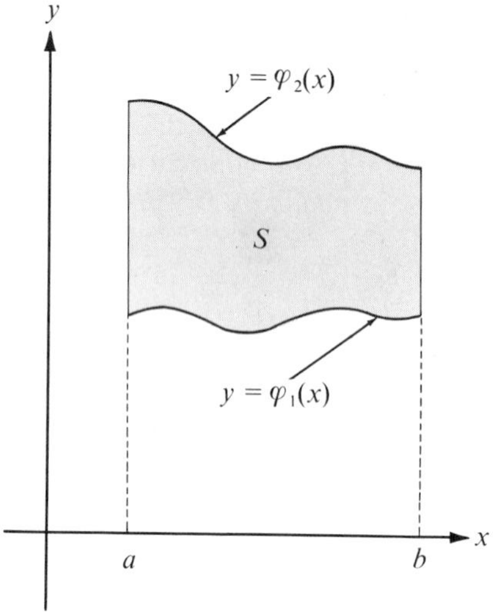

FIGURE 11.7 A region S of Type I

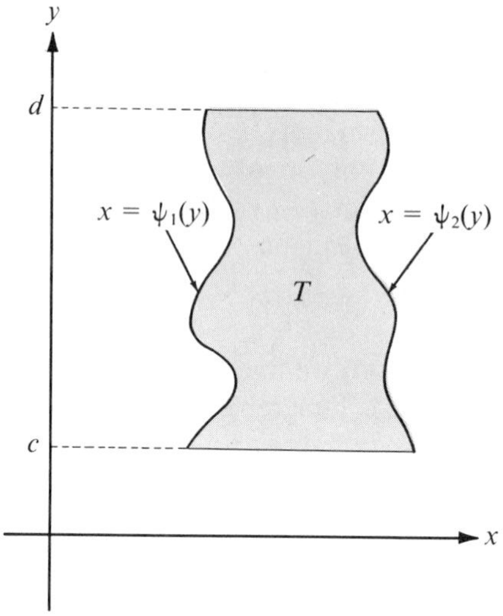

FIGURE 11.8 A region T of Type II.

Let f be defined and bounded on a region S of Type I. Enclose S in a rectangle Q and define $\tilde{f}$ on Q as indicated in Equation (11.10). The discontinuities of $\tilde{f}$ in Q will consist of the discontinuities of f in S plus those points on the boundary of S at which f is nonzero. The boundary of S consists of the graph of φ_1, the graph of φ_2, and two vertical line segments joining these graphs. Each of the line segments has content zero. The next theorem shows that each of the graphs also has content zero.

THEOREM 11.8. *Let φ be a real-valued function that is continuous on an interval $[a, b]$. Then the graph of φ has content zero.*

Proof. Let A denote the graph of φ, that is, let

$$A = \{(x, y) \mid y = \varphi(x) \quad \text{and} \quad a \leq x \leq b\}.$$

Choose any $\epsilon > 0$. We apply the small-span theorem (Theorem 3.13 of Volume I) to obtain a partition P of $[a, b]$ into a finite number of subintervals such that the span of φ in every subinterval of P is less than $\epsilon/(b - a)$. Therefore, above each subinterval of P the graph of φ lies inside a rectangle of height $\epsilon/(b - a)$. Hence the entire of graph φ lies within a finite union of rectangles, the sum of whose areas is ϵ. (An example is shown in Figure 11.9.) This proves that the graph of φ has content zero.

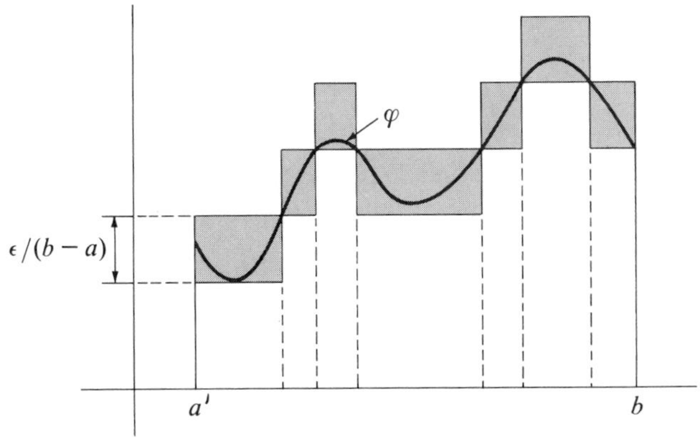

FIGURE 11.9 Proof that the graph of a continuous function φ has content zero.

The next theorem shows that the double integral $\iint_S f$ exists if f is continuous on int S, the interior of S. This is the set

$$\text{int } S = \{(x, y) \mid a < x < b \quad \text{and} \quad \varphi_1(x) < y < \varphi_2(x)\}.$$

THEOREM 11.9. *Let S be a region of Type I, between the graphs of φ_1 and φ_2. Assume that f is defined and bounded on S and that f is continuous on the interior of S. Then the double integral $\iint_S f$ exists and can be evaluated by repeated one-dimensional integration,*

$$\iint_S f(x, y)\, dx\, dy = \int_a^b \left[\int_{\varphi_1(x)}^{\varphi_2(x)} f(x, y)\, dy\right] dx. \tag{11.11}$$

Proof. Let $Q = [a, b] \times [c, d]$ be a rectangle which contains S, and let $\tilde{f}$ be defined by (11.10). The only possible points of discontinuity of $\tilde{f}$ are the boundary points of S. Since the boundary of S has content zero, $\tilde{f}$ is integrable on Q. For each fixed x in (a, b) the one-dimensional integral $\int_c^d \tilde{f}(x, y)\, dy$ exists since the integrand has at most two discontinuities on $[c, d]$. Applying the version of Theorem 11.5 given by Equation (11.7) we have

$$\iint_Q \tilde{f} = \int_a^b \left[\int_c^d \tilde{f}(x, y)\, dy\right] dx. \tag{11.12}$$

Now $\tilde{f}(x, y) = f(x, y)$ if $\varphi_1(x) \le y \le \varphi_2(x)$, and $\tilde{f}(x, y) = 0$ for the remaining values of y in $[c, d]$. Therefore

$$\int_c^d \tilde{f}(x, y)\, dy = \int_{\varphi_1(x)}^{\varphi_2(x)} f(x, y)\, dy$$

so Equation (11.12) implies (11.11).

There is, of course, an analogous theorem for a region T of Type II. If f is defined and bounded on T and continuous on the interior of T, then f is integrable on T and the formula for repeated integration becomes

$$\iint_T f(x, y)\, dx\, dy = \int_c^d \left[\int_{\psi_1(y)}^{\psi_2(y)} f(x, y)\, dx\right] dy. \tag{11.13}$$

Some regions are of both Type I and Type II. (Regions bounded by circles and ellipses are examples.) In this case the order of integration is immaterial and we may write

$$\int_a^b \left[\int_{\varphi_1(x)}^{\varphi_2(x)} f(x, y)\, dy\right] dx = \int_c^d \left[\int_{\psi_1(y)}^{\psi_2(y)} f(x, y)\, dx\right] dy.$$

In some cases one of these integrals may be much easier to compute than the other; it is usually worthwhile to examine both before attempting the actual evaluation of a double integral.

11.13 Applications to area and volume

Let S be a region of Type I given by

$$S = \{(x, y) \mid a \le x \le b \quad \text{and} \quad \varphi_1(x) \le y \le \varphi_2(x)\}.$$

Applying Theorem 11.9 with $f(x, y) = 1$ for all (x, y) in S we obtain

$$\iint_S dx\, dy = \int_a^b [\varphi_2(x) - \varphi_1(x)]\, dx.$$

By Theorem 2.1 of Volume I, the integral on the right is equal to the area of the region S. Thus, double integrals can be used to compute areas.

If f is nonnegative, the set of points (x, y, z) in 3-space such that $(x, y) \in S$ and $0 \le z \le f(x, y)$ is called the *ordinate set of f over S*. An example is shown in Figure 11.10. If f is nonnegative and continuous on S, the integral

$$\int_{\varphi_1(x)}^{\varphi_2(x)} f(x, y)\, dy$$

represents the area of a cross-section of the ordinate set cut by a plane parallel to the yz-plane (the shaded region in Figure 11.10). Formula (11.11) of Theorem 11.9 shows that the double integral of f over S is equal to the integral of the cross-sectional area. Hence

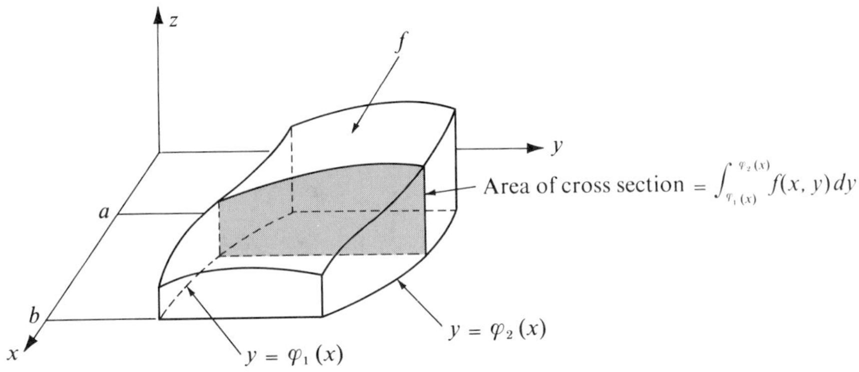

FIGURE 11.10 The integral $\int_{\varphi_1(x)}^{\varphi_2(x)} f(x, y)\,dy$ is the area of a cross section of the ordinate set. The iterated integral $\int_a^b \left[\int_{\varphi_1(x)}^{\varphi_2(x)} f(x, y)\,dy\right] dx$ is the volume of the ordinate set.

the double integral $\iint_S f$ is equal to the volume of the ordinate set of f over S. (See Theorem 2.7 of Volume I, p. 113.)

More generally, if f and g are both continuous on S with $f \le g$, then the double integral $\iint_S (g - f)$ is equal to the volume of the solid lying between the graphs of the functions f and g. Similar remarks apply, of course, to regions of Type II.

11.14 Worked examples

EXAMPLE 1. Compute the volume of the solid enclosed by the ellipsoid

$$\frac{x^2}{a^2} + \frac{y^2}{b^2} + \frac{z^2}{c^2} = 1 .$$

Solution. The solid in question lies between the graphs of two functions f and g, where

$$g(x, y) = c\sqrt{1 - x^2/a^2 - y^2/b^2} \qquad \text{and} \qquad f(x, y) = -g(x, y) .$$

Here $(x, y) \in S$, where S is the elliptical region given by

$$S = \left\{(x, y) \;\middle|\; \frac{x^2}{a^2} + \frac{y^2}{b^2} \le 1\right\}.$$

Applying Theorem 11.9 and using the symmetry we find that the volume V of the ellipsoidal solid is given by

$$V = \iint_S (g - f) = 2 \iint_S g = 8c \int_0^a \left[\int_0^{b\sqrt{1-x^2/a^2}} \sqrt{1 - x^2/a^2 - y^2/b^2}\,dy\right] dx .$$

Let $A = \sqrt{1 - x^2/a^2}$. Then the inner integral is

$$\int_0^{bA} \sqrt{A^2 - y^2/b^2}\, dy = A \int_0^{bA} \sqrt{1 - y^2/(Ab)^2}\, dy\,.$$

Using the change of variable $y = Ab \sin t$, $dy = Ab \cos t\, dt$, we find that the last integral is equal to

$$A^2 b \int_0^{\pi/2} \cos^2 t\, dt = \frac{\pi}{4} A^2 b = \frac{\pi b}{4}\left(1 - \frac{x^2}{a^2}\right).$$

Therefore

$$V = 8c \int_0^a \frac{\pi b}{4}\left(1 - \frac{x^2}{a^2}\right) dx = \frac{4}{3}\pi abc\,.$$

In the special case $a = b = c$ the solid is a sphere of radius a and the volume is $\frac{4}{3}\pi a^3$.

EXAMPLE 2. The double integral of a positive function f, $\iint_S f(x, y)\, dx\, dy$, reduces to the repeated integral

$$\int_0^1 \left[\int_{x^2}^{x} f(x, y)\, dy\right] dx\,.$$

Determine the region S and interchange the order of integration.

Solution. For each fixed x between 0 and 1, the integration with respect to y is over the interval from x^2 to x. This means that the region is of Type I and lies between the two curves $y = x^2$ and $y = x$. The region S is the set of points between these two curves and above the interval [0, 1]. (See Figure 11.11.) Since S is also of Type II we may interchange

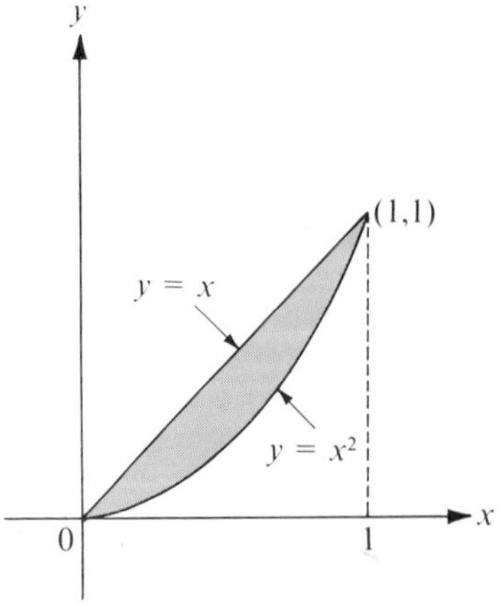

FIGURE 11.11 Example 2.

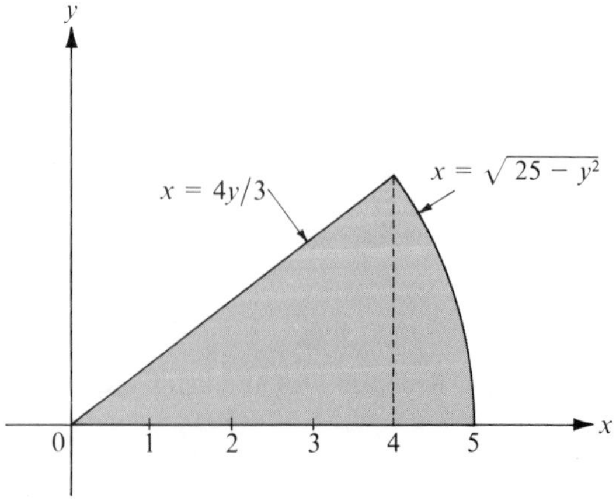

FIGURE 11.12 Example 3.

the order of integration to obtain

$$\int_0^1 \left[\int_y^{\sqrt{y}} f(x, y)\, dx\right] dy\,.$$

EXAMPLE 3. A double integral of a positive function f, $\iint_S f(x, y)\, dx\, dy$, reduces to the repeated integral:

$$\int_0^3 \left[\int_{4y/3}^{\sqrt{25-y^2}} f(x, y)\, dx\right] dy\,.$$

Determine the region S and interchange the order of integration.

Solution. For each fixed y between 0 and 3, the integration with respect to x is over the interval from $4y/3$ to $\sqrt{25 - y^2}$. Therefore region S is of Type II and lies between the two curves $x = 4y/3$ and $x = \sqrt{25 - y^2}$. This region, shown in Figure 11.12, is a sector of a circle. When the order of integration is reversed the region must be split into two regions of Type I; the result is the sum of two integrals:

$$\int_0^4 \left[\int_0^{3x/4} f(x, y)\, dy\right] dx + \int_4^5 \left[\int_0^{\sqrt{25-x^2}} f(x, y)\, dy\right] dx\,.$$

11.15 Exercises

In Exercises 1 through 5, make a sketch of the region of integration and evaluate the double integral.

1. $\iint_S x \cos (x + y)\, dx\, dy$, where S is the triangular region whose vertices are $(0, 0)$, $(\pi, 0)$, (π, π).

2. $\iint_S (1 + x) \sin y\, dx\, dy$, where S is the trapezoid with vertices $(0, 0)$, $(1, 0)$, $(1, 2)$, $(0, 1)$.

3. $\iint_S e^{x+y}\, dx\, dy$, where $S = \{(x, y) \mid |x| + |y| \le 1\}$.

4. $\iint_S x^2y^2\, dx\, dy$, where S is the bounded portion of the first quadrant lying between the two hyperbolas $xy = 1$ and $xy = 2$ and the two straight lines $y = x$ and $y = 4x$.

5. $\iint_S (x^2 - y^2)\, dx\, dy$, where S is bounded by the curve $y = \sin x$ and the interval $[0, \pi]$.

6. A pyramid is bounded by the three coordinate planes and the plane $x + 2y + 3z = 6$. Make a sketch of the solid and compute its volume by double integration.
7. A solid is bounded by the surface $z = x^2 - y^2$, the xy-plane, and the planes $x = 1$ and $x = 3$. Make a sketch of the solid and compute its volume by double integration.
8. Compute, by double integration, the volume of the ordinate set of f over S if:
   (a) $f(x, y) = x^2 + y^2$ and $S = \{(x, y) \mid |x| \le 1, |y| \le 1\}$.
   (b) $f(x, y) = 3x + y$ and $S = \{(x, y) \mid 4x^2 + 9y^2 \le 36, x > 0, y > 0\}$.
   (c) $f(x, y) = y + 2x + 20$ and $S = \{(x, y) \mid x^2 + y^2 \le 16\}$.

In Exercises 9 through 18 assume that the double integral of a positive function f extended over a region S reduces to the given repeated integral. In each case, make a sketch of the region S and interchange the order of integration.

9. $\int_0^1\left[\int_0^y f(x, y)\,dx\right] dy.$

10. $\int_0^2\left[\int_{y^2}^{2y} f(x, y)\,dx\right] dy.$

11. $\int_1^4\left[\int_{\sqrt{x}}^{2} f(x, y)\,dy\right] dx.$

12. $\int_1^2\left[\int_{2-x}^{\sqrt{2x-x^2}} f(x, y)\,dy\right] dx.$

13. $\int_{-6}^2\left[\int_{(x^2-4)/4}^{2-x} f(x, y)\,dy\right] dx.$

14. $\int_1^e\left[\int_0^{\log x} f(x, y)\,dy\right] dx.$

15. $\int_{-1}^1\left[\int_{-\sqrt{1-x^2}}^{1-x^2} f(x, y)\,dy\right] dx.$

16. $\int_0^1\left[\int_{x^3}^{x^2} f(x, y)\,dy\right] dx.$

17. $\int_0^\pi\left[\int_{-\sin(x/2)}^{\sin x} f(x, y)\,dy\right] dx.$

18. $\int_0^4\left[\int_{-\sqrt{4-y}}^{(y-4)/2} f(x, y)\,dx\right] dy.$

19. When a double integral was set up for the volume V of the solid under the paraboloid $z = x^2 + y^2$ and above a region S of the xy-plane, the following sum of iterated integrals was obtained:

$$V = \int_0^1\left[\int_0^y (x^2 + y^2)\,dx\right] dy + \int_1^2\left[\int_0^{2-y} (x^2 + y^2)\,dx\right] dy.$$

Sketch the region S and express V as an iterated integral in which the order of integration is reversed. Also, carry out the integration and compute V.

20. When a double integral was set up for the volume V of the solid under the surface $z = f(x, y)$ and above a region S of the xy-plane, the following sum of iterated integrals was obtained:

$$V = \int_0^{a\sin c}\left[\int_{\sqrt{a^2-y^2}}^{\sqrt{b^2-y^2}} f(x, y)\,dx\right] dy + \int_{a\sin c}^{b\sin c}\left[\int_{y\cot c}^{\sqrt{b^2-y^2}} f(x, y)\,dx\right] dy.$$

Given that $0 < a < b$ and $0 < c < \pi/2$, sketch the region S, giving the equations of all curves which form its boundary.

21. When a double integral was set up for the volume V of the solid under the surface $z = f(x, y)$ and above a region S of the xy-plane, the following sum of iterated integrals was obtained:

$$V = \int_1^2\left[\int_x^{x^3} f(x, y)\,dy\right] dx + \int_2^8\left[\int_x^8 f(x, y)\,dy\right] dx.$$

(a) Sketch the region S and express V as an iterated integral in which the order of integration is reversed.

(b) Carry out the integration and compute V when $f(x, y) = e^x(x/y)^{1/2}$.

22. Let $A = \int_0^1 e^{-t^2}\,dt$ and $B = \int_0^{1/2} e^{-t^2}\,dt$. Evaluate the iterated integral

$$I = 2\int_{-1/2}^{1}\left[\int_0^x e^{-y^2}\,dy\right] dx$$

in terms of A and B. These are positive integers m and n such that

$$I = mA - nB + e^{-1} - e^{-1/4}.$$

Use this fact to check your answer.

23. A solid cone is obtained by connecting every point of a plane region S with a vertex not in the plane of S. Let A denote the area of S, and let h denote the altitude of the cone. Prove that:
(a) The cross-sectional area cut by a plane parallel to the base and at a distance t from the vertex is $(t/h)^2A$, if $0 \le t \le h$.
(b) The volume of the cone is $\frac{1}{3}Ah$.
24. Reverse the order of integration to derive the formula

$$\int_0^a \left[\int_0^y e^{m(a-x)} f(x)\, dx\right] dy = \int_0^a (a - x)e^{m(a-x)} f(x)\, dx,$$

where a and m are constants, $a > 0$.

11.16 Further applications of double integrals

We have already seen that double integrals can be used to compute volumes of solids and areas of plane regions. Many other concepts such as mass, center of mass, and moment of inertia can be defined and computed with the aid of double integrals. This section contains a brief discussion of these topics. They are of special importance in physics and engineering.

Let $\boldsymbol{P}$ denote the vector from the origin to an arbitrary point in 3-space. If n positive masses $m_1, m_2, \ldots, m_n$ are located at points $\boldsymbol{P}_1, \boldsymbol{P}_2, \ldots, \boldsymbol{P}_n$, respectively, the *center of mass* of the system is defined to be the point $\boldsymbol{C}$ determined by the vector equation

$$\boldsymbol{C} = \frac{\sum_{k=1}^n m_k \boldsymbol{P}_k}{\sum_{k=1}^n m_k}.$$

The denominator, $\sum m_k$, is called the *total mass* of the system.

If each mass m_k is translated by a given vector $\boldsymbol{A}$ to a new point $\boldsymbol{Q}_k$ where $\boldsymbol{Q}_k = \boldsymbol{P}_k + \boldsymbol{A}$, the center of mass is also translated by $\boldsymbol{A}$, since we have

$$\frac{\sum m_k \boldsymbol{Q}_k}{\sum m_k} = \frac{\sum m_k(\boldsymbol{P}_k + \boldsymbol{A})}{\sum m_k} = \frac{\sum m_k \boldsymbol{P}_k}{\sum m_k} + \boldsymbol{A} = \boldsymbol{C} + \boldsymbol{A}.$$

This may also be described by saying that the location of the center of mass depends only on the points $\boldsymbol{P}_1, \boldsymbol{P}_2, \ldots, \boldsymbol{P}_n$ and the masses, and not on the origin. The center of mass is a theoretically computed quantity which represents, so to speak, a fictitious "balance point" of the system.

If the masses lie in a plane at points with coordinates $(x_1, y_1), \ldots, (x_n, y_n)$, and if the center of mass has coordinates $(\bar{x}, \bar{y})$, the vector relation which defines $\boldsymbol{C}$ can be expressed as two scalar equations

$$\bar{x} = \frac{\sum m_k x_k}{\sum m_k} \quad \text{and} \quad \bar{y} = \frac{\sum m_k y_k}{\sum m_k}.$$

In the numerator of the quotient defining $\bar{x}$, the kth term of the sum, $m_k x_k$, is called the *moment* of the mass m_k about the y-axis. If a mass m equal to the total mass of the system were placed at the center of mass, its moment about the y-axis would be equal to the moment of the system,

$$m\bar{x} = \sum_{k=1}^n m_k x_k.$$

When we deal with a system whose total mass is distributed throughout some region in the plane rather than at a finite number of discrete points, the concepts of mass, center of mass, and moment are defined by means of integrals rather than sums. For example, consider a thin plate having the shape of a plane region S. Assume that matter is distributed over this plate with a known density (mass per unit area). By this we mean that there is a nonnegative function f defined on S and that $f(x, y)$ represents the mass per unit area at the point (x, y). If the plate is made of a homogeneous material the density is constant. In this case the total mass of the plate is defined to be the product of the density and the area of the plate.

When the density varies from point to point we use the double integral of the density as the definition of the total mass. In other words, if the density function f is integrable over S, we define the total mass $m(S)$ of the plate by the equation

$$m(S) = \iint_S f(x, y)\, dx\, dy .$$

The quotient

$$\frac{\text{mass}}{\text{area}} = \frac{\iint_S f(x, y)\, dx\, dy}{\iint_S dx\, dy}$$

is called the *average density* of the plate. If S is thought of as a geometric configuration rather than as a thin plate, this quotient is called the *average* or *mean value* of the function f over the region S. In this case we do not require f to be nonnegative.

By analogy with the finite case, we define the *center of mass* of the plate to be the point $(\bar{x}, \bar{y})$ determined by the equations

$$\text{(11.14)} \qquad \bar{x}m(S) = \iint_S xf(x, y)\, dx\, dy \quad \text{and} \quad \bar{y}m(S) = \iint_S yf(x, y)\, dx\, dy .$$

The integrals on the right are called the moments of the plate about the y-axis and the x-axis, respectively. When the density is constant, say $f(x, y) = c$, a factor c cancels in each of Equations (11.14) and we obtain

$$\bar{x}a(S) = \iint_S x\, dx\, dy \quad \text{and} \quad \bar{y}a(S) = \iint_S y\, dx\, dy ,$$

where $a(S)$ is the area of S. In this case the point $(\bar{x}, \bar{y})$ is called the *centroid* of the plate (or of the region S).

If L is a line in the plane of the plate, let $\delta(x, y)$ denote the perpendicular distance from a point (x, y) in S to the line L. Then the number I_L defined by the equation

$$I_L = \iint_S \delta^2(x, y) f(x, y)\, dx\, dy$$

is called the *moment of inertia* of the plate about L. When $f(x, y) = 1$, I_L is called the moment of inertia or *second moment* of the region S about L. The moments of inertia about the x- and y-axes are denoted by I_x and I_y, respectively, and they are given by the integrals

$$I_x = \iint_S y^2 f(x, y)\, dx\, dy \quad \text{and} \quad I_y = \iint_S x^2 f(x, y)\, dx\, dy .$$

The sum of these two integrals is called the *polar moment of inertia* I_0 about the origin:

$$I_0 = I_x + I_y = \iint_S (x^2 + y^2) f(x, y)\, dx\, dy .$$

Note: The mass and center of mass of a plate are properties of the body and are independent of the location of the origin and of the directions chosen for the coordinates axes. The polar moment of inertia depends on the location of the origin but not on the directions chosen for the axes. The moments and the moments of inertia about the x- and y-axes depend on the location of the origin and on the directions chosen for the axes. If a plate of constant density has an axis of symmetry, its centroid will lie on this axis. If there are two axes of symmetry, the centroid will lie on their intersection. These facts, which can be proven from the foregoing definitions, often help to simplify calculations involving center of mass and moment of inertia.

EXAMPLE 1. A thin plate of constant density c is bounded by two concentric circles with radii a and b and center at the origin, where $0 < b < a$. Compute the polar moment of inertia.

Solution. The integral for I_0 is

$$I_0 = c \iint_S (x^2 + y^2)\, dx\, dy ,$$

where $S = \{(x, y) \mid b^2 \le x^2 + y^2 \le a^2\}$. To simplify the computations we write the integral as a difference of two integrals,

$$I_0 = c \iint_{S(a)} (x^2 + y^2)\, dx\, dy - c \iint_{S(b)} (x^2 + y^2)\, dx\, dy ,$$

where $S(a)$ and $S(b)$ are circular disks with radii a and b, respectively. We can use iterated integration to evaluate the integral over $S(a)$, and we find

$$\iint_{S(a)} (x^2 + y^2)\, dx\, dy = 4 \int_0^a \left[\int_0^{\sqrt{a^2 - x^2}} (x^2 + y^2)\, dy \right] dx = \frac{\pi a^4}{2} .$$

(We have omitted the details of the computation because this integral can be evaluated more easily with the use of polar coordinates, to be discussed in Section 11.27.) Therefore

$$I_0 = \frac{\pi c}{2} (a^4 - b^4) = \pi c (a^2 - b^2) \frac{(a^2 + b^2)}{2} = m \frac{a^2 + b^2}{2} ,$$

where $m = \pi c (a^2 - b^2)$, the mass of the plate.

EXAMPLE 2. Determine the centroid of the plane region bounded by one arch of a sine curve.

Solution. We take the region S bounded by the curve $y = \sin x$ and the interval $0 \le x \le \pi$. By symmetry, the x-coordinate of the centroid is $\bar{x} = \pi/2$. The y-coordinate, $\bar{y}$, is given by

$$\bar{y} = \frac{\iint_S y \, dx \, dy}{\iint_S dx \, dy} = \frac{\int_0^\pi [\int_0^{\sin x} y \, dy] \, dx}{\int_0^\pi \sin x \, dx} = \frac{\int_0^\pi \frac{1}{2} \sin^2 x \, dx}{2} = \frac{\pi}{8}.$$

11.17 Two theorems of Pappus

Pappus of Alexandria, who lived around 300 A.D., was one of the last geometers of the Alexandrian school of Greek mathematics. He wrote a compendium of eight books summarizing much of the mathematical knowledge of his time. (The last six and a part of the second are extant.) Pappus discovered a number of interesting properties of centroids, two of which are described in this section. The first relates the centroid of a plane region with the volume of the solid of revolution obtained by rotating the region about a line in its plane.

Consider a plane region Q lying between the graphs of two continuous functions f and g over an interval $[a, b]$, where $0 \le g \le f$. Let S be the solid of revolution generated by rotating Q about the x-axis. Let $a(Q)$ denote the area of Q, $v(S)$ the volume of S, and $\bar{y}$ the y-coordinate of the centroid of Q. As Q is rotated to generate S, the centroid travels along a circle of radius $\bar{y}$. Pappus' theorem states that *the volume of S is equal to the circumference of this circle multiplied by the area of Q;* that is,

$$v(S) = 2\pi\bar{y}a(Q). \tag{11.15}$$

To prove this formula we simply note that the volume is given by the integral

$$v(S) = \pi \int_a^b [f^2(x) - g^2(x)] \, dx$$

and that $\bar{y}$ is given by the formula

$$\bar{y}a(Q) = \iint_Q y \, dy \, dx = \int_a^b \left[\int_{g(x)}^{f(x)} y \, dy \right] dx = \int_a^b \tfrac{1}{2}[f^2(x) - g^2(x)] \, dx.$$

Comparing these two formulas we immediately obtain (11.15).

EXAMPLE 1. *Volume of a torus.* Let S be the torus generated by rotating a circular disk Q of radius R about an axis at a distance $b > R$ from the center of Q. The volume of S is easily calculated by Pappus' theorem. We have $\bar{y} = b$ and $a(Q) = \pi R^2$, so

$$v(S) = 2\pi\bar{y}a(Q) = 2\pi^2R^2b.$$

The next example shows that Pappus' theorem can also be used to calculate centroids.

EXAMPLE 2. *Centroid of a semicircular disk.* Let $\bar{y}$ denote the y-coordinate of the centroid of the semicircular disk

$$Q = \{(x, y) \mid x^2 + y^2 \le R^2, \quad y \ge 0\}.$$

The area of Q is $\frac{1}{2}\pi R^2$. When Q is rotated about the x-axis it generates a solid sphere of volume $\frac{4}{3}\pi R^3$. By Pappus' formula we have

$$\tfrac{4}{3}\pi R^3 = 2\pi\bar{y}(\tfrac{1}{2}\pi R^2),$$

so $\bar{y} = \dfrac{4R}{3\pi}$.

The next theorem of Pappus states that *the centroid of the union of two disjoint plane regions A and B lies on the line segment joining the centroid of A and the centroid of B.* More generally, let A and B be two thin plates that are either disjoint or intersect in a set of content zero. Let $m(A)$ and $m(B)$ denote their masses and let $\boldsymbol{C}_A$ and $\boldsymbol{C}_B$ denote vectors from an origin to their respective centers of mass. Then the union $A \cup B$ has mass $m(A) + m(B)$ and its center of mass is determined by the vector $\boldsymbol{C}$, where

$$\boldsymbol{C} = \frac{m(A)\boldsymbol{C}_A + m(B)\boldsymbol{C}_B}{m(A) + m(B)}. \tag{11.16}$$

The quotient for $\boldsymbol{C}$ is a linear combination of the form $a\boldsymbol{C}_A + b\boldsymbol{C}_B$, where a and b are nonnegative scalars with sum 1. A linear combination of this form is called a *convex combination* of $\boldsymbol{C}_A$ and $\boldsymbol{C}_B$. The endpoint of $\boldsymbol{C}$ lies on the line segment joining the endpoints of $\boldsymbol{C}_A$ and $\boldsymbol{C}_B$.

Pappus' formula (11.16) follows at once from the definition of center of mass given in (11.14). The proof is left as an exercise for the reader. The theorem can be extended in an obvious way to the union of three or more regions. It is especially useful in practice when a plate of constant density is made up of several pieces, each of which has geometric symmetry. We determine the centroid of each piece and then form a suitable convex combination to find the centroid of the union. Examples are given in Exercise 21 of the next section.

11.18 Exercises

In Exercises 1 through 8 a region S is bounded by one or more curves described by the given equations. In each case sketch the region S and determine the coordinates $\bar{x}$ and $\bar{y}$ of the centroid.

1. $y = x^2, \quad x + y = 2.$
2. $y^2 = x + 3, \quad y^2 = 5 - x.$
3. $x - 2y + 8 = 0, \quad x + 3y + 5 = 0, \quad x = -2, \quad x = 4.$
4. $y = \sin^2 x, \quad y = 0, \quad 0 \le x \le \pi.$
5. $y = \sin x, \quad y = \cos x, \quad 0 \le x \le \dfrac{\pi}{4}.$
6. $y = \log x, \quad y = 0, \quad 1 \le x \le a.$
7. $\sqrt{x} + \sqrt{y} = 1, \quad x = 0, \quad y = 0.$
8. $x^{2/3} + y^{2/3} = 1, \quad x = 0, \quad y = 0$, in first quadrant.

9. A thin plate is bounded by an arc of the parabola $y = 2x - x^2$ and the interval $0 \le x \le 2$. Determine its mass if the density at each point (x, y) is $(1 - y)/(1 + x)$.
10. Find the center of mass of a thin plate in the shape of a rectangle $ABCD$ if the density at any point is the product of the distances of the point from two adjacent sides AB and AD.

In Exercises 11 through 16, compute the moments of inertia I_x and I_y of a thin plate S in the xy-plane bounded by the one or more curves described by the given equations. In each case $f(x, y)$ denotes the density at an arbitrary point (x, y) of S.

11. $y = \sin^2 x, \quad y = -\sin^2 x, \quad -\pi \le x \le \pi; \quad f(x, y) = 1.$
12. $\frac{x}{a} + \frac{y}{b} = 1, \quad \frac{x}{c} + \frac{y}{b} = 1, \quad y = 0, \quad 0 < c < a, \quad b > 0; \quad f(x, y) = 1.$
13. $(x - r)^2 + (y - r)^2 = r^2, \quad x = 0, \quad y = 0, \quad 0 \le x \le r, \quad 0 \le y \le r; \quad f(x, y) = 1.$
14. $xy = 1, \quad xy = 2, \quad x = 2y, \quad y = 2x, \quad x > 0, \quad y > 0; \quad f(x, y) = 1.$
15. $y = e^x, \quad y = 0, \quad 0 \le x \le a; \quad f(x, y) = xy.$
16. $y = \sqrt{2x}, \quad y = 0, \quad 0 \le x \le 2; \quad f(x, y) = |x - y|.$
17. Let S be a thin plate of mass m, and let L_0 and L be two parallel lines in the plane of S, where L_0 passes through the center of mass of S. Prove the *parallel-axis theorem:*

$$I_L = I_{L_0} + mh^2,$$

where h is the perpendicular distance between the two lines L and L_0. [*Hint:* A careful choice of coordinates axes will simplify the work.]
18. The boundary of a thin plate is an ellipse with semiaxes a and b. Let L denote a line in the plane of the plate passing through the center of the ellipse and making an angle α with the axis of length $2a$. If the density is constant and if the mass of the plate is m, show that the moment of inertia I_L is equal to $\frac{1}{4}m(a^2 \sin^2 \alpha + b^2 \cos^2 \alpha)$.
19. Find the average distance from one corner of a square of side h to points inside the square.
20. Let δ denote the distance from an arbitrary point P inside a circle of radius r to a fixed point P_0 whose distance from the center of the circle is h. Find the average of the function δ^2 over the region enclosed by the circle.
21. Let A, B, C denote the following rectangles in the xy-plane:

$$A = [0, 4] \times [0, 1], \qquad B = [2, 3] \times [1, 3], \qquad C = [2, 4] \times [3, 4].$$

Use a theorem of Pappus to determine the centroid of each of the following figures:
(a) $A \cup B$. (c) $B \cup C$.
(b) $A \cup C$. (d) $A \cup B \cup C$.
22. An isosceles triangle T has base 1 and altitude h. The base of T coincides with one edge of a rectangle R of base 1 and altitude 2. Find a value of h so that the centroid of $R \cup T$ will lie on the edge common to R and T.
23. An isosceles triangle T has base $2r$ and altitude h. The base of T coincides with one edge of a semicircular disk D of radius r. Determine the relation that must hold between r and h so that the centroid of $T \cup D$ will lie inside the triangle.

11.19 Green's theorem in the plane

The second fundamental theorem of calculus for line integrals states that the line integral of a gradient ∇f along a path joining two points $\boldsymbol{a}$ and $\boldsymbol{b}$ may be expressed in terms of the function values $f(\boldsymbol{a})$ and $f(\boldsymbol{b})$. There is a two-dimensional analog of the second fundamental theorem which expresses a double integral over a plane region R as a line integral taken along a closed curve forming the boundary of R. This theorem is usually referred to as *Green's theorem.*† It can be stated in several ways; the most common is in the form of the identity:

$$\iint_R \left(\frac{\partial Q}{\partial x} - \frac{\partial P}{\partial y}\right) dx\, dy = \oint_C P\, dx + Q\, dy\,. \tag{11.17}$$

The curve C which appears on the right is the boundary of the region R, and the integration symbol $\oint$ indicates that the curve is to be traversed in the counterclockwise direction, as suggested by the example shown in Figure 11.13.

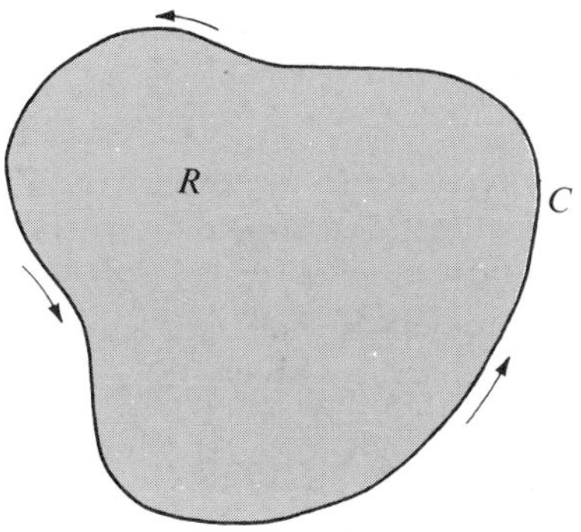

FIGURE 11.13 The curve C is the boundary of R, traversed in a counterclockwise direction.

Two types of assumptions are required for the validity of this identity. First, conditions are imposed on the functions P and Q to ensure the existence of the integrals. The usual assumptions are that P and Q are continuously differentiable on an open set S containing the region R. This implies continuity of P and Q on C as well as continuity of $\partial P/\partial y$ and $\partial Q/\partial x$ on R, although the theorem is also valid under less stringent hypotheses. Second, there are conditions of a geometric nature that are imposed on the region R and its boundary curve C. The curve C may be any *rectifiable simple closed curve.* The term "rectifiable" means, of course, that C has a finite arc length. To explain what is meant by a simple closed curve, we refer to the vector-valued function which describes the curve.

Suppose C is described by a continuous vector-valued function $\boldsymbol{\alpha}$ defined on an interval $[a, b]$. If $\boldsymbol{\alpha}(a) = \boldsymbol{\alpha}(b)$, the curve is *closed.* A closed curve such that $\boldsymbol{\alpha}(t_1) \neq \boldsymbol{\alpha}(t_2)$ for every pair of values $t_1 \neq t_2$ in the half-open interval $(a, b]$ is called a *simple* closed curve. This means that, except for the end points of the interval $[a, b]$, distinct values of t lead to distinct points on the curve. A circle is the prototype of a simple closed curve.

Simple closed curves that lie in a plane are usually called *Jordan curves* in honor of Camille Jordan (1838–1922), a famous French mathematician who did much of the pioneering work on such concepts as simple closed curves and arc length. Every Jordan curve C

† In honor of George Green (1793–1841), an English mathematician who wrote on the applications of mathematics to electricity and magnetism, fluid flow, and the reflection and refraction of light and sound. The theorem which bears Green's name appeared earlier in the researches of Gauss and Lagrange.

decomposes the plane into two disjoint open connected sets having the curve C as their common boundary. One of these regions is *bounded* and is called the *interior* (or *inner region*) of C. (An example is the shaded region in Figure 11.13.) The other is unbounded and is called the *exterior* (or *outer region*) of C. For some familiar Jordan curves such as circles, ellipses, or elementary polygons, it is intuitively evident that the curve divides the plane into an inner and an outer region, but to prove that this is true for an *arbitrary* Jordan curve is not easy. Jordan was the first to point out that this statement requires proof; the result is now known as the *Jordan curve theorem.* Toward the end of the 19th century Jordan and others published incomplete proofs. In 1905 the American mathematician Oswald Veblen (1880–1960) gave the first complete proof of this theorem. Green's theorem is valid whenever C is a rectifiable Jordan curve, and the region R is the union of C and its interior.† Since we have not defined line integrals along arbitrary rectifiable curves, we restrict our discussion here to piecewise smooth curves.

There is another technical difficulty associated with the formulation of Green's theorem. We have already remarked that, for the validity of the identity in (11.17), the curve C must be traversed in the counterclockwise direction. Intuitively, this means that a man walking along the curve in this direction always has the region R to his left. Again, for some familiar Jordan curves, such as those mentioned earlier, the meaning of the expression "traversing a curve in the counterclockwise direction" is intuitively evident. However, in a strictly rigorous treatment of Green's theorem one would have to define this expression in completely analytic terms, that is, in terms of the vector-valued function $\boldsymbol{\alpha}$ that describes the curve. One possible definition is outlined in Section 11.24.

Having pointed out some of the difficulties associated with the formulation of Green's theorem, we shall state the theorem in a rather general form and then indicate briefly why it is true for certain special regions. In this discussion the meaning of "counterclockwise" will be intuitive, so the treatment is not completely rigorous.

THEOREM 11.10. GREEN'S THEOREM FOR PLANE REGIONS BOUNDED BY PIECEWISE SMOOTH JORDAN CURVES. *Let P and Q be scalar fields that are continuously differentiable on an open set S in the xy-plane. Let C be a piecewise smooth Jordan curve, and let R denote the union of C and its interior. Assume R is a subset of S. Then we have the identity*

$$\iint_R \left(\frac{\partial Q}{\partial x} - \frac{\partial P}{\partial y}\right) dx\, dy = \oint_C P\, dx + Q\, dy, \tag{11.18}$$

where the line integral is taken around C in the counterclockwise direction.

Note: The identity in (11.18) is equivalent to the *two* formulas

$$\iint_R \frac{\partial Q}{\partial x}\, dx\, dy = \oint_C Q\, dy \tag{11.19}$$

† A proof of Green's theorem for regions of this generality can be found in Chapter 10 of the author's *Mathematical Analysis.*

and

(11.20)
$$-\iint_R \frac{\partial P}{\partial y}\,dx\,dy = \oint_C P\,dx\,.$$

In fact, if both of these are true, (11.18) follows by addition. Conversely, if (11.18) is true we may obtain (11.19) and (11.20) as special cases by taking $P = 0$ and $Q = 0$, respectively.

Proof for special regions. We shall prove (11.20) for a region R of Type I. Such a region has the form

$$R = \{(x, y) \mid a \le x \le b \quad \text{and} \quad f(x) \le y \le g(x)\},$$

where f and g are continuous on $[a, b]$ with $f \le g$. The boundary C of R consists of four parts, a lower arc C_1 (the graph of f), an upper arc C_2 (the graph of g), and two vertical line segments, traversed in the directions indicated in Figure 11.14.

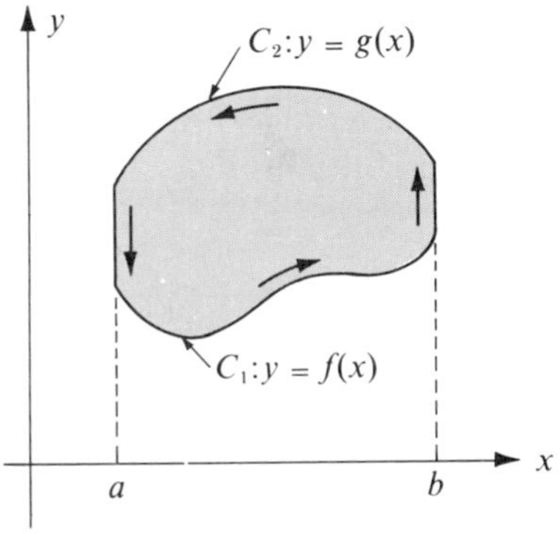

FIGURE 11.14 Proof of Green's theorem for a special region.

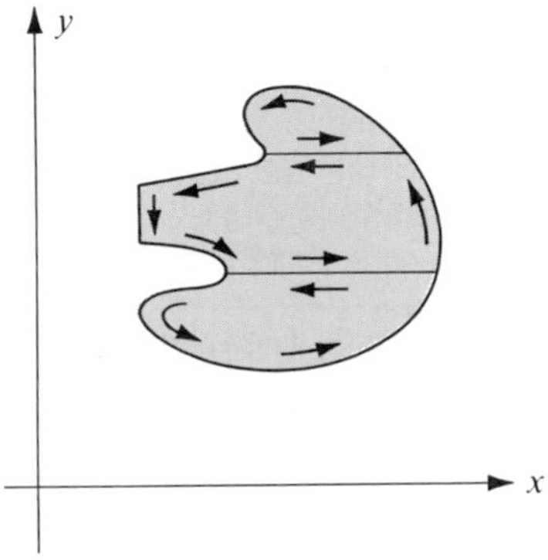

FIGURE 11.15 Proof of Green's theorem for a more general region.

First we evaluate the double integral $-\iint_R (\partial P/\partial y)\,dx\,dy$ by iterated integration. Integrating first with respect to y we have

(11.21)
$$-\iint_R \frac{\partial P}{\partial y}\,dx\,dy = -\int_a^b \left[\int_{f(x)}^{g(x)} \frac{\partial P}{\partial y}\,dy\right] dx = \int_a^b \left[\int_{g(x)}^{f(x)} \frac{\partial P}{\partial y}\,dy\right] dx$$
$$= \int_a^b P[x, f(x)]\,dx - \int_a^b P[x, g(x)]\,dx\,.$$

On the other hand, the line integral $\int_C P\,dx$ can be written as follows:

$$\int_C P\,dx = \int_{C_1} P\,dx + \int_{C_2} P\,dx\,,$$

since the line integral along each vertical segment is zero. To evaluate the integral along

C_1 we use the vector representation $\boldsymbol{\alpha}(t) = t\boldsymbol{i} + f(t)\boldsymbol{j}$ and obtain

$$\int_{C_1} P\,dx = \int_a^b P[t, f(t)]\,dt.$$

To evaluate the integral along C_2 we use the representation $\boldsymbol{\alpha}(t) = t\boldsymbol{i} + g(t)\boldsymbol{j}$ and take into account the reversal in direction to obtain

$$\int_{C_2} P\,dx = -\int_a^b P[t, g(t)]\,dt.$$

Therefore we have

$$\int_C P\,dx = \int_a^b P[t, f(t)]\,dt - \int_a^b P[t, g(t)]\,dt.$$

Comparing this equation with the formula in (11.21) we obtain (11.20).

A similar argument can be used to prove (11.19) for regions of Type II. In this way a proof of Green's theorem is obtained for regions that are of both Type I and Type II. Once this is done, the theorem can be proved for those regions R that can be decomposed into a finite number of regions that are of both types. "Crosscuts" are introduced as shown in Figure 11.15, the theorem is applied to each subregion, and the results are added together. The line integrals along the crosscuts cancel in pairs, as suggested in the figure, and the sum of the line integrals along the boundaries of the subregions is equal to the line integral along the boundary of R.

11.20 Some applications of Green's theorem

The following examples illustrate some applications of Green's theorem.

EXAMPLE 1. Use Green's theorem to compute the work done by the force field $\boldsymbol{f}(x, y) = (y + 3x)\boldsymbol{i} + (2y - x)\boldsymbol{j}$ in moving a particle once around the ellipse $4x^2 + y^2 = 4$ in the counterclockwise direction.

Solution. The work is equal to $\int_C P\,dx + Q\,dy$, where $P = y + 3x$, $Q = 2y - x$, and C is the ellipse. Since $\partial Q/\partial x - \partial P/\partial y = -2$, Green's theorem gives us

$$\int_C P\,dx + Q\,dy = \iint_R (-2)\,dx\,dy = -2a(R),$$

where $a(R)$ is the area of the region enclosed by the ellipse. Since this ellipse has semiaxes $a = 1$ and $b = 2$, its area is $\pi ab = 2\pi$ and the value of the line integral is -4π.

EXAMPLE 2. Evaluate the line integral $\int_C (5 - xy - y^2)\,dx - (2xy - x^2)\,dy$, where C is the square with vertices (0, 0), (1, 0), (1, 1), (0, 1), traversed counterclockwise.

Solution. Here $P = 5 - xy - y^2$, $Q = x^2 - 2xy$, and $\partial Q/\partial x - \partial P/\partial y = 3x$. Hence, by Green's theorem, we have

$$\int_C P\,dx + Q\,dy = 3\iint_R x\,dx\,dy = 3\bar{x},$$

where $\bar{x}$ is the x-coordinate of the centroid of the square. Since $\bar{x}$ is obviously $\frac{1}{2}$, the value of the line integral is $\frac{3}{2}$.

EXAMPLE 3. *Area expressed as a line integral.* The double integral for the area $a(R)$ of a region R can be expressed in the form

$$a(R) = \iint_R dx\,dy = \iint_R \left(\frac{\partial Q}{\partial x} - \frac{\partial P}{\partial y}\right) dx\,dy,$$

where P and Q are such that $\partial Q/\partial x - \partial P/\partial y = 1$. For example, we can take $Q(x, y) = \frac{1}{2}x$ and $P(x, y) = -\frac{1}{2}y$. If R is the region enclosed by a Jordan curve C we can apply Green's theorem to express $a(R)$ as a line integral,

$$a(R) = \int_C P\,dx + Q\,dy = \frac{1}{2}\int_C -y\,dx + x\,dy.$$

If the boundary curve C is described by parametric equations, say

$$x = X(t), \qquad y = Y(t), \qquad a \le t \le b,$$

the line integral for area becomes

$$a(R) = \frac{1}{2}\int_a^b \{-Y(t)X'(t) + X(t)Y'(t)\}\,dt = \frac{1}{2}\int_a^b \begin{vmatrix} X(t) & Y(t) \\ X'(t) & Y'(t) \end{vmatrix} dt.$$

11.21 A necessary and sufficient condition for a two-dimensional vector field to be a gradient

Let $\boldsymbol{f}(x, y) = P(x, y)\boldsymbol{i} + Q(x, y)\boldsymbol{j}$ be a vector field that is continuously differentiable on an open set S in the plane. If $\boldsymbol{f}$ is a gradient on S we have

$$\frac{\partial P}{\partial y} = \frac{\partial Q}{\partial x} \tag{11.22}$$

everywhere on S. In other words, the condition (11.22) is *necessary* for $\boldsymbol{f}$ to be a gradient. As we have already noted, this condition is not sufficient. For example, the vector field

$$\boldsymbol{f}(x, y) = \frac{-y}{x^2 + y^2}\boldsymbol{i} + \frac{x}{x^2 + y^2}\boldsymbol{j}$$

satisfies (11.22) everywhere on the set $S = \mathbf{R}^2 - \{(0, 0)\}$, but $\boldsymbol{f}$ is not a gradient on S. In Theorem 10.9 we proved that condition (11.22) is both necessary and sufficient for $\boldsymbol{f}$ to be a gradient on S if the set S is *convex*. With the help of Green's theorem we can extend this result to a more general class of plane sets known as *simply connected* sets. They are defined as follows.

DEFINITION OF A SIMPLY CONNECTED PLANE SET. *Let S be an open connected set in the plane. Then S is called simply connected if, for every Jordan curve C which lies in S, the inner region of C is also a subset of S.*

An annulus (the set of points lying between two concentric circles) is not simply connected because the inner region of a circle concentric with the bounding circles and of radius between theirs is not a subset of the annulus. Intuitively speaking, we say that S is simply connected when it has no "holes." Another way to describe simple connectedness is to say that a curve C_1 in S connecting any two points may be continuously deformed into any other curve C_2 in S joining these two points, with all intermediate curves during the deformation lying completely in S. An alternative definition, which can be shown to be equivalent to the one given here, states that an open connected set S is simply connected if its complement (relative to the whole plane) is connected. For example, an annulus is not simply connected because its complement is disconnected. An open connected set that is not simply connected is called multiply connected.

THEOREM 11.11. *Let $\boldsymbol{f}(x, y) = P(x, y)\boldsymbol{i} + Q(x, y)\boldsymbol{j}$ be a vector field that is continuously differentiable on an open simply connected set S in the plane. Then $\boldsymbol{f}$ is a gradient on S if and only if we have*

$$\frac{\partial P}{\partial y} = \frac{\partial Q}{\partial x} \qquad \textit{everywhere on } S. \tag{11.23}$$

Proof. As already noted, condition (11.23) is necessary for $\boldsymbol{f}$ to be a gradient. We shall prove now that it is also sufficient.

It can be shown that in any open connected plane set S, every pair of points $\boldsymbol{a}$ and $\boldsymbol{x}$ can be joined by a simple step-polygon, that is, by a polygon whose edges are parallel to the coordinate axes and which has no self-intersections. If the line integral of $\boldsymbol{f}$ from $\boldsymbol{a}$ to $\boldsymbol{x}$ has the same value for every simple step-polygon in S joining $\boldsymbol{a}$ to $\boldsymbol{x}$, then exactly the same argument used to prove Theorem 10.4 shows that $\boldsymbol{f}$ is a gradient on S. Therefore, we need

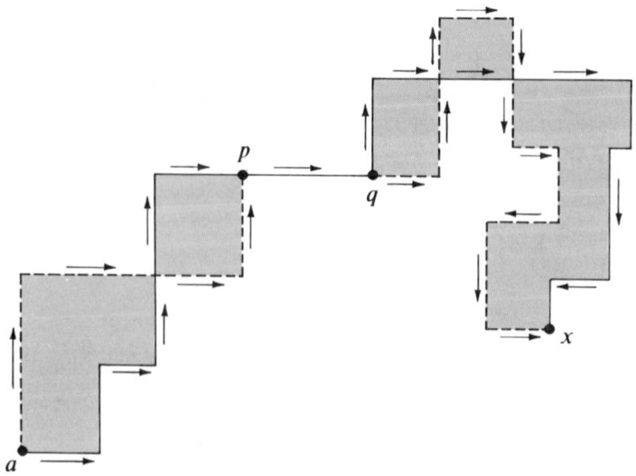

FIGURE 11.16 Independence of the path in a simply connected region.

only verify that the line integral of f from a to x has the same value for every simple step-polygon in S joining a to x.

Let C_1 and C_2 be two simple step-polygons in S joining a to x. Portions of these polygons may coincide along certain line segments. The remaining portions will intersect at most a finite number of times, and will form the boundaries of a finite number of polygonal regions, say $R_1, \ldots, R_m$. Since S is assumed to be simply connected, each of the regions R_k is a subset of S. An example is shown in Figure 11.16. The solid line represents C_1, the dotted line represents C_2, and the shaded regions represent $R_1, \ldots, R_m$. (These two particular polygons coincide along the segment pq.)

We observe next that the line integral of f from a to x along C_1 plus the integral from x back to a along C_2 is zero because the integral along the closed path is a sum of integrals taken over the line segments common to C_1 and C_2 plus a sum of integrals taken around the boundaries of the regions R_k. The integrals over the common segments cancel in pairs, since each common segment is traversed twice, in opposite directions, and their sum is zero. The integral over the boundary Γ_k of each region R_k is also zero because, by Green's theorem we may write

$$\int_{\Gamma_k} P\,dx + Q\,dy = \pm \iint_{R_k} \left(\frac{\partial Q}{\partial x} - \frac{\partial P}{\partial y}\right) dx\,dy,$$

and the integrand of the double integral is zero because of the hypothesis $\partial Q/\partial x = \partial P/\partial y$. It follows that the integral from a to x along C_1 is equal to that along C_2. As we have already noted, this implies that f is a gradient in S.

11.22 Exercises

1. Use Green's theorem to evaluate the line integral $\oint_C y^2\,dx + x\,dy$ when
   (a) C is the square with vertices $(0, 0)$, $(2, 0)$, $(2, 2)$, $(0, 2)$.
   (b) C is the square with vertices $(\pm 1, \pm 1)$.
   (c) C is the square with vertices $(\pm 2, 0)$, $(0, \pm 2)$.
   (d) C is the circle of radius 2 and center at the origin.
   (e) C has the vector equation $\alpha(t) = 2\cos^3 t\,i + 2\sin^3 t\,j$, $0 \le t \le 2\pi$.
2. If $P(x, y) = xe^{-y^2}$ and $Q(x, y) = -x^2 y\,e^{-y^2} + 1/(x^2 + y^2)$, evaluate the line integral $\oint P\,dx + Q\,dy$ around the boundary of the square of side $2a$ determined by the inequalities $|x| \le a$ and $|y| \le a$.
3. Let C be a simple closed curve in the xy-plane and let I_z denote the moment of inertia (about the z-axis) of the region enclosed by C. Show that an integer n exists such that

$$nI_z = \oint_C x^3\,dy - y^3\,dx.$$

4. Given two scalar fields u and v that are continuously differentiable on an open set containing the circular disk R whose boundary is the circle $x^2 + y^2 = 1$. Define two vector fields f and g as follows:

$$f(x, y) = v(x, y)i + u(x, y)j, \qquad g(x, y) = \left(\frac{\partial u}{\partial x} - \frac{\partial u}{\partial y}\right)i + \left(\frac{\partial v}{\partial x} - \frac{\partial v}{\partial y}\right)j.$$

Find the value of the double integral $\iint_R f \cdot g\,dx\,dy$ if it is known that on the boundary of R we have $u(x, y) = 1$ and $v(x, y) = y$.

5. If f and g are continuously differentiable in an open connected set S in the plane, show that $\oint_C f\nabla g \cdot d\boldsymbol{\alpha} = -\oint_C g\nabla f \cdot d\boldsymbol{\alpha}$ for every piecewise smooth Jordan curve C in S.
6. Let u and v be scalar fields having continuous first- and second-order partial derivatives in an open connected set S in the plane. Let R be a region in S bounded by a piecewise smooth Jordan curve C. Show that:

(a) $$\oint_C uv\,dx + uv\,dy = \iint_R \left\{ v\left(\frac{\partial u}{\partial x} - \frac{\partial u}{\partial y}\right) + u\left(\frac{\partial v}{\partial x} - \frac{\partial v}{\partial y}\right)\right\} dx\,dy.$$

(b) $$\frac{1}{2}\oint_C \left(v\frac{\partial u}{\partial x} - u\frac{\partial v}{\partial x}\right) dx + \left(u\frac{\partial v}{\partial y} - v\frac{\partial u}{\partial y}\right) dy = \iint_R \left(u\frac{\partial^2 v}{\partial x\,\partial y} - v\frac{\partial^2 u}{\partial x\,\partial y}\right) dx\,dy.$$

Normal derivatives. In Section 10.7 we defined line integrals with respect to arc length in such a way that the following equation holds:

$$\int_C P\,dx + Q\,dy = \int_C \boldsymbol{f} \cdot \boldsymbol{T}\,ds,$$

where $\boldsymbol{f} = P\boldsymbol{i} + Q\boldsymbol{j}$ and $\boldsymbol{T}$ is the unit tangent vector to C. (The dot product $\boldsymbol{f} \cdot \boldsymbol{T}$ is called the tangential component of $\boldsymbol{f}$ along C.) If C is a Jordan curve described by a continuously differentiable function $\boldsymbol{\alpha}$, say $\boldsymbol{\alpha}(t) = X(t)\boldsymbol{i} + Y(t)\boldsymbol{j}$, the unit *outer normal* $\boldsymbol{n}$ of C is defined by the equation

$$\boldsymbol{n}(t) = \frac{1}{\|\boldsymbol{\alpha}'(t)\|}\,(Y'(t)\boldsymbol{i} - X'(t)\boldsymbol{j})$$

whenever $\|\boldsymbol{\alpha}'(t)\| \neq 0$. If φ is a scalar field with gradient $\nabla\varphi$ on C, the *normal derivative* $\partial\varphi/\partial n$ is defined on C by the equation

$$\frac{\partial \varphi}{\partial n} = \nabla\varphi \cdot \boldsymbol{n}.$$

This is, of course, the directional derivative of φ in the direction of $\boldsymbol{n}$. These concepts occur in the remaining exercises of this section.

7. If $\boldsymbol{f} = Q\boldsymbol{i} - P\boldsymbol{j}$, show that
$$\int_C P\,dx + Q\,dy = \int_C \boldsymbol{f} \cdot \boldsymbol{n}\,ds.$$
(The dot product $\boldsymbol{f} \cdot \boldsymbol{n}$ is called the normal component of $\boldsymbol{f}$ along C.)
8. Let f and g be scalar fields with continuous first- and second-order partial derivatives on an open set S in the plane. Let R denote a region (in S) whose boundary is a piecewise smooth Jordan curve C. Prove the following identities, where $\nabla^2 u = \partial^2 u/\partial x^2 + \partial^2 u/\partial y^2$.

(a) $$\oint_C \frac{\partial g}{\partial n}\,ds = \iint_R \nabla^2 g\,dx\,dy.$$

(b) $$\oint_C f\frac{\partial g}{\partial n}\,ds = \iint_R (f\nabla^2 g + \nabla f \cdot \nabla g)\,dx\,dy.$$

(c) $$\oint_C \left(f\frac{\partial g}{\partial n} - g\frac{\partial f}{\partial n}\right) ds = \iint_R (f\nabla^2 g - g\nabla^2 f)\,dx\,dy.$$

The identity in (c) is known as *Green's formula*; it shows that

$$\oint_C f\frac{\partial g}{\partial n}\,ds = \oint_C g\frac{\partial f}{\partial n}\,ds$$

whenever f and g are both harmonic on R (that is, when $\nabla^2 f = \nabla^2 g = 0$ on R).

9. Suppose the differential equation

$$P(x, y)\,dx + Q(x, y)\,dy = 0$$

has an integrating factor $\mu(x, y)$ which leads to a one-parameter family of solutions of the form $\varphi(x, y) = C$. If the slope of the curve $\varphi(x, y) = C$ at (x, y) is $\tan\theta$, the unit normal vector $\boldsymbol{n}$ is taken to mean

$$\boldsymbol{n} = \sin\theta\,\boldsymbol{i} - \cos\theta\,\boldsymbol{j}.$$

There is a scalar field $g(x, y)$ such that the normal derivative of φ is given by the formula

$$\frac{\partial\varphi}{\partial n} = \mu(x, y)g(x, y),$$

where $\partial\varphi/\partial n = \nabla\varphi\cdot\boldsymbol{n}$. Find an explicit formula for $g(x, y)$ in terms of $P(x, y)$ and $Q(x, y)$.

★11.23 Green's theorem for multiply connected regions

Green's theorem can be generalized to apply to certain multiply connected regions.

THEOREM 11.12. GREEN'S THEOREM FOR MULTIPLY CONNECTED REGIONS. *Let $C_1, \ldots, C_n$ be n piecewise smooth Jordan curves having the following properties:*

(a) No two of the curves intersect.

(b) The curves $C_2, \ldots, C_n$ all lie in the interior of C_1.

(c) Curve C_i lies in the exterior of curve C_j for each $i \neq j$, $i > 1$, $j > 1$.

Let R denote the region which consists of the union of C_1 with that portion of the interior of C_1 that is not inside any of the curves $C_2, C_3, \ldots, C_n$. (An example of such a region is shown in Figure 11.17.) Let P and Q be continuously differentiable on an open set S containing R. Then we have the following identity:

$$\iint_R \left(\frac{\partial Q}{\partial x} - \frac{\partial P}{\partial y}\right) dx\,dy = \oint_{C_1} (P\,dx + Q\,dy) - \sum_{k=2}^{n} \oint_{C_k} (P\,dx + Q\,dy). \tag{11.24}$$

The theorem can be proved by introducing crosscuts which transform R into a union of a finite number of simply connected regions bounded by Jordan curves. Green's theorem is applied to each part separately, and the results are added together. We shall illustrate how this proof may be carried out when $n = 2$. The more general case may be dealt with by using induction on the number n of curves.

The idea of the proof when $n = 2$ is illustrated by the example shown in Figure 11.18, where C_1 and C_2 are two circles, C_1 being the larger circle. Introduce the crosscuts AB and CD, as shown in the figure. Let K_1 denote the Jordan curve consisting of the upper

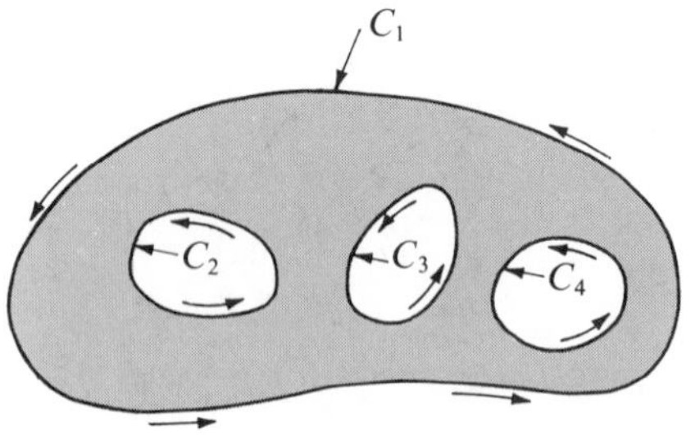

FIGURE 11.17 A multiply connected region.

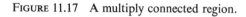

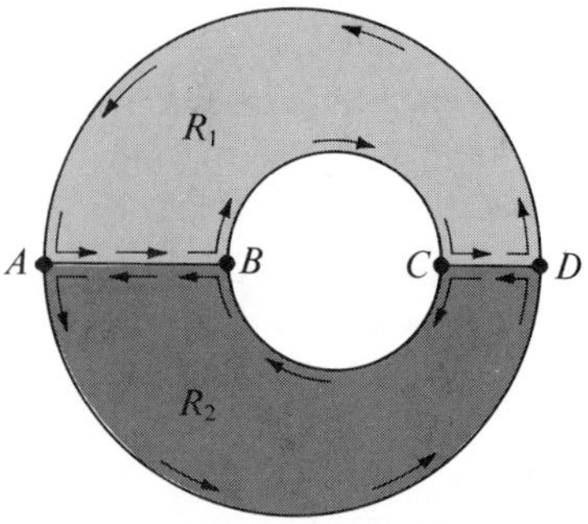

FIGURE 11.18 Proof of Green's theorem for a multiply connected region.

half of C_2, the upper half of C_1, and the segments AB and CD. Let K_2 denote the Jordan curve consisting of the lower half of C_1, the lower half of C_2, and the two segments AB and CD. Now apply Green's theorem to each of the regions bounded by K_1 and K_2 and add the two identities so obtained. The line integrals along the crosscuts cancel (since each crosscut is traversed once in each direction), resulting in the equation

$$\iint_R \left(\frac{\partial Q}{\partial x} - \frac{\partial P}{\partial y}\right) dx\, dy = \oint_{C_1} (P\, dx + Q\, dy) - \oint_{C_2} (P\, dx + Q\, dy).$$

The minus sign appears because of the direction in which C_2 is traversed. This is Equation (11.24) when $n = 2$.

For a simply connected region, the condition $\partial P/\partial y = \partial Q/\partial x$ implies that the line integral $\int P\, dx + Q\, dy$ is independent of the path (Theorem 11.11). As we have already noted, if S is *not* simply connected, the condition $\partial P/\partial y = \partial Q/\partial x$ does not necessarily imply independence of the path. However, in this case there is a substitute for independence that can be deduced from Theorem 11.12.

THEOREM 11.13. INVARIANCE OF A LINE INTEGRAL UNDER DEFORMATION OF THE PATH. *Let P and Q be continuously differentiable on an open connected set S in the plane, and assume that $\partial P/\partial y = \partial Q/\partial x$ everywhere on S. Let C_1 and C_2 be two piecewise smooth Jordan curves lying in S and satisfying the following conditions:*

(a) *C_2 lies in the interior of C_1.*

(b) *Those points inside C_1 which lie outside C_2 are in S. (Figure 11.19 shows an example.) Then we have*

$$\oint_{C_1} P\, dx + Q\, dy = \oint_{C_2} P\, dx + Q\, dy, \tag{11.25}$$

where both curves are traversed in the same direction.

Proof. Under the conditions stated, Equation (11.24) is applicable when $n = 2$. The region R consists of those points lying between the two curves C_1 and C_2 and the curves

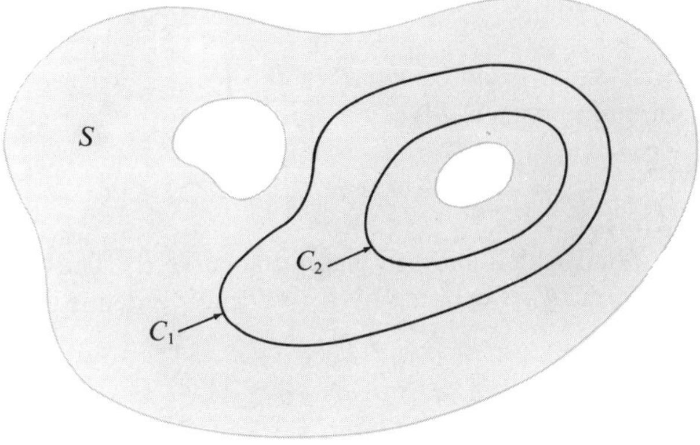

FIGURE 11.19. Invariance of a line integral under deformation of the path.

themselves. Since $\partial P/\partial y = \partial Q/\partial x$ in S, the left member of Equation (11.24) is zero and we obtain (11.25).

Theorem 11.13 is sometimes described by saying that if $\partial P/\partial y = \partial Q/\partial x$ in S the value of a line integral along a piecewise smooth simple closed curve in S is unaltered if the path is deformed to any other piecewise smooth simple closed curve in S, provided all intermediate curves remain within the set S during the deformation. The set S is assumed to be open and connected—it need not be simply connected.

★11.24 The winding number

We have seen that the value of a line integral often depends both on the curve along which the integration takes place and on the direction in which the curve is traversed. For example, the identity in Green's theorem requires the line integral to be taken in the counterclockwise direction. In a completely rigorous treatment of Green's theorem it would be necessary to describe analytically what it means to traverse a closed curve in the "counterclockwise direction." For some special curves this can be done by making specific statements about the vector-valued function $\boldsymbol{\alpha}$ which describes the curve. For example, the vector-valued function $\boldsymbol{\alpha}$ defined on the interval $[0, 2\pi]$ by the equation

$$\boldsymbol{\alpha}(t) = (a \cos t + x_0)\boldsymbol{i} + (a \sin t + y_0)\boldsymbol{j} \tag{11.26}$$

describes a circle of radius a with center at (x_0, y_0). This particular function is said to describe the circle in a *positive* or *counterclockwise* direction. On the other hand, if we replace t by $-t$ on the right of (11.26) we obtain a new function which is said to describe the circle in a *negative* or *clockwise* direction. In this way we have given a completely analytical description of "clockwise" and "counterclockwise" for a circle. However, it is not so simple to describe the same idea for an *arbitrary* closed curve. For piecewise smooth curves this may be done by introducing the concept of the *winding number*, an analytic device which gives us a mathematically precise way of counting the number of times a radius vector $\boldsymbol{\alpha}$ "winds around" a given point as it traces out a given closed curve. In this section we shall describe briefly one method for introducing the winding number.

Then we shall indicate how it can be used to assign positive and negative directions to closed curves.

Let C be a piecewise smooth closed curve in the plane described by a vector-valued function $\boldsymbol{\alpha}$ defined on an interval $[a, b]$, say

$$\boldsymbol{\alpha}(t) = X(t)\boldsymbol{i} + Y(t)\boldsymbol{j} \quad \text{if} \quad a \le t \le b.$$

Let $P_0 = (x_0, y_0)$ be a point which does not lie on the curve C. Then the winding number of $\boldsymbol{\alpha}$ with respect to the point P_0 is denoted by $W(\boldsymbol{\alpha}; P_0)$; it is defined to be the value of the following integral:

$$W(\boldsymbol{\alpha}; P_0) = \frac{1}{2\pi}\int_a^b \frac{[X(t) - x_0]Y'(t) - [Y(t) - y_0]X'(t)}{[X(t) - x_0]^2 + [Y(t) - y_0]^2}\, dt. \tag{11.27}$$

This is the same as the line integral

$$\frac{1}{2\pi}\oint_C \frac{-(y - y_0)\, dx + (x - x_0)\, dy}{(x - x_0)^2 + (y - y_0)^2}. \tag{11.28}$$

It can be shown that the value of this integral is always an *integer*, positive, negative, or zero. Moreover, if C is a *Jordan* curve (*simple* closed curve) this integer is 0 if P_0 is *outside* C and has the value $+1$ or -1 if P_0 is *inside* C. (See Figure 11.20.) Furthermore, $W(\boldsymbol{\alpha}; P_0)$

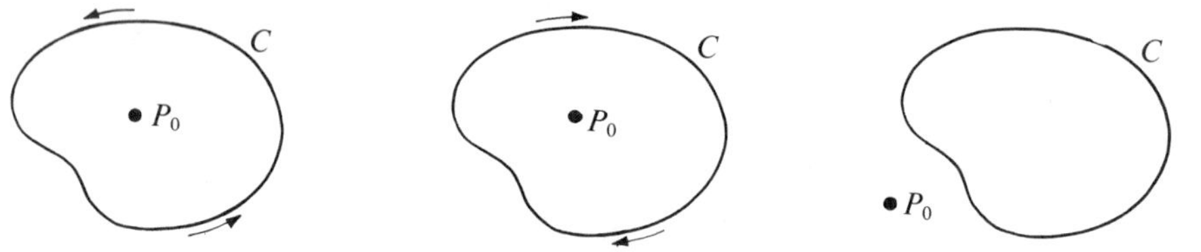

FIGURE 11.20 Illustrating the possible values of the winding number of a Jordan curve C with respect to P_0.

is either $+1$ for *every* point P_0 inside C or it is -1 for every such point. This enables us to define positive and negative orientations for C as follows: If the winding number $W(\boldsymbol{\alpha}; P_0)$ is $+1$ for every point P_0 inside C we say that $\boldsymbol{\alpha}$ traces out C in the *positive* or *counterclockwise direction*. If the winding number is -1 we say that $\boldsymbol{\alpha}$ traces out C in the *negative* or *clockwise direction*. [An example of the integral in (11.28) with $x_0 = y_0 = 0$ was encountered earlier in Example 2 of Section 10.16.]

To prove that the integral for the winding number is always $+1$ or -1 for a simple closed curve enclosing (x_0, y_0) we use Theorem 11.13. Let S denote the open connected region consisting of all points in the plane except (x_0, y_0). Then the line integral in (11.28) may be written as $\int_C P\, dx + Q\, dy$, and it is easy to verify that $\partial P/\partial y = \partial Q/\partial x$ everywhere in S. Therefore, if (x_0, y_0) is inside C, Theorem 11.13 tells us that we may replace the curve C by a circle with center at (x_0, y_0) without changing the value of the integral. Now we verify that for a circle the integral for the winding number is either $+1$ or -1, depending on whether the circle is positively or negatively oriented. For a positively oriented circle we may use the representation in Equation (11.26). In this case we have

$$X(t) = a \cos t + x_0, \qquad Y(t) = a \sin t + y_0,$$

and the integrand in (11.27) is identically equal to 1. Therefore we obtain

$$W(\boldsymbol{\alpha}; P_0) = \frac{1}{2\pi}\int_0^{2\pi} 1\, dt = 1.$$

By a similar argument we find that the intergal is -1 when C is negatively oriented. This proves that the winding number is either $+1$ or -1 for a simple closed curve enclosing the point (x_0, y_0).

★11.25 Exercises

1. Let $S = \{(x, y) \mid x^2 + y^2 > 0\}$, and let

$$P(x, y) = \frac{y}{x^2 + y^2}, \qquad Q(x, y) = \frac{-x}{x^2 + y^2}$$

if $(x, y) \in S$. Let C be a piecewise smooth Jordan curve lying in S.
(a) If $(0, 0)$ is inside C, show that the line integral $\int_C P\,dx + Q\,dy$ has the value $\pm 2\pi$, and explain when the plus sign occurs.
(b) Compute the value of the line integral $\int_C P\,dx + Q\,dy$ when $(0, 0)$ is outside C.

2. If $\boldsymbol{r} = x\boldsymbol{i} + y\boldsymbol{j}$ and $r = \|\boldsymbol{r}\|$, let

$$f(x, y) = \frac{\partial(\log r)}{\partial y}\boldsymbol{i} - \frac{\partial(\log r)}{\partial x}\boldsymbol{j}$$

for $r > 0$. Let C be a piecewise smooth Jordan curve lying in the annulus $1 < x^2 + y^2 < 25$, and find all possible values of the line integral of f along C.

3. A connected plane region with exactly one "hole" is called *doubly connected.* (The annulus $1 < x^2 + y^2 < 2$ is an example.) If P and Q are continuously differentiable on an open doubly connected region R, and if $\partial P/\partial y = \partial Q/\partial x$ everywhere in R, how many distinct values are possible for line integrals $\int_C P\,dx + Q\,dy$ taken around piecewise smooth Jordan curves in R?

4. Solve Exercise 3 for triply connected regions, that is, for connected plane regions with exactly two holes.

5. Let P and Q be two scalar fields which have continuous derivatives satisfying $\partial P/\partial y = \partial Q/\partial x$ everywhere in the plane except at three points. Let C_1, C_2, C_3 be three nonintersecting circles having centers at these three points, as shown in Figure 11.21, and let $I_k = \oint_{C_k} P\,dx + Q\,dy$. Assume that $I_1 = 12$, $I_2 = 10$, $I_3 = 15$.
(a) Find the value of $\int_C P\,dx + Q\,dy$, where C is the figure-eight curve shown.
(b) Draw another closed curve Γ along which $\int P\,dx + Q\,dy = 1$. Indicate on your drawing the direction in which Γ is traversed.
(c) If $I_1 = 12$, $I_2 = 9$, and $I_3 = 15$, show that there is no closed curve Γ along which $\int P\,dx + Q\,dy = 1$.

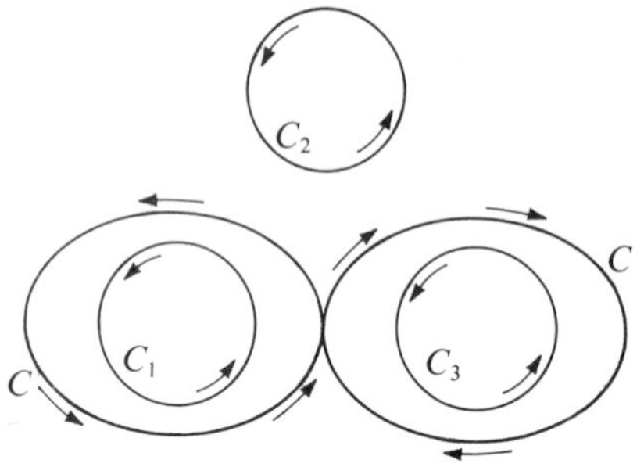

FIGURE 11.21 Exercise 5.

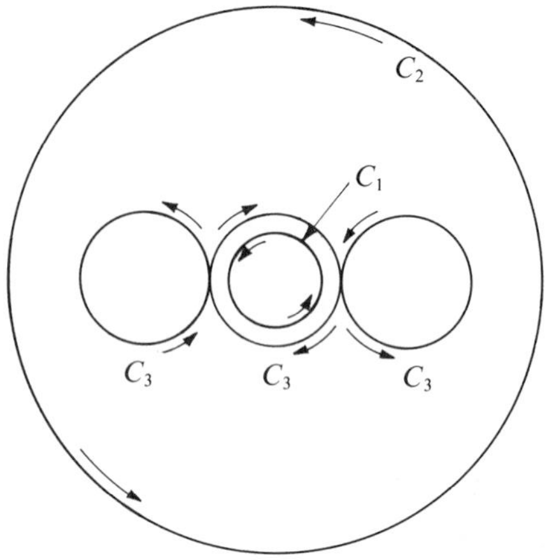

FIGURE 11.22 Exercise 6.

6. Let $I_k = \oint_{C_k} P\,dx + Q\,dy$, where

$$P(x, y) = -y\left[\frac{1}{(x-1)^2 + y^2} + \frac{1}{x^2 + y^2} + \frac{1}{(x+1)^2 + y^2}\right]$$

and

$$Q(x, y) = \frac{x-1}{(x-1)^2 + y^2} + \frac{x}{x^2 + y^2} + \frac{x+1}{(x+1)^2 + y^2}.$$

In Figure 11.22, C_1 is the smallest circle, $x^2 + y^2 = \frac{1}{8}$ (traced counterclockwise), C_2 is the largest circle, $x^2 + y^2 = 4$ (traced counterclockwise), and C_3 is the curve made up of the three intermediate circles $(x-1)^2 + y^2 = \frac{1}{4}$, $x^2 + y^2 = \frac{1}{4}$, and $(x+1)^2 + y^2 = \frac{1}{4}$ traced out as shown. If $I_2 = 6\pi$ and $I_3 = 2\pi$, find the value of I_1.

11.26 Change of variables in a double integral

In one-dimensional integration theory the method of substitution often enables us to evaluate complicated integrals by transforming them into simpler ones or into types that can be more easily recognized. The method is based on the formula

$$\int_a^b f(x)\,dx = \int_c^d f[g(t)]g'(t)\,dt, \tag{11.29}$$

where $a = g(c)$ and $b = g(d)$. We proved this formula (in Volume I) under the assumptions that g has a continuous derivative on an interval $[c, d]$ and that f is continuous on the set of values taken by $g(t)$ as t runs through the interval $[c, d]$.

There is a two-dimensional analogue of (11.29) called the formula for making a change of variables in a double integral. It transforms an integral of the form $\iint_S f(x, y)\,dx\,dy$, extended over a region S in the xy-plane, into another double integral $\iint_T F(u, v)\,du\,dv$, extended over a new region T in the uv-plane. The exact relationship between the regions S and T and the integrands $f(x, y)$ and $F(u, v)$ will be discussed presently. The method of

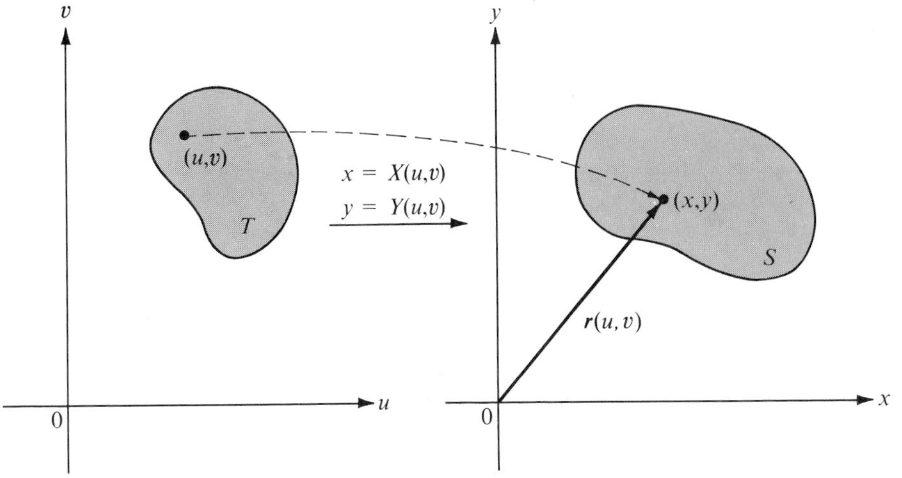

FIGURE 11.23 The mapping defined by the vector equation $\boldsymbol{r}(u, v) = X(u, v)\boldsymbol{i} + Y(u, v)\boldsymbol{j}$.

substitution for double integrals is more elaborate than in the one-dimensional case because there are two formal substitutions to be made, one for x and another for y. This means that instead of the one function g which appears in Equation (11.29), we now have two functions, say X and Y, which connect x, y with u, v as follows:

(11.30) $$x = X(u, v), \qquad y = Y(u, v).$$

The two equations in (11.30) define a mapping which carries a point (u, v) in the uv-plane into an image point (x, y) in the xy-plane. A set T of points in the uv-plane is mapped onto another set S in the xy-plane, as suggested by Figure 11.23. The mapping can also be described by means of a vector-valued function. From the origin in the xy-plane we draw the radius vector $\boldsymbol{r}$ to a general point (x, y) of S, as shown in Figure 11.23. The vector $\boldsymbol{r}$ depends on both u and v and can be considered a vector-valued function of two variables defined by the equation

(11.31) $$\boldsymbol{r}(u, v) = X(u, v)\boldsymbol{i} + Y(u, v)\boldsymbol{j} \quad \text{if} \quad (u, v) \in T.$$

This equation is called a *vector equation* of the mapping. As (u, v) runs through the points of T the endpoint of $\boldsymbol{r}(u, v)$ traces out the points of S.

Sometimes the two equations in (11.30) can be solved for u and v in terms of x and y. When this is possible we may express the result in the form

$$u = U(x, y), \qquad v = V(x, y).$$

These equations define a mapping from the xy-plane to the uv-plane, called the *inverse mapping* of the one defined by (11.30), since it carries points of S back to T. The so-called *one-to-one mappings* are of special importance. These carry *distinct* points of T onto *distinct* points of S; in other words, no two distinct points of T are mapped onto the same point of S by a one-to-one mapping. Each such mapping establishes a one-to-one correspondence between the points in T and those in S and enables us (at least in theory) to go back from S to T by the inverse mapping (which, of course, is also one-to-one).

We shall consider mappings for which the functions X and Y are continuous and have continuous partial derivatives $\partial X/\partial u$, $\partial X/\partial v$, $\partial Y/\partial u$, and $\partial Y/\partial v$ on S. Similar assumptions are made for the functions U and V. These are not serious restrictions since they are satisfied by most functions encountered in practice.

The formula for transforming double integrals may be written as

$$\iint_S f(x, y)\, dx\, dy = \iint_T f[X(u, v), Y(u, v)]\, |J(u, v)|\, du\, dv. \tag{11.32}$$

The factor $J(u, v)$ which appears in the integrand on the right plays the role of the factor $g'(t)$ which appears in the one-dimensional Formula (11.29). This factor is called the *Jacobian determinant* of the mapping defined by (11.30); it is equal to

$$J(u, v) = \begin{vmatrix} \dfrac{\partial X}{\partial u} & \dfrac{\partial Y}{\partial u} \\ \dfrac{\partial X}{\partial v} & \dfrac{\partial Y}{\partial v} \end{vmatrix}.$$

Sometimes the symbol $\partial(X, Y)/\partial(u, v)$ is used instead of $J(u, v)$ to represent the Jacobian determinant.

We shall not discuss the most general conditions under which the transformation formula (11.32) is valid. It can be shown† that (11.32) holds if, in addition to the continuity assumptions on X, Y, U, and V mentioned above, we assume that the mapping from T to S is one-to-one and that the Jacobian $J(u, v)$ is never zero. The formula is also valid if the mapping fails to be one-to-one on a subset of T of content zero, or if the Jacobian vanishes on a subset of content zero.

In Section 11.30 we show how the transformation formula (11.32) can be derived as a consequence of one of its special cases, namely, the case in which S is a rectangle and the function f has the constant value 1 at each point of S. In this special case (11.32) becomes

$$\iint_S dx\, dy = \iint_T |J(u, v)|\, du\, dv. \tag{11.33}$$

Even for this case a proof is not simple. In Section 11.29 a proof of (11.33) is given with the aid of Green's theorem. The remainder of this section will present a simple geometric argument which explains why a formula like (11.33) should hold.

Geometric motivation for Equation (11.33): Take a region T in the uv-plane, as shown in Figure 11.23, and let S denote the set of points in the xy-plane onto which T is mapped by the vector function $\boldsymbol{r}$ given by (11.31). Now introduce two new vector-valued functions $\boldsymbol{V}_1$ and $\boldsymbol{V}_2$ which are obtained by taking the partial derivatives of the components of $\boldsymbol{r}$

† See Theorem 10-30 of the author's *Mathematical Analysis*.

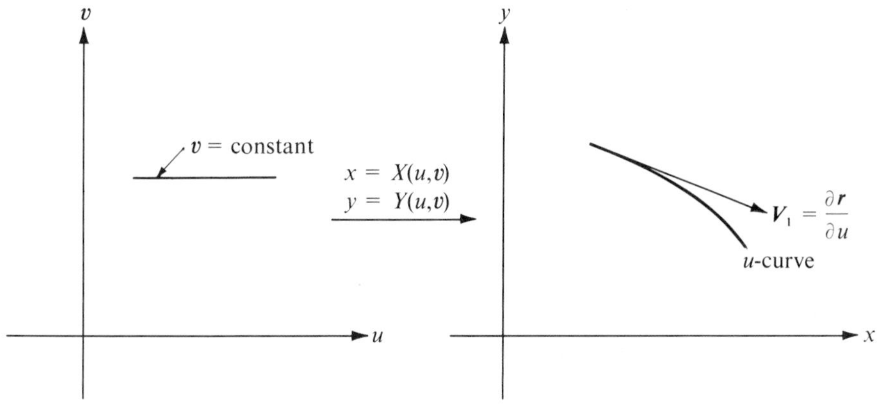

FIGURE 11.24 A u-curve and a corresponding velocity vector.

with respect to u and v, respectively. That is, define

$$V_1 = \frac{\partial \boldsymbol{r}}{\partial u} = \frac{\partial X}{\partial u}\boldsymbol{i} + \frac{\partial Y}{\partial u}\boldsymbol{j} \qquad \text{and} \qquad V_2 = \frac{\partial \boldsymbol{r}}{\partial v} = \frac{\partial X}{\partial v}\boldsymbol{i} + \frac{\partial Y}{\partial v}\boldsymbol{j}.$$

These vectors may be interpreted geometrically as follows: Consider a horizontal line segment in the uv-plane (v is constant on such a segment). The vector function $\boldsymbol{r}$ maps this segment onto a curve (called a u-curve) in the xy-plane, as suggested in Figure 11.24. If we think of u as a parameter representing time, the vector V_1 represents the velocity of the position $\boldsymbol{r}$ and is therefore tangent to the curve traced out by the tip of $\boldsymbol{r}$. In the same way, each vector V_2 represents the velocity vector of a v-curve obtained by setting u = constant. A u-curve and a v-curve pass through each point of the region S.

Consider now a small rectangle with dimensions Δu and Δv, as shown in Figure 11.25. If Δu is the length of a small time interval, then in time Δu a point of a u-curve moves along the curve a distance approximately equal to the product $\|V_1\| \Delta u$ (since $\|V_1\|$ represents the speed and Δu the time). Similarly, in time Δv a point on a v-curve moves a

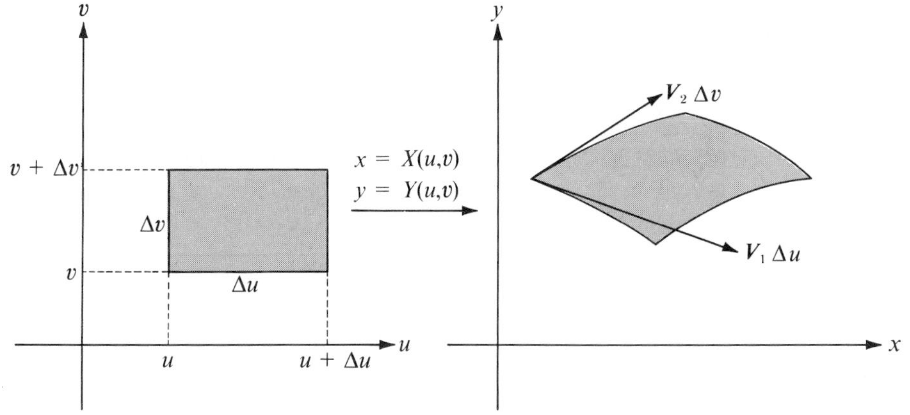

FIGURE 11.25 The image of a rectangular region in the uv-plane is a curvilinear parallelogram in the xy-plane.

distance nearly equal to $\|V_2\|\,\Delta v$. Hence the rectangular region with dimensions Δu and Δv in the uv-plane is traced onto a portion of the xy-plane that is nearly a parallelogram, whose sides are the vectors $V_1\,\Delta u$ and $V_2\,\Delta v$, as suggested by Figure 11.25. The area of this parallelogram is the magnitude of the cross product of the two vectors $V_1\,\Delta u$ and $V_2\,\Delta v$; this is equal to

$$\|(V_1\,\Delta u) \times (V_2\,\Delta v)\| = \|V_1 \times V_2\|\,\Delta u\,\Delta v\,.$$

If we compute the cross product $V_1 \times V_2$ in terms of the components of V_1 and V_2 we find

$$V_1 \times V_2 = \begin{vmatrix} \boldsymbol{i} & \boldsymbol{j} & \boldsymbol{k} \\ \dfrac{\partial X}{\partial u} & \dfrac{\partial Y}{\partial u} & 0 \\ \dfrac{\partial X}{\partial v} & \dfrac{\partial Y}{\partial v} & 0 \end{vmatrix} = \begin{vmatrix} \dfrac{\partial X}{\partial u} & \dfrac{\partial Y}{\partial u} \\ \dfrac{\partial X}{\partial v} & \dfrac{\partial Y}{\partial v} \end{vmatrix} \boldsymbol{k} = J(u, v)\boldsymbol{k}\,.$$

Therefore the magnitude of $V_1 \times V_2$ is exactly $|J(u, v)|$ and the area of the curvilinear parallelogram in Figure 11.25 is nearly equal to $|J(u, v)|\,\Delta u\,\Delta v$.

If $J(u, v) = 1$ for all points in T, then the "parallelogram" has the same area as the rectangle and the mapping preserves areas. Otherwise, to obtain the area of the parallelogram we must multiply the area of the rectangle by $|J(u, v)|$. This suggests that the Jacobian may be thought of as a "magnification factor" for areas.

Now let P be a partition of a large rectangle R enclosing the entire region T and consider a typical subrectangle of P of, say, dimensions Δu and Δv. If Δu and Δv are small, the Jacobian function J is nearly constant on the subrectangle and hence J acts somewhat like a step function on R. (We define J to be zero outside T.) If we think of J as an actual step function, then the double integral of $|J|$ over R (and hence over T) is a sum of products of the form $|J(u, v)|\,\Delta u\,\Delta v$ and the above remarks suggest that this sum is nearly equal to the area of S, which we know to be the double integral $\iint_S dx\,dy$.

This geometric discussion, which merely suggests why we might expect an equation like (11.33) to hold, can be made the basis of a rigorous proof, but the details are lengthy and rather intricate. As mentioned above, a proof of (11.33), based on an entirely different approach, will be given in a later section.

If $J(u, v) = 0$ at a particular point (u, v), the two vectors V_1 and V_2 are parallel (since their cross product is the zero vector) and the parallelogram degenerates into a line segment. Such points are called *singular points* of the mapping. As we have already mentioned, transformation formula (11.32) is also valid whenever there are only a finite number of such singular points or, more generally, when the singular points form a set of content zero. This is the case for all the mappings we shall use. In the next section we illustrate the use of formula (11.32) in two important examples.

11.27 Special cases of the transformation formula

EXAMPLE 1. *Polar coordinates.* In this case we write r and θ instead of u and v and describe the mapping by the two equations:

$$x = r\cos\theta\,, \qquad y = r\sin\theta\,.$$

That is, $X(r, \theta) = r \cos \theta$ and $Y(r, \theta) = r \sin \theta$. To obtain a one-to-one mapping we keep $r > 0$ and restrict θ to lie in an interval of the form $\theta_0 \leq \theta < \theta_0 + 2\pi$. For example, the mapping is one-to-one on any subset of the rectangle $(0, a] \times [0, 2\pi)$ in the $r\theta$-plane. The Jacobian determinant of this mapping is

$$J(r, \theta) = \begin{vmatrix} \dfrac{\partial X}{\partial r} & \dfrac{\partial Y}{\partial r} \\ \dfrac{\partial X}{\partial \theta} & \dfrac{\partial Y}{\partial \theta} \end{vmatrix} = \begin{vmatrix} \cos \theta & \sin \theta \\ -r \sin \theta & r \cos \theta \end{vmatrix} = r(\cos^2 \theta + \sin^2 \theta) = r .$$

Hence the transformation formula in (11.32) becomes

$$\iint_S f(x, y)\, dx\, dy = \iint_T f(r \cos \theta, r \sin \theta)\, r\, dr\, d\theta .$$

The r-curves are straight lines through the origin and the θ-curves are circles centered at the origin. The image of a rectangle in the $r\theta$-plane is a "parallelogram" in the xy-plane

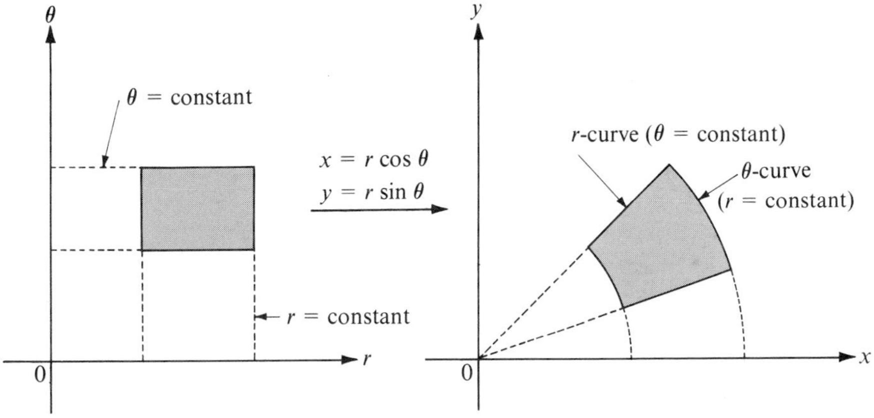

FIGURE 11.26 Transformation by polar coordinates.

bounded by two radial lines and two circular arcs, as shown in Figure 11.26. The Jacobian determinant vanishes when $r = 0$, but this does not affect the validity of the transformation formula because the set of points with $r = 0$ has content zero.

Since $V_1 = \cos \theta\, \boldsymbol{i} + \sin \theta\, \boldsymbol{j}$, we have $\|V_1\| = 1$, so there is no distortion of distances along the r-curves. On the other hand, we have

$$V_2 = -r \sin \theta\, \boldsymbol{i} + r \cos \theta\, \boldsymbol{j}, \qquad \|V_2\| = r,$$

so distances along the θ-curves are multiplied by the factor r.

Polar coordinates are particularly suitable when the region of integration has boundaries along which r or θ is constant. For example, consider the integral for the volume of one octant of a sphere of radius a,

$$\iint_S \sqrt{a^2 - x^2 - y^2}\, dx\, dy\,,$$

where the region S is the first quadrant of the circular disk $x^2 + y^2 \le a^2$. In polar coordinates the integral becomes

$$\iint_T \sqrt{a^2 - r^2}\, r\, dr\, d\theta\,,$$

where the region of integration T is now a rectangle $[0, a] \times [0, \frac{1}{2}\pi]$. Integrating first with respect to θ and then with respect to r we obtain

$$\iint_T \sqrt{a^2 - r^2}\, r\, dr\, d\theta = \frac{\pi}{2}\int_0^a r\sqrt{a^2 - r^2}\, dr = \frac{\pi}{2}\cdot\frac{(a^2 - r^2)^{3/2}}{-3}\Bigg|_0^a = \frac{\pi a^3}{6}\,.$$

The same result can be obtained by integrating in rectangular coordinates but the calculation is more complicated.

EXAMPLE 2. *Linear transformations.* Consider a linear transformation defined by a pair of equations of the form

$$x = Au + Bv\,, \qquad y = Cu + Dv\,, \tag{11.34}$$

where A, B, C, D are given constants. The Jacobian determinant is

$$J(u, v) = AD - BC\,,$$

and in order to have an inverse we assume that $AD - BC \neq 0$. This assures us that the two linear equations in (11.34) can be solved for u and v in terms of x and y.

Linear transformations carry parallel lines into parallel lines. Therefore the image of a rectangle in the uv-plane is a parallelogram in the xy-plane, and its area is that of the rectangle multiplied by the factor $|J(u, v)| = |AD - BC|$. Transformation formula (11.32) becomes

$$\iint_S f(x, y)\, dx\, dy = |AD - BC| \iint_T f(Au + Bv, Cu + Dv)\, du\, dv\,.$$

To illustrate an example in which a linear change of variables is useful, let us consider the integral

$$\iint_S e^{(y-x)/(y+x)}\, dx\, dy\,,$$

where S is the triangle bounded by the line $x + y = 2$ and the two coordinate axes. (See Figure 11.27.) The presence of $y - x$ and $y + x$ in the integrand suggests the change of variables

$$u = y - x\,, \qquad v = y + x\,.$$

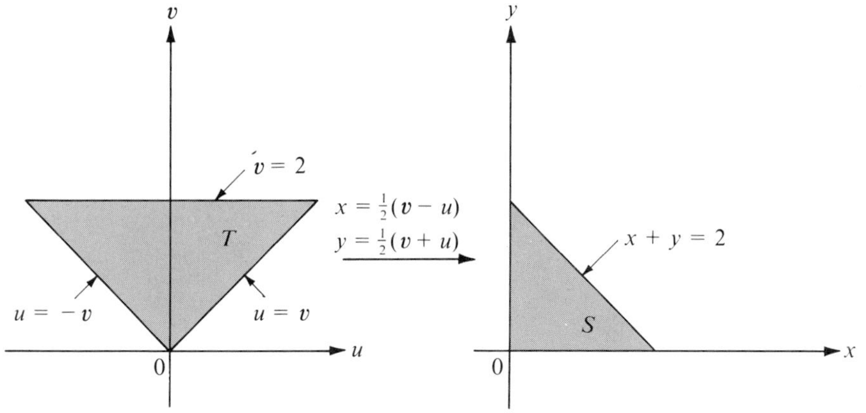

FIGURE 11.27 Mapping by a linear transformation.

Solving for x and y we find

$$x = \frac{v - u}{2} \qquad \text{and} \qquad y = \frac{v + u}{2}.$$

The Jacobian determinant is $J(u, v) = -\frac{1}{2}$. To find the image T of S in the uv-plane we note that the lines $x = 0$ and $y = 0$ map onto the lines $u = v$ and $u = -v$, respectively; the line $x + y = 2$ maps onto the line $v = 2$. Points inside S satisfy $0 < x + y < 2$ and these are carried into points of T satisfying $0 < v < 2$. Therefore the new region of integration T is a triangular region, as shown in Figure 11.27. The double integral in question becomes

$$\iint_S e^{(y-x)/(y+x)}\, dx\, dy = \frac{1}{2} \iint_T e^{u/v}\, du\, dv.$$

Integrating first with respect to u we find

$$\frac{1}{2} \iint_T e^{u/v}\, du\, dv = \frac{1}{2} \int_0^2 \left[\int_{-v}^{v} e^{u/v}\, du \right] dv = \frac{1}{2} \int_0^2 v\left(e - \frac{1}{e}\right) dv = e - \frac{1}{e}.$$

11.28 Exercises

In each of Exercises 1 through 5, make a sketch of the region S and express the double integral $\iint_S f(x, y)\, dx\, dy$ as an iterated integral in polar coordinates.

1. $S = \{(x, y) \mid x^2 + y^2 \le a^2\}$, where $a > 0$.
2. $S = \{(x, y) \mid x^2 + y^2 \le 2x\}$.
3. $S = \{(x, y) \mid a^2 \le x^2 + y^2 \le b^2\}$, where $0 < a < b$.
4. $S = \{(x, y) \mid 0 \le y \le 1 - x, 0 \le x \le 1\}$.
5. $S = \{(x, y) \mid x^2 \le y \le 1, -1 \le x \le 1\}$.

In each of Exercises 6 through 9, transform the integral to polar coordinates and compute its value. (The letter a denotes a positive constant.)

6. $\int_0^{2a}\left[\int_0^{\sqrt{2ax-x^2}}(x^2+y^2)\,dy\right]dx.$

7. $\int_0^{a}\left[\int_0^{x}\sqrt{x^2+y^2}\,dy\right]dx.$

8. $\int_0^{1}\left[\int_{x^2}^{x}(x^2+y^2)^{-1/2}\,dy\right]dx.$

9. $\int_0^{a}\left[\int_0^{\sqrt{a^2-y^2}}(x^2+y^2)\,dx\right]dy.$

In Exercises 10 through 13, transform each of the given integrals to one or more iterated integrals in polar coordinates.

10. $\int_0^{1}\left[\int_0^{1}f(x,y)\,dy\right]dx.$

11. $\int_0^{2}\left[\int_x^{x\sqrt{3}}f(\sqrt{x^2+y^2})\,dy\right]dx.$

12. $\int_0^{1}\left[\int_{1-x}^{\sqrt{1-x^2}}f(x,y)\,dy\right]dx.$

13. $\int_0^{1}\left[\int_0^{x^2}f(x,y)\,dy\right]dx.$

14. Use a suitable linear transformation to evaluate the double integral

$$\iint_S (x-y)^2\sin^2(x+y)\,dx\,dy$$

where S is the parallelogram with vertices $(\pi, 0)$, $(2\pi, \pi)$, $(\pi, 2\pi)$, $(0, \pi)$.

15. A parallelogram S in the xy-plane has vertices $(0, 0)$, $(2, 10)$, $(3, 17)$, and $(1, 7)$.
(a) Find a linear transformation $u = ax + by$, $v = cx + dy$, which maps S onto a rectangle R in the uv-plane with opposite vertices $(0, 0)$ and $(4, 2)$. The vertex $(2, 10)$ should map onto a point on the u-axis.
(b) Calculate the double integral $\iint_S xy\,dx\,dy$ by transforming it into an equivalent integral over the rectangle R of part (a).

16. If $r > 0$, let $I(r) = \int_{-r}^{r} e^{-u^2}\,du$.
(a) Show that $I^2(r) = \iint_R e^{-(x^2+y^2)}\,dx\,dy$, where R is the square $R = [-r, r] \times [-r, r]$.
(b) If C_1 and C_2 are the circular disks inscribing and circumscribing R, show that

$$\iint_{C_1} e^{-(x^2+y^2)}\,dx\,dy < I^2(r) < \iint_{C_2} e^{-(x^2+y^2)}\,dx\,dy.$$

(c) Express the integrals over C_1 and C_2 in polar coordinates and use (b) to deduce that $I(r) \to \sqrt{\pi}$ as $r \to \infty$. This proves that $\int_0^\infty e^{-u^2}\,du = \sqrt{\pi}/2$.
(d) Use part (c) to deduce that $\Gamma(\frac{1}{2}) = \sqrt{\pi}$, where Γ is the gamma function.

17. Consider the mapping defined by the equations

$$x = u + v, \qquad y = v - u^2.$$

(a) Compute the Jacobian determinant $J(u, v)$.
(b) A triangle T in the uv-plane has vertices $(0, 0)$, $(2, 0)$, $(0, 2)$. Describe, by means of a sketch, its image S in the xy-plane.
(c) Calculate the area of S by a double integral extended over S and also by a double integral extended over T.
(d) Evaluate $\iint_S (x - y + 1)^{-2}\,dx\,dy$.

18. Consider the mapping defined by the two equations $x = u^2 - v^2$, $y = 2uv$.
(a) Compute the Jacobian determinant $J(u, v)$.

(b) Let T denote the rectangle in the uv-plane with vertices $(1, 1)$, $(2, 1)$, $(2, 3)$, $(1, 3)$. Describe, by means of a sketch, the image S in the xy-plane.

(c) Evaluate the double integral $\iint_C xy\, dx\, dy$ by making the change of variables $x = u^2 - v^2$, $y = 2uv$, where $C = \{(x, y) \mid x^2 + y^2 \leq 1\}$.

19. Evaluate the double integral

$$I(p, r) = \iint_R \frac{dx\, dy}{(p^2 + x^2 + y^2)^p}$$

over the circular disk $R = \{(x, y) \mid x^2 + y^2 \leq r^2\}$. Determine those values of p for which $I(p, r)$ tends to a limit as $r \to +\infty$.

In Exercises 20 through 22, establish the given equations by introducing a suitable change of variables in each case.

20. $\displaystyle\iint_S f(x + y)\, dx\, dy = \int_{-1}^{1} f(u)\, du$, where $S = \{(x, y) \mid |x| + |y| \leq 1\}$.

21. $\displaystyle\iint_S f(ax + by + c)\, dx\, dy = 2 \int_{-1}^{1} \sqrt{1 - u^2}\, f(u\sqrt{a^2 + b^2} + c)\, du$,

where $S = \{(x, y) \mid x^2 + y^2 \leq 1\}$ and $a^2 + b^2 \neq 0$.

22. $\displaystyle\iint_S f(xy)\, dx\, dy = \log 2 \int_1^2 f(u)\, du$, where S is the region in the first quadrant bounded by the curves $xy = 1$, $xy = 2$, $y = x$, $y = 4x$.

11.29 Proof of the transformation formula in a special case

As mentioned earlier, the transformation formula

$$\iint_S f(x, y)\, dx\, dy = \iint_T f[X(u, v), Y(u, v)]\, |J(u, v)|\, du\, dv \tag{11.35}$$

can be deduced as a consequence of the special case in which S is a rectangle and f is identically 1. In this case the formula simplifies to

$$\iint_R dx\, dy = \iint_{R^*} |J(u, v)|\, du\, dv. \tag{11.36}$$

Here R denotes a rectangle in the xy-plane and R^* denotes its image in the uv-plane (see Figure 11.28) under a one-to-one mapping

$$u = U(x, y), \qquad v = V(x, y).$$

The inverse mapping is given by

$$x = X(u, v), \qquad y = Y(u, v),$$

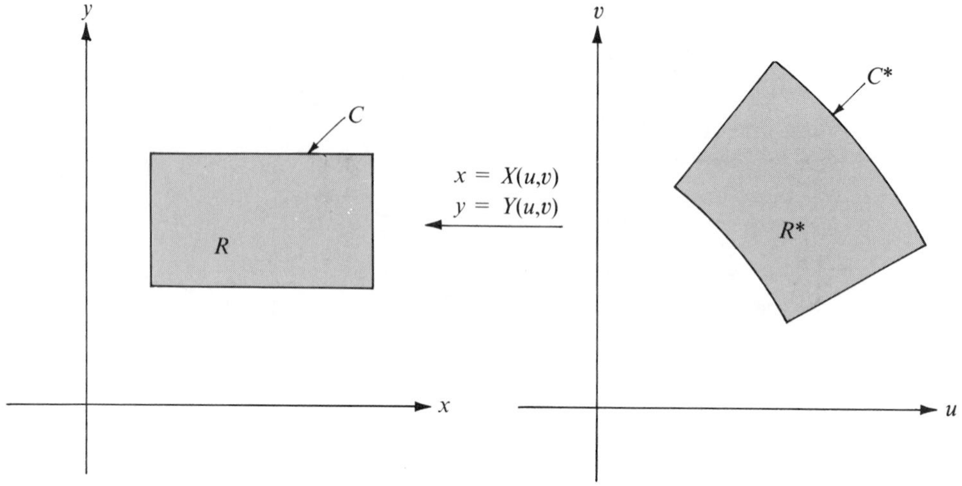

FIGURE 11.28 The transformation law for double integrals derived from Green's theorem.

and $J(u, v)$ denotes the Jacobian determinant,

$$J(u, v) = \begin{vmatrix} \dfrac{\partial X}{\partial u} & \dfrac{\partial X}{\partial v} \\ \dfrac{\partial Y}{\partial u} & \dfrac{\partial Y}{\partial v} \end{vmatrix}.$$

In this section we use Green's theorem to prove (11.36), and in the next section we deduce the more general formula (11.35) from the special case in (11.36).

For the proof we assume that the functions X and Y have continuous second-order partial derivatives and that the Jacobian is never 0 in R^*. Then $J(u, v)$ is either positive everywhere or negative everywhere. The significance of the sign of $J(u, v)$ is that when a point (x, y) traces out the boundary of R in the counterclockwise direction, the image point (u, v) traces out the boundary of R^* in the counterclockwise direction if $J(u, v)$ is positive and in the opposite direction if $J(u, v)$ is negative. In the proof we shall assume that $J(u, v) > 0$.

The idea of the proof is to express each double integral in (11.36) as a line integral, using Green's theorem. Then we verify equality of the two line integrals by expressing each in parametric form.

We begin with the double integral in the xy-plane, writing

$$\iint_R dx\,dy = \iint_R \left(\frac{\partial Q}{\partial x} - \frac{\partial P}{\partial y}\right) dx\,dy,$$

where $Q(x, y) = x$ and $P(x, y) = 0$. By Green's theorem this double integral is equal to the line integral

$$\int_C P\,dx + Q\,dy = \int_C x\,dy.$$

Here C is the boundary of R traversed in a counterclockwise direction.

Similarly, we transform the double integral in the uv-plane into a line integral around the boundary C^* of R^*. The integrand, $J(u, v)$, can be written as follows:

$$\begin{aligned} J(u, v) &= \frac{\partial X}{\partial u}\frac{\partial Y}{\partial v} - \frac{\partial X}{\partial v}\frac{\partial Y}{\partial u} = \frac{\partial X}{\partial u}\frac{\partial Y}{\partial v} + X\frac{\partial^2 Y}{\partial u\,\partial v} - X\frac{\partial^2 Y}{\partial v\,\partial u} - \frac{\partial X}{\partial v}\frac{\partial Y}{\partial u} \\ &= \frac{\partial}{\partial u}\left(X\frac{\partial Y}{\partial v}\right) - \frac{\partial}{\partial v}\left(X\frac{\partial Y}{\partial u}\right). \end{aligned}$$

Applying Green's theorem to the double integral over R^* we find

$$\iint_{R^*} J(u, v)\,du\,dv = \int_{C^*}\left(X\frac{\partial Y}{\partial u}\,du + X\frac{\partial Y}{\partial v}\,dv\right).$$

Therefore, to complete the proof of (11.36) we need only verify that

$$\int_C x\,dy = \int_{C^*}\left(X\frac{\partial Y}{\partial u}\,du + X\frac{\partial Y}{\partial v}\,dv\right). \tag{11.37}$$

We introduce a parametrization of C^* and use this to find a representation of C. Suppose C^* is described by a function $\boldsymbol{\alpha}$ defined on an interval $[a, b]$, say

$$\boldsymbol{\alpha}(t) = U(t)\boldsymbol{i} + V(t)\boldsymbol{j}.$$

Let

$$\boldsymbol{\beta}(t) = X[U(t), V(t)]\boldsymbol{i} + Y[U(t), V(t)]\boldsymbol{j}.$$

Then as t varies over the interval $[a, b]$, the vector $\boldsymbol{\alpha}(t)$ traces out the curve C^* and $\boldsymbol{\beta}(t)$ traces out C. By the chain rule, the derivative of $\boldsymbol{\beta}$ is given by

$$\boldsymbol{\beta}'(t) = \left[\frac{\partial X}{\partial u}U'(t) + \frac{\partial X}{\partial v}V'(t)\right]\boldsymbol{i} + \left[\frac{\partial Y}{\partial u}U'(t) + \frac{\partial Y}{\partial v}V'(t)\right]\boldsymbol{j}.$$

Hence

$$\int_C x\,dy = \int_a^b X[U(t), V(t)]\left(\frac{\partial Y}{\partial u}U'(t) + \frac{\partial Y}{\partial v}V'(t)\right)dt.$$

The last integral over $[a, b]$ is also obtained by parametrizing the line integral over C^* in (11.37). Therefore the two line integrals in (11.37) are equal, which proves (11.36).

11.30 Proof of the transformation formula in the general case

In this section we deduce the general transformation formula

$$\iint_S f(x, y)\,dx\,dy = \iint_T f[X(u, v), Y(u, v)]\,|J(u, v)|\,du\,dv \tag{11.38}$$

from the special case treated in the foregoing section,

$$\iint_R dx\,dy = \iint_{R^*} |J(u, v)|\,du\,dv\,, \tag{11.39}$$

where R is a rectangle and R^* is its image in the uv-plane.

First we prove that we have

$$\iint_R s(x, y)\,dx\,dy = \iint_{R^*} s[X(u, v), Y(u, v)]\,|J(u, v)|\,du\,dv\,, \tag{11.40}$$

where s is any step function defined on R. For this purpose, let P be a partition of R into mn subrectangles R_{ij} of dimensions Δx_i and Δy_j, and let c_{ij} be the constant value that s takes on the open subrectangle R_{ij}. Applying (11.39) to the rectangle R_{ij} we find

$$\Delta x_i\,\Delta y_j = \iint_{R_{ij}} dx\,dy = \iint_{R_{ij}^*} |J(u, v)|\,du\,dv\,.$$

Multiplying both sides by c_{ij} and summing on i and j we obtain

$$\sum_{i=1}^{n}\sum_{j=1}^{m} c_{ij}\,\Delta x_i\,\Delta y_j = \sum_{i=1}^{n}\sum_{j=1}^{m} c_{ij} \iint_{R_{ij}^*} |J(u, v)|\,du\,dv\,. \tag{11.41}$$

Since s is a step function, this is the same as

$$\sum_{i=1}^{n}\sum_{j=1}^{m} c_{ij}\,\Delta x_i\,\Delta y_j = \sum_{i=1}^{n}\sum_{j=1}^{m} \iint_{R_{ij}^*} s[X(u, v), Y(u, v)]\,|J(u, v)|\,du\,dv\,. \tag{11.42}$$

Using the additive property of double integrals we see that (11.42) is the same as (11.40). Thus, (11.40) is a consequence of (11.39).

Next we show that the step function s in (11.40) can be replaced by any function f for which both sides of (11.40) exist. Let f be integrable over a rectangle R and choose step functions s and t satisfying the inequalities

$$s(x, y) \le f(x, y) \le t(x, y)\,, \tag{11.43}$$

for all points (x, y) in R. Then we also have

$$s[X(u, v),\; Y(u, v)] \le f[X(u, v),\; Y(u, v)] \le t[X(u, v),\; Y(u, v)] \tag{11.44}$$

for every point (u, v) in the image R^*. For brevity, write $S(u, v)$ for $s[X(u, v),\; Y(u, v)]$ and define $F(u, v)$ and $T(u, v)$ similarly. Multiplying the inequalities in (11.44) by $|J(u, v)|$ and integrating over R^* we obtain

$$\iint_{R^*} S(u, v)\,|J(u, v)|\,du\,dv \le \iint_{R^*} F(u, v)\,|J(u, v)|\,du\,dv \le \iint_{R^*} T(u, v)\,|J(u, v)|\,du\,dv\,.$$

Because of (11.40), the foregoing inequalities are the same as

$$\iint_R s(x, y)\, dx\, dy \le \iint_{R^*} F(u, v)\, |J(u, v)|\, du\, dv \le \iint_R t(x, y)\, dx\, dy .$$

Therefore $\iint_{R^*} F(u, v)\, |J(u, v)|\, du\, dv$ is a number which lies between the integrals $\iint_R s(x, y)\, dx\, dy$ and $\iint_R t(x, y)\, dx\, dy$ for all choices of step functions s and t satisfying (11.43). Since f is integrable, this implies that

$$\iint_R f(x, y)\, dx\, dy = \iint_{R^*} F(u, v)\, |J(u, v)|\, du\, dv$$

and hence (11.38) is valid for integrable functions defined over rectangles.

Once we know that (11.38) is valid for rectangles we can easily extend it to more general regions S by the usual procedure of enclosing S in a rectangle R and extending the function f to a new function $\tilde{f}$ which agrees with f on S and has the value 0 outside S. Then we note that

$$\iint_S f = \iint_R \tilde{f} = \iint_{R^*} \tilde{f}[X(u, v), Y(u, v)]\, |J(u, v)|\, du\, dv = \iint_T F(u, v)\, |J(u, v)|\, du\, dv$$

and this proves that (11.38) is, indeed, a consequence of (11.39).

11.31 Extensions to higher dimensions

The concept of multiple integral can be extended from 2-space to n-space for any $n \ge 3$. Since the development is entirely analogous to the case $n = 2$ we merely sketch the principal results.

The integrand is a scalar field f defined and bounded on a set S in n-space. The integral of f over S, called an n-fold integral, is denoted by the symbols

$$\int \cdots \int_S f, \qquad \text{or} \qquad \int \cdots \int_S f(x_1, \ldots, x_n)\, dx_1 \cdots dx_n ,$$

with n integral signs, or more simply with one integral sign, $\int_S f(\boldsymbol{x})\, d\boldsymbol{x}$, where $\boldsymbol{x} = (x_1, \ldots, x_n)$. When $n = 3$ we write (x, y, z) instead of (x_1, x_2, x_3) and denote triple integrals by

$$\iiint_S f, \qquad \text{or} \qquad \iiint_S f(x, y, z)\, dx\, dy\, dz .$$

First we define the n-fold integral for a step function defined on an n-dimensional interval. We recall that an n-dimensional closed interval $[\boldsymbol{a}, \boldsymbol{b}]$ is the Cartesian product of n closed one-dimensional intervals $[a_k, b_k]$, where $\boldsymbol{a} = (a_1, \ldots, a_n)$ and $\boldsymbol{b} = (b_1, \ldots, b_n)$. An n-dimensional open interval $(\boldsymbol{a}, \boldsymbol{b})$ is the Cartesian product of n open intervals (a_k, b_k). The volume of $[\boldsymbol{a}, \boldsymbol{b}]$ or of $(\boldsymbol{a}, \boldsymbol{b})$ is defined to be the product of the lengths of the component intervals,

$$(b_1 - a_1) \cdots (b_n - a_n) .$$

If $P_1, \ldots, P_n$ are partitions of $[a_1, b_1], \ldots, [a_n, b_n]$, respectively, the Cartesian product $P = P_1 \times \cdots \times P_n$ is called a partition of $[\boldsymbol{a}, \boldsymbol{b}]$. A function f defined on $[\boldsymbol{a}, \boldsymbol{b}]$ is called a step function if it is constant on each of the open subintervals determined by some partition P. The n-fold integral of such a step function is defined by the formula

$$\int \cdots \int_{[\boldsymbol{a},\boldsymbol{b}]} f = \sum_i c_i v_i ,$$

where c_i is the constant value that f takes on the ith open subinterval and v_i is the volume of the ith subinterval. The sum is a finite sum extended over all the subintervals of P.

Having defined the n-fold integral for step functions, we define the integral for more general bounded functions defined on intervals, following the usual procedure. Let s and t denote step functions such that $s \le f \le t$ on $[\boldsymbol{a}, \boldsymbol{b}]$. If there is one and only one number I such that

$$\int \cdots \int_{[\boldsymbol{a},\boldsymbol{b}]} s \le I \le \int \cdots \int_{[\boldsymbol{a},\boldsymbol{b}]} t$$

for all choices of s and t satisfying $s \le f \le t$, then f is said to be integrable on $[\boldsymbol{a}, \boldsymbol{b}]$, and the number I is called the n-fold integral of f,

$$I = \int \cdots \int_{[\boldsymbol{a},\boldsymbol{b}]} f .$$

As in the two-dimensional case, the integral exists if f is continuous on $[\boldsymbol{a}, \boldsymbol{b}]$. It also exists if f is bounded on $[\boldsymbol{a}, \boldsymbol{b}]$ and if the set of discontinuities of f has n-dimensional content 0. A bounded set S has n-dimensional content 0 if for every $\epsilon > 0$ there is a finite collection of n-dimensional intervals whose union includes S and the sum of whose volumes does not exceed ϵ.

To define the n-fold integral of a bounded function f over a more general bounded set S, we extend f to a new function $\tilde{f}$ which agrees with f on S and has the value zero outside S; the integral of f over S is defined to be the integral of $\tilde{f}$ over some interval containing S.

Some multiple integrals can be calculated by using iterated integrals of lower dimension. For example, suppose S is a set in 3-space described as follows:

(11.45) $$S = \{(x, y, z) \mid (x, y) \in Q \quad \text{and} \quad \varphi_1(x, y) \le z \le \varphi_2(x, y)\} ,$$

where Q is a two-dimensional region, called the projection of S on the xy-plane, and φ_1, φ_2 are continuous on S. (An example is shown in Figure 11.29.) Sets of this type are bounded by two surfaces with Cartesian equations $z = \varphi_1(x, y)$, and $z = \varphi_2(x, y)$ and (perhaps) a portion of the cylinder generated by a line moving parallel to the z-axis along the boundary of Q. Lines parallel to the z-axis intersect this set in line segments joining the lower surface to the upper one. If f is continuous on the interior of S we have the iteration formula

(11.46) $$\iiint_S f(x, y, z)\, dx\, dy\, dz = \iint_Q \left[\int_{\varphi_1(x,y)}^{\varphi_2(x,y)} f(x, y, z)\, dz\right] dx\, dy .$$

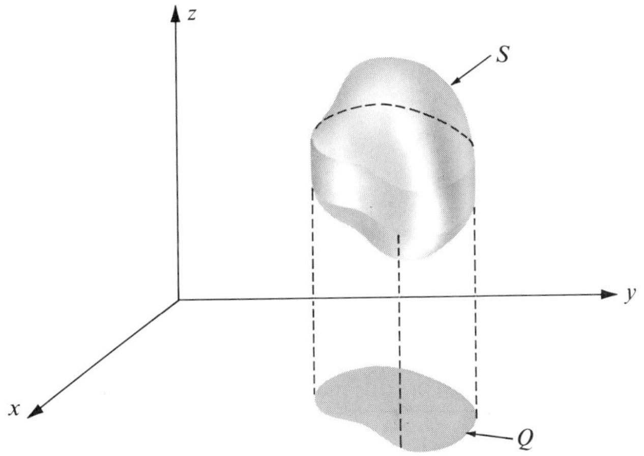

FIGURE 11.29 A solid S and its projection Q in the xy-plane.

That is, for fixed x and y, the first integration is performed with respect to z from the lower boundary surface to the upper one. This reduces the calculation to a double integral over the projection Q, which can be treated by the methods discussed earlier.

There are two more types of sets analogous to those described by (11.45) in which the x- and y-axes play the role of the z-axis, with the projections taken in the yz- or xz-planes, respectively. Triple integrals over such sets can be computed by iteration, with formulas analogous to (11.46). Most 3-dimensional sets that we shall encounter are either of one of the three types just mentioned or they can be split into a finite number of pieces, each of which is of one of these types.

Many iteration formulas exist for n-fold integrals when $n > 3$. For example, if Q is a k-dimensional interval and R an m-dimensional interval, then an $(m + k)$-fold integral over $Q \times R$ is the iteration of an m-fold integral with a k-fold integral,

$$\int \cdots \int_{Q\times R} f = \int \cdots \int_{Q} \left[\int \cdots \int_{R} f \, dx_1 \cdots dx_m \right] dx_{m+1} \cdots dx_{m+k} ,$$

provided all the multiple integrals in question exist. Later in this chapter we illustrate the use of iterated multiple integrals in computing the volume of an n-dimensional sphere.

11.32 Change of variables in an *n*-fold integral

The formula for making a change of variables in a double integral has a direct extension for n-fold integrals. We introduce new variables $u_1, \ldots, u_n$ related to $x_1, \ldots, x_n$ by n equations of the form

$$x_1 = X_1(u_1, \ldots, u_n), \qquad \ldots, \qquad x_n = X_n(u_1, \ldots, u_n).$$

Let $\boldsymbol{x} = (x_1, \ldots, x_n)$, $\boldsymbol{u} = (u_1, \ldots, u_n)$, and $\boldsymbol{X} = (X_1, \ldots, X_n)$. Then these equations define a vector-valued mapping

$$\boldsymbol{X}: T \rightarrow S$$

from a set T in n-space to another set S in n-space. We assume the mapping $\boldsymbol{X}$ is one-to-one and continuously differentiable on T. The transformation formula for n-fold integrals assumes the form

$$\int_S f(\boldsymbol{x})\, d\boldsymbol{x} = \int_T f[\boldsymbol{X}(\boldsymbol{u})]\, |\det D\boldsymbol{X}(\boldsymbol{u})|\, d\boldsymbol{u}, \tag{11.47}$$

where $D\boldsymbol{X}(\boldsymbol{u}) = [D_j X_k(\boldsymbol{u})]$ is the Jacobian matrix of the vector field $\boldsymbol{X}$. In terms of components we have

$$D\boldsymbol{X}(\boldsymbol{u}) = \begin{bmatrix} D_1X_1(\boldsymbol{u}) & D_2X_1(\boldsymbol{u}) & \cdots & D_nX_1(\boldsymbol{u}) \\ \cdot & & & \cdot \\ \cdot & & & \cdot \\ \cdot & & & \cdot \\ D_1X_n(\boldsymbol{u}) & D_2X_n(\boldsymbol{u}) & \cdots & D_nX_n(\boldsymbol{u}) \end{bmatrix}.$$

As in the two-dimensional case, the transformation formula is valid if $\boldsymbol{X}$ is one-to-one on T and if the Jacobian determinant $J(\boldsymbol{u}) = \det D\boldsymbol{X}(\boldsymbol{u})$ is never zero on T. It is also valid if the mapping fails to be one-to-one on a subset of T having n-dimensional content zero, or if the Jacobian determinant vanishes on such a subset.

For the three-dimensional case we write (x, y, z) for (x_1, x_2, x_3), (u, v, w) for (u_1, u_2, u_3), and (X, Y, Z) for (X_1, X_2, X_3). The transformation formula for triple integrals takes the form

$$\iiint_S f(x, y, z)\, dx\, dy\, dz = \iiint_T f[X(u, v, w), Y(u, v, w), Z(u, v, w)]\, |J(u, v, w)|\, du\, dv\, dw, \tag{11.48}$$

where $J(u, v, w)$ is the Jacobian determinant,

$$J(u, v, w) = \begin{vmatrix} \dfrac{\partial X}{\partial u} & \dfrac{\partial Y}{\partial u} & \dfrac{\partial Z}{\partial u} \\ \dfrac{\partial X}{\partial v} & \dfrac{\partial Y}{\partial v} & \dfrac{\partial Z}{\partial v} \\ \dfrac{\partial X}{\partial w} & \dfrac{\partial Y}{\partial w} & \dfrac{\partial Z}{\partial w} \end{vmatrix}.$$

In 3-space the Jacobian determinant can be thought of as a magnification factor for volumes. In fact, if we introduce the vector-valued function $\boldsymbol{r}$ defined by the equation

$$\boldsymbol{r}(u, v, w) = X(u, v, w)\boldsymbol{i} + Y(u, v, w)\boldsymbol{j} + Z(u, v, w)\boldsymbol{k},$$

and the vectors

$$V_1 = \frac{\partial r}{\partial u} = \frac{\partial X}{\partial u} i + \frac{\partial Y}{\partial u} j + \frac{\partial Z}{\partial u} k,$$

$$V_2 = \frac{\partial r}{\partial v} = \frac{\partial X}{\partial v} i + \frac{\partial Y}{\partial v} j + \frac{\partial Z}{\partial v} k,$$

$$V_3 = \frac{\partial r}{\partial w} = \frac{\partial X}{\partial w} i + \frac{\partial Y}{\partial w} j + \frac{\partial Z}{\partial w} k,$$

an argument similar to that given in Section 11.26 suggests that a rectangular parallelepiped of dimensions Δu, Δv, Δw in uvw-space is carried onto a solid which is nearly a curvilinear "parallelepiped" in xyz-space determined by the three vectors $V_1 \Delta u$, $V_2 \Delta v$, and $V_3 \Delta w$. (See Figure 11.30.) The boundaries of this solid are surfaces obtained by setting $u =$ constant, $v =$ constant, and $w =$ constant, respectively. The volume of a parallelepiped is equal to the absolute value of the scalar triple product of the three vectors which determine it, so the volume of the curvilinear parallelepiped is nearly equal to

$$|(V_1 \Delta u) \cdot (V_2 \Delta v) \times (V_3 \Delta w)| = |V_1 \cdot V_2 \times V_3| \, \Delta u \, \Delta v \, \Delta w = |J(u, v, w)| \, \Delta u \, \Delta v \, \Delta w .$$

11.33 Worked examples

Two important special cases of (11.48) are discussed in the next two examples.

EXAMPLE 1. *Cylindrical coordinates.* Here we write r, θ, z for u, v, w and define the mapping by the equations

$$x = r\cos\theta, \qquad y = r\sin\theta, \qquad z = z. \tag{11.49}$$

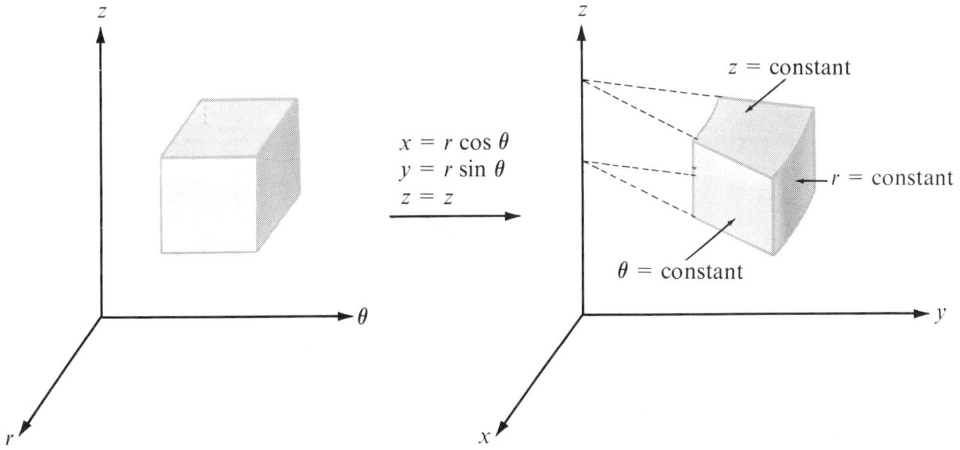

FIGURE 11.30 Transformation by cylindrical coordinates.

In other words, we replace x and y by their polar coordinates in the xy-plane and leave z unchanged. Again, to get a one-to-one mapping we must keep $r > 0$ and restrict θ to be in an interval of the form $\theta_0 \le \theta < \theta_0 + 2\pi$. Figure 11.30 shows what happens to a rectangular parallelepiped in the $r\theta z$-space.

The Jacobian determinant of the mapping in (11.49) is

$$J(r, \theta, z) = \begin{vmatrix} \cos\theta & \sin\theta & 0 \\ -r\sin\theta & r\cos\theta & 0 \\ 0 & 0 & 1 \end{vmatrix} = r(\cos^2\theta + \sin^2\theta) = r,$$

and therefore the transformation formula in (11.48) becomes

$$\iiint_S f(x, y, z)\,dx\,dy\,dz = \iiint_T f(r\cos\theta, r\sin\theta, z)r\,dr\,d\theta\,dz.$$

The Jacobian determinant vanishes when $r = 0$, but this does not affect the validity of the transformation formula because the set of points with $r = 0$ has 3-dimensional content 0.

EXAMPLE 2. *Spherical coordinates.* In this case the symbols ρ, θ, φ are used instead of u, v, w and the mapping is defined by the equations

$$x = \rho\cos\theta\sin\varphi, \qquad y = \rho\sin\theta\sin\varphi, \qquad z = \rho\cos\varphi.$$

The geometric meanings of ρ, θ, and φ are shown in Figure 11.31. To get a one-to-one mapping we keep $\rho > 0$, $0 \le \theta < 2\pi$, and $0 \le \varphi < \pi$. The surfaces ρ = constant are spheres centered at the origin, the surfaces θ = constant are planes passing through the z-axis, and the surfaces φ = constant are circular cones with their axes along the z-axis. Therefore a rectangular box in $\rho\theta\varphi$-space is mapped onto a solid of the type shown in Figure 11.31.

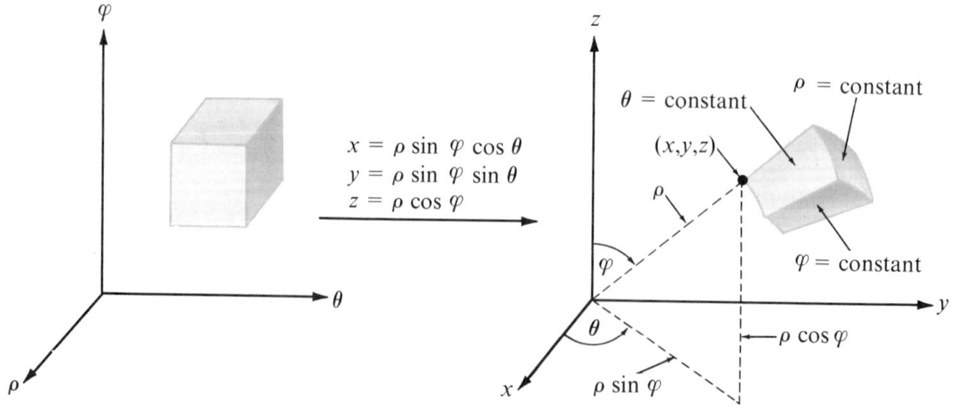

FIGURE 11.31 Transformation by spherical coordinates.

The Jacobian determinant of the mapping is

$$J(\rho, \theta, \varphi) = \begin{vmatrix} \cos\theta \sin\varphi & \sin\theta\sin\varphi & \cos\varphi \\ -\rho\sin\theta\sin\varphi & \rho\cos\theta\sin\varphi & 0 \\ \rho\cos\theta\cos\varphi & \rho\sin\theta\cos\varphi & -\rho\sin\varphi \end{vmatrix} = -\rho^2\sin\varphi .$$

Since $\sin\varphi \geq 0$ if $0 \leq \varphi < \pi$, we have $|J(\rho, \theta, \varphi)| = \rho^2 \sin\varphi$ and the formula for transforming triple integrals becomes

$$\iiint\limits_S f(x, y, z)\, dx\, dy\, dz = \iiint\limits_T F(\rho, \theta, \varphi)\rho^2 \sin\varphi\, d\rho\, d\theta\, d\varphi ,$$

where $F(\rho, \theta, \varphi) = f(\rho\cos\theta\sin\varphi, \rho\sin\theta\sin\varphi, \rho\cos\varphi)$. Although the Jacobian determinant vanishes when $\varphi = 0$ the transformation formula is still valid because the set of points with $\varphi = 0$ has 3-dimensional content 0.

The concept of volume can be extended to certain classes of sets (called measurable sets) in n-space in such a way that if S is measurable then its volume is equal to the integral of the constant function 1 over S. That is, if $v(S)$ denotes the volume of S, we have

$$v(S) = \int\limits_S \cdots \int dx_1 \cdots dx_n .$$

We shall not attempt to describe the class of sets for which this formula is valid. Instead, we illustrate how the integral can be calculated in some special cases.

EXAMPLE 3. *Volume of an n-dimensional interval.* If S is an n-dimensional interval, say $S = [a_1, b_1] \times \cdots \times [a_n, b_n]$, the multiple integral for $v(S)$ is the product of n one-dimensional integrals,

$$v(S) = \int_{a_1}^{b_1} dx_1 \cdots \int_{a_n}^{b_n} dx_n = (b_1 - a_1) \cdots (b_n - a_n).$$

This agrees with the formula given earlier for the volume of an n-dimensional interval.

EXAMPLE 4. *The volume of an n-dimensional sphere.* Let $S_n(a)$ denote the n-dimensional solid sphere (or n-ball) of radius a given by

$$S_n(a) = \{(x_1, \ldots, x_n) \mid x_1^2 + \cdots + x_n^2 \leq a^2\},$$

and let

$$V_n(a) = \int\limits_{S_n(a)} \cdots \int dx_1 \cdots dx_n ,$$

the volume of $S_n(a)$. We shall prove that

$$V_n(a) = \frac{\pi^{n/2}}{\Gamma(\frac{1}{2}n + 1)} a^n , \tag{11.50}$$

where Γ is the gamma function. For $n = 1$ the formula gives $V_1(a) = 2a$, the length of the interval $[-a, a]$. For $n = 2$ it gives $V_2(a) = \pi a^2$, the area of a circular disk of radius a. We shall prove (11.50) for $n \geq 3$.

First we prove that for every $a > 0$ we have

$$V_n(a) = a^n V_n(1). \tag{11.51}$$

In other words, the volume of a sphere of radius a is a^n times the volume of a sphere of radius 1. To prove this we use the linear change of variable $\boldsymbol{x} = a\boldsymbol{u}$ to map $S_n(1)$ onto $S_n(a)$. The mapping has Jacobian determinant a^n. Hence

$$V_n(a) = \int_{S_n(a)} \cdots \int dx_1 \cdots dx_n = \int_{S_n(1)} \cdots \int a^n \, du_1 \cdots du_n = a^n V_n(1).$$

This proves (11.51). Therefore, to prove (11.50) it suffices to prove that

$$V_n(1) = \frac{\pi^{n/2}}{\Gamma(\frac{1}{2}n + 1)}. \tag{11.52}$$

First we note that $x_1^2 + \cdots + x_n^2 \leq 1$ if and only if

$$x_1^2 + \cdots + x_{n-2}^2 \leq 1 - x_{n-1}^2 - x_n^2 \qquad \text{and} \qquad x_{n-1}^2 + x_n^2 \leq 1.$$

Therefore we can write the integral for $V_n(1)$ as the iteration of an $(n-2)$-fold integral and a double integral, as follows:

$$V_n(1) = \iint_{x_{n-1}^2 + x_n^2 \leq 1} \left[\int_{x_1^2 + \cdots + x_{n-2}^2 \leq 1 - x_{n-1}^2 - x_n^2} \cdots \int dx_1 \cdots dx_{n-2} \right] dx_{n-1} \, dx_n. \tag{11.53}$$

The inner integral is extended over the $(n-2)$-dimensional sphere $S_{n-2}(R)$, where $R = \sqrt{1 - x_{n-1}^2 - x_n^2}$, so it is equal to

$$V_{n-2}(R) = R^{n-2} V_{n-2}(1) = (1 - x_{n-1}^2 - x_n^2)^{n/2-1} V_{n-2}(1).$$

Now we write x for x_{n-1} and y for x_n. Then (11.53) becomes

$$V_n(1) = V_{n-2}(1) \iint_{x^2+y^2 \leq 1} (1 - x^2 - y^2)^{n/2-1} \, dx \, dy.$$

We evaluate the double integral by transforming it to polar coordinates and obtain

$$V_n(1) = V_{n-2}(1) \int_0^{2\pi} \int_0^1 (1 - r^2)^{n/2-1} r \, dr \, d\theta = V_{n-2}(1) \frac{2\pi}{n}.$$

In other words, the numbers $V_n(1)$ satisfy the recursion formula

$$V_n(1) = \frac{2\pi}{n} V_{n-2}(1) \qquad \text{if } n \geq 3.$$

But the sequence of numbers $\{f(n)\}$ defined by

$$f(n) = \frac{\pi^{n/2}}{\Gamma(\frac{1}{2}n + 1)}$$

satisfies the same recursion formula because $\Gamma(s + 1) = s\Gamma(s)$. Also, $\Gamma(\frac{1}{2}) = \sqrt{\pi}$ (see Exercise 16, Section 11.28), so $\Gamma(\frac{3}{2}) = \frac{1}{2}\sqrt{\pi}$ and $f(1) = V_1(1) = 2$. Also, $f(2) = V_2(1) = \pi$, hence we have $f(n) = V_n(1)$ for all $n \geq 1$. This proves (11.52).

11.34 Exercises

Evaluate each of the triple integrals in Exercises 1 through 5. Make a sketch of the region of integration in each case. You may assume the existence of all the integrals encountered.

1. $\iiint_S xy^2z^3 \, dx \, dy \, dz$, where S is the solid bounded by the surface $z = xy$ and the planes $y = x$, $x = 1$, and $z = 0$.
2. $\iiint_S (1 + x + y + z)^{-3} \, dx \, dy \, dz$, where S is the solid bounded by the three coordinate planes and the plane $x + y + z = 1$.
3. $\iiint_S xyz \, dx \, dy \, dz$, where $S = \{(x, y, z) \mid x^2 + y^2 + z^2 \leq 1, x \geq 0, y \geq 0, z \geq 0\}$.
4. $\iiint_S \left(\frac{x^2}{a^2} + \frac{y^2}{b^2} + \frac{z^2}{c^2}\right) dx \, dy \, dz$, where S is the solid bounded by the ellipsoid
$$\frac{x^2}{a^2} + \frac{y^2}{b^2} + \frac{z^2}{c^2} = 1.$$
5. $\iiint_S \sqrt{x^2 + y^2} \, dx \, dy \, dz$, where S is the solid formed by the upper nappe of the cone $z^2 = x^2 + y^2$ and the plane $z = 1$.

In Exercises 6, 7, and 8, a triple integral $\iiint_S f(x, y, z) \, dx \, dy \, dz$ of a positive function reduces to the iterated integral given. In each case describe the region of integration S by means of a sketch, showing its projection on the xy-plane. Then express the triple integral as one or more iterated integrals in which the first integration is with respect to y.

6. $\int_0^1 \left(\int_0^{1-x} \left[\int_0^{x+y} f(x, y, z) \, dz \right] dy \right) dx.$
7. $\int_{-1}^1 \left(\int_{-\sqrt{1-x^2}}^{\sqrt{1-x^2}} \left[\int_{\sqrt{x^2-y^2}}^1 f(x, y, z) \, dz \right] dy \right) dx.$
8. $\int_0^1 \left(\int_0^1 \left[\int_0^{x^2+y^2} f(x, y, z) \, dz \right] dy \right) dx.$
9. Show that:
$$\int_0^x \left(\int_0^v \left[\int_0^u f(t) \, dt \right] du \right) dv = \frac{1}{2} \int_0^x (x - t)^2 f(t) \, dt.$$

Evaluate the integrals in Exercises 10, 11, and 12 by transforming to cylindrical coordinates. You may assume the existence of all integrals encountered.

10. $\iiint_S (x^2 + y^2) \, dx \, dy \, dz$, where S is the solid bounded by the surface $x^2 + y^2 = 2z$ and the plane $z = 2$.

11. $\iiint_S dx\,dy\,dz$, where S is the solid bounded by the three coordinate planes, the surface $z = x^2 + y^2$, and the plane $x + y = 1$.
12. $\iiint_S (y^2 + z^2)\,dx\,dy\,dz$, where S is a right circular cone of altitude h with its base, of radius a, in the xy-plane and its axis along the z-axis.

Evaluate the integrals in Exercises 13, 14, and 15 by transforming to spherical coordinates.

13. $\iiint_S dx\,dy\,dz$, where S is a solid sphere of radius a and center at the origin.
14. $\iiint_S dx\,dy\,dz$, where S is the solid bounded by two concentric spheres of radii a and b, where $0 < a < b$, and the center is at the origin.
15. $\iiint_S [(x - a)^2 + (y - b)^2 + (z - c)^2]^{-1/2}\,dx\,dy\,dz$, where S is a solid sphere of radius R and center at the origin, and (a, b, c) is a fixed point outside this sphere.
16. Generalized spherical coordinates may be defined by the following mapping:

$$x = a\rho \cos^m \theta \sin^n \varphi, \qquad y = b\rho \sin^m \theta \sin^n \varphi, \qquad z = c\rho \cos^n \varphi,$$

where a, b, c, m, and n are positive constants. Show that the Jacobian determinant is equal to

$$-abcmn\rho^2 \cos^{m-1} \theta \sin^{m-1} \theta \cos^{n-1} \varphi \sin^{2n-1} \varphi.$$

Triple integrals can be used to compute volume, mass, center of mass, moment of inertia, and other physical concepts associated with solids. If S is a solid, its volume V is given by the triple integral

$$V = \iiint_S dx\,dy\,dz.$$

If the solid is assigned a density $f(x, y, z)$ at each of its points (x, y, z) (mass per unit volume), its mass M is defined to be

$$M = \iiint_S f(x, y, z)\,dx\,dy\,dz,$$

and its center of mass the point $(\bar{x}, \bar{y}, \bar{z})$ with coordinates

$$\bar{x} = \frac{1}{M} \iiint_S x f(x, y, z)\,dx\,dy\,dz,$$

and so on. The moment of inertia I_{xy} about the xy-plane is defined by the equation

$$I_{xy} = \iiint_S z^2 f(x, y, z)\,dx\,dy\,dz$$

and similar formulas are used to define I_{yz} and I_{zx}. The moment of inertia I_L about a line L is defined to be

$$I_L = \iiint_S \delta^2(x, y, z) f(x, y, z)\,dx\,dy\,dz,$$

where $\delta(x, y, z)$ denotes the perpendicular distance from a general point (x, y, z) of S to the line L.

17. Show that the moments of inertia about the coordinate axes are

$$I_x = I_{xy} + I_{xz}, \qquad I_y = I_{yx} + I_{yz}, \qquad I_z = I_{zx} + I_{zy}.$$

18. Find the volume of the solid bounded above by the sphere $x^2 + y^2 + z^2 = 5$ and below by the paraboloid $x^2 + y^2 = 4z$.
19. Find the volume of the solid bounded by the xy-plane, the cylinder $x^2 + y^2 = 2x$, and the cone $z = \sqrt{x^2 + y^2}$.
20. Compute the mass of the solid lying between two concentric spheres of radii a and b, where $0 < a < b$, if the density at each point is equal to the square of the distance of this point from the center.
21. A homogeneous solid right circular cone has altitude h. Prove that the distance of its centroid from the base is $h/4$.
22. Determine the center of mass of a right circular cone of altitude h if its density at each point is proportional to the distance of this point from the base.
23. Determine the center of mass of a right circular cone of altitude h if its density at each point is proportional to the distance of this point from the axis of the cone.
24. A solid is bounded by two concentric hemispheres of radii a and b, where $0 < a < b$. If the density is constant, find the center of mass.
25. Find the center of mass of a cube of side h if its density at each point is proportional to the square of the distance of this point from one corner of the base. (Take the base in the xy-plane and place the edges on the coordinate axes.)
26. A right circular cone has altitude h, radius of base a, constant density, and mass M. Find its moment of inertia about an axis through the vertex parallel to the base.
27. Find the moment of inertia of a sphere of radius R and mass M about a diameter if the density is constant.
28. Find the moment of inertia of a cylinder of radius a and mass M if its density at each point is proportional to the distance of this point from the axis of the cylinder.
29. The stem of a mushroom is a right circular cylinder of diameter 1 and length 2, and its cap is a hemisphere of radius R. If the mushroom is a homogeneous solid with axial symmetry, and if its center of mass lies in the plane where the stem joins the cap, find R.
30. A new space satellite has a smooth unbroken skin made up of portions of two circular cylinders of equal diameters D whose axes meet at right angles. It is proposed to ship the satellite to Cape Kennedy in a cubical packing box of inner dimension D. Prove that one-third of the box will be waste space.
31. Let $S_n(a)$ denote the following set in n-space, where $a > 0$:

$$S_n(a) = \{(x_1, \ldots, x_n) \mid |x_1| + \cdots + |x_n| \le a\}.$$

When $n = 2$ the set is a square with vertices at $(0, \pm a)$ and $(\pm a, 0)$. When $n = 3$ it is an octahedron with vertices at $(0, 0, \pm a)$, $(0, \pm a, 0)$, and $(\pm a, 0, 0)$. Let $V_n(a)$ denote the volume of $S_n(a)$, given by

$$V_n(a) = \int \cdots \int_{S_n(a)} dx_1 \cdots dx_n.$$

(a) Prove that $V_n(a) = a^n V_n(1)$.
(b) For $n \ge 2$, express the integral for $V_n(1)$ as an iteration of a one-dimensional integral and an $(n - 1)$-fold integral and show that

$$V_n(1) = V_{n-1}(1) \int_{-1}^{1} (1 - |x|)^{n-1}\, dx = \frac{2}{n} V_{n-1}(1).$$

(c) Use parts (a) and (b) to deduce that $V_n(a) = \dfrac{2^n a^n}{n!}$.

32. Let $S_n(a)$ denote the following set in n-space, where $a > 0$ and $n \geq 2$:

$$S_n(a) = \{(x_1, \ldots, x_n) \mid |x_i| + |x_n| \leq a \quad \text{for each } i = 1, \ldots, n-1\}.$$

(a) Make a sketch of $S_n(1)$ when $n = 2$ and when $n = 3$.
(b) Let $V_n(a) = \int \cdots \int_{S_n(a)} dx_1 \cdots dx_n$, and show that $V_n(a) = a^n V_n(1)$.
(c) Express the integral for $V_n(1)$ as an iteration of a one-dimensional integral and an $(n-1)$-fold integral and deduce that $V_n(a) = 2^n a^n/n$.

33. (a) Refer to Example 4, p. 411. Express the integral for $V_n(1)$, the volume of the n-dimensional unit sphere, as the iteration of an $(n-1)$-fold integral and a one-dimensional integral and thereby prove that

$$V_n(1) = 2V_{n-1}(1) \int_0^1 (1 - x^2)^{(n-1)/2}\, dx.$$

(b) Use part (a) and Equation (11.52) to deduce that

$$\int_0^{\pi/2} \cos^n t\, dt = \frac{\sqrt{\pi}}{2} \frac{\Gamma\left(\dfrac{n+1}{2}\right)}{\Gamma\left(\dfrac{n}{2} + 1\right)}.$$

12

SURFACE INTEGRALS

12.1 Parametric representation of a surface

This chapter discusses surface integrals and their applications. A surface integral can be thought of as a two-dimensional analog of a line integral where the region of integration is a surface rather than a curve. Before we can discuss surface integrals intelligently, we must agree on what we shall mean by a surface.

Roughly speaking, a surface is the locus of a point moving in space with two degrees of freedom. In our study of analytic geometry in Volume I we discussed two methods for describing such a locus by mathematical formulas. One is the *implicit representation* in which we describe a surface as a set of points (x, y, z) satisfying an equation of the form $F(x, y, z) = 0$. Sometimes we can solve such an equation for one of the coordinates in terms of the other two, say for z in terms of x and y. When this is possible we obtain an *explicit representation* given by one or more equations of the form $z = f(x, y)$. For example, a sphere of radius 1 and center at the origin has the implicit representation $x^2 + y^2 + z^2 - 1 = 0$. When this equation is solved for z it leads to two solutions, $z = \sqrt{1 - x^2 - y^2}$ and $z = -\sqrt{1 - x^2 - y^2}$. The first gives an explicit representation of the upper hemisphere and the second of the lower hemisphere.

A third method for describing surfaces is more useful in the study of surface integrals; this is the *parametric* or *vector* representation in which we have three equations expressing x, y, and z in terms of two parameters u and v:

$$x = X(u, v), \qquad y = Y(u, v), \qquad z = Z(u, v). \tag{12.1}$$

Here the point (u, v) is allowed to vary over some two-dimensional connected set T in the uv-plane, and the corresponding points (x, y, z) trace out a surface in xyz-space. This method for describing a surface is analogous to the representation of a space curve by three parametric equations involving one parameter. The presence of the two parameters in (12.1) makes it possible to transmit two degrees of freedom to the point (x, y, z), as suggested by Figure 12.1. Another way of describing the same idea is to say that a surface is the image of a plane region T under the mapping defined by (12.1).

If we introduce the radius vector $\boldsymbol{r}$ from the origin to a general point (x, y, z) of the surface, we may combine the three parametric equations in (12.1) into one vector equation

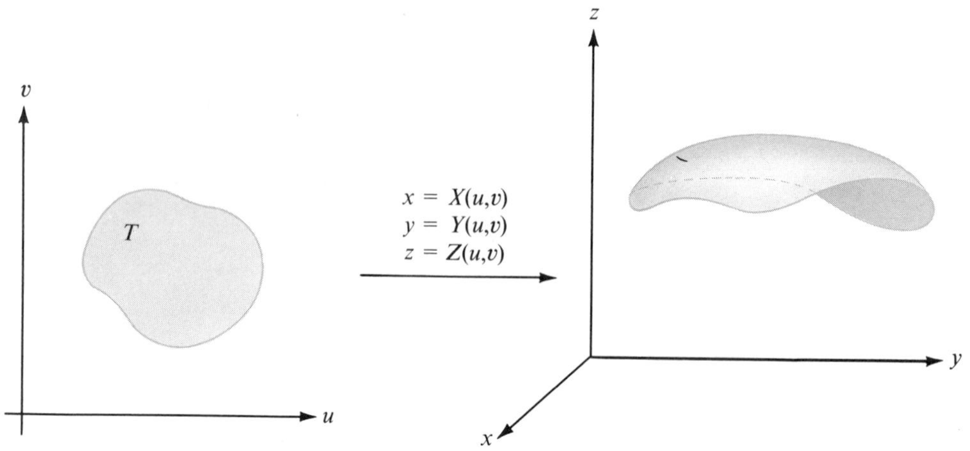

FIGURE 12.1 Parametric representation of a surface.

of the form

$$\boldsymbol{r}(u, v) = X(u, v)\boldsymbol{i} + Y(u, v)\boldsymbol{j} + Z(u, v)\boldsymbol{k}, \qquad \text{where} \quad (u, v) \in T. \tag{12.2}$$

This is called a *vector equation* for the surface.

There are, of course, many parametric representations for the same surface. One of these can always be obtained from an explicit form $z = f(x, y)$ by taking $X(u, v) = u$, $Y(u, v) = v$, $Z(u, v) = f(u, v)$. On the other hand, if we can solve the first two equations in (12.1) for u and v in terms of x and y and substitute in the third—we obtain an explicit representation $z = f(x, y)$.

EXAMPLE 1. *A parametric representation of a sphere.* The three equations

$$x = a \cos u \cos v, \qquad y = a \sin u \cos v, \qquad z = a \sin v \tag{12.3}$$

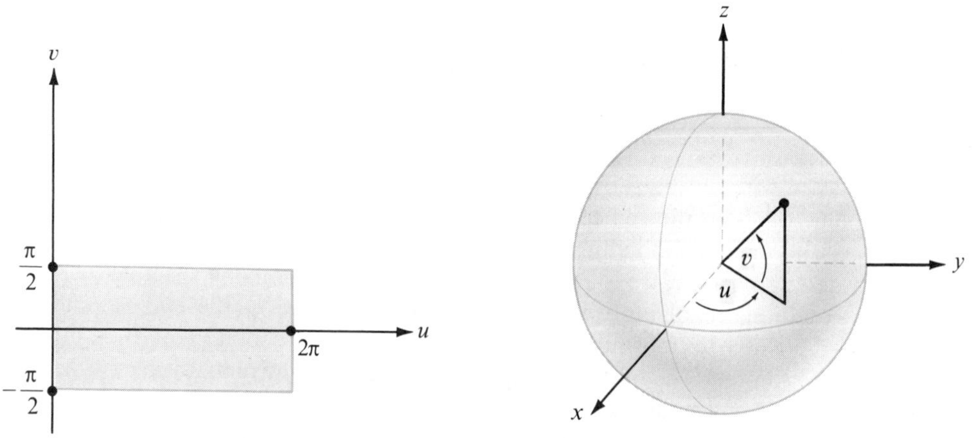

FIGURE 12.2 Parametric representation of a sphere.

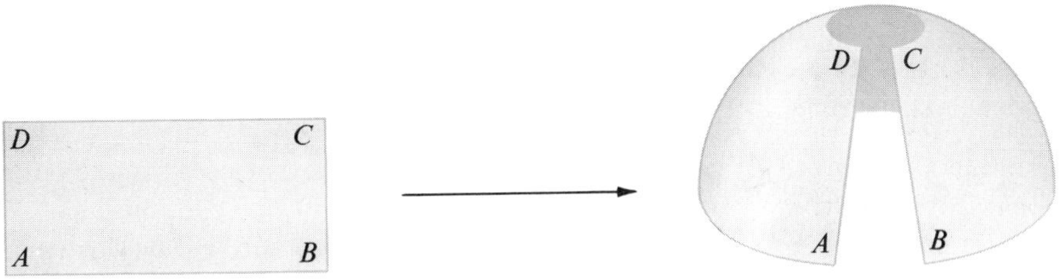

FIGURE 12.3 Deformation of a rectangle into a hemisphere.

serve as parametric equations for a sphere of radius a and center at the origin. If we square and add the three equations in (12.3) we find $x^2 + y^2 + z^2 = a^2$, and we see that every point (x, y, z) satisfying (12.3) lies on the sphere. The parameters u and v in this example may be interpreted geometrically as the angles shown in Figure 12.2. If we let the point (u, v) vary over the rectangle $T = [0, 2\pi] \times [-\frac{1}{2}\pi, \frac{1}{2}\pi]$, the points determined by (12.3) trace out the whole sphere. The upper hemisphere is the image of the rectangle $[0, 2\pi] \times [0, \frac{1}{2}\pi]$ and the lower hemisphere is the image of $[0, 2\pi] \times [-\frac{1}{2}\pi, 0]$. Figure 12.3 gives a concrete idea of how the rectangle $[0, 2\pi] \times [0, \frac{1}{2}\pi]$ is mapped onto the upper hemisphere. Imagine that the rectangle is made of a flexible plastic material capable of being stretched or shrunk. Figure 12.3 shows the rectangle being deformed into a hemisphere. The base AB eventually becomes the equator, the opposite edges AD and BC are brought into coincidence, and the upper edge DC shrinks to a point (the North Pole).

EXAMPLE 2. *A parametric representation of a cone.* The vector equation

$$\boldsymbol{r}(u, v) = v \sin \alpha \cos u\, \boldsymbol{i} + v \sin \alpha \sin u\, \boldsymbol{j} + v \cos \alpha\, \boldsymbol{k}$$

represents the right circular cone shown in Figure 12.4, where α denotes half the vertex angle. Again, the parameters u and v may be given geometric interpretations; v is the distance from the vertex to the point (x, y, z) on the cone, and u is the polar-coordinate angle. When (u, v) is allowed to vary over a rectangle of the form $[0, 2\pi] \times [0, h]$, the corresponding points (x, y, z) trace out a cone of altitude $h \cos \alpha$. A plastic rectangle

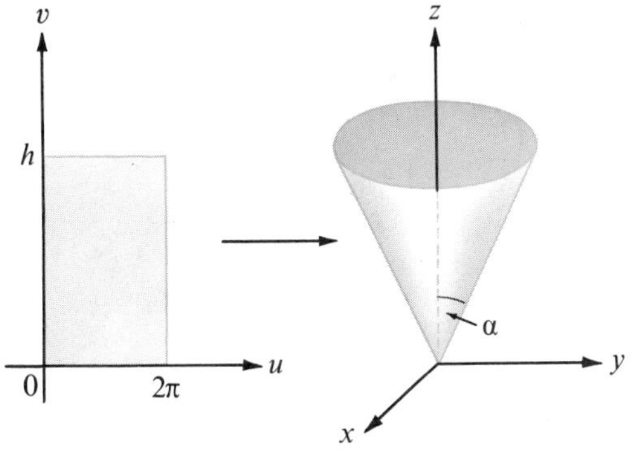

FIGURE 12.4 Parametric representation of a cone.

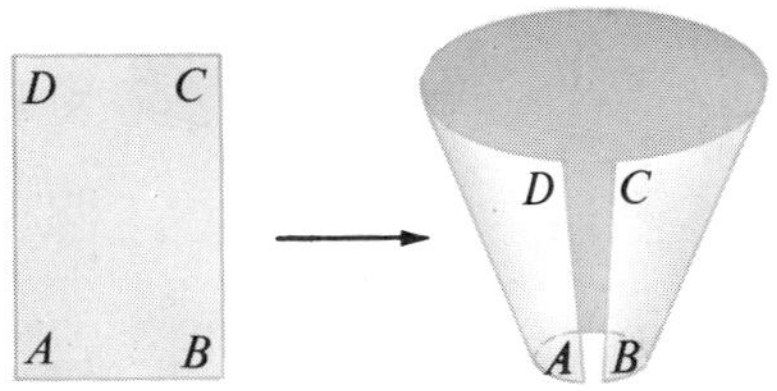

FIGURE 12.5 Deformation of a rectangle into a cone.

may be physically deformed into the cone by bringing the edges AD and BC into coincidence, as suggested by Figure 12.5, and letting the edge AB shrink to a point (the vertex of the cone). The surface in Figure 12.5 shows an intermediate stage of the deformation.

In the general study of surfaces, the functions X, Y, and Z that occur in the parametric equations (12.1) or in the vector equation (12.2) are assumed to be continuous on T. The image of T under the mapping $\boldsymbol{r}$ is called a *parametric surface* and will be denoted by the symbol $\boldsymbol{r}(T)$. In many of the examples we shall discuss, T will be a rectangle, a circular disk, or some other simply connected set bounded by a simple closed curve. If the function $\boldsymbol{r}$ is one-to-one on T, the image $\boldsymbol{r}(T)$ will be called a *simple parametric surface.* In such a case, distinct points of T map onto distinct points of the surface. In particular, every simple closed curve in T maps onto a simple closed curve lying on the surface.

A parametric surface $\boldsymbol{r}(T)$ may degenerate to a point or to a curve. For example, if all three functions X, Y, and Z are constant, the image $\boldsymbol{r}(T)$ is a single point. If X, Y, and Z are independent of v, the image $\boldsymbol{r}(T)$ is a curve. Another example of a degenerate surface occurs when $X(u, v) = u + v$, $Y(u, v) = (u + v)^2$, and $Z(u, v) = (u + v)^3$, where $T = [0, 1] \times [0, 1]$. If we write $t = u + v$ we see that the surface degenerates to the space curve having parametric equations $x = t$, $y = t^2$, and $z = t^3$, where $0 \le t \le 2$. Such degeneracies can be avoided by placing further restrictions on the mapping function $\boldsymbol{r}$, as described in the next section.

12.2 The fundamental vector product

Consider a surface described by the vector equation

$$\boldsymbol{r}(u, v) = X(u, v)\boldsymbol{i} + Y(u, v)\boldsymbol{j} + Z(u, v)\boldsymbol{k}, \qquad \text{where} \quad (u, v) \in T.$$

If X, Y, and Z are differentiable on T we consider the two vectors

$$\frac{\partial \boldsymbol{r}}{\partial u} = \frac{\partial X}{\partial u}\boldsymbol{i} + \frac{\partial Y}{\partial u}\boldsymbol{j} + \frac{\partial Z}{\partial u}\boldsymbol{k}$$

and

$$\frac{\partial \boldsymbol{r}}{\partial v} = \frac{\partial X}{\partial v}\boldsymbol{i} + \frac{\partial Y}{\partial v}\boldsymbol{j} + \frac{\partial Z}{\partial v}\boldsymbol{k}.$$

The cross product of these two vectors $\partial \boldsymbol{r}/\partial u \times \partial \boldsymbol{r}/\partial v$ will be referred to as the *fundamental vector product* of the representation $\boldsymbol{r}$. Its components can be expressed as Jacobian determinants. In fact, we have

$$(12.4) \quad \frac{\partial \boldsymbol{r}}{\partial u} \times \frac{\partial \boldsymbol{r}}{\partial v} = \begin{vmatrix} \boldsymbol{i} & \boldsymbol{j} & \boldsymbol{k} \\ \frac{\partial X}{\partial u} & \frac{\partial Y}{\partial u} & \frac{\partial Z}{\partial u} \\ \frac{\partial X}{\partial v} & \frac{\partial Y}{\partial v} & \frac{\partial Z}{\partial v} \end{vmatrix} = \begin{vmatrix} \frac{\partial Y}{\partial u} & \frac{\partial Z}{\partial u} \\ \frac{\partial Y}{\partial v} & \frac{\partial Z}{\partial v} \end{vmatrix} \boldsymbol{i} + \begin{vmatrix} \frac{\partial Z}{\partial u} & \frac{\partial X}{\partial u} \\ \frac{\partial Z}{\partial v} & \frac{\partial X}{\partial v} \end{vmatrix} \boldsymbol{j} + \begin{vmatrix} \frac{\partial X}{\partial u} & \frac{\partial Y}{\partial u} \\ \frac{\partial X}{\partial v} & \frac{\partial Y}{\partial v} \end{vmatrix} \boldsymbol{k}$$

$$= \frac{\partial(Y, Z)}{\partial(u, v)}\boldsymbol{i} + \frac{\partial(Z, X)}{\partial(u, v)}\boldsymbol{j} + \frac{\partial(X, Y)}{\partial(u, v)}\boldsymbol{k}.$$

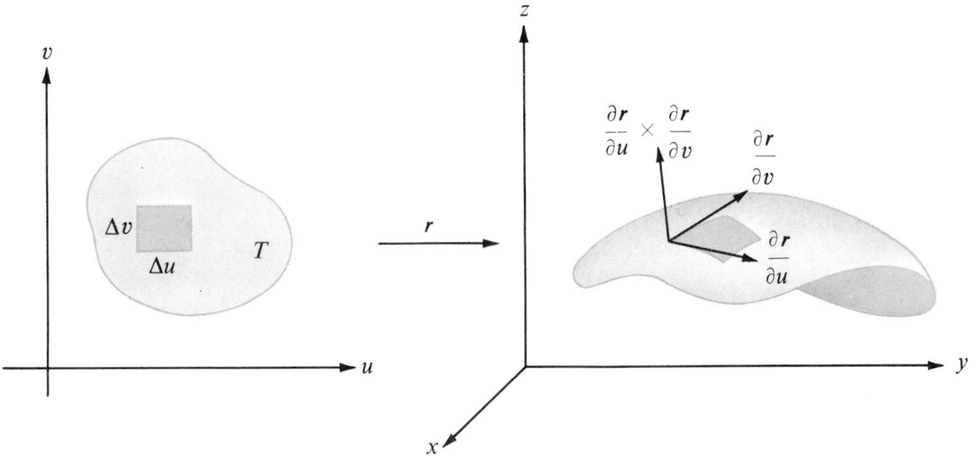

FIGURE 12.6 Geometric interpretation of the vectors $\frac{\partial \boldsymbol{r}}{\partial u}$, $\frac{\partial \boldsymbol{r}}{\partial v}$, and $\frac{\partial \boldsymbol{r}}{\partial u} \times \frac{\partial \boldsymbol{r}}{\partial v}$.

If (u, v) is a point in T at which $\partial \boldsymbol{r}/\partial u$ and $\partial \boldsymbol{r}/\partial v$ are continuous and the fundamental vector product is nonzero, then the image point $\boldsymbol{r}(u, v)$ is called a *regular point* of $\boldsymbol{r}$. Points at which $\partial \boldsymbol{r}/\partial u$ or $\partial \boldsymbol{r}/\partial v$ fails to be continuous or $\partial \boldsymbol{r}/\partial u \times \partial \boldsymbol{r}/\partial v = \boldsymbol{O}$ are called *singular points* of $\boldsymbol{r}$. A surface $\boldsymbol{r}(T)$ is called *smooth* if all its points are regular points. Every surface has more than one parametric representation. Some of the examples discussed below show that a point of a surface may be a regular point for one representation but a singular point for some other representation. The geometric significance of regular and singular points can be explained as follows:

Consider a horizontal line segment in T. Its image under $\boldsymbol{r}$ is a curve (called a u-curve) lying on the surface $\boldsymbol{r}(T)$. For fixed v, think of the parameter u as representing time. The vector $\partial \boldsymbol{r}/\partial u$ is the velocity vector of this curve. When u changes by an amount Δu, a point originally at $\boldsymbol{r}(u, v)$ moves along a u-curve a distance approximately equal to $\|\partial \boldsymbol{r}/\partial u\| \, \Delta u$ since $\|\partial \boldsymbol{r}/\partial u\|$ represents the speed along the u-curve. Similarly, for fixed u a point of a v-curve moves in time Δv a distance nearly equal to $\|\partial \boldsymbol{r}/\partial v\| \, \Delta v$. A rectangle in T having area $\Delta u \, \Delta v$ is traced onto a portion of $\boldsymbol{r}(T)$ which we shall approximate by the parallelogram determined by the vectors $(\partial \boldsymbol{r}/\partial u) \, \Delta u$ and $(\partial \boldsymbol{r}/\partial v) \, \Delta v$. (See Figure 12.6.) The area of the parallelogram spanned by $(\partial \boldsymbol{r}/\partial u) \, \Delta u$ and $(\partial \boldsymbol{r}/\partial v) \, \Delta v$ is the magnitude of their cross product,

$$\left\| \frac{\partial \boldsymbol{r}}{\partial u} \Delta u \times \frac{\partial \boldsymbol{r}}{\partial v} \Delta v \right\| = \left\| \frac{\partial \boldsymbol{r}}{\partial u} \times \frac{\partial \boldsymbol{r}}{\partial v} \right\| \Delta u \, \Delta v \, .$$

Therefore the length of the fundamental vector product may be thought of as a local magnification factor for areas. At the points at which this vector product is zero the parallelogram collapses to a curve or point, and degeneracies occur. At each regular point the vectors $\partial \boldsymbol{r}/\partial u$ and $\partial \boldsymbol{r}/\partial v$ determine a plane having the vector $\partial \boldsymbol{r}/\partial u \times \partial \boldsymbol{r}/\partial v$ as a normal. In the next section we shall prove that $\partial \boldsymbol{r}/\partial u \times \partial \boldsymbol{r}/\partial v$ is normal to every smooth curve on the surface; for this reason the plane determined by $\partial \boldsymbol{r}/\partial u$ and $\partial \boldsymbol{r}/\partial v$ is called the *tangent plane* of the surface. Continuity of $\partial \boldsymbol{r}/\partial u$ and $\partial \boldsymbol{r}/\partial v$ implies continuity of $\partial \boldsymbol{r}/\partial u \times \partial \boldsymbol{r}/\partial v$; this, in turn, means that the tangent plane varies continuously on a smooth surface.

Thus we see that continuity of $\partial \boldsymbol{r}/\partial u$ and $\partial \boldsymbol{r}/\partial v$ prevents the occurrence of sharp edges or corners on the surface; the nonvanishing of $\partial \boldsymbol{r}/\partial u \times \partial \boldsymbol{r}/\partial v$ prevents degeneracies.

EXAMPLE 1. *Surfaces with an explicit representation*, $z = f(x, y)$. For a surface with an explicit representation of the form $z = f(x, y)$, we can use x and y as the parameters, which gives us the vector equation

$$\boldsymbol{r}(x, y) = x\boldsymbol{i} + y\boldsymbol{j} + f(x, y)\boldsymbol{k}.$$

This representation always gives a simple parametric surface. The region T is called the projection of the surface on the xy-plane. (An example is shown in Figure 12.7, p. 425.) To compute the fundamental vector product we note that

$$\frac{\partial \boldsymbol{r}}{\partial x} = \boldsymbol{i} + \frac{\partial f}{\partial x}\boldsymbol{k} \quad \text{and} \quad \frac{\partial \boldsymbol{r}}{\partial y} = \boldsymbol{j} + \frac{\partial f}{\partial y}\boldsymbol{k},$$

if f is differentiable. This gives us

$$\text{(12.5)} \qquad \frac{\partial \boldsymbol{r}}{\partial x} \times \frac{\partial \boldsymbol{r}}{\partial y} = \begin{vmatrix} \boldsymbol{i} & \boldsymbol{j} & \boldsymbol{k} \\ 1 & 0 & \dfrac{\partial f}{\partial x} \\ 0 & 1 & \dfrac{\partial f}{\partial y} \end{vmatrix} = -\frac{\partial f}{\partial x}\boldsymbol{i} - \frac{\partial f}{\partial y}\boldsymbol{j} + \boldsymbol{k}.$$

Since the z-component of $\partial \boldsymbol{r}/\partial x \times \partial \boldsymbol{r}/\partial y$ is 1, the fundamental vector product is never zero. Therefore the only singular points that can occur for this representation are points at which at least one of the partial derivatives $\partial f/\partial x$ or $\partial f/\partial y$ fails to be continuous.

A specific case is the equation $z = \sqrt{1 - x^2 - y^2}$, which represents a hemisphere of radius 1 and center at the origin, if $x^2 + y^2 \le 1$. The vector equation

$$\boldsymbol{r}(x, y) = x\boldsymbol{i} + y\boldsymbol{j} + \sqrt{1 - x^2 - y^2}\,\boldsymbol{k}$$

maps the unit disk $T = \{(x, y) \mid x^2 + y^2 \le 1\}$ onto the hemisphere in a one-to-one fashion. The partial derivatives $\partial \boldsymbol{r}/\partial x$ and $\partial \boldsymbol{r}/\partial y$ exist and are continuous everywhere in the interior of this disk, but they do not exist on the boundary. Therefore every point on the equator is a singular point of this representation.

EXAMPLE 2. We consider the same hemisphere as in Example 1, but this time as the image of the rectangle $T = [0, 2\pi] \times [0, \frac{1}{2}\pi]$ under the mapping

$$\boldsymbol{r}(u, v) = a \cos u \cos v\, \boldsymbol{i} + a \sin u \cos v\, \boldsymbol{j} + a \sin v\, \boldsymbol{k}.$$

The vectors $\partial \boldsymbol{r}/\partial u$ and $\partial \boldsymbol{r}/\partial v$ are given by the formulas

$$\frac{\partial \boldsymbol{r}}{\partial u} = -a \sin u \cos v\, \boldsymbol{i} + a \cos u \cos v\, \boldsymbol{j},$$

$$\frac{\partial \boldsymbol{r}}{\partial v} = -a \cos u \sin v\, \boldsymbol{i} - a \sin u \sin v\, \boldsymbol{j} + a \cos v\, \boldsymbol{k}.$$

An easy calculation shows that their cross product is equal to

$$\frac{\partial \boldsymbol{r}}{\partial u} \times \frac{\partial \boldsymbol{r}}{\partial v} = a \cos v\, \boldsymbol{r}(u, v).$$

The image of T is not a simple parametric surface because this mapping is not one-to-one on T. In fact, every point on the line segment $v = \frac{1}{2}\pi$, $0 \le u \le 2\pi$, is mapped onto the point $(0, 0, a)$ (the North Pole). Also, because of the periodicity of the sine and cosine, $\boldsymbol{r}$ takes the same values at the points $(0, v)$ and $(2\pi, v)$, so the right and left edges of T are mapped onto the same curve, a circular arc joining the North Pole to the point $(a, 0, 0)$ on the equator. (See Figure 12.3.) The vectors $\partial \boldsymbol{r}/\partial u$ and $\partial \boldsymbol{r}/\partial v$ are continuous everywhere in T. Since $\|\partial \boldsymbol{r}/\partial u \times \partial \boldsymbol{r}/\partial v\| = a^2 \cos v$, the only singular points of this representation occur when $\cos v = 0$. The North Pole is the only such point.

12.3 The fundamental vector product as a normal to the surface

Consider a smooth parametric surface $\boldsymbol{r}(T)$, and let C^* be a smooth curve in T. Then the image $C = \boldsymbol{r}(C^*)$ is a smooth curve lying on the surface. We shall prove that at each point of C the vector $\partial \boldsymbol{r}/\partial u \times \partial \boldsymbol{r}/\partial v$ is normal to C, as illustrated in Figure 12.6.

Suppose that C^* is described by a function $\boldsymbol{\alpha}$ defined on an interval $[a, b]$, say

$$\boldsymbol{\alpha}(t) = U(t)\boldsymbol{i} + V(t)\boldsymbol{j}.$$

Then the image curve C is described by the composite function

$$\boldsymbol{\rho}(t) = \boldsymbol{r}[\boldsymbol{\alpha}(t)] = X[\boldsymbol{\alpha}(t)]\boldsymbol{i} + Y[\boldsymbol{\alpha}(t)]\boldsymbol{j} + Z[\boldsymbol{\alpha}(t)]\boldsymbol{k}.$$

We wish to prove that the derivative $\boldsymbol{\rho}'(t)$ is perpendicular to the vector $\partial \boldsymbol{r}/\partial u \times \partial \boldsymbol{r}/\partial v$ when the partial derivatives $\partial \boldsymbol{r}/\partial u$ and $\partial \boldsymbol{r}/\partial v$ are evaluated at $(U(t), V(t))$. To compute $\boldsymbol{\rho}'(t)$ we differentiate each component of $\boldsymbol{\rho}(t)$ by the chain rule (Theorem 8.8) to obtain

$$\text{(12.6)} \qquad \boldsymbol{\rho}'(t) = \nabla X \cdot \boldsymbol{\alpha}'(t)\boldsymbol{i} + \nabla Y \cdot \boldsymbol{\alpha}'(t)\boldsymbol{j} + \nabla Z \cdot \boldsymbol{\alpha}'(t)\boldsymbol{k},$$

where the gradient vectors ∇X, ∇Y, and ∇Z are evaluated at $(U(t), V(t))$. Equation (12.6) can be rewritten as

$$\boldsymbol{\rho}'(t) = \frac{\partial \boldsymbol{r}}{\partial u} U'(t) + \frac{\partial \boldsymbol{r}}{\partial v} V'(t),$$

where the derivatives $\partial \boldsymbol{r}/\partial u$ and $\partial \boldsymbol{r}/\partial v$ are evaluated at $(U(t), V(t))$. Since $\partial \boldsymbol{r}/\partial u$ and $\partial \boldsymbol{r}/\partial v$ are each perpendicular to the cross product $\partial \boldsymbol{r}/\partial u \times \partial \boldsymbol{r}/\partial v$, the same is true of $\boldsymbol{\rho}'(t)$. This proves that $\partial \boldsymbol{r}/\partial u \times \partial \boldsymbol{r}/\partial v$ is normal to C, as asserted. For this reason, the vector product $\partial \boldsymbol{r}/\partial u \times \partial \boldsymbol{r}/\partial v$ is said to be *normal* to the surface $\boldsymbol{r}(T)$. At each regular point P of $\boldsymbol{r}(T)$ the vector $\partial \boldsymbol{r}/\partial u \times \partial \boldsymbol{r}/\partial v$ is nonzero; the plane through P having this vector as a normal is called the *tangent plane* to the surface at P.

12.4 Exercises

In Exercises 1 through 6, eliminate the parameters u and v to obtain a Cartesian equation, thus showing that the given vector equation represents a portion of the surface named. Also, compute the fundamental vector product $\partial \boldsymbol{r}/\partial u \times \partial \boldsymbol{r}/\partial v$ in terms of u and v.

1. *Plane:*
$\boldsymbol{r}(u, v) = (x_0 + a_1 u + b_1 v)\boldsymbol{i} + (y_0 + a_2 u + b_2 v)\boldsymbol{j} + (z_0 + a_3 u + b_3 v)\boldsymbol{k}$.
2. *Elliptic paraboloid:*
$\boldsymbol{r}(u, v) = au \cos v\, \boldsymbol{i} + bu \sin v\, \boldsymbol{j} + u^2 \boldsymbol{k}$.
3. *Ellipsoid:*
$\boldsymbol{r}(u, v) = a \sin u \cos v\, \boldsymbol{i} + b \sin u \sin v\, \boldsymbol{j} + c \cos u\, \boldsymbol{k}$.
4. *Surface of revolution:*
$\boldsymbol{r}(u, v) = u \cos v\, \boldsymbol{i} + u \sin v\, \boldsymbol{j} + f(u)\boldsymbol{k}$.
5. *Cylinder:*
$\boldsymbol{r}(u, v) = u\boldsymbol{i} + a \sin v\, \boldsymbol{j} + a \cos v\, \boldsymbol{k}$.
6. *Torus:*
$\boldsymbol{r}(u, v) = (a + b \cos u) \sin v\, \boldsymbol{i} + (a + b \cos u) \cos v\, \boldsymbol{j} + b \sin u\, \boldsymbol{k}$, where $0 < b < a$. What are the geometric meanings of a and b?

In Exercises 7 through 10 compute the magnitude of the vector product $\partial \boldsymbol{r}/\partial u \times \partial \boldsymbol{r}/\partial v$.

7. $\boldsymbol{r}(u, v) = a \sin u \cosh v\, \boldsymbol{i} + b \cos u \cosh v\, \boldsymbol{j} + c \sinh v\, \boldsymbol{k}$.
8. $\boldsymbol{r}(u, v) = (u + v)\boldsymbol{i} + (u - v)\boldsymbol{j} + 4v^2\boldsymbol{k}$.
9. $\boldsymbol{r}(u, v) = (u + v)\boldsymbol{i} + (u^2 + v^2)\boldsymbol{j} + (u^3 + v^3)\boldsymbol{k}$.
10. $\boldsymbol{r}(u, v) = u \cos v\, \boldsymbol{i} + u \sin v\, \boldsymbol{j} + \frac{1}{2}u^2 \sin 2v\, \boldsymbol{k}$.

12.5 Area of a parametric surface

Let $S = \boldsymbol{r}(T)$ be a parametric surface described by a vector-valued function $\boldsymbol{r}$ defined on a region T in the uv-plane. In Section 12.2 we found that the length of the fundamental vector product $\partial \boldsymbol{r}/\partial u \times \partial \boldsymbol{r}/\partial v$ could be interpreted as a local magnification factor for areas. (See Figure 12.6.) A rectangle in T of area $\Delta u\, \Delta v$ is mapped by $\boldsymbol{r}$ onto a curvilinear parallelogram on S with area nearly equal to

$$\left\| \frac{\partial \boldsymbol{r}}{\partial u} \times \frac{\partial \boldsymbol{r}}{\partial v} \right\| \Delta u\, \Delta v .$$

This observation suggests the following definition.

DEFINITION OF AREA OF A PARAMETRIC SURFACE. *The area of S, denoted by $a(S)$, is defined by the double integral*

$$a(S) = \iint_T \left\| \frac{\partial \boldsymbol{r}}{\partial u} \times \frac{\partial \boldsymbol{r}}{\partial v} \right\| du\, dv . \tag{12.7}$$

In other words, to determine the area of S we first compute the fundamental vector product $\partial \boldsymbol{r}/\partial u \times \partial \boldsymbol{r}/\partial v$ and then integrate its length over the region T. When $\partial \boldsymbol{r}/\partial u \times \partial \boldsymbol{r}/\partial v$ is expressed in terms of its components, by means of Equation (12.4), we have

$$a(S) = \iint_T \sqrt{\left(\frac{\partial(Y, Z)}{\partial(u, v)}\right)^2 + \left(\frac{\partial(Z, X)}{\partial(u, v)}\right)^2 + \left(\frac{\partial(X, Y)}{\partial(u, v)}\right)^2}\, du\, dv. \tag{12.8}$$

Written in this form, the integral for surface area resembles the integral for computing the arc length of a curve.†

If S is given explicitly by an equation of the form $z = f(x, y)$ we may use x and y as the parameters. The fundamental vector product is given by Equation (12.5), so we have

$$\left\| \frac{\partial \boldsymbol{r}}{\partial x} \times \frac{\partial \boldsymbol{r}}{\partial y} \right\| = \left\| -\frac{\partial f}{\partial x}\boldsymbol{i} - \frac{\partial f}{\partial y}\boldsymbol{j} + \boldsymbol{k} \right\| = \sqrt{1 + \left(\frac{\partial f}{\partial x}\right)^2 + \left(\frac{\partial f}{\partial y}\right)^2}.$$

In this case the integral for surface area becomes

$$a(S) = \iint_T \sqrt{1 + \left(\frac{\partial f}{\partial x}\right)^2 + \left(\frac{\partial f}{\partial y}\right)^2}\, dx\, dy, \tag{12.9}$$

where the region T is now the projection of S on the xy-plane, as illustrated in Figure 12.7.

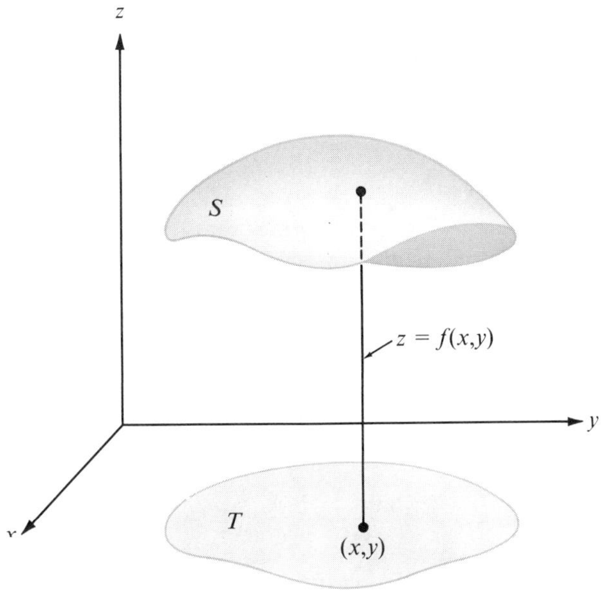

FIGURE 12.7 A surface S with an explicit representation, $z = f(x, y)$. The region T is the projection of S on the xy-plane.

† Since the integral in (12.7) involves $\boldsymbol{r}$, the area of a surface will depend on the function used to describe the surface. When we discuss surface integrals we shall prove (in Section 12.8) that under certain general conditions the area is independent of the parametric representation. The result is analogous to Theorem 10.1, in which we discussed the invariance of line integrals under a change of parameter.

When S lies in a plane parallel to the xy-plane, the function f is constant, so $\partial f/\partial x = \partial f/\partial y = 0$, and Equation (12.9) becomes

$$a(S) = \iint_T dx\,dy\,.$$

This agrees with the usual formula for areas of plane regions.

Equation (12.9) can be written in another form that gives further insight into its geometric significance. At each point of S, let γ denote the angle between the normal vector $\boldsymbol{N} = \partial \boldsymbol{r}/\partial x \times \partial \boldsymbol{r}/\partial y$ and the unit coordinate vector $\boldsymbol{k}$. (See Figure 12.8.) Since the z-component of $\boldsymbol{N}$ is 1, we have

$$\cos \gamma = \frac{\boldsymbol{N} \cdot \boldsymbol{k}}{\|\boldsymbol{N}\|\ \|\boldsymbol{k}\|} = \frac{1}{\|\boldsymbol{N}\|} = \frac{1}{\left\| \dfrac{\partial \boldsymbol{r}}{\partial x} \times \dfrac{\partial \boldsymbol{r}}{\partial y} \right\|}\,,$$

and hence $\|\partial \boldsymbol{r}/\partial x \times \partial \boldsymbol{r}/\partial y\| = 1/\cos \gamma$. Therefore Equation (12.9) becomes

(12.10) $$a(S) = \iint_T \frac{1}{\cos \gamma}\, dx\,dy\,.$$

Suppose now that S lies in a plane not perpendicular to the xy-plane. Then γ is constant and Equation (12.10) states that the area of S = (area of T)/$\cos \gamma$, or that

(12.11) $$a(T) = a(S) \cos \gamma\,.$$

Equation (12.11) is sometimes referred to as the *area cosine principle.* It tells us that if a region S in one plane is projected onto a region T in another plane, making an angle γ with the first plane, the area of T is $\cos \gamma$ times that of S. This formula is obviously true

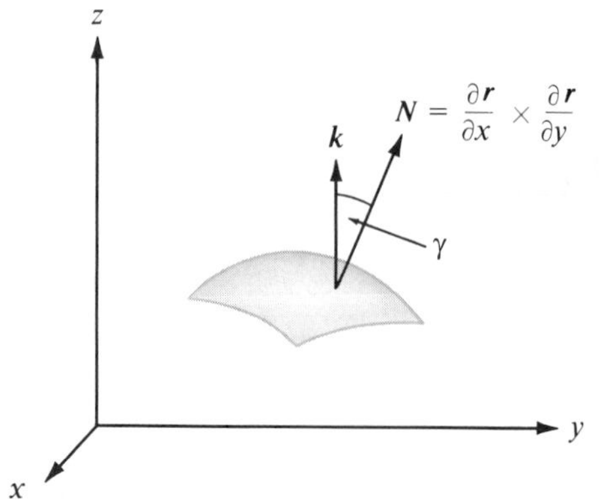

FIGURE 12.8 The length of $\dfrac{\partial \boldsymbol{r}}{\partial x} \times \dfrac{\partial \boldsymbol{r}}{\partial y}$ is $1/\cos \gamma$.

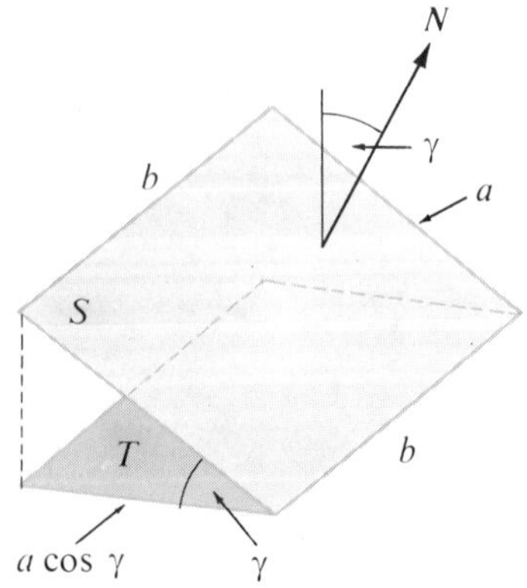

FIGURE 12.9 The area cosine principle for a rectangle.

when S is the rectangle shown in Figure 12.9, because distances in one direction are shortened by the factor $\cos \gamma$ while those in a perpendicular direction are unaltered by projection. Equation (12.11) extends this property to any plane region S having an area.

Suppose now that S is given by an implicit representation $F(x, y, z) = 0$. If S can be projected in a one-to-one fashion on the xy-plane, the equation $F(x, y, z) = 0$ defines z as a function of x and y, say $z = f(x, y)$, and the partial derivatives $\partial f/\partial x$ and $\partial f/\partial y$ are related to those of F by the equations

$$\frac{\partial f}{\partial x} = -\frac{\partial F/\partial x}{\partial F/\partial z} \quad \text{and} \quad \frac{\partial f}{\partial y} = -\frac{\partial F/\partial y}{\partial F/\partial z}$$

for those points at which $\partial F/\partial z \neq 0$. Substituting these quotients in (12.9), we find

$$a(S) = \iint_T \frac{\sqrt{(\partial F/\partial x)^2 + (\partial F/\partial y)^2 + (\partial F/\partial z)^2}}{|\partial F/\partial z|}\, dx\, dy. \tag{12.12}$$

EXAMPLE 1. *Area of a hemisphere.* Consider a hemisphere S of radius a and center at the origin. We have at our disposal the implicit representation $x^2 + y^2 + z^2 = a^2$, $z \geq 0$; the explicit representation $z = \sqrt{a^2 - x^2 - y^2}$; and the parametric representation

$$\boldsymbol{r}(u, v) = a \cos u \cos v\, \boldsymbol{i} + a \sin u \cos v\, \boldsymbol{j} + a \sin v\, \boldsymbol{k}. \tag{12.13}$$

To compute the area of S from the implicit representation we refer to Equation (12.12) with

$$F(x, y, z) = x^2 + y^2 + z^2 - a^2.$$

The partial derivatives of F are $\partial F/\partial x = 2x$, $\partial F/\partial y = 2y$, $\partial F/\partial z = 2z$. The hemisphere S projects in a one-to-one fashion onto the circular disk $D = \{(x, y) \mid x^2 + y^2 \leq a^2\}$ in the xy-plane. We cannot apply Equation (12.12) directly because the partial derivative $\partial F/\partial z$ is zero on the boundary of D. However, the derivative $\partial F/\partial z$ is nonzero everywhere in the interior of D, so we can consider the smaller concentric disk $D(R)$ of radius R, where $R < a$. If $S(R)$ denotes the corresponding portion of the upper hemisphere, Equation (12.12) is now applicable and we find

$$\text{area of } S(R) = \iint_{D(R)} \frac{\sqrt{(2x)^2 + (2y)^2 + (2z)^2}}{|2z|}\, dx\, dy$$

$$= \iint_{D(R)} \frac{a}{z}\, dx\, dy = a \iint_{D(R)} \frac{1}{\sqrt{a^2 - x^2 - y^2}}\, dx\, dy.$$

The last integral can be easily evaluated by the use of polar coordinates, giving us

$$\text{area of } S(R) = a \int_0^{2\pi} \left[\int_0^R \frac{1}{\sqrt{a^2 - r^2}}\, r\, dr \right] d\theta = 2\pi a(a - \sqrt{a^2 - R^2}).$$

When $R \to a$ this approaches the limit $2\pi a^2$.

We can avoid the limiting process in the foregoing calculation by using the parametric representation in (12.13). The calculations of Example 2 in Section 12.2 show that

$$\left\| \frac{\partial \boldsymbol{r}}{\partial u} \times \frac{\partial \boldsymbol{r}}{\partial v} \right\| = \|a \cos v\, \boldsymbol{r}(u, v)\| = a^2 |\cos v| .$$

Therefore we can apply Equation (12.7), taking for the region T the rectangle $[0, 2\pi] \times [0, \frac{1}{2}\pi]$. We find

$$a(S) = a^2 \iint_T |\cos v|\, du\, dv = a^2 \int_0^{2\pi} \left[\int_0^{\pi/2} \cos v\, dv \right] du = 2\pi a^2 .$$

EXAMPLE 2. *Another theorem of Pappus.* One of the theorems of Pappus states that a surface of revolution, obtained by rotating a plane curve of length L about an axis in the plane of the curve, has area $2\pi Lh$, where h is the distance from the centroid of the curve to the axis of rotation. We shall use Equation (12.7) to prove this theorem.

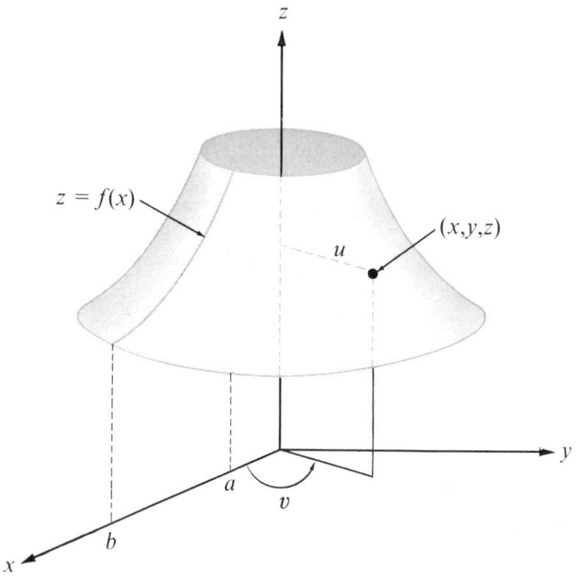

FIGURE 12.10 Area of a surface of revolution determined by Pappus' theorem.

Suppose a curve C, initially in the xz-plane, is rotated about the z-axis. Let its equation in the xz-plane be $z = f(x)$, where $a \le x \le b$, $a \ge 0$. The surface of revolution S so generated can be described by the vector equation

$$\boldsymbol{r}(u, v) = u \cos v\, \boldsymbol{i} + u \sin v\, \boldsymbol{j} + f(u)\boldsymbol{k},$$

where $(u, v) \in [a, b] \times [0, 2\pi]$. The parameters u and v can be interpreted as the radius and angle of polar coordinates, as illustrated in Figure 12.10. If $a \le u \le b$, all points (x, y, z) at a given distance u from the z-axis have the same z-coordinate, $f(u)$, so they all lie on the surface. The fundamental vector product of this representation is

$$\frac{\partial \boldsymbol{r}}{\partial u} \times \frac{\partial \boldsymbol{r}}{\partial v} = \begin{vmatrix} \boldsymbol{i} & \boldsymbol{j} & \boldsymbol{k} \\ \cos v & \sin v & f'(u) \\ -u \sin v & u \cos v & 0 \end{vmatrix} = -uf'(u) \cos v\, \boldsymbol{i} - uf'(u) \sin v\, \boldsymbol{j} + u\boldsymbol{k},$$

and hence

$$\left\| \frac{\partial \boldsymbol{r}}{\partial u} \times \frac{\partial \boldsymbol{r}}{\partial v} \right\| = u\sqrt{1 + [f'(u)]^2}\,.$$

Therefore Equation (12.7) becomes

$$a(S) = \int_0^{2\pi} \left[\int_a^b u\sqrt{1 + [f'(u)]^2}\, du \right] dv = 2\pi \int_a^b u\sqrt{1 + [f'(u)]^2}\, du\,.$$

The last integral can be expressed as $\int_C x\, ds$, a line integral with respect to arc length along the curve C. As such, it is equal to $\bar{x}L$, where $\bar{x}$ is the x-coordinate of the centroid of C and L is the length of C. (See Section 10.8.) Therefore the area of S is $2\pi L\bar{x}$. This proves the theorem of Pappus.

12.6 Exercises

1. Let S be a parallelogram not parallel to any of the coordinate planes. Let S_1, S_2, and S_3 denote the areas of the projections of S on the three coordinate planes. Show that the area of S is $\sqrt{S_1^2 + S_2^2 + S_3^2}$.
2. Compute the area of the region cut from the plane $x + y + z = a$ by the cylinder $x^2 + y^2 = a^2$.
3. Compute the surface area of that portion of the sphere $x^2 + y^2 + z^2 = a^2$ lying within the cylinder $x^2 + y^2 = ay$, where $a > 0$.
4. Compute the area of that portion of the surface $z^2 = 2xy$ which lies above the first quadrant of the xy-plane and is cut off by the planes $x = 2$ and $y = 1$.
5. A parametric surface S is described by the vector equation

$$\boldsymbol{r}(u, v) = u \cos v\, \boldsymbol{i} + u \sin v\, \boldsymbol{j} + u^2\boldsymbol{k},$$

where $0 \le u \le 4$ and $0 \le v \le 2\pi$.
(a) Show that S is a portion of a surface of revolution. Make a sketch and indicate the geometric meanings of the parameters u and v on the surface.
(b) Compute the fundamental vector product $\partial \boldsymbol{r}/\partial u \times \partial \boldsymbol{r}/\partial v$ in terms of u and v.
(c) The area of S is $\pi(65\sqrt{65} - 1)/n$, where n is an integer. Compute the value of n.
6. Compute the area of that portion of the conical surface $x^2 + y^2 = z^2$ which lies above the xy-plane and is cut off by the sphere $x^2 + y^2 + z^2 = 2ax$.
7. Compute the area of that portion of the conical surface $x^2 + y^2 = z^2$ which lies between the two planes $z = 0$ and $x + 2z = 3$.
8. Compute the area of that portion of the paraboloid $x^2 + z^2 = 2ay$ which is cut off by the plane $y = a$.
9. Compute the area of the torus described by the vector equation

$$\boldsymbol{r}(u, v) = (a + b \cos u) \sin v\, \boldsymbol{i} + (a + b \cos u) \cos v\, \boldsymbol{j} + b \sin u\, \boldsymbol{k},$$

where $0 < b < a$ and $0 \le u \le 2\pi$, $0 \le v \le 2\pi$. [*Hint:* Use the theorem of Pappus.]

10. A sphere is inscribed in a right circular cylinder. The sphere is sliced by two parallel planes perpendicular to the axis of the cylinder. Show that the portions of the sphere and cylinder lying between these planes have equal surface areas.
11. Let T be the unit disk in the uv-plane, $T = \{(u, v) \mid u^2 + v^2 \le 1\}$, and let

$$\boldsymbol{r}(u, v) = \frac{2u}{u^2 + v^2 + 1}\,\boldsymbol{i} + \frac{2v}{u^2 + v^2 + 1}\,\boldsymbol{j} + \frac{u^2 + v^2 - 1}{u^2 + v^2 + 1}\,\boldsymbol{k}.$$

(a) Determine the image of each of the following sets under $\boldsymbol{r}$: the unit circle $u^2 + v^2 = 1$; the interval $-1 \le u \le 1$; that part of the line $u = v$ lying in T.
(b) The surface $S = \boldsymbol{r}(T)$ is a familiar surface. Name and sketch it.
(c) Determine the image of the uv-plane under $\boldsymbol{r}$. Indicate by a sketch in the xyz-space the geometric meanings of the parameters u and v.

12.7 Surface integrals

Surface integrals are, in many respects, analogous to line integrals; the integration takes place along a surface rather than along a curve. We defined line integrals in terms of a parametric representation for the curve. Similarly, we shall define surface integrals in terms of a parametric representation for the surface. Then we shall prove that under certain general conditions the value of the integral is independent of the representation.

DEFINITION OF A SURFACE INTEGRAL. *Let $S = \boldsymbol{r}(T)$ be a parametric surface described by a differentiable function $\boldsymbol{r}$ defined on a region T in the uv-plane, and let f be a scalar field defined and bounded on S. The surface integral of f over S is denoted by the symbol* $\iint_{\boldsymbol{r}(T)} f\, dS$ *[or by* $\iint_S f(x, y, z)\, dS$*], and is defined by the equation*

$$\iint_{\boldsymbol{r}(T)} f\, dS = \iint_T f[\boldsymbol{r}(u, v)] \left\| \frac{\partial \boldsymbol{r}}{\partial u} \times \frac{\partial \boldsymbol{r}}{\partial v} \right\| du\, dv \tag{12.14}$$

whenever the double integral on the right exists.

The following examples illustrate some applications of surface integrals.

EXAMPLE 1. *Surface area.* When $f = 1$, Equation (12.14) becomes

$$\iint_{\boldsymbol{r}(T)} dS = \iint_T \left\| \frac{\partial \boldsymbol{r}}{\partial u} \times \frac{\partial \boldsymbol{r}}{\partial v} \right\| du\, dv.$$

The double integral on the right is that used earlier in Section 12.5 to define surface area. Thus, the area of S is equal to the surface integral $\iint_{\boldsymbol{r}(T)} dS$. For this reason, the symbol dS is sometimes referred to as an "element of surface area," and the surface integral $\iint_{\boldsymbol{r}(T)} f\, dS$ is said to be an integral of f with respect to the element of surface area, extended over the surface $\boldsymbol{r}(T)$.

EXAMPLE 2. *Center of mass. Moment of inertia.* If the scalar field f is interpreted as the density (mass per unit area) of a thin material in the shape of the surface S, the total mass m of the surface is defined by the equation

$$m = \iint_S f(x, y, z)\, dS .$$

Its center of mass is the point $(\bar{x}, \bar{y}, \bar{z})$ determined by the equations

$$\bar{x}m = \iint_S xf(x, y, z)\, dS , \qquad \bar{y}m = \iint_S yf(x, y, z)\, dS , \qquad \bar{z}m = \iint_S zf(x, y, z)\, dS .$$

The moment of inertia I_L of S about an axis L is defined by the equation

$$I_L = \iint_S \delta^2(x, y, z) f(x, y, z)\, dS ,$$

where $\delta(x, y, z)$ denotes the perpendicular distance from a general point (x, y, z) of S to the line L.

To illustrate, let us determine the center of mass of a uniform hemispherical surface of radius a. We use the parametric representation

$$\boldsymbol{r}(u, v) = a \cos u \cos v\, \boldsymbol{i} + a \sin u \cos v\, \boldsymbol{j} + a \sin v\, \boldsymbol{k} ,$$

where $(u, v) \in [0, 2\pi] \times [0, \frac{1}{2}\pi]$. This particular representation was discussed earlier in Example 2 of Section 12.2, where we found that the magnitude of the fundamental vector product is $a^2 |\cos v|$. In this example the density f is constant, say $f = c$, and the mass m is $2\pi a^2 c$, the area of S times c. Because of symmetry, the coordinates $\bar{x}$ and $\bar{y}$ of the center of mass are 0. The coordinate $\bar{z}$ is given by

$$\begin{aligned} \bar{z}m &= c \iint_S z\, dS = c \iint_T a \sin v \cdot a^2 |\cos v|\, du\, dv \\ &= 2\pi a^3 c \int_0^{\pi/2} \sin v \cos v\, dv = \pi a^3 c = \frac{a}{2} m , \end{aligned}$$

so $\bar{z} = a/2$.

EXAMPLE 3. *Fluid flow through a surface.* We consider a fluid as a collection of points called *particles*. At each particle (x, y, z) we attach a vector $\boldsymbol{V}(x, y, z)$ which represents the velocity of that particular particle. This is the velocity field of the flow. The velocity field may or may not change with time. We shall consider only *steady-state* flows, that is, flows for which the velocity $\boldsymbol{V}(x, y, z)$ depends only on the position of the particle and not on time.

We denote by $\rho(x, y, z)$ the density (mass per unit volume) of the fluid at the point (x, y, z). If the fluid is incompressible the density ρ will be constant throughout the fluid. For a compressible fluid, such as a gas, the density may vary from point to point. In any case, the density is a scalar field associated with the flow. The product of the density and the velocity we denote by F; that is,

$$F(x, y, z) = \rho(x, y, z)V(x, y, z).$$

This is a vector field called the *flux density* of the flow. The vector $F(x, y, z)$ has the same direction as the velocity, and its length has the dimensions

$$\frac{\text{mass}}{\text{unit volume}} \cdot \frac{\text{distance}}{\text{unit time}} = \frac{\text{mass}}{(\text{unit area})(\text{unit time})}.$$

In other words, the flux density vector $F(x, y, z)$ tells us how much mass of fluid per unit area per unit time is flowing in the direction of $V(x, y, z)$ at the point (x, y, z).

Let $S = r(T)$ be a simple parametric surface. At each regular point of S let n denote the unit normal having the same direction as the fundamental vector product. That is, let

(12.15)
$$n = \frac{\dfrac{\partial r}{\partial u} \times \dfrac{\partial r}{\partial v}}{\left\| \dfrac{\partial r}{\partial u} \times \dfrac{\partial r}{\partial v} \right\|}.$$

The dot product $F \cdot n$ represents the component of the flux density vector in the direction of n. The mass of fluid flowing through S in unit time in the direction of n is defined to be the surface integral

$$\iint_{r(T)} F \cdot n \, dS = \iint_T F \cdot n \left\| \frac{\partial r}{\partial u} \times \frac{\partial r}{\partial v} \right\| du \, dv.$$

12.8 Change of parametric representation

We turn now to a discussion of the independence of surface integrals under a change of parametric representation. Suppose a function r maps a region A in the uv-plane onto a parametric surface $r(A)$. Suppose also that A is the image of a region B in the st-plane under a one-to-one continuously differentiable mapping G given by

(12.16)
$$G(s, t) = U(s, t)i + V(s, t)j \quad \text{if} \quad (s, t) \in B.$$

Consider the function R defined on B by the equation

(12.17)
$$R(s, t) = r[G(s, t)].$$

(See Figure 12.11.) Two functions r and R so related will be called *smoothly equivalent.* Smoothly equivalent functions describe the same surface. That is, $r(A)$ and $R(B)$ are identical as point sets. (This follows at once from the one-to-one nature of G.) The next theorem describes the relationship between their fundamental vector products.

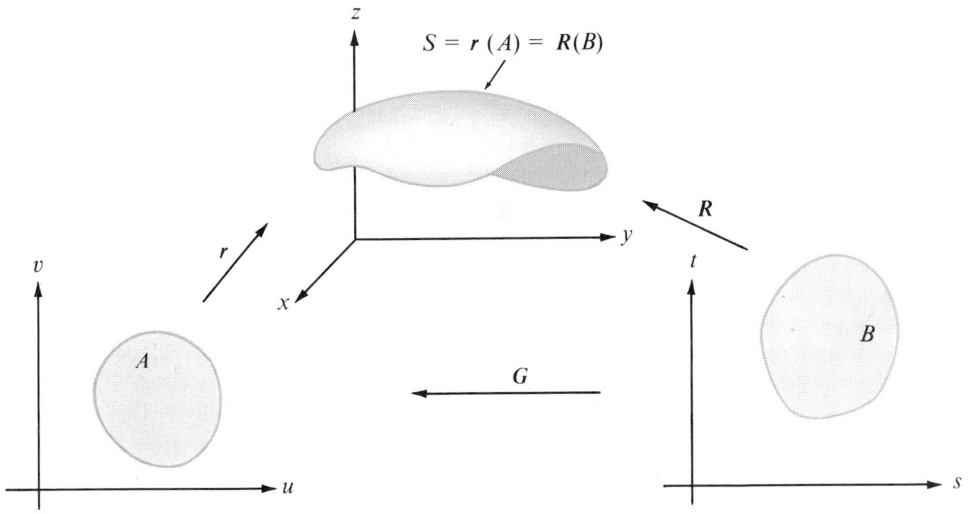

FIGURE 12.11 Two parametric representations of the same surface.

THEOREM 12.1. *Let* **r** *and* **R** *be smoothly equivalent functions related by Equation* (12.17), *where* $\boldsymbol{G} = U\boldsymbol{i} + V\boldsymbol{j}$ *is a one-to-one continuously differentiable mapping of a region B in the st-plane onto a region A in the uv-plane given by Equation* (12.16). *Then we have*

$$\frac{\partial \boldsymbol{R}}{\partial s} \times \frac{\partial \boldsymbol{R}}{\partial t} = \left(\frac{\partial \boldsymbol{r}}{\partial u} \times \frac{\partial \boldsymbol{r}}{\partial v}\right) \frac{\partial(U, V)}{\partial(s, t)}, \tag{12.18}$$

where the partial derivatives $\partial \boldsymbol{r}/\partial u$ *and* $\partial \boldsymbol{r}/\partial v$ *are to be evaluated at the point* $(U(s, t), V(s, t))$. *In other words, the fundamental vector product of* **R** *is equal to that of* **r**, *times the Jacobian determinant of the mapping* **G**.

Proof. The derivatives $\partial \boldsymbol{R}/\partial s$ and $\partial \boldsymbol{R}/\partial t$ can be computed by differentiation of Equation (12.17). If we apply the chain rule (Theorem 8.8) to each component of **R** and rearrange terms, we find that

$$\frac{\partial \boldsymbol{R}}{\partial s} = \frac{\partial \boldsymbol{r}}{\partial u}\frac{\partial U}{\partial s} + \frac{\partial \boldsymbol{r}}{\partial v}\frac{\partial V}{\partial s} \quad \text{and} \quad \frac{\partial \boldsymbol{R}}{\partial t} = \frac{\partial \boldsymbol{r}}{\partial u}\frac{\partial U}{\partial t} + \frac{\partial \boldsymbol{r}}{\partial v}\frac{\partial V}{\partial t},$$

where the derivatives $\partial \boldsymbol{r}/\partial u$ and $\partial \boldsymbol{r}/\partial v$ are evaluated at $(U(s, t), V(s, t))$. Now we cross multiply these two equations and, noting the order of the factors, we obtain

$$\frac{\partial \boldsymbol{R}}{\partial s} \times \frac{\partial \boldsymbol{R}}{\partial t} = \left(\frac{\partial \boldsymbol{r}}{\partial u} \times \frac{\partial \boldsymbol{r}}{\partial v}\right)\left(\frac{\partial U}{\partial s}\frac{\partial V}{\partial t} - \frac{\partial U}{\partial t}\frac{\partial V}{\partial s}\right) = \left(\frac{\partial \boldsymbol{r}}{\partial u} \times \frac{\partial \boldsymbol{r}}{\partial v}\right)\frac{\partial(U, V)}{\partial(s, t)}.$$

This completes the proof.

The invariance of surface integrals under smoothly equivalent parametric representations is now an easy consequence of Theorem 12.1.

THEOREM 12.2. *Let* $\boldsymbol{r}$ *and* $\boldsymbol{R}$ *be smoothly equivalent functions, as described in Theorem 12.1. If the surface integral* $\iint_{\boldsymbol{r}(A)} f\,dS$ *exists, the surface integral* $\iint_{\boldsymbol{R}(B)} f\,dS$ *also exists and we have*

$$\iint_{\boldsymbol{r}(A)} f\,dS = \iint_{\boldsymbol{R}(B)} f\,dS\,.$$

Proof. By the definition of a surface integral we have

$$\iint_{\boldsymbol{r}(A)} f\,dS = \iint_{A} f[\boldsymbol{r}(u, v)] \left\| \frac{\partial \boldsymbol{r}}{\partial u} \times \frac{\partial \boldsymbol{r}}{\partial v} \right\| du\,dv\,.$$

Now we use the mapping $\boldsymbol{G}$ of Theorem 12.1 to transform this into a double integral over the region B in the st-plane. The transformation formula for double integrals states that

$$\iint_{A} f[\boldsymbol{r}(u, v)] \left\| \frac{\partial \boldsymbol{r}}{\partial u} \times \frac{\partial \boldsymbol{r}}{\partial v} \right\| du\,dv = \iint_{B} f\{\boldsymbol{r}[\boldsymbol{G}(s, t)]\} \left\| \frac{\partial \boldsymbol{r}}{\partial u} \times \frac{\partial \boldsymbol{r}}{\partial v} \right\| \left| \frac{\partial(U, V)}{\partial(s, t)} \right| ds\,dt\,,$$

where the derivatives $\partial \boldsymbol{r}/\partial u$ and $\partial \boldsymbol{r}/\partial v$ on the right are to be evaluated at $(U(s, t), V(s, t))$. Because of Equation (12.18), the integral over B is equal to

$$\iint_{B} f[\boldsymbol{R}(s, t)] \left\| \frac{\partial \boldsymbol{R}}{\partial s} \times \frac{\partial \boldsymbol{R}}{\partial t} \right\| ds\,dt\,.$$

This, in turn, is the definition of the surface integral $\iint_{\boldsymbol{R}(B)} f\,dS$. The proof is now complete.

12.9 Other notations for surface integrals

If $S = \boldsymbol{r}(T)$ is a parametric surface, the fundamental vector product $\boldsymbol{N} = \partial \boldsymbol{r}/\partial u \times \partial \boldsymbol{r}/\partial v$ is normal to S at each regular point of the surface. At each such point there are *two* unit normals, a unit normal $\boldsymbol{n}_1$ which has the same direction as $\boldsymbol{N}$, and a unit normal $\boldsymbol{n}_2$ which has the opposite direction. Thus,

$$\boldsymbol{n}_1 = \frac{\boldsymbol{N}}{\|\boldsymbol{N}\|} \qquad \text{and} \qquad \boldsymbol{n}_2 = -\boldsymbol{n}_1\,.$$

Let $\boldsymbol{n}$ be one of the two normals $\boldsymbol{n}_1$ or $\boldsymbol{n}_2$. Let $\boldsymbol{F}$ be a vector field defined on S and assume the surface integral $\iint_S \boldsymbol{F} \cdot \boldsymbol{n}\,dS$ exists. Then we can write

(12.19)
$$\iint_S \boldsymbol{F} \cdot \boldsymbol{n}\,dS = \iint_T \boldsymbol{F}[\boldsymbol{r}(u, v)] \cdot \boldsymbol{n}(u, v) \left\| \frac{\partial \boldsymbol{r}}{\partial u} \times \frac{\partial \boldsymbol{r}}{\partial v} \right\| du\,dv$$
$$= \pm \iint_T \boldsymbol{F}[\boldsymbol{r}(u, v)] \cdot \frac{\partial \boldsymbol{r}}{\partial u} \times \frac{\partial \boldsymbol{r}}{\partial v}\,du\,dv\,,$$

where the $+$ sign is used if $\boldsymbol{n} = \boldsymbol{n}_1$ and the $-$ sign is used if $\boldsymbol{n} = \boldsymbol{n}_2$.

Suppose now we express $\boldsymbol{F}$ and $\boldsymbol{r}$ in terms of their components, say

$$\boldsymbol{F}(x, y, z) = P(x, y, z)\boldsymbol{i} + Q(x, y, z)\boldsymbol{j} + R(x, y, z)\boldsymbol{k}$$

and

$$\boldsymbol{r}(u, v) = X(u, v)\boldsymbol{i} + Y(u, v)\boldsymbol{j} + Z(u, v)\boldsymbol{k}.$$

Then the fundamental vector product of $\boldsymbol{r}$ is given by

$$\boldsymbol{N} = \frac{\partial \boldsymbol{r}}{\partial u} \times \frac{\partial \boldsymbol{r}}{\partial v} = \frac{\partial(Y, Z)}{\partial(u, v)}\boldsymbol{i} + \frac{\partial(Z, X)}{\partial(u, v)}\boldsymbol{j} + \frac{\partial(X, Y)}{\partial(u, v)}\boldsymbol{k}.$$

If $\boldsymbol{n} = \boldsymbol{n}_1$, Equation (12.19) becomes

$$\text{(12.20)} \quad \iint_S \boldsymbol{F} \cdot \boldsymbol{n}\, dS = \iint_T P[\boldsymbol{r}(u, v)] \frac{\partial(Y, Z)}{\partial(u, v)}\, du\, dv + \iint_T Q[\boldsymbol{r}(u, v)] \frac{\partial(Z, X)}{\partial(u, v)}\, du\, dv + \iint_T R[\boldsymbol{r}(u, v)] \frac{\partial(X, Y)}{\partial(u, v)}\, du\, dv\,;$$

if $\boldsymbol{n} = \boldsymbol{n}_2$, each double integral on the right must be replaced by its negative. The sum of the double integrals on the right is often written more briefly as

$$\text{(12.21)} \quad \iint_S P(x, y, z)\, dy \wedge dz + \iint_S Q(x, y, z)\, dz \wedge dx + \iint_S R(x, y, z)\, dx \wedge dy,$$

or even more briefly as

$$\text{(12.22)} \quad \iint_S P\, dy \wedge dz + Q\, dz \wedge dx + R\, dx \wedge dy.$$

The integrals which appear in (12.21) and (12.22) are also referred to as surface integrals. Thus, for example, the surface integral $\iint_S P\, dy \wedge dz$ is defined by the equation

$$\text{(12.23)} \quad \iint_S P\, dy \wedge dz = \iint_T P[\boldsymbol{r}(u, v)] \frac{\partial(Y, Z)}{\partial(u, v)}\, du\, dv.$$

This notation is suggested by the formula for changing variables in a double integral.

Despite similarity in notation, the integral on the left of (12.23) is *not* a double integral. First of all, P is a function of three variables. Also, we must take into consideration the order in which the symbols dy and dz appear in the surface integral, because

$$\frac{\partial(Y, Z)}{\partial(u, v)} = -\frac{\partial(Z, Y)}{\partial(u, v)}$$

and hence

$$\iint_S P\, dy \wedge dz = -\iint_S P\, dz \wedge dy.$$

In this notation, formula (12.20) becomes

(12.24)
$$\iint_S \boldsymbol{F}\cdot\boldsymbol{n}\,dS = \iint_S P\,dy\wedge dz + Q\,dz\wedge dx + R\,dx\wedge dy$$

if $\boldsymbol{n} = \boldsymbol{n}_1$. If $\boldsymbol{n} = \boldsymbol{n}_2$ the integral on the right must be replaced by its negative. This formula resembles the following formula for line integrals:

$$\int_C \boldsymbol{F}\cdot d\boldsymbol{\alpha} = \int_C P\,dx + Q\,dy + R\,dz\,.$$

If the unit normal $\boldsymbol{n}$ is expressed in terms of its direction cosines, say

$$\boldsymbol{n} = \cos\alpha\,\boldsymbol{i} + \cos\beta\,\boldsymbol{j} + \cos\gamma\,\boldsymbol{k}\,,$$

then $\boldsymbol{F}\cdot\boldsymbol{n} = P\cos\alpha + Q\cos\beta + R\cos\gamma$, and we can write

$$\iint_S \boldsymbol{F}\cdot\boldsymbol{n}\,dS = \iint_S (P\cos\alpha + Q\cos\beta + R\cos\gamma)\,dS\,.$$

This equation holds when $\boldsymbol{n}$ is either $\boldsymbol{n}_1$ or $\boldsymbol{n}_2$. The direction cosines will depend on the choice of the normal. If $\boldsymbol{n} = \boldsymbol{n}_1$ we can use (12.24) to write

(12.25)
$$\iint_S (P\cos\alpha + Q\cos\beta + R\cos\gamma)\,dS = \iint_S P\,dy\wedge dz + Q\,dz\wedge dx + R\,dx\wedge dy\,.$$

If $\boldsymbol{n} = \boldsymbol{n}_2$ we have, instead,

(12.26)
$$\iint_S (P\cos\alpha + Q\cos\beta + R\cos\gamma)\,dS = -\iint_S P\,dy\wedge dz + Q\,dz\wedge dx + R\,dx\wedge dy\,.$$

12.10 Exercises

1. Let S denote the hemisphere $x^2 + y^2 + z^2 = 1$, $z \geq 0$, and let $\boldsymbol{F}(x, y, z) = x\boldsymbol{i} + y\boldsymbol{j}$. Let $\boldsymbol{n}$ be the unit outward normal of S. Compute the value of the surface integral $\iint_S \boldsymbol{F}\cdot\boldsymbol{n}\,dS$, using:

   (a) the vector representation $\boldsymbol{r}(u, v) = \sin u\cos v\,\boldsymbol{i} + \sin u\sin v\,\boldsymbol{j} + \cos u\,\boldsymbol{k}$,

   (b) the explicit representation $z = \sqrt{1 - x^2 - y^2}$.

2. Show that the moment of inertia of a homogeneous spherical shell about a diameter is equal to $\frac{2}{3}ma^2$, where m is the mass of the shell and a is its radius.
3. Find the center of mass of that portion of the homogeneous hemispherical surface $x^2 + y^2 + z^2 = a^2$ lying above the first quadrant in the xy-plane.

4. Let S denote the plane surface whose boundary is the triangle with vertices at $(1, 0, 0)$, $(0, 1, 0)$, and $(0, 0, 1)$, and let $\boldsymbol{F}(x, y, z) = x\boldsymbol{i} + y\boldsymbol{j} + z\boldsymbol{k}$. Let $\boldsymbol{n}$ denote the unit normal to S having a nonnegative z-component. Evaluate the surface integral $\iint_S \boldsymbol{F} \cdot \boldsymbol{n}\, dS$, using:
   (a) the vector representation $\boldsymbol{r}(u, v) = (u + v)\boldsymbol{i} + (u - v)\boldsymbol{j} + (1 - 2u)\boldsymbol{k}$,
   (b) an explicit representation of the form $z = f(x, y)$.
5. Let S be a parametric surface described by the explicit formula $z = f(x, y)$, where (x, y) varies over a plane region T, the projection of S in the xy-plane. Let $\boldsymbol{F} = P\boldsymbol{i} + Q\boldsymbol{j} + R\boldsymbol{k}$ and let $\boldsymbol{n}$ denote the unit normal to S having a nonnegative z-component. Use the parametric representation $\boldsymbol{r}(x, y) = x\boldsymbol{i} + y\boldsymbol{j} + f(x, y)\boldsymbol{k}$ and show that

$$\iint_S \boldsymbol{F} \cdot \boldsymbol{n}\, dS = \iint_T \left(-P\frac{\partial f}{\partial x} - Q\frac{\partial f}{\partial y} + R\right) dx\, dy,$$

where each of P, Q, and R is to be evaluated at $(x, y, f(x, y))$.
6. Let S be as in Exercise 5, and let φ be a scalar field. Show that:

(a) $$\iint_S \varphi(x, y, z)\, dS = \iint_T \varphi[x, y, f(x, y)]\sqrt{1 + \left(\frac{\partial f}{\partial x}\right)^2 + \left(\frac{\partial f}{\partial y}\right)^2}\, dx\, dy.$$

(b) $$\iint_S \varphi(x, y, z)\, dy \wedge dz = -\iint_T \varphi[x, y, f(x, y)]\frac{\partial f}{\partial x}\, dx\, dy.$$

(c) $$\iint_S \varphi(x, y, z)\, dz \wedge dx = -\iint_T \varphi[x, y, f(x, y)]\frac{\partial f}{\partial y}\, dx\, dy.$$

7. If S is the surface of the sphere $x^2 + y^2 + z^2 = a^2$, compute the value of the surface integral

$$\iint_S xz\, dy \wedge dz + yz\, dz \wedge dx + x^2\, dx \wedge dy.$$

Choose a representation for which the fundamental vector product points in the direction of the outward normal.
8. The cylinder $x^2 + y^2 = 2x$ cuts out a portion of a surface S from the upper nappe of the cone $x^2 + y^2 = z^2$. Compute the value of the surface integral

$$\iint_S (x^4 - y^4 + y^2z^2 - z^2x^2 + 1)\, dS.$$

9. A homogeneous spherical shell of radius a is cut by one nappe of a right circular cone whose vertex is at the center of the sphere. If the vertex angle of the cone is α, where $0 < \alpha < \pi$, determine (in terms of a and α) the center of mass of the portion of the spherical shell that lies inside the cone.
10. A homogeneous paper rectangle of base $2\pi a$ and altitude h is rolled to form a circular cylindrical surface S of radius a. Calculate the moment of inertia of S about an axis through a diameter of the circular base.

11. Refer to Exercise 10. Calculate the moment of inertia of S about an axis which is in the plane of the base and is tangent to the circular edge of the base.
12. A fluid flow has flux density vector $\boldsymbol{F}(x, y, z) = x\boldsymbol{i} - (2x + y)\boldsymbol{j} + z\boldsymbol{k}$. Let S denote the hemisphere $x^2 + y^2 + z^2 = 1$, $z \geq 0$, and let $\boldsymbol{n}$ denote the unit normal that points out of the sphere. Calculate the mass of fluid flowing through S in unit time in the direction of $\boldsymbol{n}$.
13. Solve Exercise 12 if S also includes the planar base of the hemisphere. On the lower base the unit normal is $-\boldsymbol{k}$.
14. Let S denote the portion of the plane $x + y + z = t$ cut off by the unit sphere $x^2 + y^2 + z^2 = 1$. Let $\varphi(x, y, z) = 1 - x^2 - y^2 - z^2$ if (x, y, z) is inside this sphere, and let $\varphi(x, y, z)$ be 0 otherwise. Show that

$$\iint_S \varphi(x, y, z)\, dS = \begin{cases} \dfrac{\pi}{18}(3 - t^2)^2 & \text{if } |t| \leq \sqrt{3}, \\ 0 & \text{if } |t| > \sqrt{3}. \end{cases}$$

[*Hint:* Introduce new coordinates (x_1, y_1, z_1) with the z_1-axis normal to the plane $x + y + z = t$. The use the polar coordinates in the $x_1 y_1$-plane as parameters for S.]

12.11 The theorem of Stokes

The rest of this chapter is devoted primarily to two generalizations of the second fundamental theorem of calculus involving surface integrals. They are known, respectively, as *Stokes' theorem*† and the *divergence theorem.* This section treats Stokes' theorem. The divergence theorem is discussed in Section 12.19.

Stoke's theorem is a direct extension of Green's theorem which states that

$$\iint_S \left(\frac{\partial Q}{\partial x} - \frac{\partial P}{\partial y}\right) dx\, dy = \int_C P\, dx + Q\, dy,$$

where S is a plane region bounded by a simple closed curve C traversed in the positive (counterclockwise) direction. Stokes' theorem relates a surface integral to a line integral and can be stated as follows.

THEOREM 12.3 STOKES' THEOREM. *Assume that S is a smooth simple parametric surface, say $S = \boldsymbol{r}(T)$, where T is a region in the uv-plane bounded by a piecewise smooth Jordan curve Γ. (See Figure 12.12.) Assume also that $\boldsymbol{r}$ is a one-to-one mapping whose components have continuous second-order partial derivatives on some open set containing $T \cup \Gamma$. Let C denote the image of Γ under $\boldsymbol{r}$, and let P, Q, and R be continuously differentiable scalar fields on S. Then we have*

$$\iint_S \left(\frac{\partial R}{\partial y} - \frac{\partial Q}{\partial z}\right) dy \wedge dz + \left(\frac{\partial P}{\partial z} - \frac{\partial R}{\partial x}\right) dz \wedge dx + \left(\frac{\partial Q}{\partial x} - \frac{\partial P}{\partial y}\right) dx \wedge dy = \int_C P\, dx + Q\, dy + R\, dz. \tag{12.27}$$

† In honor of G. G. Stokes (1819–1903), an Irish mathematician who made many fundamental contributions to hydrodynamics and optics.

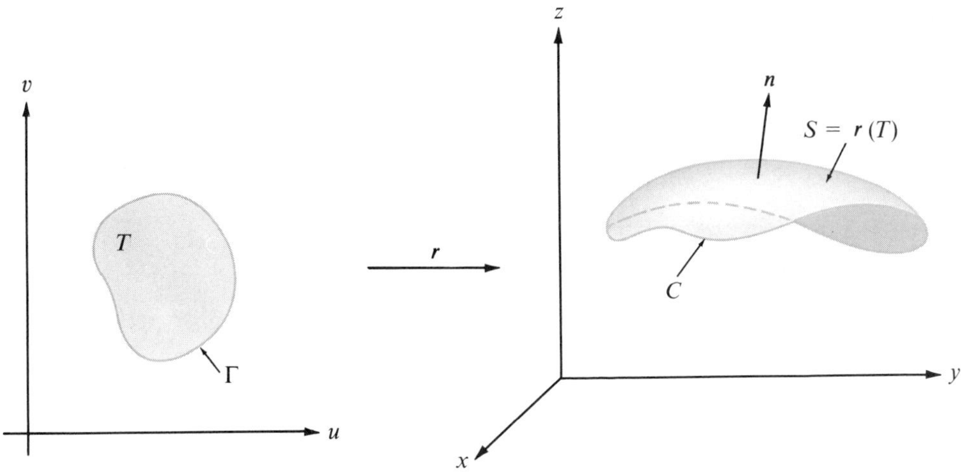

FIGURE 12.12 An example of a surface to which Stokes' theorem is applicable.

The curve Γ *is traversed in the positive (counterclockwise) direction and the curve C is traversed in the direction inherited from* Γ *through the mapping function* **r**.

Proof. To prove the theorem it suffices to establish the following three formulas,

$$\int_C P\,dx = \iint_S \left(-\frac{\partial P}{\partial y}\,dx \wedge dy + \frac{\partial P}{\partial z}\,dz \wedge dx\right), \tag{12.28}$$

$$\int_C Q\,dy = \iint_S \left(-\frac{\partial Q}{\partial z}\,dy \wedge dz + \frac{\partial Q}{\partial x}\,dx \wedge dy\right),$$

$$\int_C R\,dz = \iint_S \left(-\frac{\partial R}{\partial x}\,dz \wedge dx + \frac{\partial R}{\partial y}\,dy \wedge dz\right).$$

Addition of these three equations gives the formula (12.27) in Stokes' theorem. Since the three are similar, we prove only Equation (12.28).

The plan of the proof is to express the surface integral as a double integral over T. Then we use Green's theorem to express the double integral over T as a line integral over Γ. Finally, we show that this line integral is equal to $\int_C P\,dx$.

We write

$$\boldsymbol{r}(u, v) = X(u, v)\boldsymbol{i} + Y(u, v)\boldsymbol{j} + Z(u, v)\boldsymbol{k}$$

and express the surface integral over S in the form

$$\iint_S \left(-\frac{\partial P}{\partial y}\,dx \wedge dy + \frac{\partial P}{\partial z}\,dz \wedge dx\right) = \iint_T \left\{-\frac{\partial P}{\partial y}\frac{\partial(X, Y)}{\partial(u, v)} + \frac{\partial P}{\partial z}\frac{\partial(Z, X)}{\partial(u, v)}\right\} du\,dv.$$

Now let p denote the composite function given by

$$p(u, v) = P[X(u, v),\ Y(u, v), Z(u, v)].$$

The last integrand can be written as

$$-\frac{\partial P}{\partial y}\frac{\partial(X, Y)}{\partial(u, v)} + \frac{\partial P}{\partial z}\frac{\partial(Z, X)}{\partial(u, v)} = \frac{\partial}{\partial u}\left(p\frac{\partial X}{\partial v}\right) - \frac{\partial}{\partial v}\left(p\frac{\partial X}{\partial u}\right). \tag{12.29}$$

The verification of (12.29) is outlined in Exercise 13 of Section 12.13. Applying Green's theorem to the double integral over T we obtain

$$\iint_T \left\{\frac{\partial}{\partial u}\left(p\frac{\partial X}{\partial v}\right) - \frac{\partial}{\partial v}\left(p\frac{\partial X}{\partial u}\right)\right\} du\, dv = \int_\Gamma p\frac{\partial X}{\partial u}\, du + p\frac{\partial X}{\partial v}\, dv,$$

where Γ is traversed in the positive direction. We parametrize Γ by a function $\boldsymbol{\gamma}$ defined on an interval $[a, b]$ and let

$$\boldsymbol{\alpha}(t) = \boldsymbol{r}[\boldsymbol{\gamma}(t)] \tag{12.30}$$

be a corresponding parametrization of C. Then by expressing each line integral in terms of its parametric representation we find that

$$\int_\Gamma p\frac{\partial X}{\partial u}\, du + p\frac{\partial X}{\partial v}\, dv = \int_C P\, dx,$$

which completes the proof of (12.28).

12.12 The curl and divergence of a vector field

The surface integral which appears in Stokes' theorem can be expressed more simply in terms of the *curl* of a vector field. Let $\boldsymbol{F}$ be a differentiable vector field given by

$$\boldsymbol{F}(x, y, z) = P(x, y, z)\boldsymbol{i} + Q(x, y, z)\boldsymbol{j} + R(x, y, z)\boldsymbol{k}.$$

The curl of $\boldsymbol{F}$ is another vector field defined by the equation

$$\operatorname{curl} \boldsymbol{F} = \left(\frac{\partial R}{\partial y} - \frac{\partial Q}{\partial z}\right)\boldsymbol{i} + \left(\frac{\partial P}{\partial z} - \frac{\partial R}{\partial x}\right)\boldsymbol{j} + \left(\frac{\partial Q}{\partial x} - \frac{\partial P}{\partial y}\right)\boldsymbol{k}. \tag{12.31}$$

The components of curl $\boldsymbol{F}$ are the functions appearing in the surface integral in Stokes' formula (12.27). Therefore, this surface integral can be written as

$$\iint_S (\operatorname{curl} \boldsymbol{F}) \cdot \boldsymbol{n}\, dS,$$

where $\boldsymbol{n}$ is the unit normal vector having the same direction as the fundamental vector product of the surface; that is,

$$\boldsymbol{n} = \frac{\dfrac{\partial \boldsymbol{r}}{\partial u} \times \dfrac{\partial \boldsymbol{r}}{\partial v}}{\left\| \dfrac{\partial \boldsymbol{r}}{\partial u} \times \dfrac{\partial \boldsymbol{r}}{\partial v} \right\|}.$$

The line integral in Stokes' formula (12.27) can be written as $\int_C \boldsymbol{F} \cdot d\boldsymbol{\alpha}$, where $\boldsymbol{\alpha}$ is the representation of C given by (12.30). Thus, Stokes' theorem takes the simpler form

$$\iint_S (\operatorname{curl} \boldsymbol{F}) \cdot \boldsymbol{n}\, dS = \int_C \boldsymbol{F} \cdot d\boldsymbol{\alpha}.$$

For the special case in which S is a region in the xy-plane and $\boldsymbol{n} = \boldsymbol{k}$, this formula reduces to Green's theorem,

$$\iint_S \left(\frac{\partial Q}{\partial x} - \frac{\partial P}{\partial y}\right) dx\, dy = \int_C P\, dx + Q\, dy.$$

Equation (12.31) defining the curl can be easily remembered by writing it as an expansion of a 3×3 determinant,

$$\operatorname{curl} \boldsymbol{F} = \begin{vmatrix} \boldsymbol{i} & \boldsymbol{j} & \boldsymbol{k} \\ \dfrac{\partial}{\partial x} & \dfrac{\partial}{\partial y} & \dfrac{\partial}{\partial z} \\ P & Q & R \end{vmatrix} = \left(\frac{\partial R}{\partial y} - \frac{\partial Q}{\partial z}\right)\boldsymbol{i} + \left(\frac{\partial P}{\partial z} - \frac{\partial R}{\partial x}\right)\boldsymbol{j} + \left(\frac{\partial Q}{\partial x} - \frac{\partial P}{\partial y}\right)\boldsymbol{k}.$$

This determinant is to be expanded in terms of first-row minors, but each "product" such as $\partial/\partial y$ times R is to be interpreted as a partial derivative $\partial R/\partial y$. We can also write this formula as a cross product,

$$\operatorname{curl} \boldsymbol{F} = \nabla \times \boldsymbol{F},$$

where the symbol ∇ is treated as though it were a vector,

$$\nabla = \frac{\partial}{\partial x}\boldsymbol{i} + \frac{\partial}{\partial y}\boldsymbol{j} + \frac{\partial}{\partial z}\boldsymbol{k}.$$

If we form the "dot product" $\nabla \cdot \boldsymbol{F}$ in a purely formal way, again interpreting products such as $\partial/\partial x$ times P as $\partial P/\partial x$, we find that

(12.32)
$$\nabla \cdot \boldsymbol{F} = \frac{\partial P}{\partial x} + \frac{\partial Q}{\partial y} + \frac{\partial R}{\partial z}.$$

Equation (12.32) defines a scalar field called the *divergence* of $\boldsymbol{F}$, also written as div $\boldsymbol{F}$.

We have already used the symbol $\nabla\varphi$ to denote the gradient of a scalar field φ, given by

$$\nabla\varphi = \frac{\partial\varphi}{\partial x}\boldsymbol{i} + \frac{\partial\varphi}{\partial y}\boldsymbol{j} + \frac{\partial\varphi}{\partial z}\boldsymbol{k}.$$

This formula can be interpreted as formal multiplication of the symbolic vector ∇ by the scalar field φ. Thus, the gradient, the divergence, and the curl can be represented symbolically by the three products $\nabla\varphi$, $\nabla \cdot \boldsymbol{F}$, and $\nabla \times \boldsymbol{F}$, respectively.

Some of the theorems proved earlier can be expressed in terms of the curl. For example, in Theorem 10.9 we proved that a vector field $\boldsymbol{f} = (f_1, \ldots, f_n)$, continuously differentiable on an open convex set S in n-space, is a gradient on S if and only if the partial derivatives of the components of $\boldsymbol{f}$ satisfy the relations

$$D_k f_j(\boldsymbol{x}) = D_j f_k(\boldsymbol{x}) \qquad (j, k = 1, 2, \ldots, n). \tag{12.33}$$

In the 3-dimensional case Theorem 10.9 can be restated as follows.

THEOREM 12.4. *Let $\boldsymbol{F} = P\boldsymbol{i} + Q\boldsymbol{j} + R\boldsymbol{k}$ be a continuously differentiable vector field on an open convex set S in 3-space. Then $\boldsymbol{F}$ is a gradient on S if and only if we have*

$$\text{curl } \boldsymbol{F} = \boldsymbol{O} \quad \textit{on} \quad S. \tag{12.34}$$

Proof. In the 3-dimensional case the relations (12.33) are equivalent to the statement that curl $\boldsymbol{F} = \boldsymbol{O}$.

12.13 Exercises

In each of Exercises 1 through 4, transform the surface integral $\iint_S (\text{curl } \boldsymbol{F}) \cdot \boldsymbol{n}\, dS$ to a line integral by the use of Stokes' theorem, and then evaluate the line integral.

1. $\boldsymbol{F}(x, y, z) = y^2\boldsymbol{i} + xy\boldsymbol{j} + xz\boldsymbol{k}$, where S is the hemisphere $x^2 + y^2 + z^2 = 1$, $z \geq 0$, and $\boldsymbol{n}$ is the unit normal with a nonnegative z-component.
2. $\boldsymbol{F}(x, y, z) = y\boldsymbol{i} + z\boldsymbol{j} + x\boldsymbol{k}$, where S is the portion of the paraboloid $z = 1 - x^2 - y^2$ with $z \geq 0$, and $\boldsymbol{n}$ is the unit normal with a nonnegative z-component.
3. $\boldsymbol{F}(x, y, z) = (y - z)\boldsymbol{i} + yz\boldsymbol{j} - xz\boldsymbol{k}$, where S consists of the five faces of the cube $0 \leq x \leq 2$, $0 \leq y \leq 2$, $0 \leq z \leq 2$ not in the xy-plane. The unit normal $\boldsymbol{n}$ is the outward normal.
4. $\boldsymbol{F}(x, y, z) = xz\boldsymbol{i} - y\boldsymbol{i} + x^2y\boldsymbol{k}$, where S consists of the three faces not in the xz-plane of the tetrahedron bounded by the three coordinate planes and the plane $3x + y + 3z = 6$. The normal $\boldsymbol{n}$ is the unit normal pointing out of the tetrahedron.

In Exercises 5 through 10, use Stokes' theorem to show that the line integrals have the values given. In each case, explain how to traverse C to arrive at the given answer.

5. $\int_C y\,dx + z\,dy + x\,dz = \pi a^2\sqrt{3}$, where C is the curve of intersection of the sphere $x^2 + y^2 + z^2 = a^2$ and the plane $x + y + z = 0$.
6. $\int_C (y + z)\,dx + (z + x)\,dy + (x + y)\,dz = 0$, where C is the curve of intersection of the cylinder $x^2 + y^2 = 2y$ and the plane $y = z$.
7. $\int_C y^2\,dx + xy\,dy + xz\,dz = 0$, where C is the curve of Exercise 6.

8. $\int_C (y - z)\,dx + (z - x)\,dy + (x - y)\,dz = 2\pi a(a + b)$, where C is the intersection of the cylinder $x^2 + y^2 = a^2$ and the plane $x/a + z/b = 1$, $a > 0$, $b > 0$.
9. $\int_C (y^2 + z^2)\,dx + (x^2 + z^2)\,dy + (x^2 + y^2)\,dz = 2\pi ab^2$, where C is the intersection of the hemisphere $x^2 + y^2 + z^2 = 2ax$, $z > 0$, and the cylinder $x^2 + y^2 = 2bx$, where $0 < b < a$.
10. $\int_C (y^2 - z^2)\,dx + (z^2 - x^2)\,dy + (x^2 - y^2)\,dz = 9a^3/2$, where C is the curve cut from the boundary of the cube $0 \le x \le a$, $0 \le y \le a$, $0 \le z \le a$ by the plane $x + y + z = 3a/2$.
11. If $\boldsymbol{r} = x\boldsymbol{i} + y\boldsymbol{j} + z\boldsymbol{k}$ and $P\boldsymbol{i} + Q\boldsymbol{j} + R\boldsymbol{k} = \boldsymbol{a} \times \boldsymbol{r}$, where $\boldsymbol{a}$ is a constant vector, show that $\int_C P\,dx + Q\,dy + R\,dz = 2 \iint_S \boldsymbol{a} \cdot \boldsymbol{n}\,dS$, where C is a curve bounding a parametric surface S and $\boldsymbol{n}$ is a suitable normal to S.
12. Let $\boldsymbol{F} = P\boldsymbol{i} + Q\boldsymbol{j} + R\boldsymbol{k}$, where $P = -y/(x^2 + y^2)$, $Q = x/(x^2 + y^2)$, $R = z$, and let D be the torus generated by rotating the circle $(x - 2)^2 + z^2 = 1$, $y = 0$, about the z-axis. Show that curl $\boldsymbol{F} = \boldsymbol{O}$ in D but that $\int_C P\,dx + Q\,dy + R\,dz$ is not zero if the curve C is the circle $x^2 + y^2 = 4$, $z = 0$.
13. This exercise outlines a proof of Equation (12.29), used in the proof of Stokes' theorem.
   (a) Use the formula for differentiating a product to show that

$$\frac{\partial}{\partial u}\left(p\frac{\partial X}{\partial v}\right) - \frac{\partial}{\partial v}\left(p\frac{\partial X}{\partial u}\right) = \frac{\partial p}{\partial u}\frac{\partial X}{\partial v} - \frac{\partial p}{\partial v}\frac{\partial X}{\partial u}.$$

   (b) Now let $p(u, v) = P[X(u, v), Y(u, v), Z(u, v)]$. Compute $\partial p/\partial u$ and $\partial p/\partial v$ by the chain rule and use part (a) to deduce Equation (12.29),

$$\frac{\partial}{\partial u}\left(P\frac{\partial X}{\partial v}\right) - \frac{\partial}{\partial v}\left(P\frac{\partial X}{\partial u}\right) = -\frac{\partial P}{\partial y}\frac{\partial(X, Y)}{\partial(u, v)} + \frac{\partial P}{\partial z}\frac{\partial(Z, X)}{\partial(u, v)}.$$

12.14 Further properties of the curl and divergence

The curl and divergence of a vector field are related to the Jacobian matrix. If $\boldsymbol{F} = P\boldsymbol{i} + Q\boldsymbol{j} + R\boldsymbol{k}$, the Jacobian matrix of $\boldsymbol{F}$ is

$$\boldsymbol{D}\boldsymbol{F}(x, y, z) = \begin{bmatrix} \dfrac{\partial P}{\partial x} & \dfrac{\partial P}{\partial y} & \dfrac{\partial P}{\partial z} \\[2ex] \dfrac{\partial Q}{\partial x} & \dfrac{\partial Q}{\partial y} & \dfrac{\partial Q}{\partial z} \\[2ex] \dfrac{\partial R}{\partial x} & \dfrac{\partial R}{\partial y} & \dfrac{\partial R}{\partial z} \end{bmatrix}.$$

The trace of this matrix (the sum of its diagonal elements) is the divergence of $\boldsymbol{F}$.

Every real matrix A can be written as a sum of a symmetric matrix, $\frac{1}{2}(A + A^t)$, and a skew-symmetric matrix, $\frac{1}{2}(A - A^t)$. When A is the Jacobian matrix $\boldsymbol{D}\boldsymbol{F}$, the skew-symmetric part becomes

$$\frac{1}{2}\begin{bmatrix} 0 & \dfrac{\partial P}{\partial y} - \dfrac{\partial Q}{\partial x} & \dfrac{\partial P}{\partial z} - \dfrac{\partial R}{\partial x} \\[2ex] \dfrac{\partial Q}{\partial x} - \dfrac{\partial P}{\partial y} & 0 & \dfrac{\partial Q}{\partial z} - \dfrac{\partial R}{\partial y} \\[2ex] \dfrac{\partial R}{\partial x} - \dfrac{\partial P}{\partial z} & \dfrac{\partial R}{\partial y} - \dfrac{\partial Q}{\partial z} & 0 \end{bmatrix}. \tag{12.35}$$

The nonzero elements of this matrix are the component of curl $\boldsymbol{F}$ and their negatives. If the Jacobian matrix $D\boldsymbol{F}$ is symmetric, each entry in (12.35) is zero and curl $\boldsymbol{F} = \boldsymbol{O}$.

EXAMPLE 1. Let $\boldsymbol{F}(x, y, z) = x\boldsymbol{i} + y\boldsymbol{j} + z\boldsymbol{k}$. Then we have

$$P(x, y, z) = x, \qquad Q(x, y, z) = y, \qquad R(x, y, z) = z,$$

and the corresponding Jacobian matrix is the 3×3 identity matrix. Therefore

$$\operatorname{div} \boldsymbol{F} = 3 \qquad \text{and} \qquad \operatorname{curl} \boldsymbol{F} = \boldsymbol{O}.$$

More generally, if $\boldsymbol{F}(x, y, z) = f(x)\boldsymbol{i} + g(y)\boldsymbol{j} + h(z)\boldsymbol{k}$, the Jacobian matrix has the elements $f'(x), g'(y), h'(z)$ on the main diagonal and zeros elsewhere, so

$$\operatorname{div} \boldsymbol{F} = f'(x) + g'(y) + h'(z) \qquad \text{and} \qquad \operatorname{curl} \boldsymbol{F} = \boldsymbol{O}.$$

EXAMPLE 2. Let $\boldsymbol{F}(x, y, z) = xy^2z^2\boldsymbol{i} + z^2 \sin y\,\boldsymbol{j} + x^2e^y\boldsymbol{k}$. The Jacobian matrix is

$$\begin{bmatrix} y^2z^2 & 2xyz^2 & 2xy^2z \\ 0 & z^2 \cos y & 2z \sin y \\ 2xe^y & x^2e^y & 0 \end{bmatrix}.$$

Therefore,

$$\operatorname{div} \boldsymbol{F} = y^2z^2 + z^2 \cos y$$

and

$$\operatorname{curl} \boldsymbol{F} = (x^2e^y - 2z \sin y)\boldsymbol{i} + (2xy^2z - 2xe^y)\boldsymbol{j} - 2xyz^2\boldsymbol{k}.$$

EXAMPLE 3. *The divergence and curl of a gradient.* Suppose $\boldsymbol{F}$ is a gradient, say $\boldsymbol{F} = \operatorname{grad} \varphi = \partial\varphi/\partial x\,\boldsymbol{i} + \partial\varphi/\partial y\,\boldsymbol{j} + \partial\varphi/\partial z\,\boldsymbol{k}$. The Jacobian matrix is

(12.36)
$$\begin{bmatrix} \dfrac{\partial^2\varphi}{\partial x^2} & \dfrac{\partial^2\varphi}{\partial y\,\partial x} & \dfrac{\partial^2\varphi}{\partial z\,\partial x} \\[2ex] \dfrac{\partial^2\varphi}{\partial x\,\partial y} & \dfrac{\partial^2\varphi}{\partial y^2} & \dfrac{\partial^2\varphi}{\partial z\,\partial y} \\[2ex] \dfrac{\partial^2\varphi}{\partial x\,\partial z} & \dfrac{\partial^2\varphi}{\partial y\,\partial z} & \dfrac{\partial^2\varphi}{\partial z^2} \end{bmatrix}.$$

Therefore,

$$\operatorname{div} \boldsymbol{F} = \frac{\partial^2\varphi}{\partial x^2} + \frac{\partial^2\varphi}{\partial y^2} + \frac{\partial^2\varphi}{\partial z^2}.$$

The expression on the right is called the *Laplacian* of φ and is often written more briefly as $\nabla^2\varphi$. Thus, the divergence of a gradient $\nabla\varphi$ is the Laplacian of φ. In symbols, this is written

(12.37)
$$\operatorname{div}(\nabla\varphi) = \nabla^2\varphi.$$

When $\nabla^2\varphi = 0$, the function φ is called *harmonic*. Equation (12.37) shows that the gradient of a harmonic function has zero divergence. When the mixed partial derivatives in matrix (12.36) are continuous, the matrix is symmetric and curl $\boldsymbol{F}$ is zero. In other words,

$$\text{curl}\,(\text{grad}\,\varphi) = \boldsymbol{O}$$

for every scalar field φ with continuous second-order mixed partial derivatives. This example shows that the condition curl $\boldsymbol{F} = \boldsymbol{O}$ is necessary for a continuously differentiable vector field $\boldsymbol{F}$ to be a gradient. In other words, if curl $\boldsymbol{F} \neq \boldsymbol{O}$ on an open set S, then $\boldsymbol{F}$ is not a gradient on S. We know also, from Theorem 12.4 that if curl $\boldsymbol{F} = \boldsymbol{O}$ on an open *convex* set S, then $\boldsymbol{F}$ is a gradient on S. A field with zero curl is called *irrotational.*

EXAMPLE 4. *A vector field with zero divergence and zero curl.* Let S be the set of all $(x, y) \neq (0, 0)$, and let

$$\boldsymbol{F}(x, y) = -\frac{y}{x^2 + y^2}\boldsymbol{i} + \frac{x}{x^2 + y^2}\boldsymbol{j}$$

if $(x, y) \in S$. From Example 2 in Section 10.16 we know that $\boldsymbol{F}$ *is not a gradient on* S (although $\boldsymbol{F}$ is a gradient on every rectangle not containing the origin). The Jacobian matrix is

$$D\boldsymbol{F}(x, y) = \begin{bmatrix} \dfrac{2xy}{(x^2 + y^2)^2} & \dfrac{y^2 - x^2}{(x^2 + y^2)^2} & 0 \\[2ex] \dfrac{y^2 - x^2}{(x^2 + y^2)^2} & \dfrac{-2xy}{(x^2 + y^2)^2} & 0 \\[2ex] 0 & 0 & 0 \end{bmatrix},$$

and we see at once that div $\boldsymbol{F} = 0$ and curl $\boldsymbol{F} = \boldsymbol{O}$ on S.

EXAMPLE 5. *The divergence and curl of a curl.* If $\boldsymbol{F} = P\boldsymbol{i} + Q\boldsymbol{j} + R\boldsymbol{k}$, the curl of $\boldsymbol{F}$ is a new vector field and we can compute *its* divergence and curl. The Jacobian matrix of curl $\boldsymbol{F}$ is

$$\begin{bmatrix} \dfrac{\partial^2 R}{\partial x\,\partial y} - \dfrac{\partial^2 Q}{\partial x\,\partial z} & \dfrac{\partial^2 R}{\partial y^2} - \dfrac{\partial^2 Q}{\partial y\,\partial z} & \dfrac{\partial^2 R}{\partial z\,\partial y} - \dfrac{\partial^2 Q}{\partial z^2} \\[2ex] \dfrac{\partial^2 P}{\partial x\,\partial z} - \dfrac{\partial^2 R}{\partial x^2} & \dfrac{\partial^2 P}{\partial y\,\partial z} - \dfrac{\partial^2 R}{\partial y\,\partial x} & \dfrac{\partial^2 P}{\partial z^2} - \dfrac{\partial^2 R}{\partial z\,\partial x} \\[2ex] \dfrac{\partial^2 Q}{\partial x^2} - \dfrac{\partial^2 P}{\partial x\,\partial y} & \dfrac{\partial^2 Q}{\partial y\,\partial x} - \dfrac{\partial^2 P}{\partial y^2} & \dfrac{\partial^2 Q}{\partial z\,\partial x} - \dfrac{\partial^2 P}{\partial z\,\partial y} \end{bmatrix}.$$

If we assume that all the mixed partial derivatives are continuous, we find that

$$\text{div}\,(\text{curl}\,\boldsymbol{F}) = 0$$

and

$$\text{curl}\,(\text{curl}\,\boldsymbol{F}) = \text{grad}\,(\text{div}\,\boldsymbol{F}) - \nabla^2\boldsymbol{F}, \tag{12.38}$$

where $\nabla^2 \boldsymbol{F}$ is defined by the equation

$$\nabla^2 \boldsymbol{F} = (\nabla^2 P)\boldsymbol{i} + (\nabla^2 Q)\boldsymbol{j} + (\nabla^2 R)\boldsymbol{k}.$$

The identity in (12.38) relates all four operators, gradient, curl, divergence, and Laplacian. The verification of (12.38) is requested in Exercise 7 of Section 12.15.

The curl and divergence have some general properties in common with ordinary derivatives. First, they are *linear operators.* That is, if a and b are constants, we have

$$\text{(12.39)} \qquad \operatorname{div}(a\boldsymbol{F} + b\boldsymbol{G}) = a \operatorname{div} \boldsymbol{F} + b \operatorname{div} \boldsymbol{G},$$

and

$$\text{(12.40)} \qquad \operatorname{curl}(a\boldsymbol{F} + b\boldsymbol{G}) = a \operatorname{curl} \boldsymbol{F} + b \operatorname{curl} \boldsymbol{G}.$$

They also have a property analogous to the formula for differentiating a product:

$$\text{(12.41)} \qquad \operatorname{div}(\varphi \boldsymbol{F}) = \varphi \operatorname{div} \boldsymbol{F} + \nabla\varphi \cdot \boldsymbol{F},$$

and

$$\text{(12.42)} \qquad \operatorname{curl}(\varphi \boldsymbol{F}) = \varphi \operatorname{curl} \boldsymbol{F} + \nabla\varphi \times \boldsymbol{F},$$

where φ is any differentiable scalar field. These properties are immediate consequences of the definitions of curl and divergence; their proofs are requested in Exercise 6 of Section 12.15.

If we use the symbolic vector

$$\nabla = \frac{\partial}{\partial x}\boldsymbol{i} + \frac{\partial}{\partial y}\boldsymbol{j} + \frac{\partial}{\partial z}\boldsymbol{k}$$

once more, each of the formulas (12.41) and (12.42) takes a form which resembles more closely the usual rule for differentiating a product:

$$\nabla \cdot (\varphi \boldsymbol{F}) = \varphi \nabla \cdot \boldsymbol{F} + \nabla\varphi \cdot \boldsymbol{F}$$

and

$$\nabla \times (\varphi \boldsymbol{F}) = \varphi \nabla \times \boldsymbol{F} + \nabla\varphi \times \boldsymbol{F}.$$

In Example 3 the Laplacian of a scalar field, $\nabla^2\varphi$, was defined to be $\partial^2\varphi/\partial x^2 + \partial^2\varphi/\partial y^2 + \partial^2\varphi/\partial z^2$. In Example 5 the Laplacian $\nabla^2 \boldsymbol{F}$ of a vector field was defined by components. We get correct formulas for both $\nabla^2\varphi$ and $\nabla^2 \boldsymbol{F}$ if we interpret ∇^2 as the symbolic operator

$$\nabla^2 = \frac{\partial^2}{\partial x^2} + \frac{\partial^2}{\partial y^2} + \frac{\partial^2}{\partial z^2}.$$

This formula for ∇^2 also arises by dot multiplication of the symbolic vector ∇ with itself.

Thus, we have $\nabla^2 = \nabla \cdot \nabla$ and we can write

$$\nabla^2\varphi = (\nabla \cdot \nabla)\varphi \qquad \text{and} \qquad \nabla^2 \boldsymbol{F} = (\nabla \cdot \nabla)\boldsymbol{F}.$$

Now consider the formula $\nabla \cdot \nabla\varphi$. This can be read as $(\nabla \cdot \nabla)\varphi$, which is $\nabla^2\varphi$; or as $\nabla \cdot (\nabla\varphi)$, which is div $(\nabla\varphi)$. In Example 3 we showed that div $(\nabla\varphi) = \nabla^2\varphi$, so we have

$$(\nabla \cdot \nabla)\varphi = \nabla \cdot (\nabla\varphi);$$

hence we can write $\nabla \cdot \nabla\varphi$ for either of these expressions without danger of ambiguity. This is not true, however, when φ is replaced by a vector field $\boldsymbol{F}$. The expression $(\nabla \cdot \nabla)\boldsymbol{F}$ is $\nabla^2\boldsymbol{F}$, which has been defined. However, $\nabla \cdot (\nabla \boldsymbol{F})$ is meaningless because $\nabla \boldsymbol{F}$ is not defined. Therefore the expression $\nabla \cdot \nabla \boldsymbol{F}$ is meaningful only when it is interpreted as $(\nabla \cdot \nabla)\boldsymbol{F}$. These remarks illustrate that although symbolic formulas sometimes serve as a convenient notation and memory aid, care is needed in manipulating the symbols.

12.15 Exercises

1. For each of the following vector fields determine the Jacobian matrix and compute the curl and divergence.
   (a) $\boldsymbol{F}(x, y, z) = (x^2 + yz)\boldsymbol{i} + (y^2 + xz)\boldsymbol{j} + (z^2 + xy)\boldsymbol{k}$.
   (b) $\boldsymbol{F}(x, y, z) = (2z - 3y)\boldsymbol{i} + (3x - z)\boldsymbol{j} + (y - 2x)\boldsymbol{k}$.
   (c) $\boldsymbol{F}(x, y, z) = (z + \sin y)\boldsymbol{i} - (z - x\cos y)\boldsymbol{j}$.
   (d) $\boldsymbol{F}(x, y, z) = e^{xy}\boldsymbol{i} + \cos xy\,\boldsymbol{j} + \cos xz^2\boldsymbol{k}$.
   (e) $\boldsymbol{F}(x, y, z) = x^2 \sin y\,\boldsymbol{i} + y^2 \sin xz\,\boldsymbol{j} + xy \sin(\cos z)\boldsymbol{k}$.
2. If $\boldsymbol{r} = x\boldsymbol{i} + y\boldsymbol{j} + z\boldsymbol{k}$ and $r = \|\boldsymbol{r}\|$, compute curl $[f(r)\boldsymbol{r}]$, where f is a differentiable function.
3. If $\boldsymbol{r} = x\boldsymbol{i} + y\boldsymbol{j} + z\boldsymbol{k}$ and $\boldsymbol{A}$ is a constant vector, show that curl $(\boldsymbol{A} \times \boldsymbol{r}) = 2\boldsymbol{A}$.
4. If $\boldsymbol{r} = x\boldsymbol{i} + y\boldsymbol{j} + z\boldsymbol{k}$ and $r = \|\boldsymbol{r}\|$, find all integers n for which div $(r^n\boldsymbol{r}) = 0$.
5. Find a vector field whose curl is $x\boldsymbol{i} + y\boldsymbol{j} + z\boldsymbol{k}$ or prove that no such vector field exists.
6. Prove the elementary properties of curl and divergence in Equations (12.39) through (12.42).
7. Prove that curl (curl $\boldsymbol{F}$) = grad (div $\boldsymbol{F}$) $- \nabla^2\boldsymbol{F}$ if the components of $\boldsymbol{F}$ have continuous mixed partial derivatives of second order.
8. Prove the identity

$$\nabla \cdot (\boldsymbol{F} \times \boldsymbol{G}) = \boldsymbol{G} \cdot (\nabla \times \boldsymbol{F}) - \boldsymbol{F} \cdot (\nabla \times \boldsymbol{G}),$$

   where $\boldsymbol{F}$ and $\boldsymbol{G}$ are differentiable vector fields.
9. A vector field $\boldsymbol{F}$ will not be the gradient of a potential unless curl $\boldsymbol{F} = \boldsymbol{O}$. However, it may be possible to find a nonzero scalar field μ such that $\mu\boldsymbol{F}$ is a gradient. Prove that if such a μ exists, $\boldsymbol{F}$ is always perpendicular to its curl. When the field is two-dimensional, say $\boldsymbol{F} = P\boldsymbol{i} + Q\boldsymbol{j}$, this exercise gives us a necessary condition for the differential equation $P\,dx + Q\,dy = 0$ to have an integrating factor. (The converse is also true. That is, if $\boldsymbol{F} \cdot$ curl $\boldsymbol{F} = 0$ in a suitable region, a nonzero μ exists such that $\mu\boldsymbol{F}$ is a gradient. The proof of the converse is not required.)
10. Let $\boldsymbol{F}(x, y, z) = y^2z^2\boldsymbol{i} + z^2x^2\boldsymbol{j} + x^2y^2\boldsymbol{k}$. Show that curl $\boldsymbol{F}$ is not always zero, but that $\boldsymbol{F} \cdot$ curl $\boldsymbol{F} = 0$. Find a scalar field μ such that $\mu\boldsymbol{F}$ is a gradient.
11. Let $\boldsymbol{V}(x, y) = y^c\boldsymbol{i} + x^c\boldsymbol{j}$, where c is a positive constant, and let $\boldsymbol{r}(x, y) = x\boldsymbol{i} + y\boldsymbol{j}$. Let R be a plane region bounded by a piecewise smooth Jordan curve C. Compute div $(\boldsymbol{V} \times \boldsymbol{r})$ and curl $(\boldsymbol{V} \times \boldsymbol{r})$, and use Green's theorem to show that

$$\oint_C \boldsymbol{V} \times \boldsymbol{r} \cdot d\boldsymbol{\alpha} = 0,$$

   where $\boldsymbol{\alpha}$ describes C.

12. Show that Green's theorem can be expressed as follows:

$$\iint_R (\text{curl } \boldsymbol{V}) \cdot \boldsymbol{k}\, dx\, dy = \oint_C \boldsymbol{V} \cdot \boldsymbol{T}\, ds,$$

where $\boldsymbol{T}$ is the unit tangent to C and s denotes arc length.

13. A plane region R is bounded by a piecewise smooth Jordan curve C. The moments of inertia of R about the x- and y-axes are known to be a and b, respectively. Compute the line integral

$$\oint_C \nabla(r^4) \cdot \boldsymbol{n}\, ds$$

in terms of a and b. Here $r = \|x\boldsymbol{i} + y\boldsymbol{j}\|$, $\boldsymbol{n}$ denotes the unit outward normal of C, and s denotes arc length. The curve is traversed counterclockwise.

14. Let $\boldsymbol{F}$ be a two-dimensional vector field. State a definition for the vector-valued line integral $\int_C \boldsymbol{F} \times d\boldsymbol{\alpha}$. Your definition should be such that the following formula is a consequence of Green's theorem:

$$\int_C \boldsymbol{F} \times d\boldsymbol{\alpha} = \boldsymbol{k} \iint_R (\text{div } \boldsymbol{F})\, dx\, dy,$$

where R is a plane region bounded by a simple closed curve C.

★12.16 Reconstruction of a vector field from its curl

In our study of the gradient we learned how to determine whether or not a given vector field is a gradient. We now ask a similar question concerning the curl. Given a vector field $\boldsymbol{F}$, is there a $\boldsymbol{G}$ such that curl $\boldsymbol{G} = \boldsymbol{F}$? Suppose we write $\boldsymbol{F} = P\boldsymbol{i} + Q\boldsymbol{j} + R\boldsymbol{k}$ and $\boldsymbol{G} = L\boldsymbol{i} + M\boldsymbol{j} + N\boldsymbol{k}$. To solve the equation curl $\boldsymbol{G} = \boldsymbol{F}$ we must solve the system of partial differential equations

$$\frac{\partial N}{\partial y} - \frac{\partial M}{\partial z} = P, \qquad \frac{\partial L}{\partial z} - \frac{\partial N}{\partial x} = Q, \qquad \frac{\partial M}{\partial x} - \frac{\partial L}{\partial y} = R \tag{12.43}$$

for the three unknown functions L, M, and N when P, Q, and R are given.

It is not always possible to solve such a system. For example, we proved in Section 12.14 that the divergence of a curl is always zero. Therefore, for the system (12.43) to have a solution in some open set S it is necessary to have

$$\frac{\partial P}{\partial x} + \frac{\partial Q}{\partial y} + \frac{\partial R}{\partial z} = 0 \tag{12.44}$$

everywhere in S. As it turns out, this condition is also sufficient for system (12.43) to have a solution if we suitably restrict the set S in which (12.44) is satisfied. We shall prove now that condition (12.44) suffices when S is a three-dimensional interval.

THEOREM 12.5 *Let* $\boldsymbol{F}$ *be continuously differentiable on an open interval* S *in* 3-*space. Then there exists a vector field* $\boldsymbol{G}$ *such that* curl $\boldsymbol{G} = \boldsymbol{F}$ *if and only if* div $\boldsymbol{F} = 0$ *everywhere in* S.

Proof. The necessity of the condition $\operatorname{div} \boldsymbol{F} = 0$ has already been established, since the divergence of a curl is always zero. To establish the sufficiency we must exhibit three functions L, M, and N that satisfy the three equations in (12.43). Let us try to make a choice with $L = 0$. Then the second and third equations in (12.43) become

$$\frac{\partial N}{\partial x} = -Q \quad \text{and} \quad \frac{\partial M}{\partial x} = R.$$

This means that we must have

$$N(x, y, z) = -\int_{x_0}^{x} Q(t, y, z)\, dt + f(y, z)$$

and

$$M(x, y, z) = \int_{x_0}^{x} R(t, y, z)\, dt + g(y, z),$$

where each integration is along a line segment in S and the "constants of integration" $f(y, z)$ and $g(y, z)$ are independent of x. Let us try to find a solution with $f(y, z) = 0$. The first equation in (12.43) requires

$$\frac{\partial N}{\partial y} - \frac{\partial M}{\partial z} = P. \tag{12.45}$$

For the choice of M and N just described we have

$$\frac{\partial N}{\partial y} - \frac{\partial M}{\partial z} = -\frac{\partial}{\partial y}\int_{x_0}^{x} Q(t, y, z)\, dt - \frac{\partial}{\partial z}\int_{x_0}^{x} R(t, y, z)\, dt - \frac{\partial g}{\partial z}. \tag{12.46}$$

At this stage we interchange the two operations of partial differentiation and integration, using Theorem 10.8. That is, we write

$$\frac{\partial}{\partial y}\int_{x_0}^{x} Q(t, y, z)\, dt = \int_{x_0}^{x} D_2Q(t, y, z)\, dt \tag{12.47}$$

and

$$\frac{\partial}{\partial z}\int_{x_0}^{x} R(t, y, z)\, dt = \int_{x_0}^{x} D_3R(t, y, z)\, dt. \tag{12.48}$$

Then Equation (12.46) becomes

$$\frac{\partial N}{\partial y} - \frac{\partial M}{\partial z} = \int_{x_0}^{x} [-D_2Q(t, y, z) - D_3R(t, y, z)]\, dt - \frac{\partial g}{\partial z}. \tag{12.49}$$

Using condition (12.44) we may replace the integrand in (12.49) by $D_1P(t, y, z)$; Equation (12.49) becomes

$$\frac{\partial N}{\partial y} - \frac{\partial M}{\partial z} = \int_{x_0}^{x} D_1P(t, y, z)\, dt - \frac{\partial g}{\partial z} = P(x, y, z) - P(x_0, y, z) - \frac{\partial g}{\partial z}.$$

Therefore (12.45) will be satisfied if we choose g so that $\partial g/\partial z = -P(x_0, y, z)$. Thus, for example, we may take

$$g(y, z) = -\int_{z_0}^{z} P(x_0, y, u)\, du .$$

This argument leads us to consider the vector field $\boldsymbol{G} = L\boldsymbol{i} + M\boldsymbol{j} + N\boldsymbol{k}$, where $L(x, y, z) = 0$ and

$$M(x, y, z) = \int_{x_0}^{x} R(t, y, z)\, dt - \int_{z_0}^{z} P(x_0, y, u)\, du , \qquad N(x, y, z) = -\int_{x_0}^{x} Q(t, y, z)\, dt .$$

For this choice of L, M, and N it is easy to verify, with the help of (12.47) and (12.48), that the three equations in (12.43) are satisfied, giving us curl $\boldsymbol{G} = \boldsymbol{F}$, as required.

It should be noted that the foregoing proof not only establishes the existence of a vector field $\boldsymbol{G}$ whose curl is $\boldsymbol{F}$, but also provides a straightforward method for determining $\boldsymbol{G}$ by integration involving the components of $\boldsymbol{F}$.

For a given $\boldsymbol{F}$, the vector field $\boldsymbol{G}$ that we have constructed is not the only solution of the equation curl $\boldsymbol{G} = \boldsymbol{F}$. If we add to this $\boldsymbol{G}$ any continuously differentiable gradient $\nabla\varphi$ we obtain another solution because

$$\text{curl}\,(\boldsymbol{G} + \nabla\varphi) = \text{curl}\,\boldsymbol{G} + \text{curl}\,(\nabla\varphi) = \text{curl}\,\boldsymbol{G} = \boldsymbol{F},$$

since curl $(\nabla\varphi) = \boldsymbol{O}$. Moreover, it is easy to show that *all* continuously differentiable solutions must be of the form $\boldsymbol{G} + \nabla\varphi$. Indeed, if $\boldsymbol{H}$ is another solution, then curl $\boldsymbol{H} =$ curl $\boldsymbol{G}$, so curl $(\boldsymbol{H} - \boldsymbol{G}) = \boldsymbol{O}$. By Theorem 10.9 it follows that $\boldsymbol{H} - \boldsymbol{G} = \nabla\varphi$ for some continuously differentiable gradient $\nabla\varphi$; hence $\boldsymbol{H} = \boldsymbol{G} + \nabla\varphi$, as asserted.

A vector field $\boldsymbol{F}$ for which div $\boldsymbol{F} = 0$ is sometimes called *solenoidal.* Theorem 12.5 states that a vector field is solenoidal on an open interval S in 3-space if and only if it is the curl of another vector field on S.

The following example shows that this statement is not true for *arbitrary* open sets.

EXAMPLE. *A solenoidal vector field that is not a curl.* Let D be the portion of 3-space between two concentric spheres with center at the origin and radii a and b, where $0 < a < b$. Let $\boldsymbol{V} = \boldsymbol{r}/r^3$, where $\boldsymbol{r} = x\boldsymbol{i} + y\boldsymbol{j} + z\boldsymbol{k}$ and $r = \|\boldsymbol{r}\|$. It is easy to verify that div $\boldsymbol{V} = 0$ everywhere in D. In fact, we have the general formula

$$\text{div}\,(r^n\boldsymbol{r}) = (n + 3)r^n ,$$

and in this example $n = -3$. We shall use Stokes' theorem to prove that this $\boldsymbol{V}$ is *not* a curl in D (although it is a curl on every open three-dimensional interval not containing the origin). To do this we assume there is a vector field $\boldsymbol{U}$ such that $\boldsymbol{V} =$ curl $\boldsymbol{U}$ in D and obtain a contradiction. By Stokes' theorem we can write

$$\iint_S (\text{curl}\,\boldsymbol{U}) \cdot \boldsymbol{n}\, dS = \oint_C \boldsymbol{U} \cdot d\boldsymbol{\alpha}, \tag{12.50}$$

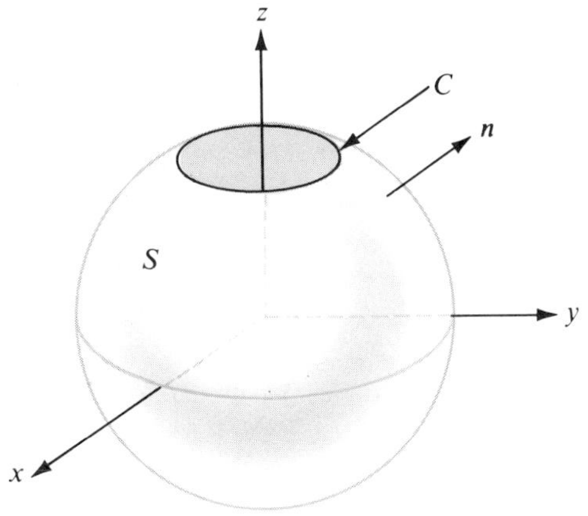

FIGURE 12.13 The surface S and curve C in Equation (12.50).

where S and C are the surface and curve shown in Figure 12.13. To construct S, we take a spherical surface of radius R concentric with the boundaries of D, where $a < R < b$, and we remove a small "polar cap," as indicated in the figure. The portion that remains is the surface S. The curve C is the circular edge shown. Let $\boldsymbol{n}$ denote the unit outer normal of S, so that $\boldsymbol{n} = \boldsymbol{r}/r$. Since curl $\boldsymbol{U} = \boldsymbol{V} = \boldsymbol{r}/r^3$, we have

$$(\text{curl } \boldsymbol{U}) \cdot \boldsymbol{n} = \frac{\boldsymbol{r}}{r^3} \cdot \frac{\boldsymbol{r}}{r} = \frac{1}{r^2}.$$

On the surface S this dot product has the constant value $1/R^2$. Therefore we have

$$\iint_S (\text{curl } \boldsymbol{U}) \cdot \boldsymbol{n} \, dS = \frac{1}{R^2} \iint_S dS = \frac{\text{area of } S}{R^2}.$$

When the polar cap shrinks to a point, the area of S approaches $4\pi R^2$ (the area of the whole sphere) and, therefore, the value of the surface integral in (12.50) approaches 4π.

Next we examine the line integral in (12.50). It is easy to prove that for any line integral $\int_C \boldsymbol{U} \cdot d\boldsymbol{\alpha}$ we have the inequality

$$\left| \int_C \boldsymbol{U} \cdot d\boldsymbol{\alpha} \right| \leq M \cdot (\text{length of } C),$$

where M is a constant depending on $\boldsymbol{U}$. (In fact, M can be taken to be the maximum value of $\|\boldsymbol{U}\|$ on C.) Therefore, as we let the polar cap shrink to a point, the length of C and the value of the line integral both approach zero. Thus we have a contradiction; the surface integral in (12.50) can be made arbitrarily close to 4π, and the corresponding line integral to which it is equal can be made arbitrarily close to 0. Therefore a function $\boldsymbol{U}$ whose curl is $\boldsymbol{V}$ cannot exist in the region D.

The difficulty here is caused by the geometric structure of the region D. Although this region is simply connected (that is, any simple closed curve in D is the edge of a parametric surface lying completely in D) there are closed *surfaces* in D that are not the complete boundaries of solids lying entirely in D. For example, no sphere about the origin is the complete boundary of a solid lying entirely in D. If the region D has the property that *every* closed surface in D is the boundary of a solid lying entirely in D, it can be shown that a vector field $\boldsymbol{U}$ exists such that $\boldsymbol{V} = \operatorname{curl} \boldsymbol{U}$ in D if and only if $\operatorname{div} \boldsymbol{V} = 0$ everywhere in D. The proof of this statement is difficult and will not be given here.

★12.17 Exercises

1. Find a vector field $\boldsymbol{G}(x, y, z)$ whose curl is $2\boldsymbol{i} + \boldsymbol{j} + 3\boldsymbol{k}$ everywhere in 3-space. What is the most general continuously differentiable vector field with this property?
2. Show that the vector field $\boldsymbol{F}(x, y, z) = (y - z)\boldsymbol{i} + (z - x)\boldsymbol{j} + (x - y)\boldsymbol{k}$ is solenoidal, and find a vector field $\boldsymbol{G}$ such that $\boldsymbol{F} = \operatorname{curl} \boldsymbol{G}$ everywhere in 3-space.
3. Let $\boldsymbol{F}(x, y, z) = -z\boldsymbol{i} + xy\boldsymbol{k}$. Find a continuously differentiable vector field $\boldsymbol{G}$ of the form $\boldsymbol{G}(x, y, z) = L(x, y, z)\boldsymbol{i} + M(x, y, z)\boldsymbol{j}$ such that $\boldsymbol{F} = \operatorname{curl} \boldsymbol{G}$ everywhere, in 3-space. What is the most general $\boldsymbol{G}$ of this form?
4. If two vector fields $\boldsymbol{U}$ and $\boldsymbol{V}$ are both irrotational, show that the vector field $\boldsymbol{U} \times \boldsymbol{V}$ is solenoidal.
5. Let $\boldsymbol{r} = x\boldsymbol{i} + y\boldsymbol{j} + z\boldsymbol{k}$ and let $r = \|\boldsymbol{r}\|$. Show that $n = -3$ is the only value of n for which $r^n\boldsymbol{r}$ is solenoidal for $r \neq 0$. For this n, choose a 3-dimensional interval S not containing the origin and express $r^{-3}\boldsymbol{r}$ as a curl in S. *Note:* Although $r^{-3}\boldsymbol{r}$ is a curl in every such S, it is *not* a curl on the set of all points different from $(0, 0, 0)$.
6. Find the most general continuously differentiable function f of one real variable such that the vector field $f(r)\boldsymbol{r}$ will be solenoidal, where $\boldsymbol{r} = x\boldsymbol{i} + y\boldsymbol{j} + z\boldsymbol{k}$ and $r = \|\boldsymbol{r}\|$.
7. Let $\boldsymbol{V}$ denote a vector field that is continuously differentiable on some open interval S in 3-space. Consider the following two statements about $\boldsymbol{V}$:
   (i) $\operatorname{curl} \boldsymbol{V} = \boldsymbol{O}$ and $\boldsymbol{V} = \operatorname{curl} \boldsymbol{U}$ for some continuously differentiable vector field $\boldsymbol{U}$ (everywhere on S).
   (ii) A scalar field φ exists such that $\nabla\varphi$ is continuously differentiable and such that
   $$\boldsymbol{V} = \operatorname{grad} \varphi \qquad \text{and} \qquad \nabla^2\varphi = 0 \quad \text{everywhere on } S.$$
   (a) Prove that (i) implies (ii). In other words, a vector field that is both irrotational and solenoidal in S is the gradient of a harmonic function in S.
   (b) Prove that (ii) implies (i), or give a counterexample.
8. Assume continuous differentiability of all vector fields involved, on an open interval S. Suppose $\boldsymbol{H} = \boldsymbol{F} + \boldsymbol{G}$, where $\boldsymbol{F}$ is solenoidal and $\boldsymbol{G}$ is irrotational. Then there exists a vector field $\boldsymbol{U}$ such that $\boldsymbol{F} = \operatorname{curl} \boldsymbol{U}$ and a scalar field φ such that $\boldsymbol{G} = \nabla\varphi$ in S. Show that $\boldsymbol{U}$ and φ satisfy the following partial differential equations in S:
   $$\nabla^2\varphi = \operatorname{div} \boldsymbol{H}, \qquad \operatorname{grad}(\operatorname{div} \boldsymbol{U}) - \nabla^2\boldsymbol{U} = \operatorname{curl} \boldsymbol{H}.$$
   Note: This exercise has widespread applications, because it can be shown that every continuously differentiable vector field $\boldsymbol{H}$ on S can be expressed in the form $\boldsymbol{H} = \boldsymbol{F} + \boldsymbol{G}$, where $\boldsymbol{F}$ is solenoidal and $\boldsymbol{G}$ is irrotational.
9. Let $\boldsymbol{H}(x, y, z) = x^2y\boldsymbol{i} + y^2z\boldsymbol{j} + z^2x\boldsymbol{k}$. Find vector fields $\boldsymbol{F}$ and $\boldsymbol{G}$, where $\boldsymbol{F}$ is a curl and $\boldsymbol{G}$ is a gradient, such that $\boldsymbol{H} = \boldsymbol{F} + \boldsymbol{G}$.
10. Let u and v be scalar fields that are continuously differentiable on an open interval R in 3-space.
    (a) Show that a vector field $\boldsymbol{F}$ exists such that $\nabla u \times \nabla v = \operatorname{curl} \boldsymbol{F}$ everywhere in R.

(b) Determine whether or not any of the following three vector fields may be used for $\boldsymbol{F}$ in part (a): (i) $\nabla(uv)$; (ii) $u\nabla v$; (iii) $v\nabla u$.
(c) If $u(x, y, z) = x^3 - y^3 + z^2$ and $v(x, y, z) = x + y + z$, evaluate the surface integral

$$\iint_S \nabla u \times \nabla v \cdot \boldsymbol{n}\, dS,$$

where S is the hemisphere $x^2 + y^2 + z^2 = 1$, $z \geq 0$, and $\boldsymbol{n}$ is the unit normal with a non-negative z-component.

12.18 Extensions of Stokes' theorem

Stokes' theorem can be extended to more general simple smooth surfaces. If T is a multiply connected region like that shown in Figure 12.14 (with a finite number of holes), the one-to-one image $S = \boldsymbol{r}(T)$ will contain the same number of holes as T. To extend

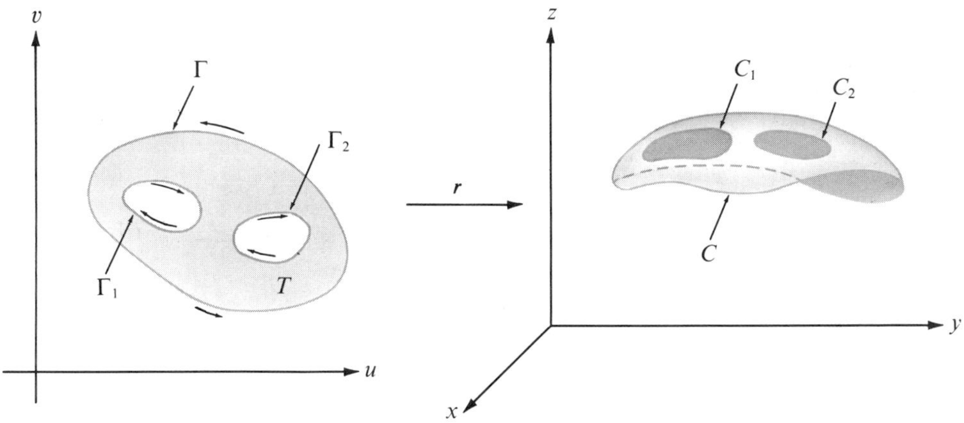

FIGURE 12.14 Extension of Stokes' theorem to surfaces that are one-to-one images of multiply connected regions.

Stokes' theorem to such surfaces we use exactly the same type of argument as in the proof of Stokes' theorem, except that we employ Green's theorem for multiply connected regions (Theorem 11.12). In place of the line integral which appears in Equation (12.27) we need a sum of line integrals, with appropriate signs, taken over the images of the curves forming the boundary of T. For example, if T has two holes, as in Figure 12.14, and if the boundary curves Γ, Γ_1, and Γ_2 are traversed in the directions shown, the identity in Stokes' theorem takes the form

$$\iint_S (\text{curl } \boldsymbol{F}) \cdot \boldsymbol{n}\, dS = \oint_C \boldsymbol{F} \cdot d\boldsymbol{\rho} + \oint_{C_1} \boldsymbol{F} \cdot d\boldsymbol{\rho}_1 + \oint_{C_2} \boldsymbol{F} \cdot d\boldsymbol{\rho}_2,$$

where C, C_1, and C_2 are the images of Γ, Γ_1, and Γ_2, respectively, and $\boldsymbol{\rho}$, $\boldsymbol{\rho}_1$, and $\boldsymbol{\rho}_2$ are the composite functions $\boldsymbol{\rho}(t) = \boldsymbol{r}[\boldsymbol{\gamma}(t)]$, $\boldsymbol{\rho}_1(t) = \boldsymbol{r}[\boldsymbol{\gamma}_1(t)]$, $\boldsymbol{\rho}_2(t) = \boldsymbol{r}[\boldsymbol{\gamma}_2(t)]$. Here $\boldsymbol{\gamma}$, $\boldsymbol{\gamma}_1$, and $\boldsymbol{\gamma}_2$ are the functions that describe Γ, Γ_1, and Γ_2 in the directions shown. The curves C, C_1, and C_2 will be traversed in the directions inherited from Γ, Γ_1, and Γ_2 through the mapping function $\boldsymbol{r}$.

Stokes' theorem can also be extended to some (but not all) smooth surfaces that are not simple. We shall illustrate a few of the possibilities with examples.

Consider first the cylinder shown in Figure 12.15. This is the union of two simple smooth parametric surfaces S_1 and S_2, the images of two adjacent rectangles T_1 and T_2, under mappings $\boldsymbol{r}_1$ and $\boldsymbol{r}_2$, respectively. If γ_1 describes the positively oriented boundary Γ_1 of T_1 and γ_2 describes the positively oriented boundary Γ_2 of T_2, the functions ρ_1 and ρ_2 defined by

$$\rho_1(t) = \boldsymbol{r}_1[\gamma_1(t)], \qquad \rho_2(t) = \boldsymbol{r}_2[\gamma_2(t)]$$

describe the images C_1 and C_2 of Γ_1 and Γ_2, respectively. In this example the representations $\boldsymbol{r}_1$ and $\boldsymbol{r}_2$ can be chosen so that they agree on the intersection $\Gamma_1 \cap \Gamma_2$. If we apply

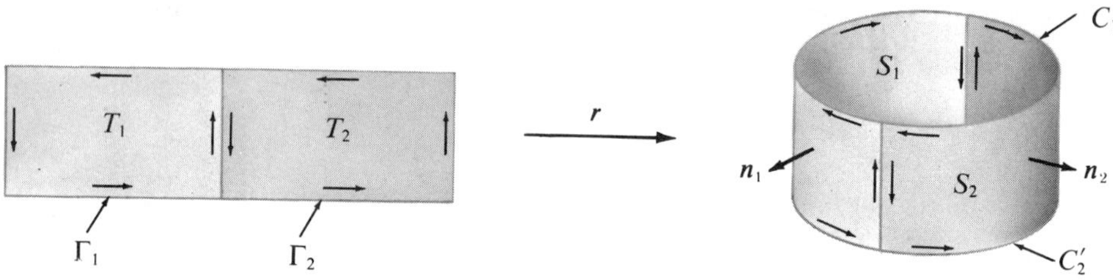

FIGURE 12.15 Extension of Stokes' theorem to a cylinder.

Stokes' theorem to each piece S_1 and S_2 and add the two identities, we obtain

$$\iint_{S_1} (\text{curl } \boldsymbol{F}) \cdot \boldsymbol{n}_1 \, dS + \iint_{S_2} (\text{curl } \boldsymbol{F}) \cdot \boldsymbol{n}_2 \, dS = \int_{C_1} \boldsymbol{F} \cdot d\rho_1 + \int_{C_2} \boldsymbol{F} \cdot d\rho_2, \tag{12.51}$$

where $\boldsymbol{n}_1$ and $\boldsymbol{n}_2$ are the normals determined by the fundamental vector products of $\boldsymbol{r}_1$ and $\boldsymbol{r}_2$, respectively.

Now let $\boldsymbol{r}$ denote the mapping of $T_1 \cup T_2$ which agrees with $\boldsymbol{r}_1$ on T_1 and with $\boldsymbol{r}_2$ on T_2, and let $\boldsymbol{n}$ be the corresponding unit normal determined by the fundamental vector product of $\boldsymbol{r}$. Since the normals $\boldsymbol{n}_1$ and $\boldsymbol{n}_2$ agree in direction on $S_1 \cap S_2$, the unit normal $\boldsymbol{n}$ is the same as $\boldsymbol{n}_1$ on S_1 and the same as $\boldsymbol{n}_2$ on S_2. Therefore the sum of the surface integrals on the left of (12.51) is equal to

$$\iint_{S_1 \cup S_2} (\text{curl } \boldsymbol{F}) \cdot \boldsymbol{n} \, dS .$$

For this example, the representations $\boldsymbol{r}_1$ and $\boldsymbol{r}_2$ can be chosen so that ρ_1 and ρ_2 determine *opposite* directions on each arc of the intersection $C_1 \cap C_2$, as indicated by the arrows in Figure 12.15. The two line integrals on the right of (12.51) can be replaced by a sum of line integrals along the two circles C_1' and C_2' forming the upper and lower edges of $S_1 \cup S_2$, since the line integrals along each arc of the intersection $C_1 \cap C_2$ cancel. Therefore, Equation (12.51) can be written as

$$\iint_{S_1 \cup S_2} (\text{curl } \boldsymbol{F}) \cdot \boldsymbol{n} \, dS = \int_{C_1'} \boldsymbol{F} \cdot d\rho_1 + \int_{C_2'} \boldsymbol{F} \cdot d\rho_2, \tag{12.52}$$

where the line integrals are traversed in the directions inherited from Γ_1 and Γ_2. The two circles C_1' and C_2' are said to form the complete boundary of $S_1 \cup S_2$. Equation (12.52) expresses the surface integral of $(\text{curl}\, \boldsymbol{F}) \cdot \boldsymbol{n}$ over $S_1 \cup S_2$ as a line integral over the complete boundary of $S_1 \cup S_2$. This equation is the extension of Stokes' theorem for a cylinder.

Suppose now we apply the same concepts to the surface shown in Figure 12.16. This surface is again the union of two smooth simple parametric surfaces S_1 and S_2, the images of two adjacent rectangles T_1 and T_2. This particular surface is called a *Möbius band*;† a

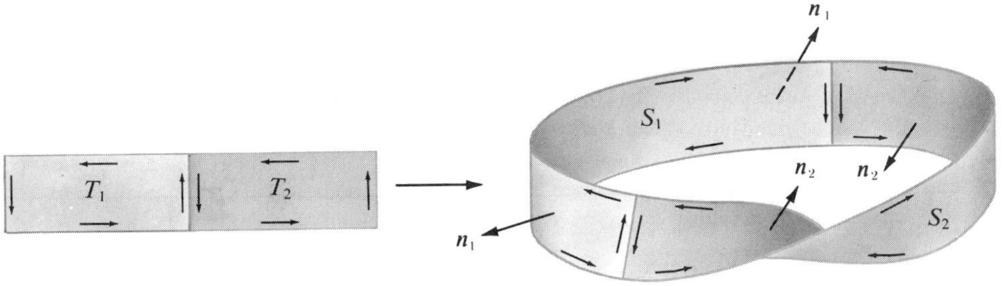

FIGURE 12.16 A Möbius band considered as the union of two simple parametric surfaces. Stokes' theorem does not extend to a Möbius band.

model can easily be constructed from a long rectangular strip of paper by giving one end a half-twist and then fastening the two ends together. We define $\boldsymbol{\rho}_1$, $\boldsymbol{\rho}_2$, C_1, and C_2 for the Möbius band as we defined them for the cylinder above. The edge of $S_1 \cup S_2$ in this case is one simple closed curve C', rather than two. This curve is called the complete boundary of the Möbius band.

If we apply Stokes' theorem to each piece S_1 and S_2, as we did for the cylinder, we obtain Equation (12.51). But if we try to consolidate the two surface integrals and the two line integrals as we did above, we encounter two difficulties. First, the two normals $\boldsymbol{n}_1$ and $\boldsymbol{n}_2$ do not agree in direction everywhere on the intersection $C_1 \cap C_2$. (See Figure 12.16.) Therefore we cannot define a normal $\boldsymbol{n}$ for the whole surface by taking $\boldsymbol{n} = \boldsymbol{n}_1$ on S_1 and $\boldsymbol{n} = \boldsymbol{n}_2$ on S_2, as we did for the cylinder. This is not serious, however, because we can define $\boldsymbol{n}$ to be $\boldsymbol{n}_1$ on S_1 and on $C_1 \cap C_2$, and then define $\boldsymbol{n}$ to be $\boldsymbol{n}_2$ everywhere else. This gives a discontinuous normal, but the discontinuities so introduced form a set of content zero in the uv-plane and do not affect the existence or the value of the surface integral

$$\iint_{S_1 \cup S_2} (\text{curl}\, \boldsymbol{F}) \cdot \boldsymbol{n}\, dS.$$

A more serious difficulty is encountered when we try to consolidate the line integrals. In this example it is not possible to choose the mappings $\boldsymbol{r}_1$ and $\boldsymbol{r}_2$ in such a way that $\boldsymbol{\rho}_1$ and $\boldsymbol{\rho}_2$ determine opposite directions on each arc of the intersection $C_1 \cap C_2$. This is illustrated by the arrows in Figure 12.16; one of these arcs is traced twice in the same direction. On this arc the corresponding line integrals will not necessarily cancel as they

† After A. F. Möbius (1790–1868), a pupil of Gauss. At the age of 26 he was appointed professor of astronomy at Leipzig, a position he held until his death. He made many contributions to celestial mechanics, but his most important researches were in geometry and in the theory of numbers.

did for the cylinder. Therefore the sum of the line integrals in (12.51) is not necessarily equal to the line integral over the complete boundary of $S_1 \cup S_2$, and Stokes' theorem cannot be extended to the Möbius band.

Note: The cylinder and the Möbius band are examples of *orientable* and *nonorientable* surfaces, respectively. We shall not attempt to define these terms precisely, but shall mention some of their differences. For an orientable surface $S_1 \cup S_2$ formed from two smooth simple parametric surfaces as described above, the mappings $\boldsymbol{r}_1$ and $\boldsymbol{r}_2$ can always be chosen so that $\boldsymbol{\rho}_1$ and $\boldsymbol{\rho}_2$ determine opposite directions on each arc of the intersection $C_1 \cap C_2$. For a nonorientable surface no such choice is possible. For a smooth orientable surface a unit normal vector can be defined in a continuous fashion over the entire surface. For a nonorientable surface no such definition of a normal is possible. A paper model of an orientable surface always has two sides that can be distinguished by painting them with two different colors. Nonorientable surfaces have only one side. For a rigorous discussion of these and other properties of orientable and nonorientable surfaces, see any book on combinatorial topology. Stokes' theorem can be extended to orientable surfaces by a procedure similar to that outlined above for the cylinder.

Another orientable surface is the sphere shown in Figure 12.17. This surface is the union of two simple parametric surfaces (hemispheres) S_1 and S_2, which we may consider

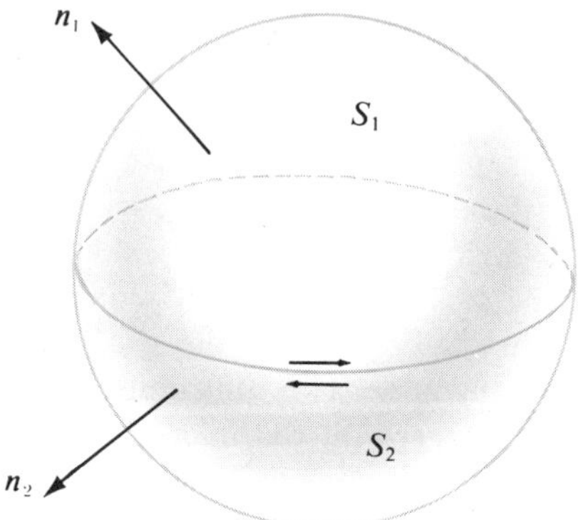

FIGURE 12.17 Extension of Stokes' theorem to a sphere.

images of a circular disk in the *xy*-plane under mappings $\boldsymbol{r}_1$ and $\boldsymbol{r}_2$, respectively. We give $\boldsymbol{r}$, $\boldsymbol{\rho}_1$, $\boldsymbol{\rho}_2$, C_1, C_2 the same meanings as in the above examples. In this case the curves C_1 and C_2 are completely matched by the mapping $\boldsymbol{r}$ (they intersect along the equator), and the surface $S_1 \cup S_2$ is said to be *closed.* Moreover, $\boldsymbol{r}_1$ and $\boldsymbol{r}_2$ can be chosen so that the directions determined by $\boldsymbol{\rho}_1$ and $\boldsymbol{\rho}_2$ are opposite on C_1 and C_2, as suggested by the arrows in Figure 12.17. (This is why $S_1 \cup S_2$ is orientable.) If we apply Stokes' theorem to each hemisphere and add the results we obtain Equation (12.51), as before. The normals $\boldsymbol{n}_1$ and $\boldsymbol{n}_2$ agree on the intersection $C_1 \cap C_2$, and we can consolidate the integrals over S_1 and S_2 into one integral over the whole sphere. The two line integrals on the right of (12.51) cancel completely, leaving us with the formula

$$\iint_{S_1 \cup S_2} (\operatorname{curl} \boldsymbol{F}) \cdot \boldsymbol{n} \, dS = 0 .$$

This holds not only for a sphere, but for any orientable closed surface.

12.19 The divergence theorem (Gauss' theorem)

Stokes' theorem expresses a relationship between an integral extended over a surface and a line integral taken over the one or more curves forming the boundary of this surface. The divergence theorem expresses a relationship between a triple integral extended over a solid and a surface integral taken over the boundary of this solid.

THEOREM 12.6. DIVERGENCE THEOREM. *Let V be a solid in 3-space bounded by an orientable closed surface S, and let $\boldsymbol{n}$ be the unit outer normal to S. If $\boldsymbol{F}$ is a continuously differentiable vector field defined on V, we have*

$$\iiint_V (\operatorname{div} \boldsymbol{F})\, dx\, dy\, dz = \iint_S \boldsymbol{F} \cdot \boldsymbol{n}\, dS. \tag{12.53}$$

Note: If we express $\boldsymbol{F}$ and $\boldsymbol{n}$ in terms of their components, say

$$\boldsymbol{F}(x, y, z) = P(x, y, z)\boldsymbol{i} + Q(x, y, z)\boldsymbol{j} + R(x, y, z)\boldsymbol{k}$$

and

$$\boldsymbol{n} = \cos \alpha\, \boldsymbol{i} + \cos \beta\, \boldsymbol{j} + \cos \gamma\, \boldsymbol{k},$$

then Equation (12.53) can be written as

$$\iiint_V \left(\frac{\partial P}{\partial x} + \frac{\partial Q}{\partial y} + \frac{\partial R}{\partial z}\right) dx\, dy\, dz = \iint_S (P \cos \alpha + Q \cos \beta + R \cos \gamma)\, dS. \tag{12.54}$$

Proof. It suffices to establish the three equations

$$\iiint_V \frac{\partial P}{\partial x}\, dx\, dy\, dz = \iint_S P \cos \alpha\, dS,$$

$$\iiint_V \frac{\partial Q}{\partial y}\, dx\, dy\, dz = \iint_S Q \cos \beta\, dS,$$

$$\iiint_V \frac{\partial R}{\partial z}\, dx\, dy\, dz = \iint_S R \cos \gamma\, dS,$$

and add the results to obtain (12.54). We begin with the third of these formulas and prove it for solids of a very special type.

Assume V is a set of points (x, y, z) satisfying a relation of the form

$$g(x, y) \le z \le f(x, y) \qquad \text{for} \quad (x, y) \text{ in } T,$$

where T is a connected region in the xy-plane, and f and g are continuous functions on T, with $g(x, y) \le f(x, y)$ for each (x, y) in T. Geometrically, this means that T is the projection of V on the xy-plane. Every line through T parallel to the z-axis intersects the solid V along a line segment connecting the surface $z = g(x, y)$ to the surface $z = f(x, y)$. The boundary surface S consists of an upper cap S_1, given by the explicit formula $z = f(x, y)$; a lower part S_2, given by $z = g(x, y)$; and (possibly) a portion S_3 of the cylinder generated by a line moving parallel to the z-axis along the boundary of T. The outer normal to S has a nonnegative z-component on S_1, has a nonpositive component on S_2, and is parallel to the xy-plane on S_3. Solids of this type will be called "xy-projectable." (An example is shown in Figure 12.18.) They include all convex solids (for example, solid spheres, ellipsoids, cubes) and many solids that are not convex (for example, solid tori with axes parallel to the z-axis).

The idea of the proof is quite simple. We express the triple integral as a double integral extended over the projection T. Then we show that this double integral has the same value

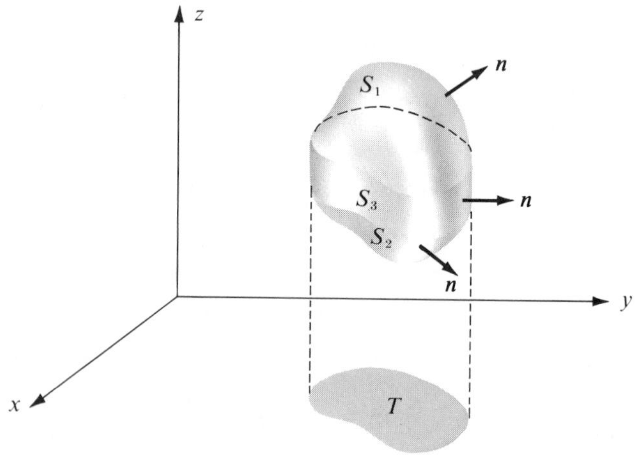

FIGURE 12.18 An example of a solid that is xy-projectable.

as the surface integral in question. We begin with the formula

$$\iiint_V \frac{\partial R}{\partial z}\,dx\,dy\,dz = \iint_T \left[\int_{g(x,y)}^{f(x,y)} \frac{\partial R}{\partial z}\,dz\right] dx\,dy\,.$$

The one-dimensional integral with respect to z may be evaluated by the second fundamental theorem of calculus, giving us

(12.55) $$\iiint_V \frac{\partial R}{\partial z}\,dx\,dy\,dz = \iint_T \{R[x, y, f(x, y)] - R[x, y, g(x, y)]\}\,dx\,dy\,.$$

For the surface integral we can write

(12.56) $$\iint_S R\cos\gamma\,dS = \iint_{S_1} R\cos\gamma\,dS + \iint_{S_2} R\cos\gamma\,dS + \iint_{S_3} R\cos\gamma\,dS\,.$$

On S_3 the normal $\boldsymbol{n}$ is parallel to the xy-plane, so $\cos\gamma = 0$ and the integral over S_3 is zero. On the surface S_1 we use the representation

$$\boldsymbol{r}(x, y) = x\boldsymbol{i} + y\boldsymbol{j} + f(x, y)\boldsymbol{k}\,,$$

and on S_2 we use the representation

$$r(x, y) = x\boldsymbol{i} + y\boldsymbol{j} + g(x, y)\boldsymbol{k}.$$

On S_1 the normal $\boldsymbol{n}$ has the same direction as the vector product $\partial\boldsymbol{r}/\partial x \times \partial\boldsymbol{r}/\partial y$, so we can write [see Equation (12.25), p. 436]

$$\iint_{S_1} R \cos \gamma \, dS = \iint_{S_1} R \, dx \wedge dy = \iint_T R[x, y, f(x, y)] \, dx \, dy .$$

On S_2 the normal $\boldsymbol{n}$ has the direction opposite to that of $\partial\boldsymbol{r}/\partial x \times \partial\boldsymbol{r}/\partial y$ so, by Equation (12.26), we have

$$\iint_{S_2} R \cos \gamma \, dS = -\iint_{S_2} R \, dx \wedge dy = -\iint_T R[x, y, g(x, y)] \, dx \, dy .$$

Therefore Equation (12.56) becomes

$$\iint_S R \cos \gamma \, dS = \iint_T \{R[x, y, f(x, y)] - R[x, y, g(x, y)]\} \, dx \, dy .$$

Comparing this with Equation (12.55) we see that

$$\iiint_V \frac{\partial R}{\partial z} \, dx \, dy \, dz = \iint_S R \cos \gamma \, dS .$$

In the foregoing proof the assumption that V is xy-projectable enabled us to express the triple integral over V as a double integral over its projection T in the xy-plane. It is clear that if V is yz-projectable we can use the same type of argument to prove the identity

$$\iiint_V \frac{\partial P}{\partial x} \, dx \, dy \, dz = \iint_S P \cos \alpha \, dS ;$$

and if V is xz-projectable we obtain

$$\iiint_V \frac{\partial Q}{\partial y} \, dx \, dy \, dz = \iint_S Q \cos \beta \, dS .$$

Thus we see that the divergence theorem is valid for all solids projectable on all three coordinate planes. In particular the theorem holds for every convex solid.

A solid torus with its axis parallel to the z-axis is xy-projectable but not xz-projectable or yz-projectable. To extend the divergence theorem to such a solid we cut the torus into four equal parts by planes through its axis parallel to the xz- and yz-planes, respectively, and we apply the divergence theorem to each part. The triple integral over the whole torus is the sum of the triple integrals over the four parts. When we add the surface integrals over the four parts we find that the contributions from the faces common to adjacent parts cancel each other, since the outward normals have opposite directions on two such faces. Therefore the sum of the surface integrals over the four parts is equal to the surface integral over the entire torus. This example illustrates how the divergence theorem can be extended to certain nonconvex solids.

12.20 Applications of the divergence theorem

The concepts of curl and divergence of a vector field $\boldsymbol{F} = P\boldsymbol{i} + Q\boldsymbol{j} + R\boldsymbol{k}$ were introduced in Section 12.12 by the formulas

$$\operatorname{div} \boldsymbol{F} = \frac{\partial P}{\partial x} + \frac{\partial Q}{\partial y} + \frac{\partial R}{\partial z} \tag{12.57}$$

and

$$\operatorname{curl} \boldsymbol{F} = \left(\frac{\partial R}{\partial y} - \frac{\partial Q}{\partial z}\right)\boldsymbol{i} + \left(\frac{\partial P}{\partial z} - \frac{\partial R}{\partial x}\right)\boldsymbol{j} + \left(\frac{\partial Q}{\partial x} - \frac{\partial P}{\partial y}\right)\boldsymbol{k}. \tag{12.58}$$

To compute div $\boldsymbol{F}$ and curl $\boldsymbol{F}$ from these formulas requires a knowledge of the components of $\boldsymbol{F}$. These components, in turn, depend on the choice of coordinate axes in 3-space. A change in the position of the coordinate axes would mean a change in the components of $\boldsymbol{F}$ and, presumably, a corresponding change in the functions div $\boldsymbol{F}$ and curl $\boldsymbol{F}$. With the help of Stokes' theorem and the divergence theorem we can obtain formulas for the divergence and curl that do not involve the components of $\boldsymbol{F}$. These formulas show that the curl and divergence represent intrinsic properties of the vector field $\boldsymbol{F}$ and do not depend on the particular choice of coordinate axes. We discuss first the formula for the divergence.

THEOREM 12.7. *Let $V(t)$ be a solid sphere of radius $t > 0$ with center at a point $\boldsymbol{a}$ in 3-space, and let $S(t)$ denote the boundary of $V(t)$. Let $\boldsymbol{F}$ be a vector field that is continuously differentiable on $V(t)$. Then if $|V(t)|$ denotes the volume of $V(t)$, and if $\boldsymbol{n}$ denotes the unit outer normal to S, we have*

$$\operatorname{div} \boldsymbol{F}(\boldsymbol{a}) = \lim_{t \to 0} \frac{1}{|V(t)|} \iint_{S(t)} \boldsymbol{F} \cdot \boldsymbol{n} \, dS. \tag{12.59}$$

Proof. Let $\varphi = \operatorname{div} \boldsymbol{F}$. If $\epsilon > 0$ is given we must find a $\delta > 0$ such that

$$\left| \varphi(\boldsymbol{a}) - \frac{1}{|V(t)|} \iint_{S(t)} \boldsymbol{F} \cdot \boldsymbol{n} \, dS \right| < \epsilon \qquad \text{whenever} \quad 0 < t < \delta. \tag{12.60}$$

Since φ is continuous at $\boldsymbol{a}$, for the given ϵ there is a 3-ball $B(\boldsymbol{a}; h)$ such that

$$|\varphi(\boldsymbol{x}) - \varphi(\boldsymbol{a})| < \frac{\epsilon}{2} \qquad \text{whenever} \quad \boldsymbol{x} \in B(\boldsymbol{a}; h).$$

Therefore, if we write $\varphi(\boldsymbol{a}) = \varphi(\boldsymbol{x}) + [\varphi(\boldsymbol{a}) - \varphi(\boldsymbol{x})]$ and integrate both sides of this equation over a solid sphere $V(t)$ of radius $t < h$, we find

$$\varphi(\boldsymbol{a})\,|V(t)| = \iiint_{V(t)} \varphi(\boldsymbol{x}) \, dx \, dy \, dz + \iiint_{V(t)} [\varphi(\boldsymbol{a}) - \varphi(x)] \, dx \, dy \, dz.$$

If we apply the divergence theorem to the first triple integral on the right and then transpose this term to the left, we obtain the relation

$$\left| \varphi(\boldsymbol{a}) \, |V(t)| - \iint_{S(t)} \boldsymbol{F} \cdot \boldsymbol{n} \, dS \right| \leq \iiint_{V(t)} |\varphi(\boldsymbol{a}) - \varphi(\boldsymbol{x})| \, dx \, dy \, dz \leq \frac{\epsilon}{2} |V(t)| < \epsilon \, |V(t)| .$$

When we divide this inequality by $|V(t)|$ we see that (12.60) holds with $\delta = h$. This proves the theorem.

In the foregoing proof we made no special use of the fact that $V(t)$ was a sphere. The same theorem holds true if, instead of spheres, we use any set of solids $V(t)$ for which the divergence theorem is valid, provided these solids contain the point $\boldsymbol{a}$ and shrink to $\boldsymbol{a}$ as $t \to 0$. For example, each $V(t)$ could be a cube inscribed in a sphere of radius t about $\boldsymbol{a}$; exactly the same proof would apply.

Theorem 12.7 can be used to give a physical interpretation of the divergence. Suppose $\boldsymbol{F}$ represents the flux density vector of a steady flow. Then the surface integral $\iint_{S(t)} \boldsymbol{F} \cdot \boldsymbol{n} \, dS$ measures the total mass of fluid flowing through S in unit time in the direction of $\boldsymbol{n}$. The quotient $\iint_{S(t)} \boldsymbol{F} \cdot \boldsymbol{n} \, dS / |V(t)|$ represents the mass per unit volume that flows through S in unit time in the direction of $\boldsymbol{n}$. As $t \to 0$, the limit of this quotient is the divergence of $\boldsymbol{F}$ at $\boldsymbol{a}$. Hence the divergence at $\boldsymbol{a}$ can be interpreted as the time rate of change of mass per unit volume per unit time at $\boldsymbol{a}$.

In some books on vector analysis, Equation (12.59) is taken as the *definition* of divergence. This makes it possible to assign a physical meaning to the divergence immediately. Also, formula (12.59) does not involve the components of $\boldsymbol{F}$. Therefore it holds true in any system of coordinates. If we choose for $V(t)$ a cube with its edges parallel to the xyz-coordinate axes and center at $\boldsymbol{a}$, we can use Equation (12.59) to deduce the formula in (12.57) which expresses div $\boldsymbol{F}$ in terms of the components of $\boldsymbol{F}$. This procedure is outlined in Exercise 14 of Section 12.21.

There is a formula analogous to (12.59) that is sometimes used as an alternative definition of the curl. It states that

(12.61)
$$\operatorname{curl} \boldsymbol{F}(\boldsymbol{a}) = \lim_{t \to 0} \frac{1}{|V(t)|} \iint_{S(t)} \boldsymbol{n} \times \boldsymbol{F} \, dS ,$$

where $V(t)$ and $S(t)$ have the same meanings as in Theorem 12.7. The surface integral that appears on the right has a vector-valued integrand. Such integrals can be defined in terms of components. The proof of (12.61) is analogous to that of Theorem 12.7.

There is another formula involving the curl that can be deduced from (12.61) or derived independently. It states that

(12.62)
$$\boldsymbol{n} \cdot \operatorname{curl} \boldsymbol{F}(\boldsymbol{a}) = \lim_{t \to 0} \frac{1}{|S(t)|} \oint_{C(t)} \boldsymbol{F} \cdot d\boldsymbol{\alpha} .$$

In this formula, $S(t)$ is a circular disk of radius t and center at $\boldsymbol{a}$, and $|S(t)|$ denotes its area. The vector $\boldsymbol{n}$ is a unit normal to $S(t)$, and $\boldsymbol{\alpha}$ is the function that traces out $C(t)$ in a direction that appears counterclockwise when viewed from the tip of $\boldsymbol{n}$. The vector field $\boldsymbol{F}$ is assumed to be continuously differentiable on $S(t)$. A proof of (12.62) can be given by the same method we used to prove (12.59). We let $\varphi(\boldsymbol{x}) = \boldsymbol{n} \cdot \operatorname{curl} \boldsymbol{F}(\boldsymbol{x})$ and argue as before, except that we use surface integrals instead of triple integrals and Stokes' theorem instead of the divergence theorem.

If $\boldsymbol{F}$ is a velocity field, the line integral over $C(t)$ is called the circulation of $\boldsymbol{F}$ along $C(t)$; the limit in (12.62) represents the circulation per unit area at the point $\boldsymbol{a}$. Thus, $\boldsymbol{n} \cdot \operatorname{curl} \boldsymbol{F}(\boldsymbol{a})$ can be regarded as a "circulation density" of $\boldsymbol{F}$ at point $\boldsymbol{a}$, with respect to a plane perpendicular to $\boldsymbol{n}$ at $\boldsymbol{a}$.

When $\boldsymbol{n}$ takes the successive values $\boldsymbol{i}$, $\boldsymbol{j}$, and $\boldsymbol{k}$, the dot products $\boldsymbol{i} \cdot \operatorname{curl} \boldsymbol{F}$, $\boldsymbol{j} \cdot \operatorname{curl} \boldsymbol{F}$, and $\boldsymbol{k} \cdot \operatorname{curl} \boldsymbol{F}$ are the components of curl $\boldsymbol{F}$ in rectangular coordinates. When Equation (12.61) is taken as the starting point for the definition of curl, the formula in (12.58) for the rectangular components of curl $\boldsymbol{F}$ can be deduced from (12.62) in exactly this manner.

12.21 Exercises

1. Let S be the surface of the unit cube, $0 \le x \le 1$, $0 \le y \le 1$, $0 \le z \le 1$, and let $\boldsymbol{n}$ be the unit outer normal to S. If $\boldsymbol{F}(x, y, z) = x^2\boldsymbol{i} + y^2\boldsymbol{j} + z^2\boldsymbol{k}$, use the divergence theorem to evaluate the surface integral $\iint_S \boldsymbol{F} \cdot \boldsymbol{n}\, dS$. Verify the result by evaluating the surface integral directly.
2. The sphere $x^2 + y^2 + z^2 = 25$ is intersected by the plane $z = 3$. The smaller portion forms a solid V bounded by a closed surface S_0 made up of two parts, a spherical part S_1 and a planar part S_2. If the unit outer normal of V is $\cos\alpha\,\boldsymbol{i} + \cos\beta\,\boldsymbol{j} + \cos\gamma\,\boldsymbol{k}$, compute the value of the surface integral
$$\iint_S (xz\cos\alpha + yz\cos\beta + \cos\gamma)\, dS$$
if (a) S is the spherical cap S_1, (b) S is the planar base S_2, (c) S is the complete boundary S_0. Solve part (c) by use of the results of parts (a) and (b), and also by use of the divergence theorem.
3. Let $\boldsymbol{n} = \cos\alpha\,\boldsymbol{i} + \cos\beta\,\boldsymbol{j} + \cos\gamma\,\boldsymbol{k}$ be the unit outer normal to a closed surface S which bounds a homogeneous solid V of the type described in the divergence theorem. Assume that the center of mass $(\bar{x}, \bar{y}, \bar{z})$ and the volume $|V|$ of V are known. Evaluate the following surface integrals in terms of $|V|$ and $\bar{x}, \bar{y}, \bar{z}$.

 (a) $\iint_S (x\cos\alpha + y\cos\beta + z\cos\gamma)\, dS$.

 (b) $\iint_S (xz\cos\alpha + 2yz\cos\beta + 3z^2\cos\gamma)\, dS$.

 (c) $\iint_S (y^2\cos\alpha + 2xy\cos\beta - xz\cos\gamma)\, dS$.

 (d) Express $\iint_S (x^2 + y^2)(x\boldsymbol{i} + y\boldsymbol{j}) \cdot \boldsymbol{n}\, dS$ in terms of the volume $|V|$ and a moment of inertia of the solid.

In Exercises 4 through 10, $\partial f/\partial n$ and $\partial g/\partial n$ denote directional derivatives of scalar fields f and g in the direction of the unit outer normal $\boldsymbol{n}$ to a closed surface S which bounds a solid V of the type described in the divergence theorem. That is, $\partial f/\partial n = \nabla f \cdot \boldsymbol{n}$ and $\partial g/\partial n = \nabla g \cdot \boldsymbol{n}$. In each of these exercises prove the given statement. You may assume continuity of all derivatives involved.

4. $\displaystyle\iint_S \frac{\partial f}{\partial n}\,dS = \iiint_V \nabla^2 f\,dx\,dy\,dz$.

5. $\displaystyle\iint_S \frac{\partial f}{\partial n}\,dS = 0$ whenever f is harmonic in V.

6. $\displaystyle\iint_S f\frac{\partial g}{\partial n}\,dS = \iiint_V f\nabla^2 g\,dx\,dy\,dz + \iiint_V \nabla f\cdot\nabla g\,dx\,dy\,dz$.

7. $\displaystyle\iint_S \left(f\frac{\partial g}{\partial n} - g\frac{\partial f}{\partial n}\right) dS = \iiint_V (f\nabla^2 g - g\nabla^2 f)\,dx\,dy\,dz$.

8. $\displaystyle\iint_S f\frac{\partial g}{\partial n}\,dS = \iint_S g\frac{\partial f}{\partial n}\,dS$ if both f and g are harmonic in V.

9. $\displaystyle\iint_S f\frac{\partial f}{\partial n}\,dS = \iiint_V |\nabla f|^2\,dx\,dy\,dz$ if f is harmonic in V.

10. $\displaystyle\nabla^2 f(\boldsymbol{a}) = \lim_{t\to 0}\frac{1}{|V(t)|}\iint_{S(t)} \frac{\partial f}{\partial n}\,dS$, where $V(t)$ is a solid sphere of radius t with center at $\boldsymbol{a}$, $S(t)$ is the surface of $V(t)$, and $|V(t)|$ is the volume of $V(t)$.

11. Let V be a convex region in 3-space whose boundary is a closed surface S and let $\boldsymbol{n}$ be the unit outer normal to S. Let $\boldsymbol{F}$ and $\boldsymbol{G}$ be two continuously differentiable vector fields such that

$$\operatorname{curl}\boldsymbol{F} = \operatorname{curl}\boldsymbol{G} \quad\text{and}\quad \operatorname{div}\boldsymbol{F} = \operatorname{div}\boldsymbol{G} \quad\text{everywhere in } V,$$

and such that

$$\boldsymbol{G}\cdot\boldsymbol{n} = \boldsymbol{F}\cdot\boldsymbol{n} \quad\text{everywhere on } S.$$

Prove that $\boldsymbol{F} = \boldsymbol{G}$ everywhere in V. [*Hint:* Let $\boldsymbol{H} = \boldsymbol{F} - \boldsymbol{G}$, find a scalar field f such that $\boldsymbol{H} = \nabla f$, and use a suitable identity to prove that $\iiint_V \|\nabla f\|^2\,dx\,dy\,dz = 0$. From this deduce that $\boldsymbol{H} = \boldsymbol{O}$ in V.]

12. Given a vector field $\boldsymbol{G}$ and two scalar fields f and g, each continuously differentiable on a convex solid V bounded by a closed surface S. Let $\boldsymbol{n}$ denote the unit outer normal to S. Prove that there is at most one vector field $\boldsymbol{F}$ satisfying the following three conditions:

$$\operatorname{curl}\boldsymbol{F} = \boldsymbol{G} \quad\text{and}\quad \operatorname{div}\boldsymbol{F} = g \quad\text{in } V, \qquad \boldsymbol{F}\cdot\boldsymbol{n} = f \quad\text{on } S.$$

13. Let S be a smooth parametric surface with the property that each line emanating from a point P intersects S once at most. Let $\Omega(S)$ denote the set of lines emanating from P and passing through S. (See Figure 12.19.) The set $\Omega(S)$ is called the *solid angle* with vertex P subtended by S. Let $\Sigma(a)$ denote the intersection of $\Omega(S)$ with the surface of the sphere of radius a centered at P. The quotient

$$\frac{\text{area of }\Sigma(a)}{a^2}$$

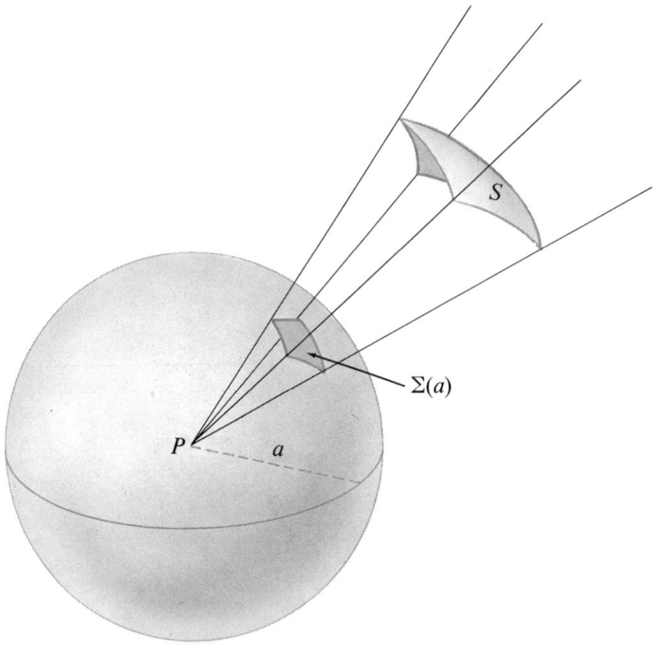

FIGURE 12.19 The solid angle $\Omega(S)$ with vertex P subtended by a surface S. It is measured by the quotient $|\Omega(S)| = \dfrac{\text{area of } \Sigma(a)}{a^2}$.

is denoted by $|\Omega(S)|$ and is used as a measure of the solid angle $\Omega(S)$.
(a) Prove that this quotient is equal to the surface integral

$$\iint_S \frac{\boldsymbol{r} \cdot \boldsymbol{n}}{r^3}\, dS,$$

where $\boldsymbol{r}$ is the radius vector from P to an arbitrary point of S, and $r = \|\boldsymbol{r}\|$. The vector $\boldsymbol{n}$ is the unit normal to S directed away from P. This shows that the quotient for $|\Omega(S)|$ is independent of the radius a. Therefore the solid angle can be measured by the area of the intersection of $\Omega(S)$ and the unit sphere about P. [*Hint:* Apply the divergence theorem to the portion of $\Omega(S)$ lying between S and $\Sigma(a)$.]
(b) Two planes intersect along the diameter of a sphere with center at P. The angle of intersection is θ, where $0 < \theta < \pi$. Let S denote the smaller portion of the surface of the sphere intercepted by the two planes. Show that $|\Omega(S)| = 2\theta$.

14. Let $V(t)$ denote a cube of edge $2t$ and center at $\boldsymbol{a}$, and let $S(t)$ denote the boundary of the cube. Let $\boldsymbol{n}$ be the unit outer normal of $S(t)$ and let $|V(t)|$ denote the volume of the cube. For a given vector field $\boldsymbol{F}$ that is continuously differentiable at $\boldsymbol{a}$, assume that the following limit exists:

$$\lim_{t\to 0} \frac{1}{|V(t)|} \iint_{S(t)} \boldsymbol{F} \cdot \boldsymbol{n}\, dS,$$

and use this limit as the definition of the divergence, div $\boldsymbol{F}(\boldsymbol{a})$. Choose xyz-coordinate axes parallel to the edges of $V(t)$ and let P, Q, and R be the components of $\boldsymbol{F}$ relative to this coordinate system. Prove that div $\boldsymbol{F}(\boldsymbol{a}) = D_1P(\boldsymbol{a}) + D_2Q(\boldsymbol{a}) + D_3R(\boldsymbol{a})$. [*Hint:* Express the surface integral as a sum of six double integrals taken over the faces of the cube. Then show that $1/|V(t)|$ times the sum of the two double integrals over the faces perpendicular to the z-axis approaches the limit $D_3R(\boldsymbol{a})$ as $t \to 0$. Argue in a similar way for the remaining terms.]

15. A scalar field φ which is never zero has the properties

$$\|\nabla\varphi\|^2 = 4\varphi \qquad \text{and} \qquad \operatorname{div}(\varphi\nabla\varphi) = 10\varphi.$$

Evaluate the surface integral

$$\iint_S \frac{\partial\varphi}{\partial n}\, dS,$$

where S is the surface of a unit sphere with center at the origin, and $\partial\varphi/\partial n$ is the directional derivative of φ in the direction of the unit outer normal to S.

PART 3

SPECIAL TOPICS

13

SET FUNCTIONS AND ELEMENTARY PROBABILITY

13.1 Historical introduction

A gambler's dispute in 1654 led to the creation of a mathematical theory of probability by two famous French mathematicians, Blaise Pascal and Pierre de Fermat. Antoine Gombaud, Chevalier de Méré, a French nobleman with an interest in gaming and gambling questions, called Pascal's attention to an apparent contradiction concerning a popular dice game. The game consisted in throwing a pair of dice 24 times; the problem was to decide whether or not to bet even money on the occurrence of at least one "double six" during the 24 throws. A seemingly well-established gambling rule led de Méré to believe that betting on a double six in 24 throws would be profitable, but his own calculations indicated just the opposite.

This problem and others posed by de Méré led to an exchange of letters between Pascal and Fermat in which the fundamental principles of probability theory were formulated for the first time. Although a few special problems on games of chance had been solved by some Italian mathematicians in the 15th and 16th centuries, no general theory was developed before this famous correspondence.

The Dutch scientist Christian Huygens, a teacher of Leibniz, learned of this correspondence and shortly thereafter (in 1657) published the first book on probability; entitled *De Ratiociniis in Ludo Aleae*, it was a treatise on problems associated with gambling. Because of the inherent appeal of games of chance, probability theory soon became popular, and the subject developed rapidly during the 18th century. The major contributors during this period were Jakob Bernoulli† (1654–1705) and Abraham de Moivre (1667–1754).

In 1812 Pierre de Laplace (1749–1827) introduced a host of new ideas and mathematical techniques in his book, *Théorie Analytique des Probabilités*. Before Laplace, probability theory was solely concerned with developing a mathematical analysis of games of chance. Laplace applied probabilistic ideas to many scientific and practical problems. The *theory of errors*, *actuarial mathematics*, and *statistical mechanics* are examples of some of the important applications of probability theory developed in the 19th century.

Like so many other branches of mathematics, the development of probability theory has been stimulated by the variety of its applications. Conversely, each advance in the theory has enlarged the scope of its influence. Mathematical statistics is one important branch of applied probability; other applications occur in such widely different fields as

† Sometimes referred to as James Bernoulli.

genetics, psychology, economics, and engineering. Many workers have contributed to the theory since Laplace's time; among the most important are Chebyshev, Markov, von Mises, and Kolmogorov.

One of the difficulties in developing a mathematical theory of probability has been to arrive at a definition of probability that is precise enough for use in mathematics, yet comprehensive enough to be applicable to a wide range of phenomena. The search for a widely acceptable definition took nearly three centuries and was marked by much controversy. The matter was finally resolved in the 20th century by treating probability theory on an axiomatic basis. In 1933 a monograph by a Russian mathematician A. Kolmogorov outlined an axiomatic approach that forms the basis for the modern theory. (Kolmogorov's monograph is available in English translation as *Foundations of Probability Theory*, Chelsea, New York, 1950.) Since then the ideas have been refined somewhat and probability theory is now part of a more general discipline known as measure theory.

This chapter presents the basic notions of modern elementary probability theory along with some of its connections to measure theory. Applications are also given, primarily to games of chance such as coin tossing, dice, and card games. This introductory account is intended to demonstrate the logical structure of the subject as a deductive science and to give the reader a feeling for probabilistic thinking.

13.2 Finitely additive set functions

The area of a region, the length of a curve, or the mass of a system of particles is a number which measures the size or content of a set. All these measures have certain properties in common. Stated abstractly, they lead to a general concept called a *finitely additive set function.* Later we shall introduce probability as another example of such a function. To prepare the way, we discuss first some properties common to all these functions.

A function $f: \mathscr{A} \to \mathbf{R}$ whose domain is a collection $\mathscr{A}$ of sets and whose function values are real numbers, is called a *set function.* If A is a set in class $\mathscr{A}$, the value of the function at A is denoted by $f(A)$.

DEFINITION OF A FINITELY ADDITIVE SET FUNCTION. *A set function $f: \mathscr{A} \to \mathbf{R}$ is said to be finitely additive if*

$$f(A \cup B) = f(A) + f(B) \tag{13.1}$$

whenever A and B are disjoint sets in $\mathscr{A}$ such that $A \cup B$ is also in $\mathscr{A}$.

Area, length, and mass are all examples of finitely additive set functions. This section discusses further consequences of Equation (13.1).

In the usual applications, the sets in $\mathscr{A}$ are subsets of a given set S, called a universal set. It is often necessary to perform the operations of union, intersection, and complementation on sets in $\mathscr{A}$. To make certain that $\mathscr{A}$ is closed under these operations we restrict $\mathscr{A}$ to be a ***Boolean algebra***, which is defined as follows.

DEFINITION OF A BOOLEAN ALGEBRA OF SETS. *A nonempty class $\mathscr{A}$ of subsets of a given universal set S is called a Boolean algebra if for every A and B in $\mathscr{A}$ we have*

$$A \cup B \in \mathscr{A} \qquad \textit{and} \qquad A' \in \mathscr{A}.$$

Here $A' = S - A$, the complement of A relative to S.

A Boolean algebra $\mathscr{A}$ is also closed under intersections and differences, since we have

$$A \cap B = (A' \cup B')' \qquad \text{and} \qquad A - B = A \cap B'.$$

This implies that the empty set $\varnothing$ belongs to $\mathscr{A}$ since $\varnothing = A - A$ for some A in $\mathscr{A}$. Also, the universal set S is in $\mathscr{A}$ since $S = \varnothing'$.

Many Boolean algebras can be formed from the subsets of a given universal set S. The smallest such algebra is the class $\mathscr{A}_0 = \{\varnothing, S\}$, which consists of only two special subsets: $\varnothing$ and S. At the other extreme is the class $\mathscr{A}_1$, which consists of *all* subsets of S. Every Boolean algebra $\mathscr{A}$ consisting of subsets of S satisfies the inclusion relations $\mathscr{A}_0 \subseteq \mathscr{A} \subseteq \mathscr{A}_1$.

The property of finite additivity of set functions in Equation (13.1) requires A and B to be disjoint sets. The next theorem drops this requirement.

THEOREM 13.1. *Let $f: \mathscr{A} \to \mathbf{R}$ be a finitely additive set function defined on a Boolean algebra $\mathscr{A}$ of sets. Then for all A and B in $\mathscr{A}$ we have*

$$f(A \cup B) = f(A) + f(B - A), \tag{13.2}$$

and

$$f(A \cup B) = f(A) + f(B) - f(A \cap B). \tag{13.3}$$

Proof. The sets A and $B - A$ are disjoint and their union is $A \cup B$. Hence, by applying (13.1) to A and $B - A$ we obtain (13.2).

To prove (13.3) we first note that $A \cap B'$ and B are disjoint sets whose union is $A \cup B$. Hence by (13.1) we have

$$f(A \cup B) = f(A \cap B') + f(B). \tag{13.4}$$

Also, $A \cap B'$ and $A \cap B$ are disjoint sets whose union is A, so (13.1) gives us

$$f(A) = f(A \cap B') + f(A \cap B). \tag{13.5}$$

Subtracting (13.5) from (13.4) we obtain (13.3).

13.3 Finitely additive measures

The set functions which represent area, length, and mass have further properties in common. For example, they are all *nonnegative* set functions. That is,

$$f(A) \geq 0$$

for each A in the class $\mathscr{A}$ under consideration.

DEFINITION OF A FINITELY ADDITIVE MEASURE. *A nonnegative set function $f: \mathscr{A} \to \mathbf{R}$ that is finitely additive is called a finitely additive measure, or simply a measure.*

Using Theorem 13.1 we immediately obtain the following further properties of measures.

THEOREM 13.2. *Let $f: \mathscr{A} \to \mathbf{R}$ be a finitely additive measure defined on a Boolean algebra $\mathscr{A}$. Then for all sets A and B in $\mathscr{A}$ we have*

(a) $f(A \cup B) \le f(A) + f(B)$.

(b) $f(B - A) = f(B) - f(A)$ *if* $A \subseteq B$.

(c) $f(A) \le f(B)$ *if* $A \subseteq B$. *(Monotone property)*

(d) $f(\varnothing) = 0$.

Proof. Part (a) follows from (13.3), and part (b) follows from (13.2). Part (c) follows from (b), and part (d) follows by taking $A = B = \varnothing$ in (b).

EXAMPLE. *The number of elements in a finite set.* Let $S = \{a_1, a_2, \ldots, a_n\}$ be a set consisting of n (distinct) elements, and let $\mathscr{A}$ denote the class of all subsets of S. For each A in $\mathscr{A}$, let $\nu(A)$ denote the number of distinct elements in A (ν is the Greek letter "nu"). It is easy to verify that this function is finitely additive on $\mathscr{A}$. In fact, if A has k elements and if B has m elements, then $\nu(A) = k$ and $\nu(B) = m$. If A and B are disjoint it is clear that the union $A \cup B$ is a subset of S with $k + m$ elements, so

$$\nu(A \cup B) = k + m = \nu(A) + \nu(B).$$

This particular set function is nonnegative, so ν is a measure.

13.4 Exercises

1. Let $\mathscr{A}$ denote the class of all subsets of a given universal set and let A and B be arbitrary sets in $\mathscr{A}$. Prove that:
   (a) $A \cap B'$ and B are disjoint.
   (b) $A \cup B = (A \cap B') \cup B$. (This formula expresses $A \cup B$ as the union of disjoint sets.)
   (c) $A \cap B$ and $A \cap B'$ are disjoint.
   (d) $(A \cap B) \cup (A \cap B') = A$. (This formula expresses A as a union of two disjoint sets.)
2. Exercise 1(b) shows how to express the union of two sets as the union of two *disjoint* sets. Express in a similar way the union of three sets $A_1 \cup A_2 \cup A_3$ and, more generally, of n sets $A_1 \cup A_2 \cup \cdots \cup A_n$. Illustrate with a diagram when $n = 3$.
3. A study of a set S consisting of 1000 college graduates ten years after graduation revealed that the "successes" formed a subset A of 400 members, the Caltech graduates formed a subset B of 300 members, and the intersection $A \cap B$ consisted of 200 members.
   (a) For each of the following properties, use set notation to describe, in terms of unions and intersections of A, B, and their complements A' and B' relative to S, the subsets consisting of those persons in S that have the property:
      (i) Neither a "success" nor a Caltech graduate.
      (i) A "success" but not a Caltech graduate.
      (iii) A "success" or a Caltech graduate, or both.

(iv) Either a "success" or a Caltech graduate, but not both.
(v) Belongs to not more than one of A or B.
(b) Determine the exact number of individuals in each of the five subsets of part (a).

4. Let f be a finitely additive set function defined on a class $\mathscr{A}$ of sets. Let $A_1, \ldots, A_n$ be n sets in $\mathscr{A}$ such that $A_i \cap A_j = \varnothing$ if $i \neq j$. (Such a collection is called a *disjoint collection* of sets.) If the union $\bigcup_{k=1}^{m} A_k$ is in class $\mathscr{A}$ for all $m \leq n$, use induction to prove that

$$f\left(\bigcup_{k=1}^{n} A_k\right) = \sum_{k=1}^{n} f(A_k).$$

In Exercises 5, 6, and 7, S denotes a finite set consisting of n distinct elements, say $S = \{a_1, a_2, \ldots, a_n\}$.

5. Let $A_1 = \{a_1\}$, the subset consisting of a_1 alone. Show that the class $\mathscr{B}_1 = \{\varnothing, A_1, A_1', S\}$ is the smallest Boolean algebra containing A_1.
6. Let $A_1 = \{a_1\}, A_2 = \{a_2\}$. Describe, in a manner similar to that used in Exercise 5, the smallest Boolean algebra $\mathscr{B}_2$ containing both A_1 and A_2.
7. Do the same as in Exercise 6 for the subsets $A_1 = \{a_1\}$, $A_2 = \{a_2\}$, and $A_3 = \{a_3\}$.
8. If $\mathscr{B}_k$ denotes the smallest Boolean algebra which contains the k subsets $A_1 = \{a_1\}$, $A_2 = \{a_2\}, \ldots, A_k = \{a_k\}$, show that $\mathscr{B}_k$ contains 2^{k+1} subsets of S if $k < n$ and 2^n subsets if $k = n$.
9. Let f be a finitely additive set function defined on the Boolean algebra of all subsets of a given universal set S. Suppose it is known that

$$f(A \cap B) = f(A)f(B)$$

for two particular subsets A and B of S. If $f(S) = 2$, prove that

$$f(A \cup B) = f(A') + f(B') - f(A')f(B').$$

10. If A and B are sets, their *symmetric difference* $A \triangle B$ is the set defined by the equation $A \triangle B = (A - B) \cup (B - A)$. Prove each of the following properties of the symmetric difference.
(a) $A \triangle B = B \triangle A$.
(b) $A \triangle A = \varnothing$.
(c) $A \triangle B \subseteq (A \triangle C) \cup (C \triangle B)$.
(d) $A \triangle B$ is disjoint from each of A and B.
(e) $(A \triangle B) \triangle C = A \triangle (B \triangle C)$.
(f) If f is a finitely additive set function defined on the Boolean algebra $\mathscr{A}$ of all subsets of a given set S, then for all A and B in $\mathscr{A}$ we have $f(A \triangle B) = f(A) + f(B) - 2f(A \cap B)$.

13.5 The definition of probability for finite sample spaces

In the language of set functions, probability is a specific kind of measure (to be denoted here by P) defined on a specific Boolean algebra $\mathscr{B}$ of sets. The elements of $\mathscr{B}$ are subsets of a universal set S. In probability theory the universal set S is called a *sample space*. We discuss the definition of probability first for finite sample spaces and later for infinite sample spaces.

DEFINITION OF PROBABILITY FOR FINITE SAMPLE SPACES. *Let $\mathscr{B}$ denote a Boolean algebra whose elements are subsets of a given finite set S. A set function P defined on $\mathscr{B}$ is called a* *probability measure if it satisfies the following three conditions:*

(a) *P is finitely additive.*

(b) *P is nonnegative.*

(c) $P(S) = 1$.

In other words, for finite sample spaces probability is simply a measure which assigns the value 1 to the whole space.

It is important to realize that a complete description of a probability measure requires three things to be specified: the sample space S, the Boolean algebra $\mathscr{B}$ formed from certain subsets of S, and the set function P. The triple $(S, \mathscr{B}, P)$ is often called a *probability space.* In most of the elementary applications the Boolean algebra $\mathscr{B}$ is taken to be the collection of *all* subsets of S.

EXAMPLE. An illustration of applied probability theory is found in the experiment of tossing a coin once. For a sample space S we take the set of all conceivable outcomes of the experiment. In this case, each outcome is either "heads" or "tails," which we label by the symbols h and t. Thus, the sample space S is $\{h, t\}$, the set consisting of h and t. For the Boolean algebra we take the collection of all subsets of S; there are four, $\varnothing$, S, H, and T, where $H = \{h\}$ and $T = \{t\}$. Next, we assign probabilities to each of these subsets. For the subsets $\varnothing$ and S we have no choice in the assignment of values. Property (c) requires that $P(S) = 1$, and, since P is a nonnegative measure, $P(\varnothing) = 0$. However, there is some freedom in assigning probabilities to the other two subsets, H and T. Since H and T are disjoint sets whose union is S, the additive property requires that

$$P(H) + P(T) = P(S) = 1 .$$

We are free to assign any nonnegative values whatever to $P(H)$ and $P(T)$ so long as their sum is 1. If we feel that the coin is unbiased so that there is no *a priori* reason to prefer heads or tails, it seems natural to assign the values

$$P(H) = P(T) = \tfrac{1}{2} .$$

If, however, the coin is "loaded," we may wish to assign different values to these two probabilities. For example, the values $P(H) = \frac{1}{3}$ and $P(T) = \frac{2}{3}$ are just as acceptable as $P(H) = P(T) = \frac{1}{2}$. In fact, for any real p in the interval $0 \le p \le 1$ we may define $P(H) = p$ and $P(T) = 1 - p$, and the resulting function P will satisfy all the conditions for a probability measure.

For a given coin, there is no mathematical way to determine what the probability p "really" is. If we choose $p = \frac{1}{2}$ we can deduce logical consequences on the assumption that the coin is fair or unbiased. The theory for unbiased coins can then be used to test the fairness of an actual coin by performing a large number of experiments with the coin and comparing the results with the predictions based on the theory. The testing of agreement between theory and empirical evidence belongs to that branch of applied probability known as *statistical inference*, and will not be discussed in this book.

The foregoing example is a typical application of the concepts of probability theory. Probability questions often arise in situations referred to as "experiments." We shall not attempt to define an experiment; instead, we shall merely mention some familiar examples: tossing one or more coins, rolling a pair of dice, dealing a bridge hand, drawing a ball from an urn, counting the number of female students at the California Institute of Technology, selecting a number from a telephone directory, recording the radiation count of a Geiger counter.

To discuss probability questions that arise in connection with such experiments, our first task is to construct a sample space S that can be used to represent all conceivable outcomes of the experiment, as we did for coin tossing. Each element of S should represent an outcome of the experiment and each outcome should correspond to one and only one element of S. Next, we choose a Boolean algebra $\mathscr{B}$ of subsets of S (usually *all* subsets of S) and then define a probability measure P on $\mathscr{B}$. The choice of the set S, the choice of $\mathscr{B}$, and the choice of P will depend on the information known about the details of the experiment and on the questions we wish to answer. The purpose of probability theory is not to discuss whether the probability space $(S, \mathscr{B}, P)$ has been properly chosen. This motivation belongs to the science or gambling game from which the experiment emanates, and only experience can suggest whether or not the choices were well made. *Probability theory is the study of logical consequences that can be derived once the probability space is given.* Making a good choice of the probability space is, strictly speaking, not probability theory — it is not even mathematics; instead, it is part of the art of applying probability theory to the real world. We shall elaborate further on these remarks as we deal with specific examples in the later sections.

If $S = \{a_1, a_2, \ldots, a_n\}$, and if $\mathscr{B}$ consists of all subsets of S, the probability function P is completely determined if we know its values on the one-element subsets, or *singletons*:

$$P(\{a_1\}),\ P(\{a_2\}),\ \ldots,\ P(\{a_n\}).$$

In fact, every subset A of S is a disjoint union of singletons, and $P(A)$ is determined by the additive property. For example, when

$$A = \{a_1\} \cup \{a_2\} \cup \cdots \cup \{a_k\},$$

the additive property requires that

$$P(A) = \sum_{i=1}^{k} P(\{a_i\}).$$

To simplify the notation and the terminology, we write $P(a_i)$ instead of $P(\{a_i\})$. This number is also called the *probability of the point* a_i. Therefore, the assignment of the point probabilities $P(x)$ for each element x in a finite set S amounts to a complete description of the probability function P.

13.6 Special terminology peculiar to probability theory

In discussions involving probability, one often sees phrases from everyday language such as "two events are equally likely," "an event is impossible," or "an event is certain to occur." Expressions of this sort have intuitive appeal and it is both pleasant and helpful to be able to employ such colorful language in mathematical discussions. Before we can do so, however, it is necessary to explain the meaning of this language in terms of the fundamental concepts of our theory.

Because of the way probability is used in practice, it is convenient to imagine that each probability space $(S, \mathscr{B}, P)$ is associated with a real or conceptual experiment. The universal set S can then be thought of as the collection of all conceivable outcomes of the experiment, as in the example of coin tossing discussed in the foregoing section. Each element of S is called an *outcome* or a *sample* and the subsets of S that occur in the Boolean algebra $\mathscr{B}$ are called *events*. The reasons for this terminology will become more apparent when we treat some examples.

Assume we have a probability space $(S, \mathscr{B}, P)$ associated with an experiment. Let A be an event, and suppose the experiment is performed and that its outcome is x. (In other words, let x be a point of S.) This outcome x may or may not belong to the set A. If it does, we say that *the event A has occurred.* Otherwise, we say that *the event A has not occurred*, in which case $x \in A'$, so the complementary event A' has occurred. An event A is called *impossible* if $A = \varnothing$, because in this case no outcome of the experiment can be an element of A. The event A is said to be *certain* if $A = S$, because then every outcome is automatically an element of A.

Each event A has a probability $P(A)$ assigned to it by the probability function P. [The actual value of $P(A)$ or the manner in which $P(A)$ is assigned does not concern us at present.] The number $P(A)$ is also called *the probability that an outcome of the experiment is one of the elements of A.* We also say that $P(A)$ is *the probability that the event A occurs* when the experiment is performed.

The impossible event $\varnothing$ must be assigned probability zero because P is a finitely additive measure. However, there may be events with probability zero that are not impossible. In other words, some of the nonempty subsets of S may be assigned probability zero. The certain event S must be assigned probability 1 by the very definition of probability, but there may be other subsets as well that are assigned probability 1. In Example 1 of Section 13.8 there are nonempty subsets with probability zero and proper subsets of S that have probability 1.

Two events A and B are said to be *equally likely* if $P(A) = P(B)$. The event A is called *more likely* than B if $P(A) > P(B)$, and *at least as likely* as B if $P(A) \geq P(B)$. Table 13.1 provides a glossary of further everyday language that is often used in probability discussions. The letters A and B represent events, and x represents an outcome of an experiment associated with the sample space S. Each entry in the left-hand column is a statement about the events A and B, and the corresponding entry in the right-hand column defines the statement in terms of set theory.

TABLE 13.1. Glossary of Probability Terms

Statement	*Meaning in set theory*
At least one of A or B occurs	$x \in A \cup B$
Both events A and B occur	$x \in A \cap B$
Neither A nor B occurs	$x \in A' \cap B'$
A occurs and B does not occur	$x \in A \cap B'$
Exactly one of A or B occurs	$x \in (A \cap B') \cup (A' \cap B)$
Not more than one of A or B occurs	$x \in (A \cap B)'$
If A occurs, so does B (A implies B)	$A \subseteq B$
A and B are mutually exclusive	$A \cap B = \varnothing$
Event A or event B	$A \cup B$
Event A and event B	$A \cap B$

13.7 Exercises

Let S be a given sample space and let A, B, and C denote arbitrary events (that is, subsets of S in the corresponding Boolean algebra $\mathscr{B}$). Each of the statements in Exercises 1 through 12 is described verbally in terms of A, B, C. Express these statements in terms of unions and intersections of A, B, C and their complements.

1. If A occurs, then B does not occur.
2. None of the events A, B, C occurs.
3. Only A occurs.
4. At least one of A, B, C occurs.
5. Exactly one of A, B, C occurs.
6. Not more than one occurs.
7. At least two of A, B, C occur.
8. Exactly two occur.
9. Not more than two occur.
10. A and C occur but not B.
11. All three events occur.
12. Not more than three occur.

13. Let A denote the event of throwing an odd total with two dice, and let B denote the event of throwing at least one 6. Give a verbal description of each of the following events:

(a) $A \cup B$,
(b) $A \cap B$,
(c) $A \cap B'$,
(d) $A' \cap B$,
(e) $A' \cap B'$,
(f) $A' \cup B$.

14. Let A and B denote events. Show that

$$P(A \cap B) \le P(A) \le P(A \cup B) \le P(A) + P(B).$$

15. Let A and B denote events and let $a = P(A)$, $b = P(B)$, $c = P(A \cap B)$. Compute, in terms of a, b, and c, the probabilities of the following events:

(a) A',
(b) B',
(c) $A \cup B$,
(d) $A' \cup B'$,
(e) $A' \cup B$,
(f) $A \cap B'$.

16. Given three events A, B, C. Prove that

$$P(A \cup B \cup C) = P(A) + P(B) + P(C) - P(A \cap B) - P(A \cap C) - P(B \cap C) + P(A \cap B \cap C).$$

13.8 Worked examples

We shall illustrate how some of the concepts of the foregoing sections may be used to answer specific questions involving probabilities.

EXAMPLE 1. What is the probability that at least one "head" will occur in two throws of a coin?

First Solution. The experiment in this case consists of tossing a coin twice; the set S of all possible outcomes may be denoted as follows:

$$S = \{hh, ht, th, tt\}.$$

If we feel that these outcomes are equally likely, we assign the point probabilities $P(x) = \frac{1}{4}$ for each x in S. The event "at least one head occurs" may be described by the subset

$$A = \{hh, ht, th\}.$$

The probability of this event is the sum of the point probabilities of its elements. Hence, $P(A) = \frac{1}{4} + \frac{1}{4} + \frac{1}{4} = \frac{3}{4}$.

Second Solution. Suppose we use the same sample space but assign the point probabilities as follows:†

$$P(hh) = 1, \qquad P(ht) = P(th) = P(tt) = 0.$$

Then the probability of the event "at least one head occurs" is

$$P(hh) + P(ht) + P(th) = 1 + 0 + 0 = 1.$$

The fact that we arrived at a different answer from that in the first solution should not alarm the reader. We began with a different set of premises. Psychological considerations might lead us to believe that the assignment of probabilities in the first solution is the more natural one. Indeed, most people would agree that this is so if the coin is unbiased. However, if the coin happens to be loaded so that heads always turns up, the assignment of probabilities in the second solution is more natural.

The foregoing example shows that we cannot expect a unique answer to the question asked. To answer such a question properly we must specify the choice of sample space and the assignment of point probabilities. Once the sample space and the point probabilities are known only one probability for a given event can be logically deduced. Different choices of the sample space or point probabilities may lead to different "correct" answers to the same question.

Sometimes the assignment of probabilities to the individual outcomes of an experiment is dictated by the language used to describe the experiment. For example, when an object is chosen "at random" from a finite set of n elements, this is intended to mean that each outcome is equally likely and should be assigned point probability $1/n$. Similarly, when we toss a coin or roll a die, if we have no *a priori* reason to feel that the coin or die is loaded, we assume that all outcomes are equally likely. This agreement will be adopted in all the exercises of this chapter.

EXAMPLE 2. If one card is drawn at random from each of two decks, what is the probability that at least one is the ace of hearts?

Solution. The experiment consists in drawing two cards, a and b, one from each deck. Suppose we denote a typical outcome as an ordered pair (a, b). The number of possible outcomes, that is, the total number of distinct pairs (a, b) in the sample space S is 52^2. We assign the probability $1/52^2$ to each such pair. The event in which we are interested is the set A of pairs (a, b), where either a or b is the ace of hearts. There are $52 + 51$ elements in A. Hence, under these assumptions we deduce that

$$P(A) = \frac{52 + 51}{52^2} = \frac{1}{26} - \frac{1}{52^2}.$$

† Note that for this assignment of probabilities there are nonempty subsets of S with probability zero and proper subsets with probability 1.

EXAMPLE 3. If two cards are drawn at random from one deck, what is the probability that one of them is the ace of hearts?

Solution. As in Example 2 we use ordered pairs (a, b) as elements of the sample space. In this case the sample space has $52 \cdot 51$ elements and the event A under consideration has $51 + 51$ elements. If we assign the point probabilities $1/(52 \cdot 51)$ to each outcome, we obtain

$$P(A) = \frac{2 \cdot 51}{52 \cdot 51} = \frac{1}{26}.$$

EXAMPLE 4. What is the probability of throwing 6 or less with three dice?

Solution. We denote each outcome of the experiment as a triple of integers (a, b, c) where a, b, and c may take any values from 1 to 6. Therefore the sample space consists of 6^3 elements and we assign the probability $1/6^3$ to each outcome. The event A in question is the set of all triples satisfying the inequality $3 \le a + b + c \le 6$. If A_n denotes the set of (a, b, c) for which $a + b + c = n$, we have

$$A = A_3 \cup A_4 \cup A_5 \cup A_6.$$

Direct enumeration shows that the sets A_n, with $n = 3, 4, 5$, and 6 contain 1, 3, 6, and 10 elements, respectively. For example, the set A_6 is given by

$$A_6 = \{(1, 2, 3), (1, 3, 2), (1, 1, 4), (1, 4, 1), (2, 1, 3), (2, 3, 1), (2, 2, 2), (3, 1, 2), (3, 2, 1), (4, 1, 1)\}.$$

Therefore A has 20 elements and

$$P(A) = \frac{20}{6^3} = \frac{5}{54}.$$

EXAMPLE 5. A die is thrown once. What is the probability that the number of points is either even or a multiple of 3?

Solution. We choose the sample space $S = \{1, 2, 3, 4, 5, 6\}$, consisting of six elements, to each of which we assign the probability $\frac{1}{6}$. The event "even" is the set $A = \{2, 4, 6\}$, the event "a multiple of 3" is $B = \{3, 6\}$. We are interested in their union, which is the set $A \cup B = \{2, 3, 4, 6\}$. Since this set contains four elements we have $P(A \cup B) = 4/6$.

This example can be solved in another way, using the formula

$$P(A \cup B) = P(A) + P(B) - P(A \cap B) = \tfrac{3}{6} + \tfrac{2}{6} - \tfrac{1}{6}.$$

13.9 Exercises

1. Let S be a finite sample space consisting of n elements. Suppose we assign equal probabilities to each of the points in S. Let A be a subset of S consisting of k elements. Prove that $P(A) = k/n$.

For each of Exercises 2 through 8, describe your choice of sample space and state how you are assigning the point probabilities. In the questions associated with card games, assume all cards have the same probability of being dealt.

2. Five counterfeit coins are mixed with nine authentic coins.
   (a) A coin is selected at random. Compute the probability that a counterfeit coin is selected.
   If two coins are selected, compute the probability that:
   (b) one is good and one is counterfeit.
   (c) both are counterfeit.
   (d) both are good.
3. Compute the probabilities of each of the events described in Exercise 13 of Section 13.7. Assign equal probabilities to each of the 36 elements of the sample space.
4. What is the probability of throwing at least one of 7, 11, or 12 with two dice?
5. A poker hand contains four hearts and one spade. The spade is discarded and one card is drawn from the remainder of the deck. Compute the probability of filling the flush—that is, of drawing a fifth heart.
6. In poker, a straight is a five-card sequence, not necessarily all of the same suit. If a poker hand contains four cards in sequence (but not A234 or JQKA) and one extra card not in the sequence, compute the probability of filling the straight. (The extra card is discarded and a new card is drawn from the remainder of the deck.)
7. A poker hand has four cards out of a five-card sequence with a gap in the middle (such as 5689), and one extra card not in the sequence. The extra card is discarded and a new one is drawn from the remainder of the deck. Compute the probability of filling the "inside straight."
8. An urn contains A white stones and B black stones. A second urn contains C white stones and D black stones. One stone is drawn at random from the first urn and transferred to the second urn. Then a stone is drawn at random from the second urn. Calculate the probability of each of the following events.
   (a) The first stone is white.
   (b) The first stone is black.
   (c) The second stone is white, given that the transferred stone was white.
   (d) The second stone is white, given that the transferred stone was black.
9. Two stones are drawn with replacement from an urn containing four red stones and two white stones. Calculate the probability of each of the following events.
   (a) Both stones are white.
   (b) Both stones are red.
   (c) Both stones are the same color.
   (d) At least one stone is red.
10. Let P_n denote the probability that exactly n of the events A and B will occur, where n takes the values 0, 1, 2. Express each of P_0, P_1, P_2 in terms of $P(A)$, $P(B)$, and $P(A \cap B)$.

Odds. Some gambling games are described in terms of "odds" rather than in terms of probabilities. For example, if we roll a fair die, the probability of the event "rolling a three" is $\frac{1}{6}$. Since there are six possible outcomes, one of which is favorable to the event "rolling a three" and five of which are unfavorable, this is often described by saying that the odds in favor of the event are 1 to 5, or the odds against it are 5 to 1. In this case the odds are related to the probability by the equation

$$\frac{1}{6} = \frac{1}{1 + 5}.$$

In general, if A is an event with probability $P(A)$ and if a and b are two real numbers such that

$$P(A) = \frac{a}{a + b}, \tag{13.6}$$

we say the *odds in favor of A* are a to b, or the *odds against A* are b to a. Since $1 - a/(a + b) = b/(a + b)$, the odds against A are the same as the odds in favor of the complementary event A'. The following exercises are devoted to further properties of odds and their relation to probabilities.

11. If $P(A) = 1$, show that (13.6) can be satisfied only when $b = 0$ and $a \neq 0$. If $P(A) \neq 1$, show that there are infinitely many choices of a and b satisfying (13.6) but that all have the same ratio a/b.
12. Compute the odds in favor of each of the events described in Exercise 2.
13. Given two events A and B. If the odds against A are 2 to 1 and those in favor of $A \cup B$ are 3 to 1, show that

$$\tfrac{5}{12} \leq P(B) \leq \tfrac{3}{4}.$$

Give an example in which $P(B) = \tfrac{5}{12}$ and one in which $P(B) = \tfrac{3}{4}$.

13.10 Some basic principles of combinatorial analysis

Many problems in probability theory and in other branches of mathematics can be reduced to problems on counting the number of elements in a finite set. Systematic methods for studying such problems form part of a mathematical discipline known as *combinatorial analysis*. In this section we digress briefly to discuss some basic ideas in combinatorial analysis that are useful in analyzing some of the more complicated problems of probability theory.

If all the elements of a finite set are displayed before us, there is usually no difficulty in counting their total number. More often than not, however, a set is described in a way that makes it impossible or undesirable to display all its elements. For example, we might ask for the total number of distinct bridge hands that can be dealt. Each player is dealt 13 cards from a 52-card deck. The number of possible distinct hands is the same as the number of different subsets of 13 elements that can be formed from a set of 52 elements. Since this number exceeds 635 billion, a direct enumeration of all the possibilities is clearly not the best way to attack this problem; however, it can readily be solved by combinatorial analysis.

This problem is a special case of the more general problem of counting the number of distinct subsets of k elements that may be formed from a set of n elements,† where $n \geq k$. Let us denote this number by $f(n, k)$. It has long been known that

$$f(n, k) = \binom{n}{k}, \tag{13.7}$$

where, as usual $\binom{n}{k}$ denotes the binomial coefficient,

$$\binom{n}{k} = \frac{n!}{k!\,(n-k)!}.$$

In the problem of bridge hands we have $f(52, 13) = \binom{52}{13} = 635{,}013{,}559{,}600$ different hands that a player can be dealt.

† When we say that a set has n elements, we mean that it has n *distinct* elements. Such a set is sometimes called an n-element set.

There are many methods known for proving (13.7). A straightforward approach is to form each subset of k elements by choosing the elements one at a time. There are n possibilities for the first choice, $n - 1$ possibilities for the second choice, and $n - (k - 1)$ possibilities for the kth choice. If we make all possible choices in this manner we obtain a total of

$$n(n-1)\cdots(n-k+1) = \frac{n!}{(n-k)!}$$

subsets of k elements. Of course, these subsets are not all distinct. For example, if $k = 3$ the six subsets

$$\{a, b, c\}, \{b, c, a\}, \{c, a, b\}, \{a, c, b\}, \{c, b, a\}, \{b, a, c\}$$

are all equal. In general, this method of enumeration counts each k-element subset exactly $k!$ times.† Therefore we must divide the number $n!/(n - k)!$ by $k!$ to obtain $f(n, k)$. This gives us $f(n, k) = \binom{n}{k}$, as asserted.

This line of argument is more or less typical of the combinatorial analysis required in the later sections. Hence it seems worthwhile to digress briefly to discuss the fundamental principles on which this analysis is based.

We often wish to count the number of elements in the Cartesian product of n finite sets $A_1, \ldots, A_n$. The Cartesian product is denoted by the symbol $A_1 \times \cdots \times A_n$ and is defined by the equation

$$A_1 \times \cdots \times A_n = \{(a_1, \ldots, a_n) \mid a_1 \in A_1, \ldots, a_n \in A_n\}.$$

That is, the Cartesian product consists of the set of all ordered n-tuples $(a_1, \ldots, a_n)$ where the kth component of the n-tuple comes from the kth set A_k.

An example with $n = 2$ is shown in Figure 13.1. Here $A_1 = \{1, 2, 4, 5\}$ and $A_2 = \{1, 3\}$. There are 4 elements in A_1, and 2 elements in A_2, giving a total of 8 elements in the Cartesian product $A_1 \times A_2$. More generally, if A_1 consists of k_1 elements and A_2 consists of k_2 elements, then $A_1 \times A_2$ consists of k_1k_2 elements. By induction on n, it follows that if A_r consists of k_r elements, then the Cartesian product $A_1 \times \cdots \times A_n$ consists of $k_1 \cdots k_n$ elements.

To express this result in the language of set functions, let $\mathscr{F}$ denote the class of all finite sets and let ν be the set function defined on $\mathscr{F}$ as follows: If $A \in \mathscr{F}$, $\nu(A)$ is the number of distinct elements in A. (For the empty set we define $\nu(\varnothing) = 0$.) Then it is easy to verify that ν is a finitely additive set function, so we may write

$$\nu\left(\bigcup_{i=1}^{n} S_i\right) = \sum_{i=1}^{n} \nu(S_i) \tag{13.8}$$

if $\{S_1, S_2, \ldots, S_n\}$ is a disjoint collection of finite sets (that is, if $S_i \cap S_j = \varnothing$ whenever $i \neq j$). The number of elements in a Cartesian product may be expressed in terms of ν as follows:

$$\nu(A_1 \times A_2 \times \cdots \times A_n) = \nu(A_1)\nu(A_2)\cdots\nu(A_n).$$

† The reason for this will become clear in Example 3 on p. 484, where we give a more detailed derivation of (13.7).

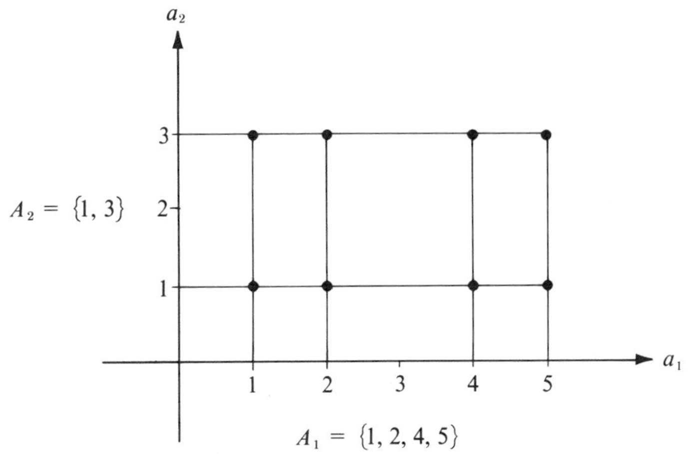

FIGURE 13.1 An example illustrating the Cartesian of two sets. The plotted points represent $A_1 \times A_2$.

A similar formula tells us how to count the number of elements in any set T of n-tuples if we know the number of possible choices for each of the successive components. For example, suppose there are k_1 possible choices for the first component x_1. Let k_2 be the number of possible choices for the second component x_2, once x_1 is known. Similarly, let k_r be the number of possible choices for the rth component x_r, once $x_1, x_2, \ldots, x_{r-1}$ are known. Then the number of n-tuples that can be formed with these choices is

$$\nu(T) = k_1 k_2 \cdots k_n .$$

This formula is often referred to as the *principle of sequential counting*. It can be proved by induction on n. In many applications the set of choices for x_r may not be easy to describe since it may not be determined until after the choices of the earlier components have been made. (This was the case when we used the principle to count bridge hands.) Fortunately, to apply the principle of sequential counting we do not need to know the actual set of choices for x_r but only the *number* of possible choices for x_r.

The additive property in formula (13.8) and the principle of sequential counting provide the key to the solution of many counting problems. The following examples show how they can be applied.

EXAMPLE 1. *Sampling with replacement.* Given a set S consisting of n elements. If $k \geq 1$, how many ordered k-tuples can be formed if each component may be an arbitrary element of S?

Note: It may be helpful to think of S as an urn containing n balls labeled $1, 2, \ldots, n$. We select a ball and record its label as the first component of our k-tuple. Replacing the ball in the urn, we again select a ball and use its label as the second component, and so on, until we have made k selections. Since we replace each ball after it is drawn, the same label may appear in different components of our k-tuple.

Solution. Each k-tuple is an element of the Cartesian product

$$T = S_1 \times \cdots \times S_k,$$

where each $S_i = S$. Conversely, each element of T is one of the k-tuples in question. Hence the number of k-tuples formed in this way is

$$\nu(T) = \nu(S_1) \cdots \nu(S_k) = n^k.$$

EXAMPLE 2. *Sampling without replacement.* Given a set S consisting of n elements. If $k \leq n$, how many ordered k-tuples can be formed if the components are chosen from S *without* replacement, that is to say, if no element of S may be used twice in any given k-tuple?

Solution. Consider any k-tuple $(x_1, x_2, \ldots, x_k)$ formed from the elements of S without replacement. For the first component x_1 there are n choices (the n elements of S). When x_1 is chosen there remain $n - 1$ ways to choose x_2. With x_2 chosen, there remain $n - 2$ ways to choose x_3, and so on, there being $n - k + 1$ ways to choose x_k. Therefore, by the principle of sequential counting, the total number of k-tuples so formed is

$$n(n-1)(n-2)\cdots(n-k+1) = \frac{n!}{(n-k)!}.$$

In particular, when $k = n$ this result tells us that $n!$ distinct n-tuples may be formed from a given set S of n elements, with no two components of any n-tuple being equal.

EXAMPLE 3. *The number of k-element subsets of an n-element set.* If $k \leq n$, how many distinct subsets of k elements can be formed from a set S consisting of n elements?

Solution. Let r denote the number of subsets in question and let us denote these subsets by

$$A_1, A_2, \ldots, A_r.$$

These sets are distinct but need not be disjoint. We shall compute r in terms of n and k by an indirect method. For this purpose, let B_i denote the collection of k-tuples that can be formed by choosing the components from the elements of A_i without replacement. The sets $B_1, B_2, \ldots, B_r$ *are* disjoint. Moreover, if we apply the result of Example 2 with $n = k$ we have

$$\nu(B_i) = k! \qquad \text{for each } i = 1, 2, \ldots, r.$$

Now let

$$T = B_1 \cup B_2 \cup \cdots \cup B_r.$$

Then T consists of all k-tuples that can be formed by choosing the components from S without replacement. From Example 2 we have

$$\nu(T) = n!/(n-k)!$$

and by additivity we also have

$$\nu(T) = \sum_{i=1}^{r} \nu(B_i) = k!\, r\,.$$

Equating the two expressions for $\nu(T)$ we obtain

$$r = \frac{n!}{k!\,(n-k)!} = \binom{n}{k}.$$

This proves formula (13.7) stated earlier in this section.

If we use the result of Example 3 to count the total number of subsets of a set S consisting of n elements, we obtain

$$\sum_{k=0}^{n} \binom{n}{k}.$$

Since this sum is also obtained when we expand $(1 + 1)^n$ by the binomial theorem, the number of subsets of S is 2^n.

13.11 Exercises

1. Let $A = \{1, 2, 3\}$. Display in roster notation the set of ordered pairs (a, b) obtained by choosing the first component from A and the second component from the *remaining* elements of A. Can this set of pairs be expressed as a Cartesian product?
2. A two-card hand can be dealt from a deck of 52 cards in $52 \cdot 51 = 2652$ ways. Determine the number of *distinct* hands, and explain your reasoning.
3. A senate committee consisting of six Democrats and four Republicans is to choose a chairman and a vice-chairman. In how many ways can this pair of officers be chosen if the chairman must be a Democrat?
4. An experiment consists of tossing a coin twice and then rolling a die. Display each outcome of this experiment as an ordered triple (a, b, c), where each of a and b is either H (heads) or T (tails) and c is the number of points on the upturned face of the die. For example, $(H, H, 3)$ means that heads came up on both tosses and 3 appeared on the die. Express the set of all possible outcomes as a Cartesian product and determine the number of possible outcomes.
5. In how many ways can a bridge deck of 52 cards be dealt into four hands, each containing 13 cards? Explain your reasoning.
6. Two dice, one red and one white, are tossed. Represent the outcome as an ordered pair (a, b), where a denotes the number of points on the red die, b the number on the white die. How many ordered pairs (a, b) are possible? How many are there for which the sum $a + b$ is:
   (a) even? (b) divisible by 3? (c) either even or divisible by 3?
7. A poker hand contains five cards dealt from a deck of 52. How many distinct poker hands can be dealt containing:
   (a) two pairs (for example, 2 kings, 2 aces, and a 3)?
   (b) a flush (five cards in a given suit)?
   (c) a straight flush (any five in sequence in a given suit, but not including ten, jack, queen, king, ace)?
   (d) a royal flush (ten, jack, queen, king, ace in a single suit)?

8. Refer to Exercise 7. Compute the probability for a poker hand to be:
   (a) a flush.
   (b) a straight flush.
   (c) a royal flush.
9. How many committees of 50 senators can be formed that contain:
   (a) exactly one senator from Alaska?
   (b) both senators from Alaska?
10. A committee of 50 senators is chosen at random. Compute the probability that both senators from Alaska are included.
11. A code group consists of four symbols in a row, each symbol being either a dot or a dash. How many distinct code groups can be formed?
12. How many k-letter words can be formed with an alphabet containing n letters?
13. Show that:

$$\text{(a)}\quad \binom{n}{0} + \binom{n}{2} + \binom{n}{4} + \cdots = \binom{n}{1} + \binom{n}{3} + \binom{n}{5} + \cdots = 2^{n-1}.$$

$$\text{(b)}\quad \binom{n}{0}^2 + \binom{n}{1}^2 + \cdots + \binom{n}{n}^2 = \binom{2n}{n}.$$

14. Suppose a set of ordered pairs (a, b) is constructed by choosing the first component from a set of k elements, say from the set $\{a_1, \ldots, a_k\}$, and the second component from a set of m elements, say from $\{b_1, \ldots, b_m\}$. There are a total of m pairs with first component a_1, namely, $(a_1, b_1), \ldots, (a_1, b_m)$. Similarly, there are m pairs $(a_i, b_1), \ldots, (a_i, b_m)$ with first component a_i. Therefore the total number of ordered pairs (a, b) is $m + m + \cdots + m$ (k summands). This sum equals km, which proves the principle of sequential counting for sets of ordered pairs. Use induction to prove the principle for sets of ordered n-tuples.

13.12 Conditional probability

An unbiased die is thrown and the result is known to be an even number. What is the probability that this number is divisible by 3? What is the probability that a child is color blind, given that it is a boy? These questions can be put in the following form: Let A and B be events of a sample space S. What is the probability that A occurs, given that B has occurred? This is not necessarily the same as asking for the probability of the event $A \cap B$. In fact, when $A = B$ the question becomes: If A occurs, what is the probability that A occurs? The answer in this case should be 1 and this may or may not be the probability of $A \cap B$. To see how to treat such problems in general, we turn to the question pertaining to rolling a die.

When we ask probability questions about rolling an unbiased die, we ordinarily use for the sample space the set $S = \{1, 2, 3, 4, 5, 6\}$ and assign point probability $\frac{1}{6}$ to each element of S. The event "divisible by 3" is the subset $A = \{3, 6\}$ and the event "even" is the subset $B = \{2, 4, 6\}$. We want the probability that an element is in A, given that it is in B. Since we are concerned only with outcomes in which the number is even, we disregard the outcomes 1, 3, 5 and use, instead of S, the set $B = \{2, 4, 6\}$ as our sample space. The event in which we are now interested is simply the singleton $\{6\}$, this being the only outcome of the new sample space that is divisible by 3. If all outcomes of B are considered equally likely, we must assign probability $\frac{1}{3}$ to each of them; hence, in particular, the probability of $\{6\}$ is also $\frac{1}{3}$.

Note that we solved the foregoing problem by employing a very elementary idea. We simply changed the sample space from S to B and provided a new assignment of probabilities. This example suggests a way to proceed in general.

Let $(S, \mathscr{B}, P)$ be a given probability space. Suppose A and B are events and consider the question: "What is the probability that A occurs, given that B has occurred?" As in the example just treated, we can change the sample space from S to B and provide a new assignment of probabilities. We are at liberty to do this in any manner consistent with the definition of probability measures. For B itself we have no choice except to assign the probability 1. Since we are interested in those elements of A which lie in the new sample space B, the problem before us is to compute the probability of the event $A \cap B$ according to the new assignment of probabilities. That is, if P' denotes the probability function associated with the new sample space B, then we must compute $P'(A \cap B)$.

We shall show now that if $P(B) \neq 0$ we can always define a probability function P' and a Boolean algebra $\mathscr{B}'$ of subsets of B such that $(B, \mathscr{B}', P')$ is a probability space. For the Boolean algebra $\mathscr{B}'$ we take the collection of all sets $T \cap B$ where T is in the original Boolean algebra $\mathscr{B}$. It is easy to verify that $\mathscr{B}'$, so defined, is indeed a Boolean algebra. One way to define a probability function P' on $\mathscr{B}'$ is simply to divide each of the old probabilities by $P(B)$. That is, if $C \in \mathscr{B}'$ we let

$$P'(C) = \frac{P(C)}{P(B)}.$$

(This is where the assumption $P(B) \neq 0$ comes in.) We are only changing the scale, with all probabilities magnified by the factor $1/P(B)$. It is easy to check that this definition of P' gives us a *bona fide* probability measure. It is obviously nonnegative and it assigns probability 1 to B. The additive property follows at once from the corresponding additive property for P.

Since each C in $\mathscr{B}'$ is of the form $A \cap B$, where A is an event in the original sample space S, we can rewrite the definition of P' as follows:

$$P'(A \cap B) = \frac{P(A \cap B)}{P(B)}.$$

This discussion suggests that the quotient $P(A \cap B)/P(B)$ provides a reasonable measure of the probability that A occurs, given that B occurs. The following definition is made with this motivation in mind:

DEFINITION OF CONDITIONAL PROBABILITY. *Let $(S, \mathscr{B}, P)$ be a probability space and let B be an event such that $P(B) \neq 0$. The conditional probability that an event A will occur, given that B has occurred, is denoted by the symbol $P(A \mid B)$ (read: "the probability of A, given B") and is defined by the equation*

$$P(A \mid B) = \frac{P(A \cap B)}{P(B)}.$$

The conditional probability $P(A \mid B)$ is not defined if $P(B) = 0$.

The following examples illustrate the use of the concept of conditional probability.

EXAMPLE 1. Let us consider once more the problem raised earlier in this section: A die is thrown and the result is known to be an even number. What is the probability that this number is divisible by 3? As a problem in conditional probabilities, we may take for the sample space the set $S = \{1, 2, 3, 4, 5, 6\}$ and assign point probabilities $\frac{1}{6}$ to each element of S. The event "even" is the set $B = \{2, 4, 6\}$ and the event "divisible by 3" is the set $A = \{3, 6\}$. Therefore we have

$$P(A \mid B) = \frac{P(A \cap B)}{P(B)} = \frac{1/6}{3/6} = \frac{1}{3}.$$

This agrees, of course, with the earlier solution in which we used B as the sample space and assigned probability $\frac{1}{3}$ to each element of B.

EXAMPLE 2. This is an example once used by the Caltech Biology Department to warn against the fallacy of superficial statistics. To "prove" statistically that the population of the U.S. contains more boys than girls, each student was asked to list the number of boys and girls in his family. Invariably, the total number of boys exceeded the total number of girls. The statistics in this case were biased because all undergraduates at Caltech were males. Therefore, the question considered here is not concerned with the probability that a child is a boy; rather, it is concerned with the conditional probability that a child is a boy, given that he comes from a family with at least one boy.

To compute the probabilities in an example of this type consider a sample of $4n$ families, each with two children. Assume that n families have 2 boys, $2n$ families one boy and one girl, and n families 2 girls. Let the sample space S be the set of all $8n$ children in these families and assign the point probability $P(x) = 1/(8n)$ to each x in S. Let A denote the event "the child is a boy" and B the event "the child comes from a family with at least one boy." The probability $P(A)$ is obviously $\frac{1}{2}$. Similarly, $P(B) = \frac{3}{4}$ since $3n$ of the $4n$ families have at least one boy. Therefore the probability that a child is a boy, given that he comes from a family with at least one boy, is the conditional probability

$$P(A \mid B) = \frac{P(A \cap B)}{P(B)} = \frac{P(A)}{P(B)} = \frac{1/2}{3/4} = \frac{2}{3}.$$

13.13 Independence

An important idea related to conditional probability is the concept of *independence of events*, which may be defined as follows:

DEFINITION OF INDEPENDENCE. *Two events A and B are called independent (or stochastically independent) if and only if*

$$P(A \cap B) = P(A)P(B). \tag{13.9}$$

If A and B are independent, then $P(A \mid B) = P(A)$ if $P(B) \neq 0$. That is, the conditional probability of A, given B, is the same as the "absolute" probability of A. This relation exhibits the significance of independence. The knowledge that B has occurred does not influence the probability that A will occur.

EXAMPLE 1. One card is drawn from a 52-card deck. Each card has the same probability of being selected. Show that the two events "drawing an ace" and "drawing a heart" are independent.

Solution. We choose a sample space S consisting of 52 elements and assign the point probability $\frac{1}{52}$ to each element. The event A, "drawing an ace," has the probability $P(A) = \frac{4}{52} = \frac{1}{13}$. The event B, "drawing a heart," has the probability $P(B) = \frac{13}{52} = \frac{1}{4}$. The event $A \cap B$ means "drawing the ace of hearts," which has probability $\frac{1}{52}$. Since $P(A \cap B) = P(A)P(B)$, Equation (13.9) is satisfied and events A and B are independent.

EXAMPLE 2. Three true dice are rolled independently, so that each combination is equally probable. Let A be the event that the sum of the digits shown is six and let B be the event that all three digits are different. Determine whether or not these two events are independent.

Solution. For a sample space S we take the set of all triples (a, b, c) with a, b, c ranging over the values 1, 2, 3, 4, 5, 6. There are 6^3 elements in S, and since they are equally probable we assign the point probability $1/6^3$ to each element. The event A is the set of all triples (a, b, c) for which $a + b + c = 6$. Direct enumeration shows that there are 10 such triples, namely:

$$(1, 2, 3), (1, 3, 2), (1, 1, 4), (1, 4, 1),$$

$$(2, 1, 3), (2, 3, 1), (2, 2, 2),$$

$$(3, 1, 2), (3, 2, 1),$$

$$(4, 1, 1).$$

The event B consists of all triples (a, b, c) for which $a \neq b$, $b \neq c$, and $a \neq c$. There are $6 \cdot 5 \cdot 4 = 120$ elements in B. Exactly six of these elements are in set A, so that $A \cap B$ has six elements. Therefore

$$P(A \cap B) = \frac{6}{6^3}, \qquad P(A) = \frac{10}{6^3}, \qquad \text{and} \qquad P(B) = \frac{120}{6^3}.$$

In this case $P(A \cap B) \neq P(A)P(B)$; therefore events A and B are not independent.

Independence for more than two events is defined as follows. A finite collection $\mathscr{A}$ of n events is said to be independent if the events satisfy the multiplicative property

$$P\left(\bigcap_{k=1}^{m} A_k\right) = \prod_{k=1}^{m} P(A_k) \tag{13.10}$$

for *every* finite subcollection $\{A_1, A_2, \ldots, A_m\}$, where m may take the values $m = 2, 3, \ldots, n$, the sets A_i being in $\mathscr{A}$.

When $\mathscr{A}$ consists of exactly three events A, B, and C, the condition of independence in (13.10) requires that

$$P(A \cap B) = P(A)P(B), \quad P(A \cap C) = P(A)P(C), \quad P(B \cap C) = P(B)P(C), \tag{13.11}$$

and

$$P(A \cap B \cap C) = P(A)P(B)P(C). \tag{13.12}$$

It might be thought that the three equations in (13.11) suffice to imply (13.12) or, in other words, that independence of three events is a consequence of independence *in pairs*. This is not true, as one can see from the following example:

Four tickets labeled a, b, c, and abc, are placed in a box. A ticket is drawn at random, and the sample space is denoted by

$$S = \{a, b, c, abc\}.$$

Define the events A, B, and C as follows:

$$A = \{a, abc\}, \qquad B = \{b, abc\}, \qquad C = \{c, abc\}.$$

In other words, the event X means that the ticket drawn contains the letter x. It is easy to verify that each of the three equations in (13.11) is satisfied so that the events A, B, and C are independent in pairs. However, (13.12) is not satisfied and hence the *three* events are not independent. The calculations are simple and are left as an exercise for the reader.

13.14 Exercises

1. Let A and B be two events with $P(A) \neq 0$, $P(B) \neq 0$. Show that

$$P(A \cap B) = P(B)P(A \mid B) = P(A)P(B \mid A). \tag{13.13}$$

Sometimes it is easier to compute the probabilities $P(A)$ and $P(B \mid A)$ directly by enumeration of cases than it is to compute $P(A \cap B)$. When this is the case, Equation (13.13) gives a convenient way to calculate $P(A \cap B)$. The next exercise is an example.

2. An urn contains seven white and three black balls. A second urn contains five white and five black balls. A ball is selected at random from the first urn and placed in the second. Then a ball is selected at random from the second. Let A denote the event "black ball on first draw" and B the event "black ball on second draw."
   (a) Compute the probabilities $P(A)$ and $P(B \mid A)$ directly by enumerating the possibilities. Use Equation (13.13) to compute $P(A \cap B)$.
   (b) Compute $P(A \cap B)$ directly by enumerating all possible pairs of drawings.
3. (a) Let A_1, A_2, A_3 be three events such that $P(A_1 \cap A_2) \neq 0$. Show that

$$P(A_1 \cap A_2 \cap A_3) = P(A_1)P(A_2 \mid A_1)P(A_3 \mid A_1 \cap A_2).$$

   (b) Use induction to generalize this result as follows: If $A_1, A_2, \ldots, A_n$ are n events ($n \geq 2$) such that $P(A_1 \cap A_2 \cap \cdots \cap A_{n-1}) \neq 0$, then

$$P(A_1 \cap A_2 \cap \cdots \cap A_n) = P(A_1)P(A_2 \mid A_1)P(A_3 \mid A_1 \cap A_2) \cdots P(A_n \mid A_1 \cap A_2 \cap \cdots \cap A_{n-1}).$$

4. A committee of 50 senators is chosen at random. Find the probability that both senators from Alaska are included, given that at least one is.

5. An urn contains five gold and seven blue chips. Two chips are selected at random (without replacement). If the first chip is gold, compute the probability that the second is also gold.
6. A deck of cards is dealt into four hands containing 13 cards each. If one hand has exactly eight spades, what is the probability that a particular one of the other hands has (a) at least one spade? (b) at least two spades? (c) a complete suit?
7. Show that $P(A \cup B \mid C) = P(A \mid C) + P(B \mid C) - P(A \cap B \mid C)$.
8. Let $A_1, A_2, \ldots, A_n$ be n disjoint events whose union is the entire sample space S. For every event E we have the equation

$$E = E \cap S = E \cap \bigcup_{i=1}^{n} A_i = \bigcup_{i=1}^{n} (E \cap A_i).$$

This equation states that E can occur only in conjunction with some A_i. Show that

(a) $P(E) = \sum_{i=1}^{n} P(E \cap A_i)$.

(b) $P(E) = \sum_{i=1}^{n} P(E \mid A_i)P(A_i)$.

This formula is useful when the conditional probabilities $P(E \mid A_i)$ are easier to compute directly than $P(E)$.
9. An unbiased coin is tossed repeatedly. It comes up heads on the first six tosses. What is the probability that it will come up heads on the seventh toss?
10. Given independent events A and B whose probabilities are neither 0 nor 1. Prove that A' and B' are independent. Is the same true if either of A or B has probability 0 or 1?
11. Given independent events A and B. Prove or disprove in each case that:
(a) A' and B are independent.
(b) $A \cup B$ and $A \cap B$ are independent.
(c) $P(A \cup B) = 1 - P(A')P(B')$.
12. If $A_1, A_2, \ldots, A_n$ are independent events, prove that

$$P\left(\bigcup_{i=1}^{n} A_i\right) + \prod_{i=1}^{n} P(A_i') = 1.$$

13. If the three events A, B, and C are independent, prove that $A \cup B$ and C are independent.

[*Hint:* Use the result of Exercise 7 to show that $P(A \cup B \mid C) = P(A \cup B)$.]

14. Let A and B be events, neither of which has probability 0. Prove or disprove the following statements:
(a) If A and B are disjoint, A and B are independent.
(b) If A and B are independent, A and B are disjoint.
15. A die is thrown twice, the sample space S consisting of the 36 possible pairs of outcomes (a, b) each assigned probability $\frac{1}{36}$. Let A, B, and C denote the following events:

$$A = \{(a, b) \mid a \text{ is odd}\}, \quad B = \{(a, b) \mid b \text{ is odd}\}, \quad C = \{(a, b) \mid a + b \text{ is odd}\}.$$

(a) Compute $P(A)$, $P(B)$, $P(C)$, $P(A \cap B)$, $P(A \cap C)$, $P(B \cap C)$, and $P(A \cap B \cap C)$.
(b) Show that A, B, and C are independent in pairs.
(c) Show that A, B, and C are not independent.

13.15 Compound experiments

We turn now to the problem of de Méré mentioned in the introduction — whether or not it is profitable to bet even money on the occurrence of at least one "double six" in 24 throws of a pair of dice. We treat the problem in a more general form: What is the probability of throwing a double six at least once in n throws of a pair of dice? Is this probability more than one-half or less than one-half when $n = 24$?

Consider first the experiment of tossing a pair of fair dice just once. The outcomes of this game can be described by ordered pairs (a, b) in which a and b range over the values 1, 2, 3, 4, 5, 6. The sample space S consists of 36 such pairs. Since the dice are fair we assign the probability $\frac{1}{36}$ to each pair in S.

Now suppose we roll the dice n times. The succession of the n experiments is one compound experiment which we wish to describe mathematically. To do this we need a new sample space and a corresponding probability measure. We consider the outcomes of the new game as ordered n-tuples $(x_1, \ldots, x_n)$, where each component x_i is one of the outcomes of the original sample space S. In other words, the sample space for the compound experiment is the n-fold Cartesian product $S \times \cdots \times S$, which we denote by S^n. The set S^n has 36^n elements, and we assign equal probabilities to each element:

$$P(x) = \frac{1}{36^n} \quad \text{if} \quad x \in S^n .$$

We are interested in the event "at least one double six in n throws." Denote this event by A. In this case it is easier to compute the probability of the complementary event A', which means "no double six in n throws." Each element of A' is an n-tuple whose components can be any element of S except (6, 6). Therefore there are 35 possible values for each component and hence $(35)^n$ n-tuples altogether in A'. Since each element of A' has probability $(\frac{1}{36})^n$, the sum of all the point probabilities in A' is $(\frac{35}{36})^n$. This gives us

$$P(A) = 1 - P(A') = 1 - (\tfrac{35}{36})^n .$$

To answer de Méré's question we must decide whether $P(A)$ is more than one-half or less than one-half when $n = 24$. The inequality $P(A) \geq \frac{1}{2}$ is equivalent to $1 - (\frac{35}{36})^n \geq \frac{1}{2}$, or $(\frac{35}{36})^n \leq \frac{1}{2}$. Taking logarithms we find

$$n \log 35 - n \log 36 \leq -\log 2, \qquad \text{or} \qquad n \geq \frac{\log 2}{\log 36 - \log 35} = 24.6+ .$$

Therefore $P(A) < \frac{1}{2}$ when $n = 24$ and $P(A) > \frac{1}{2}$ when $n \geq 25$. It is *not* advantageous to bet even money on the occurrence of at least one double six in 24 throws.

The foregoing problem suggests a general procedure for dealing with successive experiments. If an experiment is repeated two or more times, the result can be considered one compound experiment. More generally, a compound experiment may be the result of performing two or more distinct experiments successively. The individual experiments may be related to each other or they may be stochastically independent, in the sense that the probability of the outcome of any one of them is unrelated to the results of the others. For the sake of simplicity, we shall discuss how one can combine *two* independent experiments into one compound experiment. The generalization to more than two experiments will be evident.

To associate a *bona fide* probability space with a compound experiment we must explain how to define the new sample space S, the corresponding Boolean algebra $\mathscr{B}$ of subsets of S, and the probability measure P defined on $\mathscr{B}$. As in the above example, we use the concept of *Cartesian product*.

Suppose we have two probability spaces, say $(S_1, \mathscr{B}_1, P_1)$ and $(S_2, \mathscr{B}_2, P_2)$. These spaces may be thought of as associated with two experiments E_1 and E_2. By the compound experiment E we mean the one for which the sample space S is the Cartesian product $S_1 \times S_2$. An outcome of E is a pair (x, y) in S, with the first component x an outcome of E_1 and the second component y an outcome of E_2. If S_1 has n elements and if S_2 has m elements, then $S_1 \times S_2$ has nm elements.

For the new Boolean algebra $\mathscr{B}$ we take the collection of all subsets of S. Next we define the probability function P. Since S is finite we can define $P(x, y)$ for each point (x, y) in S and then use additivity to define P for subsets of S. The point probabilities $P(x, y)$ can be assigned in many ways. However, if the two experiments E_1 and E_2 are stochastically *independent*, we define P by the equation

$$(13.14) \qquad P(x, y) = P_1(x)P_2(y) \qquad \text{for each } (x, y) \text{ in } S.$$

Motivation for this definition can be given as follows. Consider two special events A and B in the new space S,

$$A = \{(x_1, y_1), (x_1, y_2), \ldots, (x_1, y_m)\}$$

and

$$B = \{(x_1, y_1), (x_2, y_1), \ldots, (x_n, y_1)\}.$$

That is, A is the set of all pairs in $S_1 \times S_2$ whose first element is x_1, and B is the set of all pairs whose second element is y_1. The intersection of the two sets A and B is the singleton $\{(x_1, y_1)\}$. If we feel that the first outcome x_1 should have no influence on the second outcome y_1 it seems reasonable to require events A and B to be independent. This means we would like to define the new probability function P in such a way that we have

$$(13.15) \qquad P(A \cap B) = P(A)P(B).$$

If we decide how to assign the probabilities $P(A)$ and $P(B)$, Equation (13.15) will tell us how to assign the probability $P(A \cap B)$, that is, the probability $P(x_1, y_1)$. Event A occurs if and only if the outcome of the first experiment is x_1. Since $P_1(x_1)$ is its probability, it seems natural to assign the value $P_1(x_1)$ to $P(A)$ as well. Similarly, we assign the value $P_2(y_1)$ to $P(B)$. Equation (13.15) then gives us

$$P(x_1, y_1) = P_1(x_1)P_2(y_1).$$

All this, of course, is merely motivation for the assignment of probabilities in (13.14). The only way to decide whether or not (13.14) is a permissible assignment of point probabilities is to check the fundamental properties of probability measures. Each number $P(x, y)$ is nonnegative, and the sum of all the point probabilities is equal to 1, since we have

$$\sum_{(x,y)\in S} P(x, y) = \sum_{x\in S_1} P_1(x) \cdot \sum_{y\in S_2} P_2(y) = 1 \cdot 1 = 1.$$

When we say that a compound experiment E is determined by two stochastically independent experiments E_1 and E_2, we mean that the probability space $(S, \mathscr{B}, P)$ is defined in the manner just described, "independence" being reflected in the fact that $P(x, y)$ is the product $P_1(x)P_2(y)$. It can be shown that the assignment of probabilities in (13.14) implies the formula

$$P(U \times V) = P_1(U)P_2(V) \tag{13.16}$$

for every pair of subsets U in $\mathscr{B}_1$ and V in $\mathscr{B}_2$. (See Exercise 12 in Section 13.23 for an outline of the proof.) We shall deduce some important consequences of this formula.

Let A be an event (in the compound experiment E) of the form

$$A = C_1 \times S_2,$$

where $C_1 \in \mathscr{B}_1$. Each outcome in A is an ordered pair (x, y) where x is restricted to be an outcome of C_1 (in the first experiment E_1) but y can be any outcome of S_2 (in the second experiment E_2). If we apply (13.16) we find

$$P(A) = P(C_1 \times S_2) = P_1(C_1)P_2(S_2) = P_1(C_1),$$

since $P_2(S_2) = 1$. Thus the definition of P assigns the same probability to A that P_1 assigns to C_1. For this reason, such an event A is said to be *determined by the first experiment* E_1. Similarly, if B is an event of E of the form

$$B = S_1 \times C_2,$$

where $C_2 \in \mathscr{B}_2$, we have

$$P(B) = P(S_1 \times C_2) = P_1(S_1)P_2(C_2) = P_2(C_2)$$

and B is said to be *determined by the second experiment* E_2. We shall now show, using (13.16), that two such events A and B are *independent*. That is, we have

$$P(A \cap B) = P(A)P(B). \tag{13.17}$$

First we note that

$$\begin{aligned} A \cap B &= \{(x, y) \mid (x, y) \in C_1 \times S_2 \text{ and } (x, y) \in S_1 \times C_2\} \\ &= \{(x, y) \mid x \in C_1 \text{ and } y \in C_2\} \\ &= C_1 \times C_2. \end{aligned}$$

Hence, by (13.16), we have

$$P(A \cap B) = P(C_1 \times C_2) = P_1(C_1)P_2(C_2). \tag{13.18}$$

Since $P_1(C_1) = P(A)$ and $P_2(C_2) = P(B)$ we obtain (13.17). Note that Equation (13.18) also shows that we can compute the probability $P(A \cap B)$ as a product of probabilities in the individual sample spaces S_1 and S_2; hence no calculations with probabilities in the compound experiment are needed.

The generalization to compound experiments determined by n experiments $E_1, E_2, \ldots, E_n$ is carried out in the same way. The points in the new sample space are n-tuples $(x_1, x_2, \ldots, x_n)$ and the point probabilities are defined as the product of the probabilities of the separate outcomes,

$$P(x_1, x_2, \ldots, x_n) = P_1(x_1)P_2(x_2)\cdots P_n(x_n). \tag{13.19}$$

When this definition of P is used we say that E is determined by *n independent experiments* $E_1, E_2, \ldots, E_n$. In the special case in which all the experiments are associated with the same probability space, the compound experiment E is said to be an example of *independent repeated trials under identical conditions.* Such an example is considered in the next section.

13.16 Bernoulli trials

An important example of a compound experiment was studied extensively by Jakob Bernoulli and is now known as a *Bernoulli sequence of trials.* This is a sequence of repeated trials executed under the same conditions, each result being stochastically independent of all the others. The experiment being repeated has just two possible outcomes, usually called "success" and "failure;" the probability of success is denoted by p and that of failure by q. Of course, $q = 1 - p$. The main result associated with Bernoulli sequences is the following theorem:

THEOREM 13.3. BERNOULLI'S FORMULA. *The probability of exactly k successes in n Bernoulli trials is*

$$\binom{n}{k} p^k q^{n-k}, \tag{13.20}$$

where $\binom{n}{k}$ denotes the binomial coefficient, $\binom{n}{k} = \dfrac{n!}{k!\,(n-k)!}$.

Proof. Denote "success" by S and "failure" by F and consider a particular sequence of n results. This may be represented by an n-tuple

$$(x_1, x_2, \ldots, x_n),$$

where each x_i is either an S or an F. The event A in which we are interested is the collection of all n-tuples that contain exactly k S's and $n - k$ F's. Let us compute the point probability of a particular n-tuple in A. The probability of each S is p, and that of each F is q. Hence, by (13.19), the probability of each particular n-tuple in A is the product of k factors equal to p with $n - k$ factors equal to q. That is,

$$P(x_1, x_2, \ldots, x_n) = p^k q^{n-k} \qquad \text{if} \quad (x_1, x_2, \ldots, x_n) \in A.$$

Therefore, to compute $P(A)$ we need only count the number of elements in A and multiply this number by $p^k q^{n-k}$. But the number of elements in A is simply the number of ways of putting exactly k S's into the n possible positions of the n-tuple. This is the same as the number of subsets of k elements that can be formed from a set consisting of n elements; we have already seen that this number is $\binom{n}{k}$. Therefore, if we add the point probabilities for all points in A we obtain

$$P(A) = \binom{n}{k} p^k q^{n-k}.$$

EXAMPLE 1. An unbiased coin is tossed 50 times. Compute the probability of exactly 25 heads.

Solution. We interpret this experiment as a sequence of 50 Bernoulli trials, in which "success" means "heads" and "failure" means "tails." Since the coin is unbiased we assign the probabilities $p = q = \frac{1}{2}$, and formula (13.20) gives us $\binom{50}{k}(\frac{1}{2})^{50}$ for the probability of exactly k heads in 50 tosses. In particular, when $k = 25$ we obtain

$$\binom{50}{25}\left(\frac{1}{2}\right)^{50} = \frac{50!}{25!\,25!}\left(\frac{1}{2}\right)^{50}.$$

To express this number as a decimal it is best to use logarithms, since tables of logarithms of factorials are readily available. If we denote the number in question by P, a table of common logarithms (base 10) gives us

$$\begin{aligned}\log P &= \log 50! - 2\log 25! - 50\log 2\\ &= 64.483 - 50.381 - 15.052 = -0.950 = 0.05 - 1.00\\ &= \log 1.12 - \log 10 = \log 0.112,\end{aligned}$$

so $P = 0.112$.

EXAMPLE 2. What is the probability of at least r successes in n Bernoulli trials?

Solution. Let A_k denote the event "exactly k successes in n trials." Then the event E in which we are interested is the union

$$E = A_r \cup A_{r+1} \cup \cdots \cup A_n.$$

Since the A_k are disjoint, we find

$$P(E) = \sum_{k=r}^{n} P(A_k) = \sum_{k=r}^{n}\binom{n}{k} p^k q^{n-k}.$$

Since

$$\sum_{k=0}^{n}\binom{n}{k} p^k q^{n-k} = (p+q)^n = 1,$$

the probability of the complementary event E' can be computed as follows:

$$P(E') = 1 - P(E) = \sum_{k=0}^{r-1} \binom{n}{k} p^k q^{n-k}.$$

This last sum gives us the probability of at most $r - 1$ successes in n trials.

13.17 The most probable number of successes in *n* Bernoulli trials

A pair of fair dice is rolled 28 times. What is the most probable number of sevens? To solve this problem we let $f(k)$ denote the probability of exactly k sevens in 28 tosses. The probability of tossing a seven is $\frac{1}{6}$. Bernoulli's formula tells us that

$$f(k) = \binom{28}{k} \left(\frac{1}{6}\right)^k \left(\frac{5}{6}\right)^{28-k}.$$

We wish to determine what value (or values) of k in the range $k = 0, 1, 2, \ldots, 28$ make $f(k)$ as large as possible. The next theorem answers this question for any sequence of Bernoulli trials.

THEOREM 13.4. *Given an integer $n \geq 1$ and a real p, $0 < p < 1$, consider the set of numbers*

$$f(k) = \binom{n}{k} p^k (1 - p)^{n-k}, \qquad \text{for} \quad k = 0, 1, \ldots, n.$$

(a) *If $(n + 1)p$ is not an integer, the largest value of $f(k)$ occurs for exactly one k:*

$$k = [(n + 1)p], \qquad \text{the greatest integer} < (n + 1)p.$$

(b) *If $(n + 1)p$ is an integer, the largest value of $f(k)$ occurs for exactly two values of k:*

$$k = (n + 1)p \qquad \text{and} \qquad k = (n + 1)p - 1.$$

Proof. To study the behavior of $f(k)$ we consider the ratio

$$r(k) = \frac{f(k)}{f(k + 1)} = \frac{k + 1}{n - k} \frac{1 - p}{p}$$

for $k = 0, 1, \ldots, n - 1$. The function $r(k)$ is strictly increasing so we have

$$0 < r(0) < r(1) < \cdots < r(n - 1).$$

We consider six cases, illustrated in Figure 13.2. In the first three cases we show that $f(k)$ takes its largest value for exactly one k. In the remaining cases $f(k)$ takes its largest value for two consecutive values of k.

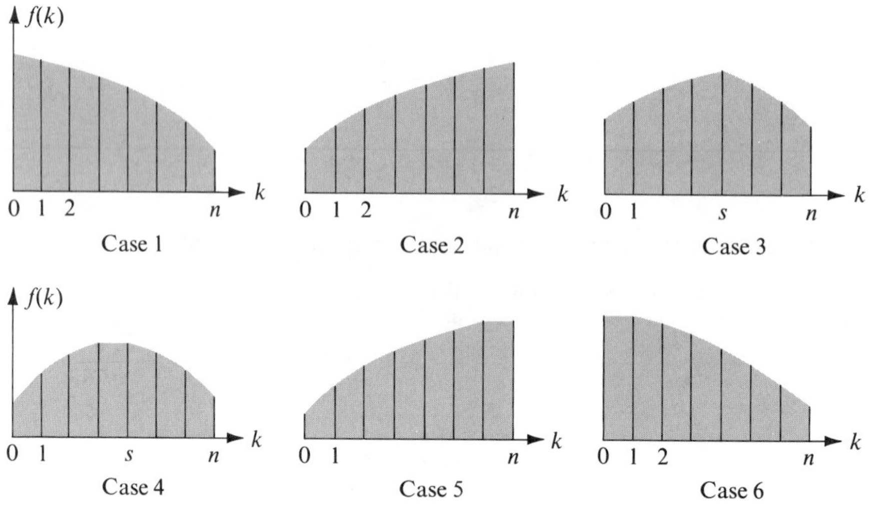

FIGURE 13.2 Calculation of the most probable number of successes in n Bernoulli trials.

CASE 1. $r(0) > 1$. In this case $r(k) > 1$ for every k so we have

$$f(0) > f(1) > \cdots > f(n).$$

Therefore the largest value of $f(k)$ occurs only for $k = 0$. Also, $r(0) = (1 - p)/(np) > 1$, so $1 - p > np$, $(n + 1)p < 1$, hence $[(n + 1)p] = 0$.

CASE 2. $r(n - 1) < 1$. In this case $r(k) < 1$ for every k so $f(0) < f(1) < \cdots < f(n)$ and the largest value of $f(k)$ occurs only for $k = n$. Since $r(n - 1) = n(1 - p)/p < 1$, we have $n - np < p$, hence $n < (n + 1)p < n + 1$, so $[(n + 1)p] = n$.

CASE 3. $r(0) < 1$, $r(n - 1) > 1$, *and* $r(k) \neq 1$ *for all* k. In this case there is a unique integer s, $0 < s < n$, such that $r(s - 1) < 1$ and $r(s) > 1$. The function $f(k)$ increases in the range $0 \leq k \leq s$ and decreases in the range $s \leq k \leq n$. Therefore $f(k)$ has a unique maximum at $k = s$. Since $r(s - 1) = s(1 - p)/(np - sp + p) < 1$ we have $s < (n + 1)p$. The inequality $r(s) > 1$ shows that $(n + 1)p < s + 1$, hence $[(n + 1)p] = s$.

Note that in each of the first three cases the maximum value of $f(k)$ occurs when $k = [(n + 1)p]$; also $(n + 1)p$ is not an integer in any of these cases.

CASE 4. $r(0) < 1$, $r(n - 1) > 1$, *and* $r(s - 1) = 1$ *for some* s, $2 \leq s < n$. In this case $f(k)$ increases for $0 \leq k \leq s - 1$ and decreases for $s \leq k \leq n$. The maximum value of $f(k)$ occurs twice, when $k = s - 1$ and when $k = s$. The equation $r(s - 1) = 1$ implies $(n + 1)p = s$.

CASE 5. $r(n - 1) = 1$. In this case $r(k) < 1$ for $k \leq n - 2$, so $f(k)$ increases in the range $0 \leq k \leq n - 1$, and $f(n - 1) = f(n)$. Hence the maximum of $f(k)$ occurs twice, when $k = n - 1$ and when $k = n$. The equation $r(n - 1) = 1$ implies $(n + 1)p = n$.

CASE 6. $r(0) = 1$. In this case $r(k) > 1$ for $k \geq 1$, so $f(k)$ decreases in the range $1 \leq k \leq n$. The maximum $f(k)$ occurs twice, when $k = 0$ and when $k = 1$. The equation $r(0) = 1$ implies $(n + 1)p = 1$.

In each of the last three cases the maximum value of $f(k)$ occurs for $k = (n + 1)p$ and for $k = (n + 1)p - 1$. This completes the proof.

EXAMPLE 1. A pair of fair dice is rolled 28 times. What is the most probable number of sevens?

Solution. We apply Theorem 13.4 with $n = 28$, $p = \frac{1}{6}$, and $(n + 1)p = \frac{29}{6}$. This is not an integer so the largest value of $f(k)$ occurs for $k = [\frac{29}{6}] = 4$.

Note: If the dice are rolled 29 times there are two solututions, $k = 4$ and $k = 5$.

EXAMPLE 2. Find the smallest n such that if a pair of fair dice is thrown n times the probability of getting exactly four sevens is at least as large as the probability of getting any other number of sevens.

Solution. We take $p = \frac{1}{6}$ in Theorem 13.4. We want the largest value of $f(k)$ to occur when $k = 4$. This requires either $[(n + 1)p] = 4$, $(n + 1)p = 4$, or $(n + 1)p - 1 = 4$. The smallest n satisfying any of these relations is $n = 23$.

13.18 Exercises

1. A coin is tossed twice, the probability of heads on the first toss being p_1 and that on the second toss p_2. Consider this a compound experiment determined by two stochastically independent experiments, and let the sample space be

$$S = \{(H, H), (H, T), (T, H), (T, T)\}.$$

(a) Compute the probability of each element of S.
(b) Can p_1 and p_2 be assigned so that

$$P(H, H) = \tfrac{1}{9}, \qquad P(H, T) = P(T, H) = \tfrac{2}{9}, \qquad P(T, T) = \tfrac{4}{9}?$$

(c) Can p_1 and p_2 be assigned so that

$$P(H, H) = P(T, T) = \tfrac{1}{3}, \qquad P(H, T) = P(T, H) = \tfrac{1}{6}?$$

(d) Consider the following four events (subsets of S):

H_1 : heads on the first toss,
H_2 : heads on the second toss,
T_1 : tails on the first toss,
T_2 : tails on the second toss.

Determine which pairs of these four events are independent.

In each of Exercises 2 through 12 describe your sample space, your assignment of probabilities, and the event whose probability you are computing.

2. A student takes a true – false examination consisting of 10 questions. He is completely unprepared so he plans to guess each answer. The guesses are to be made at random. For example, he may toss a fair coin and use the outcome to determine his guess.
(a) Compute the probability that he guesses correctly at least five times.
(b) Compute the probability that he guesses correctly at least nine times.
(c) What is the smallest n such that the probability of guessing at least n correct answers is less than $\frac{1}{2}$?
3. Ten fair dice are tossed together. What is the probability that exactly three sixes occur?
4. A fair coin is tossed five times. What is the probability of getting (a) exactly three heads? (b) at least three heads? (c) at most one head?
5. A man claims to have a divining rod which locates hidden sources of oil. The Caltech Geology Department conducts the following experiment to test his claim. He is taken into a room in which there are 10 sealed barrels. He is told that five of them contain oil and five contain water. His task is to decide which of the five contain oil and which do not.
(a) What is the probability that he locates the five oil barrels correctly just by chance?
(b) What is the probability that he locates at least three of the oil barrels correctly by chance?
6. A little old lady from Pasadena claims that by tasting a cup of tea made with milk she can tell whether the milk or the tea was added first to the cup. The lady's claim is tested by requiring her to taste and classify 10 pairs of cups of tea, each pair containing one cup of tea made by each of the two methods under consideration. Let p denote her "true" probability of classifying a pair of cups correctly. (If she is skillful, p is substantially greater than $\frac{1}{2}$; if not, $p \le \frac{1}{2}$.) Assume the 10 pairs of cups are classified under independent and identical conditions.
(a) Compute, in terms of p, the probability that she classifies correctly at least eight of the 10 pairs of cups.
(b) Evaluate this probability explicitly when $p = \frac{1}{2}$.
7. (Another problem of Chevalier de Méré.) Determine whether or not it is advantageous to bet even money on at least one 6 appearing in four throws of an unbiased die. [*Hint:* Show that the probability of at least one 6 in n throws is $1 - (\frac{5}{6})^n$.]
8. An urn contains w white balls and b black balls. If $k \le n$, compute the probability of drawing k white balls in n drawings, if each ball is replaced before the next one is drawn.
9. Two dice are thrown eight times. Compute the probability that the sum is 11 exactly three times.
10. Throw a coin 10 times or 10 coins once and count the number of heads. Find the probability of obtaining at least six heads.
11. After a long series of tests on a certain kind of rocket engine it has been determined that in approximately 5% of the trials there will be a malfunction that will cause the rocket to misfire. Compute the probability that in 10 trials there will be at least one failure.
12. A coin is tossed repeatedly. Compute the probability that the total number of heads will be at least 6 before the total number of tails reaches 5.
13. Exercise 12 may be generalized as follows: Show that the probability of at least m successes before n failures in a sequence of Bernoullian trials is

$$\sum_{k=m}^{m+n-1} \binom{m+n-1}{k} p^k q^{m+n-k-1}.$$

14. Determine all n with the following property: If a pair of fair dice is thrown n times, the probability of getting exactly ten sevens is at least as large as the probability of getting any other number of sevens.

15. A binary slot machine has three identical and independent wheels. When the machine is played the possible outcomes are ordered triples (x, y, z), where each of x, y, z can be 0 or 1. On each wheel the probability of 0 is p and the probability of 1 is $1 - p$, where $0 < p < 1$. The machine pays \$2 if the outcome is $(1, 1, 1)$ or $(0, 0, 0)$; it pays \$1 for the outcome $(1, 1, 0)$; otherwise it pays nothing. Let $f(p)$ denote the probability that the machine pays \$1 or more when it is played once.
(a) Calculate $f(p)$.
(b) Define the "payoff" to be the sum $\sum_{x \in S} g(x)P(x)$, where S is the sample space, $P(x)$ is the probability of outcome x, and $g(x)$ is the number of dollars paid by outcome x. Calculate the value of p for which the payoff is smallest.

13.19 Countable and uncountable sets

Up to now we have discussed probability theory only for finite sample spaces. We wish now to extend the theory to infinite sample spaces. For this purpose it is necessary to distinguish between two types of infinite sets, *countable* and *uncountable*. This section describes these two concepts.

To count the members of an n-element set we match the set, element by element, with the set of integers $\{1, 2, \ldots, n\}$. Comparing the sizes of two sets by matching them element by element takes the place of counting when we deal with infinite sets. The process of "matching" can be given a neat mathematical formulation by employing the function concept:

DEFINITION. *Two sets A and B are said to be in one-to-one correspondence if a function f exists with the following properties:*
(a) *The domain of f is A and the range of f is B.*
(b) *If x and y are distinct elements of A, then $f(x)$ and $f(y)$ are distinct elements of B. That is, for all x and y in A,*

$$(13.21) \qquad x \neq y \quad \textit{implies} \quad f(x) \neq f(y).$$

A function satisfying property (13.21) is said to be *one-to-one* on A. Two sets A and B in one-to-one correspondence are also said to be *equivalent*, and we indicate this by writing $A \sim B$. It is clear that every set A is equivalent to itself, since we may let $f(x) = x$ for each x in A.

A set can be equivalent to a proper subset of itself. For example, the set $P = \{1, 2, 3, \ldots\}$, consisting of all the positive integers, is equivalent to the proper subset $Q = \{2, 4, 6, \ldots\}$ consisting of the even integers. In this case a one-to-one function which makes them equivalent is given by $f(x) = 2x$ for x in P.

If $A \sim B$ we can easily show that $B \sim A$. In fact, if f is one-to-one on A and if the range of f is B, then for each b in B there is exactly one a in A such that $f(a) = b$. Therefore we can define an inverse function g on B as follows: If $b \in B$, $g(b) = a$, where a is the unique element of A such that $f(a) = b$. This g is one-to-one on B and its range is A; hence $B \sim A$. This property of equivalence is known as *symmetry*:

$$(13.22) \qquad A \sim B \quad \text{implies} \quad B \sim A.$$

It is also easy to show that equivalence has the following property, known as *transitivity*:

(13.23) $$A \sim B \quad \text{and} \quad B \sim C \qquad \text{implies} \qquad A \sim C.$$

A proof of the transitive property is requested in Exercise 2 of Section 13.20.

A set S is called *finite* and is said to contain n elements if

$$S \sim \{1, 2, \ldots, n\}.$$

The empty set is also considered to be finite. Sets which are not finite are called *infinite sets.* A set S is said to be *countably infinite* if it is equivalent to the set of all positive integers, that is, if

(13.24) $$S \sim \{1, 2, 3, \ldots\}.$$

In this case there is a function f which establishes a one-to-one correspondence between the positive integers and the elements of S; hence the set S can be displayed in roster notation as follows:

$$S = \{f(1), f(2), f(3), \ldots\}.$$

Often we use subscripts and denote $f(k)$ by a_k (or by a similar notation) and we write $S = \{a_1, a_2, a_3, \ldots\}$. The important thing here is that the correspondence in (13.24) enables us to use the positive integers as "labels" for the elements of S.

A set is said to be *countable* if it is finite or countably infinite. A set which is not countable is called *uncountable.*† (Examples will be given presently.) Many set operations when performed on countable sets produce countable sets. For example, we have the following properties:

(a) Every subset of a countable set is countable.
(b) The intersection of any collection of countable sets is countable.
(c) The union of a countable collection of countable sets is countable.
(d) The Cartesian product of a finite number of countable sets is countable.

Since we shall do very little with countably infinite sets in this book, detailed proofs of these properties will not be given.‡ Instead, we shall give a number of examples to show how these properties may be used to construct new countable sets from given ones.

EXAMPLE 1. The set S of all integers (positive, negative, or zero) is countable.

Proof. If $n \in S$, let $f(n) = 2n$ if n is positive, and let $f(n) = 2\,|n| + 1$ if n is negative or zero. The domain of f is S and its range is the set of positive integers. Since f is one-to-one on S, this shows that S is countable.

EXAMPLE 2. The set R of all rational numbers is countable.

Proof. For each fixed integer $n \geq 1$, let S_n denote the set of rational numbers of the form x/n, where x belongs to the set S of Example 1. Each set S_n is equivalent to S [take

† The words *denumerable* and *nondenumerable* are sometimes used as synonyms for *countable* and *uncountable*, respectively.

‡ Proofs are outlined in Exercises 3 through 8 of Section 13.20.

$f(t) = nt$ if $t \in S_n$] and hence each S_n is countable. Since R is the union of all the S_n, property (c) implies that R is countable.

Note. If $\mathscr{F} = \{A_1, A_2, A_3, \ldots\}$ is a countable collection of sets, the union of all sets in the family $\mathscr{F}$ is denoted by the symbols

$$\bigcup_{k=1}^{\infty} A_k \qquad \text{or} \qquad A_1 \cup A_2 \cup A_3 \cup \cdots.$$

EXAMPLE 3. Let A be a countably infinite set, say $A = \{a_1, a_2, a_3, \ldots\}$. For each integer $n \geq 1$, let $\mathscr{F}_n$ denote the family of n-element subsets of A. That is, let

$$\mathscr{F}_n = \{S \mid S \subseteq A \text{ and } S \text{ has } n \text{ elements}\}.$$

Then each $\mathscr{F}_n$ is countable.

Proof. If S is an n-element subset of A, we may write

$$S = \{a_{k_1}, a_{k_2}, \ldots, a_{k_n}\},$$

where $k_1 < k_2 < \cdots < k_n$. Let $f(S) = (a_{k_1}, a_{k_2}, \ldots, a_{k_n})$. That is, f is the function which associates with S the ordered n-tuple $(a_{k_1}, a_{k_2}, \ldots, a_{k_n})$. The domain of f is $\mathscr{F}_n$ and its range, which we denote by T_n, is a subset of the Cartesian product $C_n = A \times A \times \cdots \times A$ (n factors). Since A is countable, so is C_n [by property (d)] and hence T_n is also [by property (a)]. But $T_n \sim \mathscr{F}_n$ because f is one-to-one. This shows that $\mathscr{F}_n$ is countable.

EXAMPLE 4. The collection of all finite subsets of a countable set is countable.

Proof. The result is obvious if the given set is finite. Assume, then, that the given set (call it A) is countably infinite, and let $\mathscr{F}$ denote the class of all finite subsets of A:

$$\mathscr{F} = \{S \mid S \subseteq A \text{ and } S \text{ is finite}\}.$$

Then $\mathscr{F}$ is the union of all the families $\mathscr{F}_n$ of Example 3; hence, by property (c), $\mathscr{F}$ is countable.

EXAMPLE 5. The collection of *all* subsets of a countably infinite set is uncountable.

Proof. Let A denote the given countable set and let $\mathscr{A}$ denote the family of all subsets of A. We shall assume that $\mathscr{A}$ is countable and arrive at a contradiction. If $\mathscr{A}$ is countable, then $\mathscr{A} \sim A$ and hence there exists a one-to-one function f whose domain is A and whose range is $\mathscr{A}$. Thus for each a in A, the function value $f(a)$ is a subset of A. This subset may or may not contain the element a. We denote by B the set of elements a such that $a \notin f(a)$. Thus,

$$B = \{a \mid a \in A \text{ but } a \notin f(a)\}.$$

This B, being a subset of A, must belong to the family $\mathscr{A}$. This means that $B = f(b)$ for some b in A. Now there are only two possibilities: (i) $b \in B$, or (ii) $b \notin B$. If $b \in B$, then by the definition of B we have $b \notin f(b)$, which is a contradiction since $f(b) = B$. Therefore (i) is impossible. In case (ii), $b \notin B$, which means $b \notin f(b)$. This contradicts the definition of B, so case (ii) is also impossible. Therefore the assumption that $\mathscr{A}$ is countable leads to a contradiction and we must conclude that $\mathscr{A}$ is uncountable.

We give next an example of an uncountable set that is easier to visualize than that in Example 5.

EXAMPLE 6. The set of real x satisfying $0 < x < 1$ is uncountable.

Proof. Again, we assume the set is countable and arrive at a contradiction. If the set is countable we may display its elements as follows: $\{x_1, x_2, x_3, \ldots\}$. Now we shall construct a real number y satisfying $0 < y < 1$ which is not in this list. For this purpose we write each element x_n as a decimal:

$$x_n = 0.a_{n,1}\, a_{n,2}\, a_{n,3} \cdots,$$

where each $a_{n,i}$ is one of the integers in the set $\{0, 1, 2, \ldots, 9\}$. Let y be the real number which has the decimal expansion

$$y = 0.y_1\, y_2\, y_3 \cdots,$$

where

$$y_n = \begin{cases} 1 & \text{if } a_{n,n} \neq 1, \\ 2 & \text{if } a_{n,n} = 1. \end{cases}$$

Then no element of the set $\{x_1, x_2, x_3, \ldots\}$ can be equal to y, because y differs from x_1 in the first decimal place, differs from x_2 in the second decimal place, and in general, y differs from x_k in the kth decimal place. (A situation like $x_n = 0.249999 \cdots$ and $y = 0.250000 \cdots$ cannot occur here because of the way the y_n are chosen.) Since this y satisfies $0 < y < 1$, we have a contradiction, and hence the set of real numbers in the open interval $(0, 1)$ is uncountable.

13.20 Exercises

1. Let $P = \{1, 2, 3, \ldots\}$ denote the set of positive integers. For each of the following sets, exhibit a one-to-one function f whose domain is P and whose range is the set in question:
   (a) $A = \{2, 4, 6, \ldots\}$, the set of even positive integers.
   (b) $B = \{3, 3^2, 3^3, \ldots\}$, the set of powers of 3.
   (c) $C = \{2, 3, 5, 7, 11, 13, \ldots\}$, the set of primes. [*Note:* Part of the proof consists in showing that C is an infinite set.]
   (d) $P \times P$, the Cartesian product of P with itself.
   (e) The set of integers of the form $2^m 3^n$, where m and n are positive integers.
2. Prove the transitive property of set equivalence:

$$\text{If } A \sim B \quad \text{and} \quad B \sim C, \quad \text{then } A \sim C.$$

[*Hint:* If f makes A equivalent to B and if g makes B equivalent to C, show that the composite function $h = g \circ f$ makes A equivalent to C.]

Exercises 3 through 8 are devoted to providing proofs of the four properties (a), (b), (c), (d) of countable sets listed in Section 13.19.

3. Prove that every subset of a countable set is countable. [*Hint:* Suppose S is a countably infinite set, say $S = \{x_1, x_2, x_3, \ldots\}$, and let A be an infinite subset of S. Let $k(1)$ be the smallest positive integer m such that $x_m \in A$. Assuming $k(1), k(2), \ldots, k(n-1)$ have been defined, let $k(n)$ be the smallest positive integer $m > k(n-1)$ such that $x_m \in A$. Let $f(n) = x_{k(n)}$. Show that f is a one-to-one function whose domain is the set of positive integers and whose range is A. This proves the result when S is countably infinite. Construct a separate proof for a finite S.]
4. Show that the intersection of any collection of countable sets is countable. [*Hint:* Use the result of Exercise 3.]
5. Let $P = \{1, 2, 3, \ldots\}$ denote the set of positive integers.
(a) Prove that the Cartesian product $P \times P$ is countable. [*Hint:* Let Q denote the set of positive integers of the form $2^m 3^n$, where m and n are positive integers. Then $Q \subset P$, so Q is countable (by Exercise 3). If $(m, n) \in P \times P$, let $f(m, n) = 2^m 3^n$ and use this function to show that $P \times P \sim Q$.]
(b) Deduce from part (a) that the Cartesian product of two countable sets is countable. Then use induction to extend the result to n countable sets.
6. Let $\mathscr{B} = \{B_1, B_2, B_3, \ldots\}$ be a countable collection of *disjoint* sets ($B_i \cap B_j = \varnothing$ when $i \neq j$) such that each B_n is countable. Show that the union $\bigcup_{k=1}^{\infty} B_k$ is also countable. [*Hint:* Let $B_n = \{b_{1,n}, b_{2,n}, b_{3,n}, \ldots\}$ and $S = \bigcup_{k=1}^{\infty} B_k$. If $x \in S$, then $x = b_{m,n}$ for some unique pair (m, n) and we can define $f(x) = (m, n)$. Use this f to show that S is equivalent to a subset of $P \times P$ and deduce (by Exercise 5) that S is countable.]
7. Let $\mathscr{A} = \{A_1, A_2, A_3, \ldots\}$ be a countable collection of sets, and let $\mathscr{B} = \{B_1, B_2, B_3, \ldots\}$ be defined as follows: $B_1 = A_1$ and, for $n > 1$,
$$B_n = A_n - \bigcup_{k=1}^{n-1} A_k .$$
That is, B_n consists of those points in A_n which are not in any of the earlier sets $A_1, \ldots, A_{n-1}$. Prove that $\mathscr{B}$ is a collection of disjoint sets ($B_i \cap B_j = \varnothing$ when $i \neq j$) and that
$$\bigcup_{k=1}^{\infty} A_k = \bigcup_{k=1}^{\infty} B_k .$$
This enables us to express the union of any countable collection of sets as the union of a countable collection of *disjoint* sets.
8. If $\mathscr{F}$ is a countable collection of countable sets, prove that the union of all sets in $\mathscr{F}$ is countable. [*Hint:* Use Exercises 6 and 7.]
9. Show that the following sets are countable:
(a) The set of all intervals on the real axis with rational end points.
(b) The set of all circles in the plane with rational radii and centers having rational coordinates.
(c) Any set of disjoint intervals of positive length.
10. Show that the following sets are uncountable:
(a) The set of irrational numbers in the interval (0, 1).
(b) The set of all intervals of positive length.
(c) The set of all sequences whose terms are the integers 0 and 1. (Recall that a sequence is a function whose domain is the set of positive integers.)

13.21 The definition of probability for countably infinite sample spaces

This section extends the definition of probability to countably infinite sample spaces. Let S be a countably infinite set and let $\mathscr{B}$ be a Boolean algebra of subsets of S. We define a probability measure P on $\mathscr{B}$ as we did for the finite case, except that we require countable additivity as well as finite additivity. That is, for every countably infinite collection $\{A_1, A_2, \ldots\}$ of elements of $\mathscr{B}$, we require that

$$P\left(\bigcup_{k=1}^{\infty} A_k\right) = \sum_{k=1}^{\infty} P(A_k) \quad \text{if} \quad A_i \cap A_j = \varnothing \quad \text{whenever} \quad i \neq j. \tag{13.25}$$

Finitely additive set functions which satisfy (13.25) are said to be *countably additive* (or completely additive). Of course, this property requires assuming also that the countable union $A_1 \cup A_2 \cup A_3 \cup \cdots$ is in $\mathscr{B}$ whenever each A_k is in $\mathscr{B}$. Not all Boolean algebras have this property. Those which do are called Boolean σ-algebras. An example is the Boolean algebra of all subsets of S.

DEFINITION OF PROBABILITY FOR COUNTABLY INFINITE SAMPLE SPACES. *Let $\mathscr{B}$ denote a Boolean σ-algebra whose elements are subsets of a given countably infinite set S. A set function P is called a probability measure on $\mathscr{B}$ if it is nonnegative, countably additive, and satisfies $P(S) = 1$.*

When $\mathscr{B}$ is the Boolean algebra of all subsets of S, a probability function is completely determined by its values on the singletons (called point probabilities). Every subset A of S is either finite or countably infinite, and the probability of A is computed by adding the point probabilities for all elements in A,

$$P(A) = \sum_{x \in A} P(x).$$

The sum on the right is either a finite sum or an absolutely convergent infinite series.

The following example illustrates an experiment with a countably infinite sample space.

EXAMPLE. Toss a coin repeatedly until the first outcome occurs a second time; at this point the game ends.

For a sample space we take the collection of all possible games that can be played. This set can be expressed as the union of two countably infinite sets A and B, where

$$A = \{TT, THT, THHT, THHHT, \ldots\} \quad \text{and} \quad B = \{HH, HTH, HTTH, HTTTH, \ldots\}.$$

We denote the elements of set A (in the order listed) as $a_0, a_1, a_2, \ldots$, and those of set B as $b_0, b_1, b_2, \ldots$. We can assign arbitrary nonnegative point probabilities $P(a_n)$ and $P(b_n)$ provided that

$$\sum_{n=0}^{\infty} P(a_n) + \sum_{n=0}^{\infty} P(b_n) = 1.$$

For example, suppose the coin has probability p of coming up heads (H) and probability $q = 1 - p$ of coming up tails (T), where $0 < p < 1$. Then a natural assignment of point probabilities would be

$$P(a_n) = q^2p^n \quad \text{and} \quad P(b_n) = p^2q^n. \tag{13.26}$$

This is an acceptable assignment of probabilities because we have

$$\sum_{n=0}^{\infty} P(a_n) + \sum_{n=0}^{\infty} P(b_n) = q^2\sum_{n=0}^{\infty} p^n + p^2\sum_{n=0}^{\infty} q^n = \frac{q^2}{1-p} + \frac{p^2}{1-q} = q + p = 1.$$

Now suppose we ask for the probability that the game ends after exactly $n + 2$ tosses. This is the event $\{a_n\} \cup \{b_n\}$, and its probability is

$$P(a_n) + P(b_n) = q^2p^n + p^2q^n.$$

The probability that the game ends in at most $n + 2$ tosses is

$$\sum_{k=0}^{n} P(a_k) + \sum_{k=0}^{n} P(b_k) = q^2\frac{1-p^{n+1}}{1-p} + p^2\frac{1-q^{n+1}}{1-q} = 1 - qp^{n+1} - pq^{n+1}.$$

13.22 Exercises

The exercises in this section refer to the example in Section 13.21.

1. Using the point probabilities assigned in Equation (13.26), let $f_n(p)$ denote the probability that the game ends after exactly $n + 2$ tosses. Calculate the absolute maximum and minimum values of $f_n(p)$ on the interval $0 \le p \le 1$ for each of the values $n = 0, 1, 2, 3$.
2. Show that each of the following is an acceptable assignment of point probabilities.

   (a) $P(a_n) = P(b_n) = \dfrac{1}{2^{n+2}}$ for $n = 0, 1, 2, \ldots$.

   (b) $P(a_n) = P(b_n) = \dfrac{1}{(n+2)(n+3)}$ for $n = 0, 1, 2, \ldots$.
3. Calculate the probability that the game ends before the fifth toss, using:
   (a) the point probabilities in (13.26).
   (b) the point probabilities in Exercise 2(a).
   (c) the point probabilities in Exercise 2(b).
4. Calculate the probability that an odd number of tosses is required to terminate the game, using:
   (a) the point probabilities in (13.26).
   (b) the point probabilities in Exercise 2(a).
   (c) the point probabilities in Exercise 2(b).

13.23 Miscellaneous exercises on probability

1. What is the probability of rolling a ten with two unbiased dice?
2. Ten men and their wives are seated at random at a banquet. Compute the probability that a particular man sits next to his wife if (a) they are seated at a round table; (b) they are seated in a row.

3. A box has two drawers. Drawer number 1 contains four gold coins and two silver coins. Drawer number 2 contains three gold coins and three silver coins. A drawer is opened at random and a coin selected at random from the open drawer. Compute the probability of each of the following events:
   (a) Drawer number 2 was opened and a silver coin was selected.
   (b) A gold coin was selected from the opened drawer.
4. Two cards are picked in succession from a deck of 52 cards, each card having the same probability of being drawn.
   (a) What is the probability that at least one is a spade?
   The two cards are placed in a sack unexamined. One card is drawn from the sack and examined and found not to be a spade. (Each card has the same probability of being drawn.)
   (b) What is the probability now of having at least one spade?
   The card previously drawn is replaced in the sack and the cards mixed. Again a card is drawn and examined. No comparison is made to see if it is the same card previously drawn. The card is again replaced in the sack and the cards mixed. This is done a total of three times, including that of part (b), and each time the card examined is not a spade.
   (c) What is a sample space and a probability function for this experiment? What is the probability that one of the two original cards is a spade?
5. A man has ten pennies, 9 ordinary and 1 with two heads. He selects a penny at random, tosses it six times, and it always comes up heads. Compute the probability that he selected the double-headed penny.
6. Prove that it is impossible to load a pair of dice so that every outcome from 2 to 12 will have the same probability of occurrence.
7. A certain Caltech sophomore has an alarm clock which will ring at the appointed hour with probability 0.7. If it rings, it will wake him in time to attend his mathematics class with probability 0.8. If it doesn't ring he will wake in time to attend his class with probability 0.3. Compute the probability that he will wake in time to attend his mathematics class.
8. Three horses A, B, and C are in a horse race. The event "A beats B" will be denoted symbolically by writing AB. The event "A beats B who beats C" will be denoted by ABC, etc. Suppose it is known that
$$P(AB) = \tfrac{2}{3}, \qquad P(AC) = \tfrac{2}{3}, \qquad P(BC) = \tfrac{1}{2},$$
   and that
$$P(ABC) = P(ACB), \qquad P(BAC) = P(BCA), \qquad P(CAB) = P(CBA).$$
   (a) Compute the probability that A wins.
   (b) Compute the probability that B wins.
   (c) Compute the probability that C wins.
   (d) Are the events AB, AC, and CB independent?
9. The final step in a long computation requires the addition of three integers a_1, a_2, a_3. Assume that (a) the computations of a_1, a_2, and a_3 are stochastically independent; (b) in the computation of each a_i there is a common probability p that it is correct and that the probability of making an error of $+1$ is equal to the probability of making an error of -1; (c) no error larger than $+1$ or less than -1 can occur. Remember the possibility of compensating errors, and compute the probability that the sum $a_1 + a_2 + a_3$ is correct.
10. *The game of "odd man out."* Suppose n persons toss identical coins simultaneously and independently, where $n \geq 3$. Assume there is a probability p of obtaining heads with each coin. Compute the probability that in a given toss there will be an "odd man," that is, a person whose coin does not have the same outcome as that of any other member of the group.
11. Suppose n persons play the game "odd man out" with fair coins (as described in Exercise 10). For a given integer m compute the probability that it will take exactly m plays to conclude the game (the mth play is the first time there is an "odd man").

12. Suppose a compound experiment $(S, \mathscr{B}, P)$ is determined by two stochastically independent experiments $(S_1, \mathscr{B}_1, P_1)$ and $(S_2, \mathscr{B}_2, P_2)$, where $S = S_1 \times S_2$ and

$$P(x, y) = P_1(x)P_2(y)$$

for each (x, y) in S. The purpose of this exercise is to establish the formula

$$P(U \times V) = P_1(U)P_2(V) \tag{13.27}$$

for every pair of subsets U in $\mathscr{B}_1$ and V in $\mathscr{B}_2$. The sample spaces S_1 and S_2 are assumed to be finite.

(a) Verify that Equation (13.27) is true when U and V are singletons, and also when at least one of U or V is empty.

Suppose now that

$$U = \{u_1, u_2, \ldots, u_k\} \qquad \text{and} \qquad V = \{v_1, v_2, \ldots, v_m\}.$$

Then $U \times V$ consists of the km pairs (u_i, v_j). For each $i = 1, 2, \ldots, k$ let A_i denote the set of m pairs in $U \times V$ whose first component is u_i.

(b) Show that the A_i are disjoint sets whose union is $U \times V$.

(c) Show that

$$P(A_i) = \sum_{j=1}^{m} P(u_i, v_j) = P_1(u_i)P_2(V).$$

(d) From (b) and (c) deduce that

$$P(U \times V) = \sum_{i=1}^{k} P(A_i) = P_1(U)P_2(V).$$

14

CALCULUS OF PROBABILITIES

14.1 The definition of probability for uncountable sample spaces

A line segment is broken into two pieces, with the point of subdivision chosen at random. What is the probability that the two pieces have equal length? What is the probability that the longer segment has exactly twice the length of the shorter? What is the probability that the longer segment has at least twice the length of the shorter? These are examples of probability problems in which the sample space is uncountable since it consists of all points on a line segment. This section extends the definition of probability to include uncountable sample spaces.

If we were to use the same procedure as for countable sample spaces we would start with an arbitrary uncountable set S and a Boolean σ-algebra $\mathscr{B}$ of subsets of S and define a probability measure to be a completely additive nonnegative set function P defined on $\mathscr{B}$ with $P(S) = 1$. As it turns out, this procedure leads to certain technical difficulties that do not occur when S is countable. To attempt to describe these difficulties would take us too far afield. We shall avoid these difficulties by imposing restrictions at the outset on the set S and on the Boolean algebra $\mathscr{B}$.

First, we restrict S to be a subset of the real line $\mathbf{R}$, or of n-space $\mathbf{R}^n$. For the Boolean algebra $\mathscr{B}$ we use special subsets of S which, in the language of modern integration theory, are called *measurable* subsets of S. We shall not attempt to describe the exact meaning of a measurable set; instead, we shall mention some of the properties possessed by the class of measurable sets.

First we consider subsets of $\mathbf{R}$. The measurable subsets have the following properties:

1. If A is measurable, so is $\mathbf{R} - A$, the complement of A.
2. If $\{A_1, A_2, A_3, \ldots\}$ is a countable collection of measurable sets, then the union $A_1 \cup A_2 \cup A_3 \cup \cdots$ is also measurable.
3. Every interval (open, closed, half-open, finite, or infinite) is measurable.

Thus, the measurable sets of $\mathbf{R}$ form a Boolean σ-algebra which contains the intervals. A smallest Boolean σ-algebra exists which has this property; its members are called *Borel sets*, after the French mathematician Émile Borel (1871–1956).

Similarly, in 2-space a smallest Boolean σ-algebra exists which contains all Cartesian products of pairs of intervals; its members are the two-dimensional Borel sets. Borel sets in n-space are defined in an analogous fashion.

Henceforth, whenever we use a set S of real numbers as a sample space, or, more generally, whenever we use a set S in n-space as a sample space, we shall assume that this set is a Borel set. The Borel subsets of S themselves form a Boolean σ-algebra. These subsets are extensive enough to include all the events that occur in the ordinary applications of probability theory.

DEFINITION OF PROBABILITY FOR UNCOUNTABLE SAMPLE SPACES. *Let S be a subset of $\mathbf{R}^n$, and let $\mathscr{B}$ be the Boolean σ-algebra of Borel subsets of S. A nonnegative completely additive set function P defined on $\mathscr{B}$ with $P(S) = 1$ is called a probability measure. The triple $(S, \mathscr{B}, P)$ is called a probability space.*

14.2 Countability of the set of points with positive probability

For countable sample spaces the probability of an event A is often computed by adding the point probabilities $P(x)$ for all x in A. This process is not fruitful for uncountable sample spaces because, as the next theorem shows, most of the point probabilities are zero.

THEOREM 14.1. *Let $(S, \mathscr{B}, P)$ be a probability space and let T denote the set of all x in S for which $P(x) > 0$. Then T is countable.*

Proof. For each $n = 1, 2, 3, \ldots$, let T_n denote the following subset of S:

$$T_n = \left\{ x \;\middle|\; \frac{1}{n+1} < P(x) \le \frac{1}{n} \right\}.$$

If $P(x) > 0$ then $x \in T_n$ for some n. Conversely, if $x \in T_n$ for some n then $x \in T$. Hence $T = T_1 \cup T_2 \cup \cdots$. Now T_n contains at most n points, because if there were $n + 1$ or more points in T_n the sum of their point probabilities would exceed 1. Therefore T is countable, since it is a countable union of finite sets.

Theorem 14.1 tells us that positive probabilities can be assigned to at most a countable subset of S. The remaining points of S will have probability zero. In particular, if all the outcomes of S are equally likely, then every point in S must be assigned probability zero.

Note: Theorem 14.1 can be given a physical interpretation in terms of mass distribution which helps to illustrate its meaning. Imagine that we have an amount of mass, with the total quantity equal to 1. (This corresponds to $P(S) = 1$.) Suppose we are able to distribute this mass in any way we please along the real line, either by smearing it along the line with a uniform or perhaps a varying thickness, or by placing discrete lumps of mass at certain points, or both. (We interpret a positive amount of mass as a discrete lump.) We can place all the mass at one point. We can divide the mass equally or unequally in discrete lumps among two points, among ten points, among a million points, or among a countably infinite set of points. For example, we can put $\frac{1}{2}$ at 1, $\frac{1}{4}$ at 2, $\frac{1}{8}$ at 3, and so on, with mass $(\frac{1}{2})^n$ at each integer $n \ge 1$. Or we can smear all the mass without any concentrated lumps. Or we can smear part of it and distribute the rest in discrete lumps. Theorem 14.1 tells us that at most a countable set of points can be assigned discrete lumps of mass.

Since most (if not all) the point probabilities for an uncountable sample space will be zero, a knowledge of the point probabilities alone does not suffice to compute the probabilities of arbitrary events. Further information is required; it is best described in terms of two new concepts, *random variables* and *distribution functions*, to which we turn next. These concepts make possible the use of integral calculus in many problems with uncountable sample spaces. Integration takes the place of summation in the computation of probabilities.

14.3 Random variables

In many experiments we are interested in *numbers* associated with the outcomes of the experiment. For example, n coins are tossed simultaneously and we ask for the number of heads. A pair of dice is rolled and we ask for the sum of the points on the upturned faces. A dart is thrown at a circular target and we ask for its distance from the center. Whenever we associate a real number with each outcome of an experiment we are dealing with a *function* whose domain is the set of possible outcomes and whose range is the set of real numbers in question. Such a function is called a *random variable.* A formal definition can be given as follows:

DEFINITION OF A RANDOM VARIABLE. *Let S denote a sample space. A real-valued function defined on S is called a one-dimensional random variable. If the function values are ordered pairs of real numbers (that is, vectors in 2-space), the function is said to be a two-dimensional random variable. More generally, an n-dimensional random variable is simply a function whose domain is the given sample space S and whose range is a collection of n-tuples of real numbers (vectors in n-space).*

Thus, a random variable is nothing but a vector-valued function defined on a set. The term "random" is used merely to remind us that the set in question is a sample space.†

Because of the generality of the above definition, it is possible to have many random variables associated with a given experiment. In any particular example the experimenter must decide which random variables will be of interest and importance to him. In general, we try to work with random variables whose function values reflect, as simply as possible, the properties of the outcomes of the experiment which are really essential.

Notations. Capital letters such as X, Y, Z are ordinarily used to denote one-dimensional random variables. A typical outcome of the experiment (that is, a typical element of the sample space) is usually denoted by the Greek letter ω (omega). Thus, $X(\omega)$ denotes that real number which the random variable X associates with the outcome ω.

The following are some simple examples of random variables.

EXAMPLE 1. An experiment consists of rolling a die and reading the number of points on the upturned face. The most "natural" random variable X to consider is the one stamped on the die by the manufacturer, namely:

$$X(\omega) = \omega \qquad \text{for} \quad \omega = 1, 2, 3, 4, 5, 6.$$

† The terms "stochastic variable" and "chance variable" are also used as synonyms for "random variable." The word "stochastic" is derived from a Greek stem meaning "chance" and seems to have been invented by Jakob Bernoulli. It is commonly used in the literature of probability theory.

If we are interested in whether the number of points is even or odd, then we can consider instead the random variable Y, which is defined as follows:

$$Y(\omega) = 0 \quad \text{if} \quad \omega \text{ is even},$$
$$Y(\omega) = 1 \quad \text{if} \quad \omega \text{ is odd}.$$

The values 0 and 1 are not essential—any two distinct real numbers could be used instead. However, 0 and 1 suggest "even" and "odd," respectively, because they represent the remainder obtained when the outcome ω is divided by 2.

EXAMPLE 2. A dart is thrown at a circular target. The set of all possible outcomes is the set of all points ω on the target. If we imagine a coordinate system placed on the target with the origin at the center, we can assign various random variables to this experiment. A natural one is the two-dimensional random variable which assigns to the point ω its rectangular coordinates (x, y). Another is that which assigns to ω its polar coordinates (r, θ). Examples of one-dimensional random variables are those which assign to each ω just one of its coordinates, such as the x-coordinate or the r-coordinate (distance from the origin). In an experiment of this type we often wish to know the probability that the dart will land in a particular region of the target, for example, the first quadrant. This event can be described most simply by the random variable which assigns to each point ω its polar coordinate angle θ, so that $X(\omega) = \theta$; the event "the dart lands in the first quadrant" is the set of ω such that $0 \le X(\omega) \le \frac{1}{2}\pi$.

Abbreviations. We avoid cumbersome notation by using special abbreviations to describe certain types of events and their probabilities. For example, if t is a real number, the set of all ω in the sample space such that $X(\omega) = t$ is denoted briefly by writing

$$X = t.$$

The probability of this event is written $P(X = t)$ instead of the lengthier $P(\{\omega \mid X(\omega) = t\})$. Symbols such as $P(X = a \text{ or } X = b)$ and $P(a < X \le b)$ are defined in a similar fashion. Thus, the event "$X = a$ or $X = b$" is the union of the two events "$X = a$" and "$X = b$"; the symbol $P(X = a \text{ or } X = b)$ denotes the probability of this union. The event "$a < X \le b$" is the set of all points ω such that $X(\omega)$ lies in the half-open interval $(a, b]$, and the symbol $P(a < X \le b)$ denotes the probability of this event.

14.4 Exercises

1. Let X be a one-dimensional random variable.
   (a) If $a < b$, show that the two events $a < X \le b$ and $X \le a$ are disjoint.
   (b) Determine the union of the two events in part (a).
   (c) Show that $P(a < X \le b) = P(X \le b) - P(X \le a)$.
2. Let (X, Y) denote a two-dimensional random variable defined on a sample space S. This means that (X, Y) is a function which assigns to each ω in S a pair of real numbers $(X(\omega), Y(\omega))$. Of course, each of X and Y is a one-dimensional random variable defined on S. The notation

$$X \le a, \; Y \le b$$

stands for the set of all elements ω in S such that $X(\omega) \leq a$ and $Y(\omega) \leq b$.

(a) If $a < b$ and $c < d$, describe, in terms of elements of S, the meaning of the following notation: $a < X \leq b, c < Y \leq d$.

(b) Show that the two events "$X \leq a, Y \leq c$" and "$X \leq a, c < Y \leq d$" are disjoint. Interpret these events geometrically.

(c) Determine the union of the two events in (b).

(d) Generalize Exercise 1(c) to the two-dimensional case.

3. Two fair dice are rolled, each outcome being an ordered pair (a, b), where each of a and b is an integer from 1 to 6. Let X be the random variable which assigns the value $a + b$ to the outcome (a, b).
(a) Describe, in roster notation, the events "$X = 7$," "$X = 11$," "$X = 7$ or $X = 11$."
(b) Compute the probabilities of the events in part (a).

4. Consider an experiment in which four coins are tossed simultaneously (or one coin is tossed four times). For each coin define a random variable which assigns the value 1 to heads and the value 0 to tails, and denote these random variables by X_1, X_2, X_3, X_4. Assign the probabilities $P(X_i = 1) = P(X_i = 0) = \frac{1}{2}$ for each X_i. Consider a new random variable Y which assigns to each outcome the total number of heads among the four coins. Express Y in terms of X_1, X_2, X_3, X_4 and compute the probabilities $P(Y = 0)$, $P(Y = 1)$, and $P(Y \leq 1)$.

5. A small railroad company has facilities for transporting 100 passengers a day between two cities, at a fixed cost (to the company) of \$7 per passenger. If more than 100 passengers buy tickets in any one day the railroad is obligated to provide bus transportation for the excess at a cost of \$10 per passenger. Let X be the random variable which counts the number of passengers that buy tickets in a given day. The possible values of X are the integers $0, 1, 2, 3, \ldots$ up to a certain unknown maximum. Let Y denote the random variable which describes the total daily cost (in dollars) to the railroad for handling passengers. Express Y in terms of X.

6. A factory production line consists of two work stations A and B. At station A, X units per hour are assembled; they are immediately transported to station B, where they are inspected at the rate of Y units per hour, where $Y < X$. The possible values of X and Y are the integers 8, 9, and 10. Let Z denote the random variable which counts the number of units that come off the production line during the first hour of production.
(a) Express Z in terms of X and Y, assuming each of X and Y is constant during this hour.
(b) Describe, in a similar way, the random variable U which counts the number of units delivered in the first two consecutive hours of production. Each of X and Y is constant during each hour, but the constant values during the second hour need not be the same as those during the first.

14.5 Distribution functions

We turn now to the problem of computing the probabilities of events associated with a given random variable. Let X be a one-dimensional random variable defined on a sample space S, where S is a Borel set in n-space for some $n \geq 1$. Let P be a probability measure defined on the Borel subsets of S. For each ω in S, $X(\omega)$ is a real number, and as ω runs through the elements of S the numbers $X(\omega)$ run through a set of real numbers (the range of X). This set may be finite, countably infinite, or uncountable. For each real number t we consider the following special subset of S:

$$A(t) = \{\omega \mid X(\omega) \leq t\}.$$

If t is less than all the numbers in the range of X, the set $A(t)$ will be empty; otherwise, $A(t)$ will be a nonempty subset of S. We assume that for each t the set $A(t)$ is an *event*, that is, a Borel set. According to the convention discussed at the end of Section 14.3, we denote this event by the symbol $X \le t$.

Suppose we know the probability $P(X \le t)$ for every real t. We shall find in a moment that this knowledge enables us to compute the probabilities of many other events of interest. This is done by using the probabilities $P(X \le t)$ as a basis for constructing a new function F, called the *distribution function* of X. It is defined as follows:

DEFINITION OF A DISTRIBUTION FUNCTION. *Let X be a one-dimensional random variable. The function F defined for all real t by the equation*

$$F(t) = P(X \le t)$$

is called the distribution function of the random variable X.

Note: Sometimes the notation F_X is used to emphasize the fact that the distribution function is associated with the particular random variable X. The value of the function at t is then denoted by $F_X(t)$.

It is important to realize that the distribution function F is defined over the entire real axis, even though the range of X may be only a bounded portion of the real axis. In fact, if all numbers $X(\omega)$ lie in some finite interval $[a, b]$, then for $t < a$ the probability $P(X \le t)$ is zero (since for $t < a$ the set $X \le t$ is empty) and for $t \ge b$ the probability $P(X \le t)$ is 1 (because in this case the set $X \le t$ is the entire sample space). This means that for *bounded* random variables X whose range is within an interval $[a, b]$ we have $F(t) = 0$ for all $t < a$ and $F(t) = 1$ for all $t \ge b$.

We now proceed to derive a number of properties common to all distribution functions.

THEOREM 14.2. *Let F denote a distribution function of a one-dimensional random variable X. Then we have:*

(a) $0 \le F(t) \le 1$ *for all t.*
(b) $P(a < X \le b) = F(b) - F(a)$ *if* $a < b$.
(c) $F(a) \le F(b)$ *if* $a < b$.

Proof. Part (a) follows at once from the definition of F because probabilities always lie between 0 and 1.

To prove (b) we note that the events "$a < X \le b$" and "$X \le a$" are disjoint. Their union is the event "$X \le b$." Using additivity we obtain

$$P(a < X \le b) + P(X \le a) = P(X \le b),$$

which can also be expressed as

$$P(a < X \le b) = P(X \le b) - P(X \le a) = F(b) - F(a).$$

Part (c) follows from (b) since $P(a < X \le b) \ge 0$.

Note: Using the mass analogy, we would say that $F(t)$ represents the total amount of mass located between $-\infty$ and t (including the point t itself). The amount of mass located in a half-open interval $(a, b]$ is $F(b) - F(a)$.

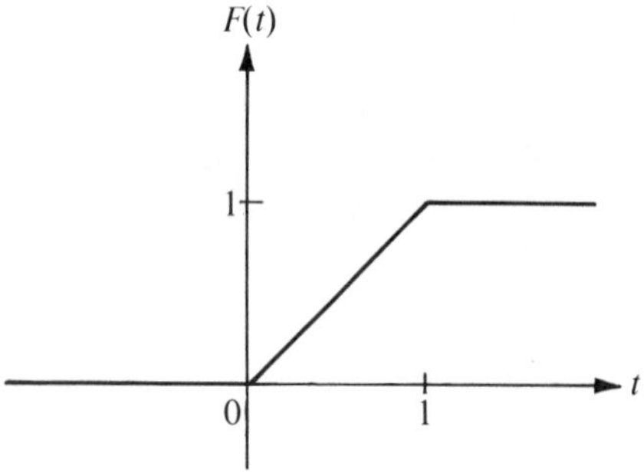

FIGURE 14.1 A distribution function of a bounded random variable.

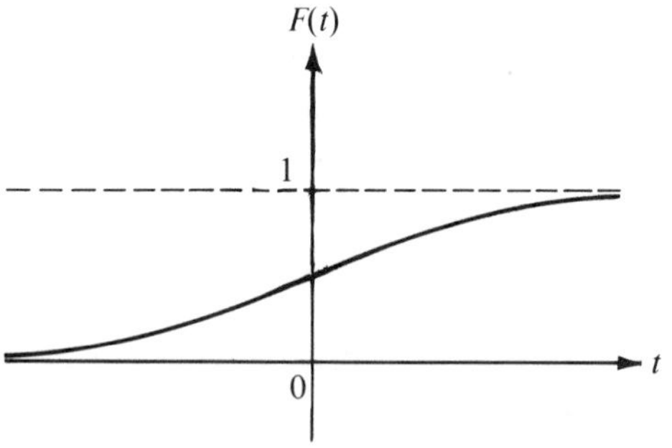

FIGURE 14.2 A distribution function of an unbounded random variable.

Figure 14.1 shows a distribution function of a bounded random variable X whose values $X(\omega)$ lie in the interval $[0, 1]$. This particular example is known as a *uniform distribution.* Here we have

$$F(t) = 0 \quad \text{for} \quad t < 0, \qquad F(t) = t \quad \text{for} \quad 0 \le t \le 1, \qquad F(t) = 1 \quad \text{for} \quad t \ge 1.$$

Figure 14.2 shows an example of a distribution function corresponding to an unbounded random variable. This example is known as a *Cauchy distribution* and its function values are given by the formula

$$F(t) = \frac{1}{2} + \frac{1}{\pi} \arctan t .$$

Experiments that lead to uniform and to Cauchy distributions will be discussed later.

Note: Using the mass analogy, we would say that in Figure 14.1 no mass has been placed to the left of the origin or to the right of the point 1. The entire mass has been distributed over the interval $[0, 1]$. The graph of F is linear over this interval because the mass is smeared with a uniform thickness. In Figure 14.2 the mass has been smeared along the entire axis. The graph is nonlinear because the mass has been smeared with a varying thickness.

Theorem 14.2(b) tells us how to compute (in terms of F) the probability that X lies in a half-open interval of the form $(a, b]$. The next theorem deals with other types of intervals.

THEOREM 14.3. *Let F be a distribution function of a one-dimensional random variable X. Then if $a < b$ we have:*

(a) $P(a \le X \le b) = F(b) - F(a) + P(X = a)$.

(b) $P(a < X < b) = F(b) - F(a) - P(X = b)$.

(c) $P(a \le X < b) = F(b) - F(a) + P(X = a) - P(X = b)$.

Proof. To prove (a) we note that the events "$a < X \le b$" and "$X = a$" are disjoint and their union is "$a \le X \le b$." Using additivity and Theorem 14.2(b) we obtain (a). Parts (b) and (c) are similarly proved.

Note that all four events

$$a < X \le b, \qquad a \le X \le b, \qquad a < X < b, \qquad \text{and} \qquad a \le X < b$$

have equal probabilities if and only if $P(X = a) = 0$ and $P(X = b) = 0$.

The examples shown in Figures 14.1 and 14.2 illustrate two further properties shared by all distribution functions. They are described in the following theorem.

THEOREM 14.4. *Let F be the distribution function of a one-dimensional random variable X. Then we have*

$$\lim_{t \to -\infty} F(t) = 0 \qquad \textit{and} \qquad \lim_{t \to +\infty} F(t) = 1. \tag{14.1}$$

Proof. The existence of the two limits in (14.1) and the fact that each of the two limits lies between 0 and 1 follow at once, since F is a monotonic function whose values lie between 0 and 1.

Let us denote the limits in (14.1) by L_1 and L_2, respectively. To prove that $L_1 = 0$ and that $L_2 = 1$ we shall use the countably additive property of probability. For this purpose we express the whole space S as a countable union of disjoint events:

$$S = \bigcup_{n=1}^{\infty} (-n < X \le -n + 1) \cup \bigcup_{n=0}^{\infty} (n < X \le n + 1).$$

Then, using additivity, we get

$$\begin{aligned} P(S) &= \sum_{n=1}^{\infty} P(-n < X \le -n + 1) + \sum_{n=0}^{\infty} P(n < X \le n + 1) \\ &= \lim_{M \to \infty} \sum_{n=1}^{M} [F(-n + 1) - F(-n)] + \lim_{N \to \infty} \sum_{n=0}^{N} [F(n + 1) - F(n)]. \end{aligned}$$

The sums on the right will telescope, giving us

$$\begin{aligned} P(S) &= \lim_{M \to \infty} [F(0) - F(-M)] + \lim_{N \to \infty} [F(N + 1) - F(0)] \\ &= F(0) - L_1 + L_2 - F(0) = L_2 - L_1. \end{aligned}$$

Since $P(S) = 1$, this proves that $L_2 - L_1 = 1$ or $L_2 = 1 + L_1$. On the other hand, we also have $L_2 \le 1$ and $L_1 \ge 0$. This implies that $L_1 = 0$ and $L_2 = 1$, as asserted.

14.6 Discontinuities of distribution functions

An example of a possible distribution function with discontinuities is shown in Figure 14.3. Using the mass analogy we would say that F has a jump discontinuity at each point which carries a positive amount of mass. As the next theorem shows, the jump is equal to the amount of mass concentrated at that particular point.

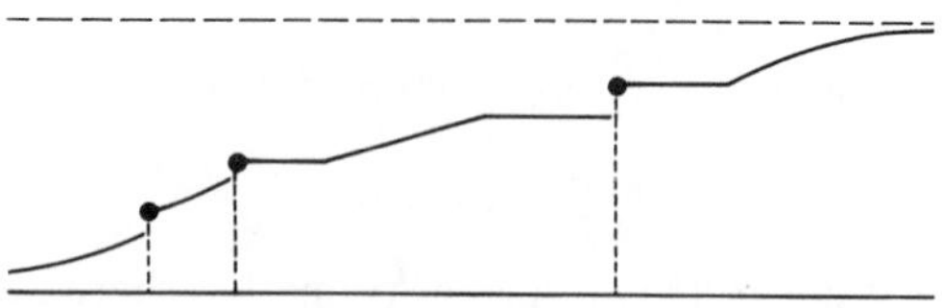

FIGURE 14.3 A possible distribution function.

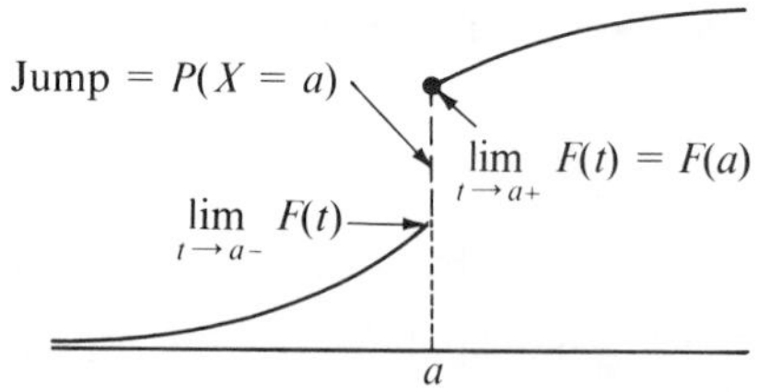

FIGURE 14.4 Illustrating a jump discontinuity of a distribution function.

THEOREM 14.5. *Let F be the distribution function of a one-dimensional random variable X. Then for each real a we have*

$$\lim_{t\to a+} F(t) = F(a) \tag{14.2}$$

and

$$\lim_{t\to a-} F(t) = F(a) - P(X = a). \tag{14.3}$$

Note: The limit relation in (14.2) tells us that F is *continuous from the right* at each point a, because $F(t) \to F(a)$ as $t \to a$ from the right. On the other hand, Equation (14.3) tells us that as $t \to a$ from the left, $F(t)$ will approach $F(a)$ if and only if the probability $P(X = a)$ is zero. When $P(X = a)$ is *not* zero, the graph of F has a jump discontinuity at a of the type shown in Figure 14.4.

Proof. The existence of the limits follows at once from the monotonicity and boundedness of F. We prove now that the limits have the values indicated. For this purpose we use part (b) of Theorem 14.2. If $t > a$ we write

$$F(t) = F(a) + P(a < X \le t); \tag{14.4}$$

if $t < a$ we write

$$F(t) = F(a) - P(t < X \le a). \tag{14.5}$$

Letting $t \to a+$ in (14.4) we find

$$\lim_{t\to a+} F(t) = F(a) + \lim_{t\to a+} P(a < X \le t),$$

whereas if $t \to a-$ in (14.5) we obtain

$$\lim_{t\to a-} F(t) = F(a) - \lim_{t\to a-} P(t < X \le a).$$

Therefore to prove (14.2) and (14.3) we must establish two equations:

$$\lim_{t\to a+} P(a < X \le t) = 0 \tag{14.6}$$

and

(14.7) $$\lim_{t \to a-} P(t < X \leq a) = P(X = a).$$

These can be justified intuitively as follows: When $t \to a+$, the half-open interval $(a, t]$ shrinks to the empty set. That is, the intersection of all half-open intervals $(a, t]$, for $t > a$, is empty. On the other hand, when $t \to a-$ the half-open interval $(t, a]$ shrinks to the point a. (The intersection of all intervals $(t, a]$ for $t < a$ is the set $\{a\}$.) Therefore, if probability behaves in a continuous fashion, Equations (14.6) and (14.7) must be valid. To convert this argument into a rigorous proof we proceed as follows:

For each integer $n \geq 1$, let

(14.8) $$p_n = P\left(a < X \leq a + \frac{1}{n}\right).$$

To prove (14.6) it suffices to show that $p_n \to 0$ as $n \to \infty$. Let S_n denote the event

$$a + \frac{1}{n+1} < X \leq a + \frac{1}{n}.$$

The sets S_n are disjoint and their union $S_1 \cup S_2 \cup S_3 \cup \cdots$ is the event $a < X \leq a + 1$. By countable additivity we have

(14.9) $$\sum_{n=1}^{\infty} P(S_n) = P(a < X \leq a + 1) = p_1.$$

On the other hand, Equation (14.8) implies that

$$p_n - p_{n+1} = P(S_n),$$

so from (14.9) we obtain the relation

(14.10) $$\sum_{n=1}^{\infty} (p_n - p_{n+1}) = p_1.$$

The convergence of the series is a consequence of (14.9). But the series on the left of (14.10) is a telescoping series with sum

$$p_1 - \lim_{n \to \infty} p_n.$$

Therefore (14.10) implies that $\lim_{n \to \infty} p_n = 0$, and this proves (14.6).

A slight modification of this argument enables us to prove (14.7) as well. Since

$$P(t < X \leq a) = P(t < X < a) + P(X = a)$$

we need only prove that

$$\lim_{t \to a-} P(t < X < a) = 0.$$

For this purpose we introduce the numbers

$$q_n = P\left(a - \frac{1}{n} < X < a\right)$$

and show that $q_n \to 0$ as $n \to \infty$. In this case we consider the events T_n given by

$$a - \frac{1}{n} < X \le a - \frac{1}{n+1}$$

for $n = 1, 2, 3, \ldots$. These are disjoint and their union is the event $a - 1 < X < a$, so we have

$$\sum_{n=1}^{\infty} P(T_n) = P(a - 1 < X < a) = q_1 .$$

We now note that $q_n - q_{n+1} = P(T_n)$, and we complete the proof as above.

The most general type of distribution is any real-valued function F that has the following properties:

(a) F is monotonically increasing on the real axis,
(b) F is continuous from the right at each point,
(c) $\lim_{t \to -\infty} F(t) = 0$ and $\lim_{t \to +\infty} F(t) = 1$.

In fact, it can be shown that for each such function F there is a corresponding set function P, defined on the Borel sets of the real line, such that P is a probability measure which assigns the probability $F(b) - F(a)$ to each half-open interval $(a, b]$. For a proof of this statement, see H. Cramér, *Mathematical Methods of Statistics*, Princeton University Press, Princeton, N.J., 1946.

There are two special types of distributions, known as *discrete* and *continuous*, that are of particular importance in practice. In the discrete case the entire mass is concentrated at a finite or countably infinite number of points, whereas in the continuous case the mass is smeared, in uniform or varying thickness, along an interval (finite or infinite). These two types of distributions will be treated in detail in the next few sections.

14.7 Discrete distributions. Probability mass functions

Let X be a one-dimensional random variable and consider a new function p, called the *probability mass function* of X. Its values $p(t)$ are defined for every real number t by the equation

$$p(t) = P(X = t).$$

That is, $p(t)$ is the probability that X takes the value t. When we want to emphasize that p is associated with X we write p_X instead of p and $p_X(t)$ instead of $p(t)$.

The set of real numbers t for which $p(t) > 0$ is either finite or countable. We denote this set by T; that is, we let

$$T = \{t \mid p(t) > 0\}.$$

The random variable X is said to be *discrete* if

$$\sum_{t \in T} p(t) = 1 .$$

In other words, X is discrete if a unit probability mass is distributed over the real line by concentrating a positive mass $p(t)$ at each point t of some finite or countably infinite set T and no mass at the remaining points. The points of T are called the *mass points* of X.

For discrete random variables a knowledge of the probability mass function enables us to compute probabilities of arbitrary events. In fact, we have the following theorem.

THEOREM 14.6. *Let A be a Borel subset of the real line* **R**, *and let $P(X \in A)$ denote the probability of the set of ω such that $X(\omega) \in A$. Then we have*

$$P(X \in A) = \sum_{x \in A \cap T} p(x), \tag{14.11}$$

where T is the set of mass points of X.

Proof. Since $A \cap T \subseteq A$ and $T - A \subseteq \mathbf{R} - A$, we have

$$\sum_{x \in A \cap T} p(x) \leq P(X \in A) \quad \text{and} \quad \sum_{x \in T - A} p(x) \leq P(X \notin A). \tag{14.12}$$

But $A \cap T$ and $T - A$ are disjoint sets whose union is T, so the second inequality in (14.12) is equivalent to

$$1 - \sum_{x \in A \cap T} p(x) \leq 1 - P(X \in A) \quad \text{or} \quad \sum_{x \in A \cap T} p(x) \geq P(X \in A).$$

Combining this with the first inequality in (14.12) we obtain (14.11).

Note: Since $p(x) = 0$ when $x \notin T$, the sum on the right of (14.11) can be written as $\sum_{x \in A} p(x)$ without danger of its being misunderstood.

When A is the interval $(-\infty, t]$, the sum in (14.11) gives the value of the distribution function $F(t)$. Thus, we have

$$F(t) = P(X \leq t) = \sum_{x \leq t} p(x).$$

If a random variable X is discrete, the corresponding distribution function F is also called discrete.

The following examples of discrete distributions occur frequently in practice.

EXAMPLE 1. *Binomial distribution.* Let p be a given real number satisfying $0 \leq p \leq 1$ and let $q = 1 - p$. Suppose a random variable X assumes the values $0, 1, 2, \ldots, n$, where n is a fixed positive integer, and suppose the probability $P(X = k)$ is given by the formula

$$P(X = k) = \binom{n}{k} p^k q^{n-k} \quad \text{for} \quad k = 0, 1, 2, \ldots, n.$$

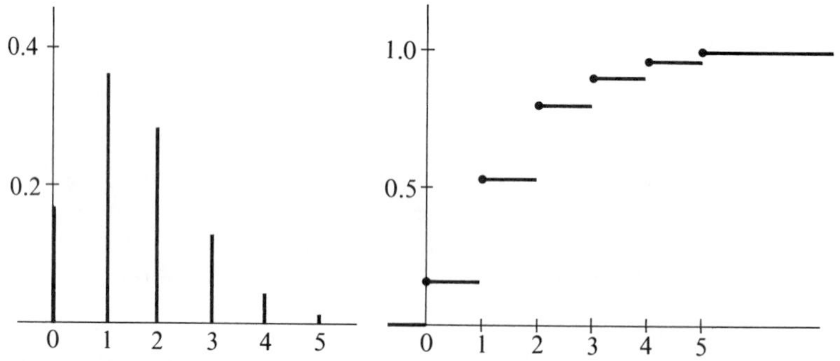

(a) The mass function (b) The distribution function

FIGURE 14.5 The probability mass function and the distribution function of a binomial distribution with parameters $n = 5$ and $p = \frac{1}{3}$.

This assignment of probabilities is permissible because the sum of all the point probabilities is

$$\sum_{k=0}^{n} P(X = k) = \sum_{k=0}^{n} \binom{n}{k} p^k q^{n-k} = (p + q)^n = 1\,.$$

The corresponding distribution function F_X is said to be a *binomial distribution* with parameters n and p. Its values may be computed by the summation formula

$$F_X(t) = \sum_{0 \le k \le t} \binom{n}{k} p^k q^{n-k}.$$

Binomial distributions arise naturally from a Bernoulli sequence of trials where p is the probability of "success" and q the probability of "failure." In fact, when the random variable X counts the number of successes in n trials, $P(X = k)$ is precisely $\binom{n}{k} p^k q^{n-k}$ because of Bernoulli's formula. (See Theorem 13.3 in Section 13.16.) Figure 14.5 shows the graphs of the probability mass function and the corresponding distribution function for a binomial distribution with parameters $n = 5$ and $p = \frac{1}{3}$.

EXAMPLE 2. *Poisson distribution.* Let λ be a positive real number and let a random variable X assume the values 0, 1, 2, 3, ... If the probability $P(X = k)$ is given by the formula

$$P(X = k) = \frac{e^{-\lambda}\lambda^k}{k!} \qquad \text{for} \quad k = 0, 1, 2, \ldots,$$

the corresponding distribution function F_X is said to be a *Poisson distribution* with parameter λ. It is so named in honor of the French mathematician S. D. Poisson (1781–1840). This assignment of probabilities is permissible because

$$\sum_{k=0}^{\infty} P(X = k) = e^{-\lambda} \sum_{k=0}^{\infty} \frac{\lambda^k}{k!} = e^{-\lambda} e^{\lambda} = 1\,.$$

The values of the distribution function are computed from the partial sums

$$F_X(t) = e^{-\lambda} \sum_{0 \le k \le t} \frac{\lambda^k}{k!}.$$

The Poisson distribution is applicable to many problems involving random events occurring in time, such as traffic accidents, connections to wrong numbers in a telephone exchange, and chromosome interchanges in cells induced by x-ray radiation. Some specific applications are discussed in the books by Feller and Parzen listed at the end of this chapter.

14.8 Exercises

1. A perfectly balanced die is rolled. For a random variable X we take the function which counts the number of points on the upturned face. Draw a graph of the corresponding distribution function F_X.
2. Two dice are rolled. Let X denote the random variable which counts the total number of points on the upturned faces. Construct a table giving the nonzero values of the probability mass function p_X and draw a graph of the corresponding distribution function F_X.
3. The distribution function F of a random variable X is given by the following formulas:

$$F(t) = \begin{cases} 0 & \text{if } t < -2, \\ \frac{1}{2} & \text{if } -2 \le t < 0, \\ \frac{3}{4} & \text{if } 0 \le t < 2, \\ 1 & \text{if } t \ge 2. \end{cases}$$

   (a) Sketch the graph of F.
   (b) Describe the probability mass function p and draw its graph.
   (c) Compute the following probabilities: $P(X = 1)$, $P(X \le 1)$, $P(X < 1)$, $P(X = 2)$, $P(X \le 2)$, $P(0 < X < 2)$, $P(0 < X \le 2)$, $P(1 \le X \le 2)$.
4. Consider a random variable X whose possible values are all rational numbers of the form $\dfrac{n}{n+1}$ and $\dfrac{n+1}{n}$, where $n = 1, 2, 3, \ldots$. If

$$P\left(X = \frac{n}{n+1}\right) = P\left(X = \frac{n+1}{n}\right) = \frac{1}{2^{n+1}},$$

   verify that this assignment of probabilities is permissible and sketch the general shape of the graph of the distribution function F_X.
5. The probability mass function p of a random variable X is zero except at points $t = 0, 1, 2$. At these points it has the values

$$p(0) = 3c^3, \qquad p(1) = 4c - 10c^2, \qquad p(2) = 5c - 1,$$

   for some $c > 0$.
   (a) Determine the value of c.
   (b) Compute the following probabilities: $P(X < 1)$, $P(X < 2)$, $P(1 < X \le 2)$, $P(0 < X < 3)$.
   (c) Describe the distribution function F and sketch its graph.
   (d) Find the largest t such that $F(t) < \frac{1}{2}$.
   (e) Find the smallest t such that $F(t) > \frac{1}{3}$.

6. A random variable X has a binomial distribution with parameters $n = 4$ and $p = \frac{1}{3}$.
   (a) Describe the probability mass function p and sketch its graph.
   (b) Describe the distribution function F and sketch its graph.
   (c) Compute the probabilities $P(1 < X \le 2)$ and $P(1 \le X \le 2)$.
7. Assume that if a thumbtack is tossed on a table, it lands either with point up or in a stable position with point resting on the table. Assume there is a positive probability p that it lands with point up.
   (a) Suppose two identical tacks are tossed simultaneously. Assuming stochastic independence, show that the probability that both land with point up is p^2.
   (b) Continuing part (a), let X denote the random variable which counts the number of tacks which land with point up (the possible values of X are 0, 1, and 2). Compute the probabilities $P(X = 0)$ and $P(X = 1)$.
   (c) Draw the graph of the distribution function F_X when $p = \frac{1}{3}$.
8. Given a random variable X whose possible values are $1, 2, \ldots, n$. Assume that the probability $P(X = k)$ is proportional to k. Determine the constant of proportionality, the probability mass function p_X, and the distribution function F_X.
9. Given a random variable X whose possible values are $0, 1, 2, 3, \ldots$. Assume that $P(X = k)$ is proportional to $c^k/k!$, where c is a fixed real number. Determine the constant of proportionality and the probability mass function p.
10. (a) A fair die is rolled. The sample space is $S = \{1, 2, 3, 4, 5, 6\}$. If the number of points on the upturned face is odd a player receives one dollar; otherwise he must pay one dollar. Let X denote the random variable which measures his financial outcome (number of dollars) on each play of the game. (The possible values of X are $+1$ and -1.) Describe the probability mass function p_X and the distribution F_X. Sketch their graphs.
    (b) A fair coin is tossed. The sample space $S = \{H, T\}$. If the outcome is heads a player receives one dollar; if it is tails he must pay one dollar. Let Y denote the random variable which measures his financial outcome (number of dollars) on each play of the game. Show that the mass function p_Y and the distribution F_Y are identical to those in part (a). This example shows that different random variables may have the same probability distribution function. Actually, there are infinitely many random variables having a given probability distribution F. (Why?) Such random variables are said to be *identically distributed.* Each theorem concerning a particular distribution function is applicable to any of an infinite collection of random variables having this distribution.
11. The number of minutes that one has to wait for a train at a certain subway station is known to be a random variable X with the following probability mass function:

$$p(t) = 0 \qquad \text{unless} \quad t = 3k/10 \qquad \text{for some} \quad k = 0, 1, 2, \ldots, 10.$$

$$p(t) = \tfrac{1}{12} \qquad \text{if} \quad t = 0, 0.3, 0.6, 0.9, 2.1, 2.4, 2.7, 3.0.$$

$$p(t) = \tfrac{1}{9} \qquad \text{if} \quad t = 1.2, 1.5, 1.8.$$

    Sketch the graph of the corresponding distribution function F. Let A be the event that one has to wait between 0 and 2 minutes (including 0 and 2), and let B be the event that one has to wait between 1 and 3 minutes (including 1 and 3). Compute the following probabilities: $P(A)$, $P(B)$, $P(A \cap B)$, $P(B \mid A)$, $P(A \cup B)$.
12. (a) If $0 < p < 1$ and $q = 1 - p$, show that

$$\binom{n}{k} p^k q^{n-k} = \frac{(np)^k}{k!}\left(1 - \frac{np}{n}\right)^n Q_n,$$

where

$$Q_n = \frac{\prod_{r=2}^{k}\left(1 - \frac{r-1}{n}\right)}{(1-p)^k}.$$

(b) Given $\lambda > 0$, let $p = \lambda/n$ for $n > \lambda$. Show that $Q_n \to 1$ as $n \to \infty$ and that

$$\binom{n}{k} p^k q^{n-k} \to \frac{\lambda^k}{k!} e^{-\lambda} \qquad \text{as} \quad n \to \infty.$$

This result suggests that for large n and small p, the binomial distribution is approximately the same as the Poisson distribution, provided the product np is nearly constant; this constant is the parameter λ of the Poisson distribution.

14.9 Continuous distributions. Density functions

Let X be a one-dimensional random variable and let F be its distribution function, so that $F(t) = P(X \le t)$ for every real t. If the probability $P(X = t)$ is zero for every t then, because of Theorem 14.5, F is continuous everywhere on the real axis. In this case F is called a *continuous distribution* and X is called a *continuous random variable.* If the derivative F' exists and is continuous on an interval $[a, t]$ we can use the second fundamental theorem of calculus to write

$$F(t) - F(a) = \int_a^t f(u)\, du, \tag{14.13}$$

where f is the derivative of F. The difference $F(t) - F(a)$ is, of course, the probability $P(a < X \le t)$, and Equation (14.13) expresses this probability as an integral.

Sometimes the distribution function F can be expressed as an integral of the form (14.13), in which the integrand f is integrable but not necessarily continuous. Whenever an equation such as (14.13) holds for all intervals $[a, t]$, the integrand f is called *a probability density function* of the random variable X (or of the distribution F) provided that f is nonnegative. In other words, we have the following definition:

DEFINITION OF A PROBABILITY DENSITY FUNCTION. *Let X be a one-dimensional random variable with a continuous distribution function F. A nonnegative function f is called a probability density of X (or of F) if f is integrable on every interval $[a, t]$ and if*

$$F(t) - F(a) = \int_a^t f(u)\, du. \tag{14.14}$$

If we let $a \to -\infty$ in (14.14) then $F(a) \to 0$ and we obtain the important formula

$$F(t) = P(X \le t) = \int_{-\infty}^t f(u)\, du, \tag{14.15}$$

valid for all real t. If we now let $t \to +\infty$ and remember that $F(t) \to 1$ we find that

$$\int_{-\infty}^{+\infty} f(u)\, du = 1. \tag{14.16}$$

For discrete random variables the sum of all the probabilities $P(X = t)$ is equal to 1. Formula (14.16) is the continuous analog of this statement. There is also a strong analogy between formulas (14.11) and (14.15). The density function f plays the same role for continuous distributions that the probability mass function p plays for discrete distributions — integration takes the place of summation in the computation of probabilities. There is one important difference, however. In the discrete case $p(t)$ is the probability that $X = t$, but in the continuous case $f(t)$ is *not* the probability that $X = t$. In fact, this probability is zero because F is continuous for every t. Of course, this also means that for a continuous distribution we have

$$P(a \le X \le b) = P(a < X < b) = P(a < X \le b) = P(a \le X < b).$$

If F has a density f each of these probabilities is equal to the integral $\int_a^b f(u)\,du$.

Note: A given distribution can have more than one density since the value of the integrand in (14.14) can be changed at a finite number of points without altering the integral. But if f is *continuous* at t then $f(t) = F'(t)$; in this case the value of the density function at t is uniquely determined by F.

Since f is nonnegative, the right-hand member of Equation (14.14) can be interpreted geometrically as the area of that portion of the ordinate set of f lying to the left of the line $x = t$. The area of the entire ordinate set is equal to 1. The area of the portion of the ordinate set above a given interval (whether it is open, closed, or half-open) is the probability that the random variable X takes on a value in that interval. Figure 14.6 shows an example of a continuous distribution function F and its density function f. The ordinate $F(t)$ in Figure 14.6(a) is equal to the area of the shaded region in Figure 14.6(b).

The next few sections describe some important examples of continuous distributions.

14.10 Uniform distribution over an interval

A one-dimensional random variable X is said to have a uniform distribution function F over a finite interval $[a, b]$ if F is given by the following formulas:

$$F(t) = \begin{cases} 0 & \text{if } t \le a, \\ \dfrac{t-a}{b-a} & \text{if } a < t < b, \\ 1 & \text{if } t \ge b. \end{cases}$$

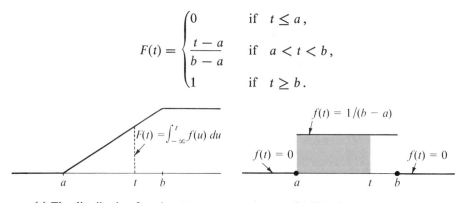

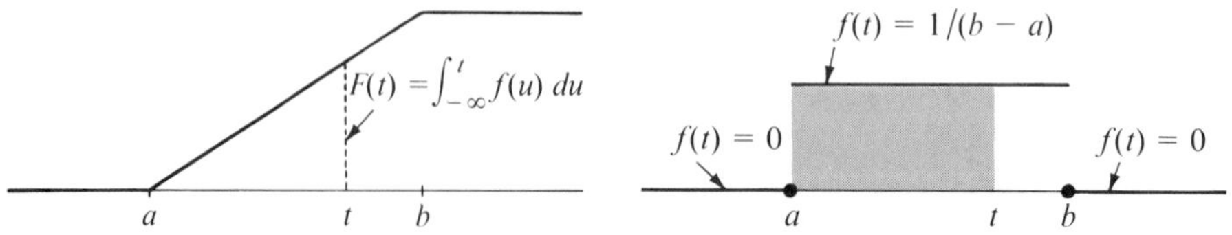

(a) The distribution function F. (b) The density function f.

FIGURE 14.6 A uniform distribution over an interval $[a, b]$ and the corresponding density function.

This is a continuous distribution whose graph is shown in Figure 14.6(a).

The derivative $F'(t)$ exists everywhere except at the points a and b, and we can write

$$F(t) = \int_{-\infty}^{t} f(u)\,du\,,$$

where f is the density function, defined as follows:

$$f(t) = \begin{cases} 1/(b-a) & \text{if } a < t < b\,, \\ 0 & \text{otherwise.} \end{cases}$$

The graph of f is shown in Figure 14.6(b).

The next theorem characterizes uniform distributions in another way.

THEOREM 14.7. *Let X be a one-dimensional random variable with all its values in a finite interval $[a, b]$, and let F be the distribution function of X. Then F is uniform over $[a, b]$ if and only if*

$$P(X \in I) = P(X \in J) \tag{14.17}$$

for every pair of subintervals I and J of $[a, b]$ having the same length, in which case we have

$$P(X \in I) = \frac{h}{b-a}\,,$$

where h is the length of I.

Proof. Assume first that X has a uniform distribution over $[a, b]$. If $[c, c + h]$ is any subinterval of $[a, b]$ of length h we have

$$P(c \le X \le c + h) = F(c + h) - F(c) = \frac{c + h - a}{b - a} - \frac{c - a}{b - a} = \frac{h}{b - a}\,.$$

This shows that $P(X \in I) = P(X \in J) = h/(b - a)$ for every pair of subintervals I and J of $[a, b]$ of length h.

To prove the converse, assume that X satisfies (14.17). First we note that $F(t) = 0$ if $t < a$ and $F(t) = 1$ if $t > b$, since X has all its values in $[a, b]$.

Introduce a new function g defined on the half-open interval $(0, b - a]$ by the equation

$$g(u) = P(a < X \le a + u) \qquad \text{if} \quad 0 < u \le b - a\,. \tag{14.18}$$

Using additivity and property (14.17) we find

$$\begin{aligned} g(u + v) &= P(a < X \le a + u + v) \\ &= P(a < X \le a + u) + P(a + u < X \le a + u + v) \\ &= g(u) + P(a < X \le a + v) = g(u) + g(v), \end{aligned}$$

provided that $0 < u + v \le b - a$. That is, g satisfies the functional equation

$$g(u + v) = g(u) + g(v)$$

for all u and v such that $u > 0$, $v > 0$, $u + v \le b - a$. This is known as *Cauchy's functional equation*. In a moment we shall prove that every nonnegative solution of Cauchy's functional equation is given by

$$g(u) = \frac{u}{b - a}\, g(b - a) \qquad \text{for} \quad 0 < u \le b - a .$$

Using this in Equation (14.18) we find that for $0 < u \le b - a$ we have

$$P(a < X \le a + u) = \frac{u}{b - a} P(a < X \le b) = \frac{u}{b - a}$$

since $P(a < X \le b) = 1$. In other words,

$$F(a + u) - F(a) = \frac{u}{b - a} \qquad \text{if} \quad 0 < u \le b - a .$$

We put $t = a + u$ and rewrite this as

$$F(t) - F(a) = \frac{t - a}{b - a} \qquad \text{if} \quad a < t \le b .$$

But $F(a) = 0$ since F is continuous from the right. Hence

$$F(t) = \frac{t - a}{b - a} \qquad \text{if} \quad a \le t \le b ,$$

which proves that F is uniform on $[a, b]$.

THEOREM 14.8. SOLUTION OF CAUCHY'S FUNCTIONAL EQUATION. *Let g be a real-valued function defined on a half-open interval* $(0, c]$ *and satisfying the following two properties:*

(a) $g(u + v) = g(u) + g(v)$ *whenever u, v, and $u + v$ are in* $(0, c]$,

and

(b) *g is nonnegative on* $(0, c]$.

Then g is given by the formula

$$g(u) = \frac{u}{c}\, g(c) \qquad \textit{for} \quad 0 < u \le c .$$

Proof. By introducing a change of scale we can reduce the proof to the special case in which $c = 1$. In fact, let

$$G(x) = g(cx) \qquad \text{for} \quad 0 < x \le 1 .$$

Then G is nonnegative and satisfies the Cauchy functional equation

$$G(x + y) = G(x) + G(y)$$

whenever x, y, and $x + y$ are in $(0, 1]$. If we prove that

$$G(x) = xG(1) \quad \text{for} \quad 0 < x \leq 1 \tag{14.19}$$

it follows that $g(cx) = xg(c)$, or that $g(u) = (u/c)g(c)$ for $0 < u \leq c$.

If x is in $(0, 1]$ then $x/2$ is also in $(0, 1]$ and we have

$$G(x) = G\left(\frac{x}{2}\right) + G\left(\frac{x}{2}\right) = 2G\left(\frac{x}{2}\right).$$

By induction, for each x in $(0, 1]$ we have

$$G(x) = nG\left(\frac{x}{n}\right) \quad \text{for} \quad n = 1, 2, 3, \ldots . \tag{14.20}$$

Similarly, if y and my are in $(0, 1]$ we have

$$G(my) = mG(y) \quad \text{for} \quad m = 1, 2, 3, \ldots .$$

Taking $y = x/n$ and using (14.20) we obtain

$$G\left(\frac{m}{n}x\right) = \frac{m}{n}G(x)$$

if x and mx/n are in $(0, 1]$. In other words, we have

$$G(rx) = rG(x) \tag{14.21}$$

for every positive rational number r such that x and rx are in $(0, 1]$.

Now take any x in the open interval $(0, 1)$ and let $\{r_n\}$ and $\{R_n\}$ be two sequences of rational numbers in $(0, 1]$ such that

$$r_n < x < R_n \qquad \text{and such that} \qquad \lim_{n\to\infty} r_n = \lim_{n\to\infty} R_n = x.$$

Cauchy's functional equation and the nonnegative property of G show that $G(x + y) \geq G(x)$ so G is monotonic increasing in $(0, 1]$. Therefore

$$G(r_n) \leq G(x) \leq G(R_n).$$

Using (14.21) we rewrite this as

$$r_nG(1) \leq G(x) \leq R_nG(1).$$

Letting $n \to \infty$ we find $xG(1) \leq G(x) \leq xG(1)$, so $G(x) = xG(1)$, which proves (14.19).

Note: Uniform distributions are often used in experiments whose outcomes are points selected at random from an interval $[a, b]$, or in experiments involving an interval $[a, b]$ as a target, where aiming is impossible. The terms "at random" and "aiming is impossible" are usually interpreted to mean that if I is any subinterval of $[a, b]$ then the probability $P(X \in I)$ depends only on the length of I and not on its location in $[a, b]$. Theorem 14.7 shows that uniform distributions are the only distributions with this property.

We turn now to the probability questions asked at the beginning of this chapter.

EXAMPLE. A line segment is broken into two pieces, with the point of subdivision chosen at random. Let X denote the random variable which measures the ratio of the length of the left-hand piece to that of the right-hand piece. Determine the probability distribution function F_X.

Solution. Use the interval $[0, 1]$ to represent the line segment and let the point of subdivision be described by the random variable $Y(\omega) = \omega$ for each ω in $(0, 1)$. Since the point of subdivision is chosen at random we assume that Y has a uniform distribution function F_Y over $[0, 1]$. Hence

$$F_Y(t) = t \qquad \text{for} \quad 0 \le t \le 1 .$$

If the segment is broken at ω, then $\omega/(1 - \omega)$ is the ratio of the length of the left-hand piece to that of the right-hand piece. Therefore $X(\omega) = \omega/(1 - \omega)$.

If $t < 0$ we have $F_X(t) = 0$ since the ratio $X(\omega)$ cannot be negative. If $t \ge 0$, the inequality $X(\omega) \le t$ is equivalent to $\omega/(1 - \omega) \le t$, which is equivalent to $\omega \le t/(1 + t)$. Therefore

$$F_X(t) = P(X \le t) = P\left(Y \le \frac{t}{1+t}\right) = F_Y\left(\frac{t}{1+t}\right) = \frac{t}{1+t}$$

since $0 \le t/(1 + t) < 1$.

Now we can calculate various probabilities. For example, the probability that the two pieces have equal length is $P(X = 1) = 0$. In fact, since F_X is a continuous distribution, the probability that X takes any particular value is zero.

The probability that the left-hand segment is at least twice as long as the right-hand segment is $P(X \ge 2) = 1 - P(X < 2) = 1 - \frac{2}{3} = \frac{1}{3}$. Similarly, the probability that the right-hand segment is at least twice as long as the left-hand segment is $P(X \le \frac{1}{2}) = \frac{1}{3}$. The probability that the longer segment is at least twice as long as the shorter segment is $P(X \ge 2) + P(X \le \frac{1}{2}) = \frac{2}{3}$.

14.11 Cauchy's distribution

A random variable X is said to have a Cauchy distribution F if

$$F(t) = \frac{1}{2} + \frac{1}{\pi} \arctan t$$

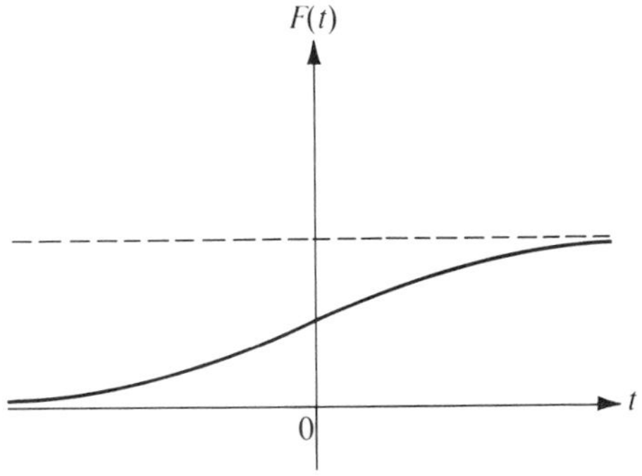

(a) The distribution function F.

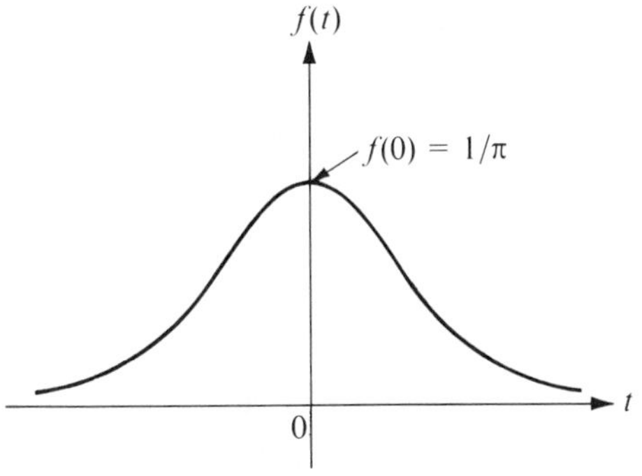

(b) The density function f.

FIGURE 14.7 Cauchy's distribution function and the corresponding density function.

for all real t. This function has a continuous derivative everywhere; a continuous density function f is given by the formula

$$f(t) = \frac{1}{\pi(1 + t^2)}.$$

The graphs of F and f are shown in Figures 14.7(a) and (b), respectively.

The following experiment leads to a Cauchy distribution. A pointer pivoted at the point $(-1, 0)$ on the x-axis is spun and allowed to come to rest. An outcome of the experiment is θ, the angle of inclination from the x-axis made by a line drawn through the pointer; θ is measured so that $-\frac{1}{2}\pi < \theta \le \frac{1}{2}\pi$. Let X be the random variable defined by $X(\theta) = \theta$, and let Y be the random variable which measures the y-intercept of the line through the pointer. If θ is the angle described above, then

$$Y(\theta) = \tan \theta .$$

We shall prove that Y has a Cauchy distribution F_Y if X has a uniform distribution over $[-\frac{1}{2}\pi, \frac{1}{2}\pi]$.

If $a < t$ let $\alpha = \arctan a$ and let $\theta = \arctan t$. Then we have

$$F_Y(t) - F_Y(a) = P(a < Y \le t) = P(\alpha < X \le \theta) = \int_\alpha^\theta f_X(u)\, du = \frac{\theta - \alpha}{\pi}.$$

Since $\alpha \to -\frac{1}{2}\pi$ as $a \to -\infty$ we find

$$F_Y(t) = \frac{\theta + \frac{1}{2}\pi}{\pi} = \frac{1}{\pi} \arctan t + \frac{1}{2}.$$

This shows that Y has a Cauchy distribution, as asserted.

14.12 Exercises

1. A random variable X has a continuous distribution function F, where

$$F(t) = \begin{cases} 0 & \text{if } t \le 0, \\ ct & \text{if } 0 \le t \le 1, \\ 1 & \text{if } t > 1. \end{cases}$$

(a) Determine the constant c and describe the density function f.
(b) Compute the probabilities $P(X = \frac{1}{3})$, $P(X < \frac{1}{3})$, $P(|X| < \frac{1}{3})$.

2. Let $f(t) = c\,|\sin t|$ for $|t| < \pi/2$ and let $f(t) = 0$ otherwise. Determine the value of the constant c so that f will be the density of a continuous distribution function F. Also, describe F and sketch its graph.
3. Solve Exercise 2 if $f(t) = c(4t - 2t^2)$ for $0 \le t \le 2$, and $f(t) = 0$ otherwise.
4. The time in minutes that a person has to wait for a bus is known to be a random variable with density function f given by the following formulas:

$$f(t) = \tfrac{1}{2} \quad \text{for} \quad 0 < t < 1, \qquad f(t) = \tfrac{1}{4} \quad \text{for} \quad 2 < t < 4, \qquad f(t) = 0 \quad \text{otherwise.}$$

Calculate the probability that the time a person has to wait is (a) more than one minute; (b) more than two minutes; (c) more than three minutes.

5. A random variable X has a continuous distribution function F and a probability density f. The density has the following properties: $f(t) = 0$ if $t < \frac{1}{4}$, $f(\frac{1}{4}) = 1$, $f(t)$ is linear if $\frac{1}{4} \le t \le \frac{1}{2}$, $f(1 - t) = f(t)$ for all t.
(a) Make a sketch of the graph of f.
(b) Give a set of formulas for determining F and sketch its graph.
(c) Compute the following probabilities: $P(X < 1)$, $P(X < \frac{3}{4})$, $P(X < \frac{1}{2})$, $P(X \le \frac{1}{4})$, $P(\frac{1}{2} < X < \frac{5}{8})$.
6. A random variable X has a uniform distribution over $[-3, 3]$.
(a) Compute $P(X = 2)$, $P(X < 2)$, $P(|X| < 2)$, $P(|X - 2| < 2)$.
(b) Find a t for which $P(X > t) = \frac{1}{3}$.
7. The Lethe Subway Company schedules a northbound train every 30 minutes at a certain station. A man enters the station at a random time. Let the random variable X count the number of minutes he has to wait for the next train. Assume X has a uniform distribution over the interval $[0, 30]$. (This is how we interpret the statement that he enters the station at "random time.")
(a) For each $k = 5, 10, 15, 20, 25, 30$, compute the probability that he has to wait at least k minutes for the next train.
(b) A competitor, the Styx Subway Company, is allowed to schedule a northbound train every 30 minutes at the same station, but at least 5 minutes must elapse between the arrivals of competitive trains. Assume the passengers come into the station at random times and always board the first train that arrives. Show that the Styx Company can arrange its schedule so that it receives five times as many passengers as its competitor.
8. Let X be a random variable with a uniform distribution F_X over the interval $[0, 1]$. Let $Y = aX + b$, where $a > 0$. Determine the distribution function F_Y and sketch its graph.
9. A roulette wheel carries the integers from 0 to 36, distributed among 37 arcs of equal length. The wheel is spun and allowed to come to rest, and the point on the circumference next to a fixed pointer is recorded. Consider this point as a random variable X with a uniform distribution. Calculate the probability that X lies in an arc containing (a) the integer 0; (b) an integer n in the interval $11 \le n \le 20$; (c) an odd integer.

10. A random variable is said to have a Cauchy distribution with parameters a and b, where $a > 0$, if its density function is given by

$$f(t) = \frac{1}{\pi} \frac{a}{a^2 + (t-b)^2}.$$

Verify that the integral of f from $-\infty$ to $+\infty$ is 1, and determine the distribution function F.

11. Let $f_1(t) = 1$ for $0 < t < 1$, and let $f_1(t) = 0$ otherwise. Define a sequence of functions $\{f_n\}$ by the recursion formula

$$f_{n+1}(x) = \int_{-\infty}^{\infty} f_1(x - t) f_n(t)\, dt.$$

(a) Prove that $f_{n+1}(x) = \int_{x-1}^{x} f_n(t)\, dt$.
(b) Make a sketch showing the graphs of f_1, f_2, and f_3.

12. Refer to Exercise 11. Prove that each function f_n is a probability density.

14.13 Exponential distributions

Let λ be a positive constant. A one-dimensional random variable X is said to have an exponential distribution F with parameter λ if

$$F(t) = \begin{cases} 1 - e^{-\lambda t} & \text{for } t \geq 0, \\ 0 & \text{for } t < 0. \end{cases}$$

A corresponding density function f is given by the formulas

$$f(t) = \begin{cases} \lambda e^{-\lambda t} & \text{for } t \geq 0, \\ 0 & \text{for } t < 0. \end{cases}$$

The graphs of F and f are like those shown in Figure 14.8.

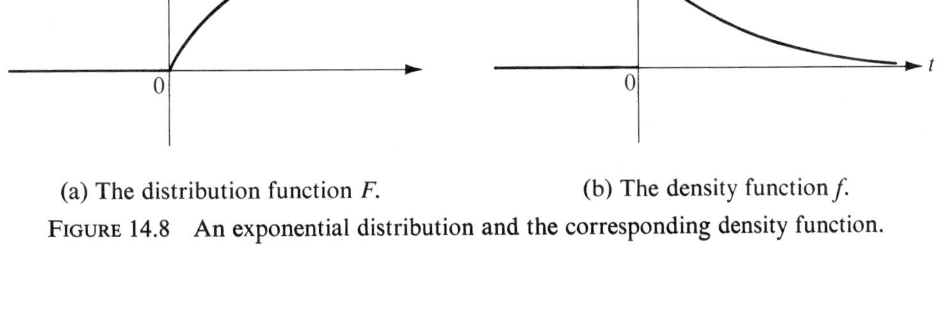

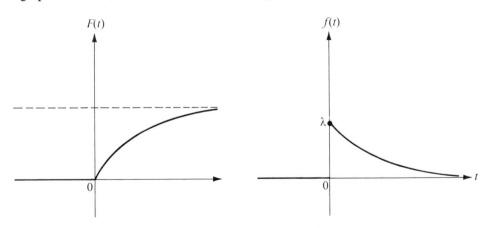

(a) The distribution function F. (b) The density function f.

FIGURE 14.8 An exponential distribution and the corresponding density function.

Exponential distributions have a characteristic property which suggests their use in certain problems involving radioactive decay, traffic accidents, and failure of electronic equipment such as vacuum tubes. This property is analogous to that which characterizes uniform distributions and can be described as follows.

Let X denote the observed waiting time until a piece of equipment fails, and let F be the distribution function of X. We assume that $F(t) = 0$ for $t \leq 0$, and for the moment we put no further restrictions on F. If $t > 0$, then $X \leq t$ is the event "failure occurs in the interval $[0, t]$." Hence $X > t$ is the complementary event, "no failure occurs in the interval $[0, t]$."

Suppose that no failure occurs in the interval $[0, t]$. What is the probability of continued survival in the interval $[t, t + s]$? This is a question in conditional probabilities. We wish to determine $P(X > t + s \mid X > t)$, the conditional probability that there is no failure in the interval $[0, t + s]$, given that there is no failure in the interval $[0, t]$.

From the definition of conditional probability we have

$$(14.22) \qquad P(X > t + s \mid X > t) = \frac{P[(X > t + s) \cap (X > t)]}{P(X > t)} = \frac{P(X > t + s)}{P(X > t)} .$$

Suppose now that F is an exponential distribution with parameter $\lambda > 0$. Then $F(t) = 1 - e^{-\lambda t}$ for $t > 0$, and $P(X > t) = 1 - P(X \leq t) = e^{-\lambda t}$. Hence Equation (14.22) becomes

$$P(X > t + s \mid X > t) = \frac{e^{-\lambda(t+s)}}{e^{-\lambda t}} = e^{-\lambda s} = P(X > s) .$$

In other words, if the piece of equipment survives in the interval $[0, t]$, then the probability of continued survival in the interval $[t, t + s]$ is equal to the probability of survival in the interval $[0, s]$ having the same length. That is, the probability of survival depends only on the length of the time interval and not on the age of the equipment. Expressed in terms of the distribution function F, this property states that

$$(14.23) \qquad \frac{1 - F(t + s)}{1 - F(t)} = 1 - F(s) \qquad \text{for all} \quad t > 0 \text{ and } s > 0 .$$

The next theorem shows that exponential distributions are the only probability distributions with this property.

THEOREM 14.9. *Let F be a probability distribution function satisfying the functional equation* (14.23), *where $F(t) < 1$ for $t > 0$. Then there is a positive constant $\lambda > 0$ such that*

$$F(t) = 1 - e^{-\lambda t} \qquad \textit{for all } t > 0 .$$

Proof. Let $g(t) = -\log [1 - F(t)]$ for $t > 0$. Then $1 - F(t) = e^{-g(t)}$, so to prove the theorem it suffices to prove that $g(t) = \lambda t$ for some $\lambda > 0$.

Now g is nonnegative and satisfies Cauchy's functional equation,

$$g(t + s) = g(t) + g(s)$$

for all $t > 0$ and $s > 0$. Therefore, applying Theorem 14.8 with $c = 1$, we deduce that $g(t) = tg(1)$ for $0 < t \leq 1$. Let $\lambda = g(1)$. Then $\lambda = -\log [1 - F(1)] > 0$, and hence $g(t) = \lambda t$ for $0 < t \leq 1$.

To prove that $g(t) = \lambda t$ for all $t > 0$, let $G(t) = g(t) - \lambda t$. The function G also satisfies Cauchy's functional equation. Moreover, G is periodic with period 1 because $G(t + 1) = G(t) + G(1)$ and $G(1) = 0$. Since G is identically 0 in (0, 1] the periodicity shows that $G(t) = 0$ for all $t > 0$. In other words, $g(t) = \lambda t$ for all $t > 0$, which completes the proof.

EXAMPLE 1. Let X be a random variable which measures the lifetime (in hours) of a certain type of vacuum tube. Assume X has an exponential distribution with parameter $\lambda = 0.001$. The manufacturer wishes to guarantee these tubes for T hours. Determine T so that $P(X > T) = 0.95$.

Solution. The distribution function is given by $F(t) = 1 - e^{-\lambda t}$ for $t > 0$, where $\lambda = 0.001$. Since $P(X > T) = 1 - F(T) = e^{-\lambda T}$, we choose T to make $e^{-\lambda T} = 0.95$. Hence $T = -(\log 0.95)/\lambda = -1000 \log 0.95 = 51.25+$.

EXAMPLE 2. Consider the random variable of Example 1, but with an unspecified value of λ. The following argument suggests a reasonable procedure for determining λ. Start with an initial number of vacuum tubes at time $t = 0$, and let $g(t)$ denote the number of tubes still functioning t hours later. The ratio $[g(0) - g(t)]/g(0)$ is the fraction of the original number that has failed in time t. Since the probability that a particular tube fails in time t is $1 - e^{-\lambda t}$, it seems reasonable to expect that the equation

$$\frac{g(0) - g(t)}{g(0)} = 1 - e^{-\lambda t} \tag{14.24}$$

should be a good approximation to reality. If we assume (14.24) we obtain

$$g(t) = g(0)e^{-\lambda t}.$$

In other words, under the hypothesis (14.24), the number $g(t)$ obeys an exponential decay law with decay constant λ. The decay constant can be computed in terms of the half-life. If t_1 is the half-life then $\frac{1}{2} = g(t_1)/g(0) = e^{-\lambda t_1}$, so $\lambda = (\log 2)/t_1$. For example, if the half-life of a large sample of tubes is known to be 693 hours, we obtain $\lambda = (\log 2)/693 = 0.001$.

14.14 Normal distributions

Let m and σ be fixed real numbers, with $\sigma > 0$. A random variable X is said to have a *normal distribution* with mean m and variance σ^2 if the density function f is given by the formula

$$f(t) = \frac{1}{\sigma\sqrt{2\pi}}\, e^{-[(t-m)/\sigma]^2/2}$$

for all real t. The corresponding distribution function F is, of course, the integral

$$F(t) = \frac{1}{\sigma\sqrt{2\pi}} \int_{-\infty}^{t} e^{-[(u-m)/\sigma]^2/2}\, du\,.$$

TABLE 14.1 Values of the standard normal distribution function

$$\Phi(t) = \frac{1}{\sqrt{2\pi}} \int_{-\infty}^{t} e^{-u^2/2}\, du\,.$$

t	0.00	0.01	0.02	0.03	0.04	0.05	0.06	0.07	0.08	0.09
0.0	0.5000	0.5040	0.5080	0.5120	0.5160	0.5199	0.5239	0.5279	0.5319	0.5359
0.1	0.5398	0.5438	0.5478	0.5517	0.5557	0.5596	0.5636	0.5675	0.5714	0.5753
0.2	0.5793	0.5832	0.5871	0.5910	0.5948	0.5987	0.6026	0.6064	0.6103	0.6141
0.3	0.6179	0.6217	0.6255	0.6293	0.6331	0.6368	0.6406	0.6443	0.6480	0.6517
0.4	0.6554	0.6591	0.6628	0.6664	0.6700	0.6736	0.6772	0.6808	0.6844	0.6879
0.5	0.6915	0.6950	0.6985	0.7019	0.7054	0.7088	0.7123	0.7157	0.7190	0.7224
0.6	0.7257	0.7291	0.7324	0.7357	0.7389	0.7422	0.7454	0.7486	0.7517	0.7549
0.7	0.7580	0.7611	0.7642	0.7673	0.7704	0.7734	0.7764	0.7794	0.7823	0.7852
0.8	0.7881	0.7910	0.7939	0.7967	0.7995	0.8023	0.8051	0.8078	0.8106	0.8133
0.9	0.8159	0.8186	0.8212	0.8238	0.8264	0.8289	0.8315	0.8340	0.8365	0.8389
1.0	0.8413	0.8438	0.8461	0.8485	0.8508	0.8531	0.8554	0.8577	0.8599	0.8621
1.1	0.8643	0.8665	0.8686	0.8708	0.8729	0.8749	0.8770	0.8790	0.8810	0.8830
1.2	0.8849	0.8869	0.8888	0.8907	0.8925	0.8944	0.8962	0.8980	0.8997	0.9015
1.3	0.9032	0.9049	0.9066	0.9082	0.9099	0.9115	0.9131	0.9147	0.9162	0.9177
1.4	0.9192	0.9207	0.9222	0.9236	0.9251	0.9265	0.9279	0.9292	0.9306	0.9319
1.5	0.9332	0.9345	0.9357	0.9370	0.9382	0.9394	0.9406	0.9418	0.9429	0.9441
1.6	0.9452	0.9463	0.9474	0.9484	0.9495	0.9505	0.9515	0.9525	0.9535	0.9545
1.7	0.9554	0.9564	0.9573	0.9582	0.9591	0.9599	0.9608	0.9616	0.9625	0.9633
1.8	0.9641	0.9649	0.9656	0.9664	0.9671	0.9678	0.9686	0.9693	0.9699	0.9706
1.9	0.9713	0.9719	0.9726	0.9732	0.9738	0.9744	0.9750	0.9756	0.9761	0.9767
2.0	0.9772	0.9778	0.9783	0.9788	0.9793	0.9798	0.9803	0.9808	0.9812	0.9817
2.1	0.9821	0.9826	0.9830	0.9834	0.9838	0.9842	0.9846	0.9850	0.9854	0.9857
2.2	0.9861	0.9864	0.9868	0.9871	0.9875	0.9878	0.9881	0.9884	0.9887	0.9890
2.3	0.9893	0.9896	0.9898	0.9901	0.9904	0.9906	0.9909	0.9911	0.9913	0.9916
2.4	0.9918	0.9920	0.9922	0.9925	0.9927	0.9929	0.9931	0.9932	0.9934	0.9936
2.5	0.9938	0.9940	0.9941	0.9943	0.9945	0.9946	0.9948	0.9949	0.9951	0.9952
2.6	0.9953	0.9955	0.9956	0.9957	0.9959	0.9960	0.9961	0.9962	0.9963	0.9964
2.7	0.9965	0.9966	0.9967	0.9968	0.9969	0.9970	0.9971	0.9972	0.9973	0.9974
2.8	0.9974	0.9975	0.9976	0.9977	0.9977	0.9978	0.9979	0.9979	0.9980	0.9981
2.9	0.9981	0.9982	0.9982	0.9983	0.9984	0.9984	0.9985	0.9985	0.9986	0.9986
3.0	0.9987	0.9987	0.9987	0.9988	0.9988	0.9989	0.9989	0.9989	0.9990	0.9990
3.1	0.9990	0.9991	0.9991	0.9991	0.9992	0.9992	0.9992	0.9992	0.9993	0.9993
3.2	0.9993	0.9993	0.9994	0.9994	0.9994	0.9994	0.9994	0.9995	0.9995	0.9995
3.3	0.9995	0.9995	0.9995	0.9996	0.9996	0.9996	0.9996	0.9996	0.9996	0.9997
3.4	0.9997	0.9997	0.9997	0.9997	0.9997	0.9997	0.9997	0.9997	0.9997	0.9998
3.5	0.9998	0.9998	0.9998	0.9998	0.9998	0.9998	0.9998	0.9998	0.9998	0.9998
3.6	0.9998	0.9998	0.9999	0.9999	0.9999	0.9999	0.9999	0.9999	0.9999	0.9999

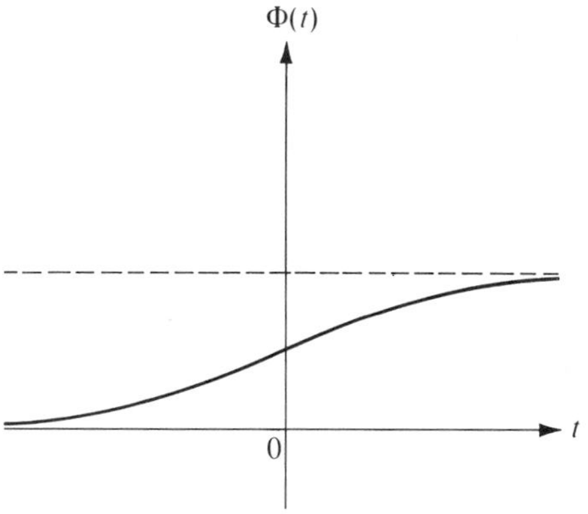

FIGURE 14.9 The standard normal distribution function: $m = 0$, $\sigma = 1$.

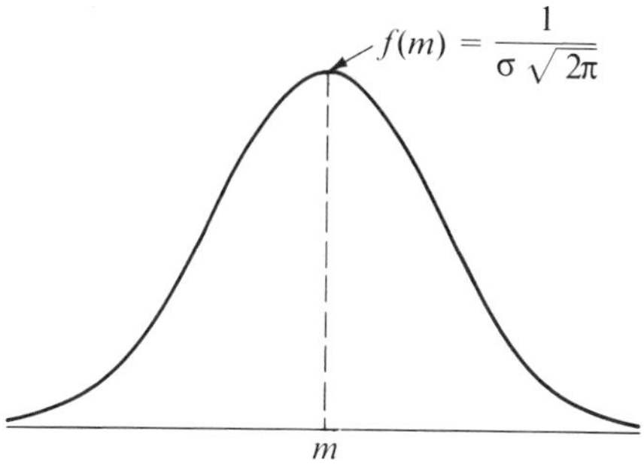

FIGURE 14.10 The density function of a normal distribution with mean m and variance σ^2.

It is clear that this function F is monotonic increasing, continuous everywhere, and tends to 0 as $t \to -\infty$. Also, it can be shown that $F(t) \to 1$ as $t \to +\infty$. (See Exercise 7 of Section 14.16.)

The special case $m = 0$, $\sigma = 1$ is called the *standard* normal distribution. In this case the function F is usually denoted by the letter Φ. Thus,

$$\Phi(t) = \frac{1}{\sqrt{2\pi}} \int_{-\infty}^{t} e^{-u^2/2}\, du .$$

The general case can be reduced to the standard case by introducing the change of variable $v = (u - m)/\sigma$ in the integral for F. This leads to the formula

$$F(t) = \Phi\left(\frac{t - m}{\sigma}\right).$$

A four-place table of values of $\Phi(t)$ for values of t spaced at intervals of length 0.01 is given in Table 14.1 for $t = 0.00$ to $t = 3.69$. The graph of Φ is shown in Figure 14.9. The graph of the density f is a famous "bell-shaped" curve, shown in Figure 14.10. The top of the bell is directly above the mean m. For large values of σ the curve tends to flatten out; for small σ it has a sharp peak, as in Figure 14.10.

Normal distributions are among the most important of all continuous distributions. Many random variables that occur in nature behave as though their distribution functions are normal or approximately normal. Examples include the measurement of the height of people in a large population, certain measurements on large populations of living organisms encountered in biology, and the errors of observation encountered when making large numbers of measurements. In physics, Maxwell's law of velocities implies that the distribution function of the velocity in any given direction of a molecule of mass M in a gas at absolute temperature T is normal with mean 0 and variance $M/(kT)$, where k is a constant (Boltzmann's constant).

The normal distribution is also of theoretical importance because it can be used to approximate the distributions of many random phenomena. One example is the binomial

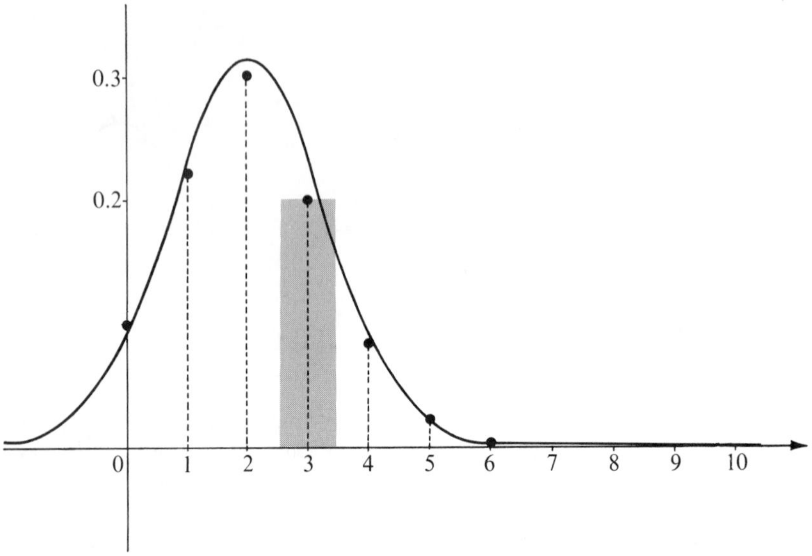

FIGURE 14.11 The density function of a normal distribution considered as an approximation to the probability mass function of a binomial distribution.

distribution with parameters n and p. If X is a random variable having a binomial distribution with parameters n and p, the probability $P(a \le X \le b)$ is given by the sum

$$\sum_{k=a}^{b} \binom{n}{k} p^k q^{n-k},$$

where $q = 1 - p$. For a large n, laborious computations are needed to evaluate this sum. In practice these computations are avoided by use of the approximate formula

$$\sum_{k=a}^{b} \binom{n}{k} p^k q^{n-k} \sim \Phi\left(\frac{b - np + \frac{1}{2}}{\sqrt{npq}}\right) - \Phi\left(\frac{a - np - \frac{1}{2}}{\sqrt{npq}}\right), \tag{14.25}$$

where the symbol $\sim$ means that the two sides of (14.25) are asymptotically equal; that is, the ratio of the left member to the right member approaches the limit 1 as $n \to \infty$. The limit relation expressed in (14.25) is a special case of the so-called *central limit theorem* of the calculus of probabilities. This theorem (discussed in more detail in Section 14.30) explains the theoretical importance of normal distributions.

Figure 14.11 illustrates approximate formula (14.25) and shows that it can be accurate even for a relatively small value of n. The dotted lines are the ordinates of the probability mass function p of a binomial distribution with parameters $n = 10$ and $p = \frac{1}{5}$. These ordinates were computed from the formula

$$p(t) = P(X = t) = \binom{10}{t}\left(\frac{1}{5}\right)^t\left(\frac{4}{5}\right)^{10-t} \qquad \text{for} \quad t = 0, 1, 2, \ldots, 10.$$

The ordinates for $t = 7$, 8, 9, and 10 are not shown because their numerical values are too near zero. For example, $p(10) = (\frac{1}{5})^{10} = 2^{10}/10^{10} = 0.0000001024$. The smooth curve is the graph of the density function f of a normal distribution (with mean $m = np = 2$ and variance $\sigma^2 = npq = 1.6$). To compute the probability $P(a \leq t \leq b)$ from the mass function p we add the function values $p(t)$ at the mass points in the interval $a \leq t \leq b$. Each value $p(t)$ may be interpreted as the area of a rectangle of height $p(t)$ located over an interval of unit length centered about the mass point t. (An example, centered about $t = 3$, is shown in Figure 14.11.) The approximate formula in (14.25) is the result of replacing the areas of these rectangles by the area of the ordinate set of f over the interval $[a - \frac{1}{2}, b + \frac{1}{2}]$.

14.15 Remarks on more general distributions

In the foregoing sections we have discussed examples of discrete and continuous distributions. The values of a discrete distribution are computed by adding the values of the corresponding probability mass function. The values of a continuous distribution with a density are computed by integrating the density function. There are, of course, distributions that are neither discrete nor continuous. Among these are the so-called "mixed" types in which the mass distribution is partly discrete and partly continuous. (An example is shown in Figure 14.3.)

A distribution function F is called *mixed* if it can be expressed as a linear combination of the form

$$F(t) = c_1 F_1(t) + c_2 F_2(t), \tag{14.26}$$

where F_1 is discrete and F_2 is continuous. The constants c_1 and c_2 must satisfy the relations

$$0 < c_1 < 1, \qquad 0 < c_2 < 1, \qquad c_1 + c_2 = 1.$$

Properties of mixed distributions may be found by studying those that are discrete or continuous and then appealing to the linearity expressed in Equation (14.26).

A general kind of integral, known as the *Riemann-Stieltjes integral*, makes possible a simultaneous treatment of the discrete, continuous, and mixed cases.† Although this integral unifies the theoretical discussion of distribution functions, in any specific problem the computation of probabilities must be reduced to ordinary summation and integration. In this introductory account we shall not attempt to describe the Riemann-Stieltjes integral. Consequently, most of the topics we discuss come in pairs, one for the discrete case and one for the continuous case. However, we shall only give complete details for one case, leaving the untreated case for the reader to work out.

Even the Riemann-Stieltjes integral is inadequate for treating the *most general* distribution functions. But a more powerful concept, called the Lebesgue-Stieltjes integral,‡ does give a satisfactory treatment of all cases. The advanced theory of probability cannot be undertaken without a knowledge of the Lebesgue-Stieltjes integral.

---

† A discussion of the Riemann-Stieltjes integral may be found in Chapter 9 of the author's *Mathematical Analysis*, Addison-Wesley Publishing Company, Reading, Mass. 1957.

‡ See any book on measure theory.

14.16 Exercises

1. Let X be a random variable which measures the lifetime (in hours) of a certain type of vacuum tube. Assume X has an exponential distribution with parameter $\lambda = 0.001$. Determine T so that $P(X > T)$ is (a) 0.90; (b) 0.99. You may use the approximate formula $-\log(1 - x) = x + x^2/2$ in your calculations.
2. A radioactive material obeys an exponential decay law with half-life 2 years. Consider the decay time X (in years) of a single atom and assume that X is a random variable with an exponential distribution. Calculate the probability that an atom disintegrates (a) in the interval $1 \le X \le 2$; (b) in the interval $2 \le X \le 3$; (c) in the interval $2 \le X \le 3$, given that it has not disintegrated in the interval $0 \le X \le 2$; (d) in the interval $2 \le X \le 3$, given that it has not disintegrated in the interval $1 \le X \le 2$.
3. The length of time (in minutes) of long distance telephone calls from Caltech is found to be a random phenomenon with probability density function

$$f(t) = \begin{cases} ce^{-t/3} & \text{for } t > 0, \\ 0 & \text{for } t \le 0. \end{cases}$$

Determine the value of c and calculate the probability that a long distance call will last (a) less than 3 minutes; (b) more than 6 minutes; (c) between 3 and 6 minutes; (d) more than 9 minutes.
4. Given real constants $\lambda > 0$ and c. Let

$$f(t) = \begin{cases} \lambda e^{-\lambda(t-c)} & \text{if } t \ge c, \\ 0 & \text{if } t < c. \end{cases}$$

Verify that $\int_{-\infty}^{\infty} f(t)\,dt = 1$, and determine a distribution function F having f as its density. This is called an exponential distribution with two parameters, a *decay parameter* λ and a *location parameter* c.
5. State and prove an extension of Theorem 14.9 for exponential distributions with two parameters λ and c.
6. A random variable X has an exponential distribution with two parameters λ and c. Let $Y = aX + b$, where $a > 0$. Prove that Y also has an exponential distribution with two parameters λ' and c', and determine these parameters in terms of a, b, c, and λ.
7. In Exercise 16 of Section 11.28 it was shown that $\int_0^\infty e^{-x^2}\,dx = \sqrt{\pi}/2$. Use this result to prove that for $\sigma > 0$ we have

$$\frac{1}{\sigma\sqrt{2\pi}} \int_{-\infty}^{\infty} \exp\left\{-\frac{1}{2}\left(\frac{u - m}{\sigma}\right)^2\right\} du = 1.$$

8. A random variable X has a standard normal distribution Φ. Prove that (a) $\Phi(-x) = 1 - \Phi(x)$; (b) $P(|X| < k) = 2\Phi(k) - 1$; (c) $P(|X| > k) = 2(1 - \Phi(k))$.
9. A random variable X has a standard normal distribution Φ. Use Table 14.1 to calculate each of the following probabilities: (a) $P(X > 0)$; (b) $P(1 < X < 2)$; (c) $P(|X| < 3)$; (d) $P(|X| > 2)$.
10. A random variable X has a standard normal distribution Φ. Use Table 14.1 to find a number c such that (a) $P(|X| > c) = \frac{1}{2}$; (b) $P(|X| > c) = 0.98$.
11. Assume X has a normal distribution function F with mean m and variance σ^2, and let Φ denote the standard normal distribution.

(a) Prove that

$$F(t) = \Phi\left(\frac{t-m}{\sigma}\right).$$

(b) Find a value of c such that $P(|X - m| > c) = \frac{1}{2}$.
(c) Find a value of c such that $P(|X - m| > c) = 0.98$.

12. A random variable X is normally distributed with mean $m = 1$ and variance $\sigma^2 = 4$. Calculate each of the following probabilities: (a) $P(-3 \le X \le 3)$; (b) $P(-5 \le X \le 3)$.
13. An architect is designing a doorway for a public building to be used by people whose heights are normally distributed, with mean $m = 5$ ft. 9 in., and variance σ^2 where $\sigma = 3$ in. How low can the doorway be so that no more than 1% of the people bump their heads?
14. If X has a standard normal distribution, prove that the random variable $Y = aX + b$ is also normal if $a \ne 0$. Determine the mean and variance of Y.
15. Assume a random variable X has a standard normal distribution, and let $Y = X^2$.

(a) Show that $F_Y(t) = \dfrac{2}{\sqrt{2\pi}} \displaystyle\int_0^{\sqrt{t}} e^{-u^2/2}\, du \quad$ if $t \ge 0$.

(b) Determine $F_Y(t)$ when $t < 0$ and describe the density function f_Y.

14.17 Distributions of functions of random variables

If φ is a real-valued function whose domain includes the range of the random variable X, we can construct a new random variable Y by the equation

$$Y = \varphi(X),$$

which means that $Y(\omega) = \varphi[X(\omega)]$ for each ω in the sample space. If we know the distribution function F_X of X, how do we find the distribution F_Y of Y? We begin with an important special case. Suppose that φ is continuous and strictly increasing on the whole real axis and takes on every real value. In this case φ has a continuous strictly increasing inverse ψ such that, for all x and y,

$$y = \varphi(x) \quad \text{if and only if} \quad x = \psi(y).$$

By the definition of F_Y we have

$$F_Y(t) = P(Y \le t) = P[\varphi(X) \le t].$$

Since φ is strictly increasing and continuous, the events "$\varphi(X) \le t$" and "$X \le \psi(t)$" are identical. Therefore $P[\varphi(X) \le t] = P[X \le \psi(t)] = F_X[\psi(t)]$. Hence the distributions F_Y and F_X are related by the equation

$$F_Y(t) = F_X[\psi(t)]. \tag{14.27}$$

When the distribution F_X and the function ψ have derivatives we can differentiate both sides of (14.27), using the chain rule on the right, to obtain

$$F_Y'(t) = F_X'[\psi(t)] \cdot \psi'(t).$$

This gives us the following equation relating the densities:

$$f_Y(t) = f_X[\psi(t)] \cdot \psi'(t).$$

EXAMPLE 1. $Y = aX + b,\ a > 0$. In this case we have

$$\varphi(x) = ax + b, \qquad \psi(y) = \frac{y - b}{a}, \qquad \psi'(y) = \frac{1}{a}.$$

Since φ is continuous and strictly increasing we may write

$$F_Y(t) = F_X\left(\frac{t - b}{a}\right) \quad \text{and} \quad f_Y(t) = \frac{1}{a} f_X\left(\frac{t - b}{a}\right).$$

EXAMPLE 2. $Y = X^2$. In this case $\varphi(x) = x^2$ and the foregoing discussion is not directly applicable because φ is not strictly increasing. However, we can use the same method of reasoning to determine F_Y and f_Y. By the definition of F_Y we have

$$F_Y(t) = P(X^2 \le t).$$

If $t < 0$ the event "$X^2 \le t$" is empty and hence $P(X^2 \le t) = 0$. Therefore $F_Y(t) = 0$ for $t < 0$. If $t > 0$ we have

$$P(X^2 \le t) = P(-\sqrt{t} \le X \le \sqrt{t}) = F_X(\sqrt{t}) - F_X(-\sqrt{t}) + P(X = -\sqrt{t}).$$

For a continuous distribution F_X we have $P(X = -\sqrt{t}) = 0$ and we obtain the following relation between F_Y and F_X:

$$F_Y(t) = \begin{cases} 0 & \text{if } t < 0, \\ F_X(\sqrt{t}) - F_X(-\sqrt{t}) & \text{if } t > 0. \end{cases}$$

For all $t < 0$ and for those $t > 0$ such that F_X is differentiable at $\sqrt{t}$ and at $-\sqrt{t}$ we have the following equation relating the densities:

$$f_Y(t) = \begin{cases} 0 & \text{if } t < 0, \\ \dfrac{f_X(\sqrt{t}) + f_X(-\sqrt{t})}{2\sqrt{t}} & \text{if } t > 0. \end{cases}$$

Further problems of this type will be discussed in Section 14.23 with the help of two-dimensional random variables.

14.18 Exercises

1. Assume X has a uniform distribution on the interval $[0, 1]$. Determine the distribution function F_Y and a probability density f_Y of the random variable Y if:
   (a) $Y = 3X + 1$,
   (b) $Y = -3X + 1$,
   (c) $Y = X^2$,
   (d) $Y = \log |X|$,
   (e) $Y = \log X^2$,
   (f) $Y = e^X$.

2. Let X be a random variable with a continuous distribution function F_X. If φ is continuous and strictly increasing on the whole real axis and if $\varphi(x) \to a$ as $x \to -\infty$ and $\varphi(x) \to b$ as $x \to +\infty$, determine the distribution function F_Y of the random variable $Y = \varphi(X)$. Also, compute a density f_Y, assuming that F_X and φ are differentiable.
3. Assume X has a standard normal distribution. Determine a probability density function of the random variable Y when
   (a) $Y = X^2$, (c) $Y = e^X$,
   (b) $Y = |X|^{1/2}$, (d) $Y = \arctan X$.

14.19 Distributions of two-dimensional random variables

The concept of a distribution may be generalized to n-dimensional random variables in a straightforward way. The treatment of the case $n = 2$ will indicate how the extension takes place.

If X and Y are two one-dimensional random variables defined on a common sample space S, (X, Y) will denote the two-dimensional random variable whose value at a typical point ω of S is given by the pair of real numbers $(X(\omega), Y(\omega))$. The notation

$$X \le a, \; Y \le b$$

is an abbreviation for the set of all elements ω in S such that $X(\omega) \le a$ and $Y(\omega) \le b$; the probability of this event is denoted by

$$P(X \le a, \; Y \le b).$$

Notations such as $a < X \le b$, $c < Y \le d$, and $P(a < X \le b, c < Y \le d)$ are similarly defined.

The set of points (x, y) such that $x \le a$ and $y \le b$ is the Cartesian product $A \times B$ of the two one-dimensional infinite intervals $A = \{x \mid x \le a\}$ and $B = \{y \mid y \le b\}$. The set $A \times B$ is represented geometrically by the infinite rectangular region shown in Figure 14.12. The number $P(X \le a, \; Y \le b)$ represents the probability that a point $(X(\omega), Y(\omega))$ lies in this region. These probabilities are the two-dimensional analogs of the one-dimensional probabilities $P(X \le a)$, and are used to define two-dimensional probability distributions.

DEFINITION. *The distribution function of the two-dimensional random variable (X, Y) is the real-valued function F defined for all real a and b by the equation*

$$F(a, b) = P(X \le a, \; Y \le b).$$

It is also known as the joint distribution of the two one-dimensional random variables X and Y.

To compute the probability that (X, Y) lies in a rectangle we use the following theorem, a generalization of Theorem 14.2(b).

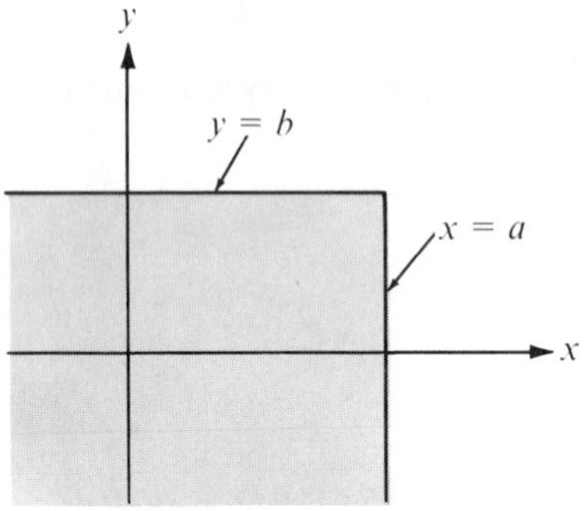

FIGURE 14.12 An infinite rectangular region $A \times B$, where $A = \{x \mid x \le a\}$ and $B = \{y \mid y \le b\}$.

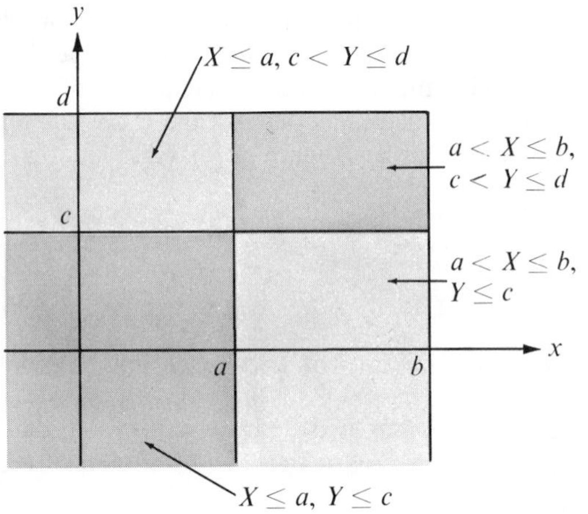

FIGURE 14.13 The event "$X \le b,\ Y \le d$" expressed as the union of four disjoint events.

THEOREM 14.10. *Let F be the distribution function of a two-dimensional random variable* (X, Y). *Then if* $a < b$ *and* $c < d$ *we have*

$$P(a < X \le b, c < Y \le d) = F(b, d) - F(a, d) - F(b, c) + F(a, c). \tag{14.28}$$

Proof. The two events "$X \le a,\ c < Y \le d$" and "$X \le a,\ Y \le c$" are disjoint, and their union is "$X \le a,\ Y \le d$." Adding probabilities we obtain $P(X \le a, c < Y \le d) + P(X \le a, Y \le c) = P(X \le a, Y \le d)$; hence

$$P(X \le a, c < Y \le d) = F(a, d) - F(a, c).$$

Similarly, we have

$$P(a < X \le b, Y \le c) = F(b, c) - F(a, c).$$

Now the four events

$$\text{“}X \le a,\ Y \le c,\text{”} \qquad \text{“}X \le a, c < Y \le d,\text{”}$$

$$\text{“}a < X \le b,\ Y \le c,\text{”} \qquad \text{“}a < X \le b, c < Y \le d\text{”}$$

are disjoint, and their union is "$X \le b,\ Y \le d$." (See Figure 14.13.) Adding the corresponding probabilities and using the two foregoing equations we obtain

$$F(a, c) + [F(a, d) - F(a, c)] + [F(b, c) - F(a, c)] + P(a < X \le b, c < Y \le d) = F(b,d),$$

which is equivalent to (14.28).

Formula (14.28) gives the probability that the random variable (X, Y) has a value in the rectangle $(a, b] \times (c, d]$. There are, of course, corresponding formulas for the rectangles, $[a, b] \times [c, d]$, $(a, b) \times (c, d)$, $[a, b) \times [c, d)$, and so forth.

Note: The analogy with mass may be extended to the two-dimensional case. Here the total mass 1 is distributed over a plane. The probability $P(a < X \le b, c < Y \le d)$ represents the total amount of mass located in the rectangle $(a, b] \times (c, d]$. The number $F(a, b)$ represents the amount in the infinite rectangular region $X \le a$, $Y \le b$. As in the one-dimensional case, the two most important types of distributions are those known as *discrete* and *continuous.* In the discrete case the entire mass is located in lumps concentrated at a finite or countably infinite number of points. In the continuous case the mass is smeared all over the plane with a uniform or varying thickness.

14.20 Two-dimensional discrete distributions

If a random variable (X, Y) is given we define a new function p, called the probability mass function of (X, Y), such that

$$p(x, y) = P(X = x, \ Y = y)$$

for every pair of real numbers (x, y). Let T denote the set of (x, y) for which $p(x, y) > 0$. It can be shown that T is either finite or countably infinite. If the sum of the $p(x, y)$ for all (x, y) in T is equal to 1, that is, if

$$\sum_{(x,y)\in T} p(x, y) = 1, \tag{14.29}$$

the random variable (X, Y) is said to be *discrete* (or *jointly discrete*). The points (x, y) in T are called the *mass points* of (X, Y).

Suppose that $x_1, x_2, x_3, \ldots$ and $y_1, y_2, y_3, \ldots$ are among the possible values of X and Y, respectively, and let

$$p_{ij} = P(X = x_i, \ Y = y_j).$$

If each p_{ij} is positive and if the sum of all the p_{ij} is 1, then the probability of an event "$(X, Y) \in E$" is the sum of all the p_{ij} taken over all x_i and y_j for which $(x_i, y_j) \in E$. We indicate this by writing

$$P[(X, Y) \in E] = \sum_{\substack{x_i \\ (x_i, y_j)\in E}} \sum_{y_j} p_{ij}.$$

In particular, since $P(X \le x, Y \le y) = F(x, y)$, the joint distribution F (which is also called discrete) is given by the double sum

$$F(x, y) = \sum_{x_i \le x} \sum_{y_j \le y} p_{ij}. \tag{14.30}$$

The numbers p_{ij} can also be used to reconstruct the probability mass functions p_X and p_Y of the one-dimensional random variables X and Y. In fact, if E_{ij} denotes the event "$X = x_i$, $Y = y_j$," the events $E_{i1}, E_{i2}, E_{i3}, \ldots$ are disjoint and their union is the event "$X = x_i$." Hence, by countable additivity, we obtain

$$P(X = x_i) = \sum_{j=1}^{\infty} P(E_{ij}) = \sum_{j=1}^{\infty} p_{ij}. \tag{14.31}$$

Similarly, we have

$$P(Y = y_j) = \sum_{i=1}^{\infty} P(E_{ij}) = \sum_{i=1}^{\infty} p_{ij}. \tag{14.32}$$

Therefore, the corresponding one-dimensional distributions F_X and F_Y can be computed from the formulas

$$F_X(t) = \sum_{x_i \le t} P(X = x_i) = \sum_{x_i \le t} \sum_{j=1}^{\infty} p_{ij}$$

and

$$F_Y(t) = \sum_{y_j \le t} P(Y = y_j) = \sum_{y_j \le t} \sum_{i=1}^{\infty} p_{ij}.$$

For finite sample spaces, of course, the infinite series are actually finite sums.

14.21 Two-dimensional continuous distributions. Density functions

As might be expected, *continuous distributions* are those that are continuous over the whole plane. For the majority of continuous distributions F that occur in practice there exists a nonnegative function f (called the *probability density* of F) such that the probabilities of most events of interest can be computed by double integration of the density. That is, the probability of an event "$(X, Y) \in Q$" is given by the integral formula

$$P[(X, Y) \in Q] = \iint_Q f. \tag{14.33}$$

When such an f exists it is also called a probability density of the random variable (X, Y), or a joint density of X and Y. We shall not attempt to describe the class of regions Q for which (14.33) is to hold, except to mention that this class should be extensive enough to include all regions that arise in the ordinary applications of probability. For example, if a joint density exists we always have

$$P(a < X \le b, c < Y \le d) = \iint_R f(x, y)\, dx\, dy, \tag{14.34}$$

where $R = [a, b] \times [c, d]$. The integrand f is usually sufficiently well behaved for the double integral to be evaluated by iterated one-dimensional integration, in which case (14.34) becomes

$$P(a < X \le b, c < Y \le d) = \int_c^d \left[\int_a^b f(x, y)\, dx\right] dy = \int_a^b \left[\int_c^d f(x, y)\, dy\right] dx.$$

In all the examples we shall consider, this formula is also valid in the limiting cases in which a and c are replaced by $-\infty$ and in which b and d are replaced by $+\infty$. Thus we have

$$(14.35) \qquad F(b, d) = \int_{-\infty}^{d} \left[\int_{-\infty}^{b} f(x, y)\, dx\right] dy = \int_{-\infty}^{b} \left[\int_{-\infty}^{d} f(x, y)\, dy\right] dx$$

for all b and d, and

$$(14.36) \qquad \int_{-\infty}^{+\infty} \left[\int_{-\infty}^{+\infty} f(x, y)\, dx\right] dy = \int_{-\infty}^{+\infty} \left[\int_{-\infty}^{+\infty} f(x, y)\, dy\right] dx = 1 .$$

Equations (14.35) and (14.36) are the continuous analogs of (14.30) and (14.29), respectively.

If a density exists it is not unique, since the integrand in (14.33) can be changed at a finite number of points without affecting the value of the integral. However, there is at most one continuous density function. In fact, at points of continuity of f we have the formulas

$$f(x, y) = D_{1,2}F(x, y) = D_{2,1}F(x, y),$$

obtained by differentiation of the integrals in (14.35).

As in the discrete case, the joint density f can be used to recover the one-dimensional densities f_X and f_Y. The formulas analogous to (14.31) and (14.32) are

$$f_X(x) = \int_{-\infty}^{+\infty} f(x, y)\, dy \qquad \text{and} \qquad f_Y(y) = \int_{-\infty}^{+\infty} f(x, y)\, dx .$$

The corresponding distributions $F_X(t)$ and $F_Y(t)$ are obtained, of course, by integrating the respective densities f_X and f_Y from $-\infty$ to t.

The random variables X and Y are called *independent* if the joint distribution $F(x, y)$ can be factored as follows,

$$F(x, y) = F_X(x)F_Y(y)$$

for all (x, y). Some consequences of independence are discussed in the next set of exercises.

EXAMPLE. Consider the function f that has the constant value 1 over the square $R = [0, 1] \times [0, 1]$, and the value 0 at all other points of the plane. A random variable (X, Y) having this density function is said to be uniformly distributed over R. The corresponding distributions function F is given by the following formulas:

$$F(x, y) = \begin{cases} xy & \text{if } (x, y) \in R, \\ x & \text{if } 0 < x < 1 \quad \text{and} \quad y > 1, \\ y & \text{if } 0 < y < 1 \quad \text{and} \quad x > 1, \\ 1 & \text{if } x \geq 1 \quad \text{and} \quad y \geq 1, \\ 0 & \text{otherwise}. \end{cases}$$

The graph of F over R is part of the saddle-shaped surface $z = xy$. At all points (x, y) not on the boundary of R the mixed partial derivatives $D_{1,2}F(x, y)$ and $D_{2,1}F(x, y)$ exist and are equal to $f(x, y)$. This distribution is the product of two one-dimensional uniform distributions F_X and F_Y. Hence X and Y are independent.

14.22 Exercises

1. Let X and Y be one-dimensional random variables with distribution functions F_X and F_Y, and let F be the joint distribution of X and Y.
(a) Prove that X and Y are independent if and only if we have
$$P(a < X \le b, c < Y \le d) = P(a < X \le b)P(c < Y \le d)$$
for all a, b, c, d, with $a < b$ and $c < d$.
(b) Consider the discrete case. Assume $x_1, x_2, \ldots$ and $y_1, y_2, \ldots$ are the mass points of X and Y, respectively. Let $a_i = P(X = x_i)$ and $b_j = P(Y = y_j)$. If $p_{ij} = P(X = x_i, Y = y_j)$, show that X and Y are independent if $p_{ij} = a_i b_j$ for all i and j.
(c) Let X and Y have continuous distributions with corresponding densities f_X and f_Y and let f denote the density of the joint distribution. Assume the continuity of all three densities. Show that the condition of independence is equivalent to the statement $f(x, y) = f_X(x)f_Y(y)$ for all (x, y). [*Hint:* Express f as a derivative of the joint distribution F.]
2. Refer to Exercise 1. Suppose that $P(X = x_1, Y = y_1) = P(X = x_2, Y = y_2) = p/2$ and that $P(X = x_1, Y = y_2) = P(X = x_2, Y = y_1) = q/2$, where p and q are nonnegative with sum 1.
(a) Determine the one-dimensional probabilities $P(X = x_i)$ and $P(Y = y_j)$ for $i = 1, 2$ and $j = 1, 2$.
(b) For what value (or values) of p will X and Y be independent?
3. If $a < b$ and $c < d$, define f as follows:
$$f(x, y) = \begin{cases} \dfrac{1}{(b - a)(d - c)} & \text{if } (x, y) \in [a, b] \times [c, d], \\ 0 & \text{otherwise.} \end{cases}$$
(a) Verify that this is the density of a continuous distribution F and determine F.
(b) Determine the one-dimensional distributions F_X and F_Y.
(c) Determine whether or not X and Y are independent.
4. If $P(Y \le b) \ne 0$, the conditional probability that $X \le a$, given that $Y \le b$, is denoted by $P(X \le a \mid Y \le b)$, and is defined by the equation
$$P(X \le a \mid Y \le b) = \frac{P(X \le a, Y \le b)}{P(Y \le b)}.$$
If $P(Y \le b) = 0$, we define $P(X \le a \mid Y \le b) = P(X \le a)$. Similarly, if $P(X \le a) \ne 0$, we define $P(Y \le b \mid X \le a) = P(X \le a, Y \le b)/P(X \le a)$. If $P(X \le a) = 0$, we define $P(Y \le b \mid X \le a) = P(Y \le b)$.
(a) Refer to Exercise 1 and describe the independence of X and Y in terms of conditional probabilities.
(b) Consider the discrete case. Assume $x_1, x_2, \ldots$ and $y_1, y_2, \ldots$ are the mass points of X and Y, respectively. Show that
$$P(X = x_i) = \sum_{j=1}^{\infty} P(Y = y_j)P(X = x_i \mid Y = y_j)$$
and
$$P(Y = y_j) = \sum_{i=1}^{\infty} P(X = x_i)P(Y = y_j \mid X = x_i).$$

5. A gambling house offers its clients the following game: A coin is tossed. If the result of the first throw is tails, the player loses and the game is over. If the first throw is heads, a second throw is allowed. If heads occur the second time the player wins \$2, but if tails comes up the player wins \$1. Let X be the random variable which is equal to 1 or 0, according to whether heads or tails occurs on the first throw. Let Y be the random variable which counts the number of dollars won by the player. Use Exercise 4 (or some other method) to compute $P(Y = 0)$, $P(Y = 1)$, and $P(Y = 2)$.
6. Refer to Exercise 4. Derive the so-called Bayes' formulas:

$$P(X = x_k \mid Y = y_j) = \frac{P(X = x_k)P(Y = y_j \mid X = x_k)}{\sum_{i=1}^{\infty} P(X = x_i)P(Y = y_j \mid X = x_i)},$$

$$P(Y = y_k \mid X = x_i) = \frac{P(Y = y_k)P(X = x_i \mid Y = y_k)}{\sum_{j=1}^{\infty} P(Y = y_j)P(X = x_i \mid Y = y_j)}.$$

7. Given two urns A and B. Urn A contains one \$5 bill and two \$10 bills. Urn B contains three \$5 bills and one \$10 bill. Draw a bill from urn A and put it in urn B. Let Y be the random variable which counts the dollar value of the bill transferred. Now draw a bill from urn B and use the random variable X to count *its* dollar value. Compute the conditional probabilities

$$P(Y = 5 \mid X = 10) \qquad \text{and} \qquad P(Y = 10 \mid X = 10).$$

[*Hint:* Use Bayes' formulas of Exercise 6.]
8. Given three identical boxes, each containing two drawers. Box number 1 has one gold piece in one drawer and one silver piece in the other. Box 2 has one gold piece in each drawer and Box 3 has one silver piece in each drawer. One drawer is opened at random and a gold piece is found. Compute the probability that the other drawer in the same box contains a silver piece. [*Hint:* Use Bayes' formulas of Exercise 6.]
9. Let Q be a plane region with positive area $a(Q)$. A continuous two-dimensional random variable (X, Y) is said to have a *uniform distribution* over Q if its density function f is given by the following formulas:

$$f(x, y) = \begin{cases} 1/a(Q) & \text{if } (x, y) \in Q, \\ 0 & \text{if } (x, y) \notin Q. \end{cases}$$

(a) If E is a subregion of Q with area $a(E)$, show that $a(E)/a(Q)$ is the probability of the event $(X, Y) \in E$.
(b) Raindrops fall at random on the square Q with vertices $(1, 0)$, $(0, 1)$, $(-1, 0)$, $(0, -1)$. An outcome is the point (x, y) in Q struck by a particular raindrop. Let $X(x, y) = x$ and $Y(x, y) = y$ and assume (X, Y) has a uniform distribution over Q. Determine the joint density function f and the one-dimensional densities f_X and f_Y. Are the random variables X and Y independent?
10. A two-dimensional random variable (X, Y) has the joint distribution function F. Let $U = X - a$, $V = Y - b$, where a and b are constants. If G denotes the joint distribution of (U, V) show that

$$G(u, v) = F(u + a, v + b).$$

Derive a similar relation connecting the density function f of (X, Y) and g of (U, V) when f is continuous.

14.23 Distributions of functions of two random variables

We turn now to the following problem: If X and Y are one-dimensional random variables with known distributions, how do we find the distribution of new random variables such as $X + Y$, XY, or $X^2 + Y^2$? This section describes a method that helps to answer questions like this. Two new random variables U and V are defined by equations of the form

$$U = M(X, Y), \qquad V = N(X, Y),$$

where $M(X, Y)$ or $N(X, Y)$ is the particular combination in which we are interested. From a knowledge of the joint distribution f of the two-dimensional random variable (X, Y) we calculate the joint distribution g of (U, V). Once g is known, the individual distributions of U and V are easily found.

To describe the method in detail, we consider a one-to-one mapping of the xy-plane onto the uv-plane defined by the pair of equations

$$u = M(x, y), \qquad v = N(x, y).$$

Let the inverse mapping be given by

$$x = Q(u, v), \qquad y = R(u, v),$$

and assume that Q and R have continuous partial derivatives. If T denotes a region in the xy-plane, let T' denote its image in the uv-plane, as suggested by Figure 14.14. Let X and Y be two one-dimensional continuous random variables having a continuous joint distribution and assume (X, Y) has a probability density function f. Define new random variables U and V by writing $U = M(X, Y)$, $V = N(X, Y)$. To determine a probability density g of the random variable (U, V) we proceed as follows:

The random variables X and Y are associated with a sample space S. For each ω in S we have $U(\omega) = M[X(\omega), Y(\omega)]$ and $V(\omega) = N[X(\omega), Y(\omega)]$. Since the mapping is one-to-one, the two sets

$$\{\omega \mid (U(\omega), V(\omega)) \in T'\} \qquad \text{and} \qquad \{\omega \mid (X(\omega), Y(\omega)) \in T\}$$

are equal. Therefore we have

$$P[(U, V) \in T'] = P[(X, Y) \in T]. \tag{14.37}$$

Since f is the density function of (X, Y) we can write

$$P[(X, Y) \in T] = \iint_T f(x, y)\, dx\, dy. \tag{14.38}$$

Using (14.37) and the formula for transforming a double integral we rewrite (14.38) as follows:

$$P[(U, Y) \in T'] = \iint_{T'} f[Q(u, v), R(u, v)] \left| \frac{\partial(Q, R)}{\partial(u, v)} \right| du\, dv.$$

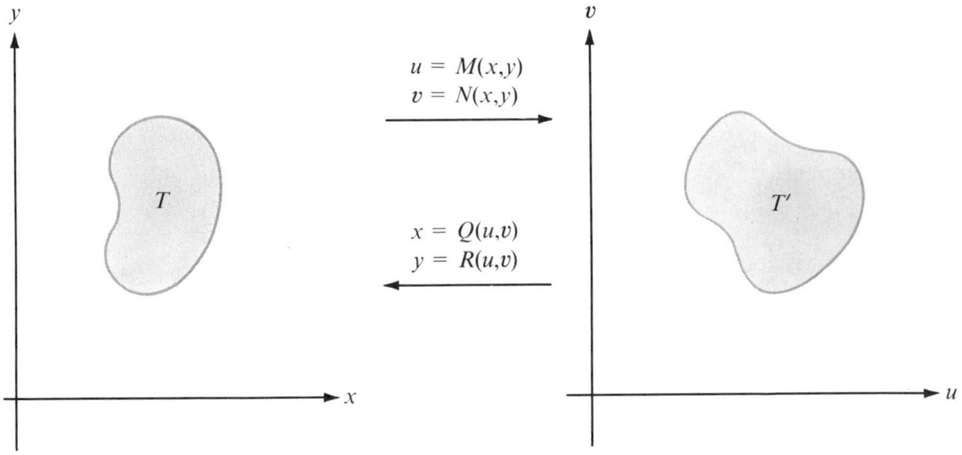

FIGURE 14.14 A one-to-one mapping of a region T in the xy-plane onto a region T' in the uv-plane.

Since this is valid for every region T' in the uv-plane a density g of (U, V) is given by the integrand on the right; that is, we have

$$g(u, v) = f[Q(u, v), R(u, v)] \left| \frac{\partial(Q, R)}{\partial(u, v)} \right| . \tag{14.39}$$

The densities f_U and f_V can now be obtained by the integration formulas

$$f_U(u) = \int_{-\infty}^{\infty} g(u, v)\, dv , \qquad f_V(v) = \int_{-\infty}^{\infty} g(u, v)\, du .$$

EXAMPLE 1. *The sum and difference of two random variables.* Given two one-dimensional random variables X and Y with joint density f, determine density functions for the random variables $U = X + Y$ and $V = X - Y$.

Solution. We use the mapping given by $u = x + y$, $v = x - y$. This is a nonsingular linear transformation whose inverse is given by

$$x = \frac{u + v}{2} = Q(u, v) , \qquad y = \frac{u - v}{2} = R(u, v) .$$

The Jacobian determinant is

$$\frac{\partial(Q, R)}{\partial(u, v)} = \begin{vmatrix} \dfrac{\partial Q}{\partial u} & \dfrac{\partial Q}{\partial v} \\ \dfrac{\partial R}{\partial u} & \dfrac{\partial R}{\partial v} \end{vmatrix} = \begin{vmatrix} \dfrac{1}{2} & \dfrac{1}{2} \\ \dfrac{1}{2} & -\dfrac{1}{2} \end{vmatrix} = -\frac{1}{2} .$$

Applying Equation (14.39) we see that a joint density g of (U, V) is given by the formula

$$g(u, v) = f\left(\frac{u+v}{2}, \frac{u-v}{2}\right) \left| \frac{\partial(Q, R)}{\partial(u, v)} \right| = \frac{1}{2} f\left(\frac{u+v}{2}, \frac{u-v}{2}\right).$$

To obtain a density $f_U = f_{X+Y}$ we integrate with respect to v and find

$$f_{X+Y}(u) = \frac{1}{2} \int_{-\infty}^{\infty} f\left(\frac{u+v}{2}, \frac{u-v}{2}\right) dv .$$

The change of variable $x = \frac{1}{2}(u + v)$, $dx = \frac{1}{2}\, dv$, transforms this to

$$f_{X+Y}(u) = \int_{-\infty}^{\infty} f(x, u - x)\, dx .$$

Similarly, we find

$$f_{X-Y}(v) = \frac{1}{2} \int_{-\infty}^{\infty} f\left(\frac{u+v}{2}, \frac{u-v}{2}\right) du = \int_{-\infty}^{\infty} f(x, x - v)\, dx .$$

An important special case occurs when X and Y are independent. In this case the joint probability density factors into a product,

$$f(x, y) = f_X(x) f_Y(y),$$

and the integrals for f_{X+Y} and f_{X-Y} become

$$f_{X+Y}(u) = \int_{-\infty}^{\infty} f_X(x) f_Y(u - x)\, dx, \qquad f_{X-Y}(v) = \int_{-\infty}^{\infty} f_X(x) f_Y(x - v)\, dx .$$

EXAMPLE 2. *The sum of two exponential distributions.* Suppose now that each of X and Y has an exponential distribution, say $f_X(t) = f_Y(t) = 0$ for $t < 0$, and

$$f_X(t) = \lambda e^{-\lambda t}, \qquad f_Y(t) = \mu e^{-\mu t} \qquad \text{for} \quad t \geq 0 .$$

Determine the density of $X + Y$ when X and Y are independent.

Solution. If $u < 0$ the integral for $f_{X+Y}(u)$ is 0 since the factor $f_X(x) = 0$ for $x < 0$, and the factor $f_Y(u - x) = 0$ for $x \geq 0$. If $u \geq 0$ the integral for $f_{X+Y}(u)$ becomes

$$f_{X+Y}(u) = \int_0^u \lambda e^{-\lambda x} \mu e^{-\mu(u-x)}\, dx = \lambda \mu e^{-\mu u} \int_0^u e^{(\mu-\lambda)x}\, dx .$$

To evaluate the last integral we consider two cases, $\mu = \lambda$ and $\mu \neq \lambda$.

If $\mu = \lambda$ the integral has the value u and we obtain

$$f_{X+Y}(u) = \lambda^2 u e^{-\lambda u} \qquad \text{for} \quad u \geq 0 .$$

If $\mu \neq \lambda$ we obtain

$$f_{X+Y}(u) = \lambda \mu e^{-\mu u} \frac{e^{(\mu-\lambda)u} - 1}{\mu - \lambda} = \lambda \mu \frac{e^{-\lambda u} - e^{-\mu u}}{\mu - \lambda} \qquad \text{for} \quad u \geq 0 .$$

EXAMPLE 3. *The maximum and minimum of two independent random variables.* Let X and Y be two independent one-dimensional random variables with densities f_X and f_Y and corresponding distribution functions F_X and F_Y. Let U and V be the random variables

$$U = \max\{X, Y\}, \qquad V = \min\{X, Y\}.$$

That is, for each ω in the sample space, $U(\omega)$ is the maximum and $V(\omega)$ is the minimum of the two numbers $X(\omega)$, $Y(\omega)$. The mapping $u = \max\{x, y\}$, $v = \min\{x, y\}$ is not one-to-one, so the procedure used to deduce Equation (14.39) is not applicable. However, in this case we can obtain the distribution functions of U and V directly from first principles.

First we note that $U \leq t$ if, and only if, $X \leq t$ and $Y \leq t$. Therefore $P(U \leq t) = P(X \leq t, Y \leq t)$. By independence this is equal to $P(X \leq t)P(Y \leq t) = F_X(t)F_Y(t)$. Thus, we have

$$F_U(t) = F_X(t)F_Y(t).$$

At each point of continuity of f_X and f_Y we can differentiate this relation to obtain

$$f_U(t) = f_X(t)F_Y(t) + F_X(t)f_Y(t).$$

Similarly, we have $V > t$ if and only if $X > t$ and $Y > t$. Therefore

$$\begin{aligned} F_V(t) &= P(V \leq t) = 1 - P(V > t) = 1 - P(X > t, Y > t) = 1 - P(X > t)P(Y > t) \\ &= 1 - (1 - F_X(t))(1 - F_Y(t)) = F_X(t) + F_Y(t) - F_X(t)F_Y(t). \end{aligned}$$

At points of continuity of f_X and f_Y we differentiate this relation to obtain

$$f_V(t) = f_X(t) + f_Y(t) - f_X(t)F_Y(t) - F_X(t)f_Y(t).$$

14.24 Exercises

1. Let X and Y be two independent one-dimensional random variables, each with a uniform distribution over the interval $[0, 1]$. Let $U = X + Y$ and let $V = X - Y$.
(a) Prove that U has a continuous density f_U given by

$$f_U(u) = \begin{cases} u & \text{if } 0 < u \leq 1, \\ 2 - u & \text{if } 1 < u < 2, \\ 0 & \text{otherwise}. \end{cases}$$

(b) Describe, in a similar way, a continuous density f_V for V.
(c) Determine whether or not U and V are independent.
2. Let X and Y be as in Exercise 1, and let $U = \max\{X, Y\}$, $V = \min\{X, Y\}$.
(a) Prove that U has a density function such that $f_U(t) = 2t$ for $0 \leq t < 1$, and $f_U(t) = 0$ otherwise.
(b) Describe a density function f_V for V.
(c) Determine whether or not U and V are independent.

3. Let X and Y be two independent one-dimensional random variables, each having an exponential distribution with parameter $\lambda = 1$, and let $f(x, y) = f_X(x)f_Y(y)$, the product of the densities of X and Y.
   (a) Let A denote the set of points in the xy-plane at which $f(x, y) > 0$. Make a sketch of A and of its image A' under the mapping defined by $u = x + y$, $v = x/(x + y)$.
   (b) Let $U = X + Y$ and $V = X/(X + Y)$ be two new random variables, and compute a probability density g of (U, V).
   (c) Compute a probability density f_U.
   (d) Compute a probability density f_V.
4. Let X and Y be two independent random variables, each with a standard normal distribution (mean = 0, variance = 1). Introduce new random variables U and V by the equations $U = X/Y$, $V = Y$.
   (a) Show that a probability density function of (U, V) is given by the formula

$$g(u, v) = -\frac{v}{2\pi} e^{-(1+u^2)v^2/2} \quad \text{if} \quad v < 0.$$

   (b) Find a similar formula for computing $g(u, v)$ when $v \geq 0$.
   (c) Determine a probability density function of U.
5. Assume X has the density function given by

$$f_X(x) = \begin{cases} \dfrac{1}{\pi\sqrt{1 - x^2}} & \text{if} \quad -1 < x < 1, \\ 0 & \text{if} \quad |x| \geq 1. \end{cases}$$

   If an independent random variable Y has density

$$f_Y(y) = \begin{cases} ye^{-y^2/2} & \text{if} \quad y \geq 0, \\ 0 & \text{if} \quad y < 0, \end{cases}$$

   find a density function of $Z = XY$.
6. Given two independent one-dimensional random variables X and Y with continuous densities f_X and f_Y. Let U and V be two random variables such that $X = U\cos V$, $Y = U\sin V$, with $U > 0$ and $-\pi < V \leq \pi$.
   (a) Prove that U has a density such that $f_U(u) = 0$ for $u < 0$, and

$$f_U(u) = u\int_{-\pi}^{\pi} f_X(u\cos v)f_Y(u\sin v)\,dv \qquad \text{for} \quad u \geq 0.$$

   (b) Determine f_U and the corresponding distribution F_U explicitly when each of X and Y has a normal distribution with mean $m = 0$ and variance σ^2.
7. (a) Assume $\sigma_1 > 0$ and $\sigma_2 > 0$. Verify the algebraic identity

$$\left(\frac{x - m_1}{\sigma_1}\right)^2 + \left(\frac{t - x - m_2}{\sigma_2}\right)^2 = \left(\frac{x - m_0}{\sigma_0}\right)^2 + \left(\frac{t - (m_1 + m_2)}{\sigma}\right)^2,$$

   where

$$\sigma^2 = \sigma_1^2 + \sigma_2^2, \qquad \sigma_0^2 = \frac{\sigma_1^2\sigma_2^2}{\sigma^2}, \qquad \text{and} \qquad m_0 = \frac{m_1\sigma_2^2 + (t - m_2)\sigma_1^2}{\sigma_1^2 + \sigma_2^2}.$$

(b) Given two independent one-dimensional random variables X and Y. Assume X has a normal distribution with mean m_1 and variance σ_1^2, and that Y has a normal distribution with mean m_2 and variance σ_2^2. Prove that $X + Y$ has a normal distribution with mean $m = m_1 + m_2$ and variance $\sigma^2 = \sigma_1^2 + \sigma_2^2$.

8. Given two one-dimensional random variables X and Y with densities f_X and f_Y and joint density f. For each fixed y, define

$$f_X(x \mid Y = y) = \frac{f(x, y)}{f_Y(y)} \qquad \text{whenever} \quad f_Y(y) > 0.$$

This is called the conditional probability density of X, given that $Y = y$. Similarly, we define the conditional probability density of Y, given that $X = x$, by the equation

$$f_Y(y \mid X = x) = \frac{f(x,y)}{f_X(x)} \qquad \text{whenever} \quad f_X(x) > 0.$$

(a) If f_Y and f_X are positive, prove that $\int_{-\infty}^{\infty} f_X(x \mid Y = y)\, dx = \int_{-\infty}^{\infty} f_Y(y \mid X = x)\, dy = 1$.
(b) If f_Y and f_X are positive, prove that

$$f_X(x) = \int_{-\infty}^{\infty} f_Y(y) f_X(x \mid Y = y)\, dy \qquad \text{and} \qquad f_Y(y) = \int_{-\infty}^{\infty} f_X(x) f_Y(y \mid X = x)\, dx.$$

9. A random variable (X, Y) is said to have a *normal bivariate distribution* if its density function f is given by the formula

$$f(x, y) = \frac{\sqrt{D}}{2\pi} e^{-Q(x,y)/2},$$

where $Q(x, y)$ is the quadratic form

$$Q(x, y) = A_{11}(x - x_0)^2 + 2A_{12}(x - x_0)(y - y_0) + A_{22}(y - y_0)^2.$$

The numbers A_{11}, A_{12}, A_{22} are constants with $A_{11} > 0$. The number $D = A_{11}A_{22} - A_{12}^2$ is called the *discriminant* of Q and is assumed to be positive. The numbers x_0 and y_0 are arbitrary.
(a) Show that $Q(x, y)$ can be expressed as a sum of squares as follows:

$$Q(x, y) = A_{11}\left(u + \frac{A_{12}}{A_{11}} v\right)^2 + \frac{D}{A_{11}} v^2, \qquad \text{where} \quad u = x - x_0,\ v = y - y_0.$$

(b) Define the "improper" double integral $\iint_{-\infty}^{+\infty} f(x, y)\, dx\, dy$ to be the limit

$$\iint_{-\infty}^{+\infty} f(x, y)\, dx\, dy = \lim_{t \to +\infty} \iint_{R(t)} f(x, y)\, dx\, dy,$$

where $R(t)$ is the square $[-t, t] \times [-t, t]$. Show that

$$\iint_{-\infty}^{+\infty} f(x, y)\, dx\, dy = 1.$$

[*Hint:* Use part (a) to transform the double integral over $R(t)$ into a double integral in the uv-plane. Then perform a linear change of variables to simplify the integral and, finally, let $t \to +\infty$.]

10. If a two-dimensional random variable (X, Y) has a normal bivariate distribution as described in Exercise 9, show that X and Y themselves are normal one-dimensional random variables with means x_0 and y_0, respectively, and with variances $\sigma^2(X) = A_{22}/D$, $\sigma^2(Y) = A_{11}/D$.
11. If (X, Y) has a normal bivariate distribution as described in Exercise 9, show that the random variable $Z = X + Y$ has a one-dimensional normal distribution with mean $x_0 + y_0$ and variance $(A_{11} - 2A_{12} + A_{22})/D$.

14.25 Expectation and variance

The mass interpretation of probability distributions may be carried a step further by introducing the concepts of *expectation* and *variance.* These play the same role in probability theory that "center of mass" and "moment of inertia" play in mechanics. Without the Stieltjes integral we must give separate definitions for the discrete and continuous cases.

DEFINITIONS OF EXPECTATION AND VARIANCE. *Let X be a one-dimensional random variable. The expectation of X and the variance of X are real numbers denoted by $E(X)$ and* Var (X) *respectively, and are defined as follows:*

(a) *For a continuous random variable with density function f_X,*

$$E(X) = \int_{-\infty}^{+\infty} t\, f_X(t)\, dt,$$

$$\operatorname{Var}(X) = \int_{-\infty}^{+\infty} [t - E(X)]^2 f_X(t)\, dt.$$

(b) *For a discrete random variable with mass points $x_1, x_2, \ldots$ having probabilities $p_k = P(X = x_k)$, we define*

$$E(X) = \sum_{k=1}^{\infty} x_k p_k,$$

$$\operatorname{Var}(X) = \sum_{k=1}^{\infty} [x_k - E(X)]^2 p_k.$$

Note: We say that $E(X)$ and Var (X) *exist* only when the integral or series in question is *absolutely convergent.* It is understood that the series is a finite sum when the sample space is finite; in this case $E(X)$ and Var (X) always exist. They also exist when f_X is 0 outside some finite interval.

The mathematical expectation $E(X)$ is a theoretically computed value associated with the random variable X. In some respects, the distribution acts as though its entire mass were concentrated at a single point, $E(X)$. The true significance of mathematical expectation in probability theory will be discussed in Section 14.29 in connection with the so-called "laws of large numbers."

In mechanics, a knowledge of the center of mass alone gives no indication of how the mass is spread or dispersed about its center. A measure of this dispersion is provided by the "second moment" or "moment of inertia." In probability theory, this second moment is the variance. It measures the tendency of a distribution to spread out from its expected value. In Section 14.28 we shall find that a small variance indicates that large deviations from the expected value are unlikely.

Although the expectation $E(X)$ may be positive or negative, the variance $\operatorname{Var}(X)$ is always nonnegative. The symbol σ^2 is also used to denote the variance. Its positive square root is called the *standard deviation* and is denoted by σ. The standard deviation is a weighted average; in fact, σ is a weighted root mean square of the distance of each value of X from the expected value $E(X)$. The analogous concept in mechanics is the "radius of gyration."

EXAMPLE 1. *Uniform distribution.* Let X have a uniform distribution over an interval $[a, b]$. Then $f(t) = 1/(b - a)$ if $a < t < b$, and $f(t) = 0$ otherwise. Therefore the expectation of X is given by

$$E(X) = \int_{-\infty}^{+\infty} tf(t)\,dt = \frac{1}{b-a}\int_a^b t\,dt = \frac{b^2 - a^2}{2(b-a)} = \frac{a+b}{2}.$$

Thus the mean is the mid-point of the interval. If we write m for $(a + b)/2$ and note that $m - a = b - m = (b - a)/2$ we find

$$\operatorname{Var}(X) = \frac{1}{b-a}\int_a^b (t-m)^2\,dt = \frac{1}{b-a}\int_{a-m}^{b-m} u^2\,du = \frac{(b-a)^2}{12}.$$

Note that the variance depends only on the *length* of the interval.

EXAMPLE 2. *Binomial distribution.* If X has a binomial distribution with parameters n and p we have

$$E(X) = \sum_{k=0}^{n} k\binom{n}{k} p^k q^{n-k},$$

where $q = 1 - p$. To evaluate this sum, let

$$f(x, y) = (x + y)^n = \sum_{k=0}^{n} \binom{n}{k} x^k y^{n-k}$$

and note that

$$\sum_{k=0}^{n} k\binom{n}{k} x^{k-1} y^{n-k} = \frac{\partial f(x, y)}{\partial x} = n(x + y)^{n-1}.$$

If we multiply both sides of this last equation by x and put $x = p$ and $y = q$, we obtain $E(X) = np$.

By a similar argument we may deduce the formula

$$\operatorname{Var}(X) = \sum_{k=0}^{n} (k - np)^2 \binom{n}{k} p^k q^{n-k} = npq.$$

Two proofs of this formula are suggested in Exercise 6 of Section 14.27.

EXAMPLE 3. *Normal distribution.* The terms "mean" and "variance" have already been introduced in connection with our description of the normal distribution in Section 14.14. These terms are justified by the formulas

$$E(X) = \frac{1}{\sigma\sqrt{2\pi}} \int_{-\infty}^{+\infty} te^{-[(t-m)/\sigma]^2/2}\, dt = m$$

and

$$\operatorname{Var}(X) = \frac{1}{\sigma\sqrt{2\pi}} \int_{-\infty}^{+\infty} (t - m)^2 e^{-[(t-m)/\sigma]^2/2}\, dt = \sigma^2.$$

Proofs of these formulas are requested in Exercise 7 of Section 14.27

Gamblers often use the concept of expectation to decide whether a given game of chance is favorable or unfavorable. As an illustration we shall consider the game of betting on "red" or "black" in roulette.

EXAMPLE 4. *Roulette.* A roulette wheel carries the numbers from 0 to 36. The number 0 appears on a gray background, half of the remaining 36 numbers on a red background, and the other half on a black background. The usual methods of betting are:

(1) Bet \$1 on a color (red or black). Possible return: \$2.
(2) Bet \$1 on a single number (0 excepted). Possible return: \$36.
(3) Bet \$1 on any dozen numbers (0 excepted). Possible return: \$3.

If 0 is the winning number the house wins and all other players lose.

Let X be the random variable which measures the financial outcome of betting by method (1). The possible values of X are $x_1 = -1$ and $x_2 = +1$. The point probabilities are $P(X = x_1) = \frac{19}{37}$, $P(X = x_2) = \frac{18}{37}$. Therefore the expectation is

$$E(X) = (-1)\tfrac{19}{37} + (+1)\tfrac{18}{37} = -\tfrac{1}{37};$$

this is usually interpreted to mean that the game is unfavorable to those who play it. The mathematical justification for this interpretation is provided by one of the *laws of large numbers*, to be discussed in Section 14.29 The reader may verify that the expectation has the same value for methods (2) and (3) as well.

EXAMPLE 5. *A coin-tossing game.* In a coin-tossing game there is a probability p that heads (H) will come up and a probability q that tails (T) will come up, where $0 \le p \le 1$ and $q = 1 - p$. The coin is tossed repeatedly until the first outcome occurs a second time; at this point the game ends. If the first outcome is H we are paid \$1 for each T that comes up until we get the next H. For example, $HTTTH$ pays \$3, but HH pays \$0. If the first outcome is T the same rules apply with H and T interchanged. The problem is to determine how much we should pay to play this game. For this purpose we shall consider the random variable which counts the number of dollars won and compute its expected value.

For the sample space we take the collection of all possible games that can be played in this manner. This set can be expressed as the union of two sets A and B, where

$$A = \{TT, THT, THHT, THHHT, \ldots\} \quad \text{and} \quad B = \{HH, HTH, HTTH, HTTTH, \ldots\}.$$

We denote the elements of set A (in the order listed) as $a_0, a_1, a_2, a_3, \ldots$ and those of set B as $b_0, b_1, b_2, b_3, \ldots$. Next, we assign the point probabilities as follows:

$$P(a_n) = p^n q^2 \qquad \text{and} \qquad P(b_n) = q^n p^2 .$$

(When $p = 0$, we put $P(a_0) = 1$ and let $P(x) = 0$ for all other x in $A \cup B$. When $q = 0$ we put $P(b_0) = 1$ and let $P(x) = 0$ for all other x.) In Section 13.21 it was shown that this is an acceptable assignment of probabilities.

The random variable X in which we are interested is defined on the sample space $A \cup B$ as follows:

$$X(a_n) = X(b_n) = n \qquad \text{for} \quad n = 0, 1, 2, \ldots .$$

The event "$X = n$" consists of the two games a_n and b_n, so we have

$$P(X = n) = p^n q^2 + q^n p^2 ,$$

where p^0 and q^0 are to be interpreted as 1 when $p = 0$ or $q = 0$. The expectation of X is given by the sum

(14.40) $$E(X) = \sum_{n=0}^{\infty} nP(X = n) = q^2 \sum_{n=0}^{\infty} np^n + p^2 \sum_{n=0}^{\infty} nq^n .$$

If either $p = 0$ or $q = 0$, we obtain $E(X) = 0$. Otherwise we may compute the sums of the series in (14.40) by noting that for $0 < x < 1$ we have

$$\sum_{n=0}^{\infty} nx^n = x \frac{d}{dx} \sum_{n=0}^{\infty} x^n = x \frac{d}{dx}\left(\frac{1}{1 - x}\right) = \frac{x}{(1 - x)^2} .$$

Using this in (14.40) with $x = p$ and $x = q$ we obtain, for $0 < p < 1$,

$$E(X) = \frac{q^2 p}{(1 - p)^2} + \frac{p^2 q}{(1 - q)^2} = p + q = 1 .$$

We interpret this result by saying that the game is unfavorable to those who pay more than \$1 to play it.

This particular example is of special interest because the expectation $E(X)$ is independent of p when $0 < p < 1$. In other words, loading the coin in favor of heads or tails does not affect the expected value except in the extreme cases in which it is so loaded that it always falls heads or always falls tails. Note that, as a function of p, the expectation $E(X)$ is discontinuous at the points $p = 0$ and $p = 1$. Otherwise it has the constant value 1. This interesting example was suggested to the author by H. S. Zuckerman.

14.26 Expectation of a function of a random variable

If a new random variable Y is related to a given one X by an equation of the form $Y = \varphi(X)$, its expectation is given (in the continuous case) by the equation

(14.41) $$E(Y) = \int_{-\infty}^{+\infty} t f_Y(t)\, dt .$$

The expectation $E(Y)$ can be computed directly in terms of the density f_X without determining the density of Y. In fact, the following formula is equivalent to (14.41):

$$E(Y) = \int_{-\infty}^{+\infty} \varphi(t) f_X(t)\, dt. \tag{14.42}$$

A proof of (14.42) in the most general case is difficult and will not be attempted here. However, for many special cases of importance the proof is simple. In one such case, φ is differentiable and strictly increasing on the whole real axis, and takes on every real value. For a continuously distributed random variable X with density f_X we have the following formula for the density function f_Y (derived in Section 14.17):

$$f_Y(t) = f_X[\psi(t)] \cdot \psi'(t),$$

where ψ is the inverse of φ. If we use this in (14.41) and make the change of variable $u = \psi(t)$ [so that $t = \varphi(u)$], we obtain

$$E(Y) = \int_{-\infty}^{+\infty} t f_Y(t)\, dt = \int_{-\infty}^{+\infty} t f_X[\psi(t)] \cdot \psi'(t)\, dt = \int_{-\infty}^{+\infty} \varphi(u) f_X(u)\, du,$$

which is the same as (14.42).

When Equation (14.42) is applied to $Y = (X - m)^2$, where $m = E(X)$, we obtain

$$E(Y) = \int_{-\infty}^{+\infty} (t - m)^2 f_X(t)\, dt = \text{Var}\,(X).$$

This shows that variance is itself an expectation. A formula analogous to (14.42) also holds, of course, in the discrete case. More generally, it can be shown that

$$E[\varphi(X, Y)] = \int_{-\infty}^{+\infty} \int_{-\infty}^{+\infty} \varphi(x, y) f(x, y)\, dx\, dy$$

if (X, Y) is a continuous random variable with joint density f.

Note: For two-dimensional random variables, expectation and variance may be defined in a manner similar to that used for the one-dimensional case, except that double integrals and double sums are employed. We shall not discuss this extension here.

14.27 Exercises

1. A die is rolled. Let X denote the number of points on the upturned face. Compute $E(X)$ and Var (X).
2. Assume that X is a continuous random variable with a probability density function. Let $Y = (X - m)/\sigma$, where $m = E(X)$ and $\sigma = \sqrt{\text{Var}\,(X)}$. Show that $E(Y) = 0$ and $E(Y^2) = 1$.
3. Derive the following general properties of expectation and variance for either the discrete or the continuous case.
   (a) $E(cX) = cE(X)$, where c is a constant.
   (b) $\text{Var}\,(cX) = c^2\, \text{Var}\,(X)$, where c is a constant.
   (c) $E(X + Y) = E(X) + E(Y)$.
   (d) $\text{Var}\,(X) = E(X^2) - [E(X)]^2$.
   (e) $\text{Var}\,(X + Y) = \text{Var}\,(X) + \text{Var}\,(Y) + 2E[(X - E(X))(Y - E(Y))]$.
   (f) $E[\varphi_1(X) + \varphi_2(Y)] = E[\varphi_1(X)] + E[\varphi_2(Y)]$. [Part (c) is a special case.]

4. If X and Y are independent random variables, show that
   (a) $\operatorname{Var}(X + Y) = \operatorname{Var}(X) + \operatorname{Var}(Y)$.
   (b) $E[\varphi(X) \cdot \psi(Y)] = E[\varphi(X)] \cdot E[\psi(Y)]$.
   (c) If $X_1, X_2, \ldots, X_n$ are independent random variables with $E(X_k) = m_k$, show that

$$\operatorname{Var}\left[\sum_{k=1}^{n} (X_k - m_k)\right] = \sum_{k=1}^{n} \operatorname{Var}(X_k - m_k) = \sum_{k=1}^{n} \operatorname{Var}(X_k).$$

5. Let $X_1, X_2, X_3, \ldots, X_n$ be n independent random variables, each having the same expectation, $E(X_k) = m$, and the same variance, $\operatorname{Var}(X_k) = \sigma^2$. Let $\bar{X}$ denote the arithmetic mean, $\bar{X} = (1/n)\sum_{i=1}^{n} X_i$. Use Exercises 3 and 4 to prove that $E(\bar{X}) = m$ and $\operatorname{Var}(\bar{X}) = \sigma^2/n$.
6. (a) If $q = 1 - p$, prove the formula

$$\sum_{k=0}^{n} (k - np)^2 \binom{n}{k} p^k q^{n-k} = npq,$$

   thereby showing that $\operatorname{Var}(X) = npq$ for a random variable X having a binomial distribution with parameters n and p. [*Hint:* $k^2 = k(k - 1) + k$.]
   (b) If X has a binomial distribution with parameters n and p, show that X can be expressed as a sum of n independent random variables $X_1, X_2, \ldots, X_n$, each assuming the possible values 0 and 1 with probabilities p and q, respectively, and each having a binomial distribution. Use this result and Exercise 5 to show that $E(X) = np$ and $\operatorname{Var}(X) = npq$.
7. Determine the expectation and variance (whenever they exist) for a random variable X having
   (a) a Poisson distribution with parameter λ.
   (b) a Cauchy distribution.
   (c) an exponential distribution with parameter λ.
   (d) a normal distribution.
8. A random variable X has a probability density function given by

$$f(t) = \frac{C(r)}{|t|^r} \quad \text{if } |t| > 1, \qquad f(t) = 0 \quad \text{if } |t| \le 1,$$

   where $r > 1$ and $C(r)$ is independent of t.
   (a) Express $C(r)$ in terms of r and make a sketch to indicate the nature of the graph of f.
   (b) Determine the corresponding distribution function F_X and make a sketch to indicate the nature of its graph.
   (c) Compute $P(X < 5)$ and $P(5 < X < 10)$ in terms of r.
   (d) For what values of r does X have a finite expectation? Compute $E(X)$ in terms of r when the expectation is finite.
   (e) For what values of r does X have a finite variance? Compute $\operatorname{Var}(X)$ in terms of r when the variance is finite.
9. A gambler plays roulette according to the following "system." He plays in sets of three games. In the first and second games he always bets \$1 on red. For the third game he proceeds as follows:
   (a) If he wins in the first and second games, he doesn't bet.
   (b) If he wins in one of the first or second and loses in the other, he bets \$1 on the color opposite to the outcome of the second game.
   (c) If he loses in both the first and second, he bets \$3 on red.
   Let X, Y, and Z denote, respectively, the financial outcomes of the first, second, and third games. Compute $E(X)$, $E(Y)$, $E(Z)$, and $E(X + Y + Z)$.

10. (*Petersburg Problem*). A player tosses a coin and wins \$1 if his first toss is heads. If he tosses heads again he wins another dollar. If he succeeds in tossing heads a third time he gets another \$2 (for a total of \$4). As long as he tosses heads in succession n times, his accumulated winnings are 2^{n-1} dollars. The game terminates when he tosses tails. Let X denote the number of dollars won in any particular game. Compute $E(X)$. In view of your result, how much would you be willing to pay Harold's Club in Reno for the privilege of playing this game?
11. (a) Assume X is a continuous random variable with probability density f_X. Let $Y = (X - m)/\sigma$, where $m = E(X)$ and $\sigma = \sqrt{\operatorname{Var}(X)}$. Prove that

$$E(e^Y) = e^{-m/\sigma} \int_{-\infty}^{+\infty} e^{t/\sigma} f_X(t)\, dt.$$

(b) Let X be a discrete random variable having a Poisson distribution with parameter λ. Define Y as in part (a) and prove that

$$E(e^Y) = e^{-\lambda G(\lambda)}, \qquad \text{where} \quad G(\lambda) = 1 + \frac{1}{\sqrt{\lambda}} - e^{1/\sqrt{\lambda}}.$$

12. A random variable X has a standard normal distribution. Compute: (a) $E(|X|)$, (b) $E(e^X)$, (c) $\operatorname{Var}(e^X)$, (d) $E(\sqrt{X^2 + Y^2})$. In part (d), Y also has a standard normal distribution but is independent of X.

14.28 Chebyshev's inequality

As mentioned earlier, a small value for the variance means that it is unlikely that a random variable X will deviate much from its expected value. To make this statement more precise we introduce the absolute value $|X - E(X)|$ which measures the actual distance between X and $E(X)$. How likely is it that this distance is more than a given amount? To answer this question we must determine the probability

$$P[|X - E(X)| > c],$$

where c is a given positive number. In the continuous case we have

$$P[|X - E(X)| > c] = 1 - P[|X - E(X)| \le c] = 1 - P[E(X) - c \le X \le E(X) + c]$$

$$= \int_{-\infty}^{+\infty} f_X(t)\, dt - \int_{E(X)-c}^{E(X)+c} f_X(t)\, dt$$

$$= \int_{-\infty}^{E(X)-c} f_X(t)\, dt + \int_{E(X)+c}^{+\infty} f_X(t)\, dt\,; \tag{14.43}$$

therefore, the calculation of this probability can be accomplished once the density function f_X is known. Of course, if f_X is *unknown* this method gives no information. However, if the *variance* is known, we can obtain an upper bound for this probability. This upper bound is provided by the following theorem of P. L. Chebyshev (1821–1894), a famous Russian mathematician who made many important contributions to probability theory and other branches of mathematics, especially the theory of numbers.

THEOREM 14.11. CHEBYSHEV'S INEQUALITY. *Let X be a one-dimensional random variable with finite expectation $E(X)$ and variance* $\operatorname{Var}(X)$. *Then for every positive number c we have*

$$P[|X - E(X)| > c] \le \frac{\operatorname{Var}(X)}{c^2}. \tag{14.44}$$

Proof. In the continuous case we have

$$\begin{aligned}
\operatorname{Var}(X) &= \int_{-\infty}^{+\infty} [t - E(X)]^2 f_X(t)\,dt \\
&\ge \int_{-\infty}^{E(X)-c} [t - E(X)]^2 f_X(t)\,dt + \int_{E(X)+c}^{+\infty} [t - E(X)]^2 f_X(t)\,dt \\
&\ge c^2\left(\int_{-\infty}^{E(X)-c} f_X(t)\,dt + \int_{E(X)+c}^{+\infty} f_X(t)\,dt\right).
\end{aligned}$$

Because of (14.43), the coefficient of c^2 on the right is $P[|X - E(X)| > c]$. Therefore, when we divide by c^2 we obtain (14.44). This completes the proof for the continuous case; the discrete case may be similarly treated.

Chebyshev's inequality tells us that the larger we make c the smaller the probability is that $|X - E(X)| > c$. In other words, it is unlikely that X will be very far from $E(X)$; it is even more unlikely if the variance $\operatorname{Var}(X)$ is small.

If we replace c by $k\sigma$, where $k > 0$ and σ denotes the standard deviation, $\sigma = \sqrt{\operatorname{Var}(X)}$, Chebyshev's inequality becomes

$$P[|X - E(X)| > k\sigma] \le \frac{1}{k^2}.$$

That is, the probability that X will differ from its expected value by more than k standard deviations does not exceed $1/k^2$. For example, when $k = 10$ this inequality tells us that the probability $P[|X - E(X)| > 10\sigma]$ does not exceed 0.010. In other words, the probability is no more than 0.010 that an observed value of X will differ from the expected value by more than ten standard deviations. Similarly, when $k = 3$ we find that the probability does not exceed $\frac{1}{9} = 0.111\ldots$ that an observed value will differ from the mean by more than three standard deviations.

Chebyshev's inequality is a general theorem that applies to all distributions. In many applications the inequality can be strengthened when more information is known about the particular distribution. For example, if X has a binomial distribution with parameters n and p it can be shown (by use of the normal approximation to the binomial distribution) that for large n the probability is about 0.003 that an observed value will differ from the mean by more than three standard deviations. (For this result, $n \ge 12$ suffices.) This is much smaller than the probability 0.111 provided by Chebyshev's inequality.

EXAMPLE. *Testing a coin for fairness.* We want to decide whether or not a particular coin is fair by tossing it 10,000 times and recording the number of heads. For a fair coin the random variable X which counts the number of heads has a binomial distribution with parameters $n = 10{,}000$ and $p = \frac{1}{2}$. The mean of X is $np = 5{,}000$ and the standard deviation is $\sigma = \sqrt{npq} = 50$. (See Example 2 in Section 14.25.) As mentioned above, the probability for a binomially distributed random variable to differ from its expected value by more than 3σ is about 0.003. Therefore, let us agree to say that a coin is *not fair* if the number of heads in 10,000 tosses differs from the mean by more than 3σ. Since $E(X) = 5{,}000$ and $3\sigma = 150$, we would say the coin is unfair if the number of heads in 10,000 tosses is less than 4,850 or more than 5,150.

14.29 Laws of large numbers

In connection with coin-tossing problems, it is often said that the probability of tossing heads with a perfectly balanced coin is $\frac{1}{2}$. This does not mean that if a coin is tossed twice it will necessarily come up heads exactly once. Nor does it mean that in 1000 tosses heads will appear exactly 500 times. Let us denote by $h(n)$ the number of heads that occur in n tosses. Experience shows that even for very large n, the ratio $h(n)/n$ is not necessarily $\frac{1}{2}$. However, experience also shows that this ratio does seem to *approach* $\frac{1}{2}$ as n increases, although it may oscillate considerably above and below $\frac{1}{2}$ in the process. This suggests that it might be possible to prove that

$$\lim_{n\to\infty} \frac{h(n)}{n} = \frac{1}{2}. \tag{14.45}$$

Unfortunately, this cannot be done. One difficulty is that the number $h(n)$ depends not only on n but also on the particular experiment being performed. We have no way of knowing in advance how $h(n)$ will vary from one experiment to another. But the real trouble is that it *is* possible (although not very likely) that in some particular experiment the ratio $h(n)/n$ may *not* tend to $\frac{1}{2}$ at all. For example, there is no reason to exclude the possibility of getting heads on *every* toss of the coin, in which case $h(n) = n$ and $h(n)/n \to 1$. Therefore, instead of trying to prove the formula in (14.45), we shall find it more reasonable (and more profitable) to ask how likely it is that $h(n)/n$ will differ from $\frac{1}{2}$ by a certain amount. In other words, given some positive number c, we seek the probability

$$P\left(\left|\frac{h(n)}{n} - \frac{1}{2}\right| > c\right).$$

By introducing a suitable random variable and using Chebyshev's inequality we can get a useful *upper bound* to this probability, a bound which does not require an explicit knowledge of $h(n)$. This leads to a new limit relation that serves as an appropriate substitute for (14.45).

No extra effort is required to treat the more general case of a Bernoullian sequence of trials, in which the probability of "success" is p and the probability of "failure" is q. (In coin tossing, "success" can mean "heads" and for p we may take $\frac{1}{2}$.) Let X denote the random variable which counts the number of successes in n independent trials. Then X has a binomial distribution with expectation $E(X) = np$ and variance $\operatorname{Var}(X) = npq$. Hence Chebyshev's inequality is applicable; it states that

$$P(|X - np| > c) \le \frac{npq}{c^2}. \tag{14.46}$$

Since we are interested in the ratio X/n, which we may call the *relative frequency* of success, we divide the inequality $|X - np| > c$ by n and rewrite (14.46) as

$$P\left(\left|\frac{X}{n} - p\right| > \frac{c}{n}\right) \leq \frac{npq}{c^2}. \tag{14.47}$$

Since this is valid for every $c > 0$, we may let c depend on n and write $c = \epsilon n$, where ϵ is a fixed positive number. Then (14.47) becomes

$$P\left(\left|\frac{X}{n} - p\right| > \epsilon\right) \leq \frac{pq}{n\epsilon^2}.$$

The appearance of n in the denominator on the right suggests that we let $n \to \infty$. This leads to the limit formula

$$\lim_{n\to\infty} P\left(\left|\frac{X}{n} - p\right| > \epsilon\right) = 0 \qquad \text{for every fixed } \epsilon > 0, \tag{14.48}$$

called the *law of large numbers for the Bernoulli distribution.* It tells us that, given any $\epsilon > 0$ (no matter how small), the probability that the relative frequency of success differs from p by more than ϵ is a function of n which tends to 0 as $n \to \infty$. This limit relation gives a mathematical justification to the assignment of the probability $\frac{1}{2}$ for tossing heads with a perfectly balanced coin.

The limit relation in (14.48) is a special case of a more general result in which the "relative frequency" X/n is replaced by the arithmetic mean of n independent random variables having the same expectation and variance. This more general theorem is usually referred to as the *weak law of large numbers*; it may be stated as follows:

THEOREM 14.12. WEAK LAW OF LARGE NUMBERS. *Let $X_1, X_2, \ldots, X_n$ be n independent random variables, each having the same expectation and the same variance, say*

$$E(X_k) = m \qquad \textit{and} \qquad \operatorname{Var}(X_k) = \sigma^2 \qquad \textit{for} \quad k = 1, 2, \ldots, n.$$

Define a new random variable $\bar{X}$ (called the arithmetic mean of $X_1, X_2, \ldots, X_n$) by the equation

$$\bar{X} = \frac{1}{n}\sum_{k=1}^{n} X_k.$$

Then, for every fixed $\epsilon > 0$, we have

$$\lim_{n\to\infty} P(|\bar{X} - m| > \epsilon) = 0. \tag{14.49}$$

An equivalent statement is

$$\lim_{n\to\infty} P(|\bar{X} - m| \leq \epsilon) = 1. \tag{14.50}$$

Proof. We apply Chebyshev's inequality to $\bar{X}$. For this we need to know the expectation and variance of $\bar{X}$. These are

$$E(\bar{X}) = m \qquad \text{and} \qquad \text{Var}\,(\bar{X}) = \frac{\sigma^2}{n}.$$

(See Exercise 5 in Section 14.27.) Chebyshev's inequality becomes $P(|\bar{X} - m| > c) \le \sigma^2/(nc^2)$. Letting $n \to \infty$ and replacing c by ϵ we obtain (14.49) and hence (14.50).

Note: To show that the limit relation in (14.48) is a special case of Theorem 14.12, we assume each X_k has the possible values 0 and 1, with probabilities $P(X_k = 1) = p$ and $P(X_k = 0) = 1 - p$. Then $\bar{X}$ is the relative frequency of success in n independent trials, $E(\bar{X}) = p$, and (14.49) reduces to (14.48).

Theorem 14.12 is called a *weak* law because there is also a *strong* law of large numbers which (under the same hypotheses) states that

$$P\left(\lim_{n\to\infty} |\bar{X} - m| = 0\right) = 1. \tag{14.51}$$

The principal difference between (14.51) and (14.50) is that the operations "limit" and "probability" are interchanged. It can be shown that the strong law implies the weak law, but not conversely.

Notice that the strong law in (14.51) seems to be closer to formula (14.45) than (14.50) is. In fact, (14.51) says that we have $\lim_{n\to\infty} \bar{X} = m$ "almost always," that is, with probability 1. When applied to coin tossing, in particular, it says that the failure of Equation (14.45) is no more likely than the chance of tossing a fair coin repeatedly and always getting heads. The strong law really shows why probability theory corresponds to experience and to our intuitive feeling of what probability "should be."

The proof of the strong law is lengthy and will be omitted. Proofs appear in the books listed as References 1, 3, 8, and 10 at the end of this chapter.

14.30 The central limit theorem of the calculus of probabilities

In many applications of probability theory, the random variables of interest are sums of other random variables. For example, the financial outcome after several plays of a game is the sum of the winnings at each play. A surprising thing happens when a large number of independent random variables are added together. Under general conditions (applicable in almost every situation that occurs in practice) the distribution of the sum tends to be normal, regardless of the distributions of the individual random variables that make up the sum. The precise statement of this remarkable fact is known as the *central limit theorem of the calculus of probabilities.* It accounts for the importance of the normal distribution in both theory and practice. A thorough discussion of this theorem belongs to the advanced study of probability theory. This section will merely describe what the theorem asserts.

Suppose we have an infinite sequence of random variables, say $X_1, X_2, \ldots,$ with finite expectations and variances. Let

$$m_k = E(X_k) \qquad \text{and} \qquad \sigma_k^2 = \text{Var}\,(X_k), \qquad k = 1, 2, \ldots.$$

We form a new random variable S_n by adding the first n differences $X_k - m_k$:

$$S_n = \sum_{k=1}^{n} (X_k - m_k). \tag{14.52}$$

We add the *differences* rather than the X_k alone so that the sum S_n will have expected value 0. The problem here is to determine the limiting form, as $n \to \infty$, of the distribution function of S_n.

If $X_1, X_2, \ldots, X_n$ are *independent*, then [by Exercise 4(c) of Section 14.27] we have

$$\operatorname{Var}(S_n) = \sum_{k=1}^{n} \operatorname{Var}(X_k - m_k) = \sum_{k=1}^{n} \operatorname{Var}(X_k) = \sum_{k=1}^{n} \sigma_k^2 .$$

Ordinarily, $\operatorname{Var}(S_n)$ will be large even though the individual variances σ_k^2 may be small. Random variables with a large variance are not fruitful objects of study because their values tend to be widely dispersed from the expected value. For this reason, a new random variable T_n is introduced by the equation

$$T_n = \frac{S_n}{\sqrt{\operatorname{Var}(S_n)}} . \tag{14.53}$$

This new variable has expectation 0 and variance 1 and is called a *standardized* random variable. The standardized variable T_n is meaningful even if the random variables X_1, $X_2, \ldots, X_n$ are not independent.

We now introduce the following definition:

DEFINITION OF THE CENTRAL LIMIT PROPERTY. *Let*

$$X_1, X_2, X_3, \ldots \tag{14.54}$$

be a sequence of random variables (not necessarily independent), where each X_k has a finite expectation m_k and a finite variance σ_k^2. Define S_n and T_n by (14.52) *and* (14.53). *The sequence in* (14.54) *is said to satisfy the central limit property if, for all a and b with $a \le b$, we have*

$$\lim_{n \to \infty} P(a \le T_n \le b) = \frac{1}{\sqrt{2\pi}} \int_a^b e^{-u^2/2}\, du . \tag{14.55}$$

In other words, the random variables in (14.54) satisfy the central limit property if the distribution of the standardized variable T_n approaches a standard normal distribution as $n \to \infty$. [Equation (14.55) is to hold also if $a = -\infty$ or $b = +\infty$.]

Laplace was the first to realize that this property is shared by many sequences of random variables, although a special case (random variables describing a Bernoullian sequence of trials) had been known earlier by DeMoivre. (Figure 14.11 shows a binomial distribution and a corresponding normal approximation.) Laplace stated a general central limit theorem which was first completely proved by the Russian mathematician A. Lyapunov in 1901. In 1922, J. W. Lindeberg generalized Laplace's result by showing that the property is satisfied if the random variables are independent and have a common distribution giving them the same expectations and variances, say $E(X_k) = m$ and $\text{Var}(X_k) = \sigma^2$ for all k. In this case the standardized variable becomes

$$T_n = \frac{\sum_{k=1}^{n} X_k - nm}{\sigma\sqrt{n}}.$$

Lindeberg realized that independence alone is not sufficient to guarantee the central limit property, but he formulated another condition (now known as the *Lindeberg condition*) which, along with independence, *is* sufficient. In 1935, W. Feller showed that the Lindeberg condition is both necessary and sufficient for independent random variables to satisfy the central limit property. We shall not discuss the Lindeberg condition here except to mention that it implies

$$\text{Var}(S_n) \to \infty \qquad \text{as} \quad n \to \infty .$$

Fortunately, many independent random variables that occur in practice automatically satisfy the Lindeberg condition and therefore also have the central limit property. Up to now, the theory for *dependent* random variables is incomplete. Only a few special cases have been treated. Much of the contemporary research in probability theory centers about the search for general theorems dealing with dependent variables.

14.31 Exercises

1. Carry out the proof of Chebyshev's inequality in the discrete case.
2. If a is any real number, prove that

$$P(|X - a| > c\lambda) \le \frac{1}{c^2}$$

for every $c > 0$, where $\lambda^2 = \int_{-\infty}^{+\infty} (t - a)^2 f_X(t)\,dt$. Chebyshev's inequality is the special case in which $a = E(X)$.

3. Let X denote the random variable which counts the number of successes in n independent trials of a Bernoullian sequence; the probability of success is p. Show that, for every $\epsilon > 0$,

$$P\left(\left|\frac{X}{n} - p\right| > \epsilon\right) \le \frac{1}{4n\epsilon^2}.$$

4. A fair coin is tossed n times; the number of heads is denoted by X. Find the smallest n for which Chebyshev's inequality implies

$$P\left(0.4 < \frac{X}{n} < 0.6\right) > 0.90 .$$

5. In a production line the number X of the defective articles manufactured in any given hour is known to have a Poisson distribution with mean $E(X) = 100$. Use Chebyshev's inequality to compute a lower bound for the probability that in a given hour there will be between 90 and 110 defective articles produced.
6. Assume that a random variable X has a standard normal distribution (mean 0 and variance 1). Let p denote the probability that X differs from its expectation $E(X)$ by more than three

times its standard deviation. Use Chebyshev's inequality to find an upper bound for p. Then use suitable tables of the normal distribution to show that there is an upper bound for p that is approximately one-fiftieth of that obtained by Chebyshev's inequality.

7. Given a sequence of independent random variables $X_1, X_2, \ldots$, each of which has a normal distribution. Let $m_k = E(X_k)$ and let $\sigma_k^2 = \text{Var}(X_k)$. Show that this sequence has the central limit property. [*Hint:* Refer to Exercise 7 in Section 14.24.]

8. Let $X_1, X_2, \ldots$ be independent random variables having the same binomial distribution. Assume each X_k takes the possible values 0 and 1 with probabilities $P(X_k = 1) = p$ and $P(X_k = 0) = q$, where $p + q = 1$. Let $Z_n = X_1 + \cdots + X_n$. The random variable Z_n counts the number of successes in n Bernoulli trials.

   (a) Show that the central limit property takes the following form:

$$\lim_{n\to\infty} P\left(\frac{Z_n - np}{\sqrt{npq}} \le t\right) = \frac{1}{\sqrt{2\pi}} \int_{-\infty}^{t} e^{-u^2/2}\, du\,.$$

   (b) Use the approximation suggested by part (a) to estimate the probability of obtaining between 45 and 55 heads if a fair coin is tossed 100 times. Refer to Table 14.1, p. 536 for the computation.

9. With the notation of Exercise 8, the central limit theorem for random variables describing a Bernoullian sequence of trials can be written in the form

$$\lim_{n\to\infty} \frac{P\left(t_1 \le \dfrac{Z_n - np}{\sqrt{npq}} \le t_2\right)}{\Phi(t_2) - \Phi(t_1)} = 1\,,$$

   where Φ is the standard normal distribution. For this particular case it can be shown that the formula is also valid when t_1 and t_2 are functions of n given by $t_1 = (a - np)/\sqrt{npq}$ and $t_2 = (b - np)/\sqrt{npq}$, where a and b are fixed positive constants, $a < b$.

   (a) Show that this relation implies the asymptotic formula

$$\sum_{k=a}^{b} \binom{n}{k} p^k q^{n-k} \sim \Phi\left(\frac{b - np + \frac{1}{2}}{\sqrt{npq}}\right) - \Phi\left(\frac{a - np - \frac{1}{2}}{\sqrt{npq}}\right) \quad \text{as} \quad n \to \infty\,.$$

   (b) An unbiased die is tossed 180 times. Use the approximation suggested in part (a) to estimate the probability that the upturned face is a six exactly 30 times. Refer to Table 14.1, p. 536 for the computation.

10. An unbiased die is tossed 100 times. Use the approximation suggested in Exercise 9(a) to estimate the probability that the upturned face is a six (a) exactly 25 times, (b) at least 25 times. Refer to Table 14.1, p. 536 for the computation.

Suggested References

1. H. Cramér, *Elements of Probability Theory*, Wiley, New York, 1955.
2. H. Cramér, *Mathematical Methods of Statistics*, Princeton Univ. Press, Princeton, N.J., 1946.
3. W. Feller, *An Introduction to Probability Theory and Its Applications*, 2nd ed., Wiley, New York, 1957.
4. B. V. Gnedenko and A. N. Kolmogorov, *Limit Distributions for Sums of Independent Random Variables*, Addison-Wesley, Reading, Mass., 1954.

5. S. Goldberg, *Probability, an Introduction*, Prentice-Hall, Englewood Cliffs, N.J., 1960.
6. H. Levy and L. Roth, *Elements of Probability*, Oxford Univ. Press, London and New York, 1936.
7. M. Loève, *Probability Theory: Foundations, Random Sequences*, Van Nostrand, New York, 1955.
8. M. E. Munroe, *Theory of Probability*, McGraw-Hill, New York, 1951.
9. J. Neyman, *First Course in Probability and Statistics*, Holt, Rinehart and Winston, New York, 1950.
10. E. Parzen, *Modern Probability Theory and Its Applications*, Wiley, New York, 1960.
11. I. Todhunter, *A History of the Mathematical Theory of Probability from the Time of Pascal to Laplace*, Chelsea, New York, 1949.
12. J. V. Uspensky, *Introduction to Mathematical Probability*, McGraw-Hill, New York, 1937.

15

INTRODUCTION TO NUMERICAL ANALYSIS

15.1 Historical introduction

The planet Uranus was discovered in 1781 by a gifted amateur astronomer, William Herschel (1738–1822), with a homemade 10-ft. telescope. With the use of Kepler's laws, the expected orbit of Uranus was quickly calculated from a few widely separated observations. It was found that the mean distance of Uranus from the sun was about twice that of Saturn and that one complete orbit would require 84 years. By 1830 the accumulated empirical data showed deviations from the scheduled orbit that could not be accounted for. Some astronomers felt that Newton's law of universal gravitation might not hold for distances as large as that of Uranus from the sun; others suspected that the perturbations were due to a hitherto undiscovered comet or more distant planet.

An undergraduate student at Cambridge University, John Couch Adams (1819–1892), was intrigued by the possibility of an undiscovered planet. He set himself the difficult task of calculating what the orbit of such a planet must be to account for the observed positions of Uranus, assuming the validity of Newton's law of gravitation. He completed his calculations in 1845 and asked the Royal Observatory at Greenwich to search for the hypothetical planet, but his request was not taken seriously.

A similar calculation was made independently and almost simultaneously by Jean Joseph Leverrier (1811–1877) of Paris, who asked Johann Galle, head of the Berlin Observatory, to confirm his prediction. The same evening that he received Leverrier's letter, Galle found the new planet, *Neptune*, almost exactly in its calculated position. This was another triumph for Newton's law of gravitation, and one of the first major triumphs of *numerical analysis*, the art and science of computation.

The history of numerical analysis goes back to ancient times. As early as 2000 B.C. the Babylonians were compiling mathematical tables. One clay tablet has been found containing the squares of the integers from 1 to 60. The Babylonians worshipped the heavenly bodies and kept elaborate astronomical records. The celebrated Alexandrian astronomer Claudius Ptolemy (circa 150 A.D.) possessed a Babylonian record of eclipses dating from 747 B.C.

In 220 B.C., Archimedes used regular polygons as approximations to a circle and deduced the inequalities $3\frac{10}{71} < \pi < 3\frac{1}{7}$. Numerical work from that time until the 17th century was centered principally around the preparation of astronomical tables. The advent of algebra in the 16th century brought about renewed activity in all branches of mathematics, including numerical analysis. In 1614, Napier published the first table of logarithms. In 1620, the logarithms of the sine and tangent functions were tabulated to seven decimal places. By 1628, fourteen-place tables of the logarithms of the numbers from 1 to 100,000 had been computed.

Computations with infinite series began to flourish near the end of the 17th century, along with the development of the calculus. Early in the 18th century Jacob Stirling and Brook Taylor laid the foundations of the *calculus of finite differences*, which now plays a central role in numerical analysis. With the prediction of the existence and location of the planet Neptune by Adams and Leverrier in 1845, the scientific importance of numerical analysis became established once and for all.

Late in the 19th century the development of automatic calculating machinery further stimulated the growth of numerical analysis. This growth has been explosive since the end of World War II because of the progress in high-speed electronic computing devices. The new machines have made possible a great many outstanding scientific achievements which previously seemed unattainable.

The art of computation (as distinct from the science of computation) lays much stress on the detailed planning required in a particular calculation. It also deals with such matters as precision, accuracy, errors, and checking. This aspect of numerical analysis will not be discussed here; it is best learned by carrying out actual numerical calculations with specific problems. For valuable advice on practical methods and techniques the reader should consult the existing books on numerical analysis, some of which are listed in the bibliography at the end of this chapter. The bibliography also contains some of the standard mathematical tables; many of them also give practical information on how to carry out a specific calculation.

This chapter provides an introduction to the *science* of computation. It contains some of the basic mathematical principles that might be required of almost anyone who uses numerical analysis, whether he works with a desk calculator or with a large-scale high-speed computing machine. Aside from its practical value, the material in this chapter is of interest in its own right, and it is hoped that this brief introduction will stimulate the reader to learn more about this important and fascinating branch of mathematics.

15.2 Approximations by polynomials

A basic idea in numerical analysis is that of using simple functions, usually polynomials, to approximate a given function f. One type of polynomial approximation was discussed in Volume I in connection with Taylor's formula (Theorem 7.1). The problem there was to find a polynomial P which agrees with a given function f and some of its derivatives at a given point. We proved that if f is a function with a derivative of order n at a point a, there is one and only one polynomial P of degree $\leq n$ which satisfies the $n + 1$ relations

$$P(a) = f(a), \qquad P'(a) = f'(a), \qquad \ldots, \qquad P^{(n)}(a) = f^{(n)}(a).$$

The solution is given by the *Taylor polynomial*,

$$P(x) = \sum_{k=0}^{n} \frac{f^{(k)}(a)}{k!}(x - a)^k.$$

We also discussed the error incurred in approximating $f(x)$ by $P(x)$ at points x other than a. This error is defined to be the difference $E_n(x) = f(x) - P(x)$, so we can write

$$f(x) = \sum_{k=0}^{n} \frac{f^{(k)}(a)}{k!}(x - a)^k + E_n(x).$$

To make further statements about the error we need more information about f. For example, if f has a continuous derivative of order $n + 1$ in some interval containing a, then for every x in this interval the error can be expressed as an integral or as an $(n + 1)$st derivative:

$$E_n(x) = \frac{1}{n!}\int_a^x (x - t)^n f^{(n+1)}(t)\,dt = \frac{f^{(n+1)}(c)}{(n + 1)!}(x - a)^{n+1},$$

where c lies between a and x. (See Sections 7.5 and 7.7 in Volume I.)

There are many other ways to approximate a given function f by polynomials, depending on the use to be made of the approximation. For example, instead of asking for a polynomial that agrees with f and some of its derivatives at a given point, we can ask for a polynomial that takes the same values as f at a number of distinct points. Specifically, if the given distinct points are $x_0, x_1, \ldots, x_n$ we seek a polynomial P satisfying the conditions

$$(15.1) \qquad P(x_0) = f(x_0), \qquad P(x_1) = f(x_1), \qquad \ldots, \qquad P(x_n) = f(x_n).$$

Since there are $n + 1$ conditions to be satisfied we try a polynomial of degree $\le n$, say

$$P(x) = \sum_{k=0}^{n} a_k x^k,$$

with $n + 1$ coefficients $a_0, a_1, \ldots, a_n$ to be determined. The $n + 1$ conditions (15.1) lead to a system of $n + 1$ linear equations for the coefficients. From the theory of linear equations it can be shown that this system has one and only one solution; hence such a polynomial always exists. If the equations are solved by Cramer's rule the coefficients a_0, $a_1, \ldots, a_n$ are expressed as quotients of determinants. In practice, however, the polynomial P is seldom determined in this manner because the calculations are extremely laborious when n is large. Simpler methods have been developed to calculate the polynomial approximation. Some of these will be discussed in later sections. The polynomial which solves the foregoing problem is called an *interpolating polynomial.*

Another common type of polynomial approximation is the so-called *least-square approximation.* Here the given function f is defined and integrable on an interval $[a, b]$ and we seek a polynomial P of degree $\le n$ such that the mean-square error

$$\int_a^b |f(x) - P(x)|^2\,dx$$

will be as small as possible. In Section 15.4 we shall prove that for a continuous f such a polynomial exists and is uniquely determined. The Legendre polynomials introduced in Section 1.14 play a fundamental role in the solution of this problem.

15.3 Polynomial approximation and normed linear spaces

All the different types of polynomial approximation described in the foregoing section can be related by one central idea which is best described in the language of linear spaces.

Let V be a linear space of functions which contains all polynomials of degree $\leq n$ and which also contains the function f to be approximated. The polynomials form a finite-dimensional subspace S, with $\dim S = n + 1$. When we speak of approximating f by a polynomial P in S, we consider the difference $f - P$, which we call the error of the approximation, and then we decide on a way to measure the size of this error.

If V is a Euclidean space, then it has an inner product (x, y) and a corresponding norm given by $\|x\| = (x, x)^{1/2}$, and we can use the norm $\|f - P\|$ as a measure of the size of the error.

Sometimes norms can be introduced in non-Euclidean linear spaces, that is, in linear spaces which do not have an inner product. These norms were introduced in Section 7.26. For convenience we repeat the definition here.

DEFINITION OF A NORM. *Let V be a linear space. A real-valued function N defined on V is called a norm if it has the following properties:*

(a) $N(f) \geq 0$ *for all f in V.*

(b) $N(cf) = |c|\, N(f)$ *for all f in V and every scalar c.*

(c) $N(f + g) \leq N(f) + N(g)$ *for all f and g in V.*

(d) $N(f) = 0$ *implies* $f = O$.

A linear space with a norm assigned to it is called a normed linear space.

The norm of f is sometimes written $\|f\|$ instead of $N(f)$. In this notation, the fundamental properties become:

(a) $\|f\| \geq 0$,

(b) $\|cf\| = |c|\, \|f\|$,

(c) $\|f + g\| \leq \|f\| + \|g\|$,

(d) $\|f\| = 0$ implies $f = O$.

A function N that satisfies properties (a), (b), and (c), but *not* (d), is called a *seminorm*. Some problems in the theory of approximation deal with seminormed linear spaces; others with normed linear spaces. The following examples will be discussed in this chapter.

EXAMPLE 1. *Taylor seminorm.* For a fixed integer $n \geq 1$, let V denote the linear space of functions having a derivative of order n at a given point a. If $f \in V$, let

$$N(f) = \sum_{k=0}^{n} |f^{(k)}(a)| .$$

It is easy to verify that the function N so defined is a seminorm. It is not a norm because $N(f) = 0$ if and only if

$$f(a) = f'(a) = \cdots = f^{(n)}(a) = 0,$$

and these equations can be satisfied by a nonzero function. For example, $N(f) = 0$ when $f(x) = (x - a)^{n+1}$.

EXAMPLE 2. *Interpolation seminorm.* Let V denote the linear space of all real-valued functions defined on an interval $[a, b]$. For a fixed set of $n + 1$ distinct points x_0, $x_1, \ldots, x_n$ in $[a, b]$, let N be defined by the equation

$$N(f) = \sum_{k=0}^{n} |f(x_k)|$$

if $f \in V$. This function N is a seminorm on V. It is not a norm because $N(f) = 0$ if and only if $f(x_0) = f(x_1) = \cdots = f(x_n) = 0$, and it is clear that these equations can be satisfied by a function f that is not zero everywhere on $[a, b]$.

EXAMPLE 3. *Square norm.* Let C denote the linear space of functions continuous on an interval $[a, b]$. If $f \in C$ define

$$N(f) = \left(\int_a^b |f(x)|^2 \, dx\right)^{1/2}. \tag{15.2}$$

This is a norm inherited from the inner product

$$(f, g) = \int_a^b f(x)\overline{g(x)} \, dx .$$

Note: Let S denote the set of functions f that are integrable on $[a, b]$. The set S is a linear space, and the function N defined by (15.2) is a seminorm on S. It is not a norm because we can have $N(f) = 0$ without f being identically zero on $[a, b]$.

EXAMPLE 4. *Max norm.* Let C denote the linear space of functions continuous on an interval $[a, b]$. If $f \in C$, define

$$N(f) = \max_{a \le x \le b} |f(x)| ,$$

where the symbol on the right stands for the absolute maximum value of $|f|$ on $[a, b]$. The verification of all four norm properties is requested in Exercise 4 of Section 15.5.

15.4 Fundamental problems in polynomial approximation

Let C be the space of functions continuous on a given interval $[a, b]$, and let S be the linear subspace consisting of all polynomials of degree $\le n$. Assume also that a norm or seminorm has been defined on C. Choose a function f in C. If there is a polynomial P in S such that

$$\|f - P\| \le \|f - Q\|$$

for all polynomials Q in S, we say that P is a *best polynomial approximation* to f with the specified degree. The term "best" is, of course, relative to the given norm (or seminorm). The best polynomial for one choice of norm need not be best for another choice of norm.

Once a norm or seminorm has been chosen, three problems immediately suggest themselves.

1. Existence. Given f in C, is there a best polynomial approximation to f with the specified degree?

2. Uniqueness. If a best polynomial approximation to f exists with the specified degree, is it uniquely determined?

3. Construction. If a best polynomial approximation to f exists with the specified degree, how can it be determined?

There are, of course, many other problems that can be considered. For example, if a unique best polynomial P_n of degree $\leq n$ exists, we may wish to obtain upper bounds for $\|f - P_n\|$ that can be used to satisfy practical requirements. Or we may ask whether $\|f - P_n\| \to 0$ as $n \to \infty$ for the given norm or possibly for some other norm. If so, we say that the polynomial approximations converge to f in this norm. In such a case arbitrarily close approximations exist relative to this norm if n is sufficiently large. These examples illustrate some of the types of problems considered in the general theory of polynomial approximation. In this introductory treatment we restrict our attention primarily to the three problems of existence, uniqueness, and construction, as described above.

For approximation by Taylor polynomials these three problems can be completely solved. If f has a derivative of order n at a point a, it is easy to prove that the best polynomial approximation of degree $\leq n$ relative to the Taylor seminorm for this n is the Taylor polynomial

$$P(x) = \sum_{k=0}^{n} \frac{f^{(k)}(a)}{k!}(x - a)^k. \tag{15.3}$$

In fact, for this polynomial we have

$$\|f - P\| = \sum_{k=0}^{n} |f^{(k)}(a) - P^{(k)}(a)| = 0,$$

so the inequality $\|f - P\| \leq \|f - Q\|$ is trivially satisfied for all polynomials Q. Therefore P is a best polynomial approximation relative to this seminorm. To establish uniqueness, we consider any polynomial Q of degree $\leq n$ such that $\|f - Q\| = 0$. This equation implies that

$$Q(a) = f(a), \qquad Q'(a) = f'(a), \qquad \ldots, \qquad Q^{(n)}(a) = f^{(n)}(a).$$

From Theorem 7.1 of Volume I we know that the Taylor polynomial in (15.3) is the only polynomial satisfying all these equations. Therefore $Q = P$. Equation (15.3) also solves the problem of construction.

All three problems can also be solved for any norm derived from an inner product. In this case, Theorem 1.16 tells us that there is a unique polynomial in S for which the norm $\|f - P\|$ is as small as possible. In fact, this P is the projection of f on S and is given by an explicit formula,

$$P(x) = \sum_{k=0}^{n} (P, e_i)e_i(x),$$

where $e_0, e_1, \ldots, e_n$ are functions forming an orthonormal basis for S.

For example, if C is the space of real functions continuous on the interval $[-1, 1]$ and if

$$(f, g) = \int_{-1}^{1} f(x)g(x)\,dx,$$

the normalized Legendre polynomials $\varphi_0, \varphi_1, \ldots, \varphi_n$ form an orthonormal basis for S, and the projection f_n of f on S is given by

$$f_n(x) = \sum_{k=0}^{n} (f, \varphi_k)\varphi_k(x), \qquad \text{where} \quad (f, \varphi_k) = \int_{-1}^{1} f(t)\varphi_k(t)\,dt.$$

We recall that the normalized Legendre polynomials are given by

$$\varphi_k(x) = \sqrt{\frac{2k+1}{2}}\, P_k(x), \qquad \text{where} \quad P_k(x) = \frac{1}{2^k k!}\frac{d^k}{dx^k}(x^2 - 1)^k.$$

The first six normalized polynomials are

$$\varphi_0(x) = \sqrt{\tfrac{1}{2}}, \qquad \varphi_1(x) = \sqrt{\tfrac{3}{2}}\,x, \qquad \varphi_2(x) = \tfrac{1}{2}\sqrt{\tfrac{5}{2}}\,(3x^2 - 1), \qquad \varphi_3(x) = \tfrac{1}{2}\sqrt{\tfrac{7}{2}}\,(5x^3 - 3x),$$

$$\varphi_4(x) = \tfrac{1}{8}\sqrt{\tfrac{9}{2}}\,(35x^4 - 30x^2 + 3), \qquad \varphi_5(x) = \tfrac{1}{8}\sqrt{\tfrac{11}{2}}\,(63x^5 - 70x^3 + 15x).$$

The corresponding problems for the interpolation seminorm will be treated next in Section 15.6. In later sections we discuss polynomial approximation relative to the max norm.

15.5 Exercises

1. Prove that each of the following collections of functions is a linear space.
   (a) All polynomials.
   (b) All polynomials of degree $\leq n$.
   (c) All functions continuous on an interval I.
   (d) All functions having a derivative at each point of I.
   (e) All functions having a derivative of order n at each point of I.
   (f) All functions having a derivative of order n at a fixed point x_0.
   (g) All functions having power-series expansions in a neighborhood of a given point x_0.
2. Determine whether or not each of the following collections of real-valued functions is a linear space.
   (a) All polynomials of degree n.
   (b) All functions defined and bounded on an interval $[a, b]$.
   (c) All step functions defined on an interval $[a, b]$.
   (d) All functions monotonic on an interval $[a, b]$.
   (e) All functions integrable on an interval $[a, b]$.
   (f) All functions that are piecewise monotonic on an interval $[a, b]$.
   (g) All functions that can be expressed in the form $f - g$, where f and g are monotonic increasing on an interval $[a, b]$.
3. Let C denote the linear space of real-valued functions continuous on an interval $[a, b]$. A function N is defined on C by the equation given. In each case, determine which of the four

properties of a norm are satisfied by N, and determine thereby whether N is a norm, a seminorm, or neither.

(a) $N(f) = f(a)$.

(b) $N(f) = |f(a)|$.

(c) $N(f) = |f(b) - f(a)|$.

(d) $N(f) = \int_a^b f(x)\,dx$.

(e) $N(f) = \left|\int_a^b f(x)\,dx\right|$.

(f) $N(f) = \int_a^b |f(x)|\,dx$.

(g) $N(f) = \int_a^b |f(x)|^2\,dx$.

(h) $N(f) = \left|\int_a^b f(x)\,dx\right|^2$.

4. Let C be the linear space of functions continuous on an interval $[a, b]$. If $f \in C$, define

$$N(f) = \max_{a \le x \le b} |f(x)|.$$

Show that N is a norm for C.

5. Let B denote the linear space of all real-valued functions that are defined and bounded on an interval $[a, b]$. If $f \in B$, define

$$N(f) = \sup_{a \le x \le b} |f(x)|,$$

where the symbol on the right stands for the supremum (least upper bound) of the set of all numbers $|f(x)|$ for x in $[a, b]$. Show that N is a norm for B. This is called the *sup norm.*

6. Refer to Exercise 3. Determine which of the given functions N have the property that $N(fg) \le N(f)N(g)$ for all f and g in C.

7. For a fixed integer $n \ge 1$, let S be the set of all functions having a derivative of order n at a fixed point x_0. If $f \in S$, let

$$N(f) = \sum_{k=0}^{n} \frac{1}{k!} |f^{(k)}(x_0)|.$$

(a) Show that N is a seminorm on S.

(b) Show that $N(fg) \le N(f)N(g)$ for all f, g in S. Prove also that the Taylor seminorm does not have this property.

8. Let f be a real continuous function on the interval $[-1, 1]$.

(a) Prove that the best quadratic polynomial approximation relative to the square norm on $[-1, 1]$ is given by

$$P(x) = \tfrac{1}{2}\int_{-1}^{1} f(t)\,dt + \tfrac{3}{2}x\int_{-1}^{1} tf(t)\,dt + \tfrac{5}{8}(3x^2 - 1)\int_{-1}^{1} (3t^2 - 1)f(t)\,dt.$$

(b) Find a similar formula for the best polynomial approximation of degree ≤ 4.

9. Calculate constants a, b, c so that the integral $\int_{-1}^{1} |e^x - (a + bx + cx^2)|^2\,dx$ will be as small as possible.

10. Let $f(x) = |x|$ for $-1 \le x \le 1$. Determine the polynomial of degree ≤ 4 that best approximates f on $[-1, 1]$ relative to the square norm.

11. Let C denote the linear space of real continuous functions on $[a, b]$ with inner product $(f, g) = \int_a^b f(x)g(x)\,dx$. Let $e_0, \ldots, e_n$ be an orthonormal basis for the subspace S of polynomials of degree $\le n$. Let P be the polynomial in S that best approximates a given f in C relative to the square norm.

(a) Prove that the square of the norm of the error is given by

$$\|f - P\|^2 = \|f\|^2 - \sum_{k=0}^{n} (f, e_k)^2.$$

(b) Calculate this error explicitly when $[a, b] = [-1, 1]$, $n = 2$, and $f(x) = |x|$.

12. Let $f(x) = 1/x$ for $x \neq 0$.
(a) Show that the constant polynomial P that best approximates f over the interval $[1, n]$ relative to the square norm is $P(x) = (\log n)/(n - 1)$. Compute $\|P - f\|^2$ for this P.
(b) Find the linear polynomial P that best approximates f over the interval $[1, n]$ relative to the square norm. Compute $\|P - f\|^2$ for this P when $n = 2$.
13. Let $f(x) = e^x$.
(a) Show that the constant polynomial P that best approximates f over the interval $[0, n]$ relative to the square norm is $P(x) = (e^n - 1)/n$. Compute $\|P - f\|^2$ for this P.
(b) Find the linear polynomial P that best approximates f over the interval $[0, 1]$ relative to the square norm. Compute $\|P - f\|^2$ for this P.
14. Let $P_0, P_1, \ldots, P_n$ be $n + 1$ polynomials orthonormal on $[a, b]$ relative to the inner product in Exercise 11. Assume also that P_k has degree k.
(a) Prove that any three consecutive polynomials in this set are connected by a recurrence relation of the form

$$P_{k+1}(x) = (a_k x + b_k)P_k(x) + c_k P_{k-1}(x)$$

for $1 \leq k \leq n - 1$, where a_k, b_k, c_k are constants.
(b) Determine this recurrence relation explicitly when the polynomials are the orthonormal Legendre polynomials.
15. Refer to Exercise 14, and let p_k denote the coefficient of x^k in $P_k(x)$.
(a) Show that $a_k = p_{k+1}/p_k$.
(b) Use the recurrence relation in Exercise 14 to derive the formula

$$\sum_{k=0}^{m} P_k(x)P_k(y) = \frac{p_m}{p_{m+1}} \frac{P_{m+1}(x)P_m(y) - P_m(x)P_{m+1}(y)}{x - y},$$

valid for $x \neq y$. Discuss also the limiting case $x = y$.

15.6 Interpolating polynomials

We turn now to approximation by interpolation polynomials. The values of a function f are known at $n + 1$ distinct points $x_0, x_1, \ldots, x_n$ and we seek a polynomial P of degree $\leq n$ that satisfies the conditions

$$(15.4) \qquad P(x_0) = f(x_0), \quad P(x_1) = f(x_1), \quad \ldots, \quad P(x_n) = f(x_n).$$

First we prove that if such a polynomial exists it is unique. Then we prove it exists by explicit construction. This polynomial minimizes the distance from f to P, measured in the interpolation seminorm for this n,

$$\|f - P\| = \sum_{k=0}^{n} |f(x_k) - P(x_k)|.$$

Since this distance is 0 if P satisfies (15.4), the interpolating polynomial P is the best approximation relative to this seminorm.

THEOREM 15.1. UNIQUENESS THEOREM. *Given $n + 1$ distinct points $x_0, x_1, \ldots, x_n$, let P and Q be two polynomials of degree $\leq n$ such that*

$$P(x_k) = Q(x_k)$$

for each $k = 0, 1, 2, \ldots, n$. Then $P(x) = Q(x)$ for all x.

Proof. Let $R(x) = P(x) - Q(x)$. The function R is a polynomial of degree $\leq n$ which has $n + 1$ distinct zeros at the points $x_0, x_1, \ldots, x_n$. The only polynomial with this property is the zero polynomial. Therefore $R(x) = 0$ for all x, so $P(x) = Q(x)$ for all x.

The interpcıating polynomial P can be constructed in many ways. We describe first a method of Lagrange. Let $A(x)$ be the polynomial given by the equation

(15.5)
$$A(x) = (x - x_0)(x - x_1) \cdots (x - x_n) = \prod_{j=0}^{n} (x - x_j).$$

This polynomial has a simple zero at each of the points x_j. Let $A_k(x)$ denote the polynomial of degree n obtained from $A(x)$ by deleting the factor $x - x_k$. That is, let

(15.6)
$$A_k(x) = \prod_{\substack{j=0 \\ j \neq k}}^{n} (x - x_j).$$

The polynomial $A_k(x)$ has a simple zero at each point $x_j \neq x_k$. At the point x_k itself we have

(15.7)
$$A_k(x_k) = \prod_{\substack{j=0 \\ j \neq k}}^{n} (x_k - x_j).$$

This is nonzero since no factor in the product is zero. Therefore the polynomial $A_k(x)/A_k(x_k)$ has the value 1 when $x = x_k$ and the value 0 when $x = x_j$ for $x_j \neq x_k$. Now let

$$P(x) = \sum_{k=0}^{n} \frac{f(x_k)A_k(x)}{A_k(x_k)}.$$

When $x = x_j$, each term in this sum vanishes except the jth term, which has the value $f(x_j)$. Therefore $P(x_j) = f(x_j)$ for each j. Since each term of this sum is a polynomial of degree n, the sum itself is a polynomial of degree $\leq n$. Thus, we have found a polynomial satisfying the required conditions. These results can be summarized by the following theorem:

THEOREM 15.2. *Given $n + 1$ distinct points $x_0, x_1, \ldots, x_n$ and $n + 1$ real numbers $f(x_0), f(x_1), \ldots, f(x_n)$, not necessarily distinct, there exists one and only one polynomial P of degree $\leq n$ such that $P(x_j) = f(x_j)$ for each $j = 0, 1, 2, \ldots, n$. This polynomial is given by the formula*

(15.8)
$$P(x) = \sum_{k=0}^{n} \frac{f(x_k)A_k(x)}{A_k(x_k)},$$

where $A_k(x)$ is the polynomial defined by (15.6).

Formula (15.8) for $P(x)$ is called *Lagrange's interpolation formula.* We can write it in the form

$$P(x) = \sum_{k=0}^{n} f(x_k)L_k(x),$$

where $L_k(x)$ is a polynomial of degree n given by

$$L_k(x) = \frac{A_k(x)}{A_k(x_k)}. \tag{15.9}$$

Thus, for each fixed x, $P(x)$ is a linear combination of the prescribed values $f(x_0), f(x_1), \ldots, f(x_n)$. The multipliers $L_k(x)$ depend only on the points $x_0, x_1, \ldots, x_n$ and not on the prescribed values. They are called *Lagrange interpolation coefficients.* If we use the formulas in (15.6) and (15.7) we can write Equation (15.9) in the form

$$L_k(x) = \prod_{\substack{j=0 \\ j\neq k}}^{n} \frac{x - x_j}{x_k - x_j}. \tag{15.10}$$

This product formula provides an efficient method for evaluating the number $L_k(x)$ for a given x.

Note: The Lagrange coefficients $L_k(x)$ are often expressed in the form

$$L_k(x) = \frac{A_k(x)}{A'(x_k)},$$

where A' is the derivative of the polynomial in (15.5). To prove this formula it suffices to show that $A'(x_k) = A_k(x_k)$. Differentiating the relation

$$A(x) = (x - x_k)A_k(x)$$

we obtain $A'(x) = (x - x_k)A_k'(x) + A_k(x)$. When $x = x_k$ this gives us $A'(x_k) = A_k(x_k)$.

EXAMPLE. Determine the polynomial of degree ≤ 3 that takes the values y_0, y_1, y_2, y_3 at the points $-2, -1, 1, 2$, respectively.

Solution. We take $x_0 = -2,\ x_1 = -1,\ x_2 = 1,\ x_3 = 2$. The polynomials $L_k(x)$ in (15.10) are given by the formulas

$$L_0(x) = \frac{(x+1)(x-1)(x-2)}{(-2+1)(-2-1)(-2-2)} = -\frac{1}{12}(x+1)(x-1)(x-2),$$

$$L_1(x) = \frac{(x+2)(x-1)(x-2)}{(-1+2)(-1-1)(-1-2)} = \frac{1}{6}(x+2)(x-1)(x-2),$$

$$L_2(x) = \frac{(x+2)(x+1)(x-2)}{(1+2)(1+1)(1-2)} = -\frac{1}{6}(x+2)(x+1)(x-2),$$

$$L_3(x) = \frac{(x+2)(x+1)(x-1)}{(2+2)(2+1)(2-1)} = \frac{1}{12}(x+2)(x+1)(x-1).$$

Therefore the required polynomial is

$$\begin{aligned} P(x) &= y_0L_0(x) + y_1L_1(x) + y_2L_2(x) + y_3L_3(x) \\ &= -\frac{y_0}{12}(x+1)(x-1)(x-2) + \frac{y_1}{6}(x+2)(x-1)(x-2) \\ &\quad - \frac{y_2}{6}(x+2)(x+1)(x-2) + \frac{y_3}{12}(x+2)(x+1)(x-1). \end{aligned}$$

To compute the value of $P(x)$ for a specific x it is usually better to leave the polynomial in this form rather than to rewrite it in increasing powers of x. For example, if $y_0 = -5$, $y_1 = 1$, $y_2 = 1$, and $y_3 = 7$, the value of $P(x)$ for $x = \frac{3}{2}$ is given by

$$\begin{aligned} P(\tfrac{3}{2}) &= \tfrac{5}{12}(\tfrac{5}{2})(\tfrac{1}{2})(-\tfrac{1}{2}) + \tfrac{1}{6}(\tfrac{7}{2})(\tfrac{1}{2})(-\tfrac{1}{2}) - \tfrac{1}{6}(\tfrac{7}{2})(\tfrac{5}{2})(-\tfrac{1}{2}) + \tfrac{7}{12}(\tfrac{7}{2})(\tfrac{5}{2})(\tfrac{1}{2}) \\ &= -\tfrac{25}{96} - \tfrac{7}{48} + \tfrac{35}{48} + \tfrac{245}{96} = \tfrac{276}{96} = 2\tfrac{7}{8}. \end{aligned}$$

15.7 Equally spaced interpolation points

In the foregoing discussion the interpolation points $x_0, x_1, \ldots, x_n$ were assumed to be distinct but otherwise arbitrary. Now we assume they are equally spaced and show that the Lagrange coefficients $L_k(x)$ can be considerably simplified. Suppose $x_0 < x_1 < x_2 < \cdots < x_n$, and let h denote the distance between adjacent points. Then we can write

$$x_j = x_0 + jh$$

for $j = 0, 1, 2, \ldots, n$. Since $x_k - x_j = (k - j)h$, Equation (15.10) becomes

$$L_k(x) = \prod_{\substack{j=0 \\ j \neq k}}^{n} \frac{x - x_0 - jh}{(k - j)h} = \prod_{\substack{j=0 \\ j \neq k}}^{n} \frac{t - j}{k - j}, \tag{15.11}$$

where

$$t = \frac{x - x_0}{h}.$$

In the last term on the right of (15.11) the product of the factors independent of t is

$$\begin{aligned} \prod_{\substack{j=0 \\ j \neq k}}^{n} \frac{1}{k - j} &= \left(\prod_{j=0}^{k-1} \frac{1}{k - j}\right)\left(\prod_{j=k+1}^{n} \frac{1}{k - j}\right) = \frac{1}{k!} \prod_{j=k+1}^{n} \frac{(-1)}{j - k} \\ &= \frac{(-1)^{n-k}}{k!\,(n - k)!} = \frac{(-1)^{n-k}}{n!} \binom{n}{k}, \end{aligned} \tag{15.12}$$

where $\binom{n}{k}$ is the binomial coefficient. Since $x = x_0 + th$, Equation (15.11) now becomes

$$L_k(x_0 + th) = \frac{(-1)^{n-k}}{n!} \binom{n}{k} \prod_{\substack{j=0 \\ j \neq k}}^{n} (t - j). \tag{15.13}$$

For each fixed n, the right member of (15.13) is a function of k and t that can be tabulated. Extensive tables of the Lagrangian coefficients for equally spaced interpolation points have been prepared by the National Bureau of Standards. (See Reference 13 in the bibliography at the end of this chapter.) If x and h are chosen so that the number $t = (x - x_0)/h$ is one for which the Lagrangian coefficients $L_k(x_0 + th)$ are tabulated, the actual calculation of $P(x_0 + th)$ is reduced to a multiplication of the $f(x_k)$ by the tabulated $L_k(x_0 + th)$, followed by addition.

15.8 Error analysis in polynomial interpolation

Let f be a function defined on an interval $[a, b]$ containing the $n + 1$ distinct points $x_0, x_1, \ldots, x_n$, and let P be the interpolation polynomial of degree $\leq n$ which agrees with f at these points. If we alter the values of f at points other than the interpolation points we do not alter the polynomial P. This shows that the function f and the polynomial P may differ considerably at points other than the interpolation points. If the given function f has certain qualities of "smoothness" throughout the interval $[a, b]$ we can expect that the interpolating polynomial P will be a good approximation to f at points other than the x_k. The next theorem gives a useful expression that enables us to study the error in polynomial interpolation when the given function has a derivative of order $n + 1$ throughout $[a, b]$.

THEOREM 15.3. *Let $x_0, x_1, \ldots, x_n$ be $n + 1$ distinct points in the domain of a function f, and let P be the interpolation polynomial of degree $\leq n$ that agrees with f at these points. Choose a point x in the domain of f and let $[\alpha, \beta]$ be any closed interval containing the points $x_0, x_1, \ldots, x_n$, and x. If f has a derivative of order $n + 1$ in the interval $[\alpha, \beta]$ there is at least one point c in the open interval (α, β) such that*

$$f(x) - P(x) = \frac{A(x)}{(n + 1)!} f^{(n+1)}(c), \tag{15.14}$$

where

$$A(x) = (x - x_0)(x - x_1) \cdots (x - x_n).$$

Note: Point c depends on both x and n.

Proof. If x is one of the interpolation points x_k, then $A(x_k) = 0$ and Equation (15.14) is trivially satisfied for any choice of c in (α, β). Suppose, then, that x is not one of the interpolation points. Keep x fixed and define a new function F on $[\alpha, \beta]$ by the equation

$$F(t) = A(x)[f(t) - P(t)] - A(t)[f(x) - P(x)]. \tag{15.15}$$

The right-hand side of this equation, as a function of t, has a derivative of order $n + 1$; hence the same is true of the left-hand side. Since $P(t)$ is a polynomial in t of degree $\leq n$, its $(n + 1)$st derivative is identically zero. The polynomial $A(t)$ has degree $n + 1$, the term of highest degree being t^{n+1}, and we have $A^{(n+1)}(t) = (n + 1)!$. Therefore, if we differentiate Equation (15.15) $n + 1$ times with respect to t we obtain the formula

$$F^{(n+1)}(t) = A(x)f^{(n+1)}(t) - (n + 1)!\,[f(x) - P(x)]. \tag{15.16}$$

From the definition in Equation (15.15) we see that F has the value zero at the $n + 1$ interpolation points $x_0, x_1, \ldots, x_n$ and *also* at the point x. Therefore $F(t) = 0$ at $n + 2$ distinct points in the interval $[\alpha, \beta]$. These points determine $n + 1$ adjacent subintervals of $[\alpha, \beta]$ and the function F vanishes at both endpoints of each of these subintervals. By Rolle's theorem, the derivative $F'(t)$ must be zero for at least one t interior to each subinterval. If we choose exactly one such t from each subinterval we obtain $n + 1$ distinct points in the open interval (α, β) at which $F'(t) = 0$. These points, in turn, determine n subintervals at whose endpoints we have $F'(t) = 0$. Applying Rolle's theorem to F' we find that the second derivative $F''(t)$ is zero for at least n distinct points in (α, β). After applying Rolle's theorem $n + 1$ times in this manner we finally find that there is at least one point c in (α, β) at which $F^{(n+1)}(c) = 0$. Substituting this value of c in Equation (15.16) we obtain

$$(n + 1)!\,[f(x) - P(x)] = A(x)f^{(n+1)}(c),$$

which is the same as (15.14). This completes the proof.

It should be noted that, as with approximation by Taylor polynomials, the error term involves the $(n + 1)$st derivative $f^{(n+1)}(c)$ evaluated at an unknown point c. If the extreme values of $f^{(n+1)}$ in $[\alpha, \beta]$ are known, useful upper and lower bounds for the error can be obtained.

Suppose now that the interpolation points are equally spaced and that $x_0 < x_1 < x_2 < \cdots < x_n$. If h denotes the spacing we can write

$$x_j = x_0 + jh \qquad \text{and} \qquad x = x_0 + th,$$

where $t = (x - x_0)/h$. Since $x - x_j = (t - j)h$, the polynomial $A(x)$ can be written as

$$A(x) = \prod_{j=0}^{n} (x - x_j) = h^{n+1} \prod_{j=0}^{n} (t - j).$$

Formula (15.14) now becomes

$$f(x) - P(x) = \frac{f^{(n+1)}(c)}{(n + 1)!} h^{n+1} \prod_{j=0}^{n} (t - j), \tag{15.17}$$

with $t = (x - x_0)/h$.

EXAMPLE. *Error in linear interpolation.* Suppose a function f with a second derivative is tabulated and we wish to estimate its value at a point x intermediate to two consecutive entries x_0 and $x_0 + h$. If we use linear interpolation we approximate the graph of f over the interval $[x_0, x_0 + h]$ by a straight line, as shown in Figure 15.1. If P denotes the linear interpolating polynomial, the error estimate in (15.17) becomes

$$f(x) - P(x) = \frac{f''(c)}{2!} h^2 t(t - 1), \tag{15.18}$$

where $t = (x - x_0)/h$. When x lies between x_0 and $x_0 + h$ we have $0 < t < 1$ and the maximum value of $|t(t - 1)|$ in this interval is $\frac{1}{4}$. Therefore (15.18) gives us the estimate

$$|f(x) - P(x)| \le \frac{|f''(c)|\, h^2}{8}.$$

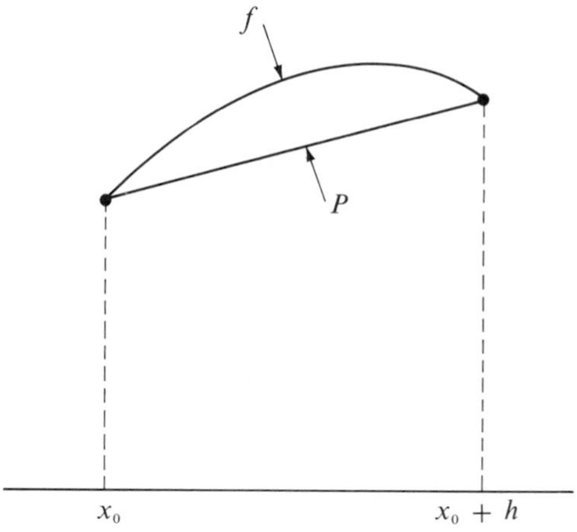

FIGURE 15.1 Linear interpolation.

The point c is an unknown point in the interval $(x_0, x_0 + h)$. If the second derivative f'' is bounded in this interval, say $|f''(x)| \le M$, the error estimate becomes

$$|f(x) - P(x)| \le \frac{Mh^2}{8}.$$

In particular, if f is a sine or cosine, then $|f''(x)| \le 1$ for all x and we have $|f(x) - P(x)| \le h^2/8$. If a table of sines or cosines has entries for every degree (one degree $= \pi/180$ radians) we have $h = \pi/180$, so

$$\frac{h^2}{8} = \frac{\pi^2}{8(180)^2} < \frac{10}{259{,}200} < \frac{1}{25{,}000} = 0.00004.$$

Since this error does not exceed $\frac{1}{2}$ in the fourth decimal place, linear interpolation would be satisfactory in a four-place table. The error estimate can be improved in portions of the table where $|f''(c)|$ is considerably less than 1.

15.9 Exercises

1. In each case find the polynomial P of lowest possible degree satisfying the given conditions.
   (a) $P(-1) = 0, \quad P(0) = 2, \quad P(2) = 7.$
   (b) $P(1) = 1, \quad P(2) = 0, \quad P(3) = 0, \quad P(4) = 1.$
   (c) $P(1) = 1, \quad P(2) = 2, \quad P(3) = 3, \quad P(0) = 1.$
   (d) $P(0) = -2, \quad P(1) = 0, \quad P(-1) = -2, \quad P(2) = 16.$
   (e) $P(-2) = 11, \quad P(-1) = -11, \quad P(0) = -5, \quad P(1) = -1.$
2. Let $f(x) = \cos(\pi x/4)$. Find the polynomial of smallest possible degree that takes the same values as f at the points $-2, -\frac{4}{3}, 0, \frac{4}{3}, 2$.
3. Let P be a polynomial of degree $\le n$ and let $A(x) = (x - x_0)(x - x_1) \cdots (x - x_n)$, where $x_0, x_1, \ldots, x_n$ are $n + 1$ distinct points.
   (a) Show that for any polynomial B the polynomial Q given by $Q(x) = P(x) + A(x)B(x)$ agrees with P at the points $x_0, x_1, \ldots, x_n$.

(b) Prove also the converse. That is, if Q is any polynomial that agrees with P at the points $x_0, x_1, \ldots, x_n$, then $Q(x) = P(x) + A(x)B(x)$ for some polynomial B.

4. (a) Find the polynomial Q of lowest possible degree that satisfies the conditions

$$Q(-2) = -5, \qquad Q(-1) = -1, \qquad Q(1) = 1, \qquad Q'(0) = -1.$$

[*Hint:* First find a polynomial P that takes the prescribed values at -2, -1, 1, and then use Exercise 3 to determine Q.]

(b) Find the polynomial Q of lowest possible degree that satisfies the conditions in part (a) with $Q'(0) = -3$ instead of $Q'(0) = -1$.

5. Let $f(x) = \log_4 x$ for $x > 0$. Compute $P(32)$, where P is the polynomial of lowest possible degree that agrees with f at the points:
(a) $x = 1, 64$.
(b) $x = 1, 16, 256$.
(c) $x = 4, 16, 64$.
(d) $x = 1, 4, 16, 64, 256$.
In each case compute the difference $f(32) - P(32)$. These examples show that the accuracy in polynomial interpolation is not necessarily improved by increasing the number of interpolation points.

6. The Lagrange interpolation coefficients $L_k(x)$ given by Equation (15.10) depend not only on x but also on the interpolation points $x_0, x_1, \ldots, x_n$. We can indicate this dependence by writing $L_k(x) = L_k(x; \boldsymbol{X})$, where $\boldsymbol{X}$ denotes the vector in $(n + 1)$-space given by $\boldsymbol{X} = (x_0, x_1, \ldots, x_n)$. For a given real number b, let $\boldsymbol{b}$ denote the vector in $(n + 1)$-space all of whose components are equal to b. If $a \neq 0$, show that

$$L_k(ax + b;\ a\boldsymbol{X} + \boldsymbol{b}) = L_k(x; \boldsymbol{X}).$$

This is called the *invariance property* of the Lagrange interpolation coefficients. The next exercise shows how this property can be used to help simplify calculations in practice.

7. Let P denote the polynomial of degree ≤ 4 that has the values

$$P(2.4) = 72, \qquad P(2.5) = 30, \qquad P(2.7) = 18, \qquad P(2.8) = 24, \qquad P(3.0) = 180.$$

(a) Introduce new interpolation points u_j related to the given points x by the equation $u_j = 10x_j - 24$. The u_j are integers. For each $k = 0, 1, 2, 3, 4$, determine the Lagrange interpolation coefficients $L_k(x)$ in terms of u, where $u = 10x - 24$.
(b) Use the invariance property of Exercise 6 to compute $P(2.6)$.

8. A table of the function $f(x) = \log x$ contains entries for $x = 1$ to $x = 10$ at intervals of 0.001. Values intermediate to each pair of consecutive entries are to be computed by linear interpolation. Assume the entries in the table are exact.
(a) Show that the error in linear interpolation will not exceed $\frac{1}{8}$ in the sixth decimal place.
(b) For what values of x will linear interpolation be satisfactory for a seven-place table?
(c) What should be the spacing of the entries in the interval $1 \leq x \leq 2$ so that linear interpolation will be satisfactory in a seven-place table?

In Exercises 9 through 15, $x_0, x_1, \ldots, x_n$ are distinct points and

$$A(x) = \prod_{j=0}^{n} (x - x_j), \qquad A_k(x) = \prod_{\substack{j=0 \\ j \neq k}}^{n} (x - x_j), \qquad L_k(x) = \frac{A_k(x)}{A_k(x_k)}.$$

9. Derive the formula $A'(x) = \sum_{k=0}^{n} A_k(x)$ by use of (a) logarithmic differentiation; (b) Lagrange's interpolation formula.
10. Prove each of the following formulas:

(a) $\displaystyle\sum_{k=0}^{n} L_k(x) = 1$ and $\displaystyle\sum_{k=0}^{n} \frac{A_k'(x)}{A'(x_k)} = 0$ for all x.

(b) $\displaystyle\sum_{k=0}^{n} \frac{1}{A'(x_k)} = 0.$ [*Hint:* Use part (a) with suitable values of x.]

11. Let P be any polynomial of degree $\le n$. Show that the coefficient of x^n is equal to

$$\sum_{k=0}^{n} \frac{P(x_k)}{A'(x_k)}.$$

12. (a) Determine a and b so that the polynomial

$$P_k(x) = \{a + b(x - x_k)\}L_k(x)^2$$

will have the following properties:

$$P_k(x_i) = 0 \quad \text{for all } i, \qquad P_k'(x_k) = 1, \qquad \text{and} \qquad P_k'(x_i) = 0 \quad \text{for} \quad i \ne k.$$

(b) Determine c and d so that the polynomial

$$Q_k(x) = \{c + d(x - x_k)\}L_k(x)^2$$

will have the following properties:

$$Q_k(x_k) = 1, \qquad Q_k(x_i) = 0 \quad \text{for} \quad i \ne k, \qquad \text{and} \qquad Q_k'(x_i) = 0 \quad \text{for all } i.$$

(c) Let $H(x) = \sum_{k=0}^{n} f(x_k)Q_k(x) + \sum_{k=0}^{n} f'(x_k)P_k(x)$, where f is a given function that is differentiable at $x_0, x_1, \ldots, x_n$. Prove that

$$H(x_i) = f(x_i) \qquad \text{and} \qquad H'(x_i) = f'(x_i) \qquad \text{for all } i.$$

Prove also that there is at most one polynomial $H(x)$ of degree $\le 2n + 1$ with this property.

13. (a) Let P and Q be two polynomials of degree $\le n$ satisfying the $n + 1$ conditions

$$P(x_0) = Q(x_0), \qquad P'(x_1) = Q'(x_1), \qquad P''(x_2) = Q''(x_2), \qquad \ldots, \qquad P^{(n)}(x_n) = Q^{(n)}(x_n).$$

Prove that $P(x) = Q(x)$ for all x.

(b) Let $B_0(x) = 1$, and for $n \ge 1$ define

$$B_n(x) = \frac{x(x - n)^{n-1}}{n!}.$$

Show that $B_n'(x) = B_{n-1}(x - 1)$ for $n \ge 1$ and deduce that

$$B_n(0) = B_n'(1) = B_n''(2) = \cdots = B_n^{(n-1)}(n - 1) = 0 \qquad \text{and} \qquad B_n^{(n)}(n) = 1.$$

(c) Show that the one and only polynomial of degree $\le n$ satisfying the conditions

$$P(0) = c_0, \qquad P'(1) = c_1, \qquad P''(2) = c_2, \qquad \ldots, \qquad P^{(n)}(n) = c_n$$

is given by

$$P(x) = \sum_{k=0}^{n} c_k B_k(x).$$

(d) If $x_k = x_0 + kh$ for $k = 0, 1, 2, \ldots, n$, where $h > 0$, generalize the results in (b) and (c).

14. Assume $x_0, x_1, \ldots, x_n$ are integers satisfying $x_0 < x_1 < \cdots < x_n$.

(a) Prove that $|A'(x_k)| \ge k!\,(n-k)!$ and deduce that

$$\sum_{k=0}^{n} \frac{1}{|A'(x_k)|} \le \frac{2^n}{n!}.$$

(b) Let P be any polynomial of degree n, with the term of highest degree equal to x^n. Let M denote the largest of the numbers $|P(x_0)|, |P(x_1)|, \ldots, |P(x_n)|$. Prove that $M \ge n!/2^n$. [*Hint:* Use part (a) and Exercise 11.]

15. Prove the following formulas. In parts (a) and (b), x is any point different from $x_0, x_1, \ldots, x_n$.

(a) $$\frac{A'(x)}{A(x)} = \sum_{j=0}^{n} \frac{1}{x - x_j}.$$

(b) $$\frac{A''(x)}{A'(x)} = \frac{A_k(x)}{A'(x)} \sum_{\substack{j=0\\ j\ne k}}^{n} \frac{1}{x - x_j} + \sum_{\substack{j=0\\ j\ne k}}^{n} \frac{1}{x - x_j} - \frac{A(x)}{A'(x)} \sum_{\substack{j=0\\ j\ne k}}^{n} \frac{1}{(x - x_j)^2}.$$

(c) $$\frac{A''(x_k)}{A'(x_k)} = 2 \sum_{\substack{j=0\\ j\ne k}}^{n} \frac{1}{x_k - x_j}.$$

16. Let $P_n(x)$ be the polynomial of degree $\le n$ that agrees with the function $f(x) = e^{ax}$ at the $n + 1$ integers $x = 0, 1, \ldots, n$. Since this polynomial depends on a we denote it by $P_n(x;\ a)$. Prove that the limit

$$\lim_{a\to 0} \frac{P_n(x;\ a) - 1}{a}$$

exists and is a polynomial in x. Determine this polynomial explicitly.

15.10 Newton's interpolation formula

Let P_n denote the interpolation polynomial of degree $\le n$ that agrees with a given function f at $n + 1$ distinct points $x_0, x_1, \ldots, x_n$. Lagrange's interpolation formula tells us that

$$P_n(x) = \sum_{k=0}^{n} L_k(x) f(x_k),$$

where $L_k(x)$ is a polynomial of degree n (the Lagrange interpolation coefficient) given by the product formula

$$L_k(x) = \prod_{\substack{j=0\\ j\ne k}}^{n} \frac{x - x_j}{x_k - x_j}, \qquad \text{for} \quad k = 0, 1, 2, \ldots, n. \tag{15.19}$$

Suppose we adjoin a new interpolation point x_{n+1} to the given points $x_0, x_1, \ldots, x_n$. To determine the corresponding polynomial P_{n+1} by Lagrange's formula it is necessary to compute a new interpolation coefficient L_{n+1} and to recompute all the earlier coefficients $L_0, L_1, \ldots, L_n$, each of which is now a polynomial of degree $n + 1$. In practice this involves considerable labor. Therefore it is desirable to have another formula for determining P_n that provides an easier transition from P_n to P_{n+1}. One such formula was discovered by Newton; we shall derive it from the following theorem.

THEOREM 15.4. *Given $n + 2$ distinct points $x_0, x_1, \ldots, x_n, x_{n+1}$. Let P_n be the polynomial of degree $\leq n$ that agrees with a given function f at $x_0, \ldots, x_n$, and let P_{n+1} be the polynomial of degree $\leq n + 1$ that agrees with f at $x_0, x_1, \ldots, x_n, x_{n+1}$. Then there is a constant c_{n+1}, uniquely determined by f and by the interpolation points $x_0, \ldots, x_{n+1}$, such that*

$$P_{n+1}(x) = P_n(x) + c_{n+1}(x - x_0) \cdots (x - x_n). \tag{15.20}$$

Proof. Let $Q(x) = P_n(x) + c(x - x_0) \cdots (x - x_n)$, where c is an unspecified constant. Then Q is a polynomial of degree $\leq n + 1$ that agrees with P_n and hence with f at each of the $n + 1$ points $x_0, \ldots, x_n$. Now we choose c to make Q agree with f also at x_{n+1}. This requires

$$f(x_{n+1}) = P_n(x_{n+1}) + c(x_{n+1} - x_0) \cdots (x_{n+1} - x_n).$$

Since the coefficient of c is nonzero, this equation has a unique solution which we call c_{n+1}. Taking $c = c_{n+1}$ we see that $Q = P_{n+1}$.

The next theorem expresses $P_n(x)$ in terms of the numbers $c_1, \ldots, c_n$.

THEOREM 15.5. NEWTON'S INTERPOLATION FORMULA. *If $x_0, \ldots, x_n$ are distinct, we have*

$$P_n(x) = f(x_0) + \sum_{k=1}^{n} c_k(x - x_0) \cdots (x - x_{k-1}). \tag{15.21}$$

Proof. We define $P_0(x) = f(x_0)$ and take $n = 0$ in (15.20) to obtain

$$P_1(x) = f(x_0) + c_1(x - x_0).$$

Now take $n = 1$ in (15.20) to get

$$P_2(x) = P_1(x) + c_2(x - x_0)(x - x_1) = f(x_0) + c_1(x - x_0) + c_2(x - x_0)(x - x_1).$$

By induction, we obtain (15.21).

The property of Newton's formula expressed in Equation (15.20) enables us to calculate P_{n+1} simply by adding one new term to P_n. This property is not possessed by Lagrange's formula.

The usefulness of Newton's formula depends, of course, on the ease with which the coefficients $c_1, c_2, \ldots, c_n$ can be computed. The next theorem shows that c_n is a linear combination of the function values $f(x_0), \ldots, f(x_n)$.

THEOREM 15.6. *The coefficients in Newton's interpolation formula are given by*

(15.22) $$c_n = \sum_{k=0}^{n} \frac{f(x_k)}{A_k(x_k)}, \qquad \text{where} \quad A_k(x_k) = \prod_{\substack{j=0 \\ j \neq k}}^{n} (x_k - x_j).$$

Proof. By Lagrange's formula we have

$$P_n(x) = \sum_{k=0}^{n} L_k(x) f(x_k),$$

where $L_k(x)$ is the polynomial of degree n given by (15.19). Since the coefficient of x^n in $L_k(x)$ is $1/A_k(x_k)$, the coefficient of x^n in $P_n(x)$ is the sum appearing in (15.22). On the other hand, Newton's formula shows that the coefficient of x^n in $P_n(x)$ is equal to c_n. This completes the proof.

Equation (15.22) provides a straightforward way for calculating the coefficients in Newton's formula. The numbers $A_k(x_k)$ also occur as factors in the denominator of the Lagrange interpolation coefficient $L_k(x)$. The next section describes an alternate method for computing the coefficients when the interpolation points are equally spaced.

15.11 Equally spaced interpolation points. The forward difference operator

In the case of equally spaced interpolation points with $x_k = x_0 + kh$ for $k = 0, 1, \ldots, n$ we can use Equation (15.12) to obtain

$$\frac{1}{A_k(x_k)} = \prod_{\substack{j=0 \\ j \neq k}}^{n} \frac{1}{x_k - x_j} = \frac{1}{h^n} \prod_{\substack{j=0 \\ j \neq k}}^{n} \frac{1}{k - j} = \frac{(-1)^{n-k}}{n!\, h^n} \binom{n}{k}.$$

In this case the formula for c_n in Theorem 15.6 becomes

(15.23) $$c_n = \frac{1}{n!\, h^n} \sum_{k=0}^{n} (-1)^{n-k} \binom{n}{k} f(x_k),$$

The sum on the right can be calculated in another way in terms of a linear operator Δ called the *forward difference operator.*

DEFINITION. *Let h be a fixed real number and let f be a given function. The function Δf defined by the equation*

$$\Delta f(x) = f(x + h) - f(x)$$

is called the first forward difference of f. It is defined at those points x for which both x and $x + h$ are in the domain of f. Higher order differences $\Delta^2 f, \Delta^3 f, \ldots$ are defined inductively as follows:

$$\Delta^{k+1} f = \Delta(\Delta^k f) \qquad \text{for} \quad k = 1, 2, 3, \ldots.$$

Note: The notations $\Delta_h f(x)$ and $\Delta f(x; h)$ are also used for $\Delta f(x)$ when it is desirable to indicate the dependence on h. It is convenient to define $\Delta^0 f = f$.

The nth difference $\Delta^n f(x)$ is a linear combination of the function values $f(x), f(x+h)$, $\ldots, f(x+nh)$. For example, we have

$$\begin{aligned}\Delta^2 f(x) &= \{f(x+2h) - f(x+h)\} - \{f(x+h) - f(x)\} \\ &= f(x+2h) - 2f(x+h) + f(x).\end{aligned}$$

In general, we have

$$\Delta^n f(x) = \sum_{k=0}^{n} (-1)^{n-k} \binom{n}{k} f(x+kh). \tag{15.24}$$

This is easily proved by induction on n, using the law of Pascal's triangle for binomial coefficients:

$$\binom{n}{k-1} + \binom{n}{k} = \binom{n+1}{k}.$$

Now suppose f is defined at $n+1$ equally spaced points $x_k = x_0 + kh$ for $k = 0, 1, \ldots,$ n. Then from (15.23) and (15.24) we obtain the formula

$$c_n = \frac{1}{h^n n!} \Delta^n f(x_0).$$

This provides a rapid method for calculating the coefficients in Newton's interpolation formula. The diagram in Table 15.1, called a *difference table*, shows how the successive differences can be systematically calculated from a tabulation of the values of f at equally spaced points. In the table we have written f_k for $f(x_k)$.

Newton's interpolation formula (15.21) now becomes

$$P_n(x) = f(x_0) + \sum_{k=1}^{n} \frac{\Delta^k f(x_0)}{k!\, h^k} \prod_{j=0}^{k-1} (x - x_j). \tag{15.25}$$

TABLE 15.1

x	$f(x)$	$\Delta f(x)$	$\Delta^2 f(x)$	$\Delta^3 f(x)$
x_0	f_0			
		$f_1 - f_0 = \Delta f(x_0)$		
x_1	f_1		$\Delta f(x_1) - \Delta f(x_0) = \Delta^2 f(x_0)$	
		$f_2 - f_1 = \Delta f(x_1)$		$\Delta^2 f(x_1) - \Delta^2 f(x_0) = \Delta^3 f(x_0)$
x_2	f_2		$\Delta f(x_2) - \Delta f(x_1) = \Delta^2 f(x_1)$	
		$f_3 - f_2 = \Delta f(x_2)$		
x_3	f_3			

If we write

$$\prod_{j=0}^{k-1}(x - x_j) = \prod_{j=0}^{k-1}(x - x_0 - jh) = h^k \prod_{j=0}^{k-1}\left(\frac{x - x_0}{h} - j\right) = h^k \prod_{j=0}^{k-1}(t - j),$$

where $t = (x - x_0)/h$, Equation (15.25) becomes

$$P_n(x) = f(x_0) + \sum_{k=1}^{n} \frac{\Delta^k f(x_0)}{k!} \prod_{j=0}^{k-1}(t - j). \tag{15.26}$$

15.12 Factorial polynomials

The product $t(t - 1) \cdots (t - k + 1)$ which appears in the sum in (15.26) is a polynomial in t of degree k called a *factorial polynomial*, or the *factorial kth power* of t. It is denoted by the symbol $t^{(k)}$. Thus, by definition,

$$t^{(k)} = \prod_{j=0}^{k-1}(t - j).$$

We also define $t^{(0)} = 1$. If we consider the forward difference operator Δ with $h = 1$, that is, $\Delta f(x) = f(x + 1) - f(x)$, we find that

$$\Delta t^{(n)} = n t^{(n-1)} \qquad \text{for} \quad n \geq 1.$$

This is analogous to the differentiation formula $Dt^n = nt^{n-1}$ for ordinary powers. Thus, the factorial power $t^{(n)}$ is related to differences much in the same way that the ordinary power t^n is related to derivatives.

With the use of factorial polynomials, Newton's interpolation formula (15.26) becomes

$$P_n(x_0 + th) = \sum_{k=0}^{n} \frac{\Delta^k f(x_0)}{k!} t^{(k)}.$$

Expressed in this form, Newton's formula resembles the Taylor formula for the polynomial of degree $\leq n$ that agrees with f and its first n derivatives at x_0. If we write

$$\binom{t}{k} = \frac{t^{(k)}}{k!} = \frac{t(t - 1) \cdots (t - k + 1)}{k!},$$

Newton's formula takes the form

$$P_n(x_0 + th) = \sum_{k=0}^{n} \binom{t}{k} \Delta^k f(x_0).$$

Further properties of factorial polynomials are developed in the following exercises.

15.13 Exercises

1. Let $\Delta f(x) = f(x+h) - f(x)$. If f is a polynomial of degree n, say

$$f(x) = \sum_{r=0}^{n} a_r x^r$$

with $a_n \neq 0$, show that (a) $\Delta^k f(x)$ is a polynomial of degree $n - k$ if $k \leq n$; (b) $\Delta^n f(x) = n!\, h^n a_n$; (c) $\Delta^k f(x) = 0$ for $k > n$.

2. Let $\Delta f(x) = f(x+h) - f(x)$. If $f(x) = \sin(ax + b)$, prove that

$$\Delta^n f(x) = \left(2 \sin \frac{ah}{2}\right)^n \sin\left(ax + b + \frac{nah + n\pi}{2}\right).$$

3. Let $\Delta f(x) = f(x+h) - f(x)$.
(a) If $f(x) = a^x$, where $a > 0$, show that $\Delta^k f(x) = (a^h - 1)^k a^x$.
(b) If $g(x) = (1 + a)^{x/h}$, where $a > 0$, show that $\Delta^k g(x) = a^k g(x)$.
(c) Show that the polynomial P_n of degree n that takes the values $P_n(k) = (1 + a)^k$ for $k = 0, 1, 2, \ldots, n$ is given by

$$P_n(x) = \sum_{k=0}^{n} \frac{a^k}{k!} x^{(k)}.$$

4. Let $x^{(n)}$ be the factorial nth power of x. Since $x^{(n)}$ is a polynomial in x of degree n with the value 0 when $x = 0$, we can write

$$x^{(n)} = \sum_{k=1}^{n} S_{k,n} x^k.$$

The numbers $S_{k,n}$ are called *Stirling numbers of the first kind.* From the definition of $x^{(n)}$ it is clear that $S_{n,n} = 1$ for $n \geq 0$.
(a) Show that $S_{n-1,n} = -n(n-1)/2$ and that $S_{1,n} = (-1)^{n-1}(n-1)!$ for $n \geq 1$.
(b) Prove that $S_{k,n+1} = S_{k-1,n} - nS_{k,n}$. Use this relation to verify the entries in Table 15.2, a table of Stirling numbers of the first kind, and construct the next three rows of the table.

TABLE 15.2

n	$S_{1,n}$	$S_{2,n}$	$S_{3,n}$	$S_{4,n}$	$S_{5,n}$	$S_{6,n}$	$S_{7,n}$
1	1						
2	−1	1					
3	2	−3	1				
4	−6	11	−6	1			
5	24	−50	35	−10	1		
6	−120	274	−225	85	−15	1	
7	720	−1764	1624	−735	175	−21	1

(c) Express the polynomial $x^{(4)} + 3x^{(3)} + 2x^{(1)} + 1$ as a linear combination of powers of x.

5. (a) Prove that

$$x = x^{(1)}, \qquad x^2 = x^{(1)} + x^{(2)}, \qquad x^3 = x^{(1)} + 3x^{(2)} + x^{(3)},$$

and that, in general,

$$x^n = \sum_{k=1}^{n} \frac{\Delta^k f(0)}{k!} x^{(k)},$$

where $f(x) = x^n$ and $\Delta f(x) = f(x + 1) - f(x)$. The numbers $T_{k,n} = \Delta^k f(0)/k!$ are called *Stirling numbers of the second kind.*

(b) Prove that

$$\Delta^k x^{n+1} = (x + k)\, \Delta^k x^n + k\, \Delta^{k-1} x^n$$

and use this to deduce that $T_{k,n+1} = T_{k-1,n} + kT_{k,n}$.

(c) Use the recursion formula in part (b) to verify the entries in Table 15.3, a table of Stirling numbers of the second kind, and construct the next three rows of the table.

TABLE 15.3

n	$T_{1,n}$	$T_{2,n}$	$T_{3,n}$	$T_{4,n}$	$T_{5,n}$	$T_{6,n}$	$T_{7,n}$
1	1						
2	1	1					
3	1	3	1				
4	1	7	6	1			
5	1	15	25	10	1		
6	1	31	90	65	15	1	
7	1	63	301	350	140	21	1

(d) Express the polynomial $x^4 + 3x^3 + 2x - 1$ as a linear combination of factorial polynomials.

6. (a) If p is a positive integer and if a and b are integers with $a < b$, prove that

$$\sum_{k=a}^{b-1} k^{(p)} = \frac{b^{(p+1)} - a^{(p+1)}}{p + 1}.$$

This formula is analogous to the integration formula for $\int_a^b x^p\, dx$. It should be noted, however, that the upper limit in the sum is $b - 1$, not b.

(b) Verify that $k(k + 3) = 4k^{(1)} + k^{(2)}$. The use part (a) to show that

$$\sum_{k=1}^{n} k(k + 3) = 4\frac{(n + 1)^{(2)}}{2} + \frac{(n + 1)^{(3)}}{3} = \frac{n(n + 1)(n + 5)}{3}.$$

(c) If $f(k)$ is a polynomial in k of degree r, prove that

$$\sum_{k=1}^{n} f(k)$$

is a polynomial in n of degree $r + 1$.

7. Use the method suggested in Exercise 6 to express each of the following sums as a polynomial in n.

(a) $\sum_{k=1}^{n} (4k^2 + 7k + 6)$.

(b) $\sum_{k=1}^{n} k^2(k + 1)$.

(c) $\sum_{k=1}^{n} k(k + 1)(k + 2)$.

(d) $\sum_{k=1}^{n} k^4$.

8. Let A denote the linear operator defined by the equation

$$A(f) = a_0 \Delta^n f + a_1 \Delta^{n-1} f + \cdots + a_{n-1} \Delta f + a_n f,$$

where $a_0, a_1, \ldots, a_n$ are constants. This is called a *constant-coefficient difference operator.* It is analogous to the constant-coefficient derivative operator described in Section 6.7. With each such A we can associate the characteristic polynomial p_A defined by

$$p_A(r) = a_0 r^n + a_1 r^{n-1} + \cdots + a_{n-1} r + a_n .$$

Conversely, with every polynomial p we can associate an operator A having this polynomial as its characteristic polynomial. If A and B are constant-coefficient difference operators and if λ is a real number, define $A + B$, AB, and λA by the same formulas used in Section 6.7 for derivative operators. Then prove that Theorem 6.6 is valid for constant-coefficient difference operators.

15.14 A minimum problem relative to the max norm

We consider a problem that arises naturally from the theory of polynomial interpolation. In Theorem 15.3 we derived the error formula

$$f(x) - P(x) = \frac{A(x)}{(n + 1)!} f^{(n+1)}(c), \tag{15.27}$$

where

$$A(x) = (x - x_0)(x - x_1) \cdots (x - x_n).$$

Here P is the unique polynomial of degree $\leq n$ that agrees with f at $n + 1$ distinct points $x_0, x_1, \ldots, x_n$ in $[a, b]$. The function f is assumed to have a derivative of order $n + 1$ on $[a, b]$, and c is an unknown point lying somewhere in $[a, b]$. To estimate the error in (15.27) we need bounds for the $(n + 1)$st derivative $f^{(n+1)}$ and for the product $A(x)$. Since A is a polynomial, its absolute value has a maximum somewhere in the interval $[a, b]$. This maximum will depend on the choice of the points $x_0, x_1, \ldots, x_n$, and it is natural to try to choose these points so the maximum will be as small as possible.

We can denote this maximum by $\|A\|$, where $\|A\|$ is the max norm, given by

$$\|A\| = \max_{a \leq x \leq b} |A(x)| .$$

The problem is to find a polynomial of specified degree that minimizes $\|A\|$. This problem was first solved by Chebyshev; its solution leads to an interesting class of polynomials that also occur in other connections. First we give a brief account of these polynomials and then return to the minimum problem in question.

15.15 Chebyshev polynomials

Let $x + iy$ be a complex number of absolute value 1. By the binomial theorem we have

$$(x + iy)^n = \sum_{k=0}^{n} \binom{n}{k} x^{n-k}(iy)^k$$

for every integer $n \geq 0$. In this formula we write $x = \cos\theta$, $y = \sin\theta$, and consider the real part of each member. Since

$$(x + iy)^n = (\cos\theta + i\sin\theta)^n = e^{in\theta} = \cos n\theta + i\sin n\theta,$$

the real part of the left member is $\cos n\theta$. The real part of the right member is the sum over even values of k. Hence we have

$$\text{(15.28)} \qquad \cos n\theta = x^n - \binom{n}{2} x^{n-2}y^2 + \binom{n}{4} x^{n-4}y^4 - + \cdots.$$

Since $y^2 = \sin^2\theta = 1 - \cos^2\theta = 1 - x^2$, the right-hand member of (15.28) is a polynomial in x of degree n. This polynomial is called the Chebyshev polynomial of the first kind and is denoted by $T_n(x)$.

DEFINITION. *The Chebyshev polynomial $T_n(x)$ is defined for all real x by the equation*

$$T_n(x) = \sum_{k=0}^{[n/2]} \binom{n}{2k} x^{n-2k}(x^2 - 1)^k.$$

From Equation (15.28) we obtain the following theorem.

THEOREM 15.7. *If $-1 \leq x \leq 1$ we have*

$$T_n(x) = \cos(n \arccos x).$$

Proof. If $\theta = \arccos x$ then $x = \cos\theta$ and $T_n(x) = \cos n\theta$.

The Chebyshev polynomials can be readily computed by taking the real part of $(x + iy)^n$ with $y^2 = 1 - x^2$, or by using the following recursion formula.

THEOREM 15.8. *The Chebyshev polynomials satisfy the recursion formula*

$$T_{n+1}(x) = 2xT_n(x) - T_{n-1}(x) \qquad \text{for} \quad n \geq 1,$$

with $T_0(x) = 1$ and $T_1(x) = x$.

Proof. First assume $-1 \leq x \leq 1$ and put $x = \cos\theta$ in the trigonometric identity

$$\cos(n+1)\theta + \cos(n-1)\theta = 2\cos\theta\cos n\theta .$$

This proves that $T_{n+1}(x) + T_{n-1}(x) = 2xT_n(x)$ for x in the interval $-1 \leq x \leq 1$. But since both members are polynomials, this relation must hold for all x.

The next five polynomials are

$$T_2(x) = 2x^2 - 1, \qquad T_3(x) = 4x^3 - 3x, \qquad T_4(x) = 8x^4 - 8x^2 + 1,$$
$$T_5(x) = 16x^5 - 20x^3 + 5x, \qquad T_6(x) = 32x^6 - 48x^4 + 18x^2 - 1.$$

The recursion formula shows that all the coefficients of $T_n(x)$ are integers; moreover, the coefficient of x^n is 2^{n-1}.

The next theorem shows that $T_n(x)$ has exactly n first order zeros and that they all lie in the interval $[-1, 1]$.

THEOREM 15.9. *If $n \geq 1$ the polynomial $T_n(x)$ has zeros at the n points*

$$x_k = \cos\frac{(2k+1)\pi}{2n}, \qquad k = 0, 1, 2, \ldots, n-1.$$

Hence $T_n(x)$ has the factorization

$$T_n(x) = 2^{n-1}(x - x_0)(x - x_1)\cdots(x - x_{n-1}) = 2^{n-1}\prod_{k=0}^{n-1}\left(x - \cos\frac{(2k+1)\pi}{2n}\right).$$

Proof. We use the formula $T_n(x) = \cos n\theta$. Since $\cos n\theta = 0$ only if $n\theta$ is an odd multiple of $\pi/2$, we have $T_n(x) = 0$ for x in $[-1, 1]$ only if $n \arccos x = (2k+1)\pi/2$ for some integer k. Therefore the zeros of T_n in the interval $[-1, 1]$ are to be found among the numbers

$$x_k = \cos\frac{2k+1}{n}\frac{\pi}{2}, \qquad k = 0, \pm 1, \pm 2, \ldots. \tag{15.29}$$

The values $k = 0, 1, 2, \ldots, n-1$ give n distinct zeros $x_0, x_1, \ldots, x_{n-1}$, all lying in the open interval $(-1, 1)$. Since a polynomial of degree n cannot have more than n zeros, these must be *all* the zeros of T_n. The remaining x_k in (15.29) are repetitions of these n.

THEOREM 15.10. *In the interval $[-1, 1]$ the extreme values of $T_n(x)$ are $+1$ and -1, taken alternately at the $n + 1$ points*

$$t_k = \cos\frac{k\pi}{n}, \qquad \textit{for} \quad k = 0, 1, 2, \ldots, n. \tag{15.30}$$

Proof. By Rolle's theorem, the relative maxima and minima of T_n must occur between successive zeros; there are $n - 1$ such points in the open interval $(-1, 1)$. From the cosine formula for T_n we see that the extreme values, ± 1, are taken at the $n - 1$ interior points

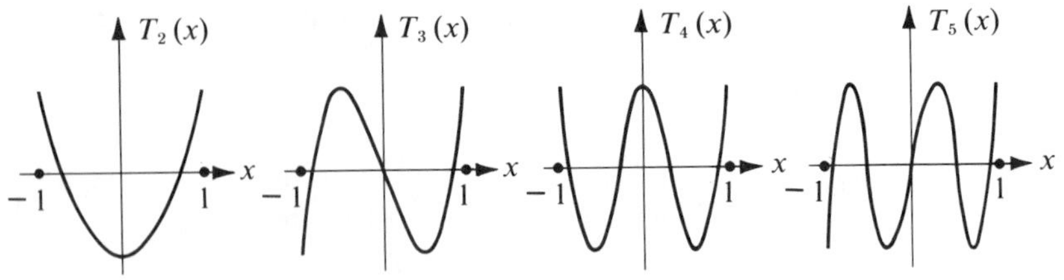

FIGURE 15.2 Graphs of Chebyshev polynomials over the interval $[-1, 1]$.

$\cos(k\pi/n)$, $k = 1, 2, \ldots, n - 1$, and also at the two endpoints $x = 1$ and $x = -1$. Therefore in the closed interval $[-1, 1]$ the extreme values $+1$ and -1 are taken *alternately* at the $n + 1$ points $t_0, t_1, \ldots, t_n$ given by $t_k = \cos(k\pi/n)$ for $k = 0, 1, 2, \ldots, n$.

Figure 15.2 shows the graphs of $T_2, \ldots, T_5$ over the interval $[-1, 1]$.

15.16 A minimal property of Chebyshev polynomials

We return now to the problem of finding a polynomial of a specified degree for which the max norm is as small as possible. The problem is solved by the following theorem.

THEOREM 15.11. *Let $p_n(x) = x^n + \cdots$ be any polynomial of degree $n \geq 1$ with leading coefficient 1, and let*

$$\|p_n\| = \max_{-1 \leq x \leq 1} |p_n(x)|.$$

Then we have the inequality

$$\|p_n\| \geq \|\tilde{T}_n\|, \tag{15.31}$$

where $\tilde{T}_n(x) = T_n(x)/2^{n-1}$. Moreover, equality holds in (15.31) if $p_n = \tilde{T}_n$.

Proof. In the interval $[-1, 1]$ the polynomial $\tilde{T}_n$ takes its extreme values, $1/2^{n-1}$ and $-1/2^{n-1}$, alternately at the $n + 1$ distinct points t_k in Equation (15.30). Therefore $\|\tilde{T}_n\| = 1/2^{n-1}$.

We show next that the inequality

$$\|p_n\| < \frac{1}{2^{n-1}} \tag{15.32}$$

leads to a contradiction. Assume, then, that p_n satisfies (15.32) and consider the difference

$$r(x) = \tilde{T}_n(x) - p_n(x).$$

At the points t_k given by (15.30) we have

$$r(t_k) = \frac{(-1)^k}{2^{n-1}} - p_n(t_k) = (-1)^k\left[\frac{1}{2^{n-1}} - (-1)^k p_n(t_k)\right].$$

Because of (15.32) the factor in square brackets is positive. Therefore $r(t_k)$ has alternating signs at the $n + 1$ points $t_0, t_1, \ldots, t_n$. Since r is continuous it must vanish at least once between consecutive sign changes. Therefore r has at least n distinct zeros. But since r is a polynomial of degree $\leq n - 1$, this means that r is identically zero. Therefore $P_n = \tilde{T}_n$, so $\|P_n\| = \|\tilde{T}_n\| = 1/2^{n-1}$, contradicting (15.32). This proves that we must have $\|p_n\| \geq 1/2^{n-1} = \|\tilde{T}_n\|$.

Although Theorem 15.11 refers to the interval $[-1, 1]$ and to a polynomial with leading coefficient 1, it can be used to deduce a corresponding result for an arbitrary interval $[a, b]$ and an arbitrary polynomial.

THEOREM 15.12. *Let $q_n(x) = c_n x^n + \cdots$ be any polynomial of degree $n \geq 1$, and let*

$$\|q_n\| = \max_{a \leq x \leq b} |q_n(x)| .$$

Then we have the inequality

$$\|q_n\| \geq |c_n| \frac{(b-a)^n}{2^{2n-1}} . \tag{15.33}$$

Moreover, equality holds in (15.33) *if*

$$q_n(x) = c_n \frac{(b-a)^n}{2^{2n-1}} T_n\left(\frac{2x - a - b}{b - a}\right).$$

Proof. Consider the transformation

$$t = \frac{2x - a - b}{b - a} .$$

This maps the interval $a \leq x \leq b$ in a one-to-one fashion onto the interval $-1 \leq t \leq 1$. Since

$$x = \frac{b - a}{2} t + \frac{b + a}{2}$$

we have

$$x^n = \left(\frac{b-a}{2}\right)^n t^n + \text{terms of lower degree},$$

hence

$$q_n(x) = c_n \left(\frac{b-a}{2}\right)^n p_n(t),$$

where $p_n(t)$ is a polynomial in t of degree n with leading coefficient 1. Applying Theorem 15.11 to p_n we obtain Theorem 15.12.

15.17 Application to the error formula for interpolation

We return now to the error formula (15.27) for polynomial interpolation. If we choose the interpolation points $x_0, x_1, \ldots, x_n$ to be the $n + 1$ zeros of the Chebyshev polynomial

T_{n+1} we can write (15.27) in the form

$$f(x) - P(x) = \frac{T_{n+1}(x)}{2^n(n+1)!} f^{(n+1)}(c).$$

The points $x_0, x_1, \ldots, x_n$ all lie in the open interval $(-1, 1)$ and are given by

$$x_k = \cos\left(\frac{2k+1}{n+1}\frac{\pi}{2}\right) \qquad \text{for} \quad k = 0, 1, 2, \ldots, n.$$

If x is in the interval $[-1, 1]$ we have $|T_{n+1}(x)| \le 1$ and the error is estimated by the inequality

$$|f(x) - P(x)| \le \frac{1}{2^n(n+1)!} |f^{(n+1)}(c)|.$$

If the interpolation takes place in an interval $[a, b]$ with the points

$$y_k = \frac{b-a}{2} x_k + \frac{b+a}{2}$$

as interpolation points, the product

$$A(x) = (x - y_0)(x - y_1) \cdots (x - y_n)$$

satisfies the inequality $|A(x)| \le (b - a)^{n+1}/2^{2n+1}$ for all x in $[a, b]$. The corresponding estimate for $f(x) - P(x)$ is

$$|f(x) - P(x)| \le \frac{(b-a)^{n+1}}{2^{2n+1}(n+1)!} |f^{(n+1)}(c)|.$$

15.18 Exercises

In this set of exercises T_n denotes the Chebyshev polynomial of degree n.

1. Prove that $T_n(-x) = (-1)^n T_n(x)$. This shows that T_n is an even function when n is even and an odd function when n is odd.
2. (a) Prove that in the open interval $(-1, 1)$ the derivative T_n' is given by the formula

$$T_n'(x) = \frac{n \sin n\theta}{\sin \theta}, \qquad \text{where} \quad \theta = \arccos x.$$

   (b) Compute $T_n'(1)$ and $T_n'(-1)$.

3. Prove that $\displaystyle\int_0^x T_n(u)\,du = \frac{1}{2}\left\{\frac{T_{n+1}(x) - T_{n+1}(0)}{n+1} - \frac{T_{n-1}(x) - T_{n-1}(0)}{n-1}\right\}$ if $n \ge 2$.

4. (a) Prove that $2T_m(x)T_n(x) = T_{m+n}(x) + T_{m-n}(x)$.
   (b) Prove that $T_{mn}(x) = T_m[T_n(x)] = T_n[T_m(x)]$.
5. If $x = \cos\theta$, prove that $\sin\theta \sin n\theta$ is a polynomial in x, and determine its degree.
6. The Chebyshev polynomial T_n satisfies the differential equation

$$(1 - x^2)y'' - xy' + n^2y = 0$$

over the entire real axis. Prove this by each of the following methods:

(a) Differentiate the relation $T_n'(x)\sin\theta = n\sin n\theta$ obtained in Exercise 2(a).

(b) Introduce the change of variable $x = \cos\theta$ in the differential equation

$$\frac{d^2(\cos n\theta)}{d\theta^2} = -n^2\cos n\theta .$$

7. Determine, in terms of Chebyshev polynomials, a polynomial $Q(x)$ of degree $\le n$ which best approximates x^{n+1} on the interval $[-1, 1]$ relative to the max norm.
8. Find a polynomial of degree ≤ 4 that best approximates the function $f(x) = x^5$ in the interval $[0, 1]$, relative to the max norm.
9. A polynomial P is called *primary* if the coefficient of the term of highest degree is 1. For a given interval $[a, b]$ let $\|P\|$ denote the maximum of $|P|$ on $[a, b]$. Prove each of the following statements:

   (a) If $b - a < 4$, for every $\epsilon > 0$ there exists a primary polynomial P with $\|P\| < \epsilon$.

   (b) If for every $\epsilon > 0$ there exists a primary polynomial P with $\|P\| < \epsilon$, then $b - a < 4$.

   In other words, primary polynomials with arbitrarily small norm exist if, and only if, the interval $[a, b]$ has length less than 4.
10. The Chebyshev polynomials satisfy the following orthogonality relations:

$$\int_{-1}^{1}\frac{T_n(x)T_m(x)}{\sqrt{1-x^2}}\,dx = \begin{cases} 0 & \text{if } n \ne m, \\ \pi & \text{if } n = m = 0, \\ \dfrac{\pi}{2} & \text{if } n = m > 0. \end{cases}$$

   Prove this by each of the following methods:

   (a) From the differential equation in Exercise 6 deduce that

$$T_m(x)\frac{d}{dx}\left(\sqrt{1-x^2}\,T_n'(x)\right) + n^2\frac{T_n(x)T_m(x)}{\sqrt{1-x^2}} = 0 .$$

   Write a corresponding formula with n and m interchanged, subtract the two equations, and integrate from -1 to 1.

   (b) Use the orthogonality relations

$$\int_0^{\pi}\cos m\theta\cos n\theta\,d\theta = \begin{cases} 0 & \text{if } n \ne m,\ n > 0,\ m > 0, \\ \pi & \text{if } n = m = 0, \\ \dfrac{\pi}{2} & \text{if } n = m > 0, \end{cases}$$

   and introduce the change of variable $x = \cos\theta$.
11. Prove that for $-1 < x < 1$ we have

$$\frac{T_n(x)}{\sqrt{1-x^2}} = (-1)^n\frac{2^n n!}{(2n)!}\frac{d^n}{dx^n}(1-x^2)^{n-1/2} .$$

12. Let $y_1, y_2, \ldots, y_n$ be n real numbers, and let

$$x_k = \cos\frac{(2k-1)\pi}{2n} \qquad \text{for } k = 1, 2, \ldots, n .$$

Let P be the polynomial of degree $\leq n - 1$ that takes the value y_k at x_k for $1 \leq k \leq n$. If x is not one of the x_k show that

$$P(x) = \frac{1}{n} \sum_{k=1}^{n} (-1)^{k-1} y_k \sqrt{1 - x_k^2} \frac{T_n(x)}{x - x_k}.$$

13. Let P be a polynomial of degree $\leq n - 1$ such that

$$\sqrt{1 - x^2}\, |P(x)| \leq 1$$

for $-1 \leq x \leq 1$. Prove that $\|P\| \leq n$, where $\|P\|$ is the maximum of $|P|$ on the interval $[-1, 1]$.

[*Hint:* Use Exercise 12. Consider three cases: $x_1 \leq x \leq 1$; $-1 \leq x \leq x_n$; $x_n \leq x \leq x_1$; in the first two use Exercise 15(a) of Section 15.9. In the third case note that $\sqrt{1 - x^2} \geq \sin(\pi/2n) > 1/n$.]

In Exercises 14 through 18, $U_n(x) = T'_{n+1}(x)/(n + 1)$ for $n = 0, 1, 2, \ldots$.

14. (a) Prove that $U_n(x) = 2xU_{n-1}(x) - U_{n-2}(x)$ for $n \geq 2$.
(b) Determine the explicit form of the polynomials $U_0, U_1, \ldots, U_5$.
(c) Prove that $|U_n(x)| \leq n + 1$ if $-1 \leq x \leq 1$.
15. Show that U_n satisfies the differential equation

$$(1 - x^2)y'' - 3xy' + n(n + 2)y = 0.$$

16. Derive the orthogonality relations

$$\int_{-1}^{1} \sqrt{1 - x^2}\, U_m(x) U_n(x)\, dx = \begin{cases} 0 & \text{if } m \neq n, \\ \dfrac{\pi}{2} & \text{if } m = n. \end{cases}$$

17. Prove that

$$\sqrt{1 - x^2}\, U_n(x) = (-1)^n \frac{2^n (n + 1)!}{(2n + 1)!} \frac{d^n}{dx^n} (1 - x^2)^{n+1/2}.$$

18. Let $y_1, y_2, \ldots, y_n$ be n real numbers and let

$$x_k = \cos \frac{k\pi}{n + 1} \qquad \text{for } k = 1, 2, \ldots, n.$$

Let P be the polynomial of degree $\leq n - 1$ that takes the value y_k at x_k for $1 \leq k \leq n$. If x is not one of the x_k show that

$$P(x) = \frac{1}{n + 1} \sum_{k=1}^{n} (-1)^{k-1} (1 - x_k^2) y_k \frac{U_n(x)}{x - x_k}.$$

15.19 Approximate integration. The trapezoidal rule

Many problems in both pure and applied mathematics lead to new functions whose properties have not been studied or whose values have not been tabulated. To satisfy certain practical needs of applied science it often becomes necessary to obtain quantitative

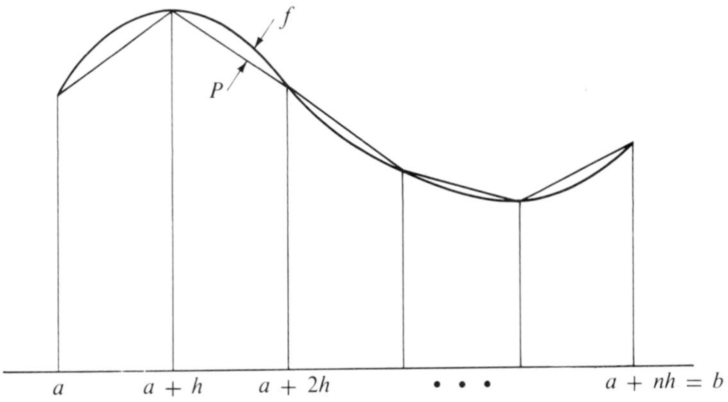

FIGURE 15.3 The trapezoidal rule obtained by piecewise linear interpolation.

information about such functions, either in graphical or numerical form. Many of these functions occur as integrals of the type

$$F(x) = \int_a^x f(t)\,dt,$$

where the integrand f is given by an explicit analytic formula or is known in part by tabular data. The remainder of this chapter describes some of the most elementary methods for finding numerical approximations to such integrals. The basic idea is very simple. We approximate the integrand f by another function P whose integral is easily computed, and then we use the integral of P as an approximation to the integral of f.

If f is nonnegative the integral $\int_a^b f(x)\,dx$ represents the area of the ordinate set of f over $[a, b]$. This geometric interpretation of the integral immediately suggests certain procedures for approximate integration. Figure 15.3 shows an example of a function f with known values at $n + 1$ equally spaced points $a, a + h, a + 2h, \ldots, a + nh = b$, where $h = (b - a)/n$. Let $x_k = a + kh$. For each $k = 0, 1, 2, \ldots, n - 1$ the graph of f over the interval $[x_k, x_{k+1}]$ has been approximated by a linear function that agrees with f at the endpoints x_k and x_{k+1}. Let P denote the corresponding piecewise linear interpolating function defined over the full interval $[a, b]$. Then we have

$$\text{(15.34)} \qquad P(x) = \frac{x_{k+1} - x}{h} f(x_k) + \frac{x - x_k}{h} f(x_{k+1}) \qquad \text{if} \quad x_k \le x \le x_{k+1}.$$

Integrating over the interval $[x_k, x_{k+1}]$ we find that

$$\int_{x_k}^{x_{k+1}} P(x)\,dx = h\,\frac{f(x_k) + f(x_{k+1})}{2}.$$

When f is positive this is the area of the trapezoid determined by the graph of P over $[x_k, x_{k+1}]$. The formula holds, of course, even if f is not positive everywhere. Adding the integrals over all subintervals $[x_k, x_{k+1}]$ we obtain

(15.35)
$$\int_a^b P(x)\,dx = \frac{h}{2}\sum_{k=0}^{n-1}[f(x_k) + f(x_{k+1})]$$
$$= \frac{h}{2}\left(f(a) + 2\sum_{k=1}^{n-1} f(a+kh) + f(b)\right).$$

To use this sum as an approximation to the integral $\int_a^b f(x)\,dx$ we need an estimate for the error, $\int_a^b f(x)\,dx - \int_a^b P(x)\,dx$. If f has a continuous second derivative on $[a, b]$ this error is given by the following theorem.

THEOREM 15.13. TRAPEZOIDAL RULE. *Assume f has a continuous second derivative f'' on $[a, b]$. If n is a positive integer, let $h = (b - a)/n$. Then we have*

(15.36)
$$\int_a^b f(x)\,dx = \frac{b-a}{2n}\left(f(a) + 2\sum_{k=1}^{n-1} f(a+kh) + f(b)\right) - \frac{(b-a)^3}{12n^2}f''(c)$$

for some c in $[a, b]$.

Note: Equation (15.36) is known as the *trapezoidal rule*. The term $-f''(c)(b-a)^3/12n^2$ represents the error in approximating $\int_a^b f(x)\,dx$ by $\int_a^b P(x)\,dx$. Once the maximum value of f'' on $[a, b]$ is known we can approximate the integral of f to any desired degree of accuracy by taking n sufficiently large. Note that no knowledge of the interpolating function P is required to use this formula. It is only necessary to know the values of f at the points $a, a + h, \ldots, a + nh$, and to have an estimate for $|f''(c)|$.

Proof. Let P be the interpolating function given by (15.34), where $x_k = a + kh$. In each subinterval $[x_k, x_{k+1}]$ we apply the error estimate for linear interpolation given by Theorem 15.3 and we find

(15.37)
$$f(x) - P(x) = (x - x_k)(x - x_{k+1})\frac{f''(c_k)}{2!}$$

for some c_k in (x_k, x_{k+1}). Let M_2 and m_2 denote the maximum and minimum, respectively, of f'' on $[a, b]$, and let

$$B(x) = (x - x_k)(x_{k+1} - x)/2.$$

Then $B(x) \geq 0$ in the interval $[x_k, x_{k+1}]$, and from (15.37) we obtain the inequalities

$$m_2 B(x) \leq P(x) - f(x) \leq M_2 B(x)$$

in this interval. Integrating, we have

(15.38)
$$m_2\int_{x_k}^{x_k+h} B(x)\,dx \leq \int_{x_k}^{x_k+h}[P(x) - f(x)]\,dx \leq M_2\int_{x_k}^{x_k+h} B(x)\,dx.$$

The integral of B is given by

$$\int_{x_k}^{x_k+h} B(x)\,dx = \frac{1}{2}\int_{x_k}^{x_k+h}(x - x_k)(x_{k+1} - x)\,dx = \frac{1}{2}\int_0^h t(h-t)\,dt = \frac{h^3}{12}.$$

Therefore the inequalities (15.38) give us

$$m_2 \le \frac{12}{h^3} \int_{x_k}^{x_k+h} [P(x) - f(x)]\,dx \le M_2 .$$

Adding these inequalities for $k = 0, 1, 2, \ldots, n - 1$ and dividing by n, we obtain

$$m_2 \le \frac{12}{nh^3} \int_a^b [P(x) - f(x)]\,dx \le M_2 .$$

Since the function f'' is continuous on $[a, b]$, it assumes every value between its minimum m_2 and its maximum M_2 somewhere in $[a, b]$. In particular, we have

$$f''(c) = \frac{12}{nh^3} \int_a^b [P(x) - f(x)]\,dx$$

for some c in $[a, b]$. In other words,

$$\int_a^b f(x)\,dx = \int_a^b P(x)\,dx - \frac{nh^3}{12} f''(c) .$$

Using (15.35) and the relation $h = (b - a)/n$ we obtain (15.36).

To derive the trapezoidal rule we used a linear polynomial to interpolate between each adjacent pair of values of f. More accurate formulas can be obtained by interpolating with polynomials of higher degree. In the next section we consider an important special case that is remarkable for its simplicity and accuracy.

15.20 Simpson's rule

The solid curve in Figure 15.4 is the graph of a function f over an interval $[a, b]$. The mid-point of the interval, $(a + b)/2$, is denoted by m. The dotted curve is the graph of a quadratic polynomial P that agrees with f at the three points a, m, and b. If we use the integral $\int_a^b P(x)\,dx$ as an approximation to $\int_a^b f(x)\,dx$ we are led to an approximate integration formula known as *Simpson's rule*.

Instead of determining P explicitly, we introduce a linear transformation that carries the interval $[a, b]$ onto the interval $[0, 2]$. If we write

$$t = \frac{x - a}{m - a}, \qquad \text{or} \qquad x = a + (m - a)t ,$$

we see that t takes the values 0, 1, 2 when x takes the values a, m, b. Now let

$$\varphi(t) = P[a + (m - a)t] .$$

Then $\varphi(t)$ is a quadratic polynomial in t that takes the values $P(a)$, $P(m)$, $P(b)$ at the points $t = 0, 1, 2$, respectively. Also, we have

$$\int_0^2 \varphi(t)\,dt = \int_0^2 P[a + (m - a)t]\,dt = \frac{1}{m - a} \int_a^b P(x)\,dx ;$$

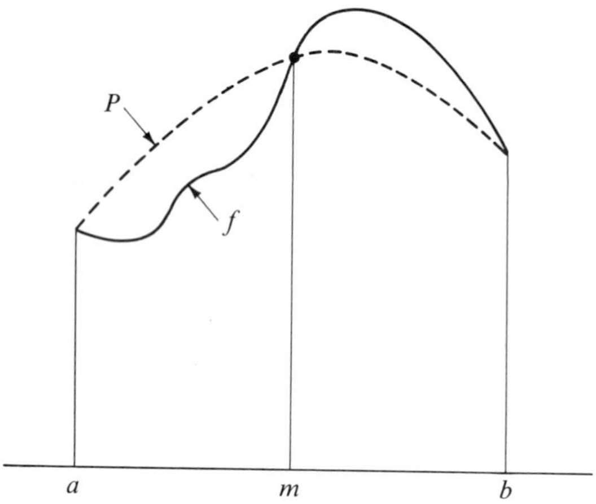

FIGURE 15.4 Interpolation by a quadratic polynomial P.

hence

$$\int_a^b P(x)\,dx = (m-a)\int_0^2 \varphi(t)\,dt = \frac{b-a}{2}\int_0^2 \varphi(t)\,dt. \tag{15.39}$$

Now we use Newton's interpolation formula to construct φ. We have

$$\varphi(t) = \varphi(0) + t\,\Delta\varphi(0) + t(t-1)\frac{\Delta^2\varphi(0)}{2!},$$

where $\Delta\varphi(t) = \varphi(t+1) - \varphi(t)$. Integrating from 0 to 2 we obtain

$$\int_0^2 \varphi(t)\,dt = 2\varphi(0) + 2\,\Delta\varphi(0) + \tfrac{1}{3}\,\Delta^2\varphi(0).$$

Since $\Delta\varphi(0) = \varphi(1) - \varphi(0)$ and $\Delta^2\varphi(0) = \varphi(2) - 2\varphi(1) + \varphi(0)$, the integral is equal to

$$\int_0^2 \varphi(t)\,dt = \tfrac{1}{3}[\varphi(0) + 4\varphi(1) + \varphi(2)] = \tfrac{1}{3}[P(a) + 4P(m) + P(b)].$$

Using (15.39) and the fact that P agrees with f at a, m, b, we obtain

$$\int_a^b P(x)\,dx = \frac{b-a}{6}[f(a) + 4f(m) + f(b)]. \tag{15.40}$$

Therefore, we may write

$$\int_a^b f(x)\,dx = \frac{b-a}{6}[f(a) + 4f(m) + f(b)] + R,$$

where $R = \int_a^b f(x)\,dx - \int_a^b P(x)\,dx$.

If f is a quadratic polynomial, P is identical to f and the error R is zero. It is a remarkable fact that we also have $R = 0$ when f is a *cubic* polynomial. To prove this property we use the error estimate for Lagrange interpolation given by Theorem 15.3, and we write

$$f(x) - P(x) = (x - a)(x - m)(x - b)\frac{f'''(c)}{3!}, \tag{15.41}$$

where $c \in (a, b)$. When f is a cubic polynomial the third derivative f''' is constant, say $f'''(x) = C$, and the foregoing formula becomes

$$f(x) - P(x) = \frac{C}{6}(x - a)(x - m)(x - b) = \frac{C}{6}(t + h)t(t - h),$$

where $t = x - m$ and $h = (b - a)/2$. Therefore

$$R = \int_a^b [f(x) - P(x)]\, dx = \frac{C}{6}\int_{-h}^{h} (t^3 - h^2 t)\, dt = 0,$$

since the last integrand is an odd function. This property is illustrated in Figure 15.5. The dotted curve is the graph of a cubic polynomial f that agrees with P at a, m, b. In this case $R = \int_a^b [f(x) - P(x)]\, dx = A_1 - A_2$, where A_1 and A_2 are the areas of the two shaded regions. Since $R = 0$ the two regions have equal areas.

We have just seen that Equation (15.40) is valid if P is a polynomial of degree ≤ 3 that agrees with f at a, m, and b. By choosing this polynomial carefully we can considerably improve the error estimate in (15.41). We have already imposed three conditions on P, namely, $P(a) = f(a)$, $P(m) = f(m)$, $P(b) = f(b)$. Now we impose a fourth condition, $P'(m) = f'(m)$. This will give P and f the same slope at $(m, f(m))$, and we can hope that this will improve the approximation of f by P throughout $[a, b]$.

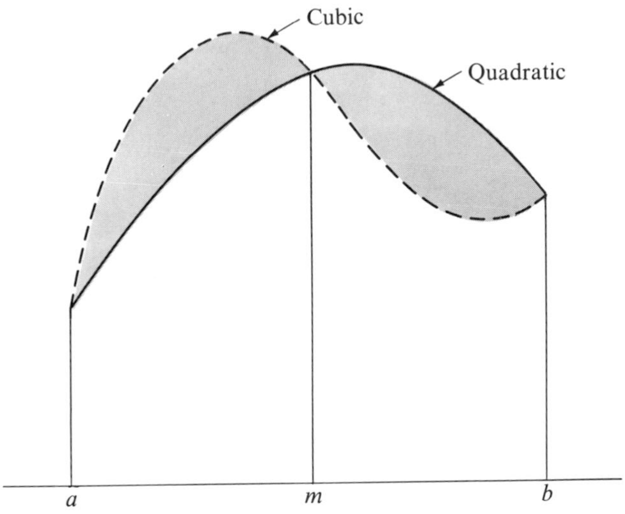

FIGURE 15.5 The two shaded regions have equal areas for every cubic interpolating polynomial.

To show that such a P can always be chosen, we let Q be the quadratic polynomial that agrees with f at a, m, b, and we let

$$P(x) = Q(x) + A(x-a)(x-m)(x-b),$$

where A is a constant to be determined. For any choice of A, this cubic polynomial P agrees with Q and hence with f at a, m, b. Now we choose A to make $P'(m) = f'(m)$. Differentiating the formula for $P(x)$ and putting $x = m$ we obtain

$$P'(m) = Q'(m) + A(m-a)(m-b).$$

Therefore if we take $A = [f'(m) - Q'(m)]/[(m-a)(m-b)]$ we also satisfy the condition $P'(m) = f'(m)$.

Next we show that for this choice of P we have

$$f(x) - P(x) = (x-a)(x-m)^2(x-b)\frac{f^{(4)}(z)}{4!} \tag{15.42}$$

for some z in (a, b), provided that the fourth derivative $f^{(4)}$ exists in $[a, b]$. To prove (15.42) we argue as in the proof of Theorem 15.3. First we note that (15.42) is trivially satisfied for any choice of z if $x = a$, m, or b. Therefore, assume $x \neq a$, $x \neq m$, $x \neq b$, keep x fixed, and introduce a new function F defined on $[a, b]$ by the equation

$$F(t) = A(x)[f(t) - P(t)] - A(t)[f(x) - P(x)],$$

where

$$A(t) = (t-a)(t-m)^2(t-b).$$

Note that $F(t) = 0$ for $t = a$, m, b, and x. By Rolle's theorem, $F'(t)$ vanishes in each of the three open intervals determined by these four points. In addition, $F'(m) = 0$ because $A'(m) = 0$ and $f'(m) = P'(m)$. Therefore $F'(t) = 0$ for at least four distinct points in (a, b). By Rolle's theorem $F''(t) = 0$ for at least three points, $F'''(t) = 0$ for at least two points, and $F^{(4)}(t) = 0$ for at least one point, say for $t = z$. From the definition of F we find

$$\begin{aligned} F^{(4)}(t) &= A(x)[f^{(4)}(t) - P^{(4)}(t)] - A^{(4)}(t)[f(x) - P(x)] \\ &= A(x)f^{(4)}(t) - 4!\,[f(x) - P(x)]. \end{aligned}$$

When we substitute $t = z$ in this equation we obtain (15.42).

Now it is a simple matter to prove Simpson's rule in the following form.

THEOREM 15.14. SIMPSON'S RULE. *Assume f has a continuous fourth derivative on $[a, b]$, and let $m = (a+b)/2$. Then we have*

$$\int_a^b f(x)\,dx = \frac{b-a}{6}[f(a) + 4f(m) + f(b)] - \frac{(b-a)^5}{2880}f^{(4)}(c) \tag{15.43}$$

for some c in $[a, b]$.

Proof. Let M_4 and m_4 denote, respectively, the maximum and minimum values of $f^{(4)}$ on $[a, b]$, and let $B(x) = -(x-a)(x-m)^2(x-b)/4!$. Since $B(x) \geq 0$ for each x in $[a, b]$, Equation (15.42) leads to the inequalities

$$m_4 B(x) \leq P(x) - f(x) \leq M_4 B(x).$$

Integrating, we find

$$m_4 \int_a^b B(x)\,dx \leq \int_a^b [P(x) - f(x)]\,dx \leq M_4 \int_a^b B(x)\,dx. \tag{15.44}$$

To evaluate the integral $\int_a^b B(x)\,dx$ we let $h = (b-a)/2$ and we have

$$\int_a^b B(x)\,dx = -\frac{1}{4!}\int_a^b (x-a)(x-m)^2(x-b)\,dx = -\frac{1}{4!}\int_{-h}^h (t+h)t^2(t-h)\,dt$$
$$= -\frac{2}{4!}\int_0^h t^2(t^2 - h^2)\,dt = \frac{1}{4!}\frac{4h^5}{15} = \frac{(b-a)^5}{2880}.$$

Therefore the inequalities in (15.44) give us

$$m_4 \leq \frac{2880}{(b-a)^5}\int_a^b [P(x) - f(x)]\,dx \leq M_4.$$

But since $f^{(4)}$ is continuous on $[a, b]$, it assumes every value between its minimum m_4 and its maximum M_4 somewhere in $[a, b]$. Therefore

$$f^{(4)}(c) = \frac{2880}{(b-a)^5}\int_a^b [P(x) - f(x)]\,dx$$

for some c in $[a, b]$. Since $\int_a^b P(x)\,dx = \frac{1}{6}(b-a)[f(a) + 4f(m) + f(b)]$, this equation gives us (15.43).

Simpson's rule is of special interest because its accuracy is greater than might be expected from a knowledge of the function f at only three points. If the values of f are known at an odd number of equally spaced points, say at $a, a+h, \ldots, a+2nh$, it is usually simpler to apply Simpson's rule successively to each of the intervals $[a, a+2h]$, $[a+2h, a+4h]$, $\ldots$, rather than to use an interpolating polynomial of degree $\leq 2n$ over the full interval $[a, a+2nh]$. Applying Simpson's rule in this manner, we obtain the following extension of Theorem 15.14.

THEOREM 15.15. EXTENDED SIMPSON'S RULE. *Assume f has a continuous fourth derivative in $[a, b]$. Let $h = (b-a)/(2n)$ and let $f_k = f(a+kh)$ for $k = 1, 2, \ldots, 2n-1$. Then we have*

$$\int_a^b f(x)\,dx = \frac{b-a}{6n}\left(f(a) + 4\sum_{k=1}^{n} f_{2k-1} + 2\sum_{k=1}^{n-1} f_{2k} + f(b)\right) - \frac{(b-a)^5}{2880n^4} f^{(4)}(\bar{c})$$

for some $\bar{c}$ in $[a, b]$.

The proof of this theorem is requested in Exercise 9 of the next section.

15.21 Exercises

1. (a) Apply the trapezoidal rule with $n = 10$ to estimate the value of the integral

$$\log 2 = \int_1^2 \frac{dx}{x}.$$

Obtain upper and lower bounds for the error. (See Exercise 10(b) to compare the accuracy with that obtained from Simpson's rule.)
(b) What is the smallest value of n that would ensure six-place accuracy in the calculation of $\log 2$ by this method?

2. (a) Show that there is a positive number c in the interval $[0, 1]$ such that the formula

$$\int_{-1}^{1} f(x)\,dx = f(c) + f(-c)$$

is exact for all polynomials of degree ≤ 3.
(b) Generalize the result of part (a) for an arbitrary interval. That is, show that constants c_1 and c_2 exist in $[a, b]$ such that the formula

$$\int_a^b f(x)\,dx = \frac{b-a}{2}\,[f(c_1) + f(c_2)]$$

is exact for all polynomials of degree ≤ 3. Express c_1 and c_2 in terms of a and b.

3. (a) Show that a positive constant c exists such that the formula

$$\int_{-1/2}^{1/2} f(x)\,dx = \tfrac{1}{3}[f(-c) + f(0) + f(c)]$$

is exact for all polynomials of degree ≤ 3.
(b) Generalize the result of part (a) for an arbitrary interval. That is, show that constants c_1 and c_2 exist in $[a, b]$ such that the formula

$$\int_a^b f(x)\,dx = \frac{b-a}{3}\left[f(c_1) + f\left(\frac{a+b}{2}\right) + f(c_2)\right]$$

is exact for all polynomials of degree ≤ 3. Express c_1 and c_2 in terms of a and b.

4. Show that positive constants a and b exist such that the formula

$$\int_0^\infty e^{-x} f(x)\,dx = \tfrac{1}{4}[af(b) + bf(a)]$$

is exact for all polynomials of degree ≤ 3.

5. Show that a positive constant c exists such that the formula

$$\int_{-\infty}^{\infty} e^{-x^2} f(x)\,dx = \frac{\sqrt{\pi}}{6}\,[f(-c) + 4f(0) + f(c)]$$

is exact for all polynomials of degree ≤ 5.

6. Let P_n be the interpolation polynomial of degree $\le n$ that agrees with f at $n + 1$ distinct points $x_0, x_1, \ldots, x_n$.
(a) Show that constants $A_0(n), A_1(n), \ldots, A_n(n)$ exist, depending only on the numbers $x_0, x_1, \ldots, x_n, a$, and b, and not on f, such that

$$\int_a^b P_n(x)\,dx = \sum_{k=0}^{n} A_k(n) f(x_k).$$

The numbers $A_k(n)$ are called *weights*. (They are sometimes called Christoffel numbers.)

(b) For a given set of distinct interpolation points and a given interval $[a, b]$, let $W_0(n)$, $W_1(n), \ldots, W_n(n)$ be $n + 1$ constants such that the formula

$$\int_a^b f(x)\,dx = \sum_{k=0}^{n} W_k(n) f(x_k)$$

is exact for all polynomials of degree $\le n$. Prove that

$$\sum_{k=0}^{n} x_k^r W_k(n) = \frac{b^{r+1} - a^{r+1}}{r+1} \quad \text{for} \quad r = 0, 1, \ldots, n.$$

This is a system of $n + 1$ linear equations that can be used to determine the weights. It can be shown that this system always has a unique solution. It can also be shown that for a suitable choice of interpolation points it is possible to make all the weights equal. When the weights are all equal the integration formula is called a Chebyshev integration formula. Exercises 2 and 3 give examples of Chebyshev integration formulas. The next exercise shows that for a proper choice of interpolation points the resulting integration formula is exact for all polynomials of degree $\le 2n + 1$.

7. In this exercise you may use properties of the Legendre polynomials stated in Sections 6.19 and 6.20. Let $x_0, x_1, \ldots, x_n$ be the zeros of the Legendre polynomial $P_{n+1}(x)$. These zeros are distinct and they all lie in the interval $[-1, 1]$. Let $f(x)$ be any polynomial in x of degree $\le 2n + 1$. Divide $f(x)$ by $P_{n+1}(x)$ and write

$$f(x) = P_{n+1}(x)Q(x) + R(x),$$

where the polynomials Q and R have degree $\le n$.

(a) Show that the polynomial R agrees with f at the zeros of P_{n+1} and that

$$\int_{-1}^{1} f(x)\,dx = \int_{-1}^{1} R(x)\,dx.$$

(b) Show that $n + 1$ weights $W_0(n), \ldots, W_n(n)$ exist (independent of f) such that

$$\int_{-1}^{1} f(x)\,dx = \sum_{k=0}^{n} W_k(n) f(x_k).$$

This gives an integration formula with $n + 1$ interpolation points that is exact for all polynomials of degree $\le 2n + 1$.

(c) Take $n = 2$ and show that the formula in part (b) becomes

$$\int_{-1}^{1} f(x)\,dx = \tfrac{5}{9} f(-\sqrt{\tfrac{3}{5}}) + \tfrac{8}{9} f(0) + \tfrac{5}{9} f(\sqrt{\tfrac{3}{5}}).$$

This is exact for all polynomials of degree ≤ 5.

(d) Introduce a suitable linear transformation and rewrite the formula in part (c) for an arbitrary interval $[a, b]$.

8. This exercise describes a method of Peano for deriving the error formula in Simpson's rule.

(a) Use integration by parts repeatedly to deduce the relation

$$\int u(t)v'''(t)\,dt = u(t)v''(t) - u'(t)v'(t) + u''(t)v(t) - \int g(t)\,dt,$$

where $g(t) = u'''(t)v(t)$.

(b) Assume φ has a continuous fourth derivative in the interval $[-1, 1]$. Take

$$v(t) = t(1 - t)^2/6, \qquad u(t) = \varphi(t) + \varphi(-t),$$

and use part (a) to show that

$$\int_{-1}^{1} \varphi(t)\,dt = \tfrac{1}{3}[\varphi(-1) + 4\varphi(0) + \varphi(1)] - \int_0^1 g(t)\,dt.$$

Then show that $\int_0^1 g(t)\,dt = \varphi^{(4)}(c)/90$ for some c in $[-1, 1]$.
(c) Introduce a suitable linear transformation to deduce Theorem 15.14 from the result of part (b).

9. (a) Let $a_1, a_2, \ldots, a_n$ be nonnegative numbers whose sum is 1. Assume φ is continuous on an interval $[a, b]$. If $c_1, c_2, \ldots, c_n$ are any n points in $[a, b]$ (not necessarily distinct), prove that there is at least one point c in $[a, b]$ such that

$$\sum_{k=1}^{n} a_k \varphi(c_k) = \varphi(c).$$

[*Hint:* Let M and m denote the maximum and minimum of φ on $[a, b]$ and use the inequality $m \le \varphi(c_k) \le M$.]

(b) Use part (a) and Theorem 15.14 to derive the extended form of Simpson's rule given in Theorem 15.15.

10. Compute $\log 2$ from the formula $\log 2 = \int_1^2 x^{-1}\,dx$ by using the extension of Simpson's rule with (a) $n = 2$; (b) $n = 5$. Give upper and lower bounds for the error in each case.

11. (a) Let $\varphi(t)$ be a polynomial in t of degree ≤ 3. Express $\varphi(t)$ by Newton's interpolation formula and integrate to deduce the formula

$$\int_0^3 \varphi(t)\,dt = \tfrac{3}{8}[\varphi(0) + 3\varphi(1) + 3\varphi(2) + \varphi(3)].$$

(b) Let P be the interpolation polynomial of degree ≤ 3 that agrees with f at the points a, $a + h$, $a + 2h$, $a + 3h$, where $h > 0$. Use part (a) to prove that

$$\int_a^{a+3h} P(x)\,dx = \frac{3h}{8}\,[f(a) + 3f(a + h) + 3f(a + 2h) + f(a + 3h)].$$

(c) Assume f has a continuous fourth derivative in $[a, b]$, and let $h = (b - a)/3$. Prove that

$$\int_a^b f(x)\,dx = \frac{b - a}{8}\,[f(a) + 3f(a + h) + 3f(a + 2h) + f(b)] - \frac{(b - a)^5}{6480} f^{(4)}(c)$$

for some c in $[a, b]$. This approximate integration formula is called *Cotes' rule*.
(d) Use Cotes' rule to compute $\log 2 = \int_1^2 x^{-1}\,dx$ and give upper and lower bounds for the error.

12. (a) Use the vector equation $\boldsymbol{r}(t) = a \sin t\, \boldsymbol{i} + b \cos t\, \boldsymbol{j}$, where $0 < b < a$, to show that the circumference L of an ellipse is given by the integral

$$L = 4a \int_0^{\pi/2} \sqrt{1 - k^2 \sin^2 t}\,dt,$$

where $k = \sqrt{a^2 - b^2}/a$.

(b) Show that Simpson's rule gives the formula

$$L = \frac{\pi}{3}\left[a + b + \sqrt{8(a^2 + b^2)}\right] - \frac{a\pi^5}{23040} f^{(4)}(c)$$

for some c in $[0, \pi/2]$, where $f(t) = \sqrt{1 - k^2 \sin^2 t}$.

15.22 The Euler summation formula

Let n be a positive integer. When the trapezoidal formula (Theorem 15.13) is applied to the interval $[0, n]$ it becomes

$$\int_0^n f(x)\,dx = \sum_{k=0}^{n-1} f(k) + \tfrac{1}{2}(f(n) - f(0)) - \frac{f''(c)n}{12}$$

for some c in $[0, n]$. If f is a quadratic polynomial then f'' is constant and hence $f''(c) = f''(0)$. In this case the formula can be rewritten in the form

$$\sum_{k=0}^{n} f(k) = \int_0^n f(x)\,dx + \frac{f(0) + f(n)}{2} + \frac{f''(0)n}{12}. \tag{15.45}$$

It is exact when f is any polynomial of degree ≤ 2.

Euler discovered a remarkable extension of this formula that is exact for any function with a continuous first derivative. It can be used to approximate integrals by sums, or, as is more often the case, to evaluate or estimate sums in terms of integrals. For this reason it is usually referred to as a "summation" formula rather than an integration formula. It can be stated as follows.

THEOREM 15.16. EULER'S SUMMATION FORMULA. *Assume f has a continuous derivative on $[0, n]$. Then we have*

$$\sum_{k=0}^{n} f(k) = \int_0^n f(x)\,dx + \frac{f(0) + f(n)}{2} + \int_0^n (x - [x] - \tfrac{1}{2})f'(x)\,dx, \tag{15.46}$$

where $[x]$ denotes the greatest integer $\leq x$.

Proof. Integration by parts gives us

$$\int_0^n (x - \tfrac{1}{2})f'(x)\,dx = (n - \tfrac{1}{2})f(n) + \tfrac{1}{2}f(0) - \int_0^n f(x)\,dx. \tag{15.47}$$

Now we consider the integral $\int_0^n [x]f'(x)\,dx$ and write it as a sum of integrals in each of which $[x]$ has a fixed value. Thus, we have

$$\begin{aligned}
\int_0^n [x]f'(x)\,dx &= \sum_{r=0}^{n-1} \int_r^{r+1} [x]f'(x)\,dx = \sum_{r=0}^{n-1} r \int_r^{r+1} f'(x)\,dx \\
&= \sum_{r=0}^{n-1} r(f(r+1) - f(r)) = \sum_{r=0}^{n-1} rf(r+1) - \sum_{r=0}^{n-1} rf(r) \\
&= -\sum_{r=0}^{n-1} f(r+1) + \sum_{r=0}^{n-1} (r+1)f(r+1) - \sum_{r=0}^{n-1} rf(r) \\
&= -\sum_{k=1}^{n} f(k) + nf(n) = -\sum_{k=0}^{n} f(k) + f(0) + nf(n).
\end{aligned}$$

Subtracting this from Equation (15.47) we obtain

$$\int_0^n (x - [x] - \tfrac{1}{2})f'(x)\,dx = \sum_{k=0}^{n} f(k) - \frac{f(0) + f(n)}{2} - \int_0^n f(x)\,dx\,,$$

which is equivalent to (15.46).

The last integral on the right of (15.46) can be written as

$$\int_0^n (x - [x] - \tfrac{1}{2})f'(x)\,dx = \int_0^n \varphi_1(x)f'(x)\,dx\,,$$

where φ_1 is the function defined by

$$\varphi_1(x) = \begin{cases} x - [x] - \frac{1}{2} & \text{if } x \text{ is not an integer,} \\ 0 & \text{if } x \text{ is an integer.} \end{cases}$$

The graph of φ_1 is shown in Figure 15.6(a). We note that $\varphi_1(x + 1) = \varphi_1(x)$, which means that φ_1 is periodic with period 1. Also, if $0 < x < 1$ we have $\varphi_1(x) = x - \frac{1}{2}$, so $\int_0^1 \varphi_1(t)\,dt = 0$.

Figure 15.6(b) shows the graph of φ_2, the indefinite integral of φ_1, given by

$$\varphi_2(x) = \int_0^x \varphi_1(t)\,dt\,.$$

It is easily verified that φ_2 is also periodic with period 1. Moreover, we have

$$\varphi_2(x) = \frac{x(x - 1)}{2} \qquad \text{if} \quad 0 \le x \le 1\,.$$

This shows that $-\frac{1}{8} \le \varphi_2(x) \le 0$ for all x. The strict inequalities $-\frac{1}{8} < \varphi_2(x) < 0$ hold except when x is an integer or half an integer.

The next theorem describes another version of Euler's summation formula in terms of the function φ_2.

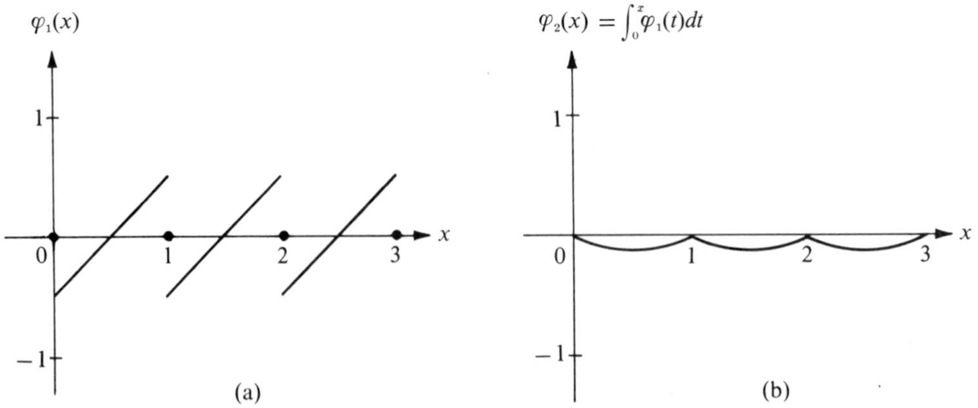

FIGURE 15.6 Graphs of the periodic functions φ_1 and φ_2.

THEOREM 15.17. *If f'' is continuous on $[0, n]$ we have*

$$\sum_{k=0}^{n} f(k) = \int_0^n f(x)\,dx + \frac{f(0) + f(n)}{2} - \int_0^n \varphi_2(x) f''(x)\,dx. \tag{15.48}$$

Proof. Since $\varphi_2'(x) = \varphi_1(x)$ at the points of continuity of φ_1, we have

$$\int_0^n \varphi_1(x) f'(x)\,dx = \int_0^n \varphi_2'(x) f'(x)\,dx.$$

Integration by parts gives us

$$\int_0^n \varphi_2'(x) f'(x)\,dx = \varphi_2(x) f'(x)\Big|_0^n - \int_0^n \varphi_2(x) f''(x)\,dx = -\int_0^n \varphi_2(x) f''(x)\,dx,$$

since $\varphi_2(n) = \varphi_2(0) = 0$. Using this in (15.46) we obtain (15.48).

Note: Although Theorems 15.16 and 15.17 refer to the interval $[0, n]$, both formulas are valid when 0 is replaced throughout by 1 or by any positive integer $< n$.

To illustrate the use of Euler's summation formula we derive the following formula for $\log n!$.

THEOREM 15.18. *For any positive integer n we have*

$$\log n! = (n + \tfrac{1}{2}) \log n - n + C + E(n), \tag{15.49}$$

where $0 < E(n) < 1/(8n)$ *and* $C = 1 + \int_1^\infty t^{-2} \varphi_2(t)\,dt$.

Proof. We take $f(x) = \log x$ and apply Theorem 15.17 to the interval $[1, n]$. This gives us

$$\sum_{k=1}^{n} \log k = \int_1^n \log x\,dx + \tfrac{1}{2} \log n + \int_1^n \frac{\varphi_2(x)}{x^2}\,dx.$$

Using the relation $\int \log t\,dt = t \log t - t$ we rewrite this equation as follows,

$$\log n! = (n + \tfrac{1}{2}) \log n - n + 1 + \int_1^n \frac{\varphi_2(t)}{t^2}\,dt. \tag{15.50}$$

Since $|\varphi_2(t)| \le \frac{1}{8}$ the improper integral $\int_1^\infty t^{-2} \varphi_2(t)\,dt$ converges absolutely and we can write

$$\int_1^n \frac{\varphi_2(t)}{t^2}\,dt = \int_1^\infty \frac{\varphi_2(t)}{t^2}\,dt - \int_n^\infty \frac{\varphi_2(t)}{t^2}\,dt.$$

Therefore Equation (15.50) becomes

$$\log n! = (n + \tfrac{1}{2}) \log n - n + C - \int_n^\infty \frac{\varphi_2(t)}{t^2}\,dt,$$

where $C = 1 + \int_1^\infty t^{-2}\varphi_2(t)\,dt$. Since we have $-\frac{1}{8} < \varphi_2(t) < 0$ except when t is an integer or half an integer, we obtain

$$0 < -\int_n^\infty \frac{\varphi_2(t)}{t^2}\,dt < \frac{1}{8n},$$

This proves (15.49), with $E(n) = -\int_n^\infty t^{-2}\varphi_2(t)\,dt$.

From Theorem 15.18 we can derive Stirling's formula for estimating $n!$.

THEOREM 15.19. STIRLING'S FORMULA. *If n is a positive integer we have*

$$\sqrt{2\pi}\, n^{n+1/2}e^{-n} < n! < \sqrt{2\pi}\, n^{n+1/2}e^{-n}\left(1 + \frac{1}{4n}\right).$$

Proof. Using Equation (15.49) and the inequalities for $E(n)$ we obtain

$$\exp\left((n + \tfrac{1}{2})\log n - n + C\right) < n! < \exp\left((n + \tfrac{1}{2})\log n - n + C + \frac{1}{8n}\right),$$

where $\exp(t) = e^t$. Using the relation $e^x < 1 + 2x$, with $x = 1/(8n)$, we can rewrite these inequalities as

$$An^{n+1/2}e^{-n} < n! < An^{n+1/2}e^{-n}\left(1 + \frac{1}{4n}\right), \tag{15.51}$$

where $A = e^C$. To complete the proof we need to show that $A = \sqrt{2\pi}$.

We shall deduce this from the inequality

$$\pi n \le \left(\frac{2^{2n}(n!)^2}{(2n)!}\right)^2 \le \frac{\pi(2n + 1)}{2}, \tag{15.52}$$

discovered by John Wallis (1616–1703). First we show how Wallis' inequality implies $A = \sqrt{2\pi}$; then we discuss the proof of (15.52).

If we let

$$A_n = \frac{n!}{n^{n+1/2}e^{-n}}$$

the inequality in (15.51) implies

$$A < A_n < A\left(1 + \frac{1}{4n}\right).$$

This shows that $A_n \to A$ as $n \to \infty$. In (15.52) we write $n! = n^{n+1/2}e^{-n}A_n$ to obtain

$$\pi n \le \left(\frac{2^{2n}n^{2n+1}e^{-2n}A_n^2}{(2n)^{2n+1/2}e^{-2n}A_{2n}}\right)^2 \le \frac{\pi(2n + 1)}{2},$$

which is equivalent to

$$\pi \le \frac{A_n^4}{2A_{2n}^2} \le \pi\,\frac{2n + 1}{2n}.$$

We let $n \to \infty$ in this last inequality. Since $A = e^C > 0$ we obtain

$$\pi \le \frac{A^4}{2A^2} \le \pi .$$

This shows that $A^2 = 2\pi$, so $A = \sqrt{2\pi}$, as asserted.

It remains to prove Wallis' inequality (15.52). For this purpose we introduce the numbers

$$I_n = \int_0^{\pi/2} \sin^n t \, dt ,$$

where n is any nonnegative integer. We note that $I_0 = \pi/2$ and $I_1 = 1$. For $0 \le t \le \pi/2$ we have $0 \le \sin t \le 1$; hence $0 \le \sin^{n+1} t \le \sin^n t$. This shows that the sequence $\{I_n\}$ is monotonic decreasing. Therefore we can write

$$\frac{1}{I_{2n}I_{2n-1}} \le \frac{1}{I_{2n}^2} \le \frac{1}{I_{2n}I_{2n+1}} . \tag{15.53}$$

Now we shall evaluate each member of this inequality; this will lead at once to Wallis' inequality.

Integration of the identity

$$\frac{d}{dt}(\cos t \sin^{n+1} t) = (n + 1) \sin^n t - (n + 2) \sin^{n+2} t$$

over the interval $[0, \pi/2]$ gives us

$$0 = (n + 1)I_n - (n + 2)I_{n+2} ,$$

or

$$I_{n+2} = \frac{n + 1}{n + 2} I_n . \tag{15.54}$$

Using this recursion formula with n replaced by $2k - 2$ we find

$$\frac{I_{2k}}{I_{2k-2}} = \frac{2k - 1}{2k} = \frac{2k(2k - 1)}{(2k)^2} .$$

Multiplying these equations for $k = 1, 2, \ldots, n$, we find

$$\prod_{k=1}^{n} \frac{I_{2k}}{I_{2k-2}} = \prod_{k=1}^{n} \frac{2k(2k - 1)}{(2k)^2} = \frac{(2n)!}{2^{2n}(n!)^2} .$$

The product on the left telescopes to I_{2n}/I_0. Since $I_0 = \pi/2$ we obtain

$$I_{2n} = \frac{(2n)!}{2^{2n}(n!)^2} \cdot \frac{\pi}{2} . \tag{15.55}$$

In a similar way we apply the recursion formula (15.54) with n replaced by $2k - 1$ and multiply the resulting equations for $k = 1, 2, \ldots, n$ to obtain

$$\prod_{k=1}^{n} \frac{I_{2k+1}}{I_{2k-1}} = \prod_{k=1}^{n} \frac{(2k)^2}{2k(2k+1)} = \frac{2^{2n}(n!)^2}{(2n+1)!} = \frac{1}{2n+1} \cdot \frac{\pi}{2} \cdot \frac{1}{I_{2n}} .$$

The product on the left telescopes to $I_{2n+1}/I_1 = I_{2n+1}$, so we get

(15.56)
$$I_{2n} I_{2n+1} = \frac{\pi}{2(2n+1)} .$$

Since $I_{2n+1} = 2nI_{2n-1}/(2n+1)$, Equation (15.56) implies

$$I_{2n} I_{2n-1} = \frac{\pi}{4n} .$$

We use this in (15.53), together with the two relations (15.55) and (15.56). Then we multiply by $\pi^2/4$ to obtain Wallis' inequality (15.52).

15.23 Exercises

1. If f is a polynomial of degree ≤ 2, show that Euler's summation formula (15.48) reduces to the trapezoidal formula, as expressed in Equation (15.45).
2. Euler's constant C is defined by the limit formula

$$C = \lim_{n \to \infty} \left(\sum_{k=1}^{n} \frac{1}{k} - \log n \right).$$

(See Section 10.17 in Volume I.) Use Euler's summation formula to prove that

$$\sum_{k=1}^{n} \frac{1}{k} = \log n + C + \frac{1}{2n} - \frac{E(n)}{n^2} ,$$

where $0 \leq E(n) \leq \frac{1}{8}$. Also, show that

$$C = 1 - \int_1^{\infty} \frac{t - [t]}{t^2} \, dt .$$

3. (a) If $s > 0, s \neq 1$, use Euler's summation formula to prove that

$$\sum_{k=1}^{n} \frac{1}{k^s} = \frac{n^{1-s}}{1-s} + C(s) + s \int_n^{\infty} \frac{t - [t]}{t^{s+1}} \, dt ,$$

where

$$C(s) = 1 + \frac{1}{s-1} - s \int_1^{\infty} \frac{t - [t]}{t^{s+1}} \, dt .$$

(b) If $s > 1$, show that $C(s) = \zeta(s)$, where ζ is the Riemann zeta function defined for $s > 1$ by the series

$$\zeta(s) = \sum_{k=1}^{\infty} \frac{1}{k^s}.$$

The series for $\zeta(s)$ diverges for $s \leq 1$. However, since the formula for $C(s)$ in part (a) is meaningful for $0 < s < 1$, it can be used to extend the definition of $\zeta(s)$ to the open interval $0 < s < 1$. Thus, for $s > 0$ and $s \neq 1$ we have the formula

$$\zeta(s) = 1 + \frac{1}{s-1} - s\int_1^{\infty} \frac{t - [t]}{t^{s+1}}\, dt.$$

This is a *theorem* if $s > 1$, and a *definition* if $0 < s < 1$.

In Exercises 4 through 6, φ_2 is the function introduced in Section 15.22.

4. (a) Use Euler's summation formula to prove that

$$\sum_{k=1}^{n} \log^2 k = (n + \tfrac{1}{2}) \log^2 n - 2n \log n + 2n - 2 + 2\int_1^n \varphi_2(x) \frac{\log x - 1}{x^2}\, dx.$$

(b) Use part (a) to deduce that for $n > e$ we have

$$\sum_{k=1}^{n} \log^2 k = (n + \tfrac{1}{2}) \log^2 n - 2n \log n + 2n + A - E(n),$$

where A is a constant and $0 < E(n) < \dfrac{\log n}{4n}$.

5. (a) Use Euler's summation formula to prove that

$$\sum_{k=1}^{n} \frac{\log k}{k} = \frac{1}{2}\log^2 n + \frac{1}{2}\frac{\log n}{n} - \int_1^n \frac{2\log x - 3}{x^3} \varphi_2(x)\, dx.$$

(b) Use part (a) to deduce that for $n > e^{3/2}$ we have

$$\sum_{k=1}^{n} \frac{\log k}{k} = \frac{1}{2}\log^2 n + \frac{1}{2}\frac{\log n}{n} + A - E(n),$$

where A is a constant and $0 < E(n) < \dfrac{\log n}{8n^2}$.

6. (a) If $n > 2$ use Euler's summation formula to prove that

$$\sum_{k=2}^{n} \frac{1}{k \log k}$$

$$= \log(\log n) + \frac{1}{2n \log n} + \frac{1}{4 \log 2} - \log(\log 2) - \int_2^n \varphi_2(x) \frac{2 + 3\log x + 2\log^2 x}{(x \log x)^3}\, dx.$$

(b) Use part (a) to deduce that for $n > 2$ we have

$$\sum_{k=2}^{n} \frac{1}{k \log k} = \log(\log n) + A + \frac{1}{2n \log n} - E(n),$$

where A is a constant and $0 < E(n) < \dfrac{1}{4n^2 \log n}$.

7. (a) If $a > 0$ and $p > 0$, use Euler's summation formula to prove that

$$\sum_{k=0}^{\infty} e^{-ak^p} = \frac{\Gamma\left(1 + \frac{1}{p}\right)}{a^{1/p}} + \frac{1}{2} - ap \int_0^{\infty} \varphi_1(x) x^{p-1} e^{-ax^p}\, dx,$$

where Γ is the gamma function.
(b) Use part (a) to deduce that

$$\sum_{k=0}^{\infty} e^{-ak^p} = \frac{\Gamma\left(1 + \frac{1}{p}\right)}{a^{1/p}} + \theta, \qquad \text{where} \quad 0 < \theta < 1.$$

8. Deduce the following limit relations with the aid of Stirling's formula and/or Wallis' inequality.

(a) $\displaystyle\lim_{n\to\infty} \frac{n}{(n!)^{1/n}} = e.$

(b) $\displaystyle\lim_{n\to\infty} \frac{(n!)^2 2^{2n}}{(2n)!\,\sqrt{n}} = \sqrt{\pi}.$

(c) $\displaystyle\lim_{n\to\infty} (-1)^n \binom{-\frac{1}{2}}{n} n = \frac{1}{\sqrt{\pi}}.$

9. Let $I_n = \int_0^{\pi/2} \sin^n t\, dt$, where n is a nonnegative integer. In Section 15.22 it was shown that the sequence $\{I_n\}$ satisfies the recursion formula

$$I_{n+2} = \frac{n+1}{n+2} I_n.$$

Let $f(n) = \frac{1}{2}\sqrt{\pi}\,\Gamma\left(\frac{n+1}{2}\right) \Big/ \Gamma\left(\frac{n}{2} + 1\right)$, where Γ is the gamma function.
(a) Use the functional equation $\Gamma(s+1) = s\Gamma(s)$ to show that

$$f(n+2) = \frac{n+1}{n+2} f(n).$$

(b) Use part (a) to deduce that

$$\int_0^{\pi/2} \sin^n t\, dt = \frac{\sqrt{\pi}}{2} \frac{\Gamma\left(\frac{n+1}{2}\right)}{\Gamma\left(\frac{n}{2} + 1\right)}.$$

SUGGESTED REFERENCES

This small list contains only a few books suggested for further reading on the general principles of numerical analysis. All these books contain further references to works of a more special nature. The list of tables given in Todd's survey (Reference 9 below) is especially recommended.

Books

1. A. D. Booth, *Numerical Methods*, Academic Press, New York, 1958; 3rd ed., Plenum Press, New York, 1966.
2. P. J. Davis, *Interpolation and Approximation*, Blaisdell, Waltham, Mass., 1963.
3. D. R. Hartree, *Numerical Analysis*, Oxford Univ. Press (Clarendon), London and New York, 1958.
4. F. B. Hildebrand, *Introduction to Numerical Analysis*, McGraw-Hill, New York, 1956.
5. A. S. Householder, *Principles of Numerical Analysis*, McGraw-Hill, New York, 1953.
6. W. E. Milne, *Numerical Calculus*, Princeton Univ. Press, Princeton, N.J., 1950.
7. J. B. Scarborough, *Numerical Mathematical Analysis*, Johns Hopkins Press, Baltimore, Md., 1958, 6th ed., 1966.
8. J. Todd, *Introduction to the Constructive Theory of Functions*, Academic Press, New York, 1963.
9. J. Todd (ed.), *Survey of Numerical Analysis*, McGraw-Hill, New York, 1962.

Tables

10. L. J. Comrie (ed.), *Chambers' Six-figure Mathematical Tables*, W. & R. Chambers, London and Edinburgh, 1949.
11. L. J. Comrie, "Interpolation and Allied Tables," 2nd rev. reprint from *Nautical Almanac* for 1937, H.M. Stationery Office, London, 1948.
12. A. J. Fletcher, J. C. P. Miller, and L. Rosenhead, *Index of Mathematical Tables*, McGraw-Hill, New York, 1946.
13. *Tables of Lagrangian Interpolation Coefficients*, Natl. Bur. Standards Columbia Press Series, Vol. 4., Columbia Univ. Press, New York, 1944.

ANSWERS TO EXERCISES

Chapter 1

1.5 Exercises (page 7)

1. Yes	8. Yes	15. Yes	22. Yes
2. Yes	9. Yes	16. Yes	23. No
3. Yes	10. Yes	17. Yes	24. Yes
4. Yes	11. No	18. Yes	25. No
5. No	12. Yes	19. Yes	26. Yes
6. Yes	13. Yes	20. Yes	27. Yes
7. Yes	14. No	21. Yes	28. Yes

31. (a) No (b) No (c) No (d) No

1.10 Exercises (page 13)

1. Yes; 2	5. Yes; 1	9. Yes; 1	13. Yes; n
2. Yes; 2	6. No	10. Yes; 1	14. Yes; n
3. Yes; 2	7. No	11. Yes; n	15. Yes; n
4. Yes; 2	8. No	12. Yes; n	16. Yes; n

17. Yes; $\dim = 1 + \frac{1}{2}n$ if n is even, $\frac{1}{2}(n + 1)$ if n is odd
18. Yes; $\dim = \frac{1}{2}n$ if n is even, $\frac{1}{2}(n + 1)$ if n is odd
19. Yes; $k + 1$
20. No
21. (a) $\dim = 3$ (b) $\dim = 3$ (c) $\dim = 2$ (d) $\dim = 2$
23. (a) If $a \neq 0$ and $b \neq 0$, set is independent, $\dim = 3$; if one of a or b is zero, set is dependent, $\dim = 2$ (b) Independent, $\dim = 2$ (c) If $a \neq 0$, independent, $\dim = 3$; if $a = 0$, dependent, $\dim = 2$ (d) Independent; $\dim = 3$ (e) Dependent; $\dim = 2$ (f) Independent; $\dim = 2$ (g) Independent; $\dim = 2$ (h) Dependent; $\dim = 2$ (i) Independent; $\dim = 2$ (j) Independent; $\dim = 2$

1.13 Exercises (page 20)

1. (a) No (b) No (c) No (d) No (e) Yes
8. (a) $\frac{1}{2}\sqrt{e^2 + 1}$ (b) $g(x) = b\left(x - \dfrac{e^2 + 1}{4}\right)$, b arbitrary
10. (b) $\dfrac{(n + 1)(2n + 1)}{6n}a + \dfrac{n + 1}{2}b$ (c) $g(t) = a\left(t - \dfrac{2n + 1}{3n}\right)$, a arbitrary
11. (c) 43 (d) $g(t) = a(1 - \frac{2}{3}t)$, a arbitrary
12. (a) No (b) No (c) No (d) No
13. (c) 1 (d) $e^2 - 1$
14. (c) $n!/2^{n+1}$

1.17 Exercises (page 30)

1. (a) and (b) $\frac{1}{3}\sqrt{3}\,(1, 1, 1)$, $\frac{1}{6}\sqrt{6}\,(1, -2, 1)$
2. (a) $\frac{1}{2}\sqrt{2}\,(1, 1, 0, 0)$, $\frac{1}{6}\sqrt{6}\,(-1, 1, 2, 0)$, $\frac{1}{6}\sqrt{3}\,(1, -1, 1, 3)$

   (b) $\frac{1}{3}\sqrt{3}\,(1, 1, 0, 1)$, $\frac{1}{\sqrt{42}}(1, -2, 6, 1)$

6. $\frac{2}{3} - \frac{1}{2}\log^2 3$
7. $e^2 - 1$
8. $\frac{1}{2}(e - e^{-1}) + \frac{3}{e}x$; $1 - 7e^{-2}$
9. $\pi - 2\sin x$
10. $\frac{3}{4} - \frac{1}{4}x$

Chapter 2

2.4 Exercises (page 35)

1. Linear; nullity 0, rank 2
2. Linear; nullity 0, rank 2
3. Linear; nullity 1, rank 1
4. Linear; nullity 1, rank 1
5. Nonlinear
6. Nonlinear
7. Nonlinear
8. Nonlinear
9. Linear; nullity 0, rank 2
10. Linear; nullity 0, rank 2
11. Linear; nullity 0, rank 2
12. Linear; nullity 0, rank 2
13. Nonlinear
14. Linear; nullity 0, rank 2
15. Nonlinear
16. Linear; nullity 0, rank 3
17. Linear; nullity 1, rank 2
18. Linear; nullity 0, rank 3
19. Nonlinear
20. Nonlinear
21. Nonlinear
22. Nonlinear
23. Linear; nullity 1, rank 2
24. Linear; nullity 0, rank $n + 1$
25. Linear; nullity 1, rank infinite
26. Linear; nullity infinite, rank 2
27. Linear; nullity 2, rank infinite
28. $N(T)$ is the set of constant sequences; $T(V)$ is the set of sequences with limit 0
29. (d) $\{1, \cos x, \sin x\}$ is a basis for $T(V)$; $\dim T(V) = 3$ (e) $N(T) = S$ (f) If $T(f) = cf$ with $c \neq 0$, then $c \in T(V)$ so we have $f(x) = c_1 + c_2 \cos x + c_3 \sin x$; if $c_1 = 0$, then $c = \pi$ and $f(x) = c_1 \cos x + c_2 \sin x$, where c_1, c_2 are not both zero but otherwise arbitrary; if $c_1 \neq 0$, then $c = 2\pi$ and $f(x) = c_1$, where c_1 is nonzero but otherwise arbitrary.

2.8 Exercises (page 42)

3. Yes; $x = v$, $y = u$
4. Yes; $x = u$, $y = -v$
5. No
6. No
7. No
8. Yes; $x = \log u$, $y = \log v$
9. No
10. Yes; $x = u - 1$, $y = v - 1$
11. Yes; $x = \frac{1}{2}(v + u)$, $y = \frac{1}{2}(v - u)$
12. Yes; $x = \frac{1}{3}(v + u)$, $y = \frac{1}{3}(2v - u)$
13. Yes; $x = w$, $y = v$, $z = u$
14. No
15. Yes; $x = u$, $y = \frac{1}{2}v$, $z = \frac{1}{3}w$
16. Yes; $x = u$, $y = v$, $z = w - u - v$
17. Yes; $x = u - 1$, $y = v - 1$, $z = w + 1$
18. Yes; $x = u - 1$, $y = v - 2$, $z = w - 3$
19. Yes; $x = u$, $y = v - u$, $z = w - v$
20. Yes; $x = \frac{1}{2}(u - v + w)$, $y = \frac{1}{2}(v - w + u)$, $z = \frac{1}{2}(w - u + v)$
25. $(S + T)^2 = S^2 + ST + TS + T^2$;
    $(S + T)^3 = S^3 + TS^2 + STS + S^2T + ST^2 + TST + T^2S + T^3$

26. (a) $(ST)(x, y, z) = (x + y + z, x + y, x)$; $(TS)(x, y, z) = (z, z + y, z + y + x)$; $(ST - TS)(x, y, z) = (x + y, x - z, -y - z)$; $S^2(x, y, z) = (x, y, z)$; $T^2(x, y, z) = (x, 2x + y, 3x + 2y + z)$; $(ST)^2(x, y, z) = (3x + 2y + z, 2x + 2y + z, x + y + z)$; $(TS)^2(x, y, z) = (x + y + z, x + 2y + 2z, x + 2y + 3z)$; $(ST - TS)^2 = (2x + y - z, x + 2y + z, -x + y + 2z)$
(b) $S^{-1}(u, v, w) = (w, v, u)$; $T^{-1}(u, v, w) = (u, v - u, w - v)$; $(ST)^{-1}(u, v, w) = (w, v - w, u - v)$; $(TS)^{-1}(u, v, w) = (w - v, v - u, u)$
(c) $(T - I)(x, y, z) = (0, x, x + y)$; $(T - I)^2(x, y, z) = (0, 0, x)$; $(T - I)^n(x, y, z) = (0, 0, 0)$ if $n \geq 3$

28. (a) $Dp(x) = 3 - 2x + 12x^2$; $Tp(x) = 3x - 2x^2 + 12x^3$; $(DT)p(x) = 3 - 4x + 36x^2$; $(TD)p(x) = -2x + 24x^2$; $(DT - TD)p(x) = 3 - 2x + 12x^2$; $(T^2D^2 - D^2T^2)p(x) = 8 - 192x$ (b) $p(x) = ax$, a an arbitrary scalar
(c) $p(x) = ax^2 + b$, a and b arbitrary scalars (d) All p in V

31. (a) $Rp(x) = 2$; $Sp(x) = 3 - x + x^2$; $Tp(x) = 2x + 3x^2 - x^3 + x^4$; $(ST)p(x) = 2 + 3x - x^2 + x^3$; $(TS)p(x) = 3x - x^2 + x^3$; $(TS)^2p(x) = 3x - x^2 + x^3$; $(T^2S^2)p(x) = -x^2 + x^3$; $(S^2T^2)p(x) = 2 + 3x - x^2 + x^3$; $(TRS)p(x) = 3x$; $(RST)p(x) = 2$ (b) $N(R) = \{p \mid p(0) = 0\}$; $R(V) = \{p \mid p \text{ is constant}\}$; $N(S) = \{p \mid p \text{ is constant}\}$; $S(V) = V$; $N(T) = \{O\}$; $T(V) = \{p \mid p(0) = 0\}$ (c) $T^{-1} = S$
(d) $(TS)^n = I - R$; $S^nT^n = I$

32. T is not one-to-one on V because it maps all constant sequences onto the same sequence

2.12 Exercises (page 50)

1. (a) The identity matrix $I = (\delta_{jk})$, where $\delta_{jk} = 1$ if $j = k$, and $\delta_{jk} = 0$ if $j \neq k$
(b) The zero matrix $O = (a_{jk})$ where each entry $a_{jk} = 0$
(c) The matrix $(c\delta_{jk})$, where (δ_{jk}) is the identity matrix of part (a)

2. (a) $\begin{bmatrix} 1 & 0 & 0 \\ 0 & 1 & 0 \end{bmatrix}$ (b) $\begin{bmatrix} 0 & 1 & 0 \\ 0 & 0 & 1 \end{bmatrix}$ (c) $\begin{bmatrix} 0 & 1 & 0 & 0 & 0 \\ 0 & 0 & 1 & 0 & 0 \\ 0 & 0 & 0 & 1 & 0 \end{bmatrix}$

3. (a) $-5\boldsymbol{i} + 7\boldsymbol{j}$, $9\boldsymbol{i} - 12\boldsymbol{j}$

(b) $\begin{bmatrix} 1 & 2 \\ 1 & -1 \end{bmatrix}$, $\begin{bmatrix} 3 & 0 \\ 0 & 3 \end{bmatrix}$ (c) $\begin{bmatrix} -\frac{7}{4} & -\frac{1}{4} \\ \frac{1}{4} & \frac{7}{4} \end{bmatrix}$, $\begin{bmatrix} 3 & 0 \\ 0 & 3 \end{bmatrix}$

4. $\begin{bmatrix} -2 & 0 \\ 0 & 2 \end{bmatrix}$, $\begin{bmatrix} 4 & 0 \\ 0 & 4 \end{bmatrix}$

5. (a) $3\boldsymbol{i} + 4\boldsymbol{j} + 4\boldsymbol{k}$; nullity 0, rank 3 (b) $\begin{bmatrix} -1 & -1 & 2 \\ 1 & -3 & 3 \\ -1 & -5 & 5 \end{bmatrix}$

6. $\begin{bmatrix} 2 & 0 & -2 \\ 1 & -1 & 1 \\ 2 & 1 & 0 \end{bmatrix}$

7. (a) $T(4\boldsymbol{i} - \boldsymbol{j} + \boldsymbol{k}) = (0, -2)$; nullity 1, rank 2 (b) $\begin{bmatrix} 0 & 1 & 1 \\ 0 & 1 & -1 \end{bmatrix}$

(c) $\begin{bmatrix} 0 & 1 & 3 \\ 0 & 0 & -2 \end{bmatrix}$ (d) $e_1 = \boldsymbol{j}$, $e_2 = \boldsymbol{k}$, $e_3 = \boldsymbol{i}$, $w_1 = (1, 1)$, $w_2 = (1, -1)$

8. (a) $(5, 0, -1)$; nullity 0, rank 2 (b) $\begin{bmatrix} 1 & -1 \\ 0 & 0 \\ 1 & 1 \end{bmatrix}$

(c) $e_1 = \boldsymbol{i}$, $e_2 = \boldsymbol{i} + \boldsymbol{j}$, $w_1 = (1, 0, 1)$, $w_2 = (0, 0, 2)$, $w_3 = (0, 1, 0)$

9. (a) $(-1, -3, -1)$; nullity 0, rank 2 (b) $\begin{bmatrix} 1 & 1 \\ 0 & 1 \\ 1 & 1 \end{bmatrix}$

(c) $e_1 = \boldsymbol{i}$, $e_2 = \boldsymbol{j} - \boldsymbol{i}$, $w_1 = (1, 0, 1)$, $w_2 = (0, 1, 0)$, $w_3 = (0, 0, 1)$

10. (a) $e_1 - e_2$; nullity 0, rank 2 (b) $\begin{bmatrix} 1 & 2 \\ 5 & 4 \end{bmatrix}$ (c) $a = 5$, $b = 4$

11. $\begin{bmatrix} 0 & -1 \\ 1 & 0 \end{bmatrix}$, $\begin{bmatrix} -1 & 0 \\ 0 & -1 \end{bmatrix}$

12. $\begin{bmatrix} 0 & 1 & 0 \\ 0 & 0 & 0 \\ 0 & 0 & 1 \end{bmatrix}$, $\begin{bmatrix} 0 & 0 & 0 \\ 0 & 0 & 0 \\ 0 & 0 & 1 \end{bmatrix}$

13. $\begin{bmatrix} 0 & 1 & 1 \\ 0 & 0 & -1 \\ 0 & 0 & 1 \end{bmatrix}$, $\begin{bmatrix} 0 & 0 & 0 \\ 0 & 0 & -1 \\ 0 & 0 & 1 \end{bmatrix}$

14. $\begin{bmatrix} 1 & 1 \\ 0 & 1 \end{bmatrix}$, $\begin{bmatrix} 1 & 2 \\ 0 & 1 \end{bmatrix}$

15. $\begin{bmatrix} 0 & -1 \\ 1 & 0 \end{bmatrix}$, $\begin{bmatrix} -1 & 0 \\ 0 & -1 \end{bmatrix}$

16. $\begin{bmatrix} 0 & -1 & 1 & 0 \\ 1 & 0 & 0 & 1 \\ 0 & 0 & 0 & -1 \\ 0 & 0 & 1 & 0 \end{bmatrix}$, $\begin{bmatrix} -1 & 0 & 0 & -2 \\ 0 & -1 & 2 & 0 \\ 0 & 0 & -1 & 0 \\ 0 & 0 & 0 & -1 \end{bmatrix}$

17. $\begin{bmatrix} 1 & -1 \\ 1 & 1 \end{bmatrix}$, $\begin{bmatrix} 0 & -2 \\ 2 & 0 \end{bmatrix}$

18. $\begin{bmatrix} 2 & -3 \\ 3 & 2 \end{bmatrix}$, $\begin{bmatrix} -5 & -12 \\ 12 & -5 \end{bmatrix}$

19. (a) $\begin{bmatrix} 0 & 0 & 0 & 0 \\ 0 & 1 & 0 & 0 \\ 0 & 0 & 2 & 0 \\ 0 & 0 & 0 & 3 \end{bmatrix}$ (b) $\begin{bmatrix} 0 & 1 & 0 & 0 \\ 0 & 0 & 4 & 0 \\ 0 & 0 & 0 & 9 \\ 0 & 0 & 0 & 0 \end{bmatrix}$ (c) $\begin{bmatrix} 0 & 0 & 0 & 0 \\ 0 & 0 & 2 & 0 \\ 0 & 0 & 0 & 6 \\ 0 & 0 & 0 & 0 \end{bmatrix}$

(d) $\begin{bmatrix} 0 & -1 & 0 & 0 \\ 0 & 0 & -2 & 0 \\ 0 & 0 & 0 & -3 \\ 0 & 0 & 0 & 0 \end{bmatrix}$ (e) $\begin{bmatrix} 0 & 0 & 0 & 0 \\ 0 & 1 & 0 & 0 \\ 0 & 0 & 4 & 0 \\ 0 & 0 & 0 & 9 \end{bmatrix}$ (f) $\begin{bmatrix} 0 & 0 & -8 & 0 \\ 0 & 0 & 0 & -48 \\ 0 & 0 & 0 & 0 \\ 0 & 0 & 0 & 0 \end{bmatrix}$

20. Choose $(x^3, x^2, x, 1)$ as a basis for V, and (x^2, x) as a basis for W. Then the matrix of TD is

$$\begin{bmatrix} 6 & 0 & 0 & 0 \\ 0 & 2 & 0 & 0 \end{bmatrix}$$

2.16 Exercises (page 57)

1. $B + C = \begin{bmatrix} 3 & 4 \\ 0 & 2 \\ 6 & -5 \end{bmatrix}, \quad AB = \begin{bmatrix} 15 & -14 \\ -15 & 14 \end{bmatrix}, \quad BA = \begin{bmatrix} -1 & 4 & -2 \\ -4 & 16 & -8 \\ 7 & -28 & 14 \end{bmatrix},$

$AC = \begin{bmatrix} 0 & 0 \\ 0 & 0 \end{bmatrix}, \quad CA = \begin{bmatrix} 0 & 0 & 0 \\ 2 & -8 & 4 \\ 4 & -16 & 8 \end{bmatrix}, \quad A(2B - 3C) = \begin{bmatrix} 30 & -28 \\ -30 & 28 \end{bmatrix}$

2. (a) $\begin{bmatrix} a & b \\ 0 & 0 \end{bmatrix}$, a and b arbitrary (b) $\begin{bmatrix} -2a & a \\ -2b & b \end{bmatrix}$, a and b arbitrary

3. (a) $a = 9, \quad b = 6, \quad c = 1, \quad d = 5$ (b) $a = 1, \quad b = 6, \quad c = 0, \quad d = -2$

4. (a) $\begin{bmatrix} -9 & -2 & -10 \\ 6 & 14 & 8 \\ -7 & 5 & -5 \end{bmatrix}$ (b) $\begin{bmatrix} -3 & 5 & -4 \\ 0 & 3 & 24 \\ 12 & -27 & 0 \end{bmatrix}$

6. $A^n = \begin{bmatrix} 1 & n \\ 0 & 1 \end{bmatrix}$

7. $A^n = \begin{bmatrix} \cos n\theta & -\sin n\theta \\ \sin n\theta & \cos n\theta \end{bmatrix}$

8. $A^n = \begin{bmatrix} 1 & n & \dfrac{n(n+1)}{2} \\ 0 & 1 & n \\ 0 & 0 & 1 \end{bmatrix}$

9. $\begin{bmatrix} 1 & 0 \\ -100 & 1 \end{bmatrix}$

10. $\begin{bmatrix} a & b \\ c & -a \end{bmatrix}$, where b and c are arbitrary, and a is any solution of the equation $a^2 = -bc$

11. (b) $\begin{bmatrix} a & 0 \\ 0 & a \end{bmatrix}$, where a is arbitrary

12. $\begin{bmatrix} 1 & 0 \\ 0 & 1 \end{bmatrix}$, $\begin{bmatrix} -1 & 0 \\ 0 & -1 \end{bmatrix}$, and $\begin{bmatrix} a & b \\ c & -a \end{bmatrix}$, where b and c are arbitrary and a is any solution of the equation $a^2 = 1 - bc$

13. $C = \begin{bmatrix} \frac{15}{2} & \frac{13}{2} \\ 8 & 7 \end{bmatrix}$, $D = \begin{bmatrix} \frac{33}{4} & \frac{19}{4} \\ \frac{43}{4} & \frac{25}{4} \end{bmatrix}$

14. (b) $(A + B)^2 = A^2 + AB + BA + B^2$; $(A + B)(A - B) = A^2 + BA - AB - B^2$
(c) For those which commute

2.20 Exercises (page 67)

1. $(x, y, z) = (\frac{8}{5}, -\frac{7}{5}, \frac{8}{5})$
2. No solution
3. $(x, y, z) = (1, -1, 0) + t(-3, 4, 1)$
4. $(x, y, z) = (1, -1, 0) + t(-3, 4, 1)$
5. $(x, y, z, u) = (1, 1, 0, 0) + t(1, 14, 5, 0)$
6. $(x, y, z, u) = (1, 8, 0, -4) + t(2, 7, 3, 0)$
7. $(x, y, z, u, v) = t_1(-1, 1, 0, 0, 0) + t_2(-1, 0, 3, -3, 1)$
8. $(x, y, z, u) = (1, 1, 1, -1) + t_1(-1, 3, 7, 0) + t_2(4, 9, 0, 7)$
9. $(x, y, z) = (\frac{4}{3}, \frac{2}{3}, 0) + t(5, 1, -3)$
10. (a) $(x, y, z, u) = (1, 6, 3, 0) + t_1(4, 11, 7, 0) + t_2(0, 0, 0, 1)$
(b) $(x, y, z, u) = (\frac{3}{11}, 4, \frac{19}{11}, 0) + t(4, -11, 7, 22)$

12. $\begin{bmatrix} -1 & 2 & 1 \\ 5 & -8 & -6 \\ -3 & 5 & 4 \end{bmatrix}$

13. $\begin{bmatrix} -\frac{5}{3} & \frac{2}{3} & \frac{4}{3} \\ -1 & 0 & 1 \\ \frac{7}{3} & -\frac{1}{3} & -\frac{5}{3} \end{bmatrix}$

14. $\begin{bmatrix} 14 & 8 & 3 \\ 8 & 5 & 2 \\ 3 & 2 & 1 \end{bmatrix}$

15. $\begin{bmatrix} 1 & -2 & 1 & 0 \\ 0 & 1 & -2 & 1 \\ 0 & 0 & 1 & -2 \\ 0 & 0 & 0 & 1 \end{bmatrix}$

16. $\begin{bmatrix} 0 & \frac{1}{2} & 0 & -1 & 0 & 1 \\ 1 & 0 & 0 & 0 & 0 & 0 \\ 0 & 0 & 0 & 1 & 0 & -1 \\ -3 & 0 & 1 & 0 & 0 & 0 \\ 0 & 0 & 0 & 0 & 0 & \frac{1}{2} \\ 9 & 0 & -3 & 0 & 1 & 0 \end{bmatrix}$

2.21 Miscellaneous exercises on matrices (page 68)

3. $P = \begin{bmatrix} 2 & 1 \\ 5 & -1 \end{bmatrix}$

4. $\begin{bmatrix} 0 & 0 \\ 0 & 0 \end{bmatrix}$, $\begin{bmatrix} 1 & 0 \\ 0 & 1 \end{bmatrix}$, and $\begin{bmatrix} a & b \\ c & 1-a \end{bmatrix}$, where b and c are arbitrary and a is any solution of the quadratic equation $a^2 - a + bc = 0$

10. (a) $\begin{bmatrix} 1 & 1 \\ -1 & 1 \end{bmatrix}$, $\begin{bmatrix} 1 & 1 \\ 1 & -1 \end{bmatrix}$, $\begin{bmatrix} -1 & 1 \\ 1 & 1 \end{bmatrix}$, $\begin{bmatrix} 1 & -1 \\ 1 & 1 \end{bmatrix}$, $\begin{bmatrix} -1 & -1 \\ 1 & -1 \end{bmatrix}$, $\begin{bmatrix} -1 & -1 \\ -1 & 1 \end{bmatrix}$, $\begin{bmatrix} 1 & -1 \\ -1 & -1 \end{bmatrix}$, $\begin{bmatrix} -1 & 1 \\ -1 & -1 \end{bmatrix}$.

Chapter 3

3.6 Exercises (page 79)

1. (a) 6 (b) 76 (c) $a^3 - 4a$
2. (a) 1 (b) 1 (c) 1
3. (b) $(b-a)(c-a)(c-b)(a+b+c)$ and $(b-a)(c-a)(c-b)(ab+ac+bc)$
4. (a) 8 (b) $(b-a)(c-a)(d-a)(c-b)(d-b)(d-c)$
   (c) $(b-a)(c-a)(d-a)(c-b)(d-b)(d-c)(a+b+c+d)$
   (d) $a(a^2-4)(a^2-16)$ (e) -160

7. $F' = \begin{vmatrix} f_1' & f_2' & f_3' \\ g_1 & g_2 & g_3 \\ h_1 & h_2 & h_3 \end{vmatrix} + \begin{vmatrix} f_1 & f_2 & f_3 \\ g_1' & g_2' & g_3' \\ h_1 & h_2 & h_3 \end{vmatrix} + \begin{vmatrix} f_1 & f_2 & f_3 \\ g_1 & g_2 & g_3 \\ h_1' & h_2' & h_3' \end{vmatrix}$

8. (b) If $F = \begin{vmatrix} f_1 & f_2 & f_3 \\ f_1' & f_2' & f_3' \\ f_1'' & f_2'' & f_3'' \end{vmatrix}$ then $F' = \begin{vmatrix} f_1 & f_2 & f_3 \\ f_1' & f_2' & f_3' \\ f_1''' & f_2''' & f_3''' \end{vmatrix}$

10 $\det A = 16$, $\det(A^{-1}) = \dfrac{1}{16}$, $A^{-1} = \begin{bmatrix} \frac{1}{2} & -\frac{3}{4} & \frac{1}{8} & \frac{1}{16} \\ 0 & \frac{1}{2} & -\frac{3}{4} & \frac{1}{8} \\ 0 & 0 & \frac{1}{2} & -\frac{3}{4} \\ 0 & 0 & 0 & \frac{1}{2} \end{bmatrix}$

3.11 Exercises (page 85)

6. $\det A = (\det B)(\det D)$
7. (a) Independent (b) Independent (c) Dependent

3.17 Exercises (page 94)

1. (a) $\begin{bmatrix} 4 & -3 \\ -2 & 1 \end{bmatrix}$ (b) $\begin{bmatrix} 2 & -1 & 1 \\ -6 & 3 & 5 \\ -4 & -2 & 2 \end{bmatrix}$ (c) $\begin{bmatrix} 109 & 113 & -41 & -13 \\ -40 & -92 & 74 & 16 \\ -41 & -79 & 7 & 47 \\ -50 & 38 & 16 & 20 \end{bmatrix}$

2. (a) $-\frac{1}{2}\begin{bmatrix} 4 & -2 \\ -3 & 1 \end{bmatrix}$ (b) $\frac{1}{8}\begin{bmatrix} 2 & -6 & -4 \\ -1 & 3 & -2 \\ 1 & 5 & 2 \end{bmatrix}$ (c) $\frac{1}{306}\begin{bmatrix} 109 & -40 & -41 & -50 \\ 113 & -92 & -79 & 38 \\ -41 & 74 & 7 & 16 \\ -13 & 16 & 47 & 20 \end{bmatrix}$

3. (a) $\lambda = 2, \quad \lambda = -3$ (b) $\lambda = 0, \quad \lambda = \pm 3$ (c) $\lambda = 3, \quad \lambda = \pm i$

5. (a) $x = 0, \quad y = 1, \quad z = 2$ (b) $x = 1, \quad y = 1, \quad z = -1$

6. (b) $\det \begin{bmatrix} x - x_1 & y - y_1 & z - z_1 \\ x_2 - x_1 & y_2 - y_1 & z_2 - z_1 \\ x_3 - x_1 & y_3 - y_1 & z_3 - z_1 \end{bmatrix} = 0; \quad \det \begin{bmatrix} x & y & z & 1 \\ x_1 & y_1 & z_1 & 1 \\ x_2 & y_2 & z_2 & 1 \\ x_3 & y_3 & z_3 & 1 \end{bmatrix} = 0$

(c) $\det \begin{bmatrix} (x - x_1)^2 + (y - y_1)^2 & (x - x_1) & (y - y_1) \\ (x_2 - x_1)^2 + (y_2 - y_1)^2 & (x_2 - x_1) & (y_2 - y_1) \\ (x_3 - x_1)^2 + (y_3 - y_1)^2 & (x_3 - x_1) & (y_3 - y_1) \end{bmatrix} = 0;$

$$\det \begin{bmatrix} x^2 + y^2 & x & y & 1 \\ x_1^2 + y_1^2 & x_1 & y_1 & 1 \\ x_2^2 + y_2^2 & x_2 & y_2 & 1 \\ x_3^2 + y_3^2 & x_3 & y_3 & 1 \end{bmatrix} = 0$$

Chapter 4

4.4 Exercises (page 101)

5. Eigenfunctions: $f(t) = Ct^{\lambda}$, where $C \neq 0$
6. The nonzero constant polynomials
7. Eigenfunctions: $f(t) = Ce^{t/\lambda}$, where $C \neq 0$
8. Eigenfunctions: $f(t) = Ce^{\frac{1}{2}t^2/\lambda}$, where $C \neq 0$
10. Eigenvectors belonging to $\lambda = 0$ are all constant sequences with limit $a \neq 0$. Eigenvectors belonging to $\lambda = -1$ are all nonconstant sequences with limit $a = 0$

4.8 Exercises (page 107)

	Eigenvalue	*Eigenvectors*	dim $E(\lambda)$
1. (a)	1, 1	$(a, b) \neq (0, 0)$	2
(b)	1, 1	$t(1, 0), t \neq 0$	1
(c)	1, 1	$t(0, 1), t \neq 0$	1
(d)	2	$t(1, 1)\ t \neq 0$	1
	0	$t(1, -1), t \neq 0$	1
2.	$1 + \sqrt{ab}$	$t(\sqrt{a}, \sqrt{b}), t \neq 0$	1
	$1 - \sqrt{ab}$	$t(\sqrt{a}, -\sqrt{b}), t \neq 0$	1

3. If the field of scalars is the set of real numbers **R**, then real eigenvalues exist only when $\sin\theta = 0$, in which case there are two equal eigenvalues, $\lambda_1 = \lambda_2 = \cos\theta$, where $\cos\theta = 1$ or -1. In this case every nonzero vector is an eigenvector, so $\dim E(\lambda_1) = \dim E(\lambda_2) = 2$.
If the field of scalars is the set of complex numbers **C**, then the eigenvalues are $\lambda_1 = \cos\theta + i\sin\theta$, $\lambda_2 = \cos\theta - i\sin\theta$. If $\sin\theta = 0$ these are real and equal. If $\sin\theta \neq 0$ they are distinct complex conjugates; the eigenvectors belonging to λ_1 are $t(i, 1)$, $t \neq 0$; those belonging to λ_2 are $t(1, i)$, $t \neq 0$; $\dim E(\lambda_1) = \dim E(\lambda_2) = 1$.

4. $\begin{bmatrix} a & b \\ c & -a \end{bmatrix}$, where b and c are arbitrary and a is any solution of the equation $a^2 = 1 - bc$.

5. Let $A = \begin{bmatrix} a & b \\ c & d \end{bmatrix}$, and let $\Delta = (a - d)^2 + 4bc$. The eigenvalues are real and distinct if $\Delta > 0$, real and equal if $\Delta = 0$, complex conjugates if $\Delta < 0$.

6. $a = b = c = d = e = f = 1$.

	Eigenvalue	*Eigenvectors*	$\dim E(\lambda)$
7. (a)	1, 1, 1	$t(0, 0, 1)$, $t \neq 0$	1
(b)	1	$t(1, -1, 0)$, $t \neq 0$	1
	2	$t(3, 3, -1)$, $t \neq 0$	1
	21	$t(1, 1, 6)$, $t \neq 0$	1
(c)	1	$t(3, -1, 3)$, $t \neq 0$	1
	2, 2	$t(2, 2, -1)$, $t \neq 0$	1

8. 1, 1, -1, -1 for each matrix

4.10 Exercises (page 112)

2. (a) Eigenvalues 1, 3; $C = \begin{bmatrix} -2c & 0 \\ c & d \end{bmatrix}$, where $cd \neq 0$

(b) Eigenvalues 6, -1; $C = \begin{bmatrix} 2a & b \\ 5a & -b \end{bmatrix}$, where $ab \neq 0$

(c) Eigenvalues 3, 3; if a nonsingular C exists then $C^{-1}AC = 3I$, so $AC = 3C$, $A = 3I$

(d) Eigenvalues 1, 1; if a nonsingular C exists then $C^{-1}AC = I$, so $AC = C$, $A = I$

3. $C = A^{-1}B$.

4. (a) Eigenvalues 1, 1, -1; eigenvectors $(1, 0, 1)$, $(0, 1, 0)$, $(1, 0, -1)$;

$$C = \begin{bmatrix} 1 & 0 & 1 \\ 0 & 1 & 0 \\ 1 & 0 & -1 \end{bmatrix}$$

(b) Eigenvalues 2, 2, 1; eigenvectors $(1, 0, -1)$, $(0, 1, -1)$, $(1, -1, 1)$;

$$C = \begin{bmatrix} 1 & 0 & 1 \\ 0 & 1 & -1 \\ -1 & -1 & 1 \end{bmatrix}$$

5. (a) Eigenvalues 2, 2; eigenvectors $t(1,0)$, $t \neq 0$. If $C = \begin{bmatrix} a & b \\ -b & 0 \end{bmatrix}$, $b \neq 0$, then

$$C^{-1}AC = \begin{bmatrix} 2 & 0 \\ 1 & 2 \end{bmatrix}$$

(b) Eigenvalues 3, 3; eigenvectors $t(1,1)$, $t \neq 0$. If $C = \begin{bmatrix} a & b \\ a+b & b \end{bmatrix}$, $b \neq 0$, then

$$C^{-1}AC = \begin{bmatrix} 3 & 0 \\ 1 & 3 \end{bmatrix}$$

6. Eigenvalues 1, 1, 1; eigenvectors $t(1, -1, -1)$, $t \neq 0$

Chapter 5

5.5 Exercises (page 118)

3. (b) T^n is Hermitian if n is even, skew-Hermitian if n is odd
7. (a) Symmetric (b) Neither (c) Symmetric (d) Symmetric
9. (d) $Q(x + ty) = Q(x) + t\bar{t}Q(y) + \bar{t}(T(x), y) + t(T(y), x)$

5.11 Exercises (page 124)

1. (a) Symmetric and Hermitian
   (b) None of the four types
   (c) Skew-symmetric
   (d) Skew-symmetric and skew-Hermitian

4. (b) $\begin{bmatrix} \cos\theta & \sin\theta \\ \sin\theta & -\cos\theta \end{bmatrix}$

5. Eigenvalues $\lambda_1 = 0$, $\lambda_2 = 25$; orthonormal eigenvectors $u_1 = \frac{1}{5}(4, -3)$, $u_2 = \frac{1}{5}(3, 4)$.

$$C = \frac{1}{5}\begin{bmatrix} 4 & 3 \\ -3 & 4 \end{bmatrix}$$

6. Eigenvalues $\lambda_1 = 2i$, $\lambda_2 = -2i$; orthonormal eigenvectors

$$u_1 = \frac{1}{\sqrt{2}}(1, -i), \quad u_2 = \frac{1}{\sqrt{2}}(1, i). \qquad C = \frac{1}{\sqrt{2}}\begin{bmatrix} 1 & 1 \\ -i & i \end{bmatrix}$$

7. Eigenvalues $\lambda_1 = 1$, $\lambda_2 = 3$, $\lambda_3 = -4$; orthonormal eigenvectors

$$u_1 = \frac{1}{\sqrt{10}}(1, 0, 3), \quad u_2 = \frac{1}{\sqrt{14}}(3, 2, -1), \quad u_3 = \frac{1}{\sqrt{35}}(3, -5, -1).$$

$$C = \begin{bmatrix} \frac{1}{\sqrt{10}} & \frac{3}{\sqrt{14}} & \frac{3}{\sqrt{35}} \\ 0 & \frac{2}{\sqrt{14}} & \frac{-5}{\sqrt{35}} \\ \frac{3}{\sqrt{10}} & \frac{-1}{\sqrt{14}} & \frac{-1}{\sqrt{35}} \end{bmatrix}$$

8. Eigenvalues $\lambda_1 = 1$, $\lambda_2 = 6$, $\lambda_3 = -4$; orthonormal eigenvectors

$$u_1 = \tfrac{1}{5}(0, 4, -3), \quad u_2 = \frac{1}{\sqrt{50}}(5, 3, 4), \quad u_3 = \frac{1}{\sqrt{50}}(5, -3, -4).$$

$$C = \frac{1}{\sqrt{50}}\begin{bmatrix} 0 & 5 & 5 \\ 4\sqrt{2} & 3 & -3 \\ -3\sqrt{2} & 4 & -4 \end{bmatrix}$$

9. (a), (b), (c) are unitary; (b), (c) are orthogonal

11. (a) Eigenvalues $\lambda_1 = ia$, $\lambda_2 = -ia$; orthonormal eigenvectors

$$u_1 = \frac{1}{\sqrt{2}}(1, i), \quad u_2 = \frac{1}{\sqrt{2}}(1, -i). \qquad \text{(b)} \quad C = \frac{1}{\sqrt{2}}\begin{bmatrix} 1 & 1 \\ i & -i \end{bmatrix}$$

5.15 Exercises (page 134)

1. (a) $A = \begin{bmatrix} 4 & 2 \\ 2 & 1 \end{bmatrix}$ (b) $\lambda_1 = 0$, $\lambda_2 = 5$ (c) $u_1 = \frac{1}{\sqrt{5}}(1, -2)$, $u_2 = \frac{1}{\sqrt{5}}(2, 1)$

   (d) $C = \frac{1}{\sqrt{5}}\begin{bmatrix} 1 & 2 \\ -2 & 1 \end{bmatrix}$

2. (a) $A = \begin{bmatrix} 0 & \frac{1}{2} \\ \frac{1}{2} & 0 \end{bmatrix}$ (b) $\lambda_1 = \frac{1}{2}$, $\lambda_2 = -\frac{1}{2}$

   (c) $u_1 = \frac{1}{\sqrt{2}}(1, 1)$, $u_2 = \frac{1}{\sqrt{2}}(1, -1)$ (d) $C = \frac{1}{\sqrt{2}}\begin{bmatrix} 1 & 1 \\ 1 & -1 \end{bmatrix}$

3. (a) $A = \begin{bmatrix} 1 & 1 \\ 1 & -1 \end{bmatrix}$ (b) $\lambda_1 = \sqrt{2}$, $\lambda_2 = -\sqrt{2}$

   (c) $u_1 = t(1 + \sqrt{2}, 1)$, $u_2 = t(-1, 1 + \sqrt{2})$, where $t = 1/\sqrt{4 + 2\sqrt{2}}$

   (d) $C = t\begin{bmatrix} 1 + \sqrt{2} & -1 \\ 1 & 1 + \sqrt{2} \end{bmatrix}$, where $t = 1/\sqrt{4 + 2\sqrt{2}}$

4. (a) $A = \begin{bmatrix} 34 & -12 \\ -12 & 41 \end{bmatrix}$ (b) $\lambda_1 = 50$, $\lambda_2 = 25$

   (c) $u_1 = \frac{1}{5}(3, -4)$, $u_2 = \frac{1}{5}(4, 3)$ (d) $C = \frac{1}{5}\begin{bmatrix} 3 & 4 \\ -4 & 3 \end{bmatrix}$

5. (a) $A = \begin{bmatrix} 1 & \frac{1}{2} & \frac{1}{2} \\ \frac{1}{2} & 0 & \frac{1}{2} \\ \frac{1}{2} & \frac{1}{2} & 0 \end{bmatrix}$ (b) $\lambda_1 = 0$, $\lambda_2 = \frac{3}{2}$, $\lambda_3 = -\frac{1}{2}$

(c) $u_1 = \frac{1}{\sqrt{3}}(1, -1, -1)$, $u_2 = \frac{1}{\sqrt{6}}(2, 1, 1)$, $u_3 = \frac{1}{\sqrt{2}}(0, 1, -1)$

(c) $C = \frac{1}{\sqrt{6}}\begin{bmatrix} \sqrt{2} & 2 & 0 \\ -\sqrt{2} & 1 & \sqrt{3} \\ -\sqrt{2} & 1 & -\sqrt{3} \end{bmatrix}$

6. (a) $A = \begin{bmatrix} 2 & 0 & 2 \\ 0 & 1 & 0 \\ 2 & 0 & -1 \end{bmatrix}$ (b) $\lambda_1 = 1$, $\lambda_2 = 3$, $\lambda_3 = -2$

(c) $u_1 = (0, 1, 0)$, $u_2 = \frac{1}{\sqrt{5}}(2, 0, 1)$, $u_3 = \frac{1}{\sqrt{5}}(1, 0, -2)$

(d) $C = \frac{1}{\sqrt{5}}\begin{bmatrix} 0 & 2 & 1 \\ \sqrt{5} & 0 & 0 \\ 0 & 1 & -2 \end{bmatrix}$

7. (a) $A = \begin{bmatrix} 3 & 2 & 4 \\ 2 & 0 & 2 \\ 4 & 2 & 3 \end{bmatrix}$ (b) $\lambda_1 = \lambda_2 = -1$, $\lambda_3 = 8$

(c) $u_1 = \frac{1}{\sqrt{2}}(1, 0, -1)$, $u_2 = \frac{1}{3\sqrt{2}}(-1, 4, -1)$, $u_3 = \frac{1}{3}(2, 1, 2)$

(d) $C = \frac{1}{3\sqrt{2}}\begin{bmatrix} 3 & -1 & 2\sqrt{2} \\ 0 & 4 & \sqrt{2} \\ -3 & -1 & 2\sqrt{2} \end{bmatrix}$

8. Ellipse; center at $(0, 0)$
9. Hyperbola; center at $(-\frac{5}{2}, -\frac{5}{2})$
10. Parabola; vertex at $(\frac{5}{16}, -\frac{15}{16})$
11. Ellipse; center at $(0, 0)$
12. Ellipse; center at $(6, -4)$
13. Parabola; vertex at $(\frac{2}{25}, \frac{11}{25})$
14. Ellipse; center at $(0, 0)$
15. Parabola; vertex at $(\frac{3}{4}, \frac{3}{4})$
16. Ellipse; center at $(-1, \frac{1}{2})$
17. Hyperbola; center at $(0, 0)$
18. Hyperbola; center at $(-1, 2)$
19. -14

5.20 Exercises (page 141)

8. $a = \pm\frac{1}{3}\sqrt{3}$
13. (a), (b), and (e)

Chapter 6

6.3 Exercises (page 144)

1. $y = e^{3x} - e^{2x}$
2. $y = \frac{2}{3}x^2 + \frac{1}{3}x^5$
3. $y = 4\cos x - 2\cos^2 x$
4. Four times the initial amount
5. $f(x) = Cx^n$, or $f(x) = Cx^{1/n}$
6. (b) $y = e^{4x} - e^{-x^3/3}$
7. $y = c_1e^{2x} + c_2e^{-2x}$
8. $y = c_1\cos 2x + c_2\sin 2x$
9. $y = e^x(c_1\cos 2x + c_2\sin 2x)$
10. $y = e^{-x}(c_1 + c_2x)$

11. $k = n^2\pi^2$; $f_k(x) = C \sin n\pi x$ $(n = 1, 2, 3, \ldots)$

13. (a) $y'' - y = 0$
(b) $y'' - 4y' + 4y = 0$
(c) $y'' + y' + \frac{5}{4}y = 0$
(d) $y'' + 4y = 0$
(e) $y'' - y = 0$

14. $y = \frac{1}{3}\sqrt{6}$, $y'' = -12y = -4\sqrt{6}$

6.9 Exercises (page 154)

1. $y = c_1 + c_2e^{-x} + c_3e^{3x}$
2. $y = c_1 + c_2e^{x} + c_3e^{-x}$
3. $y = c_1 + (c_2 + c_3x)e^{-2x}$
4. $y = (c_1 + c_2x + c_3x^2)e^{x}$
5. $y = (c_1 + c_2x + c_3x^2 + c_4x^3)e^{-x}$
6. $y = c_1e^{2x} + c_2e^{-2x} + c_3 \cos 2x + c_4 \sin 2x$
7. $y = e^{\sqrt{2}x}(c_1 \cos \sqrt{2}x + c_2 \sin \sqrt{2}x) + e^{-\sqrt{2}x}(c_3 \cos \sqrt{2}x + c_4 \sin \sqrt{2}x)$
8. $y = c_1e^{x} + e^{-x/2}(c_2 \cos \frac{1}{2}\sqrt{3}x + c_3 \sin \frac{1}{2}\sqrt{3}x)$
9. $y = e^{-x}[(c_1 + c_2x) \cos x + (c_3 + c_4x) \sin x]$
10. $y = (c_1 + c_2x) \cos x + (c_3 + c_4x) \sin x$
11. $y = c_1 + c_2x + (c_3 + c_4x) \cos \sqrt{2}x + (c_5 + c_6x) \sin \sqrt{2}x$
12. $y = c_1 + c_2x + (c_3 + c_4x) \cos 2x + (c_5 + c_6x) \sin 2x$
13. $f(x) = \dfrac{1}{2m^2}(e^{mx} - \cos mx - \sin mx)$
15. (a) $y^{(4)} - 5y'' + 4y = 0$
(b) $y''' + 6y'' + 12y' + 8y = 0$
(c) $y^{(4)} - 2y''' + y'' = 0$
(d) $y^{(4)} - 2y''' + y'' = 0$
(e) $y^{(5)} - 2y^{(4)} + y''' = 0$
(f) $y^{(4)} + 8y''' + 33y'' + 68y' + 52y = 0$
(g) $y^{(4)} - 2y'' + y = 0$
(h) $y^{(6)} + 4y'' = 0$

6.15 Exercises (page 166)

1. $y_1 = -2x - x^2 - \frac{1}{3}x^3$
2. $y_1 = \frac{1}{4}xe^{2x}$
3. $y_1 = (x - \frac{4}{3})e^{x}$
4. $y_1 = \frac{1}{3} \sin x$
5. $y_1 = \frac{1}{2}x^2e^{x} + e^{2x}$
6. $y_1 = \frac{1}{2}xe^{x}$
7. $y_1 = x \cosh x$
8. $y_1 = \frac{1}{24}x^4e^{-x}$
9. $50y_1 = (11 - 5x)e^{x} \sin 2x + (2 - 10x)e^{x} \cos 2x$
10. $y_1 = -(\frac{5}{8}x + \frac{3}{8}x^2 + \frac{1}{12}x^3)e^{-x}$
12. $y_1 = \dfrac{x^m e^{\alpha x}}{p_A^{(m)}(\alpha)}$
15. (b) $2D$ (c) $3D^2$ (d) nD^{n-1}
16. $y = Ae^{x} + Be^{-x} + \dfrac{1}{2}e^{x}\displaystyle\int \frac{e^{-x}}{x}\,dx - \frac{1}{2}e^{-x}\int \frac{e^{x}}{x}\,dx$
17. $y = (A + \frac{1}{2}x) \sin 2x + (B + \frac{1}{4} \log |\cos 2x|) \cos 2x$
18. $y = Ae^{x} + Be^{-x} + \frac{1}{2} \sec x$
19. $y = (A + Bx)e^{x} + e^{e^{x}} - xe^{x}\displaystyle\int e^{e^{x}}\,dx + e^{x}\int xe^{e^{x}}\,dx$

20. $y = -\frac{1}{8}\log|x| + \frac{1}{3}e^{x}\int \frac{e^{-x}}{x}\,dx - \frac{1}{4}e^{2x}\int \frac{e^{-2x}}{x}\,dx$

$\quad + \frac{1}{24}e^{4x}\int \frac{e^{-4x}}{x}\,dx + Ae^{x} + Be^{2x} + Ce^{4x}$

6.16 Miscellaneous exercises on linear differential equations (page 167)

1. $u(x) = 6(e^{4x} - e^{-x})/5$; $v(x) = e^{x} - e^{-5x}$
2. $u(x) = \frac{1}{2}e^{2x-\pi}\sin 5x$; $v(x) = \frac{5}{6}e^{-2x-\pi}\sin 3x$
3. $u(x) = e^{-x^2}$; $Q(x) = 4x^2 + 2$
5. $y = (A + Bx^3)e^{x} + (x^2 - 2x + 2)e^{2x}$
6. $y = Ae^{4x}\int e^{-4x-x^3/3}\,dx + Be^{4x}$
7. $y = Ax^{1/2} + Bx^{-1/2}$
8. $y = Ae^{x} + Bx^2e^{-x} - x$
9. $y = A(x^2 - 2) + B/x$
10. $y = x^{-2}[A + B(x-1)^3 + \frac{1}{9}x^3 + \frac{2}{3}x^2 - \frac{7}{6}x + \frac{1}{2} - (x-1)^3\log|x-1|]$
11. $a = 1, -1$; $y = [Ae^{g(x)} + Be^{-g(x)}]/x$

6.21 Exercises (page 177)

2. $f(x) = u_1(x)\ (\alpha = 1)$
3. (a) $A = (a-b)/2, \quad B = (a+b)/2$

   (b) $\frac{d}{dt}\left[(t^2 - 1)\frac{dy}{dt}\right] - \alpha(\alpha+1)y = 0$, where $\alpha = 1$ or -2, and $x = (t+1)/2$
4. $u_1(x) = 1 + \sum_{m=1}^{\infty} (-1)^m 2^m \frac{\alpha(\alpha-2)\cdots(\alpha-2m+2)}{(2m)!}x^{2m}$ for all x;

   $u_2(x) = x + \sum_{m=1}^{\infty} (-1)^m 2^m \frac{(\alpha-1)(\alpha-3)\cdots(\alpha-2m+1)}{(2m+1)!}x^{2m+1}$ for all x
5. $u_1(x) = 1 + \sum_{m=1}^{\infty} \frac{(-1)^m}{(3m+2)(3m-1)\cdots 8\cdot 5}x^{3m}$ for all x;

   $u_2(x) = x^{-2}\left(1 + \sum_{n=1}^{\infty} \frac{(-1)^n}{3^n n!}x^{3n}\right)$ for all $x \neq 0$
6. $y = x^2\left(\frac{1}{6} + \sum_{n=1}^{\infty} \frac{(\alpha-2)(\alpha-3)\cdots(\alpha-n-1)}{n!(n+3)!}x^{n}\right)$ for all x
11. (b) $f(x) = \frac{1}{5}P_0(x) + \frac{4}{7}P_2(x) + \frac{8}{35}P_4(x)$
15. (b) $\frac{2n}{4n^2 - 1}$

6.24 Exercises (page 188)

5. $J_{-3/2}(x) = -\left(\frac{2}{\pi x}\right)^{1/2}\left(\frac{\cos x}{x} + \sin x\right)$

9. (a) $y = x^{1/2}[c_1J_{1/3}(\frac{2}{3}x^{3/2}) + c_2J_{-1/3}(\frac{2}{3}x^{3/2})]$
(b) $y = x^{1/2}[c_1J_{1/4}(\frac{1}{2}x^2) + c_2J_{-1/4}(\frac{1}{2}x^2)]$
(c) $y = x^{1/2}[c_1J_\alpha(2\alpha x^{1+m/2}) + c_2J_{-\alpha}(2\alpha x^{1+m/2})]$, where $\alpha = 1/(m + 2)$, provided that $1/(m + 2)$ is not an integer; otherwise replace the appropriate J by K
(d) $y = x^{1/2}[c_1J_\alpha(\frac{1}{2}x^2) + c_2J_{-\alpha}(\frac{1}{2}x^2)]$, where $\alpha = \sqrt{2}/8$

10. $y = g_\alpha$ satisfies $x^2y'' + (1 - 2c)xy' + (a^2b^2x^{2b} + c^2 - \alpha^2b^2)y = 0$
(a) $y = x^{-5/2}[c_1J_5(2x^{1/2}) + c_2K_5(2x^{1/2})]$
(b) $y = x^{-5/2}[c_1J_{5/2}(x) + c_2J_{-5/2}(x)]$
(c) $y = x^{-5/2}[c_1J_1(\frac{2}{5}x^{5/2}) + c_2K_1(\frac{2}{5}x^{5/2})]$
(d) $y = x[c_1J_0(2x^{1/2}) + c_2K_0(2x^{1/2})]$

11. $a = 2, \quad c = 0$

12. $y = \sum_{n=0}^{\infty} \frac{(-1)^n}{(n!)^2} x^n; \quad y = J_0(2x^{1/2}) \quad \text{if } x > 0$

13. $b = (p_0 - a_0)/a_0, \quad c = q_0/a_0$

14. $y = x^{1/2}$

15. $t = 1: \quad y = \sum_{n=1}^{\infty} (-1)^{n-1} \frac{n!}{(2n)!} (2x)^n$

$$t = \frac{1}{2}: \quad y = x^{1/2} \sum_{n=0}^{\infty} \frac{(-1)^n}{n!} \left(\frac{x}{2}\right)^n = x^{1/2}e^{-x/2}$$

16. $u_0(x) = \cos x; \quad u_1(x) = \frac{1}{2} - \frac{1}{6}\cos x - \frac{1}{3}\cos 2x$

Chapter 7

7.4 Exercises (page 195)

3. (b) $(P^k)' = \sum_{m=0}^{k-1} P^m P' P^{k-1-m}$

7.12 Exercises (page 205)

1. (a) $A^{-1} = 2I - A, \quad A^n = nA - (n - 1)I$

(b) $$e^{tA} = e^t(1 - t)I + te^tA = e^t\begin{bmatrix} 1 & 0 \\ t & 1 \end{bmatrix}$$

2. (a) $A^{-1} = \frac{3}{2}I - \frac{1}{2}A, \quad A^n = (2^n - 1)A - (2^n - 2)I$

(b) $$e^{tA} = (2e^t - e^{2t})I + (e^{2t} - e^t)A = \begin{bmatrix} e^t & 0 \\ e^{2t} - e^t & e^{2t} \end{bmatrix}$$

3. (a) $A^{-1} = A, \quad A^n = \frac{1 + (-1)^n}{2} I + \frac{1 - (-1)^n}{2} A$

(b) $$e^{tA} = (\cosh t)I + (\sinh t)A = \begin{bmatrix} \cosh t & \sinh t \\ \sinh t & \cosh t \end{bmatrix}$$

4. (a) $A^{-1} = A, \quad A^n = \dfrac{1 + (-1)^n}{2} I + \dfrac{1 - (-1)^n}{2} A$

(b) $e^{tA} = (\cosh t)I + (\sinh t)A = \begin{bmatrix} e^{-t} & 0 \\ 0 & e^t \end{bmatrix}$

5. (b) $e^{tA} = e^{at}\begin{bmatrix} \cos bt & \sin bt \\ -\sin bt & \cos bt \end{bmatrix}$

7. $e^{A(t)} = I + (e - 1)A(t); \quad (e^{A(t)})' = (e - 1)A'(t) = \begin{bmatrix} 0 & e - 1 \\ 0 & 0 \end{bmatrix};$

$e^{A(t)}A'(t) = \begin{bmatrix} 0 & e \\ 0 & 0 \end{bmatrix}; \quad A'(t)e^{A(t)} = \begin{bmatrix} 0 & 1 \\ 0 & 0 \end{bmatrix}$

8. (a) $A^n = O$ if $n \geq 3$

(b) $e^{tA} = I + tA + \dfrac{1}{2}t^2A^2 = \begin{bmatrix} 1 & t & t + \frac{1}{2}t^2 \\ 0 & 1 & t \\ 0 & 0 & 1 \end{bmatrix}$

9. (a) $A^n = A$ if $n \geq 1$

(b) $e^{tA} = I + (e^t - 1)A = \begin{bmatrix} 1 & e^t - 1 & e^t - 1 \\ 0 & e^t & e^t - 1 \\ 0 & 0 & 1 \end{bmatrix}$

10. (a) $A^3 = 4A^2 - 5A + 2I; \quad A^n = \begin{bmatrix} 2^n & 0 & 0 \\ 0 & 1 & 0 \\ 0 & n & 1 \end{bmatrix}$

(b) $e^{tA} = \begin{bmatrix} e^{2t} & 0 & 0 \\ 0 & e^t & 0 \\ 0 & te^t & e^t \end{bmatrix}$

11. $e^{tA} = I + tA + \frac{1}{2}t^2A^2$

13. $e^Ae^B = \begin{bmatrix} e^2 & -(e - 1)^2 \\ 0 & 1 \end{bmatrix}; \quad e^Be^A = \begin{bmatrix} e^2 & (e - 1)^2 \\ 0 & 1 \end{bmatrix}; \quad e^{A+B} = \begin{bmatrix} e^2 & 0 \\ 0 & 1 \end{bmatrix}$

7.15 Exercises (page 211)

1. $e^{tA} = \frac{1}{2}(3e^t - e^{3t})I + \frac{1}{2}(e^{3t} - e^t)A$

2. $e^{tA} = (\cosh \sqrt{5}\, t)I + \dfrac{1}{\sqrt{5}}(\sinh \sqrt{5}\, t)A$

3. $e^{tA} = \frac{1}{2}e^t\{(t^2 - 2t + 2)I + (-2t^2 + 2t)A + t^2A^2\}$

4. $e^{tA} = (3e^{-t} - 3e^{-2t} + e^{-3t})I + (\frac{5}{2}e^{-t} - 4e^{-2t} + \frac{3}{2}e^{-3t})A + (\frac{1}{2}e^{-t} - e^{-2t} + \frac{1}{2}e^{-3t})A^2$

5. $e^{tA} = (4e^t - 3e^{2t} + 2te^{2t})I + (4e^{2t} - 3te^{2t} - 4e^t)A + (e^t - e^{2t} + te^{2t})A^2$

6. $e^{tA} = (4e^t - 6e^{2t} + 4e^{3t} - e^{4t})I + (-\frac{13}{3}e^t + \frac{19}{2}e^{2t} - 7e^{3t} + \frac{11}{6}e^{4t})A$
$+ (\frac{3}{2}e^t - 4e^{2t} + \frac{7}{2}e^{3t} - e^{4t})A^2 + (-\frac{1}{6}e^t + \frac{1}{2}e^{2t} - \frac{1}{2}e^{3t} + \frac{1}{6}e^{4t})A^3$

7. (b) $e^{tA} = \frac{1}{6}e^{\lambda t}\{(6 - 6\lambda t + 3\lambda^2t^2 - \lambda^3t^3)I + (6t - 6\lambda t^2 + 3\lambda^2t^3)A + (3t^2 - 3\lambda t^3)A^2 + t^3A^3\}$

8. $y_1 = c_1 \cosh \sqrt{5}\, t + \dfrac{c_1 + 2c_2}{\sqrt{5}} \sinh \sqrt{5}\, t, \quad y_2 = c_2 \cosh \sqrt{5}\, t + \dfrac{2c_1 - c_2}{\sqrt{5}} \sinh \sqrt{5}\, t$

9. $y_1 = e^t(\cos 3t - \sin 3t), \quad y_2 = e^t(\cos 3t - 3 \sin 3t)$
10. $y_1 = e^{2t} + 4te^{2t}, \quad y_2 = -2e^t + e^{2t} + 4te^{2t}, \quad y_3 = -2e^t + 4e^{2t}$
11. $y_1 = c_1 e^{2t}, \quad y_2 = c_2 e^t, \quad y_3 = (c_2 t + c_3)e^t$
12. $y_1 = 3e^{-t} - 3e^{-2t} + e^{-3t}, \quad y_2 = -3e^{-t} + 6e^{-2t} - 3e^{-3t}, \quad y_3 = 3e^{-t} - 12e^{-2t} + 9e^{-3t}$
13. $y_1 = e^{5t} + 7e^{-3t}, \quad y_2 = 2e^{5t} - 2e^{-3t}, \quad y_3 = -e^{5t} + e^{-3t}$
14. $y_1 = -\frac{1}{2}e^t + e^{2t} + \frac{1}{2}e^{3t}, \quad y_2 = e^{2t} + e^{3t}, \quad y_3 = e^{3t}, \quad y_4 = e^{4t}$
15. $y_1 = 2e^{2t} - 1, \quad y_2 = 2e^{2t} - t - 2, \quad y_3 = 2e^{2t}, \quad y_4 = e^{2t}$

7.17 Exercises (page 215)

2. (c) $y_1 = (b - 1)e^x + 2(c + 1 - b)xe^x + 1, \quad y_2 = ce^x + 2(c + 1 - b)xe^x$
4. $y_1 = -\frac{1}{3}e^t - \frac{1}{6}e^{4t} + \frac{1}{2}e^{2t}, \quad y_2 = \frac{2}{3}e^t - \frac{1}{6}e^{4t} + \frac{1}{2}e^{2t}$

5. (a) $B_0 = B, \quad B_1 = AB, \quad B_2 = \dfrac{1}{2!} A^2 B, \ldots, \quad B_m = \dfrac{1}{m!} A^m B$

   (b) $B = -m!\,(A^{-1})^{m+1} C$

6. (a) $Y(t) = \left(I + tA + \dfrac{1}{2} t^2 A^2 + \dfrac{1}{6} t^3 A^3\right) B, \quad \text{where } B = -6A^{-4}C = -\dfrac{3}{128}\begin{bmatrix} 1 \\ 1 \end{bmatrix}.$

   This gives the particular solution $y_1 = y_2 = -\frac{3}{128} - \frac{3}{32}t - \frac{3}{16}t^2 - \frac{1}{4}t^3$

   (b) $y_1 = y_2 = -\frac{3}{128} - \frac{3}{32}t - \frac{3}{16}t^2 - \frac{1}{4}t^3 + \frac{131}{128}e^{4t}$

7. $E = B, \quad F = \dfrac{1}{\alpha}(AB + C)$

8. (a) $y_1 = -\cos 2t - \frac{1}{2}\sin 2t, \quad y_2 = -\frac{1}{2}\sin 2t$

   (b) $y_1 = 2\cosh 2t + \frac{5}{2}\sinh 2t - \cos 2t - \frac{1}{2}\sin 2t, \quad y_2 = \cosh 2t + \frac{1}{2}\sinh 2t - \frac{1}{2}\sin 2t$
9. $y_1(x) = e^{2x} + e^{3x} - e^x, \quad y_2(x) = -2e^{2x} - e^{3x} + 3e^x$
10. $y_1(x) = \frac{4}{25}e^x - \frac{1}{36}e^{2x} + (c_1 - \frac{119}{900})e^{-4x} + (\frac{11}{30} - c_1 - c_2)xe^{-4x},$

    $y_2(x) = \frac{1}{25}e^x + \frac{7}{36}e^{2x} + (c_2 - \frac{211}{900})e^{-4x} + (c_1 + c_2 - \frac{11}{30})xe^{-4x}$
11. $y_1(x) = e^{-4x}(2\cos x + \sin x) + \frac{31}{26}e^x - \frac{93}{17}, \quad y_2(x) = e^{-4x}(\sin x + 3\cos x) - \frac{2}{13}e^x + \frac{6}{17}$
12. $y_1(x) = e^{-x}(x^2 + 2x + 3) + x^2 - 3x + 3, \quad y_2(x) = e^{-x}(-2x - 2) + x,$

    $y_3(x) = 2e^{-x} + x - 1$

7.20 Exercises (page 221)

4. (c) $Y(x) = e^x e^{\frac{1}{2}x^2 A} B$

5. If $A(x) = \sum_{k=0}^{\infty} x^k A_k$, then $Y(x) = B + xC + \sum_{k=2}^{\infty} x^k B_k$,

   where $(k + 2)(k + 1)B_{k+2} = \sum_{r=0}^{k} A_r B_{k-r}$ for $k \geq 0$.

7.24 Exercises (page 230)

1. (a) $Y(x) = e^x$

   (b) $Y_n(x) = 2e^x - \sum_{k=0}^{n} \dfrac{x^k}{k!}$ if n is odd; $\quad Y_n(x) = \sum_{k=0}^{n} \dfrac{x^k}{k!}$ if n is even

2. $Y_3(x) = \frac{x^2}{2} + \frac{x^5}{20} + \frac{x^8}{160} + \frac{x^{11}}{4400}$

3. $Y_3(x) = x + \frac{x^4}{4} + \frac{x^7}{14} + \frac{x^{10}}{160}$

4. $Y_3(x) = \frac{x^3}{3} + \frac{4x^7}{63} + \frac{8x^9}{405} + \frac{184x^{11}}{51975} + \frac{4x^{13}}{12285}$

5. (a) $Y_2(x) = 1 + x + x^2 + \frac{2x^3}{3} + \frac{x^4}{6} + \frac{2x^5}{15} + \frac{x^7}{63}$

(b) $M = 2;\quad c = \frac{1}{2}$

(c) $Y(x) = 1 + x + x^2 + \frac{4x^3}{3} + \frac{7x^4}{6} + \frac{6x^5}{5} + \cdots$

6. (a) $Y_4(x) = x + \frac{x^3}{3} + \frac{2x^5}{15} + \frac{17x^7}{315} + \frac{38x^9}{2835} + \frac{134x^{11}}{51975} + \frac{4x^{13}}{12285} + \frac{x^{15}}{59535}$

(d) $Y(x) = \tan x = x + \frac{x^3}{3} + \frac{2x^5}{15} + \frac{17x^7}{315} + \frac{62x^9}{2835} + \cdots$ for $|x| < \frac{\pi}{2}$

8. $Y_3(x) = 2 + x^2 + x^3 + \frac{3x^5}{20} + \frac{x^6}{10};\quad Z_3(x) = 3x^2 + \frac{3x^4}{4} + \frac{6x^5}{5} + \frac{3x^7}{28} + \frac{3x^8}{40}$

9. $Y_3(x) = 5 + x + \frac{x^4}{12} + \frac{x^6}{6} + \frac{2x^7}{63} + \frac{x^9}{72};$

$Z_3(x) = 1 + \frac{x^3}{3} + x^5 + \frac{2x^6}{9} + \frac{x^8}{8} + \frac{11x^9}{324} + \frac{7x^{11}}{264}$

10. (d) $Y_n(x) = 0;\quad \lim_{n\to\infty} Y_n(x) = 0$

(e) $Y_n(x) = \begin{cases} x^2 & \text{if } x \ge 0 \\ -x^2 & \text{if } x \le 0 \end{cases};\quad \lim_{n\to\infty} Y_n(x) = \begin{cases} x^2 & \text{if } x \ge 0 \\ -x^2 & \text{if } x \le 0 \end{cases}$

(f) $Y_n(x) = \frac{2x^n}{3^n};\quad \lim_{n\to\infty} Y_n(x) = 0$

(g) $Y_n(x) = \begin{cases} x^2 & \text{if } x \ge 0 \\ -x^2 & \text{if } x \le 0 \end{cases};\quad \lim_{n\to\infty} Y_n(x) = \begin{cases} x^2 & \text{if } x \ge 0 \\ -x^2 & \text{if } x \le 0 \end{cases}$

Chapter 8

8.3 Exercises (page 245)

2. All open except (d), (e), (h), and (j)
3. All open except (d)
5. (e) One example is the collection of all 2-balls $B(\boldsymbol{O}; 1/k)$, where $k = 1, 2, 3, \ldots$
6. (a) Both (b) Both (c) Closed (d) Open (e) Closed (f) Neither (g) Closed (h) Neither (i) Closed (j) Closed (k) Neither (l) Closed
8. (e) One example is the collection of all sets of the form $S_k = \{\boldsymbol{x} \mid \|\boldsymbol{x}\| \le 1 - 1/k\}$ for $k = 1, 2, 3, \ldots$. Their union is the open ball $B(\boldsymbol{O}; 1)$
10. No

8.5 Exercises (page 251)

1. (a) All (x, y)
   (b) All $(x, y) \neq (0, 0)$
   (c) All (x, y) with $y \neq 0$
   (d) All (x, y) with $y \neq 0$ and $\dfrac{x^2}{y} \neq \dfrac{\pi}{2} + k\pi$ $(k = 0, 1, 2, \ldots)$
   (e) All (x, y) with $x \neq 0$
   (f) All $(x, y) \neq (0, 0)$
   (g) All (x, y) with $xy \neq 1$
   (h) All $(x, y) \neq (0, 0)$
   (i) All $(x, y) \neq (0, 0)$
   (j) All (x, y) with $y \neq 0$ and $0 \leq x \leq y$ or $y \leq x \leq 0$
5. $\lim_{y \to 0} f(x, y)$ does not exist if $x \neq 0$
6. $(1 - m^2)/(1 + m^2)$; No
7. $y = \frac{1}{2}x^2$; f not continuous at $(0, 0)$
8. $f(0, 0) = 1$

8.9 Exercises (page 255)

1. $f'(\boldsymbol{x}; \boldsymbol{y}) = \boldsymbol{a} \cdot \boldsymbol{y}$
2. (a) $f'(\boldsymbol{x}; \boldsymbol{y}) = 4 \|\boldsymbol{x}\|^2 \boldsymbol{x} \cdot \boldsymbol{y}$
   (b) All points on the line $2x + 3y = \frac{3}{26}$
   (c) All points on the plane $x + 2y + 3z = 0$
3. $f'(\boldsymbol{x}; \boldsymbol{y}) = \boldsymbol{x} \cdot T(\boldsymbol{y}) + \boldsymbol{y} \cdot T(\boldsymbol{x})$
4. $\dfrac{\partial f}{\partial x} = 2x + y^3 \cos(xy)$; $\dfrac{\partial f}{\partial y} = 2y \sin(xy) + xy^2 \cos(xy)$
5. $\dfrac{\partial f}{\partial x} = x/(x^2 + y^2)^{1/2}$; $\dfrac{\partial f}{\partial y} = y/(x^2 + y^2)^{1/2}$
6. $\dfrac{\partial f}{\partial x} = y^2/(x^2 + y^2)^{3/2}$; $\dfrac{\partial f}{\partial y} = -xy/(x^2 + y^2)^{3/2}$
7. $\dfrac{\partial f}{\partial x} = -2y/(x - y)^2$; $\dfrac{\partial f}{\partial y} = 2x/(x - y)^2$
8. $D_k f(\boldsymbol{x}) = a_k$, where $\boldsymbol{a} = (a_1, \ldots, a_n)$
9. $D_k f(\boldsymbol{x}) = 2 \sum_{j=1}^{n} a_{kj} x_j$
10. $\dfrac{\partial f}{\partial x} = 4x^3 - 8xy^2$; $\dfrac{\partial f}{\partial y} = 4y^3 - 8x^2 y$
11. $\dfrac{\partial f}{\partial x} = \dfrac{2x}{x^2 + y^2}$; $\dfrac{\partial f}{\partial y} = \dfrac{2y}{x^2 + y^2}$
12. $\dfrac{\partial f}{\partial x} = -\dfrac{2x}{y} \sin(x^2)$; $\dfrac{\partial f}{\partial y} = -\dfrac{1}{y^2} \cos(x^2)$
13. $\dfrac{\partial f}{\partial x} = \dfrac{2x}{y} \sec^2 \dfrac{x^2}{y}$; $\dfrac{\partial f}{\partial y} = -\dfrac{x^2}{y^2} \sec^2 \dfrac{x^2}{y}$

14. $\dfrac{\partial f}{\partial x} = -\dfrac{y}{x^2 + y^2}$; $\dfrac{\partial f}{\partial y} = \dfrac{1}{x^2 + y^2}$

15. $\dfrac{\partial f}{\partial x} = \dfrac{1 + y^2}{1 + x^2 + y^2 + x^2y^2}$; $\dfrac{\partial f}{\partial y} = \dfrac{1 + x^2}{1 + x^2 + y^2 + x^2y^2}$

16. $\dfrac{\partial f}{\partial x} = y^2x^{y^2-1}$; $\dfrac{\partial f}{\partial y} = 2yx^{y^2}\log x$

17. $\dfrac{\partial f}{\partial x} = -\dfrac{1}{2\sqrt{x(y - x)}}$; $\dfrac{\partial f}{\partial y} = \dfrac{\sqrt{x}}{2y\sqrt{y - x}}$

18. $n = -\frac{3}{2}$
19. $a = b = 1$
22. (b) One example is $f(\boldsymbol{x}) = \boldsymbol{x} \cdot \boldsymbol{y}$, where $\boldsymbol{y}$ is a fixed nonzero vector

8.14 Exercises (page 262)

1. (a) $(2x + y^3 \cos xy)\boldsymbol{i} + (2y \sin xy + xy^2 \cos xy)\boldsymbol{j}$
   (b) $e^x \cos y\boldsymbol{i} - e^x \sin y\boldsymbol{j}$
   (c) $2xy^3z^4\boldsymbol{i} + 3x^2y^2z^4\boldsymbol{j} + 4x^2y^3z^3\boldsymbol{k}$
   (d) $2x\boldsymbol{i} - 2y\boldsymbol{j} + 4z\boldsymbol{k}$
   (e) $\dfrac{2x}{x^2 + 2y^2 - 3z^2}\boldsymbol{i} + \dfrac{4y}{x^2 + 2y^2 - 3z^2}\boldsymbol{j} - \dfrac{6z}{x^2 + 2y^2 - 3z^2}\boldsymbol{k}$
   (f) $y^zx^{y^z-1}\boldsymbol{i} + zy^{z-1}x^{y^z}\log x\boldsymbol{j} + y^zx^{y^z}\log x \log y\boldsymbol{k}$
2. (a) $-2/\sqrt{6}$
   (b) $1/\sqrt{6}$
3. $(1, 0)$, in the direction of $\boldsymbol{i}$; $(-1, 0)$, in the direction of $-\boldsymbol{i}$
4. $2\boldsymbol{i} + 2\boldsymbol{j}$; $\frac{14}{5}$
5. $(a, b, c) = (6, 24, -8)$ or $(-6, -24, 8)$
6. The set of points (x, y) on the line $5x - 3y = 6$; $\nabla f(\boldsymbol{a}) = 5\boldsymbol{i} - 3\boldsymbol{j}$
8. (c) Yes
   (d) $f(x, y, z) = \frac{1}{2}(x^2 + y^2 + z^2)$
11. (b) implies (a) and (c); (d) implies (a), (b), and (c); (f) implies (a)

8.17 Exercises (page 268)

1. (b) $F''(t) = \dfrac{\partial^2 f}{\partial x^2}[X'(t)]^2 + 2\dfrac{\partial^2 f}{\partial x\,\partial y}X'(t)Y'(t) + \dfrac{\partial^2 f}{\partial y^2}[Y'(t)]^2 + \dfrac{\partial f}{\partial x}X''(t) + \dfrac{\partial f}{\partial y}Y''(t)$

2. (a) $F'(t) = 4t^3 + 2t$; $F''(t) = 12t^2 + 2$
   (b) $F'(t) = (2\cos^2 t - 1)e^{\cos t \sin t}\cos(\cos t \sin^2 t) + (3\sin^3 t - 2\sin t)e^{\cos t \sin t}\sin(\cos t \sin^2 t)$;
   $F''(t) = (5\cos^6 t - 3\cos^4 t - 4\cos^3 t - \cos^2 t - 4\cos t)e^{\cos t \sin t}\cos(\cos t \sin^2 t)$
   $+ (14\sin^3 t - 12\sin^5 t - 4\sin t + 7\cos t - 9\cos^3 t)e^{\cos t \sin t}\sin(\cos t \sin^2 t)$

   (c) $F'(t) = \dfrac{2e^{2t}\exp(e^{2t})}{1 + \exp(e^{2t})} + \dfrac{2e^{-2t}\exp(e^{-2t})}{1 + \exp(e^{-2t})}$, where $\exp(u) = e^u$;

   $$F''(t) = \frac{4[1 + e^{2t} + \exp(e^{2t})]e^{2t}\exp(e^{2t})}{[1 + \exp(e^{2t})]^2} - \frac{4[1 + e^{-2t} + \exp(e^{-2t})]e^{-2t}\exp(e^{-2t})}{[1 + \exp(e^{-2t})]^2}$$

3. (a) $-\frac{2}{3}$
(b) $x^2 - y^2$
(c) 0
4. (a) $(1 + 3x^2 + 3y^2)(x\boldsymbol{i} + y\boldsymbol{j}) - (x^2 + y^2)^{1/2}\boldsymbol{k}$, or any scalar multiple thereof
(b) $\cos\theta = -[1 + (1 + 3(x^2 + y^2))^2]^{-1/2}$; $\cos\theta \to -\frac{1}{2}\sqrt{2}$ as $(x, y, z) \to (0, 0, 0)$
5. $U(x, y) = \frac{1}{2}\log(x^2 + y^2)$; $V(x, y) = \arctan(y/x)$
6. (b) No
8. $x/x_0 + y/y_0 + z/z_0 = 3$
9. $x + y + 2z = 4, \quad x - y - z = -1$
10. $c = \pm\sqrt{3}$

8.22 Exercises (page 275)

1. (b) $\dfrac{\partial f}{\partial x} = -2x\sin(x^2 + y^2)\cos[\cos(x^2 + y^2)]e^{\sin[\cos(x^2 + y^2)]}$

2. $\dfrac{\partial F}{\partial x} = \dfrac{1}{2}\dfrac{\partial f}{\partial u} + \dfrac{1}{2}\dfrac{\partial f}{\partial v}$; $\dfrac{\partial F}{\partial y} = -\dfrac{1}{2}\dfrac{\partial f}{\partial u} + \dfrac{1}{2}\dfrac{\partial f}{\partial v}$

3. (a) $\dfrac{\partial F}{\partial s} = \dfrac{\partial f}{\partial x}\dfrac{\partial X}{\partial s} + \dfrac{\partial f}{\partial y}\dfrac{\partial Y}{\partial s}$; $\dfrac{\partial F}{\partial t} = \dfrac{\partial f}{\partial x}\dfrac{\partial X}{\partial t} + \dfrac{\partial f}{\partial y}\dfrac{\partial Y}{\partial t}$

(c) $\dfrac{\partial^2 F}{\partial s\,\partial t} = \dfrac{\partial^2 f}{\partial x^2}\dfrac{\partial X}{\partial s}\dfrac{\partial X}{\partial t} + \dfrac{\partial^2 f}{\partial x\,\partial y}\left(\dfrac{\partial X}{\partial s}\dfrac{\partial Y}{\partial t} + \dfrac{\partial X}{\partial t}\dfrac{\partial Y}{\partial s}\right) + \dfrac{\partial^2 f}{\partial y^2}\dfrac{\partial Y}{\partial s}\dfrac{\partial Y}{\partial t} + \dfrac{\partial f}{\partial x}\dfrac{\partial^2 X}{\partial s\,\partial t} + \dfrac{\partial f}{\partial y}\dfrac{\partial^2 Y}{\partial s\,\partial t}$

4. (a) $\dfrac{\partial F}{\partial s} = \dfrac{\partial f}{\partial x} + t\dfrac{\partial f}{\partial y}$; $\dfrac{\partial F}{\partial t} = \dfrac{\partial f}{\partial x} + s\dfrac{\partial f}{\partial y}$; $\dfrac{\partial^2 F}{\partial s^2} = \dfrac{\partial^2 f}{\partial x^2} + 2t\dfrac{\partial^2 f}{\partial x\,\partial y} + t^2\dfrac{\partial^2 f}{\partial y^2}$;

$\dfrac{\partial^2 F}{\partial t^2} = \dfrac{\partial^2 f}{\partial x^2} + 2s\dfrac{\partial^2 f}{\partial x\,\partial y} + s^2\dfrac{\partial^2 f}{\partial y^2}$; $\dfrac{\partial^2 F}{\partial s\,\partial t} = \dfrac{\partial^2 f}{\partial x^2} + (s + t)\dfrac{\partial^2 f}{\partial x\,\partial y} + st\dfrac{\partial^2 f}{\partial y^2} + \dfrac{\partial f}{\partial y}$

(b) $\dfrac{\partial F}{\partial s} = t\dfrac{\partial f}{\partial x} + \dfrac{1}{t}\dfrac{\partial f}{\partial y}$; $\dfrac{\partial F}{\partial t} = s\dfrac{\partial f}{\partial x} - \dfrac{s}{t^2}\dfrac{\partial f}{\partial y}$; $\dfrac{\partial^2 F}{\partial s^2} = t^2\dfrac{\partial^2 f}{\partial x^2} + 2\dfrac{\partial^2 f}{\partial x\,\partial y} + \dfrac{1}{t^2}\dfrac{\partial^2 f}{\partial y^2}$;

$\dfrac{\partial^2 F}{\partial t^2} = s^2\dfrac{\partial^2 f}{\partial x^2} - 2\dfrac{s^2}{t^2}\dfrac{\partial^2 f}{\partial x\,\partial y} + \dfrac{s^2}{t^4}\dfrac{\partial^2 f}{\partial y^2} + \dfrac{2s}{t^3}\dfrac{\partial f}{\partial y}$; $\dfrac{\partial^2 F}{\partial s\,\partial t} = st\dfrac{\partial^2 f}{\partial x^2} - \dfrac{s}{t^3}\dfrac{\partial^2 f}{\partial y^2} + \dfrac{\partial f}{\partial x} - \dfrac{1}{t^2}\dfrac{\partial f}{\partial y}$

(c) $\dfrac{\partial F}{\partial s} = \dfrac{1}{2}\dfrac{\partial f}{\partial x} + \dfrac{1}{2}\dfrac{\partial f}{\partial y}$; $\dfrac{\partial F}{\partial t} = -\dfrac{1}{2}\dfrac{\partial f}{\partial x} + \dfrac{1}{2}\dfrac{\partial f}{\partial y}$; $\dfrac{\partial^2 F}{\partial s\,\partial t} = -\dfrac{1}{4}\dfrac{\partial^2 f}{\partial x^2} + \dfrac{1}{4}\dfrac{\partial^2 f}{\partial y^2}$;

$\dfrac{\partial^2 F}{\partial s^2} = \dfrac{1}{4}\dfrac{\partial^2 f}{\partial x^2} + \dfrac{1}{2}\dfrac{\partial^2 f}{\partial x\,\partial y} + \dfrac{1}{4}\dfrac{\partial^2 f}{\partial y^2}$; $\dfrac{\partial^2 F}{\partial t^2} = \dfrac{1}{4}\dfrac{\partial^2 f}{\partial x^2} - \dfrac{1}{2}\dfrac{\partial^2 f}{\partial x\,\partial y} + \dfrac{1}{4}\dfrac{\partial^2 f}{\partial y^2}$

5. $\dfrac{\partial^2 \varphi}{\partial r^2} = \cos^2\theta\,\dfrac{\partial^2 f}{\partial x^2} + \cos\theta\sin\theta\left(\dfrac{\partial^2 f}{\partial x\,\partial y} + \dfrac{\partial^2 f}{\partial y\,\partial x}\right) + \sin^2\theta\,\dfrac{\partial^2 f}{\partial y^2}$;

$\dfrac{\partial^2 \varphi}{\partial r\,\partial \theta} = -r\cos\theta\sin\theta\,\dfrac{\partial^2 f}{\partial x^2} + r\cos^2\theta\,\dfrac{\partial^2 f}{\partial x\,\partial y} - r\sin^2\theta\,\dfrac{\partial^2 f}{\partial y\,\partial x} + r\cos\theta\sin\theta\,\dfrac{\partial^2 f}{\partial y^2}$

$-\sin\theta\,\dfrac{\partial f}{\partial x} + \cos\theta\,\dfrac{\partial f}{\partial y}$;

6. $\dfrac{\partial F}{\partial r} = \dfrac{\partial f}{\partial x}\dfrac{\partial X}{\partial r} + \dfrac{\partial f}{\partial y}\dfrac{\partial Y}{\partial r} + \dfrac{\partial f}{\partial z}\dfrac{\partial Z}{\partial r}$; $\dfrac{\partial F}{\partial s} = \dfrac{\partial f}{\partial x}\dfrac{\partial X}{\partial s} + \dfrac{\partial f}{\partial y}\dfrac{\partial Y}{\partial s} + \dfrac{\partial f}{\partial z}\dfrac{\partial Z}{\partial s}$;

$\dfrac{\partial F}{\partial t} = \dfrac{\partial f}{\partial x}\dfrac{\partial X}{\partial t} + \dfrac{\partial f}{\partial y}\dfrac{\partial Y}{\partial t} + \dfrac{\partial f}{\partial z}\dfrac{\partial Z}{\partial t}$

7. (a) $\dfrac{\partial F}{\partial r} = \dfrac{\partial f}{\partial x} + \dfrac{\partial f}{\partial y} + 2\dfrac{\partial f}{\partial z}$; $\dfrac{\partial F}{\partial s} = \dfrac{\partial f}{\partial x} - 2\dfrac{\partial f}{\partial y} + \dfrac{\partial f}{\partial z}$; $\dfrac{\partial F}{\partial t} = \dfrac{\partial f}{\partial x} + 3\dfrac{\partial f}{\partial y} - \dfrac{\partial f}{\partial z}$

(b) $\dfrac{\partial F}{\partial r} = 2r\left(\dfrac{\partial f}{\partial x} + \dfrac{\partial f}{\partial y} + \dfrac{\partial f}{\partial z}\right)$; $\dfrac{\partial F}{\partial s} = 2s\left(\dfrac{\partial f}{\partial x} - \dfrac{\partial f}{\partial y} - \dfrac{\partial f}{\partial z}\right)$; $\dfrac{\partial F}{\partial t} = 2t\left(\dfrac{\partial f}{\partial x} - \dfrac{\partial f}{\partial y} + \dfrac{\partial f}{\partial z}\right)$

8. $\dfrac{\partial F}{\partial s} = \dfrac{\partial f}{\partial x}\dfrac{\partial X}{\partial s} + \dfrac{\partial f}{\partial y}\dfrac{\partial Y}{\partial s} + \dfrac{\partial f}{\partial z}\dfrac{\partial Z}{\partial s}$; $\dfrac{\partial F}{\partial t} = \dfrac{\partial f}{\partial x}\dfrac{\partial X}{\partial t} + \dfrac{\partial f}{\partial y}\dfrac{\partial Y}{\partial t} + \dfrac{\partial f}{\partial z}\dfrac{\partial Z}{\partial t}$

9. (a) $\dfrac{\partial F}{\partial s} = 2s\dfrac{\partial f}{\partial x} + 2s\dfrac{\partial f}{\partial y} + 2t\dfrac{\partial f}{\partial z}$; $\dfrac{\partial F}{\partial t} = 2t\dfrac{\partial f}{\partial x} - 2t\dfrac{\partial f}{\partial y} + 2s\dfrac{\partial f}{\partial z}$

(b) $\dfrac{\partial F}{\partial s} = \dfrac{\partial f}{\partial x} + \dfrac{\partial f}{\partial y} + t\dfrac{\partial f}{\partial z}$; $\dfrac{\partial F}{\partial t} = \dfrac{\partial f}{\partial x} - \dfrac{\partial f}{\partial y} + s\dfrac{\partial f}{\partial z}$

10. $\dfrac{\partial F}{\partial r} = \dfrac{\partial f}{\partial x}\dfrac{\partial X}{\partial r} + \dfrac{\partial f}{\partial y}\dfrac{\partial Y}{\partial r}$; $\dfrac{\partial F}{\partial s} = \dfrac{\partial f}{\partial x}\dfrac{\partial X}{\partial s} + \dfrac{\partial f}{\partial y}\dfrac{\partial Y}{\partial s}$; $\dfrac{\partial F}{\partial t} = \dfrac{\partial f}{\partial x}\dfrac{\partial X}{\partial t} + \dfrac{\partial f}{\partial y}\dfrac{\partial Y}{\partial t}$

11. (a) $\dfrac{\partial F}{\partial r} = \dfrac{\partial f}{\partial x}$; $\dfrac{\partial F}{\partial s} = \dfrac{\partial f}{\partial x}$; $\dfrac{\partial F}{\partial t} = \dfrac{\partial f}{\partial y}$

(b) $\dfrac{\partial F}{\partial r} = \dfrac{\partial f}{\partial x} + 2r\dfrac{\partial f}{\partial y}$; $\dfrac{\partial F}{\partial s} = \dfrac{\partial f}{\partial x} + 2s\dfrac{\partial f}{\partial y}$; $\dfrac{\partial F}{\partial t} = \dfrac{\partial f}{\partial x} + 2t\dfrac{\partial f}{\partial y}$

(c) $\dfrac{\partial F}{\partial r} = \dfrac{1}{s}\dfrac{\partial f}{\partial x}$; $\dfrac{\partial F}{\partial s} = \dfrac{-r}{s^2}\dfrac{\partial f}{\partial x} + \dfrac{1}{t}\dfrac{\partial f}{\partial y}$; $\dfrac{\partial F}{\partial t} = \dfrac{-s}{t^2}\dfrac{\partial f}{\partial y}$

13. (a) $\boldsymbol{f}(x, y, z) = x\boldsymbol{i} + y\boldsymbol{j} + z\boldsymbol{k}$, plus any constant vector
(b) $\boldsymbol{f}(x, y, z) = P(x)\boldsymbol{i} + Q(y)\boldsymbol{j} + R(z)\boldsymbol{k}$, where P, Q, R are any three functions satisfying $P' = p$, $Q' = q$, $R' = r$

14. (a) $D\boldsymbol{f}(x, y) = \begin{bmatrix} e^{x+2y} & 2e^{x+2y} \\ 2\cos(y + 2x) & \cos(y + 2x) \end{bmatrix}$; $D\boldsymbol{g}(u, v, w) = \begin{bmatrix} 1 & 4v & 9w^2 \\ -2u & 2 & 0 \end{bmatrix}$

(b) $\boldsymbol{h}(u, v, w) = e^{u+2v^2+3w^3+4v-2u^2}\boldsymbol{i} + \sin(2v - u^2 + 2u + 4v^2 + 6w^3)\boldsymbol{j}$

(c) $D\boldsymbol{h}(1, -1, 1) = \begin{bmatrix} -3 & 0 & 9 \\ 0 & -6\cos 9 & 18\cos 9 \end{bmatrix}$

15. (a) $D\boldsymbol{f}(x, y, z) = \begin{bmatrix} 2x & 1 & 1 \\ 2 & 1 & 2z \end{bmatrix}$; $D\boldsymbol{g}(u, v, w) = \begin{bmatrix} v^2w^2 & 2uvw^2 & 2uv^2w \\ 0 & w^2\cos v & 2w\sin v \\ 2ue^v & u^2e^v & 0 \end{bmatrix}$

(b) $\boldsymbol{h}(u, v, w) = (u^2v^4w^4 + w^2\sin v + u^2e^v)\boldsymbol{i} + (2uv^2w^2 + w^2\sin v + u^4e^{2v})\boldsymbol{j}$

(c) $D\boldsymbol{h}(u, 0, w) = \begin{bmatrix} 2u & w^2 + u^2 & 0 \\ 4u^3 & w^2 + 2u^4 & 0 \end{bmatrix}$

8.24 Miscellaneous exercises (page 281)

1. One example: $f(x, y) = 3x$ when $x = y$, $f(x, y) = 0$ otherwise
2. $D_1 f(0, 0) = 0$; $D_2 f(0, 0) = -1$; $D_{2,1} f(0, 0) = 0$; $D_{1,2} f(0, 0)$ does not exist
3. (a) If $\boldsymbol{a} = (a_1, a_2)$, then $f'(\boldsymbol{O}; \boldsymbol{a}) = a_2^3/a_1^2$ if $a_1 \neq 0$, and $f'(\boldsymbol{O}; \boldsymbol{a}) = 0$ if $a_1 = 0$
   (b) Not continuous at the origin
4. $\dfrac{\partial f}{\partial x} = \dfrac{1}{2} e^{-xy} x^{-1/2} y^{1/2}$; $\dfrac{\partial f}{\partial y} = \dfrac{1}{2} e^{-xy} x^{1/2} y^{-1/2}$
5. $$F'''(t) = \frac{\partial^3 f}{\partial x^3}[X'(t)]^3 + 3\frac{\partial^3 f}{\partial x^2\,\partial y}[X'(t)]^2 Y'(t) + 3\frac{\partial^3 f}{\partial x\,\partial y^2} X'(t)[Y'(t)]^2$$
$$+ \frac{\partial^3 f}{\partial y^3}[Y'(t)]^3 + 3\frac{\partial^2 f}{\partial x^2} X'(t)X''(t) + 3\frac{\partial^2 f}{\partial x\,\partial y}[X''(t)Y'(t) + X'(t)Y''(t)]$$
$$+ 3\frac{\partial^2 f}{\partial y^2} Y'(t)Y''(t) + \frac{\partial f}{\partial x} X'''(t) + \frac{\partial f}{\partial y} Y'''(t),$$
   assuming the mixed partial derivatives are independent of the order of differentiation
6. 8
7. (a) $\dfrac{\partial g}{\partial u} = \dfrac{\partial f}{\partial x} v + \dfrac{\partial f}{\partial y} u$; $\dfrac{\partial g}{\partial v} = \dfrac{\partial f}{\partial x} u - \dfrac{\partial f}{\partial y} v$; $\dfrac{\partial^2 g}{\partial u\,\partial v} = uv\dfrac{\partial^2 f}{\partial x^2} + (u^2 - v^2)\dfrac{\partial^2 f}{\partial x\,\partial y} - uv\dfrac{\partial^2 f}{\partial y^2} + \dfrac{\partial f}{\partial x}$ (b) $a = \frac{1}{2}$, $b = -\frac{1}{2}$
10. (a) $\varphi'(t) = A'(t)\int_c^{B(t)} f[A(t), y]\,dy + B'(t)\int_a^{A(t)} f[x, B(t)]\,dx$
    (b) $\varphi'(t) = 2te^{t^2}(2e^{t^2} - e^a - e^c)$
13. A sphere with center at the origin and radius $\sqrt{2}$
14. $f(x) = x^2$

Chapter 9

9.3 Exercises (page 286)

1. $f(x, y) = \sin(x - \frac{4}{3}y)$
2. $f(x, y) = e^{x+5y/2} - 1$
3. (a) $u(x, y) = x^2 y^2 e^{xy}$
   (b) $v(x, y) = 2 + \log\left|\dfrac{x}{y}\right|$
5. $A = B = C = 1$, $D = -3$; $f(x, y) = \varphi_1(3x + y) + \varphi_2(x - y)$
6. $G(x, y) = x - y$

9.8 Exercises (page 302)

1. $\partial X/\partial v = (1 + xu)/(x - y)$; $\partial Y/\partial u = (1 - yv)/(x - y)$; $\partial Y/\partial v = (1 + yu)/(y - x)$
2. $\partial X/\partial y = -(1 + xu)/(1 + u)$; $\partial V/\partial u = (1 - yv)/(1 + yu)$; $\partial V/\partial y = (1 - x)/(1 + u)$
3. $\dfrac{\partial X}{\partial v} = \dfrac{\partial(F, G)}{\partial(y, v)}\Big/\dfrac{\partial(F, G)}{\partial(x, y)}$; $\dfrac{\partial Y}{\partial u} = \dfrac{\partial(F, G)}{\partial(u, x)}\Big/\dfrac{\partial(F, G)}{\partial(x, y)}$; $\dfrac{\partial Y}{\partial v} = \dfrac{\partial(F, G)}{\partial(v, x)}\Big/\dfrac{\partial(F, G)}{\partial(x, y)}$
4. $T = \pm\dfrac{1}{\sqrt{751}}(24\boldsymbol{i} - 4\sqrt{7}\,\boldsymbol{j} + 3\sqrt{7}\,\boldsymbol{k})$

5. $2\boldsymbol{i} + \boldsymbol{j} + \sqrt{3}\,\boldsymbol{k}$, or any nonzero scalar multiple thereof
6. $\partial x/\partial u = 0, \quad \partial x/\partial v = \pi/12$
8. $n = 2$
9. $\partial f/\partial x = -1/(2y + 2z + 1)$; $\partial f/\partial y = -2(y + z)/(2y + 2z + 1)$;
$\partial^2 f/(\partial x\, \partial y) = 2/(2y + 2z + 1)^3$
10. $\partial^2 z/(\partial x\, \partial y) = [\sin(x + y)\cos^2(y + z) + \sin(y + z)\cos^2(x + y)]/\cos^3(y + z)$
11. $\dfrac{\partial f}{\partial x} = -\dfrac{D_1F + 2xD_2F}{D_1F + 2zD_2F}$; $\dfrac{\partial f}{\partial y} = -\dfrac{D_1F + 2yD_2F}{D_1F + 2zD_2F}$
12. $D_1F = f'[x + g(y)]$; $D_2F = f'[x + g(y)]g'(y)$; $D_{1,1}F = f''[x + g(y)]$;
$D_{1,2}F = f''[x + g(y)]g'(y)$; $D_{2,2}F = f''[x + g(y)][g'(y)]^2 + f'[x + g(y)]g''(y)$

9.13 Exercises (page 313)

1. Absolute minimum at (0, 1)
2. Saddle point at (0, 1)
3. Saddle point at (0, 0)
4. Absolute minimum at each point of the line $y = x + 1$
5. Saddle point at (1, 1)
6. Absolute minimum at (1, 0)
7. Saddle point at (0, 0)
8. Saddle points at (0, 6) and at $(x, 0)$, all x; relative minima at $(0, y)$, $0 < y < 6$; relative maxima at (2, 3) and at $(0, y)$ for $y < 0$ and $y > 6$
9. Saddle point at (0, 0); relative minimum at (1, 1)
10. Saddle points at $(n\pi + \pi/2, 0)$, where n is any integer
11. Absolute minimum at (0, 0); saddle point at $(-\frac{1}{4}, -\frac{1}{2})$
12. Absolute minimum at $(-\frac{1}{26}, -\frac{3}{26})$; absolute maximum at (1, 3)
13. Absolute maximum at $(\pi/3, \pi/3)$; absolute minimum at $(2\pi/3, 2\pi/3)$; relative maximum at (π, π); relative minimum at (0, 0); saddle points at $(0, \pi)$ and $(\pi, 0)$
14. Saddle point at (1, 1)
15. Absolute maximum at each point of the circle $x^2 + y^2 = 1$; absolute minimum at (0, 0)
17. (c) Relative maximum at (2, 2); no relative minima; saddle points at (0, 3), (3, 0), and (3, 3)
18. Relative maximum $\frac{1}{8}$ at $(\frac{1}{2}, \frac{1}{2})$ and $(-\frac{1}{2}, -\frac{1}{2})$; relative minimum $-\frac{1}{8}$ at $(\frac{1}{2}, -\frac{1}{2})$ and $(-\frac{1}{2}, \frac{1}{2})$; saddle points at (0, 0), $(\pm 1, 0)$, and $(0, \pm 1)$; absolute maximum 1 at $(1, -1)$ and $(-1, 1)$; absolute minimum -1 at (1, 1) and $(-1, -1)$
19. (a) $a = 1, \quad b = -\frac{1}{6}$
(b) $a = 6 \log 2 - 3\pi/2, \quad b = \pi - 3 \log 2$
21. Let $x^* = \dfrac{1}{n}\sum_{i=1}^{n} x_i$, $y^* = \dfrac{1}{n}\sum_{i=1}^{n} y_i$, $u_i = x_i - x^*$. Then $a = \left(\sum_{i=1}^{n} y_i u_i\right)\Big/\left(\sum_{i=1}^{n} u_i^2\right)$,
and $b = y^* - ax^*$
22. Let $x^* = \dfrac{1}{n}\sum_{i=1}^{n} x_i$, $y^* = \dfrac{1}{n}\sum_{i=1}^{n} y_i$ $z^* = \dfrac{1}{n}\sum_{i=1}^{n} z_i$, $u_i = x_i - x^*$, $v_i = y_i - y^*$, and let

$$\Delta = \begin{vmatrix} \sum u_i^2 & \sum u_i v_i \\ \sum u_i v_i & \sum v_i^2 \end{vmatrix}$$

, where the sums are for $i = 1, 2, \ldots, n$. Then

$$a = \frac{1}{\Delta}\begin{vmatrix} \sum u_i z_i & \sum u_i v_i \\ \sum v_i z_i & \sum v_i^2 \end{vmatrix}, \quad b = \frac{1}{\Delta}\begin{vmatrix} \sum v_i z_i & \sum u_i v_i \\ \sum u_i z_i & \sum u_i^2 \end{vmatrix}, \quad c = z^* - ax^* - by^*$$

25. Eigenvalues 4, 16, 16; relative minimum at (1, 1, 1)

9.15 Exercises (page 318)

1. Maximum value is $\frac{1}{4}$; no minimum
2. Maximum is 2; minimum is 1
3. (a) Maximum is $\dfrac{\sqrt{a^2 + b^2}}{ab}$ at $(b(a^2 + b^2)^{-1/2}, a(a^2 + b^2)^{-1/2})$; minimum is $-\dfrac{\sqrt{a^2 + b^2}}{ab}$ at $(-b(a^2 + b^2)^{-1/2}, -a(a^2 + b^2)^{-1/2})$

   (b) Minimum is $a^2b^2/(a^2 + b^2)$ at $\left(\dfrac{ab^2}{a^2 + b^2}, \dfrac{a^2b}{a^2 + b^2}\right)$; no maximum
4. Maximum is $1 + \sqrt{2}/2$ at the points $(n\pi + \pi/8, n\pi - \pi/8)$, where n is any integer; minimum is $1 - \sqrt{2}/2$ at $(n\pi + 5\pi/8, n\pi + 3\pi/8)$, where n is any integer
5. Maximum is 3 at $(\frac{1}{3}, -\frac{2}{3}, \frac{2}{3})$; minimum is -3 at $(-\frac{1}{3}, \frac{2}{3}, -\frac{2}{3})$
6. (0, 0, 1) and (0, 0, −1)
7. 1
8. (1, 0, 0), (0, 1, 0), (−1, 0, 0), (0, −1, 0)
9. $\dfrac{a^a b^b c^c}{(a + b + c)^{a+b+c}}$ at $\left(\dfrac{a}{a + b + c}, \dfrac{b}{a + b + c}, \dfrac{c}{a + b + c}\right)$
10. $abc\sqrt{3}/2$
11. $5 \log r + 3 \log \sqrt{3}$
12. $m^2 = \dfrac{A + C - \sqrt{(A - C)^2 + 4B^2}}{2(AC - B^2)}$
13. $(4 \pm \sqrt{5})/\sqrt{2}$
14. Angle is $\pi/3$; width across the bottom is $c/3$; maximum area is $c^2/(4\sqrt{3})$

Chapter 10

10.5 Exercises (page 328)

1. $-\frac{14}{15}$
2. $-2\pi a^2$
3. $\frac{1}{35}$
4. $\frac{4}{3}$
5. 0
6. 40
7. $\frac{23}{6}$
8. $\frac{5}{2}$
9. $-\frac{369}{10}$
10. -2π
11. 0
12. (a) $-2\sqrt{2}\,\pi$
    (b) $-\pi$

10.9 Exercises (page 331)

1. $\frac{23}{6}$
2. $2a^3$
3. $a = (3c/2)^{1/3}$
4. 0
5. $8\pi(\sin\theta - \cos\theta)$
6. $\pi a^3/4$
7. $-\sqrt{2}$
8. $256a^3/15$
9. $2\pi^2 a^3(1 + 2\pi^2)$
10. $[(2 + t_0^2)^{3/2} - 2\sqrt{2}]/3$
12. moment of inertia $= 4a^4$
13. $2\pi/3$
14. $\dfrac{600 - 36\sqrt{2} - 49\log(9 - 4\sqrt{2})}{64[6\sqrt{2} + \log(3 + 2\sqrt{2})]}$
15. $\bar{x} = \dfrac{6ab^2}{3a^2 + 4\pi^2 b^2}$; $\bar{y} = -\dfrac{6\pi ab^2}{3a^2 + 4\pi^2 b^2}$
16. $I_x = (a^2 + b^2)^{1/2}[\pi a^4 + (4\pi^3 - \pi/2)a^2b^2 + 32\pi^5 b^4/5]$
    $I_y = (a^2 + b^2)^{1/2}[\pi a^4 + (4\pi^3 + \pi/2)a^2b^2 + 32\pi^5 b^4/5]$

10.13 Exercises (page 336)

1. All except (f) are connected
6. (a) Not conservative
   (b) $(2e^{2\pi} - 5e^{\pi} - 5\pi - 3)/10$
7. (b) 3
8. $\frac{8}{5}$
10. $4b^2 - 8\pi b + 4$; minimum occurs when $b = \pi$

10.18 Exercises (page 345)

1. $\varphi(x, y) = \frac{1}{2}(x^2 + y^2) + C$
2. $\varphi(x, y) = x^3 y + C$
3. $\varphi(x, y) = x^2 e^y + xy - y^2 + C$
4. $\varphi(x, y) = x\sin y + y\cos x + (x^2 + y^2)/2 + C$
5. $\varphi(x, y) = x\sin(xy) + C$
6. $\varphi(x, y, z) = (x^2 + y^2 + z^2)/2 + C$
7. $\varphi(x, y, z) = x^2/2 - y^2/2 + xz - yz + C$
8. f is not a gradient
9. f is not a gradient
10. f is not a gradient
11. $\varphi(x, y, z) = y^2\sin x + xz^3 - 4y + 2z + C$
12. $\varphi(x, y, z) = x + 2x^2 y - x^3 z^2 + 2y - z^3 + C$
13. (b) $\varphi(x, y) = \dfrac{ar^{n+1}}{n + 1} + C$ if $n \neq -1$; $\varphi(x, y) = a\log r + C$ if $n = -1$
15. $\varphi(\mathbf{x}) = \dfrac{r^{p+2}}{p + 2} + C$ if $p \neq -2$; $\varphi(\mathbf{x}) = \log r + C$ if $p = -2$
16. $\varphi(\mathbf{x}) = g(r) + C$

10.20 Exercises (page 349)

1. $x^2/2 + 2xy + y^2/2 = C$
2. $x^2 y = C$

3. $x^3/3 - xy - y/2 + (\sin 2y)/4 = C$
4. $\cos 2x \sin 3y = C$
5. $x^3y + 4x^2y^2 - 12e^y + 12ye^y = C$
6. $\int Q(x)e^{\int P(x)dx}\,dx - ye^{\int P(x)dx} = C$
8. (a) $x + y = Cy^2$
   (b) $y^3/x^3 - 3\log|x| = C$
9. (a) $6(xy)^{1/2} - (y/x)^{3/2} = C$; $(x^5y)^{-1/2}$ is an integrating factor
   (b) $x + e^{-x}\sin y = C$; $e^{-x}\cos y$ is an integrating factor
10. $x^3y^4 + x^4y^5 = C$, $10x^3y^4 + x^5y^5 = C$, respectively; x^2y^3 is a common integrating factor

Chapter 11

11.9 Exercises (page 362)

1. $\frac{1}{3}$
2. 1
3. $2\sqrt{3} - \frac{38}{3}$
4. $\pi^2/4$
5. 2
6. 2π
7. 6
8. $t^{-3}(e^{t^2} - e^t) + t^{-2} - t^{-1}$
10. $\frac{1}{6}$
11. $\frac{1}{5}(\frac{21}{8} - \sqrt{2})$
12. $\pi/2$
13. $(\log 2)/6$

11.15 Exercises (page 371)

1. $-3\pi/2$
2. $\frac{3}{2} + \cos 1 + \sin 1 - \cos 2 - 2\sin 2$
3. $e - e^{-1}$
4. $\frac{7}{3}\log 2$
5. $\pi^2 - \frac{40}{9}$
6. 6
7. $\frac{80}{3}$
8. (a) $\frac{8}{3}$ (b) 2 (c) 320π

9. $\int_0^1 \left[\int_x^1 f(x, y)\,dy\right] dx$

10. $\int_0^4 \left[\int_{x/2}^{\sqrt{x}} f(x, y)\,dy\right] dx$

11. $\int_1^2 \left[\int_1^{y^2} f(x, y)\,dx\right] dy$

12. $\int_0^1 \left[\int_{2-y}^{1+\sqrt{1-y^2}} f(x, y)\,dx\right] dy$

13. $\int_{-1}^0 \left[\int_{-\sqrt{4y+4}}^{\sqrt{4y+4}} f(x, y)\,dx\right] dy + \int_0^8 \left[\int_{-\sqrt{4y+4}}^{2-y} f(x, y)\,dx\right] dy$

14. $\int_0^1 \left[\int_{e^y}^{e} f(x, y)\,dx\right] dy$

15. $\int_{-1}^0 \left[\int_{-\sqrt{1-y^2}}^{\sqrt{1-y^2}} f(x, y)\,dx\right] dy + \int_0^1 \left[\int_{-\sqrt{1-y}}^{\sqrt{1-y}} f(x, y)\,dx\right] dy$

16. $\int_0^1 \left[\int_{y^{1/2}}^{y^{1/3}} f(x, y)\,dx\right] dy$

17. $\int_{-1}^0 \left[\int_{-2\arcsin y}^{\pi} f(x, y)\,dx\right] dy + \int_0^1 \left[\int_{\arcsin y}^{\pi-\arcsin y} f(x, y)\,dx\right] dy$

18. $\int_{-2}^0 \left[\int_{2x+4}^{4-x^2} f(x, y)\,dy\right] dx$

19. $\int_0^1 \left[\int_x^{2-x} (x^2 + y^2)\, dy\right] dx = \frac{4}{3}$
20. $y = 0, \quad y = x \tan c, \quad x^2 + y^2 = a^2, \quad x^2 + y^2 = b^2$
21. (a) $\int_1^8 \left[\int_{y^{1/3}}^{y} f(x, y)\, dx\right] dy$
    (b) $4e^8 + 2e/3$
22. $m = 2; \quad n = 1$

11.18 Exercises (page 377)

1. $\bar{x} = -\frac{1}{2}, \quad \bar{y} = \frac{8}{5}$
2. $\bar{x} = 1, \quad \bar{y} = 0$
3. $\bar{x} = \frac{18}{13}, \quad \bar{y} = \frac{50}{39}$
4. $\bar{x} = \pi/2, \quad \bar{y} = \pi/8$
5. $\bar{x} = (\sqrt{2} + 1)\left(\frac{\pi\sqrt{2}}{4} - 1\right) = \frac{\pi}{2} + \frac{\pi\sqrt{2}}{4} - 1 - \sqrt{2}, \quad \bar{y} = \frac{\sqrt{2} + 1}{4}$
6. $\bar{x} = \frac{2a^2 \log a - a^2 + 1}{4(a \log a - a + 1)}, \quad \bar{y} = \frac{a(\log a)^2}{2(a \log a - a + 1)} - 1$
7. $\bar{x} = \bar{y} = \frac{1}{5}$
8. $\bar{x} = \bar{y} = 256/(315\pi)$
9. $\frac{26}{3} - \frac{15}{2} \log 3$
10. $\bar{x} = \frac{2}{3} \|\overrightarrow{AB}\|, \quad \bar{y} = \frac{2}{3} \|\overrightarrow{AD}\|$; assuming the x- and y-axes are chosen along sides AB and AD, respectively
11. $I_x = \frac{5\pi}{12}, \quad I_y = \frac{2\pi^3}{3} - \pi$
12. $I_x = \frac{1}{12}b^3(a - c), \quad I_y = \frac{1}{12}b(a^3 - c^3)$
13. $I_x = I_y = (1 - 5\pi/16)r^4$
14. $I_x = I_y = \frac{9}{8}$
15. $I_x = \frac{1}{64}[(4a - 1)e^{4a} - 1], \quad I_y = \frac{1}{32}[(a^3 - 3a^2 + 6a - 6)e^{2a} + 6]$
16. $I_x = \frac{72}{105}, \quad I_y = \frac{148}{45}$
19. $\frac{1}{3}h[\sqrt{2} + \log(1 + \sqrt{2})]$
20. $h^2 + \frac{1}{2}r^2$
21. (a) $(\frac{13}{6}, 1)$
    (b) $(\frac{7}{3}, \frac{9}{2})$
    (c) $(\frac{11}{4}, \frac{11}{4})$
    (d) $(\frac{19}{8}, \frac{13}{8})$
22. $h = 2\sqrt{3}$
23. $h > r\sqrt{2}$

11.22 Exercises (page 385)

1. (a) -4
   (b) 4
   (c) 8
   (d) 4π
   (e) $3\pi/2$

2. 0
3. $n = 3$
4. $-\pi$
9. $g(x, y) = \pm[P^2(x, y) + Q^2(x, y)]^{1/2}$

★11.25 Exercises (page 391)

1. (b) 0
2. 0, 2π, -2π
3. As many as three
4. As many as seven
5. (a) -3
6. 2π

11.28 Exercises (page 399)

1. $\int_0^{2\pi}\left[\int_0^a f(r\cos\theta, r\sin\theta)r\,dr\right]d\theta$
2. $\int_{-\pi/2}^{\pi/2}\left[\int_0^{2\cos\theta} f(r\cos\theta, r\sin\theta)r\,dr\right]d\theta$
3. $\int_0^{2\pi}\left[\int_a^b f(r\cos\theta, r\sin\theta)r\,dr\right]d\theta$
4. $\int_0^{\pi/2}\left[\int_0^{g(\theta)} f(r\cos\theta, r\sin\theta)r\,dr\right]d\theta,$ where $g(\theta) = 1/(\cos\theta + \sin\theta)$
5. $\int_0^{\pi/4}\left[\int_0^{\tan\theta\sec\theta} f(r\cos\theta, r\sin\theta)r\,dr\right]d\theta + \int_{\pi/4}^{3\pi/4}\left[\int_0^{\csc\theta} f(r\cos\theta, r\sin\theta)r\,dr\right]d\theta$
$+ \int_{3\pi/4}^{\pi}\left[\int_0^{\tan\theta\sec\theta} f(r\cos\theta, r\sin\theta)r\,dr\right]d\theta$
6. $\frac{3}{4}\pi a^4$
7. $\frac{1}{6}a^3[\sqrt{2} + \log(1 + \sqrt{2})]$
8. $\sqrt{2} - 1$
9. $\pi a^4/8$
10. $\int_0^{\pi/4}\left[\int_0^{\sec\theta} f(r\cos\theta, r\sin\theta)r\,dr\right]d\theta + \int_{\pi/4}^{\pi/2}\left[\int_0^{\csc\theta} f(r\cos\theta, r\sin\theta)r\,dr\right]d\theta$
11. $\int_{\pi/4}^{\pi/3}\left[\int_0^{2\sec\theta} f(r)r\,dr\right]d\theta$
12. $\int_0^{\pi/2}\left[\int_{g(\theta)}^{1} f(r\cos\theta, r\sin\theta)r\,dr\right]d\theta,$ where $g(\theta) = 1/(\cos\theta + \sin\theta)$
13. $\int_0^{\pi/4}\left[\int_{\tan\theta\sec\theta}^{\sec\theta} f(r\cos\theta, r\sin\theta)r\,dr\right]d\theta$
14. $\pi^4/3$
15. (a) $u = 7x - y,\quad v = -5x + y$
    (b) 60
17. (a) $1 + 2u$
    (c) $\frac{14}{3}$
    (d) $2 + \frac{2}{\sqrt{3}}\left(\arctan\frac{1}{\sqrt{3}} - \arctan\frac{5}{\sqrt{3}}\right)$

18. (a) $4(u^2 + v^2)$
(c) 0

19. $\dfrac{\pi}{1-p}[(p^2 + r^2)^{1-p} - p^{2(1-p)}]$ if $p \neq 1$; $\pi \log(1 + r^2)$ if $p = 1$.

$I(p, r)$ tends to a finite limit when $p > 1$

11.34 Exercises (page 413)

1. $\frac{1}{364}$
2. $\log\sqrt{2} - \frac{5}{16}$
3. $\frac{1}{48}$
4. $\frac{4}{5}\pi abc$
5. $\pi/6$
6. $\int_0^1 \left\{\int_0^x \left[\int_0^{1-x} f(x, y, z)\, dy\right] dz + \int_x^1 \left[\int_{z-x}^{1-x} f(x, y, z)\, dy\right] dz\right\} dx$
7. $\int_0^1 \left\{\int_{-z}^{z} \left[\int_{-\sqrt{z^2-x^2}}^{\sqrt{z^2-x^2}} f(x, y, z)\, dy\right] dx\right\} dz$
8. $\int_0^1 \left\{\int_0^{x^2} \left[\int_0^1 f(x, y, z)\, dy\right] dz + \int_{x^2}^{1+x^2} \int_{\sqrt{z-x^2}}^{1} \left[f(x, y, z)\, dy\right] dz\right\} dx$
10. $16\pi/3$
11. $\frac{1}{6}$
12. $\frac{1}{60}\pi a^2h(3a^2 + 2h^2)$
13. $\frac{4}{3}\pi a^3$
14. $\frac{4}{3}\pi(b^3 - a^3)$
15. $\frac{4}{3}\pi R^3(a^2 + b^2 + c^2)^{-1/2}$
18. $\frac{2}{3}\pi(5\sqrt{5} - 4)$
19. $\frac{32}{9}$
20. $\frac{4}{5}\pi(b^5 - a^5)$
22. On the axis at distance $\frac{2}{5}h$ from the base
23. On the axis at distance $\frac{1}{5}h$ from the base
24. On the axis of symmetry at distance $\dfrac{3}{8} \cdot \dfrac{b^4 - a^4}{b^3 - a^3}$ from the "cutting plane" of the hemispheres
25. $\bar{x} = \bar{y} = \bar{z} = \frac{7}{12}h$ (assuming the specified corner is at the origin)
26. $\frac{3}{20}M(a^2 + 4h^2)$
27. $\frac{2}{5}MR^2$
28. $\frac{3}{5}Ma^2$
29. $2^{1/4}$

Chapter 12

12.4 Exercises (page 424)

1. $(a_2b_3 - a_3b_2)(x - x_0) + (a_3b_1 - a_1b_3)(y - y_0) + (a_1b_2 - a_2b_1)(z - z_0) = 0$;

$$\frac{\partial \boldsymbol{r}}{\partial u} \times \frac{\partial \boldsymbol{r}}{\partial v} = (a_2b_3 - a_3b_2)\boldsymbol{i} + (a_3b_1 - a_1b_3)\boldsymbol{j} + (a_1b_2 - a_2b_1)\boldsymbol{k}$$

2. $x^2/a^2 + y^2/b^2 = z$; $\dfrac{\partial \boldsymbol{r}}{\partial u} \times \dfrac{\partial \boldsymbol{r}}{\partial v} = -2bu^2 \cos v\boldsymbol{i} - 2au^2 \sin v\boldsymbol{j} + abu\boldsymbol{k}$

3. $\dfrac{x^2}{a^2} + \dfrac{y^2}{b^2} + \dfrac{z^2}{c^2} = 1$; $\dfrac{\partial \boldsymbol{r}}{\partial u} \times \dfrac{\partial \boldsymbol{r}}{\partial v} = abc \sin u \left(\dfrac{\sin u \cos v}{a} \boldsymbol{i} + \dfrac{\sin u \sin v}{b} \boldsymbol{j} + \dfrac{\cos u}{c} \boldsymbol{k} \right)$

4. $z = f(\sqrt{x^2 + y^2})$; $\dfrac{\partial \boldsymbol{r}}{\partial u} \times \dfrac{\partial \boldsymbol{r}}{\partial v} = -uf'(u) \cos v \boldsymbol{i} - uf'(u) \sin v \boldsymbol{j} + u\boldsymbol{k}$

5. $\dfrac{y^2}{a^2} + \dfrac{z^2}{b^2} = 1$; $\dfrac{\partial \boldsymbol{r}}{\partial u} \times \dfrac{\partial \boldsymbol{r}}{\partial v} = b \sin v \boldsymbol{j} + a \cos v \boldsymbol{k}$

6. $(\sqrt{x^2 + y^2} - a)^2 + z^2 = b^2$;

$$\frac{\partial \boldsymbol{r}}{\partial u} \times \frac{\partial \boldsymbol{r}}{\partial v} = b(a + b \cos u)(\cos u \sin v \boldsymbol{i} + \cos u \cos v \boldsymbol{j} + \sin u \boldsymbol{k})$$

7. $|abc| \cosh v \left[\left(\dfrac{\sin^2 u}{a^2} + \dfrac{\cos^2 u}{b^2} \right) \cosh^2 v + \dfrac{\sinh^2 v}{c^2} \right]^{1/2}$

8. $\sqrt{128v^2 + 4}$

9. $|u - v| \sqrt{36u^2v^2 + 9(u + v)^2 + 4}$

10. $\sqrt{u^4 + u^2}$

12.6 Exercises (page 429)

2. $\pi a^2 \sqrt{3}$
3. $(2\pi - 4)a^2$
4. 4
5. (a) A circular paraboloid
   (b) $-2u^2 \cos v\boldsymbol{i} - 2u^2 \sin v\boldsymbol{j} + u\boldsymbol{k}$
   (c) $n = 6$
6. $\sqrt{2}\, \pi a^2/4$
7. $2\pi\sqrt{6}$
8. $2\pi a^2(3\sqrt{3} - 1)/3$
9. $4\pi^2 ab$
11. (a) A unit circle in the xy-plane; a unit semicircle in the xz-plane, with $z \le 0$; a unit semicircle in the plane $x = y$ with $z \le 0$
   (b) The hemisphere $x^2 + y^2 + z^2 = 1$, $z \le 0$
   (c) The sphere $x^2 + y^2 + z^2 = 1$ except for the North Pole; the line joining the North Pole and (x, y, z) intersects the xy-plane at $(u, v, 0)$

12.10 Exercises (page 436)

1. $4\pi/3$
3. $\bar{x} = \bar{y} = \bar{z} = a/2$
4. $\frac{1}{2}$
7. 0
8. $\pi\sqrt{2}$
9. On the axis of the cone, at a distance $\frac{1}{4}a(1 - \cos \alpha)/[1 - \cos (\alpha/2)]$ from the center of the sphere
10. $\pi a^3 h + \frac{2}{3}\pi a h^3$
11. $3\pi a^3 h + \frac{2}{3}\pi a h^3$
12. $2\pi/3$
13. $-\pi/3$

12.13 Exercises (page 442)

1. 0
2. $-\pi$
3. -4
4. $\frac{4}{3}$

12.15 Exercises (page 447)

1. (a) $\operatorname{div} \boldsymbol{F}(x, y, z) = 2x + 2y + 2z$; $\operatorname{curl} \boldsymbol{F}(x, y, z) = \boldsymbol{0}$
   (b) $\operatorname{div} \boldsymbol{F}(x, y, z) = 0$; $\operatorname{curl} \boldsymbol{F}(x, y, z) = 2\boldsymbol{i} + 4\boldsymbol{j} + 6\boldsymbol{k}$
   (c) $\operatorname{div} \boldsymbol{F}(x, y, z) = -x \sin y$; $\operatorname{curl} \boldsymbol{F}(x, y, z) = \boldsymbol{i} + \boldsymbol{j}$
   (d) $\operatorname{div} \boldsymbol{F}(x, y, z) = ye^{xy} - x \sin(xy) - 2xz \sin(xz^2)$;
   $\operatorname{curl} \boldsymbol{F}(x, y, z) = z^2 \sin(xz^2)\boldsymbol{j} - [xe^{xy} + y \sin(xy)]\boldsymbol{k}$
   (e) $\operatorname{div} \boldsymbol{F}(x, y, z) = 2x \sin y + 2y \sin(xz) - xy \sin z \cos(\cos z)$;
   $\operatorname{curl} \boldsymbol{F}(x, y, z) = [x \sin(\cos z) - xy^2 \cos(xz)]\boldsymbol{i} - y \sin(\cos z)\boldsymbol{j} + [y^2 z \cos(xz) - x^2 \cos y]\boldsymbol{k}$
2. $\boldsymbol{0}$
4. $n = -3$
5. No such vector field
10. One such field is $\mu(x, y, z) = (xyz)^{-2}$
11. $\operatorname{div}(\boldsymbol{V} \times \boldsymbol{r}) = 0$; $\operatorname{curl}(\boldsymbol{V} \times \boldsymbol{r}) = (c + 1)\boldsymbol{V}$
13. $16(a + b)$

★12.17 Exercises (page 452)

1. $(3x - 2z)\boldsymbol{j} - x\boldsymbol{k}$ is one such field
2. $(x^2/2 - xy - yz + z^2/2)\boldsymbol{j} + (x^2/2 - xz)\boldsymbol{k}$ is one such field
3. $(x^2y/2 + z^2/2)\boldsymbol{j} + \nabla f(x, y)$ for some f independent of z
5. $\boldsymbol{G}(x, y, z) = \dfrac{yz}{r(x^2 + y^2)}\boldsymbol{i} - \dfrac{xz}{r(x^2 + y^2)}\boldsymbol{j}$ satisfies $\operatorname{curl} \boldsymbol{G} = r^{-3}\boldsymbol{r}$ at all points not on the z-axis
6. $f(r) = Cr^{-3}$
9. $\boldsymbol{F}(x, y, z) = -\frac{1}{3}(z^3\boldsymbol{i} + x^3\boldsymbol{j} + y^3\boldsymbol{k})$, $\quad \boldsymbol{G}(x, y, z) = \frac{1}{3}\nabla(x^3y + y^3z + z^3x)$
10. (c) $3\pi/2$

12.21 Exercises (page 462)

1. 3
2. (a) 144π
   (b) -16π
   (c) 128π
15. 8π
3. (a) $3\,|V|$
   (b) $9\,|V|\,\bar{z}$
   (c) $|V|\,\bar{x}$
   (d) $4I_z$

Chapter 13

13.4 Exercises (page 472)

2. $A_1 \cup A_2 \cup A_3 = (A_1 \cap A_2' \cap A_3') \cup (A_2 \cap A_3') \cup A_3$; $\quad \displaystyle\bigcup_{k=1}^{n} A_k = \bigcup_{k=1}^{n-1} \Big(A_k \cap \bigcap_{j=k+1}^{n} A_j'\Big) \cup A_n$

3.

	(i)	(ii)	(iii)	(iv)	(v)
(a)	$A' \cap B'$	$A \cap B'$	$A \cup B$	$(A \cap B') \cup (A' \cap B)$	$A' \cup B'$
(b)	500	200	500	300	800

6. $\mathscr{B}_2 = \{\varnothing, A_1, A_2, A_1 \cup A_2, A_1', A_2', A_1' \cap A_2', S\}$
7. $\mathscr{B}_3 = \{\varnothing, A_1, A_2, A_3, A_1 \cup A_2, A_2 \cup A_3, A_1 \cup A_3, A_1 \cup A_2 \cup A_3, A_1', A_2', A_3', A_1' \cap A_2', A_2' \cap A_3', A_1' \cap A_3', A_1' \cap A_2' \cap A_3', S\}$ (if $n > 3$)

13.7 Exercises (page 477)

1. $A \subseteq B'$
2. $x \in A' \cap B' \cap C'$
3. $x \in A \cap B' \cap C'$
4. $x \in A \cup B \cup C$
5. $x \in (A \cap B' \cap C') \cup (A' \cap B \cap C') \cup (A' \cap B' \cap C)$
6. $x \in (A' \cap B') \cup (B' \cap C') \cup (A' \cap C')$
7. $x \in (A \cap B) \cup (A \cap C) \cup (B \cap C)$
8. $x \in (A \cap B \cap C') \cup (A \cap B' \cap C) \cup (A' \cap B \cap C)$
9. $x \in (A \cap B \cap C)'$
10. $x \in A \cap C \cap B'$
11. $x \in A \cap B \cap C$
12. $x \in A \cup B \cup C$
15. (a) $1 - a$
    (b) $1 - b$
    (c) $a + b - c$
    (d) $1 - c$
    (e) $1 - a + c$
    (f) $a - c$

13.9 Exercises (page 479)

2. (a) $\frac{5}{14}$
   (b) $\frac{45}{91}$
   (c) $\frac{10}{91}$
   (d) $\frac{36}{91}$
3. (a) $\frac{23}{36}$
   (b) $\frac{1}{6}$
   (c) $\frac{1}{3}$
   (d) $\frac{5}{36}$
   (e) $\frac{13}{36}$
   (f) $\frac{2}{3}$
4. $\frac{1}{4}$
5. $\frac{9}{47}$
6. $\frac{8}{47}$
7. $\frac{4}{47}$
8. (a) $A/(A + B)$
   (b) $B/(A + B)$
   (c) $(C + 1)/(C + D + 1)$
   (d) $C/(C + D + 1)$
9. (a) $\frac{1}{9}$
   (b) $\frac{4}{9}$
   (c) $\frac{5}{9}$
   (d) $\frac{8}{9}$
10. $P_0 = 1 - P(A) - P(B) + P(A \cap B)$, $P_1 = P(A) + P(B) - 2P(A \cap B)$, $P_2 = P(A \cap B)$
12. (a) 5 to 9
    (b) 45 to 46
    (c) 10 to 81
    (d) 36 to 55

13.11 Exercises (page 485)

1. $\{(1, 2), (1, 3), (2, 1), (2, 3), (3, 1), (3, 2)\}$
2. 1326
3. 54
4. $\{H, T\} \times \{H, T\} \times \{1, 2, 3, 4, 5, 6\}$; 24 outcomes
5. $52!/(13!)^4$
6. 36
   (a) 18
   (b) 12
   (c) 24

7. (a) $13 \cdot 12 \cdot 11 \cdot 72 = 123552$ (not including triplets or quadruplets)
   (b) 5148
   (c) 36 (not including $10JQKA$)
   (d) 4

8. (a) $4\binom{13}{5}\Big/\binom{52}{5}$ (b) $36\Big/\binom{52}{5}$ (c) $4\Big/\binom{52}{5}$

9. (a) $\dfrac{2 \cdot 98!}{(49!)^2}$ (b) $\dfrac{98!}{48! \cdot 50!}$

10. $\binom{98}{48}\Big/\binom{100}{50}$

11. 16

12. n^k

13.14 Exercises (page 490)

2. (a) $P(A) = \frac{3}{10}$; $P(B \mid A) = \frac{6}{11}$; $P(A \cap B) = \frac{9}{55}$

4. $$\frac{\binom{98}{48}}{\binom{100}{50} - \binom{98}{50}}$$

5. $\frac{4}{11}$

6. (a) $$1 - \frac{26! \cdot 34!}{21! \cdot 39!} = 1 - \frac{\binom{34}{13}}{\binom{39}{13}} = 1 - \frac{\binom{62}{5}}{\binom{39}{5}}$$
   (b) $$1 - \frac{\binom{34}{13} + 5\binom{34}{12}}{\binom{39}{13}}$$ (c) $$\frac{3\binom{26}{5}}{\binom{39}{5}\binom{39}{13}}$$

9. $\frac{1}{2}$

15. (a) $P(A) = P(B) = P(C) = \frac{1}{2}$; $P(A \cap B) = P(A \cap C) = P(B \cap C) = \frac{1}{4}$; $P(A \cap B \cap C) = 0$

13.18 Exercises (page 499)

1. (a) $P(H, H) = p_1p_2$; $P(H, T) = p_1(1 - p_2)$; $P(T, H) = (1 - p_1)p_2$; $P(T, T) = (1 - p_1)(1 - p_2)$
   (b) Yes
   (c) No
   (d) H_1 and H_2, H_1 and T_2, H_2 and T_1, T_1 and T_2
2. (a) $\frac{319}{512}$
   (b) $\frac{11}{1024}$
   (c) 6

3. $\binom{10}{3}\frac{5^7}{6^{10}} = \frac{390625}{2519424}$
4. (a) $\frac{5}{16}$
   (b) $\frac{1}{2}$
   (c) $\frac{3}{16}$
5. (a) $(5!)^2/10! = \frac{1}{252}$
   (b) $\frac{1}{2}$
6. (a) $36p^{10} - 80p^9 + 45p^8$
   (b) $\frac{7}{128}$
7. It is advantageous to bet even money
8. $\binom{n}{k}\frac{w^k b^{n-k}}{(w+b)^n}$
9. $\binom{8}{3}\frac{17^5}{18^8} = \frac{9938999}{1377495072}$
10. $\frac{193}{512}$
11. $1 - (19/20)^{10} = 0.4013$
12. $\frac{193}{512}$
14. $59 \le n \le 65$
15. (a) $f(p) = (1-p)^2 + p^3$
    (b) $(\sqrt{31} - 4)/3$

13.20 Exercises (page 504)

1. (a) $f(k) = 2k$
   (b) $f(k) = 3^k$
   (c) $f(k) = p_k$, where p_k is the kth prime ≥ 2
   (d) one such function is $f(k) = (g(k), h(k))$, where

$$g(k) = \frac{m^2(k) + 3m(k)}{2} - k + 2, \quad h(k) = k - \frac{m^2(k) + m(k)}{2},$$

and

$$m(k) = \left[\frac{\sqrt{8k-7} - 1}{2}\right]$$

where $[x]$ denotes the greatest integer $\le x$
   (e) $f(k) = 2^{g(k)}3^{h(k)}$, where $g(k)$ and $h(k)$ are as defined in part (d)

13.22 Exercises (page 507)

1. $n = 0$: max $= 1$, min $= \frac{1}{2}$
   $n = 1$: max $= \frac{1}{4}$, min $= 0$
   $n = 2$: max $= \frac{1}{8}$, min $= 0$
   $n = 3$: max $= \frac{1}{16}$, min $= 0$
3. (a) $1 - qp^3 - pq^3$
   (b) $\frac{7}{8}$
   (c) $\frac{3}{5}$
4. (a) $3pq/(pq + 2)$
   (b) $\frac{1}{3}$
   (c) $2 \log 2 - 1$

13.23 Miscellaneous exercises on probability (page 507)

1. $\frac{1}{12}$
2. (a) $\frac{2}{19}$
   (b) $\frac{1}{10}$
3. (a) $\frac{1}{4}$
   (b) $\frac{7}{12}$
4. (a) $\frac{15}{34}$
   (b) $\frac{13}{51}$
   (c) $\frac{13}{165}$
5. $\frac{64}{73}$
7. 0.65
8. (a) $\frac{5}{9}$
   (b) $\frac{2}{9}$
   (c) $\frac{2}{9}$
   (d) No
9. $p^3 + 6p\left(\frac{1-p}{2}\right)^2$
10. $np(1-p)^{n-1} + np^{n-1}(1-p)$
11. $\frac{n}{2^{n-1}}\left(1 - \frac{n}{2^{n-1}}\right)^{m-1}$

Chapter 14

14.4 Exercises (page 513)

1. (b) $X \leq b$
2. (a) $\{\omega \mid X(\omega) \in (a, b], Y(\omega) \in (c, d]\}$
   (c) $X \leq a, \quad Y \leq d$
   (d) $P(a < X \leq b, c < Y \leq d) = P(X \leq b, Y \leq d) - P(X \leq a, Y \leq d) - P(X \leq b, Y \leq c) + P(X \leq a, Y \leq c)$
3. (a) $\{(1,6), (2,5), (3,4), (4,3), (5,2), (6,1)\}$, $\{(5,6), (6,5)\}$, $\{(1,6), (2,5), (3,4), (4,3), (5,2), (6,1), (5,6), (6,5)\}$
   (b) $P(X = 7) = \frac{1}{6}$; $P(X = 11) = \frac{1}{18}$; $P(X = 7 \text{ or } X = 11) = \frac{2}{9}$
4. $Y = X_1 + X_2 + X_3 + X_4$; $P(Y = 0) = \frac{1}{16}$; $P(Y = 1) = \frac{1}{4}$; $P(Y \leq 1) = \frac{5}{16}$
5. $Y = 7X$ if $0 \leq X \leq 100$; $Y = 10X - 300$ if $X > 100$
6. (a) $Z = Y - 1$
   (b) $U = Y_1 + Y_2 - 1$

14.8 Exercises (page 523)

2.

t	2	3	4	5	6	7	8	9	10	11	12
$p_x(t)$	$\frac{1}{36}$	$\frac{1}{18}$	$\frac{1}{12}$	$\frac{1}{9}$	$\frac{5}{36}$	$\frac{1}{6}$	$\frac{5}{36}$	$\frac{1}{9}$	$\frac{1}{12}$	$\frac{1}{18}$	$\frac{1}{36}$

3. (b) $p(-2) = \frac{1}{2}$, $p(0) = p(2) = \frac{1}{4}$
   (c) $0, \frac{3}{4}, \frac{3}{4}, \frac{1}{4}, 1, 0, \frac{1}{4}, \frac{1}{4}$
5. (a) $c = \frac{1}{3}$
   (b) $\frac{1}{9}, \frac{1}{3}, \frac{2}{3}, \frac{8}{9}$
   (c) $F(t) = 0$ for $t < 0$, $F(t) = \frac{1}{9}$ for $0 \leq t < 1$, $F(t) = \frac{1}{3}$ for $1 \leq t < 2$, $F(t) = 1$ for $t \geq 2$
   (d) No such t
   (e) $t = 2$
6. (a)

k	0	1	2	3	4
$p(k)$	$\frac{16}{81}$	$\frac{32}{81}$	$\frac{8}{27}$	$\frac{8}{81}$	$\frac{1}{81}$

   $p(t) = 0$ for $t \neq 0, 1, 2, 3, 4$

   (b) $F(t) = 0$ for $t < 0$, $F(t) = \frac{16}{81}$ on $[0, 1)$, $F(t) = \frac{16}{27}$ on $[1, 2)$, $F(t) = \frac{8}{9}$ on $[2, 3)$, $F(t) = \frac{80}{81}$ on $[3, 4)$, $F(t) = 1$ for $t \geq 4$
   (c) $\frac{8}{27}, \frac{56}{81}$
7. (b) $P(X = 0) = (1 - p)^2$; $P(X = 1) = 2p(1 - p)$
8. $p_X(k) = \dfrac{2k}{n(n+1)}$; $F_X(t) = \dfrac{[t]([t]+1)}{n(n+1)}$ for $0 \leq t \leq n$, where $[t]$ denotes the greatest integer $\leq t$; $F_X(t) = 0$ for $t < 0$; $F_X(t) = 1$ for $t > n$
9. $p(k) = e^{-c}\dfrac{c^k}{k!}$; $k = 0, 1, 2, 3, \ldots$; $c \geq 0$

   $p(t) = 0$ for $t \neq 0, 1, 2, 3, \ldots$
10. (a) $p_X(t) = \frac{1}{2}$ at $t = -1$ and $t = +1$; $p_X(t) = 0$ elsewhere
    $F_X(t) = 0$ for $t < -1$; $F_X(t) = \frac{1}{2}$ for $-1 \leq t < 1$; $F_X(t) = 1$ for $t \geq 1$
11. $P(A) = \frac{2}{3}$; $P(B) = \frac{2}{3}$; $P(A \cap B) = \frac{1}{3}$; $P(B \mid A) = \frac{1}{2}$; $P(A \cup B) = 1$

14.12 Exercises (page 532)

1. (a) $c = 1$; $f(t) = 1$ if $0 \le t \le 1$; $f(t) = 0$ otherwise
   (b) $0, \frac{1}{3}, \frac{1}{3}$
2. $c = \frac{1}{2}$; $F(t) = 0$ if $t < -\pi/2$; $F(t) = \frac{1}{2}\cos t$ if $-\pi/2 \le t < 0$; $F(t) = 1 - \frac{1}{2}\cos t$ if $0 \le t < \pi/2$; $F(t) = 1$ if $t \ge \pi/2$
3. $c = \frac{3}{8}$; $F(t) = 0$ if $t < 0$; $F(t) = \frac{1}{4}(3t^2 - t^3)$ if $0 \le t \le 2$; $F(t) = 1$ if $t > 2$
4. (a) $\frac{1}{2}$
   (b) $\frac{1}{2}$
   (c) $\frac{1}{4}$
5. (a) $f(t) = 0$ if $t < \frac{1}{4}$; $f(t) = 8t - 1$ if $\frac{1}{4} \le t < \frac{1}{2}$; $f(t) = 7 - 8t$ if $\frac{1}{2} \le t < \frac{3}{4}$; $f(t) = 0$ if $t \ge \frac{3}{4}$
   (b) $F(t) = 0$ if $t < \frac{1}{4}$; $F(t) = 4t^2 - t$ if $\frac{1}{4} \le t < \frac{1}{2}$; $F(t) = -4t^2 + 7t - 2$ if $\frac{1}{2} \le t < \frac{3}{4}$; $F(t) = 1$ if $t \ge \frac{3}{4}$
   (c) $1, 1, \frac{1}{2}, 0, \frac{5}{16}$
6. (a) $0, \frac{5}{6}, \frac{2}{3}, \frac{1}{2}$
   (b) $t = 1$
7. (a)

k	5	10	15	20	25	30
$P(X \ge k)$	$\frac{5}{6}$	$\frac{2}{3}$	$\frac{1}{2}$	$\frac{1}{3}$	$\frac{1}{6}$	0

   (b) Let each Styx train arrive 5 minutes before a Lethe train
8. $F_Y(t) = 0$ if $t < b$; $F_Y(t) = (t - b)/a$ if $b \le t \le b + a$; $F_Y(t) = 1$ if $t > b + a$
9. (a) $\frac{2}{37}$
   (b) $\frac{11}{37}$
   (c) $\frac{36}{37}$
10. $F(t) = \dfrac{1}{2} + \dfrac{1}{\pi}\arctan\dfrac{t - b}{a}$

14.16 Exercises (page 540)

1. (a) 105
   (b) 10.05
2. (a) $\frac{1}{2}(\sqrt{2} - 1)$
   (b) $\frac{1}{4}(2 - \sqrt{2})$
   (c) $\frac{1}{2}(2 - \sqrt{2})$
   (d) $\frac{1}{14}(4 - \sqrt{2})$
3. (a) $1 - e^{-1}$
   (b) e^{-2}
   (c) $(e - 1)/e^2$
   (d) e^{-3}
4. $F(t) = 0$ if $t < c$; $F(t) = 1 - e^{-\lambda(t-c)}$ if $t \ge c$
6. $\lambda' = \lambda/a$, $c' = b + ac$
9. (a) 0.5000
   (b) 0.1359
   (c) 0.9974
   (d) 0.0456
10. (a) 0.675
    (b) 0.025
11. (a) 0.675σ
    (b) 0.025σ

12. (a) 0.8185
    (b) 0.8400
13. 75.98 inches
14. mean $= b$, variance $= a^2$
15. $F_Y(t) = 0$ if $t < 0$; $f_Y(t) = 0$ if $t \le 0$; $f_Y(t) = e^{-t/2}/\sqrt{2\pi t}$ if $t > 0$

14.18 Exercises (page 542)

1. (a) $F_Y(t) = 0$ if $t < 1$; $F_Y(t) = (t-1)/3$ if $1 \le t \le 4$; $F_Y(t) = 1$ if $t > 4$; $f_Y(t) = \frac{1}{3}$ if $1 \le t \le 4$; $f_Y(t) = 0$ otherwise
   (b) $F_Y(t) = 0$ if $t < -2$; $F_Y(t) = (t+2)/3$ if $-2 \le t \le 1$; $F_Y(t) = 1$ if $t > 1$; $f_Y(t) = \frac{1}{3}$ if $-2 \le t \le 1$; $f_Y(t) = 0$ otherwise
   (c) $F_Y(t) = 0$ if $t < 0$; $F_Y(t) = t^{1/2}$ if $0 \le t \le 1$; $F_Y(t) = 1$ if $t > 1$; $f_Y(t) = (2t)^{-1/2}$ if $0 \le t \le 1$; $f_Y(t) = 0$ otherwise
   (d) $F_Y(t) = e^t$ if $t \le 0$; $F_Y(t) = 1$ if $t > 0$; $f_Y(t) = e^t$ if $t \le 0$; $f_Y(t) = 0$ if $t > 0$
   (e) $F_Y(t) = e^{t/2}$ if $t \le 0$; $F_Y(t) = 1$ if $t > 0$; $f_Y(t) = \frac{1}{2}e^{t/2}$ if $t \le 0$; $f_Y(t) = 0$ if $t > 0$
   (f) $F_Y(t) = 0$ if $t < 1$; $F_Y(t) = \log t$ if $1 \le t \le e$; $F_Y(t) = 1$ if $t > e$; $f_Y(t) = 1/t$ if $1 \le t \le e$; $f_Y(t) = 0$ otherwise
2. Let ψ be the inverse of φ, defined on the open interval (a, b).
   Then $F_Y(t) = 0$ if $t \le a$; $F_Y(t) = F_X[\psi(t)]$ if $a < t < b$; $F_Y(t) = 1$ if $t \ge b$; $f_Y(t) = f_X[\psi(t)]\psi'(t)$ if $a < t < b$; $f_Y(t) = 0$ otherwise
3. (a) $f_Y(t) = 0$ if $t \le 0$; $f_Y(t) = (2\pi t)^{-1/2}e^{-t/2}$ if $t > 0$
   (b) $f_Y(t) = 0$ if $t < 0$; $f_Y(t) = 4t(2\pi)^{-1/2}e^{-t^4/2}$ if $t \ge 0$
   (c) $f_Y(t) = 0$ if $t \le 0$; $f_Y(t) = (2\pi t^2)^{-1/2}e^{-(\log t)^2/2}$ if $t > 0$
   (d) $f_Y(t) = (2\pi)^{-1/2} \sec^2 t\, e^{-(\tan^2 t)/2}$ if $|t| < \pi/2$; $f_Y(t) = 0$ if $|t| \ge \pi/2$

14.22 Exercises (page 548)

2. (a) $P(X = x_1) = P(X = x_2) = P(Y = y_1) = P(Y = y_2) = \frac{1}{2}(p + q)$
   (b) $p = q = \frac{1}{2}$
3. (a) $F(x, y) = \left(\dfrac{x-a}{b-a}\right)\left(\dfrac{y-c}{d-c}\right)$ if $a \le x \le b$ and $c \le y \le d$,

   $F(x, y) = \dfrac{x-a}{b-a}$ if $a \le x \le b$ and $y > d$, $F(x, y) = \dfrac{y-c}{d-c}$ if $x > b$ and $c \le y \le d$,

   $F(x, y) = 1$ if $x > b$ and $y > d$, $F(x, y) = 0$ otherwise
   (b) $F_X(x) = (x-a)/(b-a)$ if $a \le x \le b$; $F_X(x) = 0$ if $x < a$; $F_X(x) = 1$ if $x > b$; $F_Y(y) = (y-c)/(d-c)$ if $c \le y \le d$; $F_Y(y) = 0$ if $y < c$; $F_Y(y) = 1$ if $y > d$
   (c) X and Y are independent
5. $P(Y = 0) = \frac{1}{2}$; $P(Y = 1) = P(Y = 2) = \frac{1}{4}$
7. $\frac{1}{5}, \frac{4}{5}$
8. $\frac{1}{3}$
9. (b) $f(x, y) = \frac{1}{2}$ if $(x, y) \in Q$; $f(x, y) = 0$ if $(x, y) \notin Q$;
   $f_X(x) = 1 - |x|$ if $|x| \le 1$; $f_X(x) = 0$ if $|x| > 1$;
   $f_Y(y) = 1 - |y|$ if $|y| \le 1$; $f_Y(y) = 0$ if $|y| > 1$.
   X and Y are not independent
10. $g(u, v) = f(u + a, v + b)$

14.24 Exercises (page 553)

1. (b) $f_V(v) = 1 + v$ if $-1 \le v < 0$; $f_V(v) = 1 - v$ if $0 \le v \le 1$; $f_V(v) = 0$ if $|v| > 1$
   (c) U and V are not independent

2. (b) $f_V(t) = 2 - 2t$ if $0 \le t \le 1$; $f_V(t) = 0$ otherwise
   (c) U and V are independent
3. (b) $g(u, v) = ue^{-u}$ if $u > 0$, $0 < v < 1$; $g(u, v) = 0$ otherwise
   (c) $f_U(u) = ue^{-u}$ if $u > 0$; $f_U(u) = 0$ if $u \le 0$
   (d) $f_V(v) = 1$ if $0 < v < 1$; $f_V(v) = 0$ otherwise
4. (b) $g(u, v) = (v/2\pi)e^{-(1+u^2)v^2/2}$ if $v \ge 0$
   (c) $f_U(u) = [\pi(1 + u^2)]^{-1}$
5. $f_Z(t) = \pi^{-1/2}e^{-t^2}$
6. (b) $f_U(u) = 0$ if $u < 0$; $f_U(u) = (u/\sigma^2)\exp\left(-\dfrac{u^2}{2\sigma^2}\right)$ if $u > 0$; $F_U(t) = 0$ if $t < 0$;

$F_U(t) = 1 - \exp\left(-\dfrac{t^2}{2\sigma^2}\right)$ if $t \ge 0$

14.27 Exercises (page 560)

1. $E(X) = \frac{7}{2}$, $\text{Var}(X) = \frac{35}{12}$
7. (a) $E(X) = \text{Var}(X) = \lambda$
   (b) None
   (c) $E(X) = 1/\lambda$, $\text{Var}(X) = 1/\lambda^2$
   (d) $E(X) = m$, $\text{Var}(X) = \sigma^2$
8. (a) $C(r) = (r - 1)/2$
   (b) $F_X(t) = \frac{1}{2}|t|^{1-r}$ if $t < -1$; $F_X(t) = \frac{1}{2}$ if $-1 \le t \le 1$; $F_X(t) = 1 - \frac{1}{2}t^{1-r}$ if $t > 1$
   (c) $P(X < 5) = 1 - 5^{r-1}/2$; $P(5 < X < 10) = (5^{1-r} - 10^{1-r})/2$
   (d) X has a finite expectation when $r > 2$; $E(X) = 0$
   (e) Variance is finite for $r > 3$; $\text{Var}(X) = (r - 1)/(r - 3)$
9. $E(X) = E(Y) = -\frac{1}{37}$; $E(Z) = -1767/50653$; $E(X + Y + Z) = -4505/50653$
10. $E(X) \to \infty$ as $n \to \infty$
12. (a) $(2/\pi)^{1/2}$
    (b) $e^{1/2}$
    (c) $e^2 - e$
    (d) $(\pi/2)^{1/2}$

14.31 Exercises (page 568)

4. 251
5. 0
6. Chebyshev's inequality gives $\frac{1}{9}$; tables give 0.0027
8. (b) 0.6826
9. (b) 0.0796
10. (a) 0.0090
    (b) 0.0179

Chapter 15

15.5 Exercises (page 577)

2. (a) No
   (b) Yes
   (c) Yes
   (d) No
   (e) Yes
   (f) No
   (g) Yes
3. (a) Neither
   (b) Seminorm
   (c) Seminorm
   (d) Neither
   (e) Seminorm
   (f) Norm
   (g) Neither
   (h) Neither

6. (a), (b), (c)
8. (b) The polynomial in (a) plus $\frac{7}{8}(5x^3 - 3x)\int_{-1}^{1}(5t^3 - 3t)f(t)\,dt + \frac{9}{128}(35x^4 - 30x^2 + 3)$ $\times \int_{-1}^{1}(35t^4 - 30t^2 + 3)f(t)\,dt$
9. $a = -3e/4 + 33/(4e)$, $b = 3/e$, $c = \frac{15}{4}(e - 7/e)$
10. $\frac{15}{128}(1 + 14x^2 - 7x^4)$
11. (b) $\|f - P\|^2 = \frac{1}{96}$
12. (a) $\|P - f\|^2 = \dfrac{n-1}{n} - \dfrac{\log^2 n}{n-1}$

    (b) $P(x) = \left(\dfrac{12}{(n-1)^2} - \dfrac{6(n+1)}{(n-1)^3}\log n\right)x + \dfrac{4(n^3-1)\log n}{(n-1)^4} - \dfrac{6(n+1)}{(n-1)^2}$;

    $\|P - f\|^2 = 36\log 2 - 28\log^2 2 - \frac{23}{2} = 0.0007$ when $n = 2$
13 (a) $\|P - f\|^2 = \dfrac{1}{2n}[(n-2)e^{2n} + 4e^n - n - 2]$

    (b) $P(x) = (18 - 6e)x + 4e - 10$; $\|P - f\|^2 = 20e - \frac{7}{2}e^2 - \frac{57}{2} = 0.0038$
14. (b) $\varphi_{k+1}(x) = \dfrac{\sqrt{(2k+1)(2k+3)}}{k+1}x\varphi_k(x) - \dfrac{k}{k+1}\sqrt{\dfrac{2k+3}{2k-1}}\varphi_{k-1}(x)$, where $\varphi_k = P_k/\|P_k\|$
15. (b) $\displaystyle\sum_{k=0}^{m} P_k^2(x) = \frac{p_m}{p_{m+1}}[P_m(x)P'_{m+1}(x) - P_{m+1}(x)P'_m(x)]$

15.9 Exercises (page 585)

1. (a) $P(x) = \frac{1}{6}(x^2 + 13x + 12)$
   (b) $P(x) = \frac{1}{2}(x^2 - 5x + 6)$
   (c) $P(x) = -\frac{1}{6}(x^3 - 6x^2 + 5x - 6)$
   (d) $P(x) = 2x^3 + x^2 - x - 2$
   (e) $P(x) = -5x^3 - x^2 + 10x - 5$
2. $P(x) = \frac{1}{640}(9x^4 - 196x^2 + 640)$
4. (a) $Q(x) = 2x^3 + 3x^2 - x - 3$
   (b) $Q(x) = 4x^3 + 7x^2 - 3x - 7$
5. (a) $P(32) = \frac{31}{21}$; $f(32) - P(32) = \frac{43}{42}$
   (b) $P(32) = \frac{992}{255}$; $f(32) - P(32) = -\frac{709}{510}$
   (c) $P(32) = \frac{43}{15}$; $f(32) - P(32) = -\frac{11}{30}$
   (d) $P(32) = -\frac{403}{1530}$; $f(32) - P(32) = \frac{2114}{765}$
7. (a) $L_0(x) = \frac{1}{72}(u-1)(u-3)(u-4)(u-6)$; $L_1(x) = -\frac{1}{30}u(u-3)(u-4)(u-6)$; $L_2(x) = \frac{1}{18}u(u-1)(u-4)(u-6)$; $L_3(x) = -\frac{1}{24}u(u-1)(u-3)(u-6)$; $L_4(x) = \frac{1}{180}u(u-1)(u-3)(u-4)$
   (b) $P(2.6) = 20$
8. (b) $x \geq 1.581$
   (c) $h \leq 0.0006$
12. (a) $a = 0$, $b = 1$
   (b) $c = 1$, $d = -2L'_k(x_k)$
13. (d) Let $B_0(x) = 1$ and let $B_n(x) = (x - x_0)(x - x_0 - nh)^{n-1}/n!$ for $n \geq 1$; the one and only polynomial P of degree $\leq n$ satisfying the conditions $P(x_0) = c_0$, $P'(x_1) = c_1$, $P''(x_2) = c_2, \ldots, P^{(n)}(x_n) = c_n$ is given by $P(x) = c_0B_0(x) + \cdots + c_nB_n(x)$
16. x

15.13 Exercises (page 593)

4. (b)

8	−5040	13068	−13132	6769	−1960	322	−28	1		
9	40320	−109584	118124	−67284	22449	−4536	546	−36	1	
10	−3628800	1026576	−1172700	723680	−269325	63273	−9450	870	−45	1

(c) $1 + 2x + 2x^2 - 3x^3 + x^4$

5. (c)

8	1	127	966	1701	1050	266	28	1		
9	1	255	3025	7770	6951	2646	462	36	1	
10	1	511	9330	34105	42525	22827	5880	750	45	1

(d) $-1 + 6x^{(1)} + 16x^{(2)} + 9x^{(3)} + x^{(4)}$

7. (a) $\frac{4}{3}n^3 + \frac{11}{2}n^2 + \frac{61}{6}n$
(b) $\frac{1}{4}n^4 + \frac{5}{6}n^3 + \frac{3}{4}n^2 + \frac{1}{6}n$
(c) $\frac{1}{4}n^4 + \frac{3}{2}n^3 + \frac{11}{4}n^2 + \frac{3}{2}n$
(d) $\frac{1}{5}n^5 + \frac{1}{2}n^4 + \frac{1}{3}n^3 - \frac{1}{30}n$

15.18 Exercises (page 600)

2. (b) $T_n(1) = n^2, \quad T_n'(-1) = (-1)^{n-1}n^2$

5. $\sin\theta \sin n\theta = \dfrac{1 - x^2}{n} T_n'(x)$; degree $= n + 1$

7. $Q(x) = x^{n+1} - 2^{-n}T_{n+1}(x)$
8. $Q(x) = -\frac{5}{2}x^4 + \frac{35}{16}x^3 - \frac{25}{32}x^2 + \frac{25}{256}x - \frac{1}{512}$
14. (b) $U_0(x) = 1, \quad U_1(x) = 2x, \quad U_2(x) = 4x^2 - 1, \quad U_3(x) = 8x^3 - 4x,$
$U_4(x) = 16x^4 - 12x^2 + 1, \quad U_5(x) = 32x^5 - 32x^3 + 6x$

15.21 Exercises (page 610)

1. (a) $0.693773 - \epsilon$, where $0.000208 \le \epsilon \le 0.001667$. This gives the inequalities $0.6921 < \log 2 < 0.6936$
(b) $n = 578$
2. (a) $c = \sqrt{3}/3$

(b) $c_1 = \dfrac{a+b}{2} + \dfrac{b-a}{2}\dfrac{\sqrt{3}}{3}, \quad c_2 = \dfrac{a+b}{2} - \dfrac{b-a}{2}\dfrac{\sqrt{3}}{3}$

3. (a) $c = \sqrt{2}/2$

(b) $c_1 = \dfrac{a+b}{2} + \dfrac{b-a}{2}\dfrac{\sqrt{2}}{2}, \quad c_2 = \dfrac{a+b}{2} - \dfrac{b-a}{2}\dfrac{\sqrt{2}}{2}$

4. $a = 2 + \sqrt{2}, \quad b = 2 - \sqrt{2}$
5. $c = \sqrt{\frac{3}{2}}$

7. (d) $$\int_a^b f(x)\,dx = \frac{b-a}{18}\left[5f\left(\frac{b+a}{2} - \frac{b-a}{2}\sqrt{\frac{3}{5}}\right) + 8f\left(\frac{b+a}{2}\right) + 5f\left(\frac{b+a}{2} + \frac{b-a}{2}\sqrt{\frac{3}{5}}\right)\right]$$

10. (a) $\log 2 = 0.693254 - \epsilon$, where $0.000016 \le \epsilon \le 0.000521$; this leads to the inequalities $0.69273 < \log 2 < 0.69324$
(b) $\log 2 = 0.69315023 - \epsilon$, where $0.00000041 \le \epsilon \le 0.00001334$; this leads to the inequalities $0.693136 < \log 2 < 0.693149$

11. (d) $\log 2 = 0.693750 - \epsilon$, where $0.000115 \le \epsilon \le 0.003704$; this leads to the inequalities $0.69004 < \log 2 < 0.69364$

INDEX